Proceedings of the Seventh Conference on Probability Theory

August 29 – September 4, 1982
Brașov, Romania

These Proceedings are dedicated to the memory of Octav Onicescu (20 August 1892—19 August 1983) and Gheorghe Mihoc (7 July 1906—25 December 1981), co-founders of the Brașov Conferences on Probability Theory. May they be long remembered for their accomplishments as scientists, their wisdom and dedication as teachers and organizers, and their generosity as Men.

Proceedings of the Seventh Conference on Probability Theory

August 29 — September 4, 1982
Braşov, Romania

The Conference was organized by

THE CENTRE OF MATHEMATICAL STATISTICS
OF THE NATIONAL INSTITUTE
OF METROLOGY BUCHAREST

Edited by Marius IOSIFESCU

with the co-operation of
Şerban GRIGORESCU and Tiberiu POSTELNICU

EDITURA ACADEMIEI | VNU SCIENCE PRESS
BUCUREŞTI • ROMÂNIA | UTRECHT, THE NETHERLANDS

1985

EDITURA ACADEMIEI REPUBLICII SOCIALISTE ROMÂNIA
R-79717, București, Sectorul 1, Calea Victoriei nr. 125

VNU *SCIENCE PRESS BV*
Park Voorn 4
3454 JR De Meern
P.O. Box 2073
3500 GB Utrecht
The Netherlands

ISBN 90—6764—040—9''.

Editorial Note

The 7th Conference on Probability Theory (Braşov, 29 August — 4 September 1982) was held a few days after the 90th birthday of Professor Octav Onicescu, Member of the Romanian Academy, co-founder with Gheorghe Mihoc (1906—1981) of the Braşov Conferences. We were therefore very glad to mark the occasion in such an appropriate frame. This volume begins with the omagial address given on the opening session of the Conference by Professor Caius Jacob, President of the Mathematical Sciences Section of the Romanian Academy.

Unfortunately, Professor Onicescu died on 19th August 1983, the eve of his 91st birthday, after a short illness. Thus, to our deep regret, the present book instead of being an omagial volume has to be a memorial one. It is devoted to the memory of both Octav Onicescu and Gheorghe Mihoc. They are the co-founders of the Romanian probability school and their life-long scientific collaboration has been unique in the history of Romanian mathematics. It is certainly moving to note that the memory of Gheorghe Mihoc was omagiated at the Conference in a special session of the Probability Section, which was introduced and chaired by Octav Onicescu.

*

This volume contains the invited and contributed papers presented at the Conference, with the exceptions mentioned below. Three invited papers, namely Length bias in therapeutic observations *by H. Jesdinsky,* On the maximal inequalities of the theory of martingales and some related topics *by J. Mogyoródi, and* Information econometrics *by O. Onicescu and M. C. Botez are not included in the volume. We were unable to get the first two papers while the third one has been expanded into a book entitled* Introduction à l'econométrie informationnelle *which is in press at Editions Maloine, Paris. Instead, there have been included three invited papers (by S. K. Basu and P. K. Sen, K. J. Miescke, and A. Mukherjea) whose authors were unable to attend the Conference.*

The contributed papers by M. Altăr; V. Bally; Adriana Berechet; C. Bergthaller and M. Steinbach; D. Berşadschi; C. A. Bob; I. Bordea, Liana Leu, Sanda Popovici and Marilena Bordea; W. Breckner; I. Chiţescu, F.-C. Gheorghe and G. Obreja; G. Ciobanu, M. Stoica and Cs. Fabian; Silvia Coatu; S. Corbu; A. Cristea and C. N. Zaharia; A. Cristea, Yolanda Copelovici, N. Cajal and C. N. Zaharia; A. Cristea, M. Tîrnoveanu, C. N. Zaharia, V. Niculescu, E. Diatcu, A. Bente, Yolanda Copelovici and N. Cajal;

M. C. Demetrescu; L. Dragomirescu and M. Şerban; M. Dragomirescu; Maria Drăguţ; Monica Dumitrescu-Bad; T. Dumitrescu; V. Dumitru and Florica Luban; D. Enăchescu; O. Enchev; D. Farcaş; Argentina Filip and Aneta Muja; Liana Maria Filip; I. Florea; R. Gabriel; S. Garcea; F. Gonzacenco and G. Zbăganu; D. V. Iliescu and V. Gh. Vodă; Daniela Jarušková; Gabriela Licea; Marica Lewin; Ileana Mareş and C. Mareş; I. Maruşciac; V. Mierlea; N. V. Mihăiţă; I. Mihoc; T. Morozan; I. Niculescu and Rodica Boconcios; Şt. P. Niculescu; Elena Oancea; O. Onicescu, Gh. Oprişan and Gh. Popescu; M. Oprea and C. Marinescu; Gh. Opriş; G. Orman; A. Popescu; E. Radian; M. Rădulescu; Gh. Răuţu; Gh. Rusu; I. D. Rusu, I. Ducman and Maria Bănescu; Gh. Ruxanda, I. Smeureanu and Gh. Zaharia; Judita Samuel; B. Săvulescu and D. P. Vasiliu; S. Sburlan; B. Singer; A. Spătaru; C. Stanciu and F. Vasiliu; I. M. Stancu-Minasian · I. State; I. State and M. Tavitian; L. Stoica; V. Stoican; A. Szákats; Mihaela Şerbu and Emiliana Ursianu; Şt. Ştefănescu; A. Toia and S. Crăciunaş; Rodica Tomescu; G. Zbăganu and X. W. Zhuang will be published elsewhere.

We want to acknowledge with thanks the help our colleagues dr. C. Bergthaller, dr. E. Mărgăritescu, dr. Şt. P. Niculescu, dr. V. Preda and especially G. Zbăganu gave us with the refereeing process. Thanks are also due to Mrs. Adriana Grădinaru for her excellent typing. Last but not least, we are indebted to many colleagues from the Centre of Mathematical Statistics for their invaluable help in the proof reading of the whole volume.

THE EDITORS

CONTENTS

Invited papers

Communications

1. Probability

II. Mathematical Statistics

III. Operations Research and Mathematical Programming

IV. Biomathematics

CAIUS JACOB

L'académicien Octave Onicescu à 90 ans*

C'est un grand honneur et plaisir pour moi que de pouvoir assister aujourd'hui à l'ouverture des travaux de la 7ème Conférence sur la théorie des probabilités et de vous transmettre le salut cordial de la section des sciences mathématiques de l'Académie de la République Socialiste de Roumanie, avec les meilleurs vœux de réussite dans votre activité.

Cette réunion des spécialistes de notre pays, ainsi que de ceux de nombreux autres pays, a pour but d'examiner les directions actuelles des recherches, les problèmes modernes de cette branche si importante des sciences mathématiques, aussi bien que les progrès réalisés depuis la 6ème Conférence, tenue également en cette ville de Brașov, il y a trois ans.

Les travaux de cette Conférence s'ouvrent sous la présidence d'un grand homme de science, l'académicien Octave Onicescu, qui est le créateur de l'école roumaine de la théorie des probabilités et de statistique mathématique et qui vient d'anniverser récemment ses 90 ans.

C'est Octave Onicescu qui a donné à l'Université de Bucarest les premiers cours de spécialisation dans le domaine de la théorie des probabilités et cela dès 1928, avant que — toujours grâce à ses efforts — les leçons de théorie des probabilités soient régulièrement incluses dans le plan d'enseignement des mathématiques. Les effets de cette sage disposition ne se laissèrent d'ailleurs pas faire attendre et les leçons de calcul des probabilités du professeur Octave Onicescu s'avérèrent bientôt très utiles. Ses recherches scientifiques furent aussi de grande envergure et cela contribua essentiellement à la formation de cette école roumaine de théorie des probabilités qui s'affirme aujourd'hui avec force autant sur le plan national que sur celui international.

On doit à Octave Onicescu d'importantes recherches en théorie des probabilités en chaîne dont d'autres, plus compétents que moi, souligneront les mérites et la portée. Je voudrais seulement souligner ici que la théorie des probabilités en chaîne, au sens de Markov, et ses extensions constituaient au début de la quatrième décennie de notre siècle l'objet des préoccupations de quelques grands mathématiciens qui ont continué et élargi les recherches d'Henri Poincaré sur le battage des cartes et celles, plus générales, de Markov. Je citerai parmi ceux-ci A. N. Kolmogorov, V. I. Romanovski et B. Hostinský. Maurice Fréchet, l'un des créateurs incontestés de l'analyse fonctionnelle moderne, s'intéressa aussi au problème des probabilités en chaîne de Markov, discontinue ou continue, dans la période 1930—1934 et il donna des cours à la Sorbonne sur ce sujet.

---

* Texte de l'allocution prononcée lors de la séance d'ouverture de la 7ème Conférence

Et voilà que la période 1935—1939 enregistra une suite ininterrompue de travaux d'Octave Onicescu en collaboration avec le premier des élèves qu'il forma, Gheorghe Mihoc, qui devait d'ailleurs devenir plus tard son successeur comme chef de la chaire de théorie des probabilités à l'Université de Bucarest, avant d'être élu recteur de l'Université et ensuite président de notre Académie.

Les travaux d'Octave Onicescu et ceux qu'il réalisa en collaboration avec Gheorghe Mihoc marquèrent l'apparition de cette importante théorie des chaînes à liaisons complètes qui généralisent les chaînes de Markov.

Parmi les recherches d'Octave Onicescu, mention spéciale doit être faite de celles concernant les chaînes de mouvements discontinus, les chaînes continues à liaisons complètes, l'évolution de la répartition des grandeurs additives de la mécanique statistique, l'application de certains théorèmes de la théorie des probabilités à la statistique stellaire, l'étude des lois physiques s'exprimant par des chaînes statistiques, la définition de la probabilité et le problème de la roulette, etc.

Ces travaux d'Octave Onicescu et ceux qu'il écrivit en collaboration avec Gheorghe Mihoc ont jeté les fondements de l'école roumaine de la théorie des probabilités. Cette série de publications scientifiques, couronnée par des travaux de synthèse, tels que la monographie de 1937 (en collaboration avec Gheorghe Mihoc) sur les chaînes et familles de chaînes discontinues, celle de 1943 (aussi en collaboration avec Gheorghe Mihoc) sur les problèmes asymptotiques des chaînes de variables aléatoires, son traité de Calcul des probabilités de 1956 et celui de Calcul des probabilités et ses applications (en collaboration avec Gheorghe Mihoc et C. T. Ionescu Tulcea) de la même année, etc., toutes ces publications — dont je n'ai cité que quelques-unes — apportent des méthodes nouvelles pour l'établissement des propriétés asymptotiques des chaînes probabilistes fondées sur l'utilisation de la fonction caractéristique ainsi que d'autres résultats fondamentaux en théorie des probabilités.

L'histoire des sciences mathématiques offre rarement de pareils exemples de fructueuse collaboration scientifique que celle des deux scientifiques roumains Octave Onicescu et Gheorghe Mihoc. Dans cette collaboration chacun des deux chercheurs apportait sa contribution spécifique, propre à son caractère et à sa personnalité, à son imagination créatrice, à son esprit d'analyse, à sa technique mathématique. Ensemble, ces deux savants roumains ont eu la satisfaction de constater que leur œuvre avait porté des fruits, que notre jeunesse mathématique s'intéressait à ces recherches, qu'elle continuait leur œuvre avec un réel succès.

Ensuite, tandis que Gheorghe Mihoc se dirigeait surtout vers des problèmes d'actuariat et de statistique mathématique, de son côté Octave Onicescu reprenait ses préoccupations dans un tout autre domaine, car il était un mathématicien au large horizon, doublé d'un mécanicien, et c'est ce qui explique aussi certains aspects de son activité scientifique si vaste, englobant de nombreuses autres directions de recherches.

D'ailleurs, le professeur Octave Onicescu fut l'élève du grand mathématicien et mécanicien italien Tullio Levi-Cività et ses premières recherches, sa Thèse de 1920, furent consacrées à la géométrie différentielle et à la mécanique, notamment à l'étude des espaces d'Einstein qui admettent des groupes continus de transformations.

Il commença sa carrière didactique en enseignant la mécanique à l'Université de Bucarest dans la période d'entre les deux guerres mondiales. Aussi, quoique apparemment ses recherches et préoccupations fussent orientées plutôt vers la théorie des probabilités, l'algèbre — qu'il enseigne entre 1939 et 1954 — et l'analyse fonctionnelle, n'est-il pas étonnant qu'il fit de longues méditations dans un intervalle de temps de plus d'un quart de siècle sur les fondements et les problèmes de la mécanique, en connexion avec l'enseignement qu'il avait prodigué au début de sa carrière. Son grand ouvrage *La Mécanique*, qui parut en 1969, fit suite à une série ininterrompue de mémoires scientifiques parus à partir de 1948 et ayant trait à la fondation d'une nouvelle mécanique.

Son vif intérêt pour la mécanique se manifesta par l'analyse de l'œuvre de Galilée (en 1923) et de celle de Newton (en 1936), ainsi que par des travaux concernant spécialement les applications de l'analyse fonctionnelle à la mécanique. Suivirent ses travaux sur les mouvements discontinus, en 1948, sur le mouvement brownien, en 1950, sur l'induction de la mécanique ondulatoire de celle corpusculaire (1950), d'où résultait déjà son intention de bâtir une mécanique plus générale qui soit applicable autant dans le domaine macroscopique que dans celui microscopique. Dès lors, le professeur Octave Onicescu suivit, avec passion et avec une remarquable ténacité, la construction de cette mécanique. Son mémoire de 1954 sur une mécanique nouvelle des systèmes matériels préluda à d'autres où il examina les principes variationnels des systèmes mécaniques continus, la mécanique des systèmes inertiaux, la mécanique du solide rigide, le problème de deux ou de plusieurs corps, l'univers antiminkovskien, etc. Les idées de ce savant sont hautement imprégnées de quelques principes à caractère général d'invariance qui s'explicitèrent successivement avant d'aboutir à son traité de mécanique déjà mentionné et ensuite à sa *Mécanique invariantive et Cosmologie* (1974).

Dans ses travaux de mécanique, Octave Onicescu chercha à élargir les principes classiques de Newton-D'Alembert, Lagrange, Hamilton, Cartan en tâchant de les libérer de restrictions trop étroites, afin de permettre leur adaptation et extension au cas des vitesses et des distances très grandes.

La Mécanique invariantive d'Octave Onicescu, dont les fondements furent ainsi posés entre 1950—1970, constitue une extension de la mécanique newtonienne, à laquelle elle se réduit au cas des vitesses et distances habie tuelles. L'idée essentiellement nouvelle de la mécanique proposée par Octave Onicescu est que l'énergie, c'est-à-dire la fonction d'état d'un système de points matériels qui se meuvent sous leur interaction, est un invariant euclidien de la configuration constituée par les points matériels et leurs impulsions. Le principe de base conduisant à la loi du mouvement est alors celui de Poincaré-Cartan, selon lequel la dérivée extérieure du vecteur quadri-dimensionnel d'impulsion-énergie du point matériel est nulle. En restant donc dans le cas de l'espace-temps de la mécanique classique, Octave Onicescu réussit ainsi à retrouver dans sa mécanique la loi de variation de la masse avec la vitesse, qui fut donnée par Lorentz et Einstein, ainsi que l'expression de l'énergie donnée par Einstein. Cependant, ses considérations laissent libre voie aussi pour une mécanique dans laquelle il n'existerait pas de limitation supérieure de la vitesse.

Pour un système de points matériels, les mêmes principes conduisent Octave Onicescu à faire introduire dans l'expression des impulsions une «masse d'interaction gravitationnelle» et une «masse d'interaction Hubble». Il parvient à établir une loi de conservation de l'énergie et de l'impulsion et à donner une justification théorique à la célèbre loi empirique de Hubble concernant la récession des galaxies. Ainsi, Octave Onicescu, par sa mécanique invariantive met en évidence le fait que la gravitation et la dilatation de l'Univers ne sont que des effets inertiaux de la matière.

On ne saurait trop souligner ici le mérite de ces travaux d'Octave Onicescu, qui permettent d'intégrer dans ce nouveau domaine de la mécanique des résultats bien acquis de la Science, comme par exemple la théorie du champ électromagnétique. D'autre part, par ses développements mathématiques modernes, par ses profondes réflexions et analyses des principes de la mécanique classique, par l'élargissement du cadre de cette mécanique, l'œuvre mécanique d'Octave Onicescu incite à la méditation et replace au tout premier plan de l'actualité cette mécanique analytique classique qui, depuis Hamilton et Jacobi, Poincaré et Cartan, semblait figée dans des formes sublimes et immuables.

J'ai insisté longuement ici sur les principales contributions de l'académicien Octave Onicescu dans les domaines si importants du Calcul des probabilités et de la Mécanique. Mais son œuvre scientifique est encore plus vaste puisqu'elle s'étend de l'algèbre à la topologie et à l'analyse fonctionnelle, de l'analyse complexe à l'analyse réelle et de là à l'informatique, domaines dans lesquels il a apporté des points de vue toujours nouveaux et féconds, mais sur lesquels je n'insisterai pas ici.

Etant un mathématicien-mécanicien accompli, Octave Onicescu a su trouver dans les problèmes de la mécanique des sources d'inspiration pour des concepts nouveaux, qu'il a introduits dans d'autres domaines de recherches. Il y a une quinzaine d'années, il contribua essentiellement à la création d'un nouveau modèle mathématique qui cherche à représenter la réalité non seulement au moyen de l'analyse mathématique des données empiriques mais encore par l'intermédiaire des informations de nature probabiliste que ces données comportent. C'est à cette occasion qu'il introduisit les concepts fondamentaux, et qui s'avérèrent si utiles, de «corrélation informationnelle» et d' «énergie informationnelle».

Les travaux de cette conférence se déroulent à une date qui coïncide presque avec le 90ème anniversaire de la naissance de l'académicien Octave Onicescu. Qu'il nous soit permis de saisir cette occasion afin de lui exprimer au nom des membres de la section des sciences mathématiques de l'Académie de la République Socialiste de Roumanie et, j'en suis sûr, aussi au nom des participants à cette conférence, tous nos sentiments de profonde affection et d'admiration pour son œuvre scientifique si solide, pour ses belles réalisations, tout en lui souhaitant encore de nombreuses années de bonne santé et de travail, pour qu'il continue à nous donner l'exemple, si rare de nos jours, d'une vie orientée par le travail vers la sagesse.

Nous adressons respectueusement les mêmes voeux chaleureux de santé et bonheur à sa compagne, Madame Luiza Onicescu, qui a su assurer au professeur Onicescu des conditions si propices de travail, dans un foyer familial heureux et exemplaire.

Pour terminer, je vous adresse encore une fois mes vœux chaleureux de succès dans vos travaux.

INVITED PAPERS

SH. A. AJUPOV

Probabilistic aspects of Jordan algebras

0. Introduction

The notion of a *semifield* was introduced in [15] for an algebraic description of the space of random variables in the classical probability theory. A semifield is a commutative and associative algebra over the field of reals, which is equipped with a partial ordering consistent with the algebraic operations by means of some axioms. Elements of the semifield are interpreted as random variables, indempotents of the semifield form a complete Boolean algebra and they are considered as events, a measure on this Boolean algebra — as a probability distribution.

Every semifield with a measure can be represented as an algebra of measurable functions on a measure space. Therefore this algebraic approach is equivalent to the well-known set-theoretical approach of Kolmogorov. But its advantage is that it enables us to pass in a more natural way to the conception of the "quantum" or "non-commutative" probability theory by using the notion of ordered Jordan algebras which is a "non-associative" analogue of the notion of semifields.

Ordered Jordan algebras — OJ-*algebras*—were introduced in [1], [16] for the sake of an axiomatic approach to the quantum probability theory. Elements of an OJ-algebra E are interpreted as observables (random variables), idempotents in E form a complete orthomodular lattice (logic) [21] and they are interpreted as events, etc. (for details see [5]—[9]). Our approach is a synthesis of the above mentioned approach of Sarymsakov [15] to the classical probability theory and the algebraic approach to the quantum mechanics suggested by Jordan, von Neumann & Wigner [13]. The class of OJ-algebras is sufficiently wide and contains as particular cases semifields (the associative case), the self-adjoint parts of von Neumann algebras, and Jordan algebras which appear in the formalism of the quantum mechanics. At the same time OJ-algebras have a rich enough structure for obtaining analogues of theorems of the classical probability theory in OJ-algebras. For example, mean and pointwise ergodic theorems for Markov operators on OJ-algebras are proved in [7], [9], [16]; the convergence theorems for conditional expectations and zero-one laws — in [5], [8], [9]; strong laws of large numbers — in [6], [8], etc.

The main result of the present note — the representation theorem for OJ-algebras — gives the complete structure of these algebras.

In Section 1 the definition, examples and some basic properties of OJ-algebras are given. In particular the connection is established between OJ-algebras and Jordan Banach algebras considered in [11], [18].

In Section 2 we give general constructions of special and exceptional OJ-algebras, formulate the main result and outline its proof.

1. Ordered Jordan algebras

Let E be a real Jordan algebra [12], i.e. a commutative (but not associative in general) algebra over the field R of reals in which the associativity holds in the following weak form

$$x^2(yx) = (x^2y)x \text{ for all } x, y, \text{ in } E.$$

A subalgebra E_0 in E is called *strongly associative* if $(ac)\,b = a(cb)$ for all $a, b \in E_0$, $c \in E$. Two elements $a, b \in E$ are said to be *compatible* and denoted as $a \leftrightarrow b$ if the subalgebra $J(a, b)$ generated by these elements is strongly associative [1], [16].

A partial ordering $\geqslant$ on E is *consistent* with the algebraic operations if it satisfies the following conditions:

1) $x \geqslant y$ implies $x + z \geqslant y + z$ for all $z \in E$;

2) $x \geqslant y$ implies $\lambda x \geqslant \lambda y$ for all positive $\lambda \in R$;

3) $x \geqslant 0$, $y \geqslant 0$, $x \leftrightarrow y$, implies $xy \geqslant 0$;

4) $x^2 \geqslant 0$ for all $x \in E$;

here 0 is the null element of E.

Definition 1. A real Jordan algebra E with the unit **1** is called an *OJ-algebra* if it is equipped with a partial ordering consistent with the algebraic operations and satisfies the following two conditions.

(I) if (x_α) is an increasing bounded above net of positive elements in E, then $x = \sup(x_\alpha)$ exists and $x \leftrightarrow y$ if $x_\alpha \leftrightarrow y$ for all α (the monotone completeness);

(II) every maximal strongly associative subalgebra of E is a lattice with the induced partial ordering.

Examples and basic properties

1° Every semifield [15] and, in particular, the algebra of measurable functions on a measure space are associative OJ-algebras. Moreover every maximal strongly associative subalgebra of an OJ-algebra is a semifield. Thus the partial ordering on a Jordan algebra satisfying the conditions 1)—4), (I), (II) above is unique and defined by the cone $E^+ = E^2 = \{a^2, a \in E\}$.

2° Let $\mathfrak{U}$ be a von Neumann algebra [19] on a Hilbert space H, $M(\mathfrak{U})$— the *-algebra of measurable operators affiliated with $\mathfrak{U}$ [17], [23]. Then the family $M(\mathfrak{U})_{SA}$ of all self-adjoint operators from $M(\mathfrak{U})$ forms an OJ-algebra with respect to the natural ordering [23] and the symmetrized product $a \circ b = 1/2(ab + ba)$, where ab is the ordinary associative product of the measurable operators a and b [17].

3° Let A be a JW-*algebra* [20], i.e. a weakly closed Jordan algebra of bounded self-adjoint operators on a complex Hilbert space. Then A is an OJ-algebra. In particular, the self-adjoint part of a von Neumann algebra forms an OJ-algebra.

In order to give one more example of OJ-algebras let us recall two definitions.

Definition 2. [11]. Let B be a unital Jordan algebra and a real Banach space. B is called a JB-*algebra* (Jordan Banach algebra) if (i) $||a^2|| = ||a||^2$; (ii) $||a^2|| \leqslant ||a^2 + b^2||$ for all $a, b \in B$.

Definition 3 [18]. *A* JB-algebra B is said to be a JBW-*algebra* if it is a Banach dual space, i.e. there exists a Banach space N called *predual* for B such that $N^* = B$.

It is proved by Shultz [18, Theorem 2.3], that a JB-algebra B is a JBW-algebra iff it is monotone complete and possesses a separating set of normal states; in this case the predual space for B is unique and it can be identified with the space of all normal linear functionals on B.

Every JW-algebra is an example of a special [12] JBW-algebra. The example of an exceptional [12] JBW-algebra is given by the algebra M_3^8 of Hermitian 3×3 matrices over the Cayley numbers.

4° An arbitrary JBW-algebra B is an OJ-algebra with the positive cone $B^+ = B^2 = \{b^2, b \in B\}$ (see [11, Theorem 2.1]).

An element x in an OJ-algebra is said to be *bounded* if $-\lambda 1 \leqslant x \leqslant \lambda 1$ or some positive $\lambda \in R$. In Examples 3° and 4° OJ-algebras contain bounded elements only, but in Examples 1° and 2° they may contain unbounded elements as well.

We have the following theorem.

Theorem 1 ([2], 16]). *The set B of all bounded elements in an arbitrary* OJ-*algebra E forms a monotone complete* JB-*algebra with respect to the norm*

$$||x|| = \inf \{\lambda > 0: -\lambda 1 \leqslant x \leqslant \lambda 1\}, \quad x \in B.$$

As it is proved in [3] every monotone complete JB-algebra B has a unique central idempotent e such that

(i) the JB-algebra eB has a separating set of normal states, i.e. eB is a JBW-algebra;

(ii) the JB-algebra $(1 - e) B$ does not admit any normal state.

Since the second case is unnatural from the physical point of view, *we shall assume from now on that the set of bounded elements in an* OJ-*algebra forms a* JBW-*algebra.*

Recall also [16] that the set P of all idempotents in an arbitrary OJ-algebra forms a complete *logic* [21] and a lattice (a complete orthomodular lattice) with the induced partial ordering and the orthocomplement defined as $e^{\perp} = 1 - e$, $e \in P$; moreover the set P_0 of all central idempotents is a proper Boolean subalgebra in P_0.

One of the most useful tools in the study of OJ-algebras is the spectral theorem.

Definition 4. *A spectral family* in an OJ-algebra E is a family $(e_\lambda, \lambda \in R)$ of idempotents in E such that

(i) $e_\lambda \leqslant e_\mu$ for $\lambda \leqslant \mu$;

(ii) $\inf(e_\lambda) = 0$, $\sup(e_\lambda) = \mathbf{1}$;

(iii) $\sup\limits_{\lambda<\mu}(e_\lambda) = e_\mu$ for all $\mu \in R$.

Theorem 2. (Spectral theorem [16]). *Given any element x in an OJ-algebra E there exists a unique spectral family (e_λ) such that $x = \int_{-\infty}^{+\infty} \lambda \, de_\lambda$, and for $y \in E$ we have $y \leftrightarrow x$ iff $y \leftrightarrow e_\lambda$ for all $\lambda \in R$.*

Here the integral $\int_{-\infty}^{+\infty} \lambda \, de_\lambda$ is taken in the sense of the theory of ordered spaces [22], i.e. as the order limit of the corresponding integral sums calculated in an arbitrary maximal strongly associative subalgebra containing the element x (see [16]).

As we mentioned above, elements of the OJ-algebra E are interpreted as observables, the indempotents of E which form a logic correspond to the quantum mechanical events. Furthermore the JBW-algebra B of order bounded elements in E correspond to the algebra of bounded observables, states on B correspond to the physical states. Therefore the following theorem proved by Shultz [18, Theorem 3.9] can be considered as the representation theorem for the algebra of bounded observables.

Theorem 3. *Given any* JBW-*algebra B there exists a unique central idempotent $e \in B$ such that*

(i) *the* JBW-*algebra eB is special and isomorphic to a* JW-*algebra;*

(ii) *the* JBW-*algebra $(1 - e)B$ is "pure exceptional"* [18] *and isomorphic to the algebra $C(X, M_3^8)$ of all continuous mappings from a hyperstonean space X to the exceptional* JB-*algebra M_3^8; conversely given any hyperstonean space X the algebra $C(X, M_3^8)$ is a* JBW-*algebra.*

Our main purpose is to give a representation theorem for arbitrary OJ-algebras of observables containing unbounded elements as well.

2. Construction of OJ-algebras. The representation theorem

In order to formulate the main result we have to consider two general examples of OJ-algebras.

1° The OJ-algebra of locally measurable operators

Segal [17] introduced the notion of a measurable operator affiliated with a von Neumann algebra. Then Sankaran [14] considered the algebra $S(\mathfrak{U})$) of locally measurable operators for a von Neumann algebra $\mathfrak{U}$ with countably decomposable centre. Later Yeadon [23] has extended the definition of a locally measurable operator to the case of arbitrary von Neumann algebras and proved that the set $S(\mathfrak{U})$ of locally measurable operators form a monotone complete *-algebra.

Algebras of measurable and locally measurable operators are of great importance in the non-commutative integration theory and the quantum probability theory. Since JW-algebras are the real and non-associative counterpart of von Neumann algebras (see [20]) it is natural to try to define the notion of measurable and locally measurable operator for JW-algebras. We introduce below these notions and show that they are very useful in the representation theory of OJ-algebras.

Let A be a JW-algebra in a Hilbert space H. A projection e in A is said to be *modular* if the projection lattice $[0, e] = \{f \in A : f \leqslant e\}$ is modular, i.e.

$$(f \vee g) \wedge h = f \vee (g \wedge h) \text{ for all } f, g, h \in [0, e], f \leqslant h$$

(see [20]).

A JW-algebra A is *modular* if its projection lattice is modular, i.e. **1** is a modular projection in A.

Definition 5. A self-adjoint operator T (not bounded in general) in the Hilbert space H is said to be *affiliated* with the JW-algebra A if all its spectral projections P_λ (in the spectral resolution $T = \int_{-\infty}^{+\infty} \lambda \, dP_\lambda$) lie in A.

It is clear that if a bounded self-adjoint operator T is affiliated with A then $T \in A$.

Definition 6. A self-adjoint operator T affiliated with a JW-algebra A is called

a) *measurable*, if the projections $P_{\lambda_0}^{\perp}$ and $P_{-\lambda_0}$ are modular for some positive $\lambda_0 \in R$;

b) *locally measurable*, if there exists a sequence (q_n) of central projections in A monotone increasing to **1** such that $q_n T$ is a measurable operator for all $n = 1, 2, \ldots$.

Obviously, every measurable operator is locally measurable, and if the JW-algebra A is modular then every self-adjoint operator affiliated with A is measurable.

Note also that if the JW-algebra A happens to be the self-adjoint part of a von Neumann algebra $\mathfrak{U}$ then Definition 6 coincides with the definition of measurable and locally measurable self-adjoint operators affiliated with the von Neumann algebra $\mathfrak{U}$ [23].

Theorem 4. *There exist natural algebraic operations and a partial ordering on the set $S(A)$ of locally measurable self-adjoint operators affiliated with the* JW-*algebra A, organizing $S(A)$ to an* OJ-*algebra. Moreover the* JBW-*algebra of bounded elements of $S(A)$ coincides with the* JW-*algebra A and the set $M(A)$ of all measurable self-adjoint operators forms a solid* OJ-*subalgebra in $S(A)$.*

Recall that a Jordan subalgebra E_1 in an OJ-algebra E is called a *solid* OJ-*subalgebra* if it is an OJ-algebra with respect to the induced partial ordering and for $x, y \in E$ one has that $0 \leqslant x \leqslant y \in E_1$ implies $x \in E_1$.

An OJ-algebra E is said to be *universal* if given any spectral family (e_λ) in E there exists the integral $\int_{-\infty}^{+\infty} \lambda \, de_\lambda$. Equivalently, E is universal iff every maximal strongly associative subalgebra of E is a universal semifield [15] (i.e. an extended K-space in sense of [22]).

Corollary 1. *The* OJ-*algebra $S(A)$ is universal iff the* JW-*algebra A is modular.*

2° The exceptional OJ-algebra $S(X, M_3^8)$

Let X be a hyperstonean space, $M = M_3^8 \cup \{\infty\}$ the one-point compactification of the finite dimensional JBW-algebra M_3^8. Consider the set $S(X, M_3^8)$ of all continuous mappings $f: X \to M$ such that $f^{-1}(\infty)$ is nowhere dense in X. We are going to define the algebraic operations and a partial ordering on this set. Let $f, g \in S(X, M_3^8)$, then $Y = X \setminus \{f^{-1}(\infty) \cup g^{-1}(\infty)\}$ is an open dense subset in X, and $f(x), g(x) \in M_3^8$ for $x \in Y$. Define the mapping $f+g$ from Y to M by $(f+g)(x) = f(x) + g(x)$, $x \in Y$. Since X is a hyperstonean space, it coincides with the Čech-Stone compactification of each dense subset $Y \subset X$. Therefore $f+g$ can be uniquely extended to a continuous mapping $f+g: X \to M$ and obviously $f+g \in S(X, M_3^8)$. Similarly we can define other algebraic operations on $S(X, M_3^8)$. Since M_3^8 is a Jordan algebra, $S(X, M_3^8)$ also becomes a Jordan algebra.

Consider on $S(X, M_3^8)$ the pointwise partial ordering: $f \leqslant g$ means that $f(x) \leqslant g(x)$ for all $x \in X \setminus \{f^{-1}(\infty) \cup g^{-1}(\infty)\}$, $f, g \in S(X, M_3^8)$.

Theorem 5. *The Jordan algebra $S(X, M_3^8)$ with the above partial ordering becomes a universal* OJ*-algebra with the subalgebra of bounded elements isomorphic to the* JBW*-algebra $C(X, M_3^8)$.*

We are now in position to formulate our main result.

Theorem 6. *Let E be an arbitrary* OJ*-algebra, B the* JBW*-algebra of bounded elements in E. There exists a unique central idempotent $e \in E$ such that*

(i) *the algebra eB of bounded elements in the* OJ*-algebra eE is isomorphic to a* JW*-algebra A, and the* OJ*-algebra eE is isomorphic to a solid* OJ*-subalgebra of the* OJ*-algebra $S(A)$ of all locally measurable self-adjoint operators affiliated with A;*

(ii) *the algebra $(1-e)B$ of bounded elements in the* OJ*-algebra $(1-e)E$ is isomorphic to the* JBW*-algebra $C(X, M_3^8)$, and the* OJ*-algebra $(1-e)E$ is isomorphic to a solid* OJ*-subalgebra of the universal* OJ*-algebra $S(X, M_3^8)$.*

Remark. In the particular case when the OJ-algebra E contains only bounded elements Theorem 6 coincides with Theorem 3.

The *proof* of Theorem 6 is somewhat lengthy and its details will be given in a forthcoming paper. Here we shall outline some key moments of the proof.

The assertion of Theorem 6 concerning the subalgebras of bounded elements follows from Theorem 3. Thus the proof is needed only for the assertions about unbounded elements. Namely, we must show that if an OJ-algebra E has the subalgebra of bounded elements isomorphic to a JW-algebra (respectively to the JBW-algebra $C(X, M_3^8)$), then E is isomorphic to a solid OJ-subalgebra in $S(A)$ (respectively in $S(X, M_3^8)$).

Lemma 1. *Let E and E' be* OJ*-algebras, B and B' —* JB*-algebras of bounded elements in E and E' respectively. And let $F: B \to B'$ be a Jordan isomorphism. Suppose that the following condition is satisfied*

(*) *given any $a \in E$ with a spectral family (e_λ) $\left(i.e.\ a = \int_{-\infty}^{+\infty} \lambda \, de_\lambda\right)$ there exists an element a' in E' with the spectral family $F(e_\lambda)$ $\left((i.e.\ a' = \int_{-\infty}^{+\infty} \lambda \, dF(e_\lambda)\right)$.*

Then the mapping $F': a \to a'$ extends the isomorphism F, and F' is an isomorphism between E and the solid OJ*-subagebra $F'(E)$ in E'. Moreover if the* OJ*-algebra E is universal then E' is universal too and $F'(E) = E'$.*

It should be noted that since $F: B \to B'$ is an isomorphism it maps every spectral family from B to a spectral family in B' and therefore the condition $(*)$ is well-defined. In particular, if the OJ-algebra E' is universal then $(*)$ is satisfied in a trivial way. This remark and the universality of the OJ-algebra $S(X, M_3^8)$ imply the assertion (ii) of Theorem 6.

On the other hand the OJ-algebra $S(A)$ is not necessary universal (Corollary 1). The verification of the condition $(*)$ of Lemma 1 for the OJ-algebra $S(A)$ is the most difficult moment in the proof of Theorem 6. Therefore we formulate it as a lemma.

Lemma 2. *Let E be an* OJ-*algebra, B — the* JBW-*algebra of bounded elements in E, $F : B \to A$ an isomorphism of B onto a* JW-*algebra A on a Hilbert space H. Then given any element x in E with a spectral family (e_λ), the self-adjoint operator $T = \int_{-\infty}^{+\infty} \lambda \, dF(e_\lambda)$ in H affiliated with A is locally measurable, i.e. $T \in S(A)$. In particular, the* OJ-*algebras E and E' and the isomorphism F satisfy the condition $(*)$ of Lemma 1.*

Theorems 5, 6, Lemma 1 and Corollary 1 imply the following results.

Corollary 2. *An* OJ-*algebra E can be embedded into a universal* OJ-*algebra iff the lattice of its idempotents is modular.*

Corollary 3. *Each universal* OJ-*algebra is isomorphic to the direct sum $S(A) \oplus S(X, M_3^8)$, where A is a modular* JW-*algebra, X is a hyperstonean space.*

In the particular case, when the OJ-algebra E is associative (i.e. E is a semifield), the JBW-algebra B of bounded elements of E is an associative JBW-algebra (and therefore is special). Thus B is isometrically isomorphic to the algebra $A = L_R^\infty(\Gamma, \nu)$ of all essentially bounded locally ν-measurable real functions on a localizable measure space (Γ, ν). It is not difficult to see that $A = L_R^\infty(\Gamma, \nu)$ can be considered as an associative JW-algebra on the complex Hilbert space $L^2(\Gamma, \nu)$ of all ν-locally measurable square integrable complex functions on (Γ, ν). In this case the OJ-algebra $S(A)$ of locally measurable self-adjoint operators affiliated with the JW-algebra A is universal and coincides with the algebra (the universal semifield) $L_R^0(\Gamma, \nu)$ of all locally ν-measurable real functions on (Γ, ν).

These remarks and Theorem 6 with Corollary 3 imply the following well-known result [15] mentioned in the introduction.

Corollary 4. *Let E be an associative* OJ-*algebra (i.e. a semifield). Then there exists a localizable measure space (Γ, ν) such that E is isomorphic to a solid* OJ-*subalgebra (i.e. a solid subsemifield) of the universal associative* OJ-*algebra $L_R^0(\Gamma, \nu)$. If E is universal then the* OJ-*algebras E and $L_R^0(\Gamma, \nu)$ are isomorphic.*

REFERENCES

1. Ajupov, Sh. A., *On the theory of partially ordered Jordan algebras.* Dokl. Akad. Nauk UzSSR, *8,* 6–8(1979). (Russian)
2. Ajupov, Sh. A., *OJ-algebras of bounded elements.* Izv. Akad. Nauk UzSSR Ser. Fiz.-Mat. Nauk, 2, 3–8 (1980). (Russian)
3. Ajupov, Sh. A., *Normal states on OJB-algebras.* Izv. Akad. Nauk UzSSR Ser. Fiz.-Mat. Nauk, *3,* 9–13 (1980). (Russian)
4. Ajupov, Sh. A., *Mean ergodic theorems in Jordan algebras.* Uspehi Mat. Nauk, 36, *6(222),* 201–202 (1981). (Russian).

5. Ajupov, Sh. A., *Conditional expectations and martingales on Jordan algebras.* Dokl. Akad. Nauk UzSSR, *10*, 3–5 (1981). (Russian)
6. Ajupov, Sh. A., *Probability spaces on Jordan algebras.* Third Internat. Vilnius Conf. Probab. Theory Math. Statist.: Abstracts of Communications, **III**, pp. 3–4. Vilnius, 1981.
7. Ajupov, Sh. A., *Ergodic theorems in Jordan algebras of measurable elements.* An. Univ. Craiova Mat. Fiz.-Chim., **9**, 22–28 (1981)
8. Ajupov, Sh. A., *Martingale convergence and strong laws of large numbers in Jordan algebras.* An. Univ. Craiova Mat. Fiz.-Chim., **9**, 29–34 (1981).
9. Ajupov, Sh. A., *Independence and Markov processes in probability spaces on Jordan algebras.* In: *Limit Theorems for Stochastic Processes and Related Problems*, pp. 28–41, Tashkent, Fan, 1982. (Russian)
10. Ajupov, Sh. A., *Ergodic theorems for Markov operators on Jordan algebras. I & II.* Izv. Akad. Nauk UzSSR Ser. Fiz.-Mat. Nauk, *4* (1982). (Russian)
11. Alfsen, E. M., Shultz, F. W., Stormer, E., *A Gelfand-Neumark theorem for Jordan algebras.* Adv. in Math. **28**, 11–56 (1978).
12. Jacobson, N., *Structure and Representations of Jordan Algebras.* Providence, R. I., Amer. Math. Soc., 1968.
13. Jordan P., von Neumann, J., Wigner, E., *On an algebraic generalization of the quantum mechanical formalism.* Ann. of Math., **35**, 29–64 (1934).
14. Sankaran, S., *The*-algebra of unbounded operators.* J. London Math. Soc., **34**, 337–344(1959).
15. Sarymsakov, T. A., *Semifields and probability theory.* In: *Proc. Third Japan-USSR Sympos. Probab. Theory.* Lecture Notes in Math., Vol. 550, pp. 524–549. Berlin–Heidelberg–New York, Springer, 1976.
16. Sarymsakov, T. A., Ajupov, Sh. A., *Partially ordered Jordan algebras.* Dokl. Akad Nauk SSSR, **249**, 789–792 (1979). (Russian)
17. Segal, I., *A non commutative extension of abstract integration.* Ann. of Math., **57**, 401–457 (1953).
18. Shultz, F. W., *On normed Jordan algebras which are Banach dual spaces.* J. Funct. Anal. **31**, 360–376 (1979).
19. Takesaki, M., *Theory of Operator Algebras. I.* Berlin-Heidelberg-New York, Springer, 1979.
20. Topping, D., *Jordan algebras of self-adjoint operators.* Mem. Amer. Math. Soc., **53**, 1–48 (1965).
21. Varadarajan, V. S., *Probability in physics and a theorem on simultaneous observability.* Comm. Pure Appl. Math., **15**, 189–217 (1962).
22. Vulih, B. Z., *Introduction to the Theory of Partially Ordered Spaces.* Moscow, Fizmatgiz, 1961. (English transl. Noordhoff, 1967)

WILLEM ALBERS

Rates of convergence and expansions for rank tests: A review and some remarks on the nuisance parameter case

In the classical case of sums of independent random variables well-known refinements of the central limit theorem are Berry-Esséen-bounds and Edgeworth expansions. By now, similar refinements have been obtained for various classes of widely used rank statistics (one-sample case, two-sample case and simple linear rank statistic). In this paper a short review is given of recent developments in this area. Moreover, some attention is devoted to indicating how such results can be extended to cover cases where nuisance parameters are present.

1. Introduction

Let $X_1, \ldots, X_N$ be independent random variables (rv's) with continuous distribution functions (df's) $F_1, \ldots, F_N$. Let R_j be the rank of X_j among $(X_1, \ldots, X_N)$, for $j = 1, \ldots, N$. Then we need a vector of scores $(a(1), \ldots, a(N))$ and a vector of regression constants $(c_1, \ldots, c_N)$ to define the simple linear rank statistic

(1.1) $$T_N = \sum_{j=1}^{N} c_j a(R_j),$$

which can be used to test the hypothesis of randomness $F_1 = \ldots = F_N$. An important special case is the so-called two-sample problem, in which for some $m < N$ we have $F_1 = \ldots = F_m = F$ and $F_{m+1} = \ldots = F_N = G$. Then we choose $c_j = 0$, $j = 1, \ldots, m$, $c_j = 1$, $j = m + 1, \ldots, N$, and (1.1) reduces to the two-sample statistic

(1.2) $$T_N = \sum_{j=m+1}^{N} a(R_j).$$

Closely related to the two-sample problem is the one-sample problem. Here we have $F_1 = \ldots = F_N = G$ and the objective now is to test the hypothesis that the distribution determined by G is symmetric about zero. Let $\tilde{R}_j$ denote the rank of $|X_j|$ among $|X_1|, ..., |X_N|$ then the appropriate statistic for this case is

$$T_N = \sum_{j=1}^{N} a(\tilde{R}_j) \text{ sign } X_j. \tag{1.3}$$

The rank tests based on the statistics from (1.1)—(1.3) are widely used since these are distributionfree for the hypotheses concerned and moreover easy to apply. The choice of the scores and, in the first case, of the regression constants, is determined by the kind of alternative hypothesis one has in mind. In view of the importance of these tests, it is of considerable interest to obtain approximations to the critical values, and also to the powers under appropriate alternatives. Consequently, a lot of effort has been devoted by many authors to proving asymptotic normality of the statistics (1.1)—(1.3), under varying conditions on the F_j, $a(j)$ and c_j. A lot of information on this subject and many references are given by e.g. Hájek and Šidák (1967) and Serfling (1980). Here we shall not dwell on it any longer but move on directly to the topic which is of main interest in this review.

To that end we observe that asymptotic normality results will clearly provide us merely with first order approximations to the quantities of interest mentioned above. In view of this, it is worthwhile to try to obtain additional information. In the first place, we can try to fin dout how good normal approximations actually are, by obtaining the rate of convergence involved. Secondly, we can try to improve on the normal approximation by replacing it by a more sophisticated approximation which takes higher order terms into account. For the classical case, which concerns sums of independent rv's, this program has been carried out and the results are readily available, see e.g. Cramér (1962) and Feller (1971). It turns out that here the rate of convergence is $N^{-1/2}$ (the so-called Berry-Esséen bound), while finer approximations can be obtained by using the so-called Edgeworth expansions, which are obtained from the normal case by adding correction terms of order $N^{-k/2}$, $k = 1, 2, ..., r$. Typically the case $r = 2$ is of special interest, since it then becomes possible to evaluate deficiencies in the sense of Hodges and Lehmann (1970), which are useful in discriminating between tests that are equivalent to first order.

The objective now is to sketch the progress which has been made in extending the refinements mentioned above from the classical case to that of the statistics (1.1)—(1.3). We shall concentrate on the developments since the review paper by Bickel (1974) on expansions in nonparametric statistics. In Section 2 we consider the Berry-Esséen-type results, whereas the expansions will be dealt with in Section 3. Finally, in Section 4 we shall indicate a possible way of extending the existing results to cases involving nuisance parameters.

2. Rates of convergence

Quite a few authors have been concerned with obtaining bounds of Berry-Esséen (BE) type for the df of the simple linear rank statistic from (1.1). They typically assume that the scores are generated from a function J on (0,1) in one of the following two ways

$$a(j) = EJ(U_{j:N}) \text{ or } a(j) = J\left(\frac{j}{N+1}\right), \tag{2.1}$$

where $U_{1:N} < \dots < U_{N:N}$ are the order statistics of a sample of size N from the uniform df on (0, 1). The scores in (2.1) are called exact and approximate scores, respectively. Most authors impose rather heavy conditions on J: it is usually supposed to have a bounded derivative of a certain order (1,2 or even 4).

The simplest way of getting results on the rate of convergence is to extend the standard method of proving asymptotic normality for T_N. This method consists of writing T_N as

$$T_N = S_N + R_N, \tag{2.2}$$

where S_N is a sum of independent rv's and $R_N = o_P(|S_N|)$. Hence S_N is asymptotically normal in virtue of the classical theory, the remainder R_N is negligible and thus T_N is asymptotically normal as well. But now we realize that, as was mentioned in the previous section, the classical theory also provides a BE bound for S_N. Moreover, a closer look at R_N will typically reveal that it is actually $O_P(N^{-1/2}|S_N|)$, after which the BE bound for T_N seems to be within reach.

The first paper we mention in this connection is the one by Erickson and Koul (1976). They do arrive at $N^{-1/2}$ as the rate of convergence, but this is in L_1-norm, however. To be more precise, if F_N^* stands for the df of $T_N^* = (T_N - ET_N)/\sigma(T_N)$ and Φ stands for the standard normal df, then they show that

$$||F_N^* - \Phi||_1 = O(N^{-1/2}). \tag{2.3}$$

As $||F_N^* - \Phi||_\infty \leqslant (4||F_N^* - \Phi||_1/5)^{1/2}$, this leads to a BE bound of order $N^{-1/4}$ in the customary sup-norm interpretation, which still is rather unsatisfactory.

A first improvement over this results was obtained by Jurečková and Puri (1975), who replace $N^{-1/4}$ by $N^{-1/2+\delta}$ for $\delta > 0$ arbitrarily small. In doing so, they restrict attention to the hypothesis and contiguous alternatives $F_j(x) = F(x - \theta d_j)$, with $\sum_{j=1}^{N} d_j = 0$ and $\sum_{j=1}^{N} d_j^2 = 1$. The idea of the proof is to use an inequality like

$$F_N^*(x) \leqslant P((S_N - ET_N)/\sigma(T_N) \leqslant x + \varepsilon_N) + P(|R_N|/\sigma(T_N) > \varepsilon_N) \tag{2.4}$$

together with a similar one in the opposite direction. Then let $S_N^* = (S_N - ES_N)/\sigma(S_N)$, and show that $E(T_N^* - S_N^*)^{2k} = O(N^{-k})$, $k = 1, 2, \dots$, if S_N is chosen in a suitable way. (Under the hypothesis for example the choice $S_N = \sum_{j=1}^{N} c_j J(F(X_j))$ will do.) Application of this result and of the BE bound for S_N to (2.4) gives a bound of order $N^{-1/2} + \varepsilon_N + N^{-k}\varepsilon_N^{-2k}$ for T_N. An appropriate choice of ε_N then leads to the desired result.

The next improvement is due to Bergström and Puri (1977) who also arrive at $N^{-1/2+\delta}$, but now for general F_j. This is achieved by using the projection method due to Hájek (1968), rather than the methods for contiguous alternatives from Hájek and Šidák (1967). The role of S_N is now played by $\sum_{j=1}^{N} E(\tilde{T}_N|X_j)$, where $\tilde{T}_N$ is the following approximation to $T_N - ET_N$

$$\tilde{T}_N = \sum_{j=1}^{N} c_j\{J(\tau_{jj}) - EJ(\tau_{jj}) + (\tau_j - \tau_{jj})\, J'(\tau_{jj})\}, \tag{2.5}$$

where $\tau_j = R_j/N$ and $\tau_{jj} = E(\tau_j | X_j)$. After showing that both $E(T_N - \tilde{T}_N)^{2k}$ and $E(\tilde{T}_N - S_N)^{2k}$ are sufficiently small, the result again follows through (2.4).

The desired rate of $N^{-1/2}$ rather than $N^{-1/2+\delta}$ was achieved by Hušková (1977) for the hypothesis and contiguous alternatives. Just as in the approach by Jurečková and Puri (1975), T_N is compared to $S_N = \sum_{j=1}^{N} c_j J(F(X_j))$ under the hypothesis. However, to get rid of the last δ a crude inequality like (2.4) no longer suffices. A much more delicate analysis of $R_N = T_N - S_N$ is required, applying the same kind of techniques that are used to derive the classical BE bound. In fact, from the well-known smoothing lemma (see Esséen (1945)), it follows as a first step that for all $T > 0$

$$\text{(2.6)} \qquad \sup_x | F_N^*(x) - \tilde{F}_N(x) | = O\left(\int_{-T}^{T} |t|^{-1} | \rho_N(t) - \tilde{\rho}_N(t) | dt + T^{-1} \right),$$

where at present $\tilde{F}_N$ simply stands for Φ (in the next section, however, it will be an expansion with Φ as leading term) and $\rho_N(\tilde{\rho}_N)$ is the Fourier-Stieltjes transform of $F_N^*(\tilde{F}_N)$. Hence in particular, ρ_N is the characteristic function (cf) of T_N^*, whereas $\tilde{\rho}_N$, at least in the present section, is the cf of Φ, i.e. $\tilde{\rho}_N(t) = \exp(-t^2/2)$.

To obtain a BE bound for T_N it now clearly suffices to show that the integral in (2.6) is $O(N^{-1/2})$ for some T such that $T^{-1} = O(N^{-1/2})$. The method Hušková applies to prove this last result is closely related to the one used by Bjerve (1974) and Bickel (1974), and a very transparent sketch of this kind of approach is given by van Zwet (1977). It is observed that first the representation for T_N from (2.2) has to be replaced by one in which the remainder R_N itself is split up into a leading term Q_N and a new remainder, which will again be denoted by R_N. Hence

$$\text{(2.7)} \qquad T_N = S_N + Q_N + R_N, \text{ or } T_N^* = \bar{S}_N + \bar{Q}_N + \bar{R}_N,$$

where $\bar{S}_N = (S_N - ES_N)/\sigma(T_N)$, $\bar{Q}_N = (Q_N - EQ_N)/\sigma(T_N)$ and $\bar{R}_N = (R_N - ER_N)/\sigma(T_N)$ (Hušková for example uses $Q_N = \sum c_j J'(F(X_j)) \{R_j - E(R_j | X_j)\}/(N+1)$). From (2.7) an expansion for ρ_N follows:

$$\text{(2.8)} \qquad \rho_N(t) = E e^{it\bar{S}_N}\left(1 + \ldots + \frac{(it\bar{Q}_N)^{m-1}}{(m-1)!}\right) + O\left(\frac{|t|^m}{m!} E|\bar{Q}_N|^m + |t| E |\bar{R}_N|\right).$$

Now typically (cf. van Zwet (1977))

$$E e^{it\bar{S}_N} - e^{-t^2/2} = O(N^{-1/2}|t| e^{-t^2/4}), \quad |E e^{it\bar{S}_N}\bar{Q}_N| =$$

$$\text{(2.9)} \qquad = O(N^{-1/2} e^{-t^2/}), \quad E\bar{Q}_N^2 = O(N^{-1}), \quad E|\bar{R}_N| = O(N^{-1}),$$

in which use has been made of the fact that $\bar{S}_N$ is a sum of independent rv's and of the almost independence of Q_N and S_N. From (2.8) with $m = 2$ and (2.9), it then follows that the integral in (2.6) is indeed of order $N^{-1/2}$, as long as $T \leqslant N^{\varepsilon}$ for some $0 < \varepsilon \leqslant 1/4$.

To deal with the integral over the remaining region $N^{\varepsilon} \leqslant |t| \leqslant \delta N^{1/2}$ for some $\delta > 0$, it is noted that typically, for certain positive constants η and c and for $k = o(N)$ and $m < N$,

$$\text{(2.10)} \qquad |E e^{it\bar{S}_N}\bar{Q}_N| = O((kN)^k e^{-\eta t^2}), \quad E|\bar{Q}_N|^m = O((cm)^m N^{-m/2}).$$

The desired result now follows by applying (2.8) with $m = [\log N]$ and choosing δ sufficiently small (cf. van Zwet (1977)).

The results above are extended by Hušková (1979) to the case of general F_j by taking the paper by Bergström and Puri (1977) as a starting point, rather than the one by Jurečková and Puri (1975).

The final improvement up to date is due to Does (1982), who obtains a BE bound under the hypothesis under substantially milder conditions on J, thus including for example the normal quantile function.

3. Asymptotic expansions

From the previous section it is in principle clear how an expansion for F_N^* can be obtained. The decomposition (2.7), which replaced the crude (2.2), led to an expansion with remainder in (2.8) for ρ_N, which together with (2.9) implied that $|\rho_N(t) - \exp(-t^2/2)| = O(N^{-1/2}|t| \exp(-t^2/4))$ for $|t| \leqslant N^\varepsilon$, $0 < \varepsilon \leqslant 1/4$. If now (2.7) in its turn is replaced by a finer decomposition, a more sophisticated expansion $\tilde{\rho}_N$ plus remainder can be obtained for ρ_N, and it will e.g. follow that $|\rho_N(t) - \tilde{\rho}_N(t)| = o(N^{-1}|t| \exp(-t^2/4))$, again for $|t| \leqslant N^\varepsilon$. Then, if $\tilde{F}_N$ now is the Fourier inverse of $\tilde{\rho}_N$, the integral in (2.6) will be $o(N^{-1})$ rather than $O(N^{-1/2})$, again for $T \leqslant N^\varepsilon$.

Hence, just as in the previous section, it remains to deal with the integral over the remaining region. But here the analogy stops: the device from (2.10) breaks down beyond $N^{1/2}$ (cf. again van Zwet (1977)), while now T has to satisfy $T^{-1} = o(N^{-1})$, at least if we want to show that $\tilde{F}_N$ is an expansion to $o(N^{-1})$ for F_N^*. Consequently, a different approach is needed. As $\int |t|^{-1}|\tilde{\rho}_N(t)|\, dt$ is negligible for $|t| \geqslant N^\varepsilon$, the problem reduces to showing that

$$\int_I |t|^{1-} |\rho_N(t)|\, dt = o(N)^{-1}, \tag{3.1}$$

where $I = \{t | N^\varepsilon \leqslant |t| \leqslant T\}$ for some T such that $T^{-1} = o(N^{-1})$. (Incidentally, the problem of proving (3.1) already occurs in the classical case of independent rv's, where it is dealt with by imposing the so-called Cramér's condition (C)). Typically (3.1) will be satisfied if cases are excluded where T_N^* has a lattice distribution the span h of which does not satisfy $h = o(N^{-1})$. For then F_N^* will have jumps which are not of a smaller order than N^{-1} and approximation to $o(N^{-1})$ by means of the continuous function $\tilde{F}_N$ clearly is impossible. For the case of the simple linear rank statistic from (1.1) it is shown by van Zwet (1982) that (3.1) holds under very mild conditions on the c_j, $a(j)$ and F_j.

After the general outline above, we shall now briefly mention some of the papers which have appeared on the subject. Unlike the Berry-Esséen case, an attack of T_N from (1.1) was postponed until the special — and simpler — statistics from (1.2) and (1.3) had been dealt with. Our starting point is the one-sample problem, which was considered by Albers, Bickel and van Zwet (1976). The special case of Wilcoxon's signed rank test was also considered by Prášková (1976.) We shall be very brief about this case, as Bickel (1974) already gave a preview of the results. The basic tool is the observation that the statistic from (1.3) is a sum of independent rv's conditional on the vector $Z = (Z_1, ..., Z_N)$ of order statistics of $(|X_1|, ..., |X_N|)$. This makes it possible to use a direct approach here, without having to expand the statistic T_N^* in the manner described above (cf. (2.7)). For, by applying the classical

Edgeworth expansions for sums of independent rv's, an expansion can be established for the df of T_N^*, conditional on Z, from which an unconditional expansion then follows by taking expectations. The result is an expansion to $o(N^{-1})$, under the hypothesis and under contiguous location alternatives, under mild conditions on J, including e.g. the normal quantile function. Contiguous alternatives in general were considered by Albers (1974), and a specialization to contiguous nonparametric alternatives was given by Albers (1979 b). Finally, expansions for one-sample rank tests under restricted adaptation were considered by Albers (1979a, 1980).

The two-sample statistic from (1.2) has been studied by Bickel and van Zwet (1978). The result for the case of the hypothesis was also proved independently by Robinson (1978). Of course, the two-sample case is closely related to the one-sample case, but there are a few additional complications. In the first place, if we now condition on the order statistics $Z = (Z_1, ..., Z_N)$ of $X_1, ..., X_N$, the statistic T_N from (1.2) is distributed as a sum of independent rv's, conditional on another such sum. Hence, additional steps are required before the classical theory of expansions can be applied. The second difference lies in the fact that in the one-sample problem we are always dealing with symmetric distributions, which is not necessarily the case in the two-sample problem. Consequently, in the latter case the terms of order $N^{-1/2}$ typically do not vanish.

Before going to the simple linear rank statistic, we note in passing that the special case of Kendall's τ under the hypothesis has been considered by Prašková-Vizková (1976), also using a direct approach. Some additional remarks on this case were made by Albers (1978). As concerns the simple linear rank statistic from (1.1.), expansions for this case have been obtained by Does (1982). He uses the method involving the expansion of the statistic T_N^* itself, as outlined at the beginning of this section, and he invokes the result by van Zwet (1982) to deal with the cf for large t. His results include an expansion under the hypothesis under mild conditions on J, comparable to those used for his BE bound, and an expansion under contiguous location alternatives under stronger conditions on J, comparable to those used by the other authors mentioned in the previous section.

4. Nuisance parameters

In this last section we shall devote some attention to the question how the results of the previous sections might be extended to cases involving nuisance parameters. As a motivating example we shall use the two-sample scale problem. Thus, let $X_1, ..., X_m$ have continuous df $F(x - \mu)$ and let $X_{m+1}, ..., X_N$ have df $F((x - \nu)/\sigma)$. Then, if $\mu = \nu$, we can test H_0: $\sigma = 1$ in a standard manner by means of a two-sample rank statistic (cf. (1.2)) $T_N = \sum_{j=m+1}^{N} a_N(R_j) - \lambda \sum_{i=1}^{N} a_N(i)$, where $\lambda = 1 - m/N$. Suppose that the scores are as in (2.1). Well-known examples are the Ansari-Bradley test, which has $J(t) = \frac{1}{2} - \left| t - \frac{1}{2} \right|$, and the Klotz test, which has $J(t) = \{\Phi^{-1}(t)\}^2$.

Quite often, however, the assumption that $\mu = \nu$ cannot be trusted. But without it, no rank test can perform in a satisfactory manner (see Hájek and Šidák (1967), p. 94). A natural remedy is mentioned e.g. by Lehmann

(1975), p. 104: let $\hat{\mu}$ and $\hat{\nu}$ be suitable estimators of the nuisance parameters μ and ν, respectively, then one should replace T_N by

$$\hat{T} = \sum_{j=m+1}^{N} a_N(\hat{R}_j) - \lambda \sum_{i=1}^{N} a_N(i), \tag{4.1}$$

where $\hat{R}_1, \ldots, \hat{R}_N$ are the ranks of the adjusted variables $X_j - \hat{\mu}, j = 1, \ldots, m,$ $X_j - \hat{\nu}, j = m + 1, \ldots, N$. In view of the translation invariance of these adjusted variables we can without loss of generality assume that $\mu = \nu = 0$ in the sequel.

Now the use of $\hat{\mu}$ and ν will clearly affect the distribution of the test statistic, and the best we can hope for is that the asymptotic laws of $\hat{T}_N$ and T_N will coincide. Conditions under which this is indeed the case have been obtained by several students of Lehmann (see once more Lehmann (1975), p. 105) and by Hájek (1970). It turns out that one typically needs the following symmetries

$$F(-x) = 1 - F(x), \ -\infty < x < \infty, \ J(1-t) = J(t), \ 0 < t < 1. \tag{4.2}$$

Incidentally, note that if the first of these conditions holds, the second is automatically fulfilled for the LMP score function corresponding to F, which is given by $J(t) = -1 - F^{-1}(t) f'(F^{-1}(t))/f(F^{-1}(t))$, in which f' is the derivative of the density f of F. Jurečková (1979) has shown that under asymmetric F the asymptotic law of $\hat{T}_N$ will indeed differ from that of T_N and will moreover depend on F, even under H_0.

Having thus observed the asymptotic equivalence to first order under (4.2) of $\hat{T}_N$ and T_N, it is only natural in the context of the present paper to wonder whether e.g. a BE bound for $\hat{T}_N$, preferably of the classical order $N^{-1/2}$, will hold under (4.2) (and suitable additional regularity conditions). In trying to answer this question, we encounter the following problem. In the short review given in the previous sections, we have tried to focus on the main ideas underlying the papers discussed and we have omitted technicalities whenever possible. However, the papers involved are in fact of a highly technical nature, as the reader may easily verify by consulting some of the references. In view of this situation, the idea of attacking the "doubly" complicated statistic $\hat{T}_N$ seems rather uninviting.

For this reason, we would like to point out here a possible way of solving the problem by splitting it up into parts in a natural manner. To this end, we observe that we not only know that $\hat{T}_N - T_N$ is small, but also that $T_N - S_N$ is small for some suitably chosen classical statistic S_N, e.g. $S_N = \sum_{j=m+1}^{N} J(F(X_j)) - \lambda \sum_{i=1}^{N} J(F(X_i))$ (cf. Section 2). Moreover, for $\hat{T}_N$ a classical counterpart, say $\hat{S}_N$, can be found in a similar way. To be slightly more specific, the approximation of $J(R_j/(N+1))$ by $J(F(X_j))$ is based on the following observations. Let $u(x) = 1$ for $x \geqslant 0$ and let $u(x) = 0$ for $x < 0$, then $R_j = \sum_{k=1}^{N} u(X_j - X_k)$. Hence under H_0 we have that $E(R_j/(N+1) \mid X_j) = (1 + (N-1) F(X_j))/(N+1) \approx F(X_j)$. Next we note that

$$\hat{R}_j = \sum_{k=1}^{N} u(X_j - X_k - \delta_{jk}(\hat{\mu} - \hat{\nu})),$$

with $\delta_{jk} = 1$ for $j \leqslant m < k$, $\delta_{jk} = -1$ for $k \leqslant m < j$ and $\delta_{jk} = 0$ otherwise, which suggests approximation of $J(\hat{R}_j/(N+1))$ through $J((1-\lambda)F(X_j) + \lambda F(X_j - (\hat{\mu} - \hat{\nu})))$ for $j \leqslant m$ and by $J((1-\lambda)F(X_j + (\hat{\mu} - \hat{\nu})) + \lambda F(X_j))$ for $j > m$.

The idea now is that the difference $\hat{T}_N - T_N$ will in first order be equal to the corresponding classical difference $\hat{S}_N - S_N$, which suggests a decomposition like

$$\hat{T}_N = S_N + (T_N - S_N) + (\hat{S}_N - S_N) + R_{0N}, \tag{4.3}$$

where $R_{0N} = \{(\hat{T}_N - T_N) - (\hat{S}_N - S_N)\}$ hopefully is $O_P(N^{-1}|S_N|)$. Next we recall that in (2.7) a three-term decomposition of T_N was given. Clearly, the same can be done for $\hat{S}_N$. Therefore we shall write here $T_N = S_N + Q_{1N} + R_{1N}$ and $\hat{S}_N = S_N + Q_{2N} + R_{2N}$. Using a standardization similar to the one in (2.7), we then replace (4.3) by

$$\hat{T}_N^* = \bar{S}_N + \sum_{i=1}^{2} \bar{Q}_{iN} + \sum_{i=0}^{2} \bar{R}_{iN}. \tag{4.4}$$

Now note that in (4.4) we again have the same structure as in (2.7). Hence a BE bound for $\hat{T}_N$ will hold if conditions (2.9) and (2.10) are satisfied, with $\sum_{i=1}^{2} \bar{Q}_{iN}$ and $\sum_{i=0}^{2} \bar{R}_{iN}$ playing the roles of $\bar{Q}_N$ and $\bar{R}_N$, respectively. Inspection of these conditions readily reveals that they will be satisfied if (i) (2.9) and (2.10) hold for the case of $T_N = S_N + Q_{1N} + R_{1N}$, (ii) the same is true for $\hat{S}_N = S_N + Q_{2N} + R_{2N}$, and (iii)

$$E|\bar{R}_{0N}| = O(N^{-1}), \quad |Ee^{it\bar{S}_N}\bar{Q}_{1N}^{j}\bar{Q}_{2N}^{k-j}| = O((kN)^k e^{-\eta t^2}), \tag{4.5}$$

for $k = o(N)$ and some $\eta > 0$.

Since, as we saw in Section 2, the BE bound for T_N has already been established, the problem of obtaining such a bound for $\hat{T}_N$ now has been reduced to finding one for its simpler classical counterpart $\hat{S}_N$, provided that it is possible to show that (4.5) holds. Note that the analysis of $\hat{S}_N$ will be quite similar to that of T_N. Taylor expansion will lead to $\hat{S}_N - S_N = \hat{\mu} Z_{1N} + \hat{\nu} Z_{2N} + \ldots$, and $\hat{\mu}$ and $\hat{\nu}$ will typically satisfy $\hat{\mu} = N^{-1}Z_{3N} + \ldots$, $\hat{\nu} = N^{-1}Z_{4N} + \ldots$, . Here Z_{iN}, $i = 1, \ldots, 4$ are sums of independent rv's with zero expectation. Hence $\bar{Q}_{2N}$ will be proportional to $N^{-3/2}(Z_{1N}Z_{3N} + Z_{2N}Z_{4N})$, which is indeed $O_P(N^{-1/2})$.

As concerns (4.5), we observe that it should be possible to verify the second statement relatively easily: the cases $j = k$ and $j = 0$ have already been dealt with in obtaining the bounds for T_N and $\hat{S}_N$, respectively. Since the argument involved there is the almost independence of S_N and Q_{1N} and of S_N and Q_{2N} respectively, a similar argument should work without additional effort for the intermediate cases $0 < j < k$.

To prove the first statement in (4.5), it seems indicated to proceed as in proving that $E\bar{Q}_{1N}^2 = O(N^{-1})$. Remembering that $Q_{1N} = \sum_{j=1}^{N} c_j J'(F(X_j)) \times$

$\times (R_j - E(R_j | X_j))/(N+1)$, while $R_j = \sum_{k=1}^{N} u(X_j - X_k)$, we see that evaluation of EQ^2_{1N} involves evaluation of expectations of expressions like

$$J'(F(X_i))\, J'F(X_j))\{(u(X_i - X_k) - F(X_i)\}\{u(X_j - X_l) - F(X_l)\},\ k \neq i,\ l \neq j.$$

By conditioning on X_i and X_j it is observed that such expectations are 0 for $k \neq l$. Now for $\tilde{R}_{0N}$ a similar but more complicated approach can by tried, which involves replacing factors like $\{u(X_i - X_k) - F(X_i)\}$ by factors like $\{u(X_i - X_k - \delta_{1k}(\hat{\mu} - \hat{\nu})) - u(X_i - X_k) - F(X_i - \delta_{ik}(\hat{\mu} - \hat{\nu})) + F(X_i)\}$. Then again a conditioning argument can be applied, but now on all rv's except on X_k and X_l, after which one can proceed by using that $\hat{\mu} = N^{-1}Z_{3N} + + \ldots$ and $\hat{\nu} = N^{-1}Z_{4N} + \ldots$, together with the fact that J has bounded derivatives of a suitable order.

A final remark we add to the sketch above is that for the special case where $\hat{\mu}$ and $\hat{\nu}$ are the sample medians the desired result has actually been proved by Albers (1982). The method employed differs from the one sketched above: it is a direct approach exploiting the special structure of this case.

REFERENCES

1. Albers, W., *Asymptotic Expansions and the Deficiency Concept in Statistics.* Math. Centre Tracts, 58. Amsterdam, Math. Centrum, 1974.
2. Albers, W., Bickel P. J., van Zwet, W. R., *Asymptotic expansions for the power of distributionfree tests in the one-sample problem.* Ann. Statist., 4. 108—156 (1976).
3. Albers, W., *A note on the Edgeworth expansion for the Kendall rank correlation coefficient.* Ann. Statist., 6, 923—925 (1978).
4. Albers, W., *Asymptotic deficiences of one-sample rank tests under restricted adaptation.* Ann. Statist., 7, 944—954 (1979 a).
5. Albers, W., *Asymptotic expansions for the power of one-sample rank tests against contiguous nonparametric alternatives.* In: *Proc. 2 nd Prague Sympos. Asymptotic Statistics,* pp. 105—117. Amsterdam, North Holland, 1979 b.
6. Albers, W., *Asymptotic expansions for the power of adaptive rank tests in the one-sample problem.* In: *Statistique non Paramétrique Asymptotique.* Lecture Notes in Math., Vol. 821, pp. 108—158, Berlin—Heidelberg—New York, Springer, 1980.
7. Albers, W., *A Berry-Esséen bound for scale rank tests with nuisance medians.* Technical Report, Twente University of Technology, 1982.
8. Bergström, H., Puri, M. L., *Convergence and remainder terms in linear rank statistics.* Ann. Statist., 5, 671—680 (1977).
9. Bickel, P. J., *Edgeworth expansions in nonparametric statistics.* Ann. Statist., 2, 1—20 (1974).
10. Bickel, P. J., van Zwet, W. R., *Asymptotic expansions for the power of distributionfree tests in the two-sample problem.* Ann. Statist., 6, 937—1004 (1978).
11. Bjerve, S., *Error bounds and asymptotic expansions for linear combinations of order statistics.* Unpublished Ph. D. Thesis, University of California, Berkeley, 1974.
12. Cramér, H., *Random Variables and Probability Distributions, 2nd Ed.* Cambridge, Cambridge University Press, 1962.
13. Does, R.J.M.M., *Higher Order Asymptotics for Simple Linear Rank Statistics.* Math. Centre Tracts, 151. Amsterdam, Math. Centrum, 1982.
14. Erickson, R. V., Koul, H. L., *L_1 rates of convergence for linear rank statistics.* Ann. Statist., 4, 771—774 (1976).
15. Esséen, C. G., *Fourier analysis of distribution functions.* Acta Math., 77, 1—125 (1945).
16. Feller, W., *An Introduction to Probability Theory and Its Applications, Vol.* 2, *2nd Ed.*, New York, Wiley, 1971.
17. Hájek, J., Šidák, Z., *Theory of Rank Tests.* Prague, Academia, 1967.
18. Hájek, J., *Asymptotic normality of simple linear rank statistics under alternatives.* Ann. Math. Statist., 39, 325—346 (1968).
19. Hájek, J., *Miscellaneous problems of rank test theory.* In: Puri, M. L. (Ed.), *Nonparametric Techniques in Statistical Inference.* Cambridge, Cambridge University Press, 1970.

20. Hodges, J. L., Jr., Lehmann, E. L., *Deficiency*. Ann. Math. Statist., **41**, 783—801 (1970).
21. Hušková, M., *The rate of convergence of simple linear rank statistics under hypcthesis and alternatives*. Ann. Statist., **5**, 658—670 (1977).
22. Hušková, M., *The rate of convergence of simple linear rank statistics under alternatives*. In: *Contributions to Statistics*, pp. 99—108. Dordrecht, Reidel, 1979.
23. Jurečková, J., Puri, M. L., *Order of normal approximation for rank test statistics distribution*. Ann. Probab., **3**, 526—533 (1975).
24. Jurečková, J., *Nuisance medians in rank testing scale*. In: *Contributions to Statistics*, pp. 109—118. Dordrecht, Reidel, 1979.
25. Lehmann, E. L., *Nonparametrics: Statistical Methods Based on Ranks*. San Francisco, Holden-Day, 1975.
26. Prašková, Z., *Asymptotic expansion and a local limit theorem for the signed Wilcoxon statistic*. Comment. Math. Univ. Carolin., **17**, 335—344 (1976).
27. Prašková-Vizková, Z., *Asymptotic expansion and a local limit theorem for a function of the Kendall rank correlation coefficient*. Ann. Statist., **4**. 597—607 (1976).
28. Robinson, J., *An asymptotic expansion for samples from a finite population*. Ann.Statist., **6**, 1005—1011 (1978).
29. Serfling, R. J., *Approximation Theorems of Mathematical Statistics*. New York, Wiley, 1980.
30. van Zwet, W. R., *Asymptotic expansions for the distribution functions of linear combinations of order statistics*. In: *Statistical Decision Theory and Related Topics II*, pp. 421—437. New York, Academic Press, 1977.
31. van Zwet, W. R., *On the Edgeworth expansion for the simple linear rank statistic*. In: Gnedenko, B. V., Puri, M. L., Vincze, I. (Eds.), *Nonparametric Statistical Inference*. Colloq. Math. Soc. J. Bolyai, Vol. 32, pp. 889—909. Amsterdam, North-Holland, 1982.

SUJIT K. BASU*, PRANAB KUMAR SEN **

Asymptotically efficient estimator for the index of a stable distribution

Based on the sample characteristic function, a method of estimating the index parameter of a stable distribution is considered. Weak convergence of the empirical characteristic function (process) is incorporated in the choice of an asymptotically optimal estimator and in the study of its asymptotic properties.

1. Introduction

Although the recent years have witnessed some developments in the area of inference concerning parameters of the *stable laws*, the efforts have hardly matched their importance in many areas of applications including astronomy, business and economics. The attempts have been mostly disconcerted, in the sense that often these lack generality and apply to specific situations only. The main reason for such a state of affairs lies in the fact that while the probability density function (p.d.f.) of a stable distribution function (d.f.) always exists, it may not always be expressible in a closed form. However, the characteristic function (c.f.) $\varphi(t)$ of a stable d.f. F is representable as

$$\varphi(t) = \exp\{ita - |t\delta|^{\alpha}[1 + i\beta(\text{sgn}t)\,\omega(t, \alpha)]\}, \tag{1.1}$$

where t is real, the ***index*** (characteristic exponent) α satisfies $0 < \alpha \leqslant 2$, $\delta \geqslant 0$, $|\beta| \leqslant 1$, $-\infty < a < \infty$, and $\omega(t, \alpha) = (2/\pi)\log|t|$ or $\tan(\pi\alpha/2)$ according as $\alpha = 1$ or $\neq 1$. Press (1972) exploited this canonical representation of $\varphi(t)$ to suggest suitable estimators for the parameters α, β, δ and a; in particular, for F symmetric, these estimators were shown to have asymptotically a (multi-) normal distribution. His estimators are combinations of estimators of $\varphi(t)$ at four fixed (but arbitrary) t points, and hence, the asymptotic dispersion matrix (as well as the efficiency) may depend on the choice of these t-points. In a followup, Press (1975) has given a chronological account of the efforts

* The work of this author, who is with the Indian Institute of Management, Calcutta, India 700027, was partially carried out in the Department of Statistics, University of North Carolina, Chapel Hill, NC 27514 USA.

** Work of this author, who is with the University of North Carolina, Chapel Hill, NC 27514, USA, was partially supported by the National Heart, Lung and Blood Institute, Contract NIH-NHLBI-71-2243-L.

in this area, while, for the positive stable laws, some alternative (and efficient) methods have been considered by Brockwell and Brown (1979, 1981). For a general stable law, de Haan and Resnick (1980) proposed an estimator of the index parameter α, based on the order statistics. Unfortunately, their estimator is not generally asymptotically efficient.

The object of the present investigation is to have a deeper look into the method of Press (1972) and to locate the optimal t points (leading to asymptotically efficient estimators of the parameters). In this context, the weak convergence of the sample characteristic function, studied earlier by Feuerverger and Mureika (1977) and Csörgő (1981), has been incorporated to provide valid asymptotic solutions under minimal regularity conditions. An algorithm has also been provided. In fact, for the sake of simplicity and for the purpose of getting a straight-forward insight into the issues involved, we have considered here the case of a stable d.f. F symmetric about the origin (so that $a = \beta = 0$) with $\delta = 1$. The problem of simultaneous estimation of α, δ in such a (symmetric) case can be treated similarly at the expense of a bit more algebra, while the general case of α, β, δ and a requires a more complicated algorithm (but no new tools).

2. Empirical characteristic function based estimator of α

We are concerned with the estimation of the index parameter α $(0 < \alpha \leqslant 2)$ of a symmetric stable law for which in (1.1) $\delta = 1$ (and $a = \beta = 0$. Thus, (1.1) reduces to

(2.1) $$\varphi(t) = \exp(-|t|^{\alpha}), \text{ for all } t \in (-\infty, \infty).$$

Let $X_1, \ldots, X_n$ be a set of n independent and identically distributed random variables (i.i.d.r.v.) with a d.f. F with c.f. φ, given by (2.1). The *empirical characteristic function* (e.c.f.) is defined as

(2.2) $$\varphi_n(t) = n^{-1}\sum_{j=1}^{n} \exp\{itX_j\}, \; t \in (-\infty, \infty)$$

Note that by (2.1),

(2.3) $$\alpha = (\log|t|)^{-1} \log(-\log \varphi(t)), \; \forall \text{ real } t.$$

As such, following Press's (1972) moment-estimation method, for an arbitrary t, we consider the estimator

(2.4) $$\hat{\alpha}_n = \hat{\alpha}_n(t) = (\log|t|)^{-1} \log(-\log U_n(t)),$$

where

(2.5) $$U_n(t) = \text{Re}(\varphi_n(t)) = \text{real part of } \varphi_n(t).$$

Note that (2.4) is properly defined for all real t, excepting $t = 0$ and ± 1. Since $\hat{\alpha}_n(-t) = \hat{\alpha}_n(t)$, $\forall$ real t, in the sequel, we shall consider only positive values of $t (\neq 1)$.

Now, by Theorem 2.1 of Feuerverger and Mureika (1977), for every $T (< \infty)$,

(2.6) $$\sup_{|t| \leqslant T} |\varphi_n(t) - \varphi(t)| \to 0 \text{ almost surely, as } n \to \infty,$$

so that the real part of $\varphi_n(t)$ $(= U_n(t))$ converges a.s. to $\varphi(t)$, and hence, by (2.3) and (2.4), for every fixed $T(<\infty)$

$$\sup\{|\hat{\alpha}_n(t) - \alpha| : 0 < t(\neq 1) \leqslant T\} \to 0 \text{ a.s., as } n \to \infty. \tag{2.7}$$

Having established this (strong) consistency of $\alpha_n(t)$, one would naturally like to have an optimal choice of t. In this respect, we proceed to study first the asymptotic behaviour of $\{\sqrt{n}(\hat{\alpha}_n(t) - \alpha),\ 0 \leqslant t \leqslant T\}$ and incorporate the same in our desired solution. Let then

$$Y_n(t) = n^{1/2}\{U_n(t) - \varphi(t)\},\quad 0 \leqslant t \leqslant T. \tag{2.8}$$

Note that as F is a stable distribution and, in (2.1), $\alpha > 0$,

$$\int_{-\infty}^{\infty} |x|^p \, dF(x) < \infty,\quad 0 \leqslant p < \alpha, \tag{2.9}$$

so that by Theorem 3.1 of Feuerverger and Mureika (1977) [and an improved and generalized version of this theorem by Csörgő (1981)], as $n \to \infty$, for every fixed $T < \infty$,

$$Y_n = \{Y_n(t),\ 0 \leqslant t \leqslant T\} \underset{D}{\to} Y = \{Y(t),\ 0 \leqslant t \leqslant T\} \tag{2.10}$$

where Y is Gaussian with 0 drift and

$$EY(s)Y(t) = {}^1/_2[\varphi(s+t) + \varphi(s-t)] - \varphi(s)\varphi(t), \tag{2.11}$$

for every $(s, t) \in [0, T]^2$. Further, (2.10) ensures that

$$\sup\{|Y_n(t)| : 0 \leqslant t \leqslant T\} = O_p(1). \tag{2.12}$$

Therefore, writing $U_n(t) = \varphi(t) + n^{-1/2}Y_n(t)$ and using (2.3), (2.4) and (2.12), we obtain that for every fixed $T < \infty$,

$$\begin{aligned} -\log U_n(t) = &-\log \varphi(t)\,[1 - n^{-1/2}\, Y_n(t)\{-\varphi(t) \log \varphi(t)\}^{-1} + \\ &+ O(n^{-1}Y_n^2(t))\{-\varphi^2(t) \log \varphi(t)\}^{-1}],\quad \forall\, t \in (0, T)\backslash\{1\}, \end{aligned} \tag{2.13}$$

so that by some routine steps,

$$\begin{aligned} n^{1/2}(\hat{\alpha}_n(t) - \alpha) = &- e^{t^\alpha}\, Y_n(t)\, (t^\alpha \log t^\alpha)^{-1}\, \alpha + \\ &+ O(n^{-1/2}\, Y_n^2(t)\, e^{2t^\alpha}\, (t^\alpha \log t^\alpha)^{-1}). \end{aligned} \tag{2.14}$$

As such, by (2.11), (2.12) and (2.14), for every fixed $t(\neq 0$ or $1)$,

$$n^{1/2}(\hat{\alpha}_n(t) - \alpha) \sim N(0, \sigma^2(t)), \tag{2.15}$$

where

$$\begin{aligned} \sigma^2(t) &= {}^1/_2\, \alpha^2\, e^{2t^\alpha}\, (-t^\alpha \log t^\alpha)^{-2}\, (1 - 2\, e^{-2t^\alpha} + e^{-2t^\alpha t^\alpha}) = \\ &= {}^1/_2\, \alpha^2 h_\alpha\, (t^*);\quad t^* = t^\alpha,\ t \geqslant 0, \end{aligned} \tag{2.16}$$

and

$$h_\alpha(s) = e^{2s}(-s \log s)^{-2}\, (1 - 2\, e^{-2s} + e^{-2^\alpha s}),\quad s \geqslant 0. \tag{2.17}$$

Note that for every $\alpha \in (0, 2]$, $h_\alpha(s) \to +\infty$ as $s \downarrow 0$ or to 1 or to ∞. Further, we shall show in the Appendix that for every fixed $\alpha \in (0, 2]$, there exists a unique $s_0 \in (0, e^{-2})$, such that

$$\inf_{0<s<\infty} h_\alpha(s) = h_\alpha(s_0), \tag{2.18}$$

where $s_0(=s_{0\alpha})$ may depend on α. Let then $t_{0\alpha}$ be the solution of the equation $t_{0\alpha}^{\alpha} = s_{0\alpha}$. Then, from (2.15) through (2.18), we conclude that within the class $\{\hat{\alpha}_n(t), 0 < t\,(\neq 1) < \infty\}$ of estimators of α, an asymptotically optimal estimator is $\hat{\alpha}_n(t_{0\alpha})$. Unfortunately, $s_{0\alpha}$ as well as $t_{0\alpha}$ may generally depend on the unknown parameter α (as $h_\alpha(s)$ does so), and hence, we may not be in a position to compute $\hat{\alpha}_n(t_{0\alpha})$. For this reason, in the next section, we proceed to construct an iterative procedure yielding an alternative asymptotically efficient estimator.

3. An asymptotically optimal iterative estimator

Basically, we consider an iterative method of estimating $t_{0\alpha}$ and incorporate this in the formulation of the estimator $\hat{\alpha}_n$. The suggested procedure depends implicitly on the weak convergence result in (2.6)—(2.7). Note that by virtue of (2.10)—(2.11), for every $\varepsilon > 0$ and $\eta > 0$, there exist a $\delta > 0$ and an $n_0(= n_0(\varepsilon, \eta))$, such that for every $n \geqslant n_0$,

$$P\{\sup_{t:\,|t-t_{0\alpha}|\leqslant\delta} |Y_n(t) - Y_n(t_{0\alpha})| > \varepsilon\} < \eta. \tag{3.1}$$

Hence, using (2.16) and (3.1), we conclude that for every $\varepsilon' > 0$ and $\eta' > 0$ (and fixed $\alpha \in (0, 2]$), there exist a $\delta > 0$ and an $n_0'(= n_0'(\alpha, \varepsilon', \eta'))$, such that for every $n \geqslant n_0'$,

$$P\{\sup_{t:\,|t-t_{0\alpha}|\leqslant\delta} n^{1/2}|\hat{\alpha}_n(t) - \hat{\alpha}_n(t_{0\alpha})| > \varepsilon'\} < \eta'. \tag{3.2}$$

As a result, to find an asymptotically optimal estimator (within the class $\{\hat{\alpha}(t), t > 0\}$, it suffices to consider a (stochastic) sequence $\{\hat{t}_{nm}, m \geqslant 0\}$ of t-points, such that $\hat{t}_{nm}$ stochastically converges to $t_{0\alpha}$ as m increases and $n \to \infty$, and then to use $\hat{\alpha}_n(\hat{t}_{nm})$, for some appropriate m, as the desired estimator. Towards this objective, we consider the following iterative procedures.

As will be seen in the Appendix, $s_{0\alpha}$ is always $\leqslant e^{-2}$, for all α: $0 < \alpha \leqslant 2$. As such, at the initial stage, we choose an arbitrary $\hat{t}_{n0} \in (0,1)$, preferably close to $^1/_2$, compute $\hat{\alpha}_{n0} = \hat{\alpha}_n(\hat{t}_{n0})$ and $\hat{s}_{0\alpha}^{(0)} = s_0(\hat{\alpha}_{n0})$, by (2.4) and (2.18), respectively. In the next step, we consider

$$\hat{t}_{n1}: \hat{t}_{n1}^{\hat{\alpha}_{n0}} = \hat{s}_{0\alpha}^{(0)}, \quad \hat{\alpha}_{n1} = \hat{\alpha}_n(\hat{t}_{n1}) \text{ and } \hat{s}_{0\alpha}^{(1)} = s_0(\hat{\alpha}_{n1}). \tag{3.3}$$

In this way, at the m-th step, we define

$$\hat{t}_{nm}: \hat{t}_{nm}^{\hat{\alpha}_{n(m-1)}} = \hat{s}_{0\alpha}^{(m-1)}, \quad \hat{\alpha}_{nm} = \hat{\alpha}_n(\hat{t}_{nm}) \text{ and } \hat{s}_{0\alpha}^{(m)} = s_0(\hat{\alpha}_{nm}), \tag{3.4}$$

for every $m \geqslant 1$. The *stopping number* $M(= M_{n\varepsilon})$, corresponding to some preassigned $\varepsilon > 0$, is defined by

$$M = \min\{m \geqslant 2: \sqrt{n}\,|\hat{\alpha}_{nm} - \hat{\alpha}_{n(m-1)}| < \varepsilon\}. \tag{3.5}$$

In the Appendix, we shall show that $s_{0\alpha}$ is a continuous function of α: $0<\alpha\leqslant 2$, and hence, by (2.6), (2.7) and the definition of $s_{0\alpha}$, $\hat{t}_{nm}\to t_{0\alpha}$ a.s., as $n\to\infty$ (for every $m\geqslant 1$), and hence, $M(=M_n)$ is a.s. finite. Actually, a few iterations will lead to the desired estimator

$$\hat{\alpha}_{nM}=\hat{\alpha}_n(\hat{t}_{nM}). \tag{3.6}$$

Since $\hat{t}_{nM}$ stochastically converges to t_0, by virtue of (3.2), (3.5), (3.6) and (2.15)—(2.17), we conclude that

$$n^{1/2}(\hat{\alpha}_{nM}-\alpha)\sim N(0,\ {}^1/_2\alpha^2 h_\alpha(s_{0\alpha})), \tag{3.7}$$

which, by (2.18), reveals the desired asymptotic optimality property.

It may be noted that if, as in Press (1972), we consider an arbitrary t and $\hat{\alpha}_n(t)$, then by (2.15) and (3.7), the asymptotic relative efficiency (A.R.E.) of $\hat{\alpha}_n(t)$ with respect to $\hat{\alpha}_{nM}$ is given by

$$e\,(t,t_{0\alpha})=h_\alpha(s_{0\alpha})/h_\alpha(t^\alpha)=\{h_\alpha(t^\alpha)\}^{-1}\{\inf_s h_\alpha(s)\}\leqslant 1,\ \forall t, \tag{3.8}$$

where the equality sign holds when $t=t_{0\alpha}$. Actually, (3.8) may be quite below 1 depending on t, $t_{0\alpha}$ and α. This explains the utility of the iterative procedure considered here. We may term $\hat{\alpha}_{nM}$ an *adaptive estimator* of α also, since t_{nM} is adapted from the data set.

4. Appendix

The function $h_\alpha(x)$ in (2.17) plays a fundamental role in the procedure described in Section 3. We define $s_0=s_{0\alpha}$ as in (2.18). Then, we have the following.

Lemma 4.1. *For every (fixed) α: $0<\alpha\leqslant 2$,*

$$0<s_{0\alpha}<e^{-2}. \tag{4.1}$$

Proof. Consider first the case of $h_\alpha(x)$, $0<x<1$. Note that for every $\alpha(0<\alpha\leqslant 2)$, $e^{2x}(1-2e^{-2x}+e^{-x2^\alpha})$ is nondecreasing in $x(\geqslant 0)$, while $(-x\log x)^{-2}$ is a convex function $(0\leqslant x\leqslant 1)$ with values at 0 and 1 equal to $+\infty$ and a unique minimum at $x=e^{-1}$. Hence, by (2.17), $h_\alpha(x)$ is nondecreasing on $[e^{-1},1]$, so that for $0\leqslant x\leqslant 1$, the minimum value of $h_\alpha(x)$ occurs at some point below e^{-1}. Putting $\gamma=2-2^\alpha$ note that

$$h_\alpha(e^{-1})=e^2(e^{2/e}+e^{\gamma/e}-2)\leqslant e^2(e^{2/e}+e^{1/e}-2)\quad(\text{as }-2\leqslant\gamma\leqslant 1)$$

$$=11.74,\ \text{for all }\alpha:\ 0<\alpha<2. \tag{4.2}$$

Next, observe that for every $x\geqslant 1$, $0<\alpha\leqslant 2$,

$$h_\alpha(x)=(x\log x)^{-2}e^{2x}(1-2e^{-2x}+e^{-2^\alpha x})\geqslant(x\log x)^{-2}e^{2x}(1-2e^{-2x}+e^{-4x})$$

$$=(x\log x)^{-2}e^{2x}(1-e^{-2x})^2=\{(e^x-e^{-x})/(x\log x)\}^2=g(x),\ \text{say}. \tag{4.3}$$

Also, note that

$$(4.4)\qquad {}^1/_2\frac{d}{dx}(\log g(x)) = \frac{e^x + e^{-x}}{e^x - e^{-x}} - \frac{1}{x} - \frac{1}{x\log x},$$

where the first term on the right hand side of (4.4) is ↓ in x and converges to 1 as $x \to \infty$, while the other two go to 0 as $x \to \infty$. As such, if x_0 be defined by $x_0^{-1} + (x_0 \log x_0)^{-1} = 1$ ($x_0 \sim 2.35$), then (4.4) is positive for all $x \geqslant x_0$, so that $g'(x)$ is strictly positive for every $x \geqslant x_0$. Hence, the minimum of $g(x)$ on $(1, \infty)$ does not occur at any $x \geqslant x_0$. Let us then consider the interval $(1, x_0)$ and denote x_1 by $x_1 \log x_1 = e^{-1}$, $x_2 = 1.5$, $x_3 = 1.75$, $x_4 = 2.00$ and $x_5 = x_0$. Then, for $1 \leqslant x \leqslant x_1$, $g(x) \geqslant e^2 (e^1 - e^{-1})^2 = e^4 (1 - e^{-2})^2 \geqslant 40 \geqslant h_\alpha (e^{-1})$. Also, for $x_1 \leqslant x \leqslant x_2$, $g(x) \geqslant (x_2 \log x_2)^{-2} \times (e^{x_1} - e^{-x_1})^2 \geqslant 20 \geqslant h_\alpha (e^{-1})$, and a similar case holds for $x_2 \leqslant x \leqslant x_3$ as well as $x_3 \leqslant x \leqslant x_4$ and $x_4 \leqslant x \leqslant x_0$. Hence, $g(x) \geqslant h_\alpha(e^{-1})$, for every $x \geqslant 1$. Thus, by (4.3), $\inf\{h_\alpha(x) \colon x \geqslant 1\} \geqslant h_\alpha(e^{-1})$, $\forall \alpha \in (0, 2]$. Therefore, the global minimum of $h_\alpha(x)$ $(0 < x < \infty)$ occurs at a value of $x < e^{-1}$.

Note that by definition, for every α: $0 < \alpha \leqslant 2$,

$$h_\alpha(x) = \sum_{r=1}^{\infty} \{(2^r + \gamma^r)/r!\}\, x^{r-2}(-\log x)^{-2}$$

$$(4.5)\qquad = \sum_{r=1}^{\infty} c_r x^{r-2}(-\log x)^{-2},\quad 0 < x \leqslant e^{-1},$$

where the c_r are ll non-negative. Thus, for every $x > 0$,

$$(4.6)\qquad \frac{d}{dx} h_\alpha(x) = \sum_{r=1}^{\infty} c_r\{(r-2) + 2(-\log x)^{-1}\}\, x^{r-3}(-\log x)^{-2},$$

so that for every x: $2 \geqslant -\log x$ (i.e., $x \geqslant e^{-2}$), $(r-2) + 2(-\log x)^{-1} \geqslant r - 1 \geqslant 0$, $\forall r \geqslant 1$, and $(d/dx)h_\alpha(x) > 0$. Therefore, the minima of $h_\alpha(x)$ do not occur within the interval (e^{-2}, e^{-1}). Hence, the minima of $h_\alpha(x)$, no matter whether unique or not, occur at a point lying in the interval $(0, e^{-2})$ Q.E.D.

Note that though $h_\alpha(x)$ is the product of two convex functions, it is not itself convex everywhere. Hence, (4.2), (4.3) and (4.6) provide a convenient tool for proving (4.1).

It may be remarked that for $\alpha = 2$, $h_2(x)$ is equal to

$$\sum_{r=1}^{\infty} \{2^{2r+1}/(2r)!\}\, x^{2(r-1)}(-\log x)^{-2},\quad x \geqslant 0,$$

and hence, is ↑ in $x \in (0, 1)$, so that $s_{02} = 0$. On the other hand, for $0 < \alpha < 2$, we like to show that $0 < s_{0\alpha} < e^{-2}$ and is unique. Towards this, we have the following.

Lemma 4.2. *For every (fixed) α: $0 < \alpha < 2$, $h_\alpha(x)$ has a unique minimum on $(0, e^{-2})$.*

Proof. Note that by (4.6) and the non-negativity of $x^{-2}(-\log x)^{-3}$ (on $(0, e^{-2})$), $(d/dx)h_\alpha(x)$ and $g_\alpha^*(x) = x^2(-\log x)^3\, (d/dx)h_\alpha(x)$ both have the same sign. Thus, it suffices to show that $g_\alpha(x)$ has a unique root in $(0, e^{-2})$. Note that by (4.6)

$$g_\alpha(x) = \sum_{r=1}^{\infty} c^r\{(r-2)(-\log x) + 2\}\, x^{r-1} =$$

$$= c_1(2 + \log x) + 2c_2 x + \sum_{r=3}^{\infty} c_r x^{r-1}((r-2)(-\log x) + 2). \tag{4.7}$$

Note that $g_\alpha(0) = -\infty$, $g_\alpha(e^{-2}) > 0$, and hence, it suffices to show that $(d/dx)\, g_\alpha(x) > 0$, $\forall\, x \in (0, e^{-2})$. Towards this, we have by (4.7),

$$\frac{d}{dx} g_\alpha(x) = x^{-1}c_1 + 2c_2 + \sum_{r=3}^{\infty} c_r \left\{(r-1)\, x^{r-2}((r-2)(-\log x)+2)+\right.$$

$$\left. + x^{r-1}(r-2)\left(-\frac{1}{x}\right)\right\} = x^{-1}c_1 + 2c_2 + \sum_{r=3}^{\infty} c_r\{x^{r-2}[(r-1)(r-2)(-\log x) +$$

$$+ 2(r-1) - (r-2)]\} = x^{-1}c_1 + 2c_2 +$$

$$+\sum_{r=3}^{\infty} c_r x^{r-2}\{(r-1)(r-2)(-\log x) + r\} > 0 \tag{4.8}$$

for every $x \in (0, 1)$. Hence, $g_\alpha(x)$ has a unique root $(s_{0\alpha})$ in $(0, e^{-2})$. Q.E.D.

Note that the c_r are polynomial functions of $\gamma(= 2 - 2^\alpha)$, and hence, are continuous and differentiable functions of α. Hence, by (4.7) and (4.8), we conclude that

$$s_{0\alpha} \text{ is a continuous function of } \alpha\colon 0 < \alpha \leqslant 2. \tag{4.9}$$

We conclude this section with the remark that for $\alpha = 2$, $s_{02} = 0$ and for $\alpha \to 2$, $s_{0\alpha} \downarrow 0$. Hence, looking at (2.4), (2.14), (3.3) and (3.4), we may argue that for $\alpha = 2$ or very, very close to 2, the iterative procedure may not work out well. However, in such a case, if we let

$$\hat{s}_{0\alpha}^{(m)} = \max\{n^{-\lambda},\ s_0(\hat{\alpha}_{nm})\},\ m \geqslant 1, \tag{4.10}$$

where $\lambda (0 < \lambda < 1/4)$ is some prefixed positive number, then the procedure works out — though we may not have the asymptotically optimal $\hat{\alpha}_{nM}$.

REFERENCES

1. Brockwell, P. J., Brown, B. M., *Estimation for the positive stable laws, I.* Austral. J. Statist., **21**, 139–148 (1979).
2. Brockwell, P. J., Brown, B. M., *High-efficiency estimation for the positive stable laws.* J. Amer. Statist. Assoc., **76**, 626–631 (1981).
3. Csörgő, S., *Limit behaviour of the empirical characteristic function.* Ann. Probab., **9**, 130–144 (1981).
4. De Haan, L., Resnick, S. I., *A simple asymptotic estimate for the index of a stable distribution.* J. R. Statist. Soc. Ser. B, **42**, 83–87 (1980).
5. Feuerverger, A., Mureika, R. A., *The empirical characteristic function and its applications.* Ann. Statist., **5**, 88–97 (1977).
6. Press, S. J., *Estimation in univariate and multivariate stable distributions.* J. Amer. Statist. Assoc., **67**, 842–846(1972).
7. Press, S. J., *Stable distributions: Probability, inference and applications in finance — a survey, and a review of recent results.* In: Patil, G. P., Kotz, S., Ord, J. K. (Eds.), *A Modern Course on Statistical Distributions in Scientific Work.* Boston, Reidel, 1975.

D.A. BERRY *

Some gambling problems with nonintuitive solutions

1. Introduction

You find yourself in a gambling house with positive fortune f. You are to select one of a number of available gambles, the characteristics of which may not be completely known. The gamble chosen depends on your state of knowledge concerning the various gambles.

Taking a gamble gives a random reward (possibly negative) and also gives some information concerning the unknown aspects of the various gambles. Your fortune is incremented by this reward and your state of knowledge is adjusted. Proceeding in this way, you choose a gamble at each of a possibly unlimited number of stages. Choices are made sequentially, depending on previous gambles and rewards. Your objective is to increase your fortune to M before it becomes 0.

Following the notation and terminology of Dubins and Savage (1976), F denotes the set of fortunes and $\Gamma(f)$ is the set of gambles available at $f \in F$. A strategy $\sigma \in \Gamma$ available at f is a sequence $\sigma_0, \sigma_1, \ldots$ where $\sigma_0 \in \Gamma(f)$ and, for each positive n and each finite sequence $(f_1, \ldots, f_n)$ of elements of F, $\sigma_n(f_1, \ldots, f_n) \in \Gamma(f_n)$. The utility of a fortune or of a strategy is denoted by u. Let $U(f) = \sup u(\sigma)$ where the supremum is over all $\sigma \in \Gamma$ available at f. A strategy is ε-optimal in Γ at f if $u(\sigma) \geqslant U(f) - \varepsilon$ for $\varepsilon \geqslant 0$. A strategy is optimal if it is 0-optimal. A gamble is ε-optimal at f if it is the initial gamble of at least one ε-optimal strategy.

Consistent with the objective of reaching M before 0, the utility of a fortune is defined to be

$$u(f) = 0, \; f < M$$
$$= 1, \; f \geqslant M.$$

The utility of a strategy is then the probability that $f_n \geqslant M$ for all n sufficiently large.

Two problems are considered in this paper: red-and-black and two-armed bandits. In the first there is a single stochastic process and the problem

* Supported in part by NSF Grant MCS8102477.

is to decide how much to stake at each stage. In the second there are two processes to choose from at each stage and stakes are fixed throughout.

In red-and-black a gamble is determined by a stake s ($0 \leqslant s \leqslant f$). The fortune space is either discrete, taken to be the nonnegative integres, or continuous, taken to be the nonnegative reals. The probability of winning, thereby changing your fortune to $f+s$, is p and of losing, your fortune becoming $f-s$, is $1-p$. The parameter p is unknown. This problem has been considered by Berry, Heath, and Sudderth (1974). An ε-optimal strategy which has an unexpected characteristic is given in Section 2 for the continuous case. An example in which the optimal stake as a function of f has an unusual form is given in Section 3 for the discrete case.

In a two-armed bandit there are but two gambles: choose process 1 and choose process 2. Since the stakes are fixed throughout the fortune space is effectively discrete; the stake is taken to be 1 and the fortune space taken to be the nonnegative integers. Choosing process i increases your fortune by 1 with probability p_i and decreases it by 1 with probability $1-p_i$. Both p_1 and p_2 may be unknown. This problem has been considered by Berry and Fristedt (1980a, 1980b). Two examples with nonintuitive solutions are given in Section 4 for the case in which p_1 is known; one has $M < \infty$. Another counterintuitive example is given in Section 5 for the case in which the values of p_1 and p_2 are known, but not which goes with p_1.

2. Continuous red-and-black

In continuous red-and-black with fixed p, $M = 1$ and $F = [0, \infty)$. An available stake s is any current fortune f, $0 \leqslant s \leqslant f$, so that

$$\Gamma_p(f) = \{p\delta(f+s) + \bar{p}\delta(f-s)\colon\ 0 \leqslant s \leqslant f\}$$

for all $f \in F$, where the symbol $\delta(f)$ denotes the measure which assigns mass 1 to $\{f\}$ and $\bar{p} = 1-p$. Since gambles and stakes are equivalent, a strategy can as well be defined in terms of stakes. This interpretation has the advantage that the same stakes are available at a fortune f for every value of p.

Suppose first that p is fixed and $0 \leqslant p \leqslant {}^1/_2$. Dubins and Savage (1976) show that "bold play" is optimal: always staking $f \wedge (1-f)$ is optimal. But bold play may not be uniquely optimal. If $p = 0$ then every strategy is optimal. If $p = {}^1/_2$ then every strategy that stakes enough to eventually reach 0 or 1 (with probability 1) is optimal. If $0 < p < {}^1/_2$ and $i/2^k < f < (i+1)/2^k < 1$ for some positive integers k and i, then it is also optimal to play boldly on this binary interval — the initial stake $[f - i/2^k] \wedge [(i+1)/2^k - f]$ is optimal.

If ${}^1/_2 < p < 1$ then the opposite kind of play is optimal. To take advantage of this superfair situation, you should never risk all of your fortune. For example, Berry, Heath, and Sudderth (1974) show that always taking the "timid near 0" stake,

$$\begin{aligned} t(f) &= f^2, 0 \leqslant f < 1 \\ &= 0,\ f \geqslant 1 \end{aligned}$$

is optimal for any $p > {}^1/_2$.

Suppose p is unknown and information about p is given by probability measure π on $[0, 1]$. The outcome of each gamble may now affect future stakes for two reasons: your fortune f changes and your information regarding p changes. Your "fortune" depends on π as well as on f. The set of fortunes is $G = \{(f, \pi): f \in F, \pi \in \mathcal{P}\}$, where $\mathcal{P}$ can be taken to be the collection of countably additive probabilities defined on the Borel subsets of $[0, 1]$ or the collection of finitely additive probabilities defined on all subsets of $[0, 1]$. Your utility for $(f, \pi) \in G$ is $u(f, \pi) = u(f)$ which is again 0 or 1 according as $f < 1$ or $f \geqslant 1$.

The information present concerning p changes by Bayes' theorem depending on the outcome of a gamble

$$\pi(B) \xrightarrow[\text{win}]{} \pi_W(B) = \frac{1}{Ep} \int_B x \, d\pi(x) \quad \text{if } Ep \neq 0,$$

$$\pi(B) \xrightarrow[\text{loss}]{} \pi_L(B) = \frac{1}{\overline{Ep}} \int_B \bar{x} \, d\pi(x) \quad \text{if } \overline{Ep} \neq 0,$$

for appropriate sets B. The definition of π_W when $Ep = 0$ and of π_L when $\overline{Ep} = 0$ are immaterial since "win" has probability 0 in the first case and 1 in the second.

To stake s at (f, π) is to use the gamble $g(s, f, \pi) = (Ep)\, \delta(f + s, \pi_W) + (\overline{Ep})\, \delta(f - s, \pi_L)$. The available gambles are defined by $\Gamma(f, \pi) = \{g(s, f, \pi): 0 \leqslant s \leqslant f\}$. Letting U be the utility of the house Γ, $U(f, \pi)$ is the supremum over all strategies σ available at (f, π) in Γ of the σ-probability that 1 is reached.

For any π with support contained in $\{0\} \cup [1/2, 1]$ or $[0, 1/2] \cup \{1\}$, there are optimal strategies: timid and bold play, respectively. If, on the other hand, $\pi((0, 1/2)) > 0$ and $\pi((1/2, 1)) > 0$ then there is no optimal strategy. But there are many ε-optimal strategies in this case. For example, you can stake 0 at every stage until $\pi'((0, 1/2)) \vee \pi'((1/2, 1)) > 1 - \varepsilon$, where π' is the current measure, which happens with probability 1 for all p and therefore for all π; subsequently, you play boldly if $\pi'((0, 1/2)) > 1 - \varepsilon$ and timidly otherwise. But no finite sequence of wins and losses gives complete information concerning the sign of $p - 1/2$ in this case, so there is always a chance of playing inoptimally.

It can take an arbitrarily long time staking 0 to determine with confidence the sign of $p - 1/2$. And there may be a minimum stake which could make timid play very bad—especially if it is quite likely under π that p is small. Fortunately, there are ε-optimal strategies in which you do not play cautiously at first. You can stake all but δ of your fortune playing boldly on the chance that $p \leqslant 1/2$. Having given yourself nearly the best chance of reaching 1 in case $p < 1/2$, you now assume that $p \geqslant 1/2$ and play timidly—reaching 1 with probability 1 if $p > 1/2$. Consider the "δ-bold strategy" which always stakes $b_\delta(f)$ at (f, π),

$$\begin{aligned} b_\delta(f) &= f^2 && \text{if } 0 < f \leqslant \delta \\ &= f - \delta && \text{if } \delta < f < (1 + \delta)/2 \\ &= \bar{f} && \text{if } (1 + \delta)/2 \leqslant f < 1 \\ &= 0 && \text{if } f \geqslant 1. \end{aligned}$$

This strategy is bold for $f \in (\delta, 1)$ with δ playing the role of 0, and timid for $f \in (0, \delta]$. Despite the fact that the δ-bold strategy does not depend on π it is ε-optional for all $\pi \in \mathcal{P}$.

Theorem (Berry, Heath, Sudderth 1974). *For* $\varepsilon > 0$, *there is a* $\delta(\varepsilon) > 0$ *such that, for all* $\delta \in (0, \delta(\varepsilon))$ *and all* $(f, \pi) \in G$, *the* δ*-bold strategy at* (f, π) *is* ε*-optimal. Furthermore, for all* $(f, \pi) \in G$, $U(f, \pi) = \int U_x(f)\, d\pi(x) = \pi(p > {}^1/_2) + {}$

$+ \int_{x \leqslant {}^1/_2} B_x(f)\, d\pi(x)$, *where* B_p *is the utility of house* Γ_p *for fixed* p.

This result means that you can behave ε-optimally ignoring the information you have concerning p; in addition, you can ignore the results of your gambles (except that they affect f and your stakes do of course depend on your current f).

There are other ε-optimal strategies that, like δ-bold, do not play timidly initially (see Berry, Heath, Sudderth 1974). Among ε-optimal strategies, δ-bold has the advantage of getting to 0 or 1 quickly. But it is not best for this purpose (f^2 near 0 is rather small so δ-bold gets to 1 slowly when $p > {}^1/_2$ and to 0 slowly from $f \leqslant \delta$ when $p < {}^1/_2$). In general, the fastest ε-optimal strategy will be neither bold nor timid, and will depend on π.

The next section treats a similar problem but stakes are discrete and there is a minimum stake until $f = 0$ or M.

3. Discrete red-and-black

Now suppose M is a positive integer and $F = \{0, 1, \ldots\}$. If $f = 0$ or $f \geqslant M$ then $\Gamma_p(f) = \{\delta(f)\}$; if $f = 1, 2, \ldots, M-1$ then $\Gamma_p(f) = \{p\delta(f+s) + \bar{p}\delta(f-s) : s = 1, \ldots, f\}$. As in the previous section, $u(f) = 1$ for $f \geqslant M$ and $u(f) = 0$ for $f < M$.

For fixed $p \in [0, {}^1/_2]$ then bold play — always staking $f \wedge (M-f)$ — is optimal (Dubins and Savage 1976). For fixed $p \in [{}^1/_2, 1]$ then timid play — always staking 1 — is optimal. The latter is easy to prove using the fundamental theorem of gambling (Dubins and Savage 1976, Theorem 2.12.1). The important calculation is to show that $T_p(f)$, the utility of timid play, is *excessive:* staking s followed by timid play is not an improvement on timid play, that is, on $s = 1$ followed by timid play. To wit, for ${}^1/_2 < p < 1$:

$$pT_p(f+s) + \bar{p}T_p(f-s) = p\,\frac{1-(\bar{p}/p)^{f+s}}{1-(\bar{p}/p)^M} + \bar{p}\,\frac{1-(\bar{p}/p)^{f-s}}{1-(\bar{p}/p)^M} =$$

$$= \frac{1-(\bar{p}/p)^f\,[p(\bar{p}/p)^s + \bar{p}(p/\bar{p})^s]}{1-(\bar{p}/p)^M} \leqslant \frac{1-(\bar{p}/p)^f}{1-(\bar{p}/p)^M} = T_p(f)$$

with equality if and only if $s = 1$. Therefore, timid play is uniquely optimal. That $T_{{}^1/_2}(f) = f/M$ is excessive is trivial, but the corresponding calculation is an equality for all $s \leqslant f \wedge (M-f)$ since all such stakes are optimal.

In discrete red-and-black there do not exist strategies that are ε-optimal uniformly in p for sufficiently small ε. On the other hand, since there are only a finite number of strategies through a particular stage, there is an optimal strategy when p is unknown and it depends on the information concerning p. So the results of the gambles, and not just the current f, affect the stakes made when following an optimal strategy.

While there exists an optimal strategy for each π, a numerical iteration is necesary to find one when $\pi((0, {}^1/_2)) > 0$ and $\pi(({}^1/_2, 1)) > 0$. The difficulty is that finding $U(f, \pi)$ for $f = 1, \ldots, M-1$ one must simultaneously find $U(f, \pi')$ for all possible (f, π') that might arise using an arbitrary strategy. Unless π happens to be very special — for example, a symmetric two-point measure — then some truncation is necessary.

To minimize the number of iterations one should initialize with a good approximation, and not start with all utilities equal to 0, say. One can use the following lower bound (for all possible (f, π'))

$$U(f, \pi') \geqslant B(f, \pi') \vee T(f, \pi'),$$

where B and T are the utilities of the bold and timid strategies

$$B(f, \pi) = \int B_x(f)\, d\pi(x),$$

$$T(f, \pi) = \int T_x(f)\, d\pi(x).$$

Or with the following upper bound

$$U(f, \pi') \leqslant \int [B_x(f) \vee T_x(f)]\, d\pi'(x) = \int_{[0,1/2)} B_x(f) d\pi'(x) + \int_{[1/2,1]} T_x(f)\, d\pi'(x).$$

For definiteness, suppose

$$U_0(f, \pi') = B(f, \pi') \vee T(f, \pi').$$

For each possible (f, π'), using a convenient ordering, one calculates

$$U_i(f, \pi') = \max_{1 \leqslant s \leqslant f \wedge (M-f)} \left[\int x\, d\pi'(x)\, U_{i-1}(f+s, \pi'_W) + \right.$$

$$\left. + \int \bar{x}\, d\pi'(x)\, U_{i-1}(f-s, \pi'_L) \right]$$

for $i = 1, 2, \ldots,$ at the same time keeping track of the stakes at which the maxima obtain. The various $U_i(f, \pi)$ converge monotonically (upward in this case) to $U(f, \pi)$. However, some care must be exercised to insure that the stakes found by an iteration which is suitably truncated are indeed optimal.

Table 1 lists the optimal stakes when $M = 16$ and π is the symmetric two-point measure on $p = 1/4$ and $p = 3/4$. The calculations are enormously eased in this case because $(\pi_W)_L = (\pi_L)_W = \pi$; at most $2M - 3 = 29$ dfiferent measures are possible.

The odd values of f from 5 to 13 illustrate a curious phenomenon. If p were known to be $1/4$ then 1 is an optimal stake at these fortunes (of course, there are other optimal stakes — for example, 7 is also optimal at $f = 7$); and if p were known to be $3/4$ then 1 is the uniquely optimal stake. But with π a mixture of these two values of p, 1 is not an optimal stake! On the other hand, if $p = 1/4$

Table 1

$M = 16,\ \pi(1/4) = \pi(3/4) = 1/2$

f	Optimal Initial Stake	$U(f, \pi)$
1	1	0.333
2	1	0.444
3	1	0.481
4	1	0.494
5	2	0.501
6	4	0.516
7	2	0.531
8	1	0.556
9	7	0.611
10	1	0.625
11	5	0.645
12	4	0.688
13	3	0.723
14	2	0.781
15	1	0.840

and $f = 10$ then 2 and 6 are the only optimal stakes; 1 is optimal in the mixture.

Table 1 reports only a small part of an optimal strategy — the initial stake. One cannot deduce the optimal second stake from the table (for either win or lose) starting at $f = 7$, say, and staking 2 initially; f becomes 5 or 9 but π changes.

4. Two-armed bandit, one arm known

As in the previous section, M is a positive integer, $F = \{0, 1, \dots\}$, $u(f) = 1$ for $f \geqslant M$ and $u(f) = 0$ for $f < M$. Now, there are two processes, or arms, characterized by win probabilities p_1 and p_2. If $f = 0$ or $f \geqslant M$ then $\Gamma_{p_1,p_2}(f) = \{\delta(f)\}$; if $f = 1, 2, \dots, M-1$ then $\Gamma_{p_1,p_2}(f) = \{p_i\delta(f+1) + \bar{p}_i\delta(f-1) : i = 1, 2\}$. The problem is trivial: you should always choose the gamble with higher win probability, and $U_{p_1,p_2}(f) = \max_{i=1,2} \{T_{p_i}(f)\}$.

When p_1 and p_2 are unknown the problem is very complicated (see Berry and Fristedt (1980a, 1980b)). Let π now denote the joint measure of (p_1, p_2), again reflecting your information concerning the arms, and $u(f, \pi) = u(f)$ as above. To choose arm i is to use gamble $g_i(f, \pi) = Ep_i\delta(f+1, \pi_{W_i}) + \overline{Ep_i}\,\delta(f_i - 1, \pi_{L_i})$, in an obvious adaptation of previous notation. Again, $U(f, \pi)$ denotes the maximum over all strategies σ of the σ-probability of reaching M; the arguments in Berry and Fristedt (1980a) are easily adapted to show that optimal strategies exist since $M < \infty$.

As an example (Berry and Fristedt (1980a), Example 1.1), suppose that under π the value of p_1 is known to be 1/2 and p_2 is either 0 or 1: $\pi(p_2 = 1) = r = 1 - \pi\,(p_2 = 0)$. Then $\pi_{W_1} = \pi_{L_1} = \pi$ and $\pi_{W_2} = \delta(1)$, $\pi_{L_2} = \delta(0)$. Choosing arm 2 gives you complete information with but one observation. It is reasonable to expect that choosing arm 2 is optimal initially, especially for large r, since you can then proceed with perfect knowledge of the arms' characteristics. Actually, the opposite is true. It is optimal (uniquely optimal unless $r = 1$) to choose arm 1 for all r and $f = 2, \dots, M-1$ and for $r < 1/M$ and $f = 1$; when $r \geqslant 1/M$ and $f = 1$ then (and only then when $r \neq 1$) arm 2 is optimal. Let σ^* denote the implied strategy. Optimality follows if $f_n \to 0$ or 1 with σ^*-probability 1 and if $U^*(f, r) = u(\sigma^*)$ is excessive. The first is clear; the second is demonstrated below.

For $f = 1, 2, \dots, M-1$,

$$U^*(f, r) = \frac{f-1}{M-1} + \frac{M-f}{M-1}\, r, \quad r \geqslant 1/M$$
$$= f/M \qquad , r < 1/M.$$

Suppose either $f > 1$ and $r < 1$, or $r < 1/M$; arm 1 is chosen when using σ^*. The utility of choosing arm 2 initially and then following σ^* is

$$rU^*(f+1, 1) + \bar{r}U^*(f-1, 0) = r + \bar{r}(f-1)/M =$$

$$= \begin{cases} U^*(f, r) - (1-r)\dfrac{f-1}{M(M-1)}, & \text{if } r \geqslant 1/M \\[2ex] U^*(f, r) - \left[\dfrac{1}{M} - r + \dfrac{r}{M}(f-1)\right], & \text{if } r < 1/M \end{cases}$$

$$< U^*(f, r)$$

in the case at hand. Now suppose either $f = 1$ and $r \geqslant 1/M$ or $f > 1$ and $r = 1$; arm 2 is chosen using σ^*. If $f = 1$ the utility of choosing arm 1 followed by σ^* is

$$\frac{1}{2} U^*(f, r) = \frac{1}{2(M-1)} + \frac{M-2}{2(M-1)} r = U^*(f, r) - \frac{rM-1}{2M-2} \leqslant U^*(f, r)$$

with equality only if $r = 1/M$. The utility of choosing arm 1 followed by σ^* when $f > 1$ and $r = 1$ is $1 = U^*(f, 1)$. Therefore, σ^* is optimal. All other optimal strategies can be found by considering those situations in which equality occurs in the above calculations: both arms are optimal when $f = 1$ and $r = 1/M$ or when $f > 1$ and $r = 1$; otherwise the arm indicated by σ^* is uniquely optimal.

For the purposes of the next example suppose $M = \infty$. Now $p_1 = 6/10$ (the problem would be uninteresting unless $p_1 > {}^1/_2$) and $\pi(p_2 = 0) = 3/35 = 1 - \pi(p_2 = 5/8)$; a win with arm 2 reveals the better arm ($p_2 = 5/8$) but a loss is ambiguous ($\pi_{L_2}(p_2 = 0) = 1/5$). For an analysis of this problem see Example 5.1 of Berry and Fristedt (1980a). If f is sufficiently large you might be willing to sustain an arbitrarily large number of losses with arm 2 to establish the better arm with high probability; following an optimal strategy you would sustain exactly three! The optimal strategy is given below.

Clearly, if arm 2 ever wins then it is used forever. The only question is whether to choose arm 2 and what to do if it loses. If $f_0 = 1$ then you should choose arm 2 and there is nothing further to resolve. If $f_0 = 2$ then arm 2 is optimal. If it loses (so $f_1 = 1$) then arm 1 should be used until $f = 3$ is reached (if ever) at which time arm 2 should be used again. If arm 2 loses a second time then arm 1 is used until $f = 14$ is reached, at which time arm 2 is used again. If arm 2 loses a third time then it is never used again; after three losses the probability that $p_2 = 0$ is $16/25$.

If $f_0 = 3$ then arm 2 is optimal. If it loses (so $f_1 = 2$) arm 1 is used until $f = 3$ is reached again, when arm 2 is used for the second time. As in the previous case, $f = 14$ must be reached before a nonwinning arm 2 is used for the third time.

If $4 \leqslant f_0 \leqslant 15$ then you should choose arm 2 twice initially — and a third time if both are losses and $f = 14$ is reached using arm 1. Finally, if the initial $f \geqslant 16$ then you should choose arm 2 at the first three stages — and never again if it loses all three times, regardless of how large f becomes.

This section applies when π is concentrated on a line in the unit square with p_1 constant; in the next section π concentrates its mass on two points symmetric about $p_1 = p_2$.

5. Two-armed bandit, dependent arms

The setting is the same as in the previous section except that π now concentrates its mass on $\{(\alpha, \beta), (\beta, \alpha)\}$; set $r = \pi(\alpha, \beta) = 1 - \pi(\beta, \alpha)$ and suppose $\beta \geqslant \alpha$. It seems reasonable to expect that a myopic strategy — choosing arm 2 if $r \geqslant {}^1/_2$ and arm 1 if $r < {}^1/_2$ — is optimal. This is myopic in the sense that the arm chosen is the more likely to win: if $r > {}^1/_2$ then $Ep_2 > Ep_1$. There are many reasons to choose arm 2 in this case and little argument against it; in particular, the probability of k immediate wins is greater with arm 2 than with arm 1 for all $k > 1$ ($Ep_2^k > Ep_1^k$), and, more generally, the expectation of any increasing function is greater for p_2 than p_1. Moreover, for classical two-armed bandits, Feldman (1962) showed that myopic strategies are optimal in this setting. But, as will be seen, they may not be optimal in the current problem!

However, myopic strategies are optimal in a number of special cases — among them, $\alpha + \beta = 1$. (This situation arises, for example, when you are offered the option of bètting on heads or tails with the same coin.)

Theorem (Berry and Fristedt (1980b)). *Suppose* $0 < \alpha = \bar{\beta} < {}^1/_2$. *Arm* 1 *is optimal if* $\bar{r}/r > (\beta/\alpha)^{f-1}$, *arm* 2 *if* $\bar{r}/r < (\alpha/\beta)^{f-1}$, *and both are optimal otherwise. Moreover, these statements completely characterize optimal strategies.*

While all myopic strategies are included in the class of optimal strategies, this class includes many other strategies as well — especially when $\beta - \alpha$ is large. Again, $f = 1$ is critical: you must choose the arm most likely to have win probability β when $f = 1$ but have more flexibility for $f > 1$. An interesting feature of the class of optimal strategies for this case is that it does not depend on M; even $M = +\infty$ is allowed in the theorem. Naturally, the probability of reaching M depends on M but you may safely ignore $M - f$ in determining your behavior.

That myopic strategies are not optimal for general (α, β) is shown by the following example (a special case of Example 3.1, Berry and Fristedt (1980b)). Set $\alpha = 7/15$, $\beta = 14/15$, and $M = 3$. Consider the myopic strategy σ^m, with utility $U_m(f, r)$, which chooses arm 1 whenever $r < {}^1/_2$ and arm 2 whenever $r \geqslant {}^1/_2$. Suppose that $f_0 = 2$ and, initially, $r = 8/9$. The fact that your fortune steps back and forth between $f = 1$ and $f = 2$ when neither 0 nor 3 is reached — $f_1 = 1, f_2 = 2, f_3 = 1, f_4 = 2, \ldots$ — facilitates the analysis. Assuming 0 and 3 are not reached, the arm chosen under σ^m, the result, the new fortune, and the new value of r are as follows

	(2, loss, 1, 1/2),
(2, win, 2, 7/11),	(2, loss, 1, 2/9),
(1, win, 2, 1/9),	(1, loss, 1, 1/2),

and the process continues to repeat (until 0 or 3 is reached), alternating between two successive choices of arm 2 and two of arm 1. $U_m(2, 8/9)$ can be calculated explicitly

$$U_m(1,{}^1/_2) = ({}^1/_2)\,[\beta^2 + \beta\bar{\beta}\alpha^2 + \beta\bar{\beta}\alpha\bar{\alpha}U_m(1, {}^1/_2)] + ({}^1/_2)\,[\alpha^2 + \alpha\bar{\alpha}\beta^2 +$$
$$+ \alpha\bar{\alpha}\beta\bar{\beta}U_m(1,{}^1/_2)] = \frac{\beta^2 + \alpha^2 + \alpha\beta^2 + \alpha^2\beta - 2\alpha^2\beta^2}{2(1 - \alpha\bar{\alpha}\beta\bar{\beta})} = 0.67000,$$

so

$$U_m(2,8/9) = \frac{119}{135} + \frac{16}{135}\,U_m(1, {}^1/_2) = 0.96089.$$

Now suppose arm 1 is chosen first!, followed by arm 2 at the next five stages. Then $f_6 = 2$, assuming 0 and 3 are not reached, and the new r is again $8/9$; σ^m is followed from stage 7 onward. The utility of this strategy can be calculated in a straightforward manner from the utility of σ^m and turns out to be 0.96105, which is greater than $U_m(2, 8/9)$. Therefore, while this strategy is clearly not optimal itself, it is better than σ^m. Whether arm 1 is optimal at stages 1, 7, 13, ..., and arm 2 otherwise, is an open question. But the precise form of an optimal strategy is not very important since the context is so special. The importance of the example is that myopic strategies are not optimal.

Acknowledgement. Professor David C. Heath programmed the calculations reported in Table 1.

REFERENCES

1. Berry, D. A., Fristedt, B., *Two-armed bandits with a goal, I. One arm known.* Adv. in Appl. Probab., **12**, 775–798 (1980 a).
2. Berry, D. A., Fristedt, B., *Two-armed bandits with a goal, II. Dependent arms.* Adv. in Appl. Probab., **12**, 958–971 (1980 b).
3. Berry, D. A., Heath, D. C., Sudderth, W. D., *Red-and-black with unknown win probability.* Ann. Statist., **2**, 602–608 (1974).
4. Dubins, L. E., Savage, L. J., *Inequalities for Stochastic Processes: How to Gamble if You Must.* New York, Dover, 1976.
5. Feldman, D., *Contributions to the 'two-armed-bandit' problem.* Ann. Math. Statist., **33**, 847–856 (1962).

E. M. J. BERTIN

Some results on unimodal distribution functions

1. The notion of unimodal probability shares with many other useful mathematical notions the property that there is no way to pinpoint the origin of the concept. An early, and perhaps decisive, instance of the absence of unimodality has been recorded by the zoologist W. F. R. Weldon, taking measurements of the Naples crabs. In 1892 he writes to K. Pearson: "Out of the mouths of babes and sucklings hath he perfected praise. In the last few evenings I have wrestled with a double humped curve, and have overthrown it. Enclosed is the diagram. If you scoff at this, I shall never forgive you" [E. S. Pearson; 1970].

Pearson did not scoff at the bimodal distribution, but, like Weldon, thought of a possible mixture of normal distributions within the same race, perhaps due to evolution. The problem of the decomposition of the given distribution into two normal distributions led to the fundamental paper [Pearson; 1894].

The first paper, dealing with unimodality as an independent notion, is perhaps [Hinčin; 1938]. The intuitive notion of a distribution function F, *unimodal at* a, as a d.f. having a one humped density, is replaced by the more precise definition that F is convex on $(-\infty, a)$ and concave on (a, ∞), a being a *mode* of F.

The well known *theorem of Hinčin* asserts the equivalence of the following statements

(1) *F is unimodal at a.*

(2) *F is absolutely continuous on* $\complement\{a\}$ *and* $G(x) = F(x+a) - xF'(x+a)$, a.e., *for some d.f.* G.

(3) $F(x) = \int_0^1 G\left(\frac{x-a}{u}\right) \mathrm{d}u$, $x \in \mathbb{R}$, *for some d.f.* G.

(4) *The c.f.* χ *of* F *satisfies* $\chi(t) = \frac{e^{ita}}{t}\int_0^t \Psi(s)\, \mathrm{d}s$, $t > 0$, *for some c.f.* Ψ.

A modern proof of this theorem would obtain the d.f. G by applying the Choquet theorem on the closed convex set U_a of all d.f.'s. unimodal at a.

Since the publication of this theorem did not prevent a subsequent mathematical comedy of false theorems, wrong examples, erroneous counterexamples and vain conjectures, there might be some interest in finding other necessary and sufficient conditions for unimodality.

2. A clue for another characterisation of unimodality of the d.f. F is given by the Lévy *concentration function*

$$Q(l) = \sup \{F(x+l) - F_-(x) : x \in \mathbb{R}\}, \text{ for } l \in [0, \infty],$$
$$Q(l) = 0 \text{ for } l < 0.$$

The relations between the properties of F and those of Q are governed by the *pointer* A of F

$$A(l) = \{x \in \overline{\mathbb{R}} : Q(l) = F(x+l) - F_-(x)\}, \ l \in [0, \infty].$$

Since F is a u.s.c. function on $\overline{\mathbb{R}}$, $l \to F(x+l) - F_-(x)$ is a u.s.c. function on $\overline{\mathbb{R}}$ and A is a multivalued function from $[0, \infty]$ into $\overline{\mathbb{R}}$, with non empty compact values. Remembering that A is said to be *u.s.c. at* l_0 if, for each neighbourhood V of $A(l_0)$, we have $A<W> \subset V$ for some neighbourhood W of l_0, it is not difficult to prove

Theorem 2.1.

1. $l_0 \in [0, \infty]$, *Q continuous at $l_0 \Rightarrow A$ u.s.c. at l_0.*

2. $l \in (0, \infty) \Rightarrow A(l) \subset \mathbb{R}$.

The concentration function of any unimodal d.f. is necessarily unimodal at 0. Indeed, given l, $k > 0$ and $\lambda \in (0, 1)$, we choose $x \in A(l)$, $y \in A(k)$ and observe that $Q(\lambda l + (1-\lambda) k) \geqslant F(\lambda x + (1-\lambda) y + l + k) - F_-(\lambda x + (1-\lambda) y) \geqslant \lambda(F(x+l) - F_-(x)) + (1-\lambda)(F(y+k) - F_-(y)) = \lambda Q(l) + (1-\lambda) Q(k)$.

As another necessary condition for unimodality we mention the fact that $A(l)$ is a convex set for $l \in [0, \infty]$ whenever F is unimodal. Indeed, the set of modes of F is a compact interval $[a_-, a_+]$ and $A(l)$ is the set of maximal points in $[a_+ - l, a_+]$ of the concave function $x \to F(x+l) - F_-(x)$.

It may be shown ([1], p. 332) that neither unimodality of the concentration function, nor convex valuedness of the pointer are sufficient for the unimodality of F. However

Theorem 2.2. *Let F be a d.f., Q its concentration function and A its pointer. Then F is unimodal if and only if Q is unimodal and $A(l)$ is convex for $l > 0$.*

Indeed, let Q be unimodal and let each $A(l)$ be convex. Since Q is concave and continuous on $(0, \infty)$, it follows from Theorem 2.1 that A is u.s.c. on $(0, \infty)$ and hence that $J = A < (0, \infty) >$ is a convex set. By direct computation, it is easily shown that F is convex on J. In the same way F is concave on the interval $K = (A + I) < (0, \infty) >$. Finally, the support of F is contained in $\overline{J \cup K}$. See ([1], p. 331) for details.

3. In this section, F denotes a unimodal d.f., with mode a. Clearly, if F and A are smooth functions, we have $F'(A(l)) = F'(A(l) + l) = Q'(l)$. An extension of this useful equality to the general case is obtained by introducing the *larboard function* $b : l \to \inf A(l)$ and the *starboard function* $t : l \to \sup A(l)$. We mention the following result. The sub(super) differential of a convex (concave) function is denoted by ∂.

Theorem 3.1.

1. *b and t are decreasing and absolutely continuous on* $[0, \infty)$, $b(l) \subset [a_- - l, a_-]$ *and* $t(l) \subset [a_+ - l, a_+]$.

2. *$b + I$ and $t + I$ are increasing and absolutely continuous on* $[0, \infty)$.

3. $b(l) < a_+ \Rightarrow \partial Q(l) \subset \partial F(b(l))$; $t(l) > a_- - l \Rightarrow \partial Q(l) \subset \partial F(t(l) + l)$.

Using these relations, one may corroborate slightly the Hinčin representation. With the convention $h^-(z) = \inf h^{-1}(z)$, one obtains e. g.

Corollary 3.2. *Let G be the d.f. associated with F by the theorem of Hinčin. Then*

$$G(x) = -\int_{-\infty}^{x} s d(Q' \circ b^-(s + a) \ \textit{for} \ x < 0$$

and

$$G(x) = 1 + \int_{x}^{\infty} s d\,(Q'_- \circ (b + I)^-)\,(s + a) \ \textit{for} \ x > 0.$$

4. There are similar, but not completely analogous, results for discrete unimodality. The notions of a (discrete) interval of $\mathbb{Z}$, and of (discrete) convexity of a function $F\colon \mathbb{Z} \to \mathbb{R}$ are defined in the obvious way. Remember that a distribution F on $\mathbb{Z}$ is called *d-unimodal at a* if F is convex on $(-\infty, a]$ and concave on $[a, \infty]$. For each $G\colon\mathbb{Z} \to \mathbb{R}$, we denote by $\hat{G}$ the piecewise linear extension of G to $\mathbb{R}$. The following result is almost obvious.

Lemma 4.1. *Let F be a discrete d.f. The following assertions are equivalent*

1. *F is d-unimodal at a.*

2. *$\hat{F}$ is unimodal at a.*

3. *F is the restriction to $\mathbb{Z}$ of a d.f. G on $\mathbb{R}$, unimodal at a.*

For each d.f. F on $\mathbb{Z}$, the *concentration function* Q_F and the *pointer* A_F of F are defined as in Section 2. The following links with the continuous case lead to the main Theorem 4.4 below.

Lemma 4.2. *Let F be a discrete d.f. Then*

1. $Q_F = Q_{\hat{F}}$ *on* $\mathbb{Z}$ *and* $Q_{\hat{F}}$ *is piecewise linear (but not necessarily* $Q_{\hat{F}} = \hat{Q}_F$).

2. $A_{\hat{F}}(l) = A_F(l) \cup \{[n, n + 1]\colon n, n + 1 \in A_F(l)\}$, $l \in \mathbb{N}$.

In the continuous case, the unimodality of Q implies the upper semicontinuity of the pointer, which is essential for the proof of the main Theorem 2.2. In the discrete case, there is no automatic d-semicontinuity of the pointer

Definition 4.3. A multivalued function $A\colon \mathbb{N}^* \to \mathbb{L}$ is called *d-u.s.c.* if k, $l \in \mathbb{N}^*$, $|k - l| \leqslant 1 \Rightarrow A(l) \subset A(k) \cup (A(k) + 1) \cup (A(k) - 1)$.

However we have

Theorem 4.4. *Let F be a discrete d.f., A its pointer and Q its concentration function. The following assertions are equivalent*

1. *F is d-unimodal.*

2. *Q is d-unimodal, A is d-u.s.c. and each $A(l)$, $l \in \mathbb{N}^*$, is a discrete interval.*

The implication $1 \Rightarrow 2$ follows easily from the corresponding properties for the continuous case and Lemmas 4.1 and 4.2. The proof of the implication

$2 \Rightarrow 1$ is based on the fact that A maps a discretely connected set onto a discretely connected set. See [2] for details.

Corollary 4.5. *Let F be **d**-unimodal. Then* $Q_{\hat{F}} = \hat{Q}_F$.

5. *A* d.f. F, unimodal at a, is absolutely continuous on $\mathbf{C}\{a\}$, and hence admits a density. Intuitively, a d.f. F is unimodal at a if the density is increasing on $(-\infty, a)$ and decreasing on (a, ∞). This statement can be made exact by introducing the following notions.

Definition 5.1.

1. *A* function $f \geqslant 0$ is called an *admissible density* of the d.f. F if f is measurable and $f(x) \in [\min\{F'_-(x), F'_+(x)\}, \max\{F'_-(x), F'_+(x)\}]$, whenever these quantities exist.

2. A function $f\colon \mathbb{R} \Rightarrow (-\infty, \infty]$ is called *single humped* if there exists a partition $\{I, J\}$ of $\mathbb{R}$, with I and J intervals, I preceding J, such that f is increasing on I and decreasing on J.

In optimization theory, a single humped function is often called a *unimodal function*. By a simple topological reasoning, one sees that f is single humped if and only if each set $\{f \geqslant \alpha\}$, $\alpha < \sup f$, is convex. Moreover, from standard results on convex functions, follows

Lemma 5.2.

1. *Each unimodal d.f. F has an admissible density, and each admissible density of F is single humped.*

2. *Conversely, if the a.c. d.f. F has a single humped density, then F is unimodal.*

A weaker sufficient condition for unimodality is produced by the theory of quasi-concavity.

Definition 5.3. Let $\lambda \in (0, 1)$.

1. A function $f\colon \mathbb{R} \to (-\infty, \infty]$ is called *λ-quasi-concave* if $f(\lambda x + (1-\lambda)y) \geqslant \min\{f(x), f(y)\}$ for $x < y$.

2. A subset A of $\mathbb{R}$ is called *λ-quasi-convex* if its characteristic function is λ-quasi-concave.

3. A function f is called *quasi-concave* if *f is μ—quasi-concave for every* $\mu \in (0, 1)$.

Clearly f single humped $\Leftrightarrow f$ quasi-concave $\Rightarrow f$ λ-quasi-concave. In [Pijnappel, 1980], it is shown that the theorem of Ostrowski, stating that every measurable midpoint convex function is convex, admits an analogue for λ-quasi-concave functions. We sketch the steps of a proof.

Lemma 5.4. *Let $f \leqslant 0$ be λ-quasi concave, $a < b$ and $f(a) = f(b) = 0$. Let A denote the λ-quasi-convex hull of $\{a, b\}$. Then $A \subset [a,b] \cap f^{-1}(0)$ and $\bar{A} = [a, b]$.*

Lemma 5.5. *If, moreover, f is integrable on $[a, b]$, then*

$$(b-a)\int_a^c f(t)\,dt = (c-a)\int_a^b f(t)dt \quad \textit{for } c \in [a, b].$$

Indeed, the set of all c for which this equality holds contains the closed λ-quasi-convex hull of $\{a, b\}$.

Lemma 5.6. *Let f be an integrable and λ-quasi-concave function such that $f \leqslant$ essup f. Then f is quasi-concave.*

One has to prove the convexity of $\{f \geqslant \alpha\}$ for $\alpha < \sup f$. Let $a < b$, $f(a) \geqslant \alpha$ and $f(b) \geqslant \alpha$. One may suppose that the measure of $\{f \geqslant \alpha\} \cap [a, b]$ is strictly positive. The function g: $x \rightarrow \min\{f(x), \alpha\} - \alpha$ is λ-quasi-concave and integrable, $g^{-1}(0)$ is dense and not negligible in $[a, b]$ (Lemma 5.4), and $g(x) = (b-a)^{-1} \int_a^b g(t)dt$ a.e. on $[a, b]$ (Lemma 5.5]. In particular, $(b - a)^{-1} \int_a^b g(t)dt = 0$. Since g is λ-quasi-concave, one has $g = 0$ and $f \geqslant \alpha$ on $[a, b]$.

Summing up we have

Theorem 5.7. *Let F be an a.c. d.f. and* $\lambda \in (0,1)$. *The following assertions are equivalent*

1. *F is unimodal.*
2. *F has a* λ*-quasi-concave density.*
3. *Every admissible density of F is quasi-concave.*

Indeed, if f is a λ-quasi-concave density of F, the function $\hat{f} = \inf \{f, \operatorname{essup} f\}$ is a λ-quasi-concave density of F, satisfying $\hat{f} \leqslant \operatorname{essup} \hat{f}$.

REFERENCES

1. Bertin, E. M. J., Hengartner, W., Theodorescu, R., *Some characterizations of unimodal distribution functions.* Z. Wahrsch. Verw. Gebiete, **57**, 327–338 (1981).
2. Bertin, E. M. J., Theodorescu, R., *Some characterizations of discrete unimodality.* Preprint 173, Utrecht, 1980.
3. Bertin, E. M. J., Theodorescu, R., *Propriétés de quasiconcavité des fonctions de répartition unimodales.* Preprint 495, Bonn, 1982.
4. Hinčin, A. Y., *On unimodal distribution.* Izv. Nauchno-Issled Inst. Math. Mekh. Tomsk. Gos. Univ., **2**, 1–7 (1938). (Russian)
5. Pearson, E. S., *Some incidents in the early history of statistics,* 1890– 1894. In: Pearson, E.S., Kendall, M. G., (Eds.) *Studies in the History of Statistics and Prcbability,* pp. 323—353. London, Charles Griffin & Co, 1970.
6. Pearson, K., *Contribution to the mathematical theory of evolution.* Philos. Trans. Roy. Soc. London Ser, A, **185**, 71–110 (1894).
7. Pijnappel, W.W.F., *Halfconvexe functies.* Thesis, Nijmegen, 1980.

LOUIS H. BLAKE

Some implications of metric tightness on sub σ-fields of a probability space

1. The usual metric on sub σ-fields of a probability space

Let $(\Omega, \mathbf{A}, P)$ be a probability space. From a naive point of view, it can be asked how "close" are (two) sub σ-fields of $\mathbf{A}$. The answer must surely be dictated by how "close" the individual sets in the one σ-field are to the individual sets in the other σ-field. This gives rise to the definition of the "usual" metric on sub σ-fields.

Let $\mathbf{B}$, $\mathbf{C}$ be two sub σ-fields of $\mathbf{A}$, then

i) $d(B, \mathbf{C}) = \inf_{C \in \mathbf{C}} P(B \Delta C)$

ii) $d(\mathbf{B}, \mathbf{C}) = \sup_{B \in \mathbf{B}} d(B, \mathbf{C})$

iii) $D(\mathbf{B}, \mathbf{C}) = d(\mathbf{B}, \mathbf{C}) + d(\mathbf{C}, \mathbf{B})$.

It is clear that D is a pseudo-metric on the set of sub σ-fields of $\mathbf{A}$.

It should be stressed that $D(\mathbf{B}, \mathbf{C}) = 0$ means for each $B \in \mathbf{B}$, $B \in \mathbf{C}$ up to a P-null set in $\mathbf{A}$ and for each $C \in \mathbf{C}$, $C \in \mathbf{B}$ up to a P-null set in $\mathbf{A}$. This is discussed in [4],

In [9], Neveu introduces the idea of *relative completeness;* that is, for $\mathbf{B} \subset \mathbf{C}$, $\mathbf{B}$ is relatively complete with respect to $\mathbf{C}$ if every $\mathbf{C}$-null set is a $\mathbf{B}$ null set. If in the previous definition of the usual metric $\mathbf{B}$ and $\mathbf{C}$ are both complete relative to $\mathbf{A}$ and $D(\mathbf{B}, \mathbf{C}) = 0$, then clearly $\mathbf{B} = \mathbf{C}$.

Very closely related to a metric distance between sub σ-fields is the idea of how much smoothing the conditional expectation operator of the one σ-field does on L_1 functions of the other σ-field. More precisely, in [3] for $\mathbf{B}$, $\mathbf{C}$ sub σ-fields of $\mathbf{A}$

$$\rho(\mathbf{B}, \mathbf{C}) \equiv \inf_{\varepsilon > 0} \{\sup_{B \in \mathbf{B}} P[|E(I_B|\mathbf{C}) - I_B| > \varepsilon] \leqslant \varepsilon\}$$

represents the amount of smoothing done on $\mathbf{B}$ by $\mathbf{C}$. It is proved in [3] that for any sequence of sub σ-fields $\{\mathbf{A}_n\}_{n=1}^{\infty}$ of $\mathbf{A}$ that $\rho(\mathbf{A}, \mathbf{A}_n) \to 0$ if and only if $d(\mathbf{A}, \mathbf{A}_n) \to 0$.

A rather fundamental question addressed by Boylan in [4] is that for all $f \in L_1(\Omega, \mathbf{A}, P)$ is the conditioning on f done by $\mathbf{B}$ and $\mathbf{C}$ (by the conditional expectation operator) the "same" if $D(\mathbf{B}, \mathbf{C}) = 0$? The answer is "yes" as expected. It is worthwhile to see how the idea of smoothing makes the answer very immediate as opposed to the clever but protracted proof in [4].

If $D(\mathbf{B}, \mathbf{C}) = 0$ then for all $B \in \mathbf{B}$, $E(I_B \mid \mathbf{C}) - I_B = 0$ a.e. (and $\rho(\mathbf{B}, \mathbf{C}) = \rho(\mathbf{C}, \mathbf{B}) = 0$). Merely let $\{S_n\}_{n=1}^{\infty}$ be an increasing sequence of $\mathbf{B}$-measurable simple functions with $\lim\limits_{n\to\infty} S_n = E(f \mid \mathbf{B})$. Clearly $E(S_n \mid \mathbf{C}) = S_n$ a.e. and $\lim\limits_{n\to\infty} E(S_n \mid \mathbf{C}) = E(E(f \mid \mathbf{B}) \mid \mathbf{C})$ and also $\lim\limits_{n\to\infty} E(S_n \mid \mathbf{C}) = E(f \mid \mathbf{B})$ a.e.

Thus $E(f \mid \mathbf{B})$ is $\mathbf{C}$-measurable a.e. Next, let $C \in \mathbf{C}$. Then, there exists $\bar{C} \in \mathbf{B}$ and $P(C \,\Delta\, \bar{C}) = 0$. So,

$$\int_C E(E(f \mid \mathbf{B}) \mid \mathbf{C}) dP = \int_{\bar{C}} E(f \mid \mathbf{B}) dP = \int_{\bar{C}} f \, dP = \int_C f \, dP.$$

It has thus been established that there exists $N \in \mathbf{C}$ with $P(N) = 0$ and $E(f \mid \mathbf{B}) + I_N$ being $\mathbf{C}$-measurable and for every $C \in \mathbf{C}$, $\int_C E(f \mid \mathbf{B}) + I_N = \int_C f$. And this characterizes $E(f \mid \mathbf{C})$. That is, $E(f \mid \mathbf{B}) = E(f \mid \mathbf{C})$ a.e.

There is another characterization of tightness with respect to the usual metric in terms of "traces" of measurable spaces which is developed by Neveu in [9]. The limitation here, however, is that tightness is between two σ-fields related by containment.

Proposition (Neveu). *Let $\mathbf{B} \subset \mathbf{C}$ be two σ-fields complete relative to $\mathbf{A}$. Then there exists $C \in \mathbf{C}$ such that $C \cap \mathbf{B} = C \cap \mathbf{C}$ and $P(\sim C) \leqslant 4d(\mathbf{B}, \mathbf{C})$. Conversely, if $C \in \mathbf{C}$ and $C \cap \mathbf{B} = C \cap \mathbf{C}$, then $d(\mathbf{B}, \mathbf{C}) \leqslant P(\sim C)$.*

One of the main applications of the usual metric is an "equiconvergence" theorem for martingales.

The following definition is needed.

Definition. Let ∂ be any collection of functions in $L_1(\Omega, \mathbf{A}, P)$ with $\{E(|x| \mid \mathbf{A}_n), x \in \delta, n = 1, 2, \ldots, \infty\} \equiv F(\partial)$ a uniformly integrable family, where $\{\mathbf{A}_n\}_{n=1}^{\infty}$ is an increasing (or decreasing) sequence of sub σ-fields in $\mathbf{A}$ with $\sigma\left(\bigcup_{n=1}^{\infty} \mathbf{A}_n\right) \equiv \mathbf{A}_{\infty}$. Then ∂ has an *equiconvergence* (of martingales) theorem if $E(x \mid \mathbf{A}_n) \to E(x \mid \mathbf{A}_{\infty})$ in L_1 at a rate independent of $x \in \partial$.

The main result in this direction first proved by Martin [8] and subsequently by Boylan [4], Neveu [9] and Blake [3] is given by the

Theorem. *With an increasing (decreasing) sequence of sub σ-fields $\{\mathbf{A}_n\}_{n=1}^{\infty}$, with limit σ-field $\mathbf{A}_{\infty}$, $F(\partial)$ a uniformly integrable family and $d(\mathbf{A}_{\infty}, \mathbf{A}_n) \to 0$ as $n \to \infty$, ∂ has an equiconvergence theorem.*

It is really Neveu [9] who first produces a rate at which this equiconvergence takes place. Also in Blake [3], Lemma 4 produces the basis for extracting a rate in terms of the smoothing coefficient.

It would not be appropriate to conclude this discussion of the usual metric without drawing attention to the carefully crafted theorem of Boylan [4] which says that the usual metric is complete.

2. A conditional metric on sub σ-fields of a probability space

In [7], Hanen and Neveu establish the following useful result: with $\mathbf{B}$ a sub σ-field of $\mathbf{A}$ and $A \in \mathbf{A}$, $\{P^{\mathbf{B}}(\sim A) = 0\}$ is the largest $\mathbf{B}$-measurable set contained $P^{\mathbf{A}}$ a.e. in A. This set is unique up to P-equivalence.

In the sequel write $(A)_{\mathbf{B}} \equiv \{P^{\mathbf{B}}(\sim A) = 0\}$.

The following definition establishes a conditional metric.

Definition. $\mathbf{B}, \mathbf{B}_1, \mathbf{B}_n$ are sub σ-fields of $\mathbf{A}$.

(i) $g(\mathbf{B}, \mathbf{B}_1) = \sup\limits_{B \in \mathbf{B}} P(B \setminus (B)_{\mathbf{B}_1})$

(ii) $G(\mathbf{B}, \mathbf{B}_1) = g(\mathbf{B}, \mathbf{B}_1) + g(\mathbf{B}_1, \mathbf{B})$.

It is an easy exercise to show that if $G(\mathbf{B}, \mathbf{B}_n) \to 0$ as $n \to \infty$ then $D(\mathbf{B}, \mathbf{B}_n) \to 0$ as $n \to \infty$.

As in the case of the usual metric a relationship between the conditional metric and a smoothing index can be developed.

Definition. For any two sub σ-fields $\mathbf{B}, \mathbf{C}$ let

$$\lambda(\mathbf{B}, \mathbf{C}) \equiv \inf_{\varepsilon>0}\{(\sup_{B \in \mathbf{B}} P[E(I_B \mid \mathbf{C}) - I_{(B)_{\mathbf{C}}} > \varepsilon]) \leqslant \varepsilon\}.$$

$\lambda(\mathbf{B}, \mathbf{C})$ is the amount $\mathbf{B}$ *is smoothed to* $\mathbf{C}$.

The following relationship exists between γ and g

$$\lambda^2(\mathbf{B}, \mathbf{C}) \leqslant g(\mathbf{B}, \mathbf{C}) \leqslant 2\gamma(\mathbf{B}, \mathbf{C}).$$

(Proof to appear elsewhere.)

With this result, a direct relationship between ρ and γ can be established.

Proposition. $\rho(\mathbf{B}, \mathbf{C}) \leqslant 3\gamma(\mathbf{B}, \mathbf{C})$.

Before the proposition is proved, the following result will be needed.

Lemma. *If* $(\sup\limits_{B \in \mathbf{B}} P(\mid E(I_B \mid \mathbf{C}) - I_B \mid > \delta)) > \delta$, *then* $\rho(\mathbf{B}, \mathbf{C}) \geqslant \delta$. *The analogous result holds for* $\gamma(\mathbf{B}, \mathbf{C})$.

Proof. Suppose $\xi < \delta$ and $(\sup\limits_{B \in \mathbf{B}} P(\mid E(I_B \mid \mathbf{C}) - I_B \mid > \xi)) \leqslant \xi$. Then,

$$\delta \leqslant \sup_{B \in \mathbf{B}} P(\mid E(I_B \mid C) - I_B \mid > \delta) \leqslant \sup_{B \in \mathbf{B}} P(\mid E(I_B \mid C) - I_B \mid > \xi) \leqslant \xi$$

which is a contradiction. Thus,

$$\rho(\mathbf{B}, \mathbf{C}) = \inf_{\varepsilon>0}\{(\sup_{B \in \mathbf{B}} P(\mid E(I_B \mid \mathbf{C}) - I_B \mid > \varepsilon) \leqslant \varepsilon\} \geqslant \delta.$$

Proof of Proposition.

$$\mid E(I_B \mid \mathbf{C}) - I_B \mid \leqslant \mid E(I_B \mid \mathbf{C}) - I_{(B)_{\mathbf{C}}} \mid + \mid I_B - I_{(B)_{\mathbf{C}}} \mid.$$

So, consider any $\tau > 1$ and write

$$P(E(I_B \mid \mathbf{C}) - I_{(B)_{\mathbf{C}}} > \varepsilon/\tau) + P(I_B - I_{(B)_{\mathbf{C}}} > \frac{\tau - 1}{\tau}\varepsilon) =$$

$$= P(E(I_B \mid \mathbf{C}) - I_{(B)_{\mathbf{C}}} > \varepsilon/\tau) + P(B - (B)_{\mathbf{C}}).$$

So,

$$\sup_{B \in \mathbf{B}} P(\mid E(I_B \mid \mathbf{C}) - I_B \mid > \varepsilon) \leqslant \sup_{B \in \mathbf{B}} P(E(I_B \mid \mathbf{C}) - I_{(B)_{\mathbf{C}}} > \varepsilon/\tau) + g(\mathbf{B}, \mathbf{C}).$$

In particular write for any $\delta > 0$, $\varepsilon = \rho - \delta$. So,

$$\frac{\rho-\delta}{\tau} \leqslant \rho - \delta \leqslant \sup_{B \in \mathbf{B}} P(|E(I_B|\mathbf{C}) - I_B| > \rho - \delta) \leqslant$$

$$\sup_{B \in \mathbf{B}} P\left(E(I_B \mid \mathbf{C}) - I_{(B)\mathbf{C}} > \frac{\rho-\delta}{\tau}\right) + g(\mathbf{B}, \mathbf{C}).$$

Thus,

$$\frac{\rho-\delta}{\tau} - g \leqslant \sup_{B \in \mathbf{B}} P\left(E(I_B \mid \mathbf{C}) - \mathrm{I}_{(B)\mathbf{C}} > \frac{\rho-\delta}{\tau}\right) \leqslant$$

$$\sup_{\mathbf{B} \in B} P\left(E(I_B \mid \mathbf{C}) - I_{(B)\mathbf{C}} > \frac{\rho-\delta}{\tau} - g\right).$$

So, by the lemma

$$\lambda(\mathbf{B}, \mathbf{C}) \geqslant \frac{\rho(\mathbf{B},\mathbf{C}) - \delta}{\tau} - g(\mathbf{B}, \mathbf{C}) \geqslant \frac{\rho(\mathbf{B}, \mathbf{C}) - \delta}{\tau} - 2\,\gamma(\mathbf{B}, \mathbf{C}).$$

and when $\delta > 0$ is arbitrary and $\tau > 1$ is arbitrary, the result follows.

The following theorem is a Neveu-like characterization of tightness in the conditional metric.

Theorem. *Let* $\{\mathbf{B}_n\}_{n=1}^{\infty}$ *be a monotone sequence (say increasing) of sub σ-fields generating and complete relative to* $\mathbf{B}_\infty$. $g(\mathbf{B}_\infty, \mathbf{B}_n) \to 0$ *as* $n \to \infty$ *if and only if there exists a sequence of sets* $\{B_n\}_{n=1}^{\infty}$ *such that* $B_n \in \mathbf{B}_n$ *each* n, $P(B_n) \to 1$ *as* $n \to \infty$ *and* $B_n \cap \mathbf{B}_n = B_n \cap \mathbf{B}_\infty$.

Proof. To appear elsewhere.

It is exactly this result which facilitates the proof of the following equiconvergence

Theorem. *With an increasing (decreasing) sequence of sub σ-fields* $\{\mathbf{A}_n\}_{n=1}^{\infty}$ *with limit σ-field* $\mathbf{A}_\infty$ *and* $F(\partial)$ *a uniformly integrable family and* $\rho(\mathbf{A}_n, \mathbf{A}_\infty) \to 0$ *as* $n \to \infty$, ∂ *has an equiconvergence theorem. Moreover, the rate of equiconvergence is faster than that suggested in [9] by Neveu in the case of tightness between sub σ-fields in the usual metric.*

Proof. To appear elsewhere.

This section will close with a result and a question about the conditional metric.

Theorem. *Let* $\{\mathbf{A}_n\}_{n=1}^{\infty}$ *be an increasing (decreasing) sequence of sub σ-fields which is Cauchy in the G metric. Then* $g(\mathbf{A}_\infty, \mathbf{A}_n) \to 0$ *as* $n \to \infty$ *where* $\mathbf{A}_\infty \equiv \sigma\left(\bigcup_{n=1}^{\infty} \mathbf{A}_n\right)$. *That is, G is "order-complete".*

Proof. To appear elsewhere.

The *open question* is whether G is a complete metric.

It is of course not true in a general topological setting that a metric space with a partial ordering which is "order complete" is complete. An easy example would be the unit square under the usual Euclidean metric with the top removed. Say $(x_1, y_1) \prec (x_2, y_2)$ if $y_1 = y_2$ and $x_1 \leqslant x_2$. Obviously the metric space is "order complete" but the metric space is not complete.

3. The conditional metric and regularity of conditional probability

Let $\mathbf{B}$ be a sub σ-field of $\mathbf{A}$ and $P^{\mathbf{B}}$ be a conditional probability on $\mathbf{A}$. Let $p(\cdot, \cdot \mid \mathbf{B}, \mathbf{A})$ denote a version of $P^{\mathbf{B}}$ on $\mathbf{A}$.

Suppose $\{\mathbf{B}_n\}_{n=1}^{\infty}$ is an increasing (decreasing) sequence of sub σ-fields complete relative to $\mathbf{B}_\infty = \sigma\left(\bigcup_{n=1}^{\infty} \mathbf{B}_n\right)$. Suppose $P^{\mathbf{B}_n}$ is regular on $\mathbf{A}$ for each n. Is $P^{\mathbf{B}}$ regular on $\mathbf{A}$? The answer in general is "no".

Suppose $\{\mathbf{A}_n\}_{n=1}^{\infty}$ is an increasing (decreasing) sequence of sub σ-fields complete relative to $\mathbf{A} = \sigma\left(\bigcup_{n=1}^{\infty} \mathbf{A}_n\right)$. Suppose $P^{\mathbf{B}}$ is regular on $\mathbf{A}_n$ for each n. Is $P^{\mathbf{B}}$ regular on $\mathbf{A}$? The answer is in general "no".

The purpose of this section is to show that with sufficient tightness between the ascending sub σ-fields the answer to each of the previous questions is "yes".

First consider how the Dieudonné example [5] of a non-regular conditional probability can be used to display a "no" answer to each of the previous questions when no tightness conditions are imposed.

Consider $(\Omega, \mathbf{B}, \nu)$ as the usual Lebesgue measure space on $[0, 1] = \Omega$ Let $\mathbf{A} = \sigma(\mathbf{B}, M)$, where M is a non Borel subset of $[0, 1]$ with $\nu_*(M) = 0$ and $\nu^*(M) = 1$. P is then defined as the extension of ν from $\mathbf{B}$ to $\mathbf{A}$. More specifically, for each $A \in \mathbf{A}$ with $A = B_1M + B_2M^c$ with B_1, B_2 in $\mathbf{B}$ write $P(A) = \nu(B_1)$. For further details on this construction see [5], [6], [1] and [2].

Let $\mathbf{N}$ be the null sets of $\mathbf{B}$ where $\mathbf{B}$ is generated by $\{\mathbf{B}_n\}_{n=1}^{\infty}$, an increasing sequence of sub σ-fields with each $\mathbf{B}_n$ generated by a finite sequence of sets of non zero measure and $\mathbf{N}$. Clearly $P^{\mathbf{B}_n}$ is regular on $\mathbf{A}$ for each n but $P^{\mathbf{B}}$ is not regular on $\mathbf{A}$.

Next, let $\mathbf{N}$ be the null sets of $\mathbf{A}$ where $\mathbf{A}$ is generated by $\{\mathbf{A}_n\}_{n=1}^{\infty}$, an increasing sequence of sub σ-fields with each $\mathbf{A}_n$ generated by a finite sequence of sets of non zero measure and $\mathbf{N}$. Clearly $P^{\mathbf{B}}$ is regular on $\mathbf{A}_n$ for each n but $P^{\mathbf{B}}$ is not regular on $\mathbf{N}$.

In the following two theorems sufficient tightness between sub σ-fields is given to assure preservation of regularity of the conditional probabilities.

Theorem. *Suppose* $\{\mathbf{B}_n\}_{n=1}^{\infty}$ *is a monotone (say increasing) sequence of sub* σ*-fields with each* $\mathbf{B}_n$ *complete relative to* $\mathbf{B}_\infty$ *and* $g(\mathbf{B}_\infty, \mathbf{B}_n) \to 0$ *as* $n \to \infty$. *If* $P^{\mathbf{B}_n}$ *is regular on* $\mathbf{A}$ *for each* n, *then* $P^{\mathbf{B}_\infty}$ *is regular on* $\mathbf{A}$ *provided* $\mathbf{A}$ *is countably generated.*

Proof. By a theorem in the previous section there exists a sequence of sets $\{B_n\}_{n=1}^{\infty}$ with $B_n \in \mathbf{B}_n$, $P(B_n) \to 1$ as $n \to \infty$ and $B_n\mathbf{B}_n = B_n\mathbf{B}_\infty$. There exists a subsequence B_{n_j} such that $I_{B_{n_j}} \to I_\Omega P_{\mathbf{B}_\infty}$ a.e. For ease of notation (and without loss of generality) assume $I_{\mathbf{B}_n} \to I_\Omega$ off N, a $\mathbf{B}_\infty$ null set.

Let $p(\cdot, \cdot \mid \mathbf{B}_n, \mathbf{A})$ denote a regular version of $P^{\mathbf{B}_n}$ on $\mathbf{A}$.

Let $\mathcal{A}$ be a countable field generating $\mathbf{A}$. Let $p(\cdot, \cdot \mid \mathbf{B}_\infty, \mathbf{A})$ be an arbitrary version of $P^{\mathbf{B}_\infty}$ on $\mathcal{A}$. For each $A \in \mathcal{A}$ there exists a $\mathbf{B}_\infty$-null set N_A off which $p(\omega, A \mid \mathbf{B}_n, \mathbf{A})\, I_{\mathbf{B}_n} = p(\omega, A \mid \mathbf{B}_\infty, \mathbf{A}) I_{\mathbf{B}_n}$ for all n. It is then clear that off $N \cup \bigcup_{A \in \mathcal{A}} N_A$, $\tilde{p}(\omega, A \mid \mathbf{B}_\infty, \mathbf{A}) \equiv \lim_{n \to \infty} p(\omega, A \mid \mathbf{B}_\infty, \mathbf{A}) I_{\mathbf{B}_n}$ and $\tilde{p}(\omega, \cdot \mid \mathbf{B}_\infty, \mathbf{A})$ is countably additive on $\mathcal{A}$. $\tilde{p}(\omega, . \mid \mathbf{B}_\infty, \mathbf{A})$ can then be extended in the usual manner for $\omega \notin N \cup \bigcup_{A \in \mathcal{A}} N_A$ as a measure on $\mathbf{A}$ giving a regular version of $P^{\mathbf{B}}$ on $\mathbf{A}$.

Theorem. *Let $\{\mathbf{A}_n\}_{n=1}^{\infty}$ be a monotone (say increasing) sequence of sub σ-fields with each $\mathbf{A}_n$ complete relative to $\mathbf{A}_\infty$ and $g(\mathbf{A}_\infty, \mathbf{A}_n) \to 0$ as $n \to \infty$. If $P^{\mathbf{B}}$ is regular on $\mathbf{A}_n$ for each n, then $P^{\mathbf{B}}$ is regular on $\mathbf{A}$ provided $\mathbf{A}$ is countably generated.*

Proof. By a theorem in the previous section a sequence of sets $A'_n \in \mathbf{A}_n$ can be produced such that $A'_n\mathbf{A}_n = A'_n\mathbf{A}_\infty$ and $P(A'_n) \to 1$ as $n \to \infty$. Then there exists a sequence of sets $A_n \in \mathbf{A}_n$ such that $A_n \subset A_{n+1}$ $P^{\mathbf{A}_\infty}$ a.e. and $P(A_n) \uparrow 1$ as $n \to \infty$ and $A_n\mathbf{A}_\infty \subset \mathbf{A}_n$. (Note: $\{\mathbf{A}_n\}_{n=1}^{\infty}$ can be produced by letting $A_n = \bigcup_{k=1}^{n} A'_k$ for $n \geqslant 1$.) Let $\{p(\cdot, \cdot \mid \mathbf{B}, \mathbf{A}_n)\}_{n=1}^{\infty}$ be versions of $P^{\mathbf{B}}$ on $\mathbf{A}_n$ which are regular.

Let $\mathcal{A}$ be a countable field generating $\mathbf{A}_\infty$. A null set N_1 can be produced off which $p(\omega, A_nC \mid \mathbf{B}, \mathbf{A}_n) \leqslant p(\omega, A_{n+1}C \mid \mathbf{B}, \mathbf{A}_{n+1})$ for each $C \in \mathcal{A}$. A null set N_2 exists off which $p(\omega, C \mid \mathbf{B}, \mathbf{A}_\infty) \equiv \lim_{n\to\infty} p(\omega, A_nC \mid \mathbf{B}, \mathbf{A}_n)$ for all $C \in \mathcal{A}$. Observe that off $N_1 \cup N_2$, $p(\omega, A_nC \mid \mathbf{B}, \mathbf{A}_n) \uparrow p(\omega, C \mid \mathbf{B}, \mathbf{A}_\infty)$ for all $C \in \mathcal{A}$. Thus, consider $C = \bigcup_{i=1}^{\infty} C_i$ where $C \in \mathcal{A}$ and $\{C_i\}^{\infty}$ are in $\mathcal{A}$ and pairwise disjoint. Let $\omega \notin N_1 \cup N_2$ and $\varepsilon > 0$. Then there exists an integer M so that for all $n \geqslant M$

$$p(\omega, C \mid \mathbf{B}, \mathbf{A}_\infty) \leqslant \varepsilon + p(\omega, A_nC \mid \mathbf{B}, \mathbf{A}_n) =$$

$$\varepsilon + \sum_{i=1}^{\infty} p(\omega, A_nC_i \mid \mathbf{B}, \mathbf{A}_n) \leqslant \varepsilon + \sum_{i=1}^{\infty} p(\omega, C_i \mid \mathbf{B}, \mathbf{A}_\infty).$$

But, where ε is arbitrary $p(\omega, \cdot \mid \mathbf{B}, \mathbf{A}_\infty)$ is subadditive.

Next, there exists a null set N_3 off which $p(\omega, C \mid \mathbf{B}, \mathbf{A}_\infty)$ $p(\omega, \geqslant B \mid \mathbf{B}, \mathbf{A}_\infty)$ whenever $C \supset B$ and $C, B \in \mathcal{A}$. So, off N_3,

$$p(\omega, C \mid \mathbf{B}, \mathbf{A}_\infty) \geqslant \sum_{i=1}^{\infty} p(\omega, C_i \mid \mathbf{B}, \mathbf{A}_\infty).$$

Thus $p(\omega, \cdot \mid \mathbf{B}, \mathbf{A})$ is countably additive on $\mathbf{A}$ off $N_1 \cup N_2 \cup N_3$ and so $p(\omega, \cdot \mid \mathbf{B}, \mathbf{A}_\infty)$ can be extended in the usual manner to a measure on $\mathbf{A}_\infty$ giving a regular version of $P^{\mathbf{B}}$ on $\mathbf{A}_\infty$.

Remark. By a theorem in the previous section regarding the "order" completeness of g, it follows that $\mathbf{A}_\infty = \sigma\left(\bigcup_{n=1}^{\infty} \mathbf{A}_n\right)$ or $\mathbf{A}_\infty \supset \sigma\left(\bigcup_{n=1}^{\infty} \mathbf{A}_n\right)$ and $g\left(\mathbf{A}_\infty, \sigma\left(\bigcup_{n=1}^{\infty} \mathbf{A}_n\right)\right) = 0$. (The same can be said about $\mathbf{B}_\infty$.)

As a concluding thought, consider the following *open question:* In the previous two theorems, if tightness in the G metric is replaced by tightness in the (weaker) D metric will the respective conclusions hold?

REFERENCES

1. Blake, L. H., *Canonical extensions of measures and the extension of regularity of conditonal probabilities.* Pacific J. Math., **41**, 25–31 (1972).
2. Blake, L. H., *Simple extensions of measures and the preservation of regularity of conditional probabilities.* Pacific J. Math., **46**, 355–359 (1973).

3. Blake, L. H., *A smoothing index and equiconvergence of martingales.* J. London Math. Soc., 11, 385–388 (1975).
4. Boylan, E. S., *Equiconvergence of martingales.* Ann. Math. Statist., **42**, 552–559 (1971).
5. Dieudonné, J., *Sur le théorème de Lebesgue-Nikodym. III.* Ann. Univ. Grenoble Sect. Sci. Math. Phys. (N.S.), **25**, 25–43 (1948).
6. Halmos, P.R., *Measure Theory.* Princeton, Van Nostrand, 1950.
7. Hanen, A., Neveu, J., *Atomes conditionnels d'un espace de probabilité.* Acta Math. Acad. Sci. Hungar., **17**, 443–449 (1966).
8. Martin, N. F. G., *Uniform convergence of families of martingales.* Ann. Math. Statist., **40** 1071–1073 (1969).
9. Neveu, J., *Note on the tightness of the metric on the set of complete sub σ-algebras of a probability space.* Ann. Math. Statist., **43**, 1369–1371 (1972).

RICHARD C. BRADLEY

Some remarks on strong mixing conditions*

0. Introduction

The purpose of this note is to review some recent work done on a few special types of problems in the field of "strong mixing conditions". The reader is referred to Iosifescu [20] for a more general, comprehensive review of the field of strong mixing conditions as well as a more complete list of references.

In Section 1 we shall review several basic measures of dependence and the corresponding mixing conditions. In Section 2 the well known mixing condition based on maximal correlations will be reformulated in a simpler manner involving pairs of events. In Section 3 we shall investigate the possible limit values of the various dependence coefficients for strictly stationary sequences of random variables under the assumptions of ergodicity and mixing in the ergodic-theoretic sense. In Section 4 some counterexamples to the central limit theorem will be discussed. And in Section 5 we shall survey some limit theorems in which logarithmic mixing rates are used (in a nontrivial way).

All random variables here are understood to be real-valued. Also, if $(f_n, n = 1, 2, \ldots)$ is a sequence of real numbers, then Σf_n always means $\sum_{n=1}^{\infty} f_n$, $f_n \to c$ always means $f_n \to c$ as $n \to \infty$, $\lim f_n$ always means $\lim_{n\to\infty} f_n$, and so on.

1. Definition of the mixing conditions

Let $(\Omega, \mathcal{F}, P)$ be a probability space.

For any two σ-fields $\mathcal{A}$ and $\mathcal{B} (\subset \mathcal{F})$ define the following measures of dependence

$$\alpha(\mathcal{A}, \mathcal{B}) = \sup |P(A \cap B) - P(A)P(B)| : A \in \mathcal{A},\ B \in \mathcal{B},$$

$$\varphi(\mathcal{A}, \mathcal{B}) = \sup |P(B \mid A) - P(B)| : A \in \mathcal{A},\ B \in \mathcal{B},\ P(A) > 0,$$

$$\psi(\mathcal{A}, \mathcal{B}) = \sup \left| \frac{P(A \cap B)}{P(A)\,P(B)} - 1 \right| : A \in \mathcal{A},\ B \in \mathcal{B},\ P(A)\,P(B) > 0,$$

* This work was partially supported by NSF Grant MCS 81-01583.

$$\rho(\mathcal{A}, \mathcal{B}) = \sup \mid \mathrm{Corr}(f, g) \mid : f \in L^2(\mathcal{A})\,,\ g \in L^2(\mathcal{B}),$$

$$\beta(\mathcal{A}, \mathcal{B}) = \sup\ (1/2) \sum_{i=1}^{I} \sum_{j=1}^{J} \mid P(A_i \cap B_j) - P(A_i)\, P(B_j) \mid,$$

$$I(\mathcal{A}, \mathcal{B}) = \sup \sum_{i=1}^{I} \sum_{j=1}^{J} P(A_i \cap B_j) \log\left[\frac{P(A_i \cap B_j)}{P(A_i)\, P(B_j)}\right],$$

where in the last two definitions the sup is taken over all pairs of partitions $\{A_1, ..., A_I\}$ and $\{B_1, ..., B_J\}$ of Ω such that $A_i \in \mathcal{A}$ for each i and $B_j \in \mathcal{B}$ for each j. Here "log" means the natural logarithm. In the definition of $I(\mathcal{A}, \mathcal{B})$ it is understood that the summand is 0 if $P(A_i \cap B_j) = 0$. In the definition of $\rho(\mathcal{A}, \mathcal{B})$, $\mathrm{Corr}(f, g) = 0$ if f, or g, is constant a.s.; and the notations $L^2(\mathcal{A})$ and $L^2(\mathcal{B})$ refer only to classes of real-valued (not complex-valued) random variables.

Suppose $(X_k,\ k = ..., -1, 0, 1, ...)$ is a strictly stationary sequence of (real-valued) random variables. For $-\infty \leqslant J \leqslant L \leq \infty$ let $\mathcal{F}_J^L$ denote the σ-field of events generated by $(X_k,\ J \leqslant k \leqslant L)$. For each $n = 1, 2, ...$ define

$$\alpha(n) = \alpha(\mathcal{F}_{-\infty}^0, \mathcal{F}_n^\infty),\ \varphi(n) = \varphi(\mathcal{F}_{-\infty}^0, \mathcal{F}_n^\infty),\ \psi(n) = \psi(\mathcal{F}_{-\infty}^0, \mathcal{F}_n^\infty)$$

$$\rho(n) = \rho(\mathcal{F}_{-\infty}^0, \mathcal{F}_n^\infty),\ \beta(n) = \beta(\mathcal{F}_{-\infty}^0, \mathcal{F}_n^\infty),\ I(n) = I(\mathcal{F}_{-\infty}^0, \mathcal{F}_n^\infty)$$

The mixing conditions we shall look at are as follows

$\alpha(n) \to 0$ strong mixing (in Rosenblatt's sense),

$\varphi(n) \to 0$ φ-mixing,

$\psi(n) \to 0$ ψ-mixing,

$\rho(n) \to 0$ ρ-mixing,

$\beta(n) \to 0$ absolute regularity (weak Bernoulli),

$I(n) \to 0$ information regularity.

The general "implications" between them are as follows

$$\begin{array}{ccccc} \psi(n) \to 0 & \Rightarrow & \varphi(n) \to 0 & \Rightarrow & \rho(n) \to 0 \\ \Downarrow & & \Downarrow & & \Downarrow \\ I(n) \to 0 & \Rightarrow & \beta(n) \to 0 & \Rightarrow & \alpha(n) \to 0 \end{array}$$

For the fact that $I(n) \to 0$ implies $\beta(n) \to 0$ and the fact that $\varphi(n) \to 0$ implies $\rho(n) \to 0$, the reader is referred to [25, p. 195, Corollary 2] and [17, p. 309, Theorem 17.2.3] respectively. The other five implications in this chart are easy to verify.

2. Equivalent measures of dependence

For any two σ-fields $\mathcal{A}$ and $\mathcal{B}$ define

$$\lambda(\mathcal{A}, \mathcal{B}) = \sup \frac{\mid P(A \cap B) - P(A)\, P(B) \mid}{[P(A)\, P(B)]^{1/2}} : A \in \mathcal{A},\ B \in \mathcal{B},\ P(A)\, P(B) > 0.$$

It is easy to see that $0 \leqslant \alpha(\mathcal{A}, \mathcal{B}) \leqslant \lambda(\mathcal{A}, \mathcal{B}) \leqslant \rho(A, B) \leqslant 1$.

2.1. **Theorem** ([11, Theorem 1]). *The inequality* $\rho(\mathcal{A}, \mathcal{B}) \leqslant 14[\lambda(\mathcal{A}, \mathcal{B})]^{1/31}$ *always holds.*

Hence for strictly stationary random sequences (X_k) the "ρ-mixing" condition $\rho(n) \to 0$ is equivalent to the condition $\lambda(n) \equiv \lambda(\mathcal{F}_{-\infty}^{0}, \mathcal{F}_{n}^{\infty}) \to 0$. This may help make the ρ-mixing condition easier to verify in practice.

Remark. In an earlier, unpublished version of [11] the sharper inequality $\rho(\mathcal{A}, \mathcal{B}) \leq 4[\lambda(\mathcal{A}, \mathcal{B})]^{1/18}$ was established with a more complicated proof; and in the special case where $\mathcal{A}$ (or $\mathcal{B}$) is purely atomic with exactly two atoms a sharp inequality was established which, in simple terms, says that $\rho(\mathcal{A}, \mathcal{B}) = O(\lambda(\mathcal{A}, \mathcal{B}) \, [1 + \log(1/\lambda(\mathcal{A}, \mathcal{B}))]^{1/2})$.

3. Possible limit values of some dependence coefficients

3.1 **Theorem.** *If* (X_k) *is strictly stationary and mixing in the ergodic-theoretic sense, then the following statements hold*

(i) ([7, Theorem 1]) *Either* $\beta(n) \to 0$, *or* $\beta(n) = 1 \ \forall n$.

(ii) ([4, Theorem 1]) *Either* $\varphi(n) \to 0$, *or* $\varphi(n) = 1 \ \forall n$.

(iii) (See [8, Lemma 0.6]) *Either* $I(n) \to 0$, *or* $I(n) = \infty \ \forall n$.

(iv) ([9, Theorem 1]) $\psi(n) \to 0$, $\psi(n) \to 1$, *or* $\psi(n) = \infty \ \forall n$.

The assumption of mixing in the ergodic-theoretic sense is of course weaker than any of the six mixing conditions defined in Section 1. Statement (i) in Theorem 3.1 is a slight extension of a theorem of Volkonskii and Rozanov [25, Theorem 4.1] characterizing the stationary sequences satisfying $\beta(n) \to 0$. Statement (iii) is elementary and is probably already known. Statement (iv) is separated into two other statements in [9, Theorem 1]: under the hypothesis of Theorem 3.1,

$$\sup \{P(A \cap B)/[P(A)\,P(B)]: A \in \mathcal{F}_{-\infty}^{0}, B \in \mathcal{F}_{n}^{\infty}, P(A)\,P(B) > 0\}$$

must either $\to 1$ as $n \to \infty$ or be $= \infty \ \forall n$, and similarly the inf must either $\to 1$ or be $= 0 \ \forall n$. In all four statements in Theorem 3.1 the assumption of strict stationarity is crucial. For example, by modifying the construction given by Kesten and O'Brien for [21, Theorem 1] one can construct a *non-stationary* random sequence (X_k) such that as $n \to \infty$,

$$\alpha(n) \equiv \sup_J \alpha(\mathcal{F}_{-\infty}^{J}, \mathcal{F}_{J+n}^{\infty}) \to 0 \text{ and } \varphi(n) \equiv \sup_J \varphi(\mathcal{F}_{-\infty}^{J}, \mathcal{F}_{J+n}^{\infty})$$

converges to any given number between 0 and 1. (The notation $\mathcal{F}_{L}^{J}$ is as in Section 1.) It has been known for a long time that without further restrictions, results like Theorem 3.1 (with (X_k) strictly stationary) do not hold for $\alpha(n)$ or $\rho(n)$. For example the following statement holds instead

3.2 **Theorem** ([5, Theorem 6]). *Suppose* $(c_n, n = 1, 2, \ldots)$ *and* $(d_n, n = 1, 2, \ldots)$ *are each a non-increasing sequence of real numbers such that* $0 \leqslant 4d_n \leqslant c_n \leqslant 1 \ \forall n$. *Let* $(f_n, n = 1, 2, \ldots)$ *be an arbitrary sequence of positive numbers. Then there exists a strictly stationary sequence* (X_k) *such that*

(i) $\rho(n) = c_n \ \forall n$,

(ii) $d_n \leqslant \alpha(n) \leqslant d_n + f_n \ \forall n$, *and*

(iii) *each tail* σ*-field of* (X_k) *is trivial (i.e. contains only events of probability* 0 *or* 1).

Property (iii) here (which implies mixing in the ergodic-theoretic sense) is not part of the statement of [5, Theorem 6] but is easy to verify for the construction of (X_k) given in the proof. (Similar results on dependence coefficients can be found in [21], [4], and [9].)

What happens in Theorem 3.1 if the assumption of mixing is replaced by the weaker assumption of ergodicity? The answer is given by H.C.P. Berbee.

3.3. **Theorem** (Berbee [1, Theorem 2.1]). *If (X_k) is strictly stationary and ergodic, then $\beta(n) \to 1 - 1/p$ for some $p \in \{1, 2, 3, \dots\} \cup \{\infty\}$. If $2 \leqslant p < \infty$ then, letting T denote the usual shift operator (on events in $\mathcal{F}_{-\infty}^{\infty}$), the invariant σ-field of T^p is identical to each tail σ-field of (X_k) (up to null-sets) and is purely atomic with exactly p atoms, each having probability $1/p$, and conditional on any one of these atoms the sequence of random vectors $(Y_k, k = \dots, -1, 0, 1, \dots)$ defined by $Y_k \equiv (X_{(k-1)p+1}, X_{(k-1)p+2}, \dots, X_{kp})$ is strictly stationary and satisfies the absolute regularity (weak Bernoulli) condition $\beta(n) \to 0$.*

Berbee [1] refers to this property of (Y_k) by saying that the sequence (X_k) is "periodic weak Bernoulli with period p". His result yields, with the help of Theorem 3.1 (applied to (Y_k) when $2 \leqslant p < \infty$), the following additional properties of strictly stationary ergodic sequences (X_k)

(a) $\varphi(n) \to 1 - 1/p$ for some $p \in \{1, 2, 3, \dots, \infty\}$.

(b) $I(n) \to \log p$ for some $p \in \{1, 2, 3, \dots, \infty\}$.

(c) $\psi(n) \to p$ for some $p \in \{0, 1, 2, \dots, \infty\}$.

4. Counterexamples to the central limit theorem

Here $N(0, 1)$ denotes the standard normal distribution. For each $n = 1, 2, \dots$ let $S_n = X_1 + X_2 + \dots + X_n$.

I.A. Ibragimov conjectured that if (X_k) is strictly stationary and φ-mixing with $EX_k^2 < \infty$ and Var $S_n \to \infty$, then S_n is asymptotically normally distributed as $n \to \infty$. This problem is still unsolved; see the comments on it in Iosifescu [20]. If "φ-mixing" is replaced by "ρ-mixing" the answer becomes negative; a counterexample was given in [3, Theorem 1] and later extended.

4.1 **Theorem** ([8, Theorem 1]). *There exists a strictly stationary sequence (X_k) such that*

(i) $EX_k = 0$, Var $X_k < \infty$, *and* Var $S_n \to \infty$,

(ii) $I(n) \to 0$ (*hence* $\beta(n) \to 0$, *hence* $\alpha(n) \to 0$),

(iii) $\rho(n) \to 0$, *and*

(iv) *S_n is partially attracted to all distributions owned by r.v. 's of the form $W^{1/2}Z$ where Z is a $N(0,1)$ r.v., W is a nonnegative r.v. with an infinitely divisible distribution, and W and Z are independent.*

Ibragimov [15] proved some central limit theorems for strictly stationary sequences (X_k) satisfying the strong mixing condition $\alpha(n) \to 0$ with a certain mixing rate depending on the moments of $|X_k|$. In the case where $E|X_k|^{2+\delta} < \infty$ for some $\delta > 0$ (respectively $|X_k| < C$ a.s. for some $C < \infty$)

he used the rate $\Sigma\alpha(n)^{\delta/(2+\delta)} < \infty$ (respectively $\Sigma\alpha(n) < \infty$). (See [17, Theorems 18.5.3 and 18.5.4].) Yu. A. Davydov [14, pp. 320—324, Examples 1 and 2] showed that this rate (depending on the moments) is "natural" in the sense that if it is weakened by a small amount then S_n no longer has to be asymptotically normally distributed. These counterexamples constructed by Davydov are strictly stationary countable-state irreducible aperiodic Markov chains for which $\text{Var } S_n \sim n^d$ for some $1 < d < 2$ and S_n is attracted to a nonnormal stable law. In the case where X_k is only assumed to have finite second moments there is no mixing rate for $\alpha(n)$ short of m-dependence that will yields the central limit theorem

4.2 **Theorem** ([8, Theorem 2]). *Let* $(d_n, n = 1, 2, ...)$ *and* $(h_n, n = 1, 2,...)$ *be arbitrary sequences of positive numbers such that* $d_n \to 0$ *(arbitrarily fast) and* $h_n \to 0$ *(arbitrarily slowly). Then there exists a strictly stationary sequence* (X_k) *with* $EX_k^2 < \infty$ *such that*

(i) $\alpha(n) < d_n \ \forall n$,

(ii) $\text{Var } S_n > n^2 h_n \ \forall n$,

(iii) S_n *is partially attracted to all infinitely divisible distributions, and*

(iv) $X_k = f(Y_k) \ \forall k$ *where* (Y_k) *is a strictly stationary countable-state irreducible aperiodic Markov chain and* f *is a real-valued function on the state space of* (Y_k).

(Part (iv) comes from [8 Remark 2.1]. This theorem was given with (i) replaced by the (essentially stronger) condition $I(n) < d_n \ \forall n$; see [8, Lemma 0.2 and Theorem 2].)

In [7, Theorem 2] a strictly stationary two-state (zero-one) sequence (X_k) is constructed such that (ii), (iii), and (iv) (but not (i)) of Theorem 4.2 hold (with (ii) permitted to fail for finitely many values of n). In that example, $\beta(n) \to 0$ holds automatically by (iv), but the mixing rate is very slow.

5. Logarithmic mixing rates

In research on mixing conditions, when a mixing rate was needed in a limit theorem it has been common practice to use a polynomial or exponential rate. In 1970 Ibragimov and Rozanov [18] used a *logarithmic* rate for the ρ-mixing condition in a nontrivial way (quated in Theorem 5.1(i) below), and recently it has become apparent that logarithmic mixing rates will suffice in many cases where, previously, polynomial rates were used. To illustrate this, let us collect together some results under logarithmic ρ-mixing

5.1. **Theorem.** *Suppose* (X_k) *is strictly stationary with* $EX_k = 0$, $\text{Var } X_k < \infty$, *and* $\text{Var } S_n \to \infty$. *Suppose also that* $\Sigma\rho(2^n) < \infty$. *Then the following five statements hold*

(i) (Ibragimov and Rozanov [18], (19, p. 190, Note 2])(X_k) *has a continuous spectral density.*

(ii) (See [6, Theorem 1] or [22, Theorem 4.1]) *There exists* σ^2, $0 < \sigma^2 < \infty$, *such that* $\lim n^{-1} \text{Var } S_n = \sigma^2$.

(iii) (Ibragimov [16, Theorem 2.2]) $S_n/(n^{1/2}\sigma) \to N(0, 1)$ *in distribution as* $n \to \infty$.

(iv) (Peligrad [22, Lemma 3.5]) *The family of r.v.'s* $(n^{-1}S_n^2, n = 1, 2, \dots)$ *is uniformly integrable.*

(v) (Peligrad [22, Theorem 2.5]) *Assuming the (slightly faster) mixing rate* $\Sigma\rho^{1/2}(2^n) < \infty$, *the normalized sample paths* $(S_{[nt]}/(n^{1/2}\sigma), 0 \leqslant t \leqslant 1)$ *converge in distribution to a standard Wiener process* $(W(t), 0 \leqslant t \leqslant 1)$. *(Here* $[t]$ = *the greatest integer* $\leqslant t$, *and* $S_0 \equiv 0$.*)*

In a sense statement (ii) is implicitly a part of the ideas behind the proof of statement (i) in [18,19]. Note that $\Sigma\rho(2^n) < \infty$ is satisfied if $\rho(n) = o((\log n)^{-(1+\varepsilon)})$, for some $\varepsilon > 0$. All five statements fail to follow if the mixing rate on $\rho(n)$ is omitted and one only assumes $\rho(n) \to 0$ instead; see Theorem 4.1 and [19, pp. 179—180, Example 1]. Except for this, however, the hypothesis of Theorem 5.1 is unnecessarily strong; it was used for the sake of brevity. In the original statements of some of the results quoted in Theorem 5.1 parts of that hypothesis were either weakened or omitted.

In [10, Theorem 2] a set of (unfortunately rather restrictive) sufficient conditions is given for an array $(X_k^{(m)}, m \geqslant 1, 1 \leqslant k \leqslant k(m))$ of random variables to satisfy

$$\lim_{m\to\infty}\left[\operatorname{Var}\left(\sum_{k=1}^{k(m)} X_k^{(m)}\right)\right]\Bigg/\left[\sum_{k=1}^{k(m)} \operatorname{Var}\ X_k^{(m)}\right] = 1.$$

One of those conditions is $\Sigma\rho(2^n) < \infty$ where $\rho(n)$ is suitably redefined for this context; and (as in Theorem 5.1) it cannot be replaced by the weaker assumption $\rho(n) \to 0$ without strengthening the other conditions. The above result was used to show that if (X_k) is a strictly stationary sequence with $\Sigma\rho(2^n) < \infty$ and satisfies some additional assumptions on probability densities, then for some kernel-type estimators of probability density a multidimensional central limit theorem holds with the same normalizing constants as though the X_k's were independent ([10, Theorem 1]). The details are too cumbersome to discuss here.

In connection with the Marcinkiewicz-Zygmund strong law of large numbers, Peligrad [23, Theorem 2.1] showed that, under certain standard assumptions on positive numbers α and p, if (X_k) is strictly stationary and φ-mixing with a certain logarithmic rate, then $E\,|\,X_k\,|^p < \infty$ if and only if

$$\Sigma n^{p\alpha-2}\, P\left(\max_{j\leqslant n}|\,S_j\,| \geqslant \varepsilon n^{\alpha}\right) < \infty, \forall\varepsilon > 0.$$

This is a well known type of expression for the rate of convergence in the strong law.

Berkes and Philipp [2, Theorem 4] proved an "almost sure invariance principle" under φ-mixing with a certain logarithmic mixing rate. Dabrowski [13] gave a refined version of their argument and thereby relaxed even further the mixing rate in their result

5.2 **Theorem** (Dabrowski [13]). *Suppose* (X_k) *is strictly stationary and* $EX_k = 0$, $\operatorname{Var} X_k < \infty$, *and* $\operatorname{Var} S_n \to \infty$. *Suppose also that*

(i) $E\,|\,X_k\,|^{2+\delta} < \infty$ *for some* $0 < \delta \leq 2$, *and*

(ii) $\varphi(n) = o((\log n)^{-\lambda})$ *for some* $\lambda > 1 + 2/\delta$. *Then there exists* σ^2, $0 < \sigma^2 < \infty$, *such that* $n^{-1}\operatorname{Var} S_n \to \sigma^2$. *Without changing its probability distribution the process* $(S_n, n = 1, 2, \dots)$ *can be redefined on another probability space, together with a standard Wiener process* $(W(t), t \geqslant 0)$, *such that* $P(|\,S_n - W(\sigma^2 n)\,| = o(n^{1/2}(\log\log n)^{-1/2})) = 1$.

Under the same conditions, the law of the iterated logarithm and some other limit theorems follow automatically as corollaries (see[24, Chapter 1]). With slight modifications in Dabrowki's argument, one obtains a similar result under logarithmic strong mixing, but one has to use a faster mixing rate and a more stringent moment condition.

5.3 **Theorem** ([12, Theorem 4]). *The statement of Theorem* 5.2 *remains valid if one replaces* (i) *and* (ii) *there by the following pair of conditions*

(i) $\sup_n E\,|\,S_n\,|^{2+\delta}/(\mathrm{Var}\ S_n)^{(2+\delta)/2} < \infty$ *for some* $\delta > 0$, *and*

(ii) $\alpha(n) = o((\log n)^{-\lambda})$ *for some* $\lambda > 1 + 3/\delta$.

Remarks added April, 1983. 1. W. Bryc, in a letter to the author, has proved the following rather sharp improvement of Theorem 2.1: *for some universal constant C the inequality* $\rho(\mathcal{A}, \mathcal{B}) \leqslant C \cdot \lambda(\mathcal{A}, \mathcal{B}) \cdot [1 + \log(1/\lambda(\mathcal{A}, \mathcal{B}))]$ *always holds.* (The proof will be published later.)

2. N. Herrndorf, in a paper titled *Stationary strongly mixing sequences not satisfying the central limit theorem* (Ann. Probab., **11**, 809—813(1983)), has constructed a strictly stationary sequence of *uncorrelated* r.v. 's (X_k) (with $0 < \mathrm{Var}\ X_k < \infty$) such that (as in Theorem 4.2) $\alpha(n) \to 0$ arbitrarily fast but the CLT fails to hold.

REFERENCES

1. Berbee, H. C. P., *Weak Bernoulli processes, periodicity and renewal theory on the integers.* Report No. 167, Free University of Amsterdam, 1981.
2. Berkes, I., Philipp, W., *Approximation theorems for independent and weakly dependent random vectors.* Ann. Probab., **7**, 29—54 (1979).
3. Bradley, R. C., *A remark on the central limit question for dependent random variables.* J. Appl. Probab., **17**, 94—101 (1980).
4. Bradley, R. C., *On the φ-mixing condition for stationary random sequences.* Duke Math. J., **47**, 421—433 (1980).
5. Bradley, R. C., *Central limit theorems under weak dependence.* J. Multivariate Anal., **11**, 1—16 (1981).
6. Bradley, R. C., *A sufficient condition for linear growth of variances in a stationary random sequence.* Proc. Amer. Math. Soc., **83**, 586—589 (1981).
7. Bradley, R. C., *Absolute regularity and functions of Markov chains.* Stochastic Process. Appl., **14**, 67—77 (1983).
8. Bradley, R. C., *Information regularity and the central limit question.* Rocky Mountain J. Math., **13**, 77—97 (1983).
9. Bradley, R. C., *On the ψ-mixing condition for stationary random sequences.* Trans. Amer. Math. Soc., **276**, 55—66 (1983).
10. Bradley, R. C., *The asymptotic normality of some kernel-type estimators of probability density* Statist. Probab. Lett., **1**, 295—300 (1983).
11. Bradley, R. C., *Equivalent measures of dependence.* J. Multivariate Anal., **13**, 167—176(1983).
12. Bradley, R. C., *Approximation theorems for strongly mixing random variables.* Michigan Math. J. **30**, 69—81 (1983).
13. Dabrowski, A. R., *A note on a theorem of Berkes and Philipp for dependent sequences.* Statist. Probab. Lett., **1**, 53—55 (1982).
14. Davydov, Yu. A., *Mixing conditions for Markov chains.* Theory Probab. Appl., **18**, 312—328 (1973).
15. Ibragimov, I. A., *Some limit theorems for stationary processes.* Theory Probab. Appl., **7**, 349—382 (1962).
16. Ibragimov, I. A., *A note on the central limit theorem for dependent random variables.* Theory Probab. Appl., **20**, 135—141 (1975).

17. Ibragimov, I. A., Linnik, Yu. V., *Independent and Stationary Sequences of Random Variables.* Groningen, Wolters-Noordhoff, 1971.

18. Ibragimov, I.A., Rozanov, Yu. A., *Gaussian Random Processes.* Moscow, Nauka, 1970. (Russian)

19. Ibragimov, I. A., Rozanov, Yu, A., *Gaussian Random Processes.* Applications of Math., No. 9. Berlin—Heidelberg—New York, Springer, 1978.

20. Iosifescu, M., *Recent advances in mixing sequences of random variables.* In: *Third International Summer School on Probability Theory and Mathematical Statistics (Varna, 1978),* pp. 111–138. Sofia, Publishing House of the Bulgarian Academy of Sciences, 1980.

21. Kesten, H., O'Brien, G. L., *Examples of mixing sequences.* Duke Math. J., **43**, 405—415(1976).

22. Peligrad, M., *Invariance principles for mixing sequences of random variables.* Ann. Probab., 10(1982), 968–981.

23. Peligrad, M., *Convergence rates of the strong law for stationary φ-mixing sequences.* INCREST Preprint Series in Math., No. 112. Bucureşti, 1981.

24. Philipp, W., Stout, W. F., *Almost sure invariance principles for sums of weakly dependent random variables.* Amer. Math. Soc. Memoir No. 161 (1975).

25. Volkonskii, V. A., Rozanov, Yu. A., *Some limit theorems for random functions, Part II.* Theory Probab. Appl., **6**, 186–198 (1961).

L. BUTZ*, P. L. HAMMER, D. HAUSSMANN*

Reduction methods for the vertex packing problem

For a given graph $G = (V, E)$ several criteria are proposed for removing vertices and inserting edges such that every maximum stable set in the resulting graph $G' = (V', E')$ is also maximum stable with respect to G.

1. Introduction

The non-weighted set packing problem

$$\begin{aligned} &\max\ 1 \cdot x \\ (1) \qquad &\text{subject to } Ax \leqslant 1 \\ &x_j \in \{0,1\} \end{aligned}$$

(where A is an (n, m) zero-one matrix) can be restated as the *vertex packing problem* of determining a *stable* (i.e. pairwise non-adjacent) set of vertices with maximum cardinality in the graph $G = (V, E)$ (finite, undirected, loopless, no multiple edges) given by $V = \{1, 2, ..., n\}$ and

$$E = \left\{ \{i, j\} : \sum_{k=1}^{m} a_{ki} a_{kj} > 0;\ i \neq j;\ i, j \in V \right\}.$$

Thus a graph $G = (V, E)$ is uniquely determined by a given zero-one matrix A, and it follows readily that any maximum stable set of vertices in G corresponds to an optimal solution of (1) and vice versa.

Some papers dealing with this problem are given in the reference list.

We shall use the term *feasible operation* for altering a graph $G^\nu = (V^\nu, E^\nu)$ in such a way that every maximum stable set of vertices in the resulting graph $G^{\nu+1} = (V^{\nu+1}, E^{\nu+1})$ is also maximum stable with respect to the graph G^ν. Clearly, if the graphs $G^1, G^2, ..., G^k$ have been determined by a sequence of feasible operations from the initial graph $G = G^\circ$, then G^k has been derived from G by a feasible operation.

As one usually is satisfied by obtaining only one optimum solution of (1), it is promising to drastically reduce the number of vertices by feasible

* Institut für Okonometrie und Operations Research Universität Bonn, Bonn, West Germany.

operations in order to arrive at a problem of much smaller dimension. This is the underlying idea of looking into criteria for removing vertices and inserting edges in turn such that the implied alteration of the graph is a feasible operation. Removing a vertex i from a graph $G = (V, E)$ should always be understood as proceeding to the subgraph G' of G induced by $V\setminus\{i\}$, i.e. $G' = (V\{i\}, \{\{k, j\} \in E: \{k, j\} \subseteq V\setminus\{i\}\})$. Analogously, removing a set I of vertices is understood as removing every element of I one by one.

The following characterizations are immediate consequences of the definition of a feasible operation

(2) Removing a vertex i from the graph G is a feasible operation iff there exists a maximum stable set for G which does not contain i.

(3) Inserting an edge between two non-adjacent vertices i and j of the graph G is a feasible operation iff there exists a maximum stable set for G which does not contain both i and j.

In Sections II and III, we give criteria for feasible operations of the kind indicated by (2) and (3), respectively. Inserting new edges will serve mainly as a means for further removal of vertices. Finally, section IV deals which reduction methods obtained by partitioning the graph G and its complementary graph G into their components.

2. Removing vertices

For a vertex i of a graph $G = (V, E)$ (i.e. $i \in V$) let $N_G(i)$ denote the set of neighbors of i with respect to G,

$$N_G(i) := \{j \in V : \{i, j\} \in E\}.$$

(4) **Definition.** Let $i, j \in V$, $G = (V, E)$. Vertex i *dominates* vertex j (notation: $i \succeq_G j$) if $N_G\setminus(i)j\} \supseteq N_G(j)\setminus\{i\}$. We speak of *linked dominance* if of two linked (i.e. adjacent) vertices $i, j \in V$ one of them is dominating the other.[1]

The following lemma implies how the notion of linked dominance provides a useful reduction method for determining maximum stable sets.

(5) **Lemma.** *Let i, j be adjacent vertices of the graph $G = (V, E)$ such that i dominates j. There exists a maximum stable set S^+ of vertices in G which does not contain vertex i.*

Proof. For any stable set of vertices $S \subset V$ with $i \in S$ we have $|(S\setminus\{i\}) \cup \cup\{j\}| = |S|$, where $(S\setminus\{i\}) \cup \{j\}$ is also stable because of $i \succeq_G j$.

By means of (2) and (5), linked dominance renders a feasible operation to be applied to $G = (V, E)$.

(5) can be generalized as follows

(6) **Lemma.** *In the graph $G = (V, E)$, let $I \subseteq V$ be a set of vertices such that* (i) *each $i \in I$ dominates an adjacent vertex $j \in V$, and* (ii) *for each pair of vertices i_1, $i_2 \in I$ at most one of them dominates the other.*

---

[1] We noticed that Trotter [17] mentioned the concept of linked dominance in his doctoral dissertation.

Then there exists a maximum stable set S^+ of vertices in G such that $S^+ \cap I = \emptyset$.

Proof. It suffices to show that eliminating the vertices of the set I one after another, in an appropriate order $i_1, i_2, \ldots$, is a feasible operation.

The order of elimination is only restricted by the dominance relation, i.e. if $i \succeq_G j$, i has to be eliminated before j (this can be done as $\succeq_G$ induces a partial ordering of the set V). Now let $G^\nu = (V^\nu, E^\nu)$ be the subgraph of G induced by eliminating the vertices $i_1, i_2, \ldots, i_\nu \in I$. Then let $i_{\nu+1}$ be the next vertex to be eliminated, and choose $i_o \in \{i_1, \ldots, i^\nu\}$ arbitrarily. By (i) there is $j_0, j_{\nu+1} \in V$ such that $\{i_0, j_0\}$, $\{i_{\nu+1}, j_{\nu+1}\} \in E$ and $i_0 \succeq_G j_0$, $i_{\nu+1} \succeq_G j_{\nu+1}$. Clearly, $j_{\nu+1} \neq i_0$ because otherwise $i_{\nu+1}$ would have been eliminated before i_0. One can check easily that $\{i_{\nu+1}, j_{\nu+1}\} \in E^\nu$ and $i_{\nu+1} \succeq_{G^\nu} j_{\nu+1}$, as i_0 has been chosen arbitrarily from $\{i_1, \ldots, i_\nu\}$, hence eliminating $i_{\nu+1}$ from G^ν is a feasible operation.

Thus removing a set of dominating vertices fulfilling the assumptions of (6) is a feasible operation. The practical relevance of (6) seems to be minor, as checking for linked dominances occurring among the vertices of a given graph is usually facilitated if one applies feasible operations as soon as possible, i.e. if one simply applies (5) as often as possible. Moreover, it might be the case that repeated application of (5) results in eliminating a number of vertices with cardinality greater than that of any set I in (6).

Another criterion for a feasible operation is given by

(7) **Lemma.** *Let i be a vertex of $G = (V, E)$, let $N \subseteq V$ be the set of neighbors of i (i.e. $N = N_G(i)$) and let $T \subseteq V$ be the set of vertices at distance 2 from i (i.e. $T = \left(\bigcup_{j \in N_G(i)} N_G(j)\right) \setminus (N_G(i) \cup \{i\})$).*

If the subgraph $G_T = (T, E_T)$ induced by T contains no set S which is a maximal stable set with respect to the subgraph $G_{N \cup T} = (N \cup T, E_{N \cup T})$ induced by $N \cup T$, then removing the vertex i from G is a feasible operation.

Proof. Let S^+ be a maximum stable set in G, let $i \in S^+$ (otherwise we are done by means of (2)). As $S^+ \cap T$ is stable in G_T, there exists a maximal stable set S' in G_T containing $S^+ \cap T$ and a vertex $j \in N$ such that $S' \cup \{j\}$ is stable in $G_{N \cup T}$. By definition of N and T, $(S^+ \setminus \{i\}) \cup \{j\}$ is a maximum stable set in G. Hence by (2) removing i is a feasible operation.

(8) **Lemma.** *In $G = (V, E)$, let $i_1, \ldots, i_r$ be vertices adjacent to a vertex i_0 such that the sets of their non-neighbors are stable (i.e. $\overline{N}_\nu := V \setminus (N_G(i_\nu) \cup \{i_\nu\})$ stable, $\nu = 0, 1, \ldots, r$). Let $|\overline{N}_\nu| \leqslant |\overline{N}_0|$, $\nu = 1, \ldots, r$. Then removing the vertices $i_1, \ldots, i_r$ from G is a feasible operation.*

Proof. Again, let S^+ be a maximum stable set in G containing some i_{v^+}, $1 \leqslant v^+ \leqslant r$. Clearly, $S^+ = \overline{N}_{v^+} \cup \{i_{v^+}\}$, $i_v \notin S^0 = \overline{N}_0 \cup \{i_0\}$ for all $1 \leqslant v \leqslant r$, and $|S^0| = |S^+|$, hence S^0 is a maximum stable set in G not containing any of the i_v, $1 \leqslant v \leqslant r$.

(9) **Lemma.** *In $G = (V, E)$, let i be a vertex having exactly two neighbors j, k, and let j and k be non-adjacent (i.e. $N_G(i) = \{j, k\} \notin E$). Furthermore, assume $N_G(j) \setminus \{i\} = N_G(k) \setminus \{i\} =: N$. Let $T := \left(\bigcup_{h \in N} N_G(h)\right) \setminus (\{j, k\} \cup N)$ (i.e. $T \subseteq V$ is the set of vertices at distance 2 from j, k). If the subgraph G_T induced by T contains no set S which is a maximal stable set with respect to the subgraph*

$G_{N\cup T}$ induced by $N \cup T$, then removing the vertices j and k from G is a feasible operation.

(10) **Remark.** After removal of j and k, vertex i becomes isolated, hence it is contained in every maximum stable set of any graph G^ν obtained by further feasible operations. Therefore there is an optimum solution to problem (1) having $x_i = 1$. This of course holds for any isolated vertex occurring during the applications of feasible operations.

As will be pointed out later on (cf. (13)), the condition $N_G(j) = N_G(k)$ is implied by $N_G(i) = \{j, k\} \notin E$ for a special class of graphs.

Proof of (9). (Analogous to the proof of (7)). Let S^+ be a maximum stable set in G, let $j, k \in S^+$ ((2) applies in case of $\{j, k\} \cap S^+ = \emptyset$, and $j \in S^+$ iff $k \in S^+$, as S^+ is maximum and $N_G(j) = N_G(k)$, $\{j, k\} \notin E$). $S^+ \cap T$ is stable in G_T, let S' be a maximal stable set in G_T containing $S^+ \cap T$. There exists a vertex $l \in N$ such that $S' \cup \{l\}$ is stable in $G_{N\cup T}$. Thus $(S^+ \setminus \{j, k)\} \cup \cup \{i, l\}$ is a maximum stable set in G, hence (9) follows from (2).

(11) **Lemma.** *Let i, j be two adjacent vertices in the graph $G = (V, E)$ both having exactly two neighbors. Then there is a maximum stable set of vertices in G containing either i or j.*

Proof. Let $i_0 \neq j$ be adjacent to i and let $j_0 \neq i$ be adjacent to j. Every maximal stable set S in G containing neither i nor j must contain i_0 or j_0 because of maximality, say i_0. But then $(S/\{i_0\}) \cup \{i\}$ is a stable set having the same cardinality as S.

If two vertices i, j in $G = (V, E)$ fulfil the conditions of (11) they can be replaced by an edge connecting their neighbors i_0, j_0. After determining a maximum stable set S in the resulting graph $G' = (V \setminus \{i, j\}, (E \cup \{i_0, j_0\}) \setminus (\{i, i_0\} \cup \{j, j_0\}))$, one can easily decide whether $S^+ \cup \{i\}$ or $S^+ \cup \{j\}$ is a maximum stable set in G.

Coming to the end of this section, we want to give some brief comparisons of the efficiency of applying the feasible operations characterized in (5), (7), (8), (9), (11).

The "linked dominance" criterion of (5) seems to be by far the best, as its condition is easy to check and applies quite frequently. The criterion of (11) is even easier to check but it applies less likely. The advantage of (8) is that it might result in the removal of more than one vertex, but especially in graphs of low density it can be a lot of work to determine whether the sets $\bar{N}_v$ are stable or not.

Checking the applicability of (7) and (9) amounts in enumerating all maximal stable sets of the graphs G_T, therefore this should be considered only up to some reasonable cardinality $|T|$.

3. Inserting edges

The concept of removing vertices is a helpful tool for reducing the dimension of problem (1) and one would like to use it as often as possible. If none of the criteria given in (5), (7), (8), (9) or (11) applies, there is still a chance that inserting new edges might yield a graph which does allow feasible operations as described in Section II. In view of (3) we are therefore looking for conditions under which inserting a new edge is a feasible operation.

(12) **Lemma.** *Linking the non-adjacent vertices i, j in a graph $G = (V, E)$ is a feasible operation if there exists a vertex k which is adjacent to exactly one of the vertices i, j such that every other neighbor of k is adjacent to i or to j.*

Proof. Without loss of generality, let k be adjacent to j. Observing that every stable set in G containing i and j remains stable after exchanging j by k, the lemma follows from (3).

(13) **Remark.** Let $G = (V, E)$ be the graph resulting from applying (12) as often as possible to an initial graph $G^0 = (V, E^0)$, $E^0 \subseteq E$. Then $N_G(i) = \{j, k\} \notin E$ implies $N_G(j) = N_G(k)$.

Thus in graphs which do not allow any further application of the kind of feasible operation characterized in (12), one of the conditions of (9) is automatically fulfilled.

As a consequence, if a vertex i dominates an adjacent vertex j we know by (12) that it is a feasible operation to link i with any of its non-neighbors. Thus (5) may be regarded as a corollary to (12).

(14) **Lemma.** *Let $\{i_1, i_2\}$, $\{j_1, j_2\}$ be two pairs of non-adjacent vertices in $G=(V, E)$ such that $\{j_1, j_2\} \subseteq N_G(i_1) \cup N_G(i_2)$ and $N_G(j_1) \cup N_G(j_2) \subseteq N_G(i_1) \cup \cup N_G(i_2) \cup \{i_1, i_2\}$.*

Then linking i_1 and i_2 is a feasible operation.

Proof. Any stable set in G containing i_1 and i_2 does not contain j_1 or j_2 and remains stable after exchanging i_1, i_2 by j_1, j_2. Therefore again the lemma follows from (3).

It should be pointed out that inserting edges does not give a reduction method in itself. The only motivation for (12) and (14) is to alter a graph in such a way that reduction methods can be applied, which could not have been applied before.

4. Decomposition procedures

A quite different reduction method is given by partitioning the graph $G = (V, E)$ and its complementary graph

$$\bar{G} := (V, \{\{i, j\} \notin E : i, j \in V, i \neq j\})$$

into their connected components.

(15) **Lemma.** *Let $G_1, \ldots, G_p$ be the components of G and let $\bar{G}_1, \ldots, \bar{G}_q$ be the subgraphs of G induced by the sets of vertices of the components of the complementary graph $\bar{G}$ of G. Let S_r, $\bar{S}_t$ be maximum stable sets of G_r, $\bar{G}_t$, respectively. Then both $S := \bigcup_{r=1}^{p} S_r$ and $\bar{S} := \bar{S}_{t^+}$ where $|\bar{S}_{t^+}| = \max_{1 \leq t \leq q} |\bar{S}_t|$ are maximum stable sets of G.*

Proof. Note that every stable set of G is contained in exactly one of the subgraphs $\bar{G}_t$.

If $\bar{G}$ or G decomposes into more than one component, (15) indicates an immediate reduction of Problem (1) to lower dimensional vertex packing problems. Obviously, this leads to an iterative branching process, a useful framework for the feasible operations described in Sections II and III. In the following paragraphs we outline an implementation of this branching process.

Essentially, the procedure is to generate and examine one after another certain subgraphs of G and $\bar{G}$ which can be stored efficiently by merely reordering the list of unremoved vertices such that always the vertices of the subgraph under current consideration precede all the other vertices. When a specified subgraph has been examined completely, its vertices have to be marked and will be neglected during the further branching process.

Let the first subgraph to be considered be G. At every stage of the process one determines that component of the complementary graph of the current subgraph which contains the first unmarked vertex from our list. The component obtained this way becomes the new current subgraph, and so on. If no further reduction by this process is possible, one can try to apply feasible operations to the current subgraph (as proposed in Sections II and III) and restart the search for components. When the current subgraph cannot be reduced, neither by feasible operations nor by partitioning into components, all its vertices become marked (as many test examples showed, at this stage the current subgraph is very likely to consist only of a single vertex.) Then one backtracks to a subgraph having still unmarked vertices.

This branching process stops when all the vertices of G have been marked. If it was possible to reduce every subgraph to a single vertex during this procedure, then by another backtracking process one can easily compose a maximum stable set from these single vertices as indicated by (15).

The branching and the backtracking process can be carried out in a computationally efficient manner. Only four arrays are necessary for these processes

— a list of the vertices (with or without marks)

— a list of the number of vertices in each subgraph,

— a list that gives for each subgraph G^v the index of the subgraph from which G^v originated,

— a list noting for each subgraph G^v whether G^v or $\bar{G}^v$ decomposes into components.

REFERENCES

1. Balas, E., Samuelson, H., *Finding a minimum node cover in an arbitrary graph.* Carnegie Mellon University, M.Sc. Rep. 325, Nov. 1973.
2. Balas, E., Zemel, E., *Graph substitution and set packing polytopes.* Carnegie Mellon University, M.Sc. Rep. 384, WP 55—75—76, Jan. 1976.
3. Balas, E., Zemel, E., *Critical cutsets of graphs and canonical facets of set packing polytopes.* Carnegie Mellon University, M.Sc. Rep. 385, WP 59—75—76, Feb. 1976.
4. Balinski, M. L., *On maximum matching, minimum covering and their connections.* In: Kuhn, H. W. (Ed.), *Proc. Princeton Sympos. Math. Programming*, pp. 303—311. Princeton, 1970.
5. Berge, C., *Alternating chain methods: A survey.* In: Read, R. C. (Ed.), *Graph Theory and Computing*, pp. 1—13. New York, Academic Press, 1972.
6. Chvátal, V., *On certain polytopes associated with graphs.* J. Combin. Theory Ser. B, **18**, 138—154 (1975).
7. Chvátal, V., Hammer, P. L., *Set-packing problems and threshold graphs.* Research Report CORR 73—21, University of Waterloo, Aug. 1973.
8. Edmonds, J., *Covers and packings in a family of sets.* Bull. Amer. Math. Soc., **68**, 494—499(1962).

9. Hakimi, S. L., Frank, H., *Maximum internally stable sets of a graph.* J. Math. Anal. Appl., **25**, 296–308 (1969).

10. Houck, D., Vemuganti, R. R., *An algorithm for the vertex packing problem.* University of Baltimore, Mathematics Research Report No. 75–7, 1975.

11. Nemhauser, G. L., Trotter, L. E., *Properties of vertex packing and independence system polyhedra.* Math. Programming, **6**, 48–61 (1974).

12. Nemhauser, G. L., Trotter, L. E., *Vertex packings: Structural properties and algorithms.* Math. Programming, **8**, 232–248 (1975).

13. Nieminen, U. J., *On maximum internally stable sets of a graph.* Naval. Res. Logist. Quart., 21, 557–561 (1974).

14. Padberg, M. W., *On the facial structure of set packing polyhedra.* Math. Programming, **5**, 199–215 (1973).

15. Padberg, M. W., *On the complexity of set packing polyhedra.* New York University, GBA Working Paper No. 75–105, Nov. 1975.

16. Samuelson, H. M., *Integer programming duality for set-packing-partitioning problems.* State University of New York at Buffalo, M. Sc. Rep. 337, WP 193, Feb. 1974.

17. Trotter, L. E., *Solution characteristics and algorithms for the vertex packing problem.* Cornell University, Ithaca, Operations Research Technical Report 168, 1973.

18. Trotter, L. E., *A class of facet producing graphs for vertex packing polyhedra.* Discrete Math., **12**, 373–388 (1975).

MIKLÓS CSÖRGŐ*, SÁNDOR CSÖRGŐ[1],
LAJOS HORVÁTH[1], PÁL RÉVÉSZ[2]

On weak and strong approximations of the quantile process

An up-to-date summary of weak convergence of the empirical quantile process is given in Section 1. In Section 2 necessary and sufficient conditions are given for the weak convergence of the empirical quantile process in $||\cdot/q||$-metrics. Some new strong approximation results are proved for the quantile process in Section 3. Many problems of theoretical and statistical interest are discussed throughout.

1. Introduction

Let $X_1, X_2, \dots$ be independent identically distributed random variables (i.i.d. rv) with a continuous distribution function F. Define the *quantile function* Q of F by

$$(1.1)\qquad Q(y) = F^{-1}(y) = \inf\{x \in \mathbb{R}^1\colon F(x) > y\}$$
$$= \inf\{x \in \mathbb{R}^1\colon F(x) = y\},\ 0 \leq y \leq 1.$$

Thus $F(Q(y)) = y \in [0, 1]$, and if we put $U_1 = F(X_1)$, then U_1 is a uniformly distributed rv on $[0, 1]$. Also $U_1 = F(X_1)$, $U_2 = F(X_2)$, ... are independent uniform-$[0, 1]$ rv.

Let $X_{1:n} \leq X_{2:n} \leq \dots \leq X_{n:n}$ be the order statistics of a random sample $X_1, \dots, X_n$ on F which, in turn, induce the uniform-$[0,1]$ order statistics $U_{1:n} = F(X_{1:n}) \leq U_{2:n} = F(X_{2:n}) \leq \dots \leq U_{n:n} = F(X_{n:n})$ of $U_1, U_2, \dots, U_n$.

Define the *empirical distribution function* F_n of the first n i.i.d. rv $X_1, \dots, X_n$ by

$$(1.2)\qquad F_n(x) = \begin{cases} 0, & \text{if } X_{1:n} > x \\ k/n, & \text{if } X_{k:n} \leq x < X_{k+1:n} \\ 1, & \text{if } X_{n:n} \leq x,\ x \in \mathbb{R}^1 \end{cases}$$

---

* Research supported by a NSERC Canada grant at Carleton University, Ottawa.

[1] Szeged University, Szeged, Hungary.

[2] Mathematical Institute of the Hungarian Academy of Sciences, Budapest, Hungary.

and the thus induced *uniform empirical distribution function* E_n of $U_1, \dots, U_n$ by

$$E_n(y) = F_n(Q(y)) = \begin{cases} 0, & \text{if } U_{1:n} > y \\ k/n, & \text{if } U_{k:n} \leqslant y < U_{k+1:n} \\ 1, & \text{if } U_{n:n} \leqslant y, \ 0 \leqslant y \leqslant 1. \end{cases} \tag{1.3}$$

In terms of the latter empirical distribution functions F_n and E_n, we define now the *empirical quantile function* Q_n by

$$\begin{aligned} Q_n(y) &= F_n^{-1}(y) = \inf\{x: F_n(x) \geqslant y\} \\ &= X_{k:n}, \text{ if } (k-1)/n < y \leqslant k/n \ \ (k = 1, 2, \dots, n), \end{aligned} \tag{1.4}$$

and the *uniform empirical quantile function* U by

$$\begin{aligned} U_n(y) &= E_n^{-1}(y) = \inf\{u: F_n(Q(u)) \geqslant y\} \\ &= U_{k:n} = F(X_{k:n}), \ (k-1)/n < y \leqslant k/n \ (k = 1, 2, \dots, n), \\ &= F(Q_n(y)). \end{aligned} \tag{1.5}$$

Let $f = F'$ be the density function of the distribution function F. Define the *quantile process* ρ_n by

$$\rho_n(y) = n^{1/2} f(Q(y))\,(Q_n(y) - Q(y)), \ y \in (0, 1), \tag{1.6}$$

and the corresponding *uniform quantile process* u_n by

$$u_n(y) = n^{1/2}(U_n(y) - y). \tag{1.7}$$

M. Csörgő and P. Révész have proved

Theorem A (Csörgő and Révész (1978)). *Let $X_1, X_2, \dots$ be i.i.d. rv with a continuous distribution function F and assume*

(i) *F is twice differentiable on (a, b), where*

$$a = \sup\{x: F(x) = 0\}, \ b = \inf\{x: F(x) = 1\}, \ -\infty \leqslant a < b \leqslant \infty$$

(ii) *$F'(x) = f(x) > 0$ on (a, b),*

(iii) *for some $\gamma > 0$ we have*

$$\sup_{0<y<1} y(1-y)\frac{|f'(Q(y))|}{f^2(Q(y))} \leqslant \gamma.$$

Then, with $\delta_n = 25n^{-1}\log\log n$,

$$\sup_{\delta_n \leqslant y \leqslant 1-\delta_n} |\rho_n(y) - u_n(y)| \overset{a.s.}{=} O(n^{-1/2}\log\log n). \tag{1.8}$$

If, in addition to (i), (ii) *and* (iii), *we also assume*

(iv) $$A = \lim_{x \downarrow a} f(x) < \infty, \ B = \lim_{x \uparrow b} f(x) < \infty,$$

(v) *one of*

(v,α) $A \wedge B > 0$,

(v, β) *if $A = 0$ (respectively $B = 0$) then f is nondecreasing (respectively nonincreasing) on an interval to the right of a (respectively to the left of b),* *then, if* (v, α) *obtains,*

(1.9) $$\sup_{0\leqslant y\leqslant 1} |\rho_n(y) - u_n(y)| \stackrel{a.s.}{=} O(n^{-1/2} \log\log n),$$

and if (v, β) *obtains,*

(1.10) $$\sup_{0\leqslant y\leqslant 1} |\rho_n(y) - u_n(y)| \stackrel{a.s.}{=}$$

$$= \begin{cases} O(n^{-1/2} \log\log n), & \text{if } \gamma < 1 \\ O(n^{-1/2} (\log\log n)^2), & \text{if } \gamma = 1 \\ O(n^{-1/2} (\log\log n)^{\gamma} (\log n)^{(1+\varepsilon)(\gamma-1)}), & \text{if } \gamma > 1, \end{cases}$$

where $\varepsilon > 0$ *is arbitrary, and* γ *is as in* (iii).

The above theorem also implies approximations of ρ_n in terms of appropriate sequences of Brownian bridges $\{B_n(y),\ 0 \leqslant y \leqslant 1\}$ (cf. Csörgő and Révész (1978)) due to

Theorem B (Csörgő and Révész (1975, 1978)). *For an i.i.d. sequence of rv* $X_1, X_2, \ldots$ *there exists a probability space with a sequence of Brownian bridges* $\{B_n\}$ *on it such that*

(1.11) $$\sup_{0\leqslant y\leqslant 1} |u_n(y) - B_n(y)| \stackrel{a.s.}{=} O(n^{-1/2} \log n).$$

Naturally, from the above two theorems it follows that

(1.12) $$\rho_n(\cdot) \xrightarrow{D} B(\cdot)$$

in Skorohod's space $D[0, 1]$.

Let $q(y) \geqslant 0$ be a continuous function on $[0, 1]$ which is strictly positive on $(0, 1)$, nondecreasing on $[0, \frac{1}{2}]$ and symmetric about $y = 1/2$, and let

(1.13) $$h(y) = \left(y(1-y) \log\log \frac{1}{y(1-y)}\right)^{1/2}, \quad 0 \leqslant y \leqslant 1.$$

Define also $g(y) = q(y)/h(y)$ so that

(1.14) $$g(y) = q(y)/h(y) \to \infty \ \text{as} \ y \to 0.$$

Then $g(y) = g(1-y)$ by definition.

Shorack (1979) showed that with q and g as above, the condition (1.14) is sufficient for O'Reilly's (1974) necessary and sufficient condition on q for

(1.15) $$\sup_{1/(n+1)\leqslant y\leqslant n/(n+1)} |(u_n(y) - B_n(y))/q(y)| \xrightarrow{P} 0$$

to be true with the Brownian bridges B_n of (1.11). Similar work to that of O'Reilly's was done earlier by Chibisov (1964), and Pyke and Shorack (1968). S. Csörgő (1982) showed that on assuming conditions (i), (ii), (iii), (iv) and (v) of Theorem A and condition (1.14), we have

(1.16) $$\sup_{1/(n+1)\leqslant y\leqslant n/(n+1)} |(\rho_n(y) - B_n(y))/q(y)| \xrightarrow{P} 0$$

with B_n as in (1.11), provided that γ of condition (iii) is less than 1. Otherwise, i.e., if in (iii) $\gamma \geqslant 1$, growth conditions had to be introduced for the function $G(t) = \inf \{g(y): 0 \leqslant y \leqslant t\} \uparrow \infty$ as $t \downarrow 0$. M. Csörgő (1982, Theorem 5.1.1) verified (1.16) only under the conditions (i), (ii), (iii) of Theorem A and (1.14).

Stute (1982) proved (1.16) with $q \equiv 1$ under only conditions (iv), (v) of Theorem A as follows: we have (1.16) with $q \equiv 1$ if (v, α) obtains, or if (v, β) obtains, provided that in the latter case both

$$g(y)(f(Q(y/\lambda))/f(Q(y)) \to \infty \text{ as } y \to 0 \tag{1.17}$$

and (symmetrically)

$$g(y)(f(Q(1-y)/\lambda))/f(Q(1-y)) \to \infty \text{ as } y \to 0 \tag{1.18}$$

for each $\lambda \geqslant 1$ (note that in Stute's (1982) case $g = 1/h$ on account of $q \equiv 1$). Shorack (1982) announced the latter result with q and g as in (1.14) (for a proof we may, for example, refer to M. Csörgő (1982, Corollary 5.3.2)). All the afore-mentioned results concerning (1.16) are contained in

Theorem C (Csörgő (1982, Theorem 5.3.1)). *Let a, b be as in Theorem A and assume that F has a continuous density function $F' = f$ that is positive on (a, b), the support of F. Let q be any given O'Reilly weight function. Then, as $n \to \infty$, with the sequence of Brownian bridges $\{B_n\}$ of* (1.11) *we have* (1.16) *under* (1.14), *provided that with g of the latter the following assumption also holds true*

(1.19) *For any given $0 < \eta < 1$ and $\varepsilon > 0$ there exist $0 < c < 1$ and n_0 such that*

$$P\left\{\sup_{1/(n+1)\leqslant y\leqslant c} \frac{f(Q(y))}{f(Q(\theta_{y,n}))}\frac{1}{g(y)} > \varepsilon\right\} \leqslant \eta,$$

and similarly

$$P\left\{\sup_{1-c\leqslant y\leqslant n/(n+1)} \frac{f(Q(y))}{f(Q(\theta_{y,n}))}\frac{1}{g(y)} > \varepsilon\right\} \leqslant \eta$$

for all $n \geqslant n_0$, where $U_n(y) \wedge y < \theta_{y,n} < U_n(y) \vee y$.

All the afore quoted results concerning (1.16) can be put in terms of weak convergence on $D[0, 1]$, *provided* we redefine u_n (and hence also ρ_n) to be equal to zero on $[0, 1/(n+1))$ and $(n/(n+1), 1]$.

2. Weak approximations

First we show that the sufficient conditions of Theorem C (cf. (1.19)) for the weak approximation of ρ_n on $[1/(n+1), n/(n+1)]$ are nearly necessary as well. Then we examine the problem of weak convergence of ρ_n on $[0,1]$. A weight function q will be called an O'Reilly weight function throughout if $g = q/h$ satisfies (1.14).

Theorem 2.1. *Let a, b be as in Theorem A and assume that F has a continuous density function $F' = f$ that is positive on (a, b), the support of F. If for any given g of an O'Reilly weight function q we have that the rv*

$$\left(f\left(Q\left(\frac{1}{n}\right)\right)\Big/f(Q(\theta_{1/n,n}))\right)\Big/g\left(\frac{1}{n}\right)(\log\log n)^{1/2}, \text{ or}$$

$$\left(f\left(Q\left(\frac{n-1}{n}\right)\right)\Big/f(Q(\theta_{(n-1)/n,n}))\right)\Big/g\left(\frac{1}{n}\right)(\log\log n)^{1/2} \tag{2.1}$$

is, contrary to (1.19), *bounded away from zero in probability as $n \to \infty$, then the rv of* (1.16) *is also bounded away from zero in probability for any sequence of Brownian bridges $\{B_n\}$.*

Proof. It suffices to show for example that, given the first condition of (2.1), the rv

$$\sup_{1/(n+1)\leqslant y\leqslant 1/2} |(\rho_n(y) - B_n(y))/q(y)| \tag{2.2}$$

is bounded away from zero in probability. By the mean value theorem we have (with $\theta_{y,n}$ as in (1.19))

$$\rho_n(y)/q(y) = u_n(y)(f(Q(y))/f(Q(\theta_{y,n})))/q(y). \tag{2.3}$$

Consider

$$\sup_{1/(n+1)\leqslant y\leqslant 1/2} |(\rho_n(y) - B_n(y))/q(y)|$$

$$\geqslant n^{1/2}\left|u_n\left(\frac{1}{n}\right)\right|\left(f\left(Q\left(\frac{1}{n}\right)\right)\Big/f(Q(\theta_{1/n,n}))\right)\Big/\left[(\log\log n)^{1/2}g\left(\frac{1}{n}\right)\right] -$$

$$-\left|B_n\left(\frac{1}{n}\right)\Big/q\left(\frac{1}{n}\right)\right|.$$

Since

$$\lim_{n\to\infty} P\left\{n^{1/2}\left|u_n\left(\frac{1}{n}\right)\right| > \varepsilon\right\} = e^{-(\varepsilon+1)} + 1 - e^{-(1-\varepsilon)},\ \varepsilon > 0, \tag{2.4}$$

and $B_n\left(\frac{1}{n}\right)\Big/q\left(\frac{1}{n}\right) \xrightarrow{P} 0$ as $n\to\infty$, we get (2.2) by (2.1). This also completes the proof of Theorem 2.1.

As to the problem of weak convergence of ρ_n/q in sup-norm metric over [0, 1], we first observe that for a Brownian bridge B $\sup_{0<y\leqslant 1/n} |B(y)/q(y)| \xrightarrow{P}$ $\xrightarrow{P} 0$ and $\sup_{1-\frac{1}{n}\leqslant y<1} |B(y)/q(y)| \xrightarrow{P} 0$ as $n\to\infty$ for any O'Reilly weight function q. Hence the only way for ρ_n/q to converge weakly to B/q over [0, 1] in sup-norm metric is that we have the latter two statements holding true also with ρ_n replacing B in them. In this context then, we prove our

Theorem 2.2. *Let a, b be as in Theorem A and assume that F has a continuous density function $F' = f$ that is positive on (a, b), the support of F. Let q be any given O'Reilly weight function so that as $n\to\infty$*

$$\sup_{0\leqslant y\leqslant 1/n} \frac{n^{-1/2}}{q(y)}\,\frac{f(Q(y))}{f(Q(\theta_{y,n}))} \xrightarrow{P} 0,$$

(2.5)

$$\sup_{(n-1)/n\leqslant y\leqslant 1} \frac{n^{-1/2}}{q(y)}\,\frac{f(Q(y))}{f(Q(\theta_{y,n}))} \xrightarrow{P} 0,$$

where $\theta_{y,n}$ is as in (1.19). Then as $n\to\infty$

(2.6)

$$\sup_{0\leqslant y\leqslant 1/n} |\rho_n(y)/q(y)| \xrightarrow{P} 0,$$

$$\sup_{(n-1)/n\leqslant y\leqslant 1} |\rho_n(y)/q(y)| \xrightarrow{P} 0.$$

If, on the other hand, for the given q and g there exists a sequence of rv $\tau_n \leqslant 1/n$ so that the rv

$$n^{-1/2}(f(Q(\tau_n))/f(Q(\theta_{\tau_n,n})))/q(\tau_n), \text{ or} \tag{2.7}$$
$$n^{-1/2}(f(Q(1-\tau_n))/f(Q(\theta_{1-\tau_n,n})))/q(\tau_n)$$

with $\tau_n \wedge U_{1:n} \leqslant \theta_{\tau_n,n} \leqslant \tau_n \vee U_{1:n}$ *and* $(1-\tau_n) \wedge U_{n:n} \leqslant \theta_{1-\tau_n,n} \leqslant (1-\tau_n) \vee \vee U_{n:n}$, *is bounded away from zero in probability as* $n \to \infty$, *then so will be also the rv of* (2.6).

Proof. Consider (cf. (2.3))

$$\sup_{0 \leqslant y \leqslant 1/n} |\rho_n(y)/q(y)|$$
$$\leqslant n^{1/2} \sup_{0 \leqslant y \leqslant 1/n} |u_n(y)| \sup_{0 \leqslant y \leqslant 1/n} \frac{n^{-1/2}}{q(y)} \frac{f(Q(y))}{f(Q(\theta_{y,n}))}.$$

Since the rv $n^{1/2} \sup_{0 \leqslant y \leqslant 1/n} |u_n(y)|$ is bounded in probability as $n \to \infty$ (cf. (2.4)), by the first condition of (2.5) we get the first statement of (2.6). The second statement of (2.6) is proved in a similar way. Consider now.

$$\sup_{0 \leqslant y \leqslant 1/n} |\rho_n(y)/q(y)| \geqslant n^{1/2}|u_n(\tau_n)| \frac{f(Q(\tau_n))/f(Q(\theta_{\tau_n,n}))}{n^{1/2}q(\tau_n)}. \tag{2.8}$$

Since the rv $n^{1/2}|u_n(\tau_n)|$ is bounded away from zero in probability (cf. (2.4)) by the first condition of (2.7) so is the left hand side rv of the inequality of (2.8). The second statement of (2.7) follows similarly from its second alternative condition. This also completes the proof of Theorem 2.2.

Now we discuss some *implications* (examples) of interest which follow from Theorems C, 2.1 and 2.2.

(1) We have already noted in the Introduction that Theorem C implies the Stute (1982) -Shorack (1982) result: (1.16) holds true under the conditions (iv) and (v) of Theorem A if (v, α) obtains, or if (v, β) obtains and (1.17), (1.18) are also assumed. If (v, α) obtains then F has finite support and hence (1.16) holds immediately. If (v, β) obtains and, say, A of (iv) is zero, then for small enough c, we have

$$\sup_{1/(n+1) \leqslant y \leqslant c} \frac{f(Q(y))}{f(Q(\theta_{y,n}))} \frac{1}{g(y)} \leqslant \sup_{1/(n+1) \leqslant y \leqslant c} \frac{f(Q(y))}{f(Q(y/\lambda))} \frac{1}{g(y)}$$

for some $\lambda \geqslant 1$ and all $n \geqslant 1$ with probability arbitrarily near to one by Remark 1 in Wellner (1978). Hence (1.17) implies the first condition of (1.19) and the Stute (1982) -Shorack (1982) theorem follows from Theorem C.

(2) For any O'Reilly weight function q such that $q(y) \to 0$ as $y \to 0$ we have $\sup_{0 < y \leqslant 1/n} \frac{1}{q(y)} = \infty$. Hence in case of the uniform quantile process u_n we can choose $\{\tau_n\}$ of (2.7) such that for any given constant $K > 0$ we have $(1/n^{1/2}q(\tau_n)) > K$. Consequently, for any q with $q(0) = 0$ and for any sequence of Brownian bridges $\{B_n\}$ we have

$$P\left\{\sup_{0 \leqslant y \leqslant 1} |(u_n(y) - B_n(y))/q(y)| = \infty\right\} = 1.$$

(3) As to the *problem of having*

$$\sup_{0\leqslant y\leqslant 1} |(\rho_n(y) - B(y))/q(y)| \xrightarrow{P} 0 \tag{2.9}$$

with B_n as in (1.11), we note that choosing the g function of q appropriately for some specific distributions (2.9) might turn out to be true. For example, in case of $F(x) = 1 - e^{-x}$, $x \geqslant 0$, we may choose $g(y) = Q(y)\,a(y)\Big/\left(\log\log\dfrac{1}{1-y}\right)^{1/2}$, where now $Q(y) = \log\,(1/(1-y))$, $a(y) \to \infty$ as $y \to 1$, and then (2.9) will hold true. Naturally a similar statement holds true symmetrically for $F(x) = 1 - e^{-x}$, $x \leqslant 0$. We note also that with the same g function as above, (2.9) will hold true also for the Weibull distribution.

(4) As mentioned already in the Introduction, M. Csörgő (1982, Theorem 5.1.1) verified (1.16) only under conditions (i), (ii), (iii) of Theorem A, and noted also that Theorem C also contained the latter result. In order to see this, consider

$$\sup_{1/(n+1)\leqslant y\leqslant n/(n+1)} \frac{f(Q(y))}{f(Q(\theta_{y,n}))} \tag{2.10}$$

$$\leqslant \sup_{1/(n+1)\leqslant y\leqslant n/(n+1)} \left(\frac{U_n(y)\vee y}{U_n(y)\wedge y}\,\frac{1-(U_n(y)\wedge y)}{1-(U_n(y)\vee y)}\right)^{\gamma},$$

where λ is as in (iii) of Theorem A, and the inequality is by Lemma 1 in Csörgő and Révész (1978). Using now Lemma 2 in Wellner (1978) on the right hand side rv of the inequality of (2.10) it follows (cf. e.g., (2.8) in M. Csörgő (1982)) that

$$\lim_{\varepsilon\to\infty}\ \overline{\lim_{n\to\infty}}\ P\left\{\sup_{1/(n+1)\leqslant y\leqslant n/n+1} \frac{f(Q(y))}{f(Q(\theta_{y,n}))} > \varepsilon\right\} = 0. \tag{2.11}$$

Hence condition (1.19) of Theorem C is satisfied and consequently (1.16) holds true with *any* O'Reilly .weight function q under the conditions (i), (ii), (iii) of Theorem A.

(5) Given conditions (i), (ii) of Theorem A and replacing its condition (iii) by requiring the existence of the limits (cf. condition (vi) of Theorem 3.2)

$$\lim_{y\downarrow 0} y\,\frac{f'(Q(y))}{f^2(Q(y))} = \gamma_1, \quad \lim_{y\uparrow 1}(1-y)\,\frac{f'(Q(y))}{f^2(Q(y))} = \gamma_2,$$

where γ_1 and γ_2 are real numbers, then it can be shown (cf. Theorem 3.A in Parzen (1980), or page 7 of Seneta (1976), or Mason (1982)) that we have

$$\begin{aligned} f(Q(y)) &= y^{\gamma_1}L_1(y) \text{ as } y \downarrow 0 \\ f(Q(y)) &= (1-y)^{\gamma_2}L_2(y) \text{ as } y \uparrow 1, \end{aligned} \tag{2.12}$$

where L_1 and L_2 are slowly varying functions at 0 respectively at 1. If we simply *assume the forms of* (2.12) *for* $f(Q)$ *on the tails*, then these are weaker assumptions on f than that of (iii), for then we do not require the existence of f' on (a, b), the support of F. So let us *assume that $f(Q)$ is as in* (2.12) *on the tails*, and consider its first statement (regarding the second one, similar conclusions will hold true). It follows from Corollary on page 274 in Feller (1966) that for a slowly varying function L_1 we have: for any $\varepsilon > 0$ there exist some positive constants K_1, K_2 and $0 < y_0 < 1$ such that we have

$$K_1 y^{\varepsilon} < L_1(y) < K_2 y^{-\varepsilon} \text{ for all } 0 < y \leqslant y_0. \tag{2.13}$$

Consider now (cf. first statement of (2.12) with $\gamma_1 \equiv \gamma$)

$$\begin{aligned}
&\sup_{1/(n+1)\leqslant y\leqslant c} \frac{f(Q(y))}{f(Q(\theta_{y,n}))} \frac{1}{g(y)} \\
&= \sup_{1/(n+1)\leqslant y\leqslant c} \left(\frac{y}{\theta_{y,n}}\right)^{\gamma} \frac{L_1(y)}{L_1(\theta_{y,n})} \frac{1}{g(y)} \\
&\leqslant \sup_{1/(n+1)\leqslant y\leqslant c} \left(\frac{y}{\theta_{y,n}}\right)^{\gamma} \frac{K_2 y^{-\varepsilon}}{K_1 \theta^{\varepsilon}_{y,n}} \frac{1}{g(y)} \\
&= \sup_{1/(n+1)\leqslant y\leqslant c} \left(\frac{y}{\theta_{y,n}}\right)^{\gamma+\varepsilon} \frac{K_2}{K_1} \frac{1}{y^{2\varepsilon}} \frac{1}{g(y)}.
\end{aligned} \tag{2.14}$$

It follows from Remark 1 of Wellner (1978) that for both $\gamma + \varepsilon > 0$ and $\gamma + \varepsilon < 0$ we have

$$\lim_{a\to\infty} \overline{\lim_{n\to\infty}} P\left\{ \sup_{1/(n+1)\leqslant y\leqslant c} \left(\frac{y}{\theta_{y,n}}\right)^{\gamma+\varepsilon} > a \right\} = 0. \tag{2.15}$$

Hence choosing $g(y) = y^{-\delta}$, $\delta > 0$ and then $\varepsilon > 0$ of (2.13) so that $\varepsilon < \delta/2$, then the first condition of (1.19) holds true by (2.14) and (2.15) combined. This means that having assumed (2.12), which is weaker than (i), (ii), (iii) of Theorem A combined, the statement of (1.16) holds only with $g(y) = y^{-\delta}$ for $\delta > 0$, arbitrary otherwise (cf. the last sentence of our example (4)).

(6) We note that the conditions discussed in (4) and (5) for the validity of (1.16) cannot, in general, insure also the validity of (2.9). Namely we have the following

Remark. *If* $\overline{\lim}_{y\to 0} f(Q(y))/q(y) = \infty$ *or* $\overline{\lim}_{y\to 1} f(Q(y))/q(y) = \infty$, *then*

$$P\left\{ \sup_{0\leqslant y\leqslant 1} |\rho_n(y)/q(y)| = \infty \right\} = 1. \tag{2.16}$$

Proof. If a of the support (a, b) of F is $-\infty$, then there is a $\delta > 0$ such that $Q(y) < 0$ and $Q(U_{1:n}) > Q(y)$, $0 < y < \delta$. Hence

$$|Q(U_{1:n}) - Q(y)| = -Q(y) + Q(U_{1:n}) > -(1/2)Q(y),$$

and therefore

$$\sup_{0\leqslant y\leqslant 1} |\rho_n(y)/q(y)| \geqslant (n^{1/2}/2) \sup_{0\leqslant y\leqslant \delta} \frac{f(Q(y))}{q(y)} |Q(y)| = \infty.$$

If a of the support (a, b) of F is larger than $-\infty$, then $Q(U_{1:n}) > a$ and we can choose a constant $\delta > 0$ such that

$$|Q(U_{1:n}) - Q(y)| \geqslant \frac{1}{2}|Q(y)|, \quad 0 < y < \delta,$$

and get

$$\sup_{0 \leqslant y \leqslant 1} |\rho_n(y)/q(y)| \geqslant (n^{1/2}/2) Q(U_{1:n}) \sup_{0<y<\delta} \frac{f(Q(y))}{q(y)} = \infty.$$

Now a similar argument on the right hand side of (a, b) completes the proof of our Remark.

3. Strong approximations

First we recall that conditions (i), (ii), (iii) of Theorem A imply (1.8), and given also the tail monotonicity assumptions of (iv), (v) we have also (1.9) and (1.10). We have just seen in Examples (4) and (5) of our previous section that conditions (i), (ii), (iii) of Theorem A alone, or the somewhat weaker assumptions of (2.12) for the tail behaviour of $f(Q)$, imply (1.16) (in (4) without any further restrictions on q, while in (5) with $g(y)$ of the form $y^{-\delta}$ $(\delta > 0)$ only). The initial aim of the present section is to examine the possibility of extending the statement of (1.8) over a wider range than $[\delta_n, 1 - \delta_n]$, $\delta_n = 25n^{-1} \log\log n$, but using only the assumptions (i), (ii), (iii) of Theorem A and not those of its conditions (iv), (v) which, when combined with (i), (ii), (iii), made (1.9) and (1.10) possible. We have

Theorem 3.1. *Assume conditions* (i), (ii), (iii) *of Theorem A. Then*

$$\sup_{1/(n+1) \leqslant y \leqslant n/(n+1)} |\rho_n(y) - u_n(y)|$$

(3.1)
$$\overset{\text{a.s.}}{=} \begin{cases} O(n^{-1/2}(\log\log n)^{1+\gamma}), & \text{if } \gamma \leqslant 1 \\ O(n^{-1/2}(\log\log n)^{\gamma}(\log n)^{(1+\varepsilon)(\gamma-1)}), & \text{if } \gamma > 1, \end{cases}$$

where $\varepsilon > 0$ *is arbitrary, and* γ *is as in condition* (iii).

Proof. We have (1.8). Hence by (3.10) and respectively (4.5.17) in Csörgő and Révész (1978, respectively 1981), in order to verify (3.1), it suffices to show that the rv

$$\sup_{1/(n+1) \leqslant y \leqslant \delta_n} |\rho_n(y)| \quad \text{and} \quad \sup_{1-\delta_n \leqslant y \leqslant n/(n+1)} |\rho_n(y)|$$

are almost surely negligible with the indicated $O(\cdot)$ rates of (3.1). We demonstrate this for the first one of the just mentioned rv. Consider

$$\sup_{1/(n+1) \leqslant y \leqslant \delta_n} |\rho_n(y)|$$

(3.2)
$$\leqslant \sup_{0 \leqslant y \leqslant \delta_n,\, U_{[ny]+1:n} < y} |\rho_n(y)| + \sup_{1/(n+1) \leqslant y \leqslant \delta_n,\, U_{[ny]+1:n} \geqslant y} |\rho_n(y)|.$$

It is shown by Csörgő and Révész (cf. (3.14) in [4] respectively (4.5.21) in [5]) that given only conditions (i), (ii), (iii)

$$(3.3) \qquad \sup_{0 \leqslant y \leqslant \delta_n,\, U_{[ny]+1:n} < y} |\rho_n(y)| \overset{a.s.}{=} O(\cdot) \text{ rates of } (1.10),$$

i.e., in the proof of (3.3) the tail monotonicity assumptions (iv), (v) are not used. Hence, in order to verify (3.1), it suffices to show that

$$(3.4) \qquad \sup_{1/(n+1) \leqslant y \leqslant \delta_n,\, U_{[ny]+1:n} \geqslant y} |\rho_n(y)| \overset{a.s.}{=} O(n^{-1/2}(\log\log n)^{\gamma+1})$$

for all $\gamma > 0$. We therefore consider

$$(3.5) \qquad |\rho_n(y)| = n^{1/2} \int_y^{U_{[ny]+1:n}} \frac{f(Q(y))}{f(Q(u))} du$$

$$\leqslant 2^\gamma n^{1/2} \int_y^{U_{[ny]+1:n}} \left(\frac{u}{y}\right)^\gamma du, \text{ (cf. Lemma 1 in Csörgö and Révész (1978))}$$

$$= \frac{2^\gamma}{\gamma+1} n^{1/2} \frac{1}{y^\gamma} ((U_{[ny]+1:n})^{\gamma+1} - y^{\gamma+1})$$

$$= \frac{2^\gamma}{\gamma+1} n^{1/2} \left(\left(\frac{U_{[ny]+1:n}}{y}\right)^\gamma U_{[ny]+1:n} - y\right)$$

$$= \frac{2^\gamma}{\gamma+1} n^{1/2} (U_{[ny]+1:n} - y) \left(\frac{U_{[ny]+1:n}}{y}\right)^\gamma + \left(\left(\frac{U_{[ny]+1:n}}{y}\right)^\gamma - 1\right) y n^{1/2} \frac{2^\gamma}{\gamma+1}.$$

We have (cf. (3.12) respectively (4.5.19) in Csörgő and Révész [4], respectively [5])

$$(3.6) \quad \left(\frac{U_{[ny]+1:n} - y}{y} + 1\right)^\gamma \leqslant \left(\frac{46 n^{-1} \log\log n}{1/(n+1)} + 1\right)^\gamma \leqslant (47(\log\log n) + 1)^\gamma$$

almost surely for n large whenever $U_{[ny]:n} \geqslant y$ and $1/(n+1) \leqslant y \leqslant \delta_n$.

Hence (cf. again (3.12) respectively (4.5.19) in Csörgő and Révész [4], respectively [5])

$$(3.7) \qquad \sup_{1/(n+1) \leqslant y \leqslant \delta_n,\, U_{[ny]+1:n} \geqslant y} \frac{2^\gamma}{\gamma+1} n^{1/2} (U_{[ny]+1:n} - y) \left(\frac{U_{[ny]+1:n}}{y}\right)^\gamma$$

$$\leqslant \frac{2^\gamma}{\gamma+1} 46\, n^{-1/2} \log\log n(47 \log\log n + 1), \text{ a.s. for } n \text{ large,}$$

and

$$(3.8) \qquad \sup_{1/(n+1) \leqslant y \leqslant \delta_n,\, U_{[ny]+1;n} \geqslant y} \left(\left(\frac{U_{[ny]+1:n}}{y}\right)^\gamma - 1\right) y n^{1/2} \frac{2^\gamma}{\gamma+1}$$

$$\leqslant ((47 \log\log n + 1)^\gamma - 1)\, 25 n^{-1/2} \log\log n,$$

almost surely for large n.

Combining now (3.3), (3.5), (3.7) and (3.8) we get (3.1).

It is clear from Theorem 3.1 that, when proving (1.9) and (1.10) conditions (iv) and (v) of Theorem A come into play only because of the tail regions $[0, 1/(n+1)), (n/(n+1), 1]$. Having replaced δ_n of (1.8) by $1/(n+1)$ in (3.1), we have only paid the price of slightly weakened rates of convergence. While they render (1.9) and (1.10) true, the extra conditions (iv) and (v) of Theorem A are somewhat disjoint from that of (iii). Next we modify the latter somewhat for the sake of seeking further insight into the effect of the tail behaviour of the density-quantile function $f(Q)$ on a statement like (3.1). We are going to do this over the interval $[0, 1/2]$ only and note that similar statements can be made over $[1/2, 1]$.

Theorem 3.2. *Assume conditions* (i), (ii), (iii) *of Theorem A and, instead of its conditions* (iv), (v), *we assume now that*

$$\text{(vi)} \qquad \lim_{y \downarrow 0} y \frac{f'(Q(y))}{f^2(Q(y))} = \gamma_1.$$

Then, if $\gamma_1 > 0$, *we have* (1.10), *and if* $\gamma_1 < 0$ *we have*

$$\text{(3.9)} \qquad \sup_{n^{-\alpha} \leqslant y \leqslant 1/2} |\rho_n(y) - u_n(y)| \xrightarrow{a.s.} 0, \quad \text{as } n \to \infty,$$

provided that $\alpha < 1 + 1/(2|\gamma_1|)$. *On the other hand, when* $\gamma_1 < 0$, *for* $\alpha > 1 + 1/(2|\gamma_1|)$ *there exist positive constants* $K = K(\alpha)$ *and* $\lambda = \lambda(\alpha)$ *such that*

$$\text{(3.10)} \qquad \varliminf_{n \to \infty} n^{-\lambda} \sup_{n^{-\alpha} \leqslant y \leqslant 1/2} |\rho_n(y)| \geqslant K \ a.s.$$

Proof. If $\gamma_1 > 0$ then $f' > 0$ near zero, and condition (v, β) of Theorem A holds true. Consequently in the latter case we have also (1.10). In order to show that (3.9) holds true, we first note that we have (3.1) under (i), (ii), (iii). Hence and by (3.10) respectively (4.5.17) in Csörgő and Révész [4] respectively [5] it suffices to show that the rv $\sup_{n^{-\alpha} \leqslant y \leqslant 1/n} |\rho_n(y)| \xrightarrow{a.s.} 0$. Consider

$$\text{(3.11)} \qquad \sup_{n^{-\alpha} \leqslant y \leqslant 1/n} |\rho_n(y)| \leqslant \sup_{0 \leqslant y \leqslant 1/n,\, U_{1:n} < y} |\rho_n(y)| + \sup_{n^{-\alpha} \leqslant y \leqslant 1/n,\, U_{1:n} \geqslant y} |\rho_n(y)|.$$

By (3.3), $\sup_{0 \leqslant y \leqslant 1/n,\, U_{1:n} < y} |\rho_n(y)| \xrightarrow{a.s.} 0$ under conditions (i), (ii), (iii) of Theorem A. Hence it suffices to show that under (i), (ii), (iii) *and* (vi) we have

$$\text{(3.12)} \qquad \sup_{n^{-\alpha} \leqslant y \leqslant 1/n,\, U_{1:n} \geqslant y} |\rho_n(y)| \xrightarrow{a.s.} 0.$$

Condition (vi) implies that for any $\delta > 0$ there exists a $y_0 > 0$ such that

$$\text{(3.13)} \qquad \sup_{0 \leqslant y \leqslant y_0} \left| y \frac{f'(Q(y))}{f^2(Q(y))} \right| < |\gamma_1| + \delta.$$

Hence, by Lemma 1 in Csörgő and Révész (1978), we have

$$\frac{f(Q(y_1))}{f(Q(y_2))} \leqslant \left(\frac{y_1 \vee y_2}{y_1 \wedge y_2} \, \frac{1-(y_1 \wedge y_2)}{1-(y_1 \vee y_2)}\right)^{|\gamma_1|+\delta} \text{ for all } y_1, y_2 \in (0, y_0].$$

Computations like those of (3.5) for $U_{1:n} \geqslant y$ result in

$$(3.14) \qquad |\rho_n(y)| = \frac{2^{|\gamma_1|+\delta}}{|\gamma_1|+\delta+1} \left\{ n^{1/2}(U_{1:n}-y)\left(\frac{U_{1:n}}{y}\right)^{|\gamma_1|+\delta} + \left(\left(\frac{U_{1:n}}{y}\right)^{|\gamma_1|+\delta} - 1\right) y n^{1/2} \right\}.$$

By (3.6) for $n^{-\alpha} \leqslant y \leqslant 1/n$, $U_{1:n} \geqslant y$, we get a.s. for n large that

$$(3.15) \qquad \left(\frac{U_{1:n}}{y}\right)^{|\gamma_1|+\delta} \leqslant \left(\frac{46n^{-1}\log\log n}{n^{-\alpha}} + 1\right)^{|\gamma_1|+\delta} = O(n^{(\alpha-1)(|\gamma_1|+\delta)}(\log\log n)^{|\gamma_1|+\delta}).$$

Now we choose $\delta > 0$ of (3.13) so that $(\alpha - 1)(|\gamma_1| + \delta) < 1/2$ and computations like those of (3.7) and (3.8) result in (3.12), which in turn implies (3.9).

Now we turn to the proof of (3.10). First we note that in proving (3.10), only condition (vi) will be used. Given the latter condition we have (cf. Theorem 3A in Parzen (1980), or page 7 of Seneta (1976))

$$(3.16) \qquad f(Q(y)) = y^{\gamma_1} L_1(y) \text{ as } y \downarrow 0,$$

where L_1 is a slowly varying function at $y = 0$. By Corollary on page 274 in Feller (1966), for any $\varepsilon > 0$ there exist some positive constants K_1, K_2 and $0 < y_0 < 1$ such that we have (cf. (2.13))

$$(3.17) \qquad K_1 y^{\gamma_1+\varepsilon} \leqslant f(Q(y)) \leqslant K_2 y^{\gamma_1-\varepsilon} \text{ for all } 0 < y \leqslant y_0.$$

Hence on taking reciprocals in (3.17) and integrating, we get

$$(3.18) \qquad C_2 y^{-\gamma_1+\varepsilon+1} \leqslant Q(y) - Q(0) \leqslant C_1 y^{-(\gamma_1+\varepsilon)+1} \text{ for all } 0 < y \leqslant y_0,$$

where C_1, C_2 are some positive constants if $\varepsilon < |\gamma_1|/2$, i.e., said integration can be performed and (3.18) holds true, provided $-(\gamma_1 + \varepsilon) + 1 > 0$. We note also that by (3.18) $Q(0) > -\infty$. Next consider

$$(3.19) \qquad \sup_{n^{-\alpha} \leqslant y \leqslant 1/2} |\rho_n(y)| \geqslant |\rho_n(n^{-\alpha})| = n^{1/2} f(Q(n^{-\alpha}))|Q(U_{1:n}) - Q(n^{-\alpha})|$$

$$\geqslant n^{1/2} f(Q(n^{-\alpha}))|Q(U_{1:n}) - Q(0)| - n^{1/2} f(Q(n^{-\alpha}))|Q(0) - Q(n^{-\alpha})|$$

$$\geqslant n^{1/2} K_1 n^{-\alpha(\gamma_1+\varepsilon)} C_2 (U_{1:n})^{-\gamma_1+\varepsilon+1} - n^{1/2} K_2 n^{-\alpha(\gamma_1-\varepsilon)} C_1 n^{-\alpha(-(\gamma_1+\varepsilon)+1)}$$

$$\geqslant K_1 C_2 n^{1/2-\alpha(\gamma_1+\varepsilon)+\gamma_1-\varepsilon-1} (\log n)^{(1+\varepsilon)(-\gamma_1+\varepsilon+1)} - K_2 C_1 n^{1/2-\alpha(\gamma_1-\varepsilon)+\alpha(\gamma_1+\varepsilon)-\alpha} \text{ a.s.},$$

where the last a.s. inequality is in view of $\underline{\lim}_{n\to\infty} U_{1:n}(n(\log n)^{1+\varepsilon}) = \infty$ for every $\varepsilon > 0$, while the inequality preceding the last one is via (3.17) and (3.18). Now (3.10) will be proved, provided we can show that the first term on the rigth hand side of the last inequality above goes to ∞ and the second one of the same inequality goes to zero as $n \to \infty$ for $\alpha > 1 + 1/(2|\gamma_1|)$. Indeed, given the latter condition on α, and choosing $\varepsilon > 0$ of (3.17) small enough, the exponent of n in the said first term can be made positive, while that of n in the second term can be made negative. This also completes the proof of (3.10), and Theorem 3.2 is also proved.

One of the interesting consequences of Theorem A is the following law of iterated logarithm (LIL) for ρ_n

$$\overline{\lim_{n\to\infty}} \left(\frac{2}{\log\log n}\right)^{1/2} \sup_{0\leqslant y\leqslant 1} |\rho_n(y)| \overset{a.s.}{=} 1. \tag{3.20}$$

An interesting consequence of (3.9) of Theorem 3.2 is that it throws new light on the latter LIL. We have

Corollary 3.1. *Assume conditions* (i), (ii), (iii) *of Theorem A and condition* (vi) *of Theorem* 3.2. *Then if* $\gamma_1 > 0$ *we have* (3.20), *and if* $\gamma_1 < 0$ *we have*

$$\overline{\lim_{n\to\infty}} \left(\frac{2}{\log\log n}\right)^{1/2} \sup_{n^{-\alpha}\leqslant y\leqslant 1/2} |\rho_n(y)| \overset{a.s.}{=} \begin{cases} 1 & \textit{if } \alpha < 1 + 1/(2|\gamma_1|) \\ \infty & \textit{if } \alpha > 1 + 1/(2|\gamma_1|). \end{cases} \tag{3.21}$$

Proof. We have $\overline{\lim_{n\to\infty}} \left(\frac{2}{\log\log n}\right)^{1/2} \sup_{n^{-\alpha}\leqslant y\leqslant 1/2} |u_n(y)| \overset{a.s.}{=} 1$. Now the first statement of (3.21) follows from the latter combined with (3.9). The second statement of (3.21) is by (3.10).

We note that Corollary 3.1 implies the non existence of LIL for ρ_n over [0, 1] under (i), (ii), (iii) and (vi) if $\gamma_1 < 0$. On the other hand it follows from Theorem 3 in Mason (1982) that if we replace the weight function $f(Q)$ in $n^{-1/2}\rho_n$ by y^ε, $\varepsilon > 0$, then

$$\lim_{n\to\infty} \sup_{0\leqslant y\leqslant 1/2} y^\varepsilon |Q_n(y) - Q(y)| \overset{a.s.}{=} 0$$

for every $\varepsilon > 0$, given conditions (i), (ii), (iii) and (vi), i.e., under the latter conditions we always have a Mason type Glivenko-Cantelli theorem for $(Q_n(y) - Q(y))$.

Summarizing the main features of the problem of strong approximation of ρ_n by u_n over [0, 1] in general, we have seen so far that under conditions (i), (ii), (iii), (iv), (v) of Theorem A we have (1.9) and (1.10), on dropping conditions (iv) and (v) we have (3.1), and when we replace conditions (iv), (v) by that of (vi) we have (3.9) and (3.10). Now we give an example which will amount to saying that for results like (1.9) and (1.10) neither conditions (iv), (v), nor condition (vi) are necessary. This example is due to Parzen (1979, page 116). Continuing the numbering of examples of our Section 2, we now have

(7) Parzen's example (1979). Let

$$1 - F(x) = \exp(-x - C\sin x), \quad x \geqslant 0, \; 0.5 < C < 1.$$

Letting $x = Q(y)$ $(0 \leqslant y \leqslant 1)$ we get

$$-\log(1 - y) = Q(y) + C\sin Q(y)$$

and

$$f(Q(y)) = (1-y)(1+C\cos Q(y)).$$

Hence

$$Q(y) \leqslant |\log(1-y)| + C \tag{3.22}$$

and

$$f(Q(y)) \leqslant (1-y)(1+C). \tag{3.23}$$

Also

$$f'(Q(y)) = -(1-y)((1+C\cos Q(y))^2 + C\sin Q(y)).$$

Clearly then

$$\sup_{0<x<\infty} F(x)(1-F(x))\frac{|f'(x)|}{f^2(x)} = \sup_{0<y<1} y(1-y)\frac{|f'(Q(y))|}{f^2(Q(y))} \leqslant \frac{1}{1-C},$$

i.e., *conditions* (i), (iii), (iii) *of Theorem A are satisfied. Hence by Theorem* 3.1 *we have* (3.1) *with* $\gamma = 1/(1-C)$. On the other hand *as* $y \to 1$, $Q(y) \to \infty$ *and* $f(Q(y))$ *oscillates.* Hence *conditions* (iv), (v) of Theorem A *are not satisfied.* Also *as* $y \to 1$

$$(1-y)\frac{f'(Q(y))}{f^2(Q(y))} \text{ oscillates,}$$

i.e., the right tail version limit requirement of *condition* (vi) *is also not satisfied. Nevertheless* we show now that *in case of this example we have*

$$\sup_{0\leqslant y\leqslant 1} |\rho_n(y) - u_n(y)| \overset{a.s.}{=} O\left(n^{-1/2}(\log\log n)^{1/(1-C)}(\log n)^{(1+\varepsilon)\frac{C}{1-C}}\right) \tag{3.24}$$

where $C \in (0.5, 1)$ *as above.* In view of having (3.1) and because of $\sup_{1-1/n\leqslant y\leqslant 1} |u_n(y)| \overset{a.s.}{=} O(n^{-1/2}\log\log n)$ (cf. e.g., (4.5.17) in Csörgő and Révész (1981)), for the sake of verifying (3.24) it will suffice to show that

$$\sup_{1-1/n\leqslant y\leqslant 1} |\rho_n(y)| \overset{a.s.}{=} O(n^{-1/2}\log n). \tag{3.25}$$

Hence consider (cf. (3.22), (3.23))

$$\sup_{1-1/n\leqslant y\leqslant 1} |\rho_n(y)| \leqslant n^{1/2} \sup_{1-1/n\leqslant y\leqslant 1} f(Q(y))\,Q(y) + n^{1/2} \sup_{1-1/n\leqslant y\leqslant 1} f(Q(y))\,Q(U_{n:n})$$

$$\leqslant n^{1/2} \sup_{1-1/n\leqslant y\leqslant 1} (1-y)(1+C)(|\log(1-y)| + C)$$

$$+ n^{1/2} \sup_{1-1/n\leqslant y\leqslant 1} (1-y)(1+C)(|\log(1-U_{n:n})| + C).$$

As to the first term of the latter inequality we have the asymptotic upper bound $O(n^{-1/2}\log n)$, while for the second one we have the same asymptotic upper bound almost surely, due to $|\log(1-U_{n:n})| \overset{a.s.}{=} O(\log n)$. Hence (3.25) holds true and (3.24) also follows.

REFERENCES

1. Chibisov, D., *Some theorems on the limiting behaviour of empirical distribution function.* In: *Selected Translations in Mathematical Statistics and Probability, Vol. 6*, pp. 147–156. Providence, R.I., Amer. Math. Soc., 1964.
2. Csörgő, M., *Quantile Processes with Statistical Applications.* Philadelphia, S.I.A.M., 1982.
3. Csörgő, M., Révész, P., *Some notes on the empirical distribution function and the quantile process.* In: Révész, P. (Ed.) *Limit Theorems of Probability Theory.* Colloq. Math. Soc. J. Bolyai, Vol. 11, pp. 59–71. Amsterdam, North-Holland, 1975.
4. Csörgő, M., Révész, P., *Strong approximations of the quantile process* Ann., Statist., **6**, 882–894 (1978).
5. Csörgő, M., Révész, P., *Strong Approximations in Probability and Statistics.* New York, Academic Press & Budapest, Akadémiai Kiadó, 1981.
6. Csörgő, S., *On general quantile processes in weighted sup-norm metrics.* Stochastic Process. Appl., **12**, 215–220 (1982).
7. Feller, W., *An Introduction to Probability Theory and Its Applications, Vol. II.* New York, Wiley, 1966.
8. Mason, D. M., *Some characterizations of almost sure bounds for weighted multidimensional empirical distributions and a Glivenko-Cantelli theorem for sample quantiles.* Z. Wahrsch. Verw. Gebiete, **59**, 505–513 (1982).
9. O'Reilly, N., *On the weak convergence of empirical processes in sup-norm metrics.* Ann. Probab., **2**, 642–651 (1974).
10. Parzen, E., *Nonparametric statistical data modelling.* J. Amer. Statist. Assoc., **58**, 13–30 (1979).
11. Parzen, E., *Quantile functions, convergence in quantile, and extreme value distribution theory.* Tech. Report No. B-3, Texas A&M Univ., 1980.
12. Pyke, R., Shorack, G. R., *Weak convergence of a two-sample empirical process and a new approach to Chernoff-Savage theorems.* Ann. Math. Statist., **39**, 755–771 (1968).
13. Seneta, E., *Regularly Varying Functions.* Lecture Notes in Math., Vol. 508. Berlin–Heidelberg–New York, Springer, 1976.
14. Shorack, G. R., *Weak convergence of empirical and quantile processes in sup-norm metrics via KMT-construction.* Stochastic Process. Appl., **9**, 95–98 (1979).
15. Shorack, G. R., *Weak convergence of the general quantile process in* $||.//q||$ *-metrics.* Inst. Math. Statist. Bull., **11**, Abstract 82t-2 (1982).
16. Stute, W., *The oscillation behaviour of empirical processes.* Ann. Probab. **10**, 86–107 (1982).
17. Wellner, J. A., *Limit theorems for the ratio of the empirical distribution function to the true distribution function.* Z. Wahrsch. Verw. Gebiete, **45**, 73–88 (1978).

I. CUCULESCU

Strong Markov property for a class of transition semigroups

We defined in [7] and [8] a class of transition semigroups (with fictitious states and cemetery set) and proved that an operation, defined also in [7] and [8], when applied to a finite number or to a sequence of such transition semigroups, leads to one in the same class. Here we introduce a concept of such a strong Markov semigroup, weaker than the usual one, and show that the above mentioned operation, when applied to such strong Markov semigroups, gives also a strong Markov one.

1. Introduction

In [7] and [8] we introduced the class of fc-semigroups. If F is a metrisable space, if $E \in \mathbf{B}_F$ (the elements in $F \setminus E$ will be named "fictitious") and if $\{\Delta, \mathbf{D}\}$ is a measurable space (the "cemetery set"), then a fc-semigroup on $E \subset F$, $\{\Delta, \mathbf{D}\}$ is a transition semigroup (Q_t) on $\{E \oplus \Delta, \mathbf{B}_E \oplus \mathbf{D}\}$ satisfying the conditions stated below. Let $\Omega(Q)$ be the set of all sample paths $f : [0, \infty) \to F \oplus \Delta$ for which there is a $\zeta(f) \in [0, \infty]$ (its "lifetime") such that $f(t) \in F$ and is right continuous for $t < \zeta(f)$ and $f(t) \in \Delta$ and is constant for $t \geqslant \zeta(f)$ Let $x_{Q,t}(f) = f(t)$, $\mathbf{K}(Q) = \mathbf{B}(x_{Q,t} | t \in [0, \infty))$. The first condition is that for every probability μ on $E \oplus \Delta$ there is a probability P_Q^μ on $\mathbf{K}(Q)$ for which $P_Q^\mu(x_{Q,t} \in F \setminus E) = 0$ for all $t \geqslant 0$ and for which $(x_{Q,t})_{t \in [0, \infty)}$ is, in an obvious sense, a (Q_t)-process; the second condition is that, defining $x_{Q,\infty}(f) = x_{Q,t}(f)$ for $\zeta(f) < \infty$, $t \geqslant \zeta(f)$ and $x_{Q.\infty}(f) = \xi$ for $\zeta(f) = \infty$, the conditional distribution of ζ with respect to $(x_{Q,0}, x_{Q.\infty})$ charges no $\{t\}$ with $t \in (0, \infty)$.

Let us recall from [4] that, if we introduce also $\mathbf{F}_{Q,t} = \mathbf{B}(x_{Q,s} | s \leqslant t)$, then there exists a transition probability G_Q from E to $(0, \infty] \times (\Delta \oplus \{\xi\})$, defined by $G_Q(x, (0, t] \times D) = Q_t(x, D)$ for $D \in \mathbf{D}$, $G_Q(x, \{\infty\} \times \Delta) = 0$, such that the conditional distribution of $(\zeta - t, x_{Q,\infty})$ with respect to $\mathbf{F}_{Q,t}$ is, on $(\zeta > t)$, $G_Q(x_{Q,t},\qquad)$.

The first idea in order to introduce the concept of a strong Markov fc-semigroup (Q_t) is to require that for every $(\mathbf{F}_{Q,t+})$-stopping time τ and every probability μ on $E \oplus \Delta$, for which $P_Q^\mu(x_{Q,\tau} \in F \setminus E, \tau < \infty) = 0$ the strong Markov property is valid for τ and μ, i.e. $M_{P_Q^\mu}((g \circ x_{Q,\tau+t})\, \chi_{(\tau<\infty)} \mid \mathbf{F}_{Q,\tau+}) = ((Q_t\, g) \circ x_{Q,\tau})\, \chi_{(\tau<\infty)}$ for all measurable $g : E \oplus \Delta \to [0, \infty)$ (where we consider g and $Q_t g$ to be null on $F \setminus E$).

But this is not the definition which we have in view, as will be explained in the following section.

2. An example

Let us recall the following facts ([7], [8]). Let us denote by $\mathbf{E}(Q)$, where (Q_t) is a fc-semigroup on $E \cup F$, Δ, the set of all entrance laws $\mathbf{I}$ with respect to (Q_t) for which there exists a probability $P_R^{\mathbf{I}}$ on $\mathbf{K}(Q)$ such that $P_Q^{\mathbf{I}}(x_{Q,t} \in \in F \setminus E) = 0$ for all $t \in (0, \infty)$ (only!) and $(x_{Q,t})_{t \in (0,\infty)}$ is a (Q_t)-process with the entrance law $\mathbf{I}$.

Let us define a fc-(Q_t)-process as one with time set $[0, \infty)$, with sample paths in $\Omega(Q)$ and which transports (via the sample path mapping) the probability from its underlying space into a P_Q^{μ}. In a similar manner, a fc$_+$-(Q_t)-process will be one in which the result of that transport is a $P_Q^{\mathbf{I}}$ with $\mathbf{I} \in \mathbf{E}(Q)$.

Now, if (Q_t^1), (Q_t^2) are two fc-semigroups and if T is a measurable mapping from the cemetery set of the first to $\mathbf{E}(Q^2)$ then there exists an unique fc-semigroup (Q_t^{12}), named "the result of combining (Q_t^1), (Q_t^2) using T", having as state space the direct sum of the two state spaces without the cemetery set of (Q_t^1), coinciding with (Q_t^2) on its state space and having also the following property. Let, on the same probability space, (x_t^1) be a fc-(Q_t^1)-process, with initial distribution noncharging the cemetery set, with life time ζ^1 and (x_t^2) be a fc$_+$-(Q_t^2)-process. Suppose that the family consisting of all x_t^1, x_s^2, in which every x_t^1 appears before every x_s^2, is Markov and that T is a "conditional entrance law" of (x_t^2) with respect to x_∞^1 on $(\zeta^1 < \infty)$ (see [7] for details). Then $x_t^{12} = x_t^1$ for $t < \zeta^1$, $x_t^{12} = x_{t-\zeta^1}^2$ for $t \geqslant \zeta^1$ define a fc-(Q_t^{12})-process.

A similar result is true for a finite or infinite sequence of fc-semigroups and measurable mappings; the second condition at the beginning of §1 above is essential in the proof of the "infinite" case. We shall use in this paper only the consequence $P_Q^{\mu}(\zeta = t) = 0$ for $t \in (0, \infty)$ of that "second condition".

We did not express in [7] explicitely (Q_t^{12}); the proof there shows, besides $Q_t^{12}(x, \quad) = Q_t^2(x, \quad)$ for x in the state space of (Q_t^2), $Q_t^{12}(x, \quad) = Q_t^1(x, \quad)$ on the state space of (Q_t^1) minus its cemetery set for x not in the state space of (Q_t^2), that ,if we write $T(\delta) = (T_t(\delta, \quad))_{t \in (0,\infty)}$, we have $Q_t^{12}(x, A) =$

$$= \int \chi_{(0,t)\times\Delta}(s, \delta)\, T_{t-s}(\delta, A)\, dG_{Q^1}(x, (s, \delta)) \text{ for every measurable subset } A$$

of the state space of (Q_t^2) and every x in the state space of (Q_t^1) but not in its cemetery set.

Let us consider now the following example. Let (Q_t^1) be the minimal semigroup on $\{-2, -1, 0, 1, 2\}$ coresponding to a transition matrix (q_{ij}) defined by $q_{01} = q_{0,-1} = \frac{1}{2}$, $q_{12} = q_{-1,-2} = q_{22} = q_{-2,-2} = 1$ and arbitrary nonnul parameteres for the holding times; transform, in its state space, $\{-2, 2\}$ into a cemetery set. Let (Q_t^2) be defined, on the real line R, by $Q_t^2(x, \quad) = \varepsilon_{x+t}$ for $x \geqslant 0$ and $Q_t^2(x, \quad) = \varepsilon_{x-t}$ for $x < 0$ (thus there are no fictitious states in the considered semigroups, and the second has void cemetery set). It is easy to see that both are fc-semigroups and that they are strong Markov in the sense proposed in § 1 above. Let $\mathbf{I}_+, \mathbf{I}_- \in \mathbf{E}(Q^2)$ be defined as $\mathbf{I}_+ = (\varepsilon_t)_{t \in (0,\infty)}$, $\mathbf{I}_- = (\varepsilon_{-t})_{t \in (0,\infty)}$; consider $T(-2) = \mathbf{I}_-$, $T(2) = \mathbf{I}_+$. The resulting (Q_t^{12}) is a fc-semigroup on $\{-1, 0, 1\} \oplus$ R, having no fictitious states and void cemetery set.

Let us show that (Q_t^{12}) is not a strong Markov fc-semigroup in the sense in § 1. Let us denote by $-2'$, etc. the elements in $\{-2, -1, 0, 1, 2\}$ in order to mark that they are different of -2, etc. in R, in the state spaces appearing in our reasonment. Consider the first visit σ of R, defined on $\Omega(Q^{12})$; it is a $(\mathbf{F}_{Q^{12}, t+})$-stopping time. Consider also the events $A \in \mathbf{F}_{Q^{12}, \sigma+}$ defined as "1′ is visited before σ" and $B \in \mathbf{B}(x_{Q^{12}, \sigma+t} \mid t \in [0, \infty))$ defined as "$(-\infty, 0)$ is visited".

Let us recall the description above of a fc-(Q_t^{12})-process starting from 0′. In it σ corresponds to ζ^1, A equals $(x^1_\infty = 2')$. Since $P^{\mathbf{I}_+}_{Q^2}$ and $P^{\mathbf{I}_-}_{Q^2}$ are concentrated on $(x_{Q^2, 0} = 0)$ it follows that $x^{12}_{\zeta^1} = 0$ a.e., hence the strong Markov property with σ requires the independence of A and B (we have to remark also that $\zeta^1 < \infty$ a.e.). The probability of A is obviously $\frac{1}{2}$. Since $\complement A = (x^1_\infty = -2')$, $T(-2') = \mathbf{I}_-$ and $P^{\mathbf{I}_-}_{Q^2}$ is concentrated on $(x_{Q^2, t} < 0)$ for every $t > 0$, it follows that $B \supset \complement A$ up to a null set. A similar reasonment, based on the facts $T(2') = \mathbf{I}_+$, $P^{\mathbf{I}_+}_{Q^2}(x_{Q^2, t} < 0) = 0$ for all t, shows that the probability of $A \cap B$ is null, hence the probability of B is $\frac{1}{2}$ and A and B cannot be independent.

Our example may be easily transformed in order to arrive to an analogous situation in which both semigroups have nonvoid sets of fictitious states and nonvoid cemetery sets.

We shall define in the following section a concept of strong Markov fc-semigroup by imposing the strong Markov property for a more restrictive class of stopping times, but still sufficiently rich. The main purpose of this definition will be to make true the fact that the result of combining two (or a finite or infinite sequence of) strong Markov fc-semigroups is also a strong Markov fc-semigroup.

We remain at this elementary level and do not try to exploit the possibilities of "regularising" (Q_t^2) above by changing the topology on its state space (splitting 0 into "itself" and a fictitious state, etc.). This may lead to more involved examples for the "regularisation methods".

3. Strong Markov fc-semigroups

Let us define, as usual, for $s \geq 0$, $\theta_s : \Omega(Q) \to \Omega(Q)$ by $(\theta_s f)(t) = f(s + t)$; we have $\theta_s^{-1}(\mathbf{F}_{Q, t}) \subset \mathbf{F}_{Q, s+t}$, etc.

Definition 3.1. Let (Q_t) be a fc-semigroup on $E \subset F$, Δ.

A set $M \subset (0, \infty) \times \Omega(Q)$ will be called a sm-set if the following facts are valid.

a) M is progressively measurable with respect to $(\mathbf{F}_{Q, t+})$.

b) $P^\mu_Q(M_t) = 1$ for every probability μ on $E \oplus \Delta$ and every $t > 0$ (where, as usual, M_t means $\{f \mid (t, f) \in M\}$ and it belongs to $\mathbf{F}_{Q, t+}$).

c) $(t, f) \in M$ implies $f(t) \notin F \setminus E$.

d) For every $s < t$, $(t, f) \in M$ and $(t - s, \theta_s(f)) \in M$ are equivalent.

Definition 3.2. We say that a fc-semigroup (Q_t) is a strong Markov one if there exists a sm-set $M \subset (0, \infty) \times \Omega(Q)$ such that, whatever be the $(\mathbf{F}_{Q, t+})$-stopping time τ and the probability μ on $E \oplus \Delta$ for which $P^\mu_Q(\{f \mid 0 < \tau(f) < \infty, (\tau(f), f) \notin M\}) = 0$ (we say it shortly "τ is a (μ, M)-stopping time") we have (the following will be called shortly "strong Markov property for τ, μ") for all $t > 0$ and measurable $g \geqslant 0$ $M_{P_Q^\mu}((g \circ x_{Q,\tau+t})\, \chi_{(\tau<\infty)} \mid \mathbf{F}_{Q,\tau+}) = ((Q_t g) \circ x_{Q,\tau})\, \chi_{(\tau<\infty)}$.

Remarks. *a.* Of course, the above is inspired from [2], part **II**, § 9.

b. $f \to (\tau(f), f)$, defined on $(\tau < \infty)$, is measurable with respect to the progressive Borel field and $\mathbf{F}_{Q,\tau+}$ (see, say, [9]).

c. We may state the situation in Definition 3.2. more precisely as "(Q_t) is a M-strong Markov fc-semigroup".

d. We get a definition equivalent to 3.2 above if we require also that $f(t) \in \Delta$ implies $(t, f) \in M$. This follows from the fact (which will be used also in a proof below) that, if $A \in \mathbf{F}_{Q,\tau+}$ and if the strong Markov property is true for (τ_A, μ) and $(\tau_{\complement A}, \mu)$, then it is also true for (τ, μ) (where, as usual, $\tau_A = \tau$ on A and $\tau_A = \infty$ on $\complement A$). In our case $A = (\tau < \zeta)$ (ζ is a stopping time, the strong Markov property for $\tau_{\complement A}$ is obvious).

Proposition 3.3. *Let (Q_t) be a M-strong Markov* fc-*semigroup, let σ be a $(\mathbf{F}_{Q,t+})$-stopping time and τ be $[0, \infty]$-valued $\mathbf{F}_{Q,\sigma+}$-measurable, $\tau \geqslant \sigma$, such that $\sigma(f) < \infty$ implies $\tau(f) > \sigma(f)$. Then τ is a M-stopping time for all the probabilities μ on the state space.*

Proof. It will be sufficient to find a sequence (A_n) of sets in $\mathbf{F}_{Q,\tau+}$ having as union $\Omega(Q)$ such that τ_{A_n} is, for every n, a (μ, M) stopping time for all μ. We choose as one term of the sequence $(\tau = \infty)$, for which the property is obvious, and the others as $(\sigma < s < \tau)$, where s runs over the rationals.

The above reasonment showed that we may restrict ourselves to the case $\sigma = s$ (constant), τ $\mathbf{F}_{Q,s}$-measurable, $\tau > s$.

Let us consider the mapping $j_s : \Omega(Q) \to \Omega(Q) \times \Omega(Q)$, defined by $j_s(f) = (f, \theta_s(f))$, where $\Omega(Q) \times \Omega(Q)$ is endowed with $\mathbf{F}_{Q,s} \otimes \mathbf{K}(Q)$.

The Markov property shows that the probability $P = P_Q^\mu \circ j_s^{-1}$ equals $P_Q^\mu \mid_{\mathbf{F}_{Q,s}} \otimes P'$ where P' is the transition probability defined by $P'(f) = P_Q^{\varepsilon_y}$, $y = x_{Q,s}(f)$.

According to property d in Definition 3.1, $\{f \mid \tau(f) < \infty,\ (\tau(f), f) \notin M\}$ equals $j_s^{-1}(\{(f, g) \mid \tau(f) < \infty, (\tau(f) - s, g) \notin M\})$ and the set under j_s^{-1} is in the Borel field on which P is defined due to the progressive measurability of M and to the $\mathbf{F}_{Q,s}$-measurability of τ.

We obtain

$$P_Q^\mu(\{f \mid \tau(f) < \infty,\ (\tau(f), f) \notin M\}) = \int \chi_{(\tau<\infty)}(f) P'(f, \complement M_{\tau(f)-s})\, dP_Q^\mu(f) = 0.$$

Corollary 3.4. *If (Q_t) is a M-strong Markov* fc-*semigroup and τ is a (μ, M)-stopping time for $(\mathbf{F}_{Q,t+})$ then, for all $\mathbf{K}(Q)$-measurable $g \geqslant 0$ we have*

$$M_{P_Q^\mu}((g \circ \theta_\tau)\, \chi_{(\tau<\infty)} \mid \mathbf{F}_{Q,\tau+}) = \left(\int g\, dP_Q^{\varepsilon_{x_{Q,\tau}}}\right) \chi_{(\tau<\infty)}.$$

The proof goes along the usual lines (see [1]); the above proposition shows that $\tau + t$ are (μ, M)-stopping times.

Corollary 3.5. *Let, in the hypothesis of* 3.4, *moreover, $\tau(f) < \infty$ imply $\tau(f) < \zeta(f)$. Then, on $(\tau < \infty)$, the conditional distribution of $(\zeta - \tau, x_{Q,\infty})$ with respect to $\mathbf{F}_{Q,\tau+}$ and P_Q^μ is $G_Q(x_{Q,\tau},\ \ )$.*

The proof consists in an application of 3.4, in the remark that $\zeta \circ \theta_{\tau} = \zeta - \tau$ on $(\tau < \infty)$, in the definition of G_Q in § 1 above and in its consequence: the distribution of $(\zeta, \ x_{Q, \infty})$ with respect to P^{ε_x} is $G_Q(x, \quad)$.

Proposition 3.6. *If* (Q_t) *is a M-strong Markov* fc-*semigroup, if* $\mathbf{I} \in \mathbf{E}(Q)$ *and if* τ *is a* $(\mathbf{F}_{Q, t+})$*-stopping time for which* $\tau > 0$ $P^{\mathbf{I}}_Q$-a.e. *and* $P^{\mathbf{I}}_Q(\{f \mid \tau(f) < \infty, (\tau(f), f) \notin M\}) = 0$ *then the strong Markov property is true for* τ *and* $\mathbf{I}$, i.e., *for every* $t > 0$ *and measurable* $g \geqslant 0$, $M_{P^{\mathbf{I}}_Q}(g \circ \chi_{Q, \tau+t}) \chi_{(\tau<\infty)} \mid \mathbf{F}_{Q, \tau+}) = ((Q_t g) \circ x_{Q, \tau}) \chi_{(\tau>\infty)}$.

Proof. According to Remark *d* after Definition 3.2, it will be sufficient to find a sequence (A_n) of sets in $\mathbf{F}_{Q, \tau+}$, having as union a set with $P^{\mathbf{I}}_Q$-null complementary, and to prove the statement for all τ_{A_n}. Taking A_n as $(\tau > s_n)$, where s_n runs over all positive rationals, we see that it will be enough to prove the proposition in the supplementary hypothesis $\tau > s$, with $s > 0$.

In order to do it, let us consider the situation in the proof of 3.3, concerning j_s, in which we replace P^{μ}_Q there by $P^{\mathbf{I}}_Q$ here.

We have $\overset{-1}{j_s}(\mathbf{F}_{Q, s} \otimes \mathbf{F}_{Q, t}) = \mathbf{F}_{Q, s+t}$; since $\tau - s$ is a $(\mathbf{F}_{Q, s+t+})_{t \in [0, \infty)}$ — stopping time, a well known lemma (stated in 3.7 below) shows that $\tau = s + \sigma \circ j_s$, σ being a $(\mathbf{G}_{t+})$-stopping time, where $\mathbf{G}_t = \mathbf{F}_{Q, s} \otimes \mathbf{F}_{Q, t}$, and that $\mathbf{F}_{Q, \tau+} = \overset{-1}{j_s}(\mathbf{G}_{\sigma+})$.

Moreover $x_{Q, s+t} = x_{Q, t} \circ \mathrm{pr}_2 \circ j_s$ and, denoting $y_t = x_{Q, t} \circ \mathrm{pr}_2$, we arrive to $x_{Q, \tau+t} = y_{\sigma+t} \circ j_s$. Hence it will be sufficient to show that
(*) $M_P((g \circ y_{\sigma+t}) \chi_{(\sigma<\infty)} \mid \mathbf{G}_{\sigma+}) = ((Q_t t) \circ y_\sigma) \chi_{(\sigma<\infty)}$.

The right member of the relation (*) is $\mathbf{G}_{\sigma+}$-measurable.

Let us remark that, for all $f \in \Omega(Q)$, $\sigma(f, \quad)$ is a $(\mathbf{F}_{Q, t+})$-stopping time. Moreover, $\mathbf{G}_{\sigma+} \subset \mathbf{H}$, where $\mathbf{H} = \{A \mid A \in \mathbf{F}_{Q, s} \otimes \mathbf{K}(Q),\ A_f \in \mathbf{F}_{Q, \sigma(f, \quad)+}$ for all $f \in \Omega(Q)\}$.

Exactly as in the proof of 3.3 we obtain $P(\{(f_1, f_2) \mid \sigma(f_1, f_2) < \infty, (\sigma(f_1, f_2), f_2) \notin M\}) = 0$ and then that, for $P^{\mathbf{I}}_Q|_{\mathbf{F}_{Q,s}}$-almost all f, $P^{\varepsilon_{x_{Q,s}(f)}}(\{f_2 \mid \sigma(f, f_2) < \infty, (\sigma(f, f_2), f_2) \notin M\}) = 0$.

For such f we know that $M((g \circ x_{Q, \sigma(f, \quad)+t}) \chi_{(\sigma, (f, \quad)<\infty)} \mid \mathbf{F}_{Q, \sigma(f, \quad)+} = ((Q_t g) \circ x_{Q, \sigma(f, \quad)}) \chi_{(\sigma(f, \quad)<\infty)}$, the conditional mean value being with respect to P^{ε_z}, $z = x_{Q, s}(f)$.

Now Lemma 3.8 below shows the truth of (*) with $\mathbf{H}$ in place of $\mathbf{G}_{\sigma+}$, etc.

Lemma 3.7. *If* $\varphi : X \to Y$, *if* $(\mathbf{F}_t)_{t \in [0, \infty)}$ *is an ascending family of Borel fields on* Y, $\mathbf{G}_t = \varphi^{-1}(\mathbf{F}_t)$ *and* $\tau : Y \to [0, \infty]$, *then* τ *is a* $(\mathbf{F}_{t+})$*-stopping time iff* $\tau \circ \varphi$ *is a* $(\mathbf{G}_{t+})$*-one. In this situation* $\varphi^{-1}(\mathbf{F}_{\tau+}) = \mathbf{G}_{(\tau \circ \varphi)+}$.

Lemma 3.8. *Let* P *be a probability on* $\{E, \mathbf{K}\}$, Q *be a transition probability from* $\{E, \mathbf{K}\}$ *to* $\{F, \mathbf{L}\}$, *let, for every* $\omega \in E$, $\mathbf{H}_\omega \subset \mathbf{L}$ *be a Borel field and let* $f \geqslant 0$ *be* $\mathbf{K} \otimes \mathbf{L}$*-measurable.*

Consider the Borel field $\mathbf{H} = \{A \mid A \in \mathbf{K} \otimes \mathbf{L},\ A_\omega \in \mathbf{H}_\omega$ *for all* $\omega\}$.

Suppose that h *is* $\mathbf{H}$*-measurable and that, for* P*-almost all* $\omega \in E$, $h(\omega, \quad) = M_{Q(\omega, \quad)}(f(\omega, \quad) \mid \mathbf{H}_\omega \quad)$.

Then $h = M(f \mid \mathbf{H})$.

4. Combining two strong Markov fc-semigroups

We begin with two lemmas.

Lemma 4.1. *If* $(\mathbf{F}_t)_{t\in[0,\infty)}$, $(\mathbf{G}_t)_{t\in[0,\infty)}$ *are ascending families of Borel fields on X, Y respectively and if* $\varphi : X \to Y$ *is such that* $\varphi^{-1}(\mathbf{G}_s) \subset \mathbf{F}_t$ *for* $s < t$ *then* $\psi : (0, \infty) \times X \to (0, \infty) \times Y$ *defined by* $\psi(t, \omega) = (t, \varphi(\omega))$ *is measurable with respect to the progressive Borel fields of the families* $(\mathbf{F}_{s+})$, $(\mathbf{G}_{t+})$.

Proof. If follows that $\varphi^{-1}(\mathbf{G}_{t+}) \subset \mathbf{F}_{t+}$ and then that we have to show that for every $u < t$ and $(\mathbf{G}_{t+})$-progressively measurable z the mapping $(s, \omega) \to \to z(s, \varphi(\omega))$ is, on $(0, u] \times X$, measurable with respect to $\mathbf{B}_{(0,u]} \otimes \mathbf{F}_{t+}$. Let $v \in (u, t)$. Then $(s, \omega) \to (s, \varphi(\omega))$ is $(\mathbf{B}_{(0,u]} \otimes \mathbf{F}_t, \mathbf{B}_{(0,u]} \otimes \mathbf{G}_v)$-measurable and z is $\mathbf{B}_{(0,u]} \otimes \mathbf{G}_v$-measurable.

Lemma 4.2. *Let* (Q_t) *be a* fc-*semigroup and let* σ *be a* $(\mathbf{F}_{Q,t+})$-*stopping time. Then* $S_\sigma = \{(t, f) \mid \sigma(f) < t\}$ *is progressive with respect to* $(\mathbf{F}_{Q,t+})$ *(since its* χ *is adapted and left continuous in t) and the mapping* $\alpha(t,f) = (t - -\sigma(f), \theta_{\sigma(f)}(f))$, *defined on* S_σ, *is measurable with respect to the* $(\mathbf{F}_{Q,t+})$-*progressive Borel field* $\mathbf{P}$ (i.e. $\alpha^{-1}(\mathbf{P}) \subset \mathbf{P}$).

Proof. We have to show that, if z is $(\mathbf{F}_{Q,t+})$-progressively measurable, then $(t,f) \to z(t - \sigma(f), \theta_{\sigma(f)}(f))$ is, on $S_\sigma \cap ((0, s - \varepsilon) \times \Omega(Q))$, measurable with respect to $\mathbf{B}_{(0,s-\varepsilon)} \otimes \mathbf{F}_{Q,s}$ and this even in the supplementary hypothesis that $n = \dfrac{s}{\varepsilon}$ is an integer.

Let us split $((0, s - \varepsilon) \times \Omega(Q)) \cap S_\sigma$ into $\bigcup_{k=1}^{n-1} D_k$, where $D_k = ((0, s - \varepsilon) \times ((k-1)\,\varepsilon \leqslant \sigma < k\varepsilon)) \cap S_\sigma \in \mathbf{B}_{(0,s-\varepsilon)} \otimes \mathbf{F}_{Q,s}$.

The image $\alpha(D_k)$ is contained in $(0, s - k\varepsilon) \times \Omega(Q)$, set on which the progressive Borel field is contained in $\mathbf{B}_{(0,s-k\varepsilon)} \otimes \mathbf{F}_{Q,s-k\varepsilon}$. Hence the proof will be finished if we show that, on D_k, α is measurable with respect to $\mathbf{B}_{(0,s-\varepsilon)} \otimes \mathbf{F}_{Q,s}$ on D_k and $\mathbf{B}_{(0,s-k\varepsilon)} \otimes \mathbf{F}_{Q,s-k\varepsilon}$ on the image.

Taking into account the definition of α, the arguments proving the last stated fact are the following.

The first is that, on $((k-1)\,\varepsilon \leqslant \sigma < k\varepsilon)$, $\mathbf{F}_{Q,\sigma+}$ is contained in $\mathbf{F}_{Q,k\varepsilon} \subset \mathbf{F}_{Q,s}$, thus σ is $\mathbf{F}_{Q,s}$-measurable on this set, etc.

The second starts with the progressive measurability of $(t,f) \to x_{Q,t}(f)$ which gives the $\mathbf{F}_{Q,(\sigma+r)+}$-measurability of $f \to x_{Q,\sigma(f)+r}(f)$ on $\{f \mid \sigma(f) < \infty\}$ and then the $\mathbf{F}_{Q,s}$-measurability of the same mapping on $(\sigma + r < s) = (\sigma < s - r)$, continues with the obvious remarks $x_{Q,r} \circ \theta_\sigma = x_{Q,\sigma+r}$ and $((k-1)\,\varepsilon \leqslant \sigma < k\varepsilon) \subset (\sigma < s - r)$ for $k\varepsilon \leqslant s - r$ i.e. for $r \leqslant s - k\varepsilon$ which lead to the $\mathbf{F}_{Q,s}$-measurability of $x_{Q,r} \circ \theta_\sigma$ on $((k-1)\varepsilon \leqslant \sigma < k\varepsilon)$ for every fixed k and all $r \leqslant s - k\varepsilon$ and finishes with the measurability on D_k of $(t,f) \to \theta_{\sigma(f)}(f)$ with respect to $\mathbf{F}_{k,s-k\varepsilon}$ on the image and $\mathbf{B}_{(0,s-\varepsilon)} \otimes \mathbf{F}_{Q,s}$ on D_k.

Definition 4.3. Let (Q_t) be a fc-semigroup on $E \subset F$, Δ.

If $f, g \in \Omega(Q)$, $t > 0$, we define $f \overset{t}{\sim} g$ if there is an $\varepsilon > 0$ such that $f(s) = g(s)$ for all $s \in (t - \varepsilon, t + \varepsilon) \cap (0, \infty)$.

This is obviously an equivalence relation; let $\mathbf{G}_t(Q)$ be the set of equivalence classes. If $f \in \Omega(Q)$, its class in $\mathbf{G}_t(Q)$ will be named "the germ defined by f in t".

Let us remark that, for $0<s<t$, $f \overset{t}{\sim} g$ is equivalent to $\theta_{t-s}f \overset{t}{\sim} \theta_{t-s}g$. In this way we arrive to a bijection $b_{st} : \mathbf{G}_t(Q) \to \mathbf{G}_s(Q)$; we have obviously $b_{rs}b_{st} = b_{rt}$ for $r<s<t$. Thus we may consider all $\mathbf{G}_t(Q)$ as "equal to the same set", which we denote by $\mathbf{G}(Q)$.

Since $f \overset{t}{\sim} g$ implies $f(t) = g(t)$, we may say that the germ defined by f in t "passes by $f(t)$".

We denote by $\mathbf{G}^{nf}(Q)$ the set of germs in $\mathbf{G}(Q)$ passing by a nonfictitious state, i.e. by a state in $E \oplus \Delta$, and by $\mathbf{G}^r(Q)$ the set of germs which pass by states in E.

Lemma 4.4. *Let* (Q_t) *be a* fc-*semigroup and* $M \subset (0, \infty) \times \Omega(Q)$ *be a* sm-*set (or, more generally, a set having the properties* a, c, d *in Definition* 3.1*). Let* $\mathbf{G}_M$ *be the set of all* $a \in \mathbf{G}(Q)$ *for which there is a* $(t, f) \in M$ *such that* a *is the germ defined by* f *in* t. *Let* $f \in \Omega(Q)$, $t > 0$.

Then $(t, f) \in M$ *iff the germ defined by* f *in* t *is in* $\mathbf{G}_M$.

Proof. The "only if" is obvious. If the germ defined by f in t is in $\mathbf{G}_M$, then there is a $(s, g) \in M$ and an $\varepsilon \in (0, \min(t, s))$ such that $f(t-\eta) = g(s-\eta)$ for all $\eta \in [-\varepsilon, \varepsilon]$.

It follows from d in Definition 3.1 that $(\varepsilon, \theta_{s-\varepsilon} g) \in M$, hence $\theta_{s-\varepsilon} g \in \mathbf{F}_{Q,2\varepsilon}$ (progressive measurability of M). On the other hand $(\theta_{t-\varepsilon}f)(r) = (\theta_{s-\varepsilon} g)(r)$ for all $r \in [0, 2\varepsilon]$. The "easy part" of the Galmarino lemma shows that $\theta_{t-\varepsilon} f \in M_\varepsilon$. We deduce $(\varepsilon, \theta_{t-\varepsilon} f) \in M$, $(t, f) \in M$, according to d in Definition 3.1.

Remarks. *a.* In the situation in Lemma 4.4 we have $\mathbf{G}_M \subset \mathbf{G}^{nf}(Q)$.

b. If $0 < s < t$, $f \in \Omega(Q)$, then the germ defined by f in t is the same as that defined by $\theta_{t-s} f$ in s.

c. If $\mathbf{H} \subset \mathbf{G}^{nf}(Q)$ is an arbitrary set and if we define M as the set of all (t, f) with $t>0$, $f \in \Omega(Q)$ for which the germ defined by f in t is in $\mathbf{H}$ then we get, due to *b* above, a subset of $(0, \infty) \times \Omega(Q)$ satisfying c, d in Definition 3.1.

Lemma 4.4 shows that every sm-set may be defined in this way. But for an arbitrary $\mathbf{H}$ it appears as a first question if the set M defined above is progressively measurable. Lemma 4.5 below is a result in this sense.

Lemma 4.5. *Let* (Q_t) *be a* fc-*semigroup on* $E \subset F$, Δ, *let* σ *be a* $(\mathbf{F}_{Q,t+})$-*stopping time such that* $\sigma \circ \theta_s = \sigma - s$ *on* $(\sigma > s)$, *whatever be* s. *Let* $N \subset (0, \infty) \times \Omega(Q)$ *be a set having the properties:*

a. N *is* $(\mathbf{F}_{Q,t+})$-*progressively measurable.*

b'. $(t, f) \in N$ *implies* $\sigma(f) > t$.

c. $(t, f) \in N$ *implies* $f(t) \notin F \setminus E$.

d'. *If* $0 < s < t\sigma < (t)$ *and* $(t, f) \in N$ *then* $(t-s, \theta_s f) \in N$.

Let $\mathbf{G}'$ *be the set of all germs defined by all* f *in all* t *for which* $(t, f) \in N$. *Let* M *be the set of all* (t, f) *for which the germ defined by* f *in* t *is in* $\mathbf{G}'$.

Then M *has the properties* a, c, d *in Definition* 3.1 *and is the smallest such set containing* N.

Proof. According to Lemma 4.4 and to Remark *c* above, we have to prove only the progressive measurability of M.

Les $\rho_s : (s, \infty) \times \Omega(Q) \to (0, \infty) \times \Omega(Q)$ be defined by $\rho_s(t, f) = (t-s, \theta_s f)$. It is measurable with respect to the progressive Borel field (Lemma 4.2). It will be sufficient to show that $M = N \cup \bigcup_{r \in \Gamma} \rho_r^{-1}(N)$, where

Γ is the set of positive rationals. The inclusion $\supset$ is immediate, in view of Remark *b* above.

In order to establish the opposite inclusion, let $(t, f) \in M$. It follows that there is a $(s, g) \in N$ and an $\varepsilon > 0$ such that $f(t + \eta) = g(s + \eta)$ for all $\eta \in$ $\in [-\varepsilon, \varepsilon]$; we may suppose $\varepsilon < \min(s, t)$. Let us choose a rational $r \in (t - \varepsilon, t)$, $r = t - \delta$. We deduce first that $(\theta_r f)(u) = (\theta_{t-\delta} f)(u) = (\theta_{s-\delta} g)(u)$ for all $u \in [0, \delta + \varepsilon]$. On the other hand condition b' gives $\sigma(g) > s$ and then d' gives $(\delta, \theta_{s-\delta} g) \in N$. It follows that $\theta_{s-\delta} g \in N_\delta \in \mathbf{F}_{Q,\delta+}$ (progressive measurability of N) and the Galmarino lemma implies $\theta_r f \in N_\delta$, hence $(t - r, \theta_r f) =$ $= (\delta, \theta_r f) \in N$ and finally $(t, f) \in \rho_r^{-1}(N)$.

Remark. *d.* If (Q_t) is a fc-semigroup, if σ is a $(\mathbf{F}_{Q, t+})$-stopping time such that $\sigma = s + \sigma \circ \theta_s$ on $(\sigma > s)$ and if $\mathbf{H} \subset \mathbf{G}^{nf}(Q)$ is an arbitrary set, then the set of all (t, f) with $0 < t < \sigma(f)$ for which the germ defined by f in t is in $\mathbf{H}$ has the properties b', c, d' in the statement of Proposition 4.5 (see Remark *b* above, $0 < s < t < \sigma(f)$ implies $0 < t - s < \sigma(\theta_s f)$, etc.).

Theorem 4.6. *The result of combining two strong Markov* fc-*semigroups is a strong Markov one.*

Proof. Let (Q_t^1), (Q_t^2) be M^1, M^2, respectively, strong Markov fc-semigroups, $E^1 \subset F^1$, Δ^1 and $E^2 \subset F^2$, Δ^2 be the components of their state spaces, $T : \Delta^1 \to \mathbf{E}(Q^2)$ be the measurable mapping occurring in the combining construction, (Q_t^{12}) be the resulting fc-semigroup.

a. Since F^1 and $F^2 \oplus \Delta^2$ are parts of the state space of (Q^{12}), etc., we may consider that $\mathbf{G}^r(Q^1)$ and $\mathbf{G}^{nf}(Q^2)$ are subsets of $\mathbf{G}^{nf}(Q^{12})$. Let us introduce (as in Remark *c* above) the set $M^{12} \subset (0, \infty) \times \Omega(Q^{12})$ defined as the set of all (t, f) for which the germ defined by f in t is in $(\mathbf{G}_{M^1} \cap \mathbf{G}^r(Q^1)) \cup \mathbf{G}_{M^2}$; this last set will be denoted by $\mathbf{H}$. M^{12} has the properties c, d in Definition 3.1.

b. We prove now that M^{12} is progressively measurable by applying Lemma 4.5 to the $(\mathbf{F}_{Q^{12}, t+})$-stopping time $\sigma = \sigma_1 + \sigma_2 \circ \theta_{\sigma_1}$, where σ_1 is the first visit in $F^2 \oplus \Delta^2$ and σ_2 is the first visit in F^1, and to the set N defined as in Remark *d* above: N is the set of all (t, f) with $0 < t < \sigma(f)$ for which the germ defined by f in t is in $\mathbf{H}$ above. Hence N has the properties b', c, d' in Lemma 4.5.

$\sigma > 0$ everywhere, $\sigma = s + \sigma \circ \theta_s$ on $(\sigma > s)$, every germ in $\mathbf{H}$ "appears" in N, hence the construction in Lemma 4.5 applied to N gives M^{12}. It remains to show that N is progressively measurable.

Consider, with this purpose, the sets $L_1 = \{(t, f) \mid 0 < t < \sigma_1(f)\}$, $L_2 = \{(t, f) \mid 0 < t < \sigma_2(f)\}$. Both are $(\mathbf{F}_{Q^{12}, t+})$-progressively measurable. Fix $\delta_0 \in \Delta^1$, $a_0 \in F^2 \oplus \Delta^2$; let for $i = 1, 2$. $\varphi_i : \Omega(Q^{12}) \to \Omega(Q^i)$ be defined as $(\varphi_i(f))(s) = f(s)$ for $s < \sigma_i(f)$, $(\varphi_1(f))(s) = \delta_0$, $(\varphi_2(f))(s) = a_0$ for $s \geqslant \sigma_i(f)$. We define also $\psi_i : (0, \infty) \times \Omega(Q^{12}) \to (0, \infty) \times \Omega(Q^i)$ by $\psi_i(t, f) = (t, \varphi_i(f))$. Consider also α: $\{(t, f) \mid f \in \Omega(Q^{12}), \sigma_1(f) < t < \infty\} \to (0, \infty) \times \Omega(Q^{12})$, $\alpha(t, f)$ $= (t - \sigma_1(f), \theta_{\sigma_1(f)}(f))$.

According to Lemmas 4.1, 4.2, ψ_i and α are measurable with respect to the corresponding progressive Borel fields.

We have (once more Remark *b* above) $N = (\psi_1^{-1}(M^1) \cap L^1) \cup$ $\cup\, \alpha^{-1}(L^2 \cap \varphi_2^{-1}(M^2))$, hence it is progressively measurable and (4.5) so is M^{12}.

Let us mention especially that $(\sigma_1(f), f) \notin M^{12}$ either because $\sigma_1(f) = 0$ or ∞, or because the germ in $\sigma_1(f)$ corresponds to a jump from F^1 to $F^2 \oplus \Delta^2$.

c. We have to prove, for every probability μ on the state space of (Q_t^{12}), the relation $P_{Q^{12}}^{\mu}(M_t^{12}) = 1$ for all $t > 0$ and the strong Markov property for every (μ, M^{12})-stopping time. μ is a convex combination of a probability on E^1 and of one on $E^2 \oplus \Delta^2$, the corresponding $P^{\mu}_{Q^{12}}$ are related in the same way, etc., hence it will be suffcient to consider only the cases $\mu(E^1) = 0$ and $\mu(E^2 \oplus \Delta^2) = 0$.

In the first case μ appears also as a probability on $E^2 \oplus \Delta^2$, $P^{\mu}_{Q^{12}} = = P^{\mu}_{Q^2} \circ k^{-1}$, where $k: \Omega(Q^2) \to \Omega(Q^{12})$ is defined by $(k(f))(s) = f(s)$ for all s, i.e. by $x_{Q^{12},t} \circ k = x_{Q^2,t}$. We have $\sigma_1(k(f)) = 0$, $\sigma(k(f)) = \sigma_2(k(f)) = \infty$ for all f, $(t, k(f)) \in M^{12}$ iff $(t, f) \in M^2$, $k^{-1}(\mathbf{F}_{Q^{12},t}) = \mathbf{F}_{Q^2,t}$, etc. and the desired facts follow easily (see also Lemma 3.7).

d. In order to treat the second case, in which μ appears also as a probability on E^1, consider $\Omega = \Omega(Q^1) \times \Omega(Q^2)$, $\mathbf{K} = \mathbf{K}(Q^1) \otimes \mathbf{K}(Q^2)$, $P^{\mu} = = P^{\mu}_{Q^1} \otimes P'$, where P' is the transition probability defined by $P'(f, \quad) = = P^{\mathbf{I}}_{Q^2}$, $\mathbf{I} = T(x_{Q^1,\infty}(f), \quad)$, $T(\xi, \quad)$ being arbitrarily defined. Let $j : \Omega \to \Omega(Q^{12})$ be defined by $(j(f_1, f_2))(s) = f_1(s)$ for $s < \zeta(f_1)$, $(j(f_1, f_2))(s) = f_2(s - \zeta(f_1))$ for $s \geqslant \zeta(f_1)$. Then $P^{\mu}_{Q^{12}} = P^{\mu} \circ j^{-1}$ (see the introduction or, better, [7] § 5, proof of the Theorem).

Moreover, $\sigma_1(j(f_1, f_2)) = \zeta(f_1)$, $\sigma(j(f_1, f_2)) = \infty$ for all f_1, f_2.

Let us prove now that, in this case, for $t > 0$, $P^{\mu}_{Q^{12}}(M^1_{t2}) = 1$, i.e. t at $P^{\mu}(j^{-1}(M^{12}_t)) = 1$. The set $j^{-1}(M^{12}_t)$ consists of all (f_1, f_2) for which $(t, f(f_1, f_2)) \in \in M^{12}$; recalling the definition of M^{12}, it follows the disjoint union of two sets. The first is the set of all (f_1, f_2) for which $\zeta(f_1) > t$ and the germ defined by f_1 in t is in $\mathbf{G}_{M^1}$, hence it is nothing but $((\zeta > t) \cap M^1_t) \times \Omega(Q^2)$. The second consists of all (f_1, f_2) for which $\zeta(f_1) < t$ and the germ defined by f_2 in $t - \zeta(f_1)$ is in $\mathbf{G}_{M^2}$. Hence $P^{\mu}_{Q^{12}}(M^{12}_t) = P^{\mu}_{Q^1}((\zeta > t) \cap M^1_t) + \int_{(\zeta<t)} P'(f_1, M^2_{t-\zeta(f_1)})\, dP_{Q^1}(f)$.

The first term equals $P^{\mu}_{Q^1}(\zeta > t)$ due to condition b in Definition 3.1. The integrand of the second, for a fixed f_1, is a $P^{\mathbf{I}}_{Q^2}(M^2_r) = P^{\mathbf{I}}_{Q^2}(\theta^{-1}_{r-s}(M^2_s))$ due to Remark *b* above; the Markov property, introducing the notati on $\nu = = P^{\mathbf{I}}_{Q^2} \circ x^{-1}_{Q^2, r-s}$, shows that it equals $P^{\nu}_{Q^2}(M^2_s) = 1$ (Definition 3.1, b).

We obtain $P^{\mu}_{Q^{12}}(M^{12}_t) = P^{\mu}_{Q^1}(\zeta > t) + P^{\mu}_{Q^1}(\zeta < t)$ and its value is 1 due to $P^{\mu}_{Q^1}(\zeta = t) = 0$ (see [7] § 5, point c in the proof of the theorem).

The fact that M^{12} is a sm-set is completely proved.

e. It remains to prove the strong Markov property for (μ, M^{12})-stopping times τ with respect to $(\mathbf{F}_{Q^{12},t+})$ in the case $\mu(E^2 \oplus \Delta^2) = 0$. A transport via j above shows that it will be sufficient to prove that, denoting $x^{12}_t = x_{Q^{12},t} \circ j$, $\mathbf{F}_t = \mathbf{B}(x^{12}_s \mid s \leqslant t)$, for every $(\mathbf{F}_{t+})$-stopping time τ and every probability μ on E^1, for which, for P^{μ}-almost all f, the germ defined by $t \to x^{12}_t(f)$ (equal to $j(f)$) in $\tau(f)$ is in $\mathbf{G}_{M^{12}}$($\mathbf{H}$ in *a* above), we have $M_{P^{\mu}}((g \circ x^{12}_{\tau+t})\, \chi_{(\tau<\infty)} \mid \mathbf{F}_{\tau+}) = = \chi_{(\tau<\infty)}((Q^{12}_t g) \circ x^{12}_{\tau})$, for all $t > 0$ and measurable $g \geqslant 0$.

Let us remark that $\mathbf{F}_t$ is included in the direct sum $\mathbf{F}'_t$ of $\mathrm{pr}_1^{-1}((\mathbf{F}_{Q^1,t})_{(\zeta>t)})$ and of the Borel field on $(\zeta \leqslant t) \times \Omega(Q^2)$ of all its subsets $A \in \mathbf{K}$ for which $A_{f_1} \in \mathbf{F}_{Q^2, t-\zeta(f_1)}$ for all f_1. Hence it will be sufficient to prove the same relation with $\mathbf{F}'_{\tau+}$ in place of $\mathbf{F}_{\tau+}$.

Moreover, let us remark that, since the germ defined by $t \to x^{12}_t(f)$ in $\zeta(\mathrm{pr}_1 f)$ is "a jump from F^1 to $F^2 \oplus \Delta^2$ " hence is never in $\mathbf{H}$, we have $P^{\mu}(\tau = = \zeta \circ \mathrm{pr}_1 < \infty) = 0$ in our hypothesis. According to Remark *d* after 3.2, it will be sufficient to consider the subcases "$\tau(f) < \infty$ implies $\tau(f) > \zeta(\mathrm{pr}_1 f)$" and "$\tau(f) < \infty$ implies $\tau(f) < \zeta(\mathrm{pr}_1 f)$".

f. Let us prove the strong Markov property in the first subcase. If $\zeta(f_1) < \infty$, $\tau(f_1, \quad) - \zeta(f_1)$ follows a $(\mathbf{F}_{Q^2, t+})$-stopping time and $\mathbf{F}'_{\tau+}$ equals the set of all $A \in \mathbf{K}$ for which $\zeta(f_1) < \infty$ implies $A_{f_1} \in \mathbf{F}_{Q^2, (\tau(f_1, \quad) - \zeta(f_1))+}$.

The germ defined by $j(f_1, f_2)$ in $\tau(f_1, f_2)$ is, if $\zeta(f_1) < \infty$, the germ defined by f_2 in $\tau(f_1, f_2) - \zeta(f_1)$. Since the first is P^μ - a.e. in $\mathbf{H}$, it follows that for $P^\mu_{Q^1}$-almost all f_1 for which $\zeta(f_1) < \infty$ we have $P'(f_1, \{f_2 \mid \tau(f_1, f_2) < \infty, (\tau(f_1, f_2) - \zeta(f_1), f_2) \notin M^2\}) = 0$, hence the strong Markov property is valid for $P'(f_1, \quad)$ and $\tau(f_1, \quad) - \zeta(f_1)$, according to Proposition 3.6.

An application of Lemma 3.8, recalling that $Q^{12}_t(x, \quad) = Q^2_t(x, \quad)$ for $x \in E^2 \oplus \Delta^2$, finishes the proof in this subcase.

g. The treatment of the second subcase at the end of f begins by considering $(\tau < t < \zeta \circ \mathrm{pr}_1)$, set on which τ is $\mathbf{F}'_t$-measurable, $\mathbf{F}'_{\tau+} \subset \mathbf{F}'_t$(the set itself is not in $\mathbf{F}'_{\tau+}$) and $\mathbf{F}'_t \subset \mathrm{pr}_1^{-1}(\mathbf{F}_{Q^1, t})$. It follows (varying t, etc.) that $\tau = \rho \circ \mathrm{pr}_1$ where ρ is $(\mathbf{F}_{Q^1, t+})$-optional and that $(\mathbf{F}'_{\tau+})_{(\tau<\infty)} = \mathrm{pr}_1^{-1}((\mathbf{F}_{Q^1, \rho+})_{(\rho<\infty)})$.

We know from Proposition 3.3 that for P^μ-almost all f the germ defined by $j(f)$ in $\tau(f) + t$, when the last is finite, is in $\mathbf{H}$, hence, as in e above, $P^\mu(\tau + t = \zeta \circ \mathrm{pr}_1 < \infty) = 0$.

The germ defined by $j(f_1, f_2)$ in $\tau(f_1, f_2)$, when the last is finite, is the germ defined by f_1 in $\rho(f_1)$; it follows that ρ is a (μ, M^1)-stopping time.

Let $T(\delta, \quad) = (T_t(\delta, \quad))_{t \in (0, \infty)}$.

In order to prove the relation expressing the strong Markov property for τ, we consider separately the situations $g = 0$ on $F^2 + \Delta^2$, $g = 0$ on F^1; $g = 0$ also on Δ^1 in the first, during the proof.

In the first $(g \circ x^{12}_{\tau+t})\, \chi_{(\tau<\infty)} = ((g \circ x_{Q^1, \rho+t})\, \chi_{(\rho<\infty)}) \circ \mathrm{pr}_1$, etc. and the strong Markov property for τ follows from that for (ρ, μ), recalling that $\chi_{E^1} Q^{12}_t(x, \quad) = \chi_{E^1} Q^1_t(x, \quad)$ for $x \in E^1$.

The second situation is handled by writing $((g \circ x^{12}_{\tau+t})\, \chi_{(\tau<\infty)})\,(f_1, f_2) = \chi_{(\infty>\rho+\tau>\zeta)}(f_1)\, g(x_{Q^2, \rho(f_1)+t-\zeta(f_1)}(f_2))$ P^μ-a.e.; conditioning first by $\mathrm{pr}_1^{-1}(\mathbf{K}(Q^1))$ we obtain $\chi_{(\infty>\rho+t>\zeta)}(T_{\rho+t-\zeta}\, g)\; (x_{Q^1, \infty}) = \chi_{(\rho<\infty)}\, \chi_{(\zeta-\rho<t)}(T_{t-(\zeta-\rho)}\, g)\; (x_{Q^1, \infty})$ composed with pr_1.

Conditioning the result with respect to $\mathrm{pr}_1^{-1}(\mathbf{F}_{Q^1, \rho+})$, we get, due to the property of ρ pointed out before and to Corollary 3.5,

$$\chi_{(\rho<\infty)} \int_{(0, t) \times \Delta^1} (T_{t-s}\, g)\, (\delta)\, \mathrm{d}G_{Q^1}\, (x_{Q^1, \rho}, (s, \delta)) = \chi_{(\rho<\infty)}\, (Q^{12}_t g)\; (x_{Q^1, \rho}),$$

composed with pr_1 (see also § 2) and the proof is complete.

Remark. If, for all δ, $T(\delta) = (\nu(\delta)\, Q^2_t)$, ν being a transition probability, etc., we may take a larger $\mathbf{H}$, namely we may add to it all the germs corresponding to jumps from F^1 to $F^2 \oplus \Delta^2$, passing by elements in $E^2 \oplus \Delta^2$. The proof works in the same way.

However, if $T(\delta) = (\nu(\delta)\, Q^2_t)$ is valid only for δ in a subset of Δ^1, an analogous enlargement of $\mathbf{H}$ is generally impossible, since if a sample path defines in t a germ corresponding to a jump as above, we cannot infer if it came from a (f_1, f_2) with $x_{Q^1, \infty}(f_1)$ in that subset or not.

Theorem 4.7. *The result of combining a finite or infinite sequence of strong Markov* fc-*semigroups is a strong Markov one.*

Proof. Induction for a finite one. According to Remark *d* after 3.2, it will be sufficient (the sm-set being defined in a way analogous to that in a, proof of 4.6) to consider the case of a stopping time which, when finite, is smaller than the first visit in the nth component of the state space (n depending on the stopping time). But it reduces to the case of "two fc-semigroups" in 4.6, the first being the result of combining the fc-semigroups at the beginning of the sequence, up to the $(n - 1)$th and the second being the result of combining the nth, the $(n + 1)$th, ..., etc.

Remark. We meet more difficulties if we try to state and prove an analogous result for a fc-semigroup obtained by repeating (see 5.2 in [8]) a strong Markov one.

REFERENCES

1. Blumenthal, R. M., Getoor, R. K., *Markov Processes and Potential Theory.* New York—London, Academic Press, 1968.
2. Chung, K. L., *Markov Chains with Stationary Transition Probabilities.* Berlin—Göttingen—Heidelberg, Springer, 1960.
3. Courrège, Ph., Priouret, P., *Temps d'arrêt d'une fonction aléatoire: Relations d'équivalence associées et propriétés de décomposition.* Publ. Inst. Statist. Univ. Paris, **14**, 245—274 (1965).
4. Cuculescu, I., *Some constructions of Markov processes (I).* Rev. Roumaine Math. Pures Appl., **22**, 1397—1410 (1977).
5. Cuculescu, I., *Some constructions of Markov processes (II).* Rev. Roumaine Math. Pures Appl., **24**, 207—214 (1979).
6. Cuculescu, I., *Some constructions of Markov processes (III).* Rev. Roumaine Math. Pures Appl., **24**, 357—372 (1979).
7. Cuculescu, I., *Constructions involving Markov processes with time set* $(0, \infty)$ *(I).* In: Demetrescu, M. C., Iosifescu, M. (Eds.), *Studies in Probability and Related Topics. Papers in Honour of Octav Onicescu on His 90th Birthday*, pp. 131—143. Roma, Editrice Nagard, 1983.
8. Cuculescu, I., *Constructions involving Markov processes with time set* $(0, \infty)$ *(II).* Rev. Roumaine Math. Pures Appl., **30** (1985).
9. Dellacherie, C., *Capacités et Processus Stochastiques.* Berlin—Heidelberg—New York, Springer, 1972.

PAUL DEHEUVELS

Asymptotic results for the pseudo-prime sequence generated by Hawkins' random sieve: Twin primes and Riemann's hypothesis

We solve the twin primes problem in the case of the pseudo-prime sequence generated by Hawkins' random sieve, by showing that there exists almost surely an infinity of pairs of pseudo-primes whose increments take any a priori given positive integer value. Some extensions are given, and recent investigations on the asymptotic theory of the Hawkins pseudo-primes are discussed. A new random sieve is presented, giving a natural extension of Hawkins' sieve.

1. Introduction

The random sieve introduced by Hawkins (1958) produces an increasing random integer sequence with asymptotic properties very analogous to those of prime numbers.

It is constructed as follows. Let $A_1 = \{2, 3, 4, \dots\}$. Next, if A_n is given for some $n \geqslant 1$, put $X_n = \min A_n$ and successively delete with probability X_n^{-1} or keep with probability $1 - X_n^{-1}$ each number (independently of the others) in the set $A_n - \{X_n\}$. The remaining elements in $A_n - \{X_n\}$ constitute the set A_{n+1}.

The comparison of Hawkins' sieve and Erathosthenes' sieve which generates prime numbers is straightforward, since in Erathosthenes' sieve, one would delete in $A_n - \{X_n\}$ the multiples of X_n instead of using a random selection procedure, giving on the whole identical global rates of deletion.

Because of this analogy, it is not a surprize that many asymptotic results which are known to be true for the prime number sequence $\{p_1, p_2, \dots\}$ are also true with probability one for the random "pseudo-prime" sequence $\{X_1, X_2, \dots\}$. In particular we may cite the following properties.

Wunderlich (1974, 1976) has obtained that if

$$(1) \qquad Y_n = \prod_{1 \leqslant k \leqslant n} (1 - X_k^{-1})^{-1}, \quad y_n = \prod_{1 \leqslant k \leqslant n} (1 - p_k^{-1})^{-1},$$

then

$$(2) \qquad \lim_{n \to \infty} (n \log n)^{-1} X_n = 1 \text{ and } \lim_{n \to \infty} (\log n)^{-1} Y_n = 1 \text{ a.s.,}$$

whereas (Prime Number Theorem and Mertens' Theorem)

(3) $$\lim (n \log n)^{-1} p_n = 1 \text{ and } \lim (\log n)^{-1} y_n = e^{\gamma},$$

where γ is the Euler constant.

If li(.) is the logarithm-integral function,

(4) $$\mathrm{li}\,(x) = \lim_{\delta \downarrow 0} \left(\int_0^{1-\delta} + \int_{1+\delta}^{x} \right) \frac{dz}{\log z},$$

then, the Riemann hypothesis about the zeros of the Riemann zeta function is equivalent to

(5) $$\mathrm{li}\,(p_n) = n + O\left(n^{\frac{1}{2}+\varepsilon}\right) \text{ for any } \varepsilon > 0.$$

The corresponding result for Hawkins' sieve is due to Neudecker and Williams (1974), who showed that

(6) $$L\,\mathrm{li}\,(X_n L^{-1}) = n + O\left(n^{\frac{1}{2}+\varepsilon}\right) \text{ a.s. for any } \varepsilon > 0,\ n \to \infty,$$

where L denotes the (random) limit

(7) $$L = \lim_{n \to \infty} X_n \exp(-Y_n) \text{ a.s.},$$

which they prove to exist and to be non zero a.s.

Following that idea, it appears logical to investigate systematically the limit theorems true for the Hawkins "pseudo-primes", in order to deduce possible conjectures for prime numbers. This has been done first with the Riemann hypothesis by Heyde (1978) and Foster and Williams (1978), who showed that

(7) $$\limsup_{n \to \infty} (2n \log \log n)^{-1/2} (L\,\mathrm{li}\,(X_n L^{-1}) - n) = 1 \text{ a.s., and}$$
$$\liminf_{n \to \infty} (2n \log \log n)^{-1/2} (L\,\mathrm{li}\,(X_n L^{-1}) - n) = -1 \text{ a.s..}$$

In Deheuvels (1982), a complete characterization of the upper and lower class of the sequence $\{L\mathrm{li}(X_n L^{-1}) - n\}$ is given, proving in particular that

(8) $$P\left(L\,\mathrm{li}\,(X_n L^{-1}) - n \leqslant \left(2n\left\{\log_2 n + \frac{3}{2}\log_3 n + \log_4 n + \dots + (1+\varepsilon)\log_p n\right\}\right)^{1/2} \text{ i.o.}\right) = P\left(L\,\mathrm{li}\,(X_n L^{-1}) - n \geqslant -\left(2n\left\{\log_2 n + \frac{3}{2}\log_3 n + \log_4 n + \dots + (1+\varepsilon)\log_p n\right\}\right)^{1/2} \text{ i.o.}\right) = 0 \text{ or } 1,$$

according as $\varepsilon > 0$ or $\varepsilon \leqslant 0$ (where $\log_j$ is the j times iterated logarithm and $p \geqslant 4$).

The question of establishing optimal asymptotic bounds for prime numbers is open, except for the first term given in (3) by the prime number theorem. For the Hawkins "pseudo-primes", an asymptotic development has been obtained by Deheuvels (1982) yielding, for any $N \geqslant 1$, as $n \to \infty$,

$$X_n = n \log n + n \log_2 n + n(C - 1) + n \sum_{k=1}^{N} \frac{P_k(\log_2 n + C - 1)}{(\log n)^k} +$$

$$(9) \qquad + \frac{O(1)}{(\log n)^N},$$

where $C = -\log L$ is defined by (7), and $P_1, P_2, \ldots$ is a sequence of polynomials of degrees 1, 2,. ..., defined by

$$(10) \qquad \log\left(1 + \xi z + \sum_{k=1}^{N-1} z^{k+1} P_k(\xi)\right) = \sum_{k=1}^{N} z^k P_k(\xi) + o(z^N), \ z \to 0,$$

for any $N = 2, 3, \ldots$.

Another question which deserves interest is the famous twin primes problem, concerning the number (it is unknown wether it is finite or not) of primes (p_n, p_{n+1}) such that $p_{n+1} - p_n = 2$.

We intend here to investigate the following general problem for the Hawkins "pseudo-prime" sequence $\{X_n, n > 1\}$: what is the number of pairs (X_n, X_{n+1}) such that $X_{n+1} - X_n = k$ (where $k = 1, 2, \ldots$ is an arbitrary positive integer)?

The answer is given in the following theorem

Theorem 1. *For any fixed $k \geqslant 1$, there exists almost surely an infinite set of indices $n \geqslant 1$ such that $X_{n+1} - X_n = k$.*

It has to be noted here that there is an important discrepancy between the prime sequence $\{p_n, n \geqslant 1\}$ and the Hawkins "pseudo-prime" sequence $\{X_n, n \geqslant 1\}$, precisely in the fact that nothing prevents X_{n+1} to be equal to $X_n + 1$, whereas this is impossible for p_n (except for $p_1 = 2$, $p_2 = 3$).

To fill the gaps between the results obtained from Hawkins' sieve and the results one should expect from the sieve of Eratosthenes, one should define and study other kinds of random sieves, trying to get each time closer to the exact generating system of the prime sequence. To do so, we can start from Lemma 4 in Deheuvels (1982) where it is shown for the Hawkins "pseudo-primes" $\{X_n, n \geqslant 1\}$, that there exists an i.i.d. sequence $\{\omega_n, n \geqslant 1\}$ of exponentially $E(1)$ distributed random variables, such that, for any $n \geqslant 1$,

$$(11) \qquad X_{n+1} - X_n = \left[\frac{\omega_n}{-\log(1 - Y_n^{-1})}\right] + 1,$$

where Y_n is defined by (1). A more sophisticated version of (11) is given by

$$(12) \qquad X_{n+1} - X_n = Y_n \omega_n + \left(1 - \frac{1}{2}\omega_n - \zeta_n\right) + (\omega_n - 1)\,\varphi(Y_n^{-1}) +$$

$$+ \{\psi(\zeta_n, Y_n^{-1}) + \varphi(Y_n^{-1})\},$$

where $\{\zeta_n, n \geqslant 1\}$ is an i.i.d. sequence of uniformly distributed on (0, 1) random variables, dependent on $\{\omega_n, n \geqslant 1\}$ by the formula (see Deheuvels (1982, 1984))

$$(13) \qquad \zeta_n = \frac{1 - \exp(-\omega_n + (X_{n+1} - X_n - 1)\log(1 - Y_n^{-1}))}{Y_n^{-1}},$$

and

$$(14) \qquad \psi(\zeta, z) = -\frac{\log(1-\zeta z)}{\log(1-z)}, \quad \varphi(z) = -\frac{1}{\log(1-z)} - \frac{1}{z} + \frac{1}{2}.$$

With respect to the preceding arguments, the following family of random sequences appear as generalizations of the Hawkins "pseudo-primes"

Definition. Let $X_1 = 2$. If, for $n > 1$, $X_1, \ldots, X_n$ are given, put

$$Y_n = \prod_{1 \leqslant k \leqslant n} (1 - X_k^{-1})^{-1}.$$

Let now $\{\theta_n, n \geqslant 1\}$ be an i.i.d. sequence of non-negative random variables, and let $L(y, \theta)$ be an arbitrary non-negative function of $y \geqslant 2$, and $\theta > 0$. Define X_{n+1} as

$$(15) \qquad X_{n+1} = X_n + [L(Y_n, \theta_n)] + 1.$$

It can be seen from (11) and (15) that $\{X_n, n \geqslant 1\}$ is the Hawkins sequence if $L(y, \theta) = \theta / \{-\log(1 - y^{-1})\}$, and $\{\theta_n, n \geqslant 1\}$ is an i.i.d. sequence of exponentially $E(1)$ distributed random variables.

Among the conditions which would appear as natural for $L(y, \theta)$, one could cite

$$(16) \qquad L(y, \theta) \sim y\theta \quad \text{as} \quad y \to \infty.$$

The study of such sequences is an open problem (except for the Hawkins "pseudo-primes" for which much has been done). The adequacy with primes should be helped if one could answer the question

Is there a limiting repartition for $(p_{n+1} - p_n)/y_n$? (In other words, does the proportion of indices $k : 1 \leqslant k \leqslant n$, such that $(p_{k+1} - p_k)/y_k \leqslant x$ converge to a limit $F(x)$ as n increases to infinity?) If it is the case, then one should choose $F(x)$ as the distribution function of θ_n, $n = 1, 2, \ldots$, noting that in the case of the Hawkins random sieve, $F(x) = 1 - e^{-x}$.

2. The twin primes problem for Hawkins' random sequence

Consider first the case where $k = 1$ in Theorem 1; clearly then, by (11), $X_{n+1} - X_n = 1$ each time that (for some $\varepsilon > 0$)

$$(17) \qquad \varepsilon_n \leqslant (1 - \varepsilon)(-\log(1 - Y_n^{-1})).$$

But, by (2), $Y_n \sim \log n$ a.s. as $n \to \infty$. It follows from (17) that there exists almost surely an infinite set of indices n such that $X_{n+1} - X_n = 1$ if the event

$$\omega_n \leqslant (1 - \varepsilon)/\log n$$

happens infinitely often with probability one. Clearly this is true by Borel-Cantelli:

$$\sum_{n=1}^{+\infty} P(\omega_n \leqslant (1 - \varepsilon)/\log n) = +\infty.$$

Likewise, if k is arbitrary ($k \geq 1$), then by (11), $X_{n+1} - X_n = k$ each time that

$$(18) \qquad (k - 1 + \varepsilon)(-\log(1 - Y_n^{-1})) \leq \varepsilon_n \leq (k - \varepsilon)(-\log(1 - Y_n^{-1})),$$

In turn, (18) happens infinitely often with probability one if (for some $\varepsilon > 0$) the event

$$(k - 1 + \varepsilon)/\log n \leq \omega_n < (k - \varepsilon)/\log n$$

does, which is true like in the case $k = 1$ by the same arguments.

This proves Theorem 1. The simplicity of the argument enables to ask the question whether there are double twins, that is, indices n such that

$$X_{n+1} - X_n = X_{n+2} - X_{n+1} = k.$$

The same argument used on events such as

$$A_n^p = \{(k - 1 + \varepsilon)/\log n \leq \omega_{n+1}, \ldots, \omega_{n+p} \leq (k - \omega)/\log n\}$$

and using the fact that

$$\sum_{k=1}^{+\infty} P(A_{pk}^p) = +\infty,$$

enables to show

Theorem 2. *For any $p \geq 1$, for any $k \geq 1$, there is almost surely an infinite set of indices n such that*

$$(19) \qquad X_{n+1} - X_n = X_{n+2} - X_{n+1} = \ldots = X_{n+p} - X_{n+p-1} = k.$$

As far as the Hawkins "pseudo-prime" sequence is near to the real prime sequence, (19) would seem to imply that it is very likely that there exists an infinity of twin primes, and even much better than that! In fact, one should be careful about such statements, and the comparison between (2) and (3) shows that some cautiousness is necessary before making the transposition.

An interesting related question concerns the series

$$S_k = \sum_{n=1}^{+\infty} (X_n^{-1} + X_{n+1}^{-1})\, I(X_{n+1} - X_n = k),$$

where $I(A)$ is the indicator function of the event A. It is known for primes that the series

$$s_2 = \sum_{n=1}^{+\infty} (p_n^{-1} + p_{n+1}^{-1})\, I(p_{n+1} - p_n = 2) =$$

$$= \left(\frac{1}{3} + \frac{1}{5}\right) + \left(\frac{1}{5} + \frac{1}{7}\right) + \left(\frac{1}{11} + \frac{1}{13}\right) + \left(\frac{1}{17} + \frac{1}{19}\right) + \cdots$$

is convergent. This is sometime used as an argument to support the idea that the total number of twin primes is finite.

It happens that his phenomenon occurs also for "pseudo-primes", since

Theorem 3. *For any $p \geq 1$, for any $k \geq 1$, the series*

$$(20) \qquad \sum_{n=1}^{+\infty} X_n^{-1} I(X_{n+1} - X_n = \ldots = X_{n+p} - X_{n+p-1} = k)$$

is almost surely convergent.

Proof. Note that in (20) we could change without harm X_n^{-1} into $X_n^{-1} + + \dots + X_{n+p}^{-1}$. If we use (2), we can change, without altering the convergence or divergence of (20), X_n^{-1} into $(n \log n)^{-1}$. Hence, we have to consider the series

$$\sum_{n=2}^{+\infty} (n \log n)^{-1} I(X_{n+1} - X_n = \dots = X_{n+p} - X_{n+p-1} = k).$$

Next, by (11) and (18), note that the event $\{X_{n+1} - X_n = \dots = X_{n+p} - - X_{n+p-1} = k\}$ is included in $\{X_{n+1} - X_n = k\}$, which in turn is included in

$$B_n^k = \{(k-2)/\log n \leqslant \omega_n \leqslant (k+1)/\log n\},$$

but clearly $P(B_n^k) \sim 3/\log n$ as $n \to \infty$. It follows that $\sum (n \log n)^{-1} P(B_n^k) < \infty$, which proves the convergence of (21) and (22).

Even if no conclusion can be derived from Theorem 3 in the case of prime numbers, it gives at least support to reject the argument that, since s_2 converges, it would be likely that the number of twin primes should be finite.

3. The maximal gap between successive pseudo-primes

We investigate now the almost sure upper class of $X_{n+1} - X_n$. It is characterized by the following result.

Theorem 4. *For any* $q \geqslant 4$,

(21) $$P(X_{n+1} - X_n) \geqslant (\log n)(\log n + 2\log_2 n + \log_3 n + \dots$$
$$\dots + (1+\varepsilon)\log_q n) \text{ i.o.}) = 0 \text{ or } 1,$$

according as $\varepsilon > 0$ *or* $\varepsilon \leqslant 0$.

Proof. We first note (see Deheuvels (1982), (36)) that

$$Y_n = \log n + \log_2 n + C + o(1) \text{ a.s. as } n \to \infty,$$

and by Borel-Cantelli, that

$$P(\omega_n \geqslant \log n + \log_2 n + \log_3 n + \dots + (1+\varepsilon)\log_q n \text{ i.o.}) = 0 \text{ or } 1,$$

according as $\varepsilon > 0$ or $\varepsilon \leqslant 0$. It follows that (by multiplying both expressions and deleting the unnecessary terms) (21) holds, using either (11) or (12).

The following corollary would be worthwile proving for prime numbers(!):

Corollary 1. *The difference* $X_{n+1} - X_n$ *between the* $(n+1)$*-th and the nth "pseudo-prime" of Hawkins' sequence takes infinitely often every value* $k = 2, 3, \dots,$ *and is infinitely often higher than* $\log^2 n$, *but not exceeding more than finitely often with probability one* $(1+\varepsilon)\log^2 n$, *for any given* $\varepsilon > 0$.

4. Further comments

The main technique which enabled us to obtain in Deheuvels (1982) results such as (8) and (9) is a strong approximation method which is developed in a more general setting in Deheuvels (1981, 1984). A comprehensive analysis of the Hawkins pseudo-primes can be found in Hall and Heyde (1980, p. 226—234).

Aknowledgements. The author would like to express his thanks to Professor Marius Iosifescu for suggesting him to investigate the problem of twin primes in the case of sequences generated by random sieves.

REFERENCES

1. Deheuvels, P., *The strong approximation of extremal processes.* Z. Wahrsch. Verw. Gebiete, **58**, 1–6 (1981).
2. Deheuvels, P., *Asymptotic expansions for prime numbers and improvements of the Riemann hypothesis based on Hawkins random sieve* (submitted) (1982).
3. Deheuvels, P., *Strong approximation in extreme value theory and applications.* In: Révész, P. (Ed.), *Limit Theorems in Probability and Statistics.* Colloq. Math. Soc. J. Bolyai, Vol. 36. Amsterdam, North-Holland, 1984.
4. Foster, D., Williams, D., *The Hawkins sieve and Brownian motion.* Compositio Math., **37**, 277–289 (1978).
5. Hall, P., Heyde, C. C., *Martingale Limit Theory and Its Application.* New York, Academic Press, 1980.
6. Hawkins, D., *The random sieve.* Math. Mag., **31**, 1–3 (1958).
7. Heyde, C. C., *A loglog improvement to the "Riemann hypcthesis" for the Hawkins randcm sieve.* Ann. Probab. **6**, 870–875 (1978).
8. Neudecker, W., Williams, D., *The "Riemann hypothesis" for the Hawkins random sieve.* Compositio Math., **29**, 197–200 (1974).
9. Wunderlich, M. C., *A probabilistic setting for prime number theory.* Acta Arith., **26**, 59–81 (1974).
10. Wunderlich, M. C., *The prime number theorem for random sequences.* J. Number Theory, **8**, 369–371 (1976).

DANIEL DUGUÉ

Rôles des matrices en statistique et problèmes qu'elles posent

Au maître à qui est dédiée cette conférence je désire apporter mon hommage personnel et celui de mes collègues statisticiens français. Le professeur Octav Onicescu figure parmi les savants roumains qui ont toujours entretenu des liens étroits avec mon pays, où il a été l'élève des plus grands mathématiciens parmi lesquels je dois citer Paul Montel, très attaché aux mathématiques roumaines.

J'ai rencontré le Professeur Onicescu il y a quarante-cinq ans à Genève au cours de journées probabilistes et statistiques où étaient réunis avec de modestes étudiants dont j'étais, toutes les gloires de l'époque: Borel, Fréchet, Darmois, Cramér, Sir Ronald Fisher, Heisenberg, Paul Lévy et lui-même bien sûr. En 1967 je participais au colloque de Cluj dédié à son 75ème anniversaire.

Très sensible au grand honneur qu'il m'a fait en m'invitant ici je lui souhaite une longue et heureuse continuation de son activité.

Le but de cette conférence est de commenter l'ensemble des publications faites par l'auteur sur ce sujet (voir les références).

Fixons tout d'abord quelques notations. Elles seront utilisées dans la suite de l'exposé. $\mathbf{A}_n$ est une matrice $n \times n$ où $a_{ij} = 0$ si $j > i$ et $a_{ij} = 1$ si $j \leqslant i$. $\mathbf{B}_n$ sera égal à ${}^t\mathbf{A}_n\mathbf{A}_n$ avec naturellement $b_{ij} = \min(n - i + 1, n - j + 1)$ et $\mathbf{M}_n$ sera une matrice dont tous les éléments sont égaux à l'unité.

Dans le mémoire [3] je donne les valeurs propres de $\mathbf{B}_n$ et de $\mathbf{B}_n - (1/n)\,\mathbf{M}_n$.

Après quelques calculs simples utilisant les suites de Fibonacci on trouve comme valeurs propres de $\mathbf{B}_n$:

$$z_k = \left[4 \sin^2 \frac{2k-1}{2n+1} \frac{\pi}{2}\right]^{-1}, \quad k = 1, 2, \dots, n,$$

et celles de $\mathbf{B}_n - (1/n)\,\mathbf{M}_n$:

$$z'_k = \left[4 \sin^2 \frac{k}{n} \frac{\pi}{2}\right]^{-1}, \quad k = 1, 2, \dots, n-1.$$

Ces formules dérivent de la façon dont on peut écrire les déterminants:

$$P_n = \det(\mathbf{I}_n - z\mathbf{B}_n) \quad \text{et} \quad Q_n = \det\{\mathbf{I}_n - z[\mathbf{B}_n - (1/n)\,\mathbf{M}_n]\}.$$

En posant $z = 2(1 - \cos\theta)$ on a:

$$P_n = \cos\left(n + \frac{1}{2}\right)\theta \Big/ \cos\frac{\theta}{2} \quad \text{et} \quad Q_n = \sin n\theta / n \sin\theta.$$

Ces deux suites de valeurs propres sont naturellement liées aux valeurs propres des noyaux browniens $\min(x, y)$ et $\min(x, y) - xy$. Une autre suite de nombres s'est révélée très utile en particulier dans la note [2] et dans le mémoire [4] qui contient les démonstrations de [2].

Soit $\mathbf{D}_n$ la forme diagonale de la matrice $\mathbf{B}_n$. On a donc $\mathbf{T}_n\mathbf{B}_n\mathbf{T}_n^{-1} = \mathbf{D}_n$ et posons $\mathbf{T}_n\mathbf{M}_n\mathbf{T}_n^{-1} = \mathbf{N}_n$ dont les éléments diagonaux seront n_k, $k=1, 2, \ldots, n$. On obtient ainsi:

$$n_k = \frac{1}{(2n+1)\,\mathrm{tg}^2 \dfrac{2k-1}{2n+1}\dfrac{\pi}{2}}.$$

Dans [4] utilisant les résultats de [1] j'établis le résultat suivant qui donne la fonction caractéristique de von Mises-Smirnoff pour plusieurs dimensions. Appelons $Y_n^{(p)}$ l'intégrale

$$n\int_{\mathbf{R}^p} [I_n(x_1, x_2, \ldots, x_p) - F_1(x_1)\,F_2(x_2) \ldots F_p(x_p)]^2 \,\mathrm{d}F_1(x_1)\,\mathrm{d}F_2(x_2) \ldots \mathrm{d}F_p(x_p),$$

I_n étant la distribution empirique sur n résultats d'une variable p-dimensionnelle dont les coordonnées sont indépendantes et ont pour loi de probabilité $F_i(x_i)$. On a

$$\lim_{n\to\infty} \mathrm{E}\,(\exp iuY_{(p)}^n) =$$

$$\left\{ i\,2^{p-1}\frac{\mathrm{d}}{\mathrm{d}u} \prod_{n_1=1}^{\infty}\prod_{n_2=1}^{\infty} \prod_{p=1}^{\infty}\left[1 - \frac{2\,iu}{\left(n_1 - \frac{1}{2}\right)^2\left(n_2 - \frac{1}{2}\right)^2 \cdots \left(n_p - \frac{1}{2}\right)^2 \pi^{2p}}\right]\right\}^{-\frac{1}{2}}.$$

M. Csörgő m'a fait l'honneur d'utiliser ce résultat pour construire les tables de lois de probabilité correspondantes. Dans un travail ultérieur, dans le but de donner la loi liée de l'intégrale de von Mises-Smirnoff et de l'intégrale de Paul Lévy je donnerai la forme du déterminant

$$R_n = \det\,[\mathbf{I}_n - u\,{}^t\mathbf{A}\mathbf{A} - v({}^t\mathbf{A} - \mathbf{A})].$$

En posant

$$\cos\theta = \frac{1 - \dfrac{u}{2}}{\sqrt{1 - v^2}}$$

soit $u = 2(1 - \sqrt{1 - v^2}\cos\theta)$ on trouve

$$R_n = (1 - v^2)^{\frac{n}{2}}\left[\frac{\cos\left(n + \frac{1}{2}\right)\theta}{\cos\dfrac{\theta}{2}} + \frac{\sin n\theta}{\sin\theta}\left(1 - \frac{1}{\sqrt{1 - v^2}}\right)\right].$$

Ce sera je pense une partie de ma communication à la session de Madrid de l'Institut International de Statistique.

Dans [10] un calcul analogue donne les valeurs propres de la matrice ${}^t\mathbf{A}_n - \mathbf{A}_n$ soit $\left[\mathrm{i}\,\mathrm{tg}\left(k+\frac{1}{2}\right)\frac{\pi}{n}\right]^{-1}$ (k variant de $-p$ à $p-1$, n étant égal à $2p$ ou $2p+1$) et de la matrice ${}^t\mathbf{A}_n - \mathbf{A}_n - (1/n)\,[{}^t\mathbf{A}_n - \mathbf{A}_n]\,\mathbf{M}_n$ qui sont $\left[\mathrm{i}\,\mathrm{tg}\,k\frac{\pi}{n}\right]^{-1}$, k variant de $-p$ à $+p$ (0 excepté), n étant égal à $2p+1$ ou $2p+2$. De plus $\mathrm{i}\,({}^t\mathbf{A}_n - \mathbf{A}_n)$ étant une matrice hermitienne peut être diagonalisée par une matrice unitaire $\mathbf{U}_n$. En posant ${}^t\mathbf{U}_n\mathbf{M}_n\mathbf{U}_n = \mathbf{N}'_n$, les éléments diagonaux n'_k de $\mathbf{N}'_n$ seront donnés par la formule

$$n'_k = \frac{1}{n\sin^2\left(k+\frac{1}{2}\right)\frac{\pi}{n}}$$

si n est pair et une formule analogue si n est impair.

Ces trois formules sont utilisées dans [10] pour généraliser des résultats de Paul Lévy dans le domaine multidimensionnel. Ces méthodes sont exactement analogues à celles qui ont permis la généralisation multidimensionnelle de la loi de Kolmogoroff-Smirnoff.

Donnons maintenant une démonstration qui m'a été obligeamment envoyée par mon collègue Benzécri.

Soit une matrice $n\times n$ symétrique $\mathbf{H}_n$ de valeurs propres $\lambda_1, \dots, \lambda_n$ et posons $\mathbf{H}_n = {}^t\mathbf{U}_n\mathbf{D}_n\mathbf{U}_n$, $\mathbf{D}_n$ étant diagonale et $\mathbf{U}_n$ une matrice unitaire, il est important pour de nombreux calculs statistiques de connaître le jacobien

$$\frac{D(\{\mathbf{H}_n\})}{D(\{\mathbf{D}_n\}, \{\mathbf{U}_n\})}.$$

$\{\mathbf{H}_n\}$, l'ensemble des éléments de $\mathbf{H}_n$ est un $\mathbb{R}_{n(n+1)/2}$, $\{\mathbf{D}_n\}$ un $\mathbb{R}_n$ et $\{\mathbf{U}_n\}$ un $\mathbb{R}_{n(n-1)/2}$. Il s'agit donc de trouver le déterminant fonctionnel de l'application

$$\mathbb{R}_{n(n+1)/2} \to \mathbb{R}_n \times \mathbb{R}_{n(n-1)/2}.$$

$\mathbf{H}_n$, matrice symétrique $n\times n$ définie sur un espace orthonormé E_n, peut être considérée comme un tenseur symétrique sur le produit tensoriel $E_n \otimes E_n$.

Appelons $E_n \underset{s}{\otimes} E_n$ l'espace des tenseurs symétriques et $E_n \underset{a}{\otimes} E_n$ l'espace des tenseurs antisymétriques.

On peut donner à $E_n \underset{s}{\otimes} E_n$ la base orthonormée

$$\left\{(\mathbf{e}_i \otimes \mathbf{e}_i,\ i = 1, 2, \dots, n) \cup \left(\frac{\mathbf{e}_i \otimes \mathbf{e}_{i'} + \mathbf{e}_{i'} \otimes \mathbf{e}_i}{\sqrt{2}},\quad 1 \leqslant i < i' \leqslant n\right)\right\}$$

et à $E_n \underset{a}{\otimes} E_n$ la base orthonormée

$$\left\{\frac{\mathbf{e}_i \otimes \mathbf{e}_{i'} - \mathbf{e}_{i'} \otimes \mathbf{e}_i}{\sqrt{2}},\quad 1 \leqslant i < i' \leqslant n\right\},$$

$\{\mathbf{e}_i,\ i = 1, 2, \dots, n\}$ étant naturellement une base orthonormée de E_n.

L'élément de volume de E_n sera $\bigwedge_{i=1}^{n} \mathrm{d}x_i \mathbf{e}_i$, celui de $E_n \underset{s}{\otimes} E_n$ sera:

$$\bigwedge_{i=1}^{n} \mathrm{d}x_{ii}\, \mathbf{e}_i \otimes \mathbf{e}_i \bigwedge_{i=1}^{n} \bigwedge_{j=i+1}^{n} \mathrm{d}x_{ij} \frac{\mathbf{e}_i \otimes \mathbf{e}_j + \mathbf{e}_j \otimes \mathbf{e}_i}{\sqrt{2}}$$

et celui de $E_n \underset{a}{\otimes} E_n$

$$\bigwedge_{i=1}^{n} \bigwedge_{j=i+1}^{n} \mathrm{d}x_{ij} \frac{\mathbf{e}_i \otimes \mathbf{e}_j - \mathbf{e}_j \otimes \mathbf{e}_i}{\sqrt{2}}.$$

Ces volumes seront évidemment invariants dans un changement de base orthonormée. Si l'on repère le tenseur $\mathbf{H}_n$ par rapport à ses vecteurs propres unitaires, $\mathbf{H}_n$ s'écrira $\mathbf{H}_n = \sum_{i=1}^{n} \lambda_i \mathbf{a}_i \otimes \mathbf{a}_i$, λ_i étant une valeur propre et $\mathbf{a}_i$ un vecteur propre.

On aura

$$\mathrm{d}\mathbf{H}_n = \sum_{i=1}^{n} \mathrm{d}\lambda_i\, \mathbf{a}_i \otimes \mathbf{a}_i + \sum_{i=1}^{n} \lambda_i (\mathrm{d}\mathbf{a}_i \otimes \mathbf{a}_i + \mathbf{a}_i \otimes \mathrm{d}\mathbf{a}_i).$$

Posons: $\mathbf{a}_i \mathrm{d}\mathbf{a}_j = \mathrm{d}\omega_{ij}$; on voit facilement à cause de $\mathbf{a}_i \mathbf{a}_i = 1$ et $\mathbf{a}_i \mathbf{a}_j = 0$ que $\mathrm{d}\omega_{ii} = 0$ et $\mathrm{d}\omega_{ij} = -\mathrm{d}\omega_{ji}$. La matrice $\{\mathrm{d}\omega_{ij}\}$ est antisymétrique.

Donc:

$$\mathrm{d}\mathbf{H}_n = \sum_{i=1}^{n} \mathrm{d}\lambda_i \mathbf{a}_i \otimes \mathbf{a}_i + \sum_{i,j=1 (i \neq j)}^{n} \lambda_i \mathrm{d}\omega_{ij} (\mathbf{a}_j \otimes \mathbf{a}_i + \mathbf{a}_i \otimes \mathbf{a}_j),$$

ce qui peut s'écrire

$$\mathrm{d}\mathbf{H}_n = \sum_{i=1}^{n} \mathrm{d}\lambda_i\, \mathbf{a}_i \otimes \mathbf{a}_i + \sum_{i=1}^{n} \sum_{j=i+1}^{n} (\lambda_i - \lambda_j) \sqrt{2}\, \mathrm{d}\omega_{ij} \frac{\mathbf{a}_j \otimes \mathbf{a}_i + \mathbf{a}_i \otimes \mathbf{a}_j}{\sqrt{2}}.$$

L'élément de volume de $E_n \underset{s}{\otimes} E_n$ exprimé dans l'espace des $\{\mathbf{D}_n\}$ et $\{\mathbf{U}_n\}$ est donc

$$\bigvee_{i=1}^{n} \mathrm{d}\lambda_i\, \mathbf{a}_i \otimes \mathbf{a}_i \bigwedge_{i=1}^{n} \bigwedge_{j=i+1}^{n} (\lambda_i - \lambda_j)\, 2^{\frac{n(n-1)}{4}}\, \mathrm{d}\omega_{ij} \frac{\mathbf{a}_j \otimes \mathbf{a}_i + \mathbf{a}_i \otimes \mathbf{a}_j}{\sqrt{2}} =$$

$$= \prod_{i=1}^{n} \prod_{j=i+1}^{n} (\lambda_i - \lambda_j) \prod_{i=1}^{n} \mathrm{d}\lambda_i \prod_{i=1}^{n} \prod_{j=i+1}^{n} 2^{\frac{n(n-1)}{4}}\, \mathrm{d}\omega_{ij}.$$

Les éléments différentiels $\sqrt{2}\, \mathrm{d}\omega_{ij}$ sont les coordonnées d'un tenseur antisymétrique

$$\sum_{i=1}^{n} \sum_{j=i+1}^{n} \sqrt{2}\, \mathrm{d}\omega_{ij} \frac{\mathbf{a}_i \otimes \mathbf{a}_j - \mathbf{a}_j \otimes \mathbf{a}_i}{\sqrt{2}}.$$

L'élément de volume de l'espace $\{\mathbf{D}_n\}$ est donc: $\prod_{i=1}^{n} \mathrm{d}\lambda_i$. Celui de l'espace $\{\mathbf{U}_n\}$ sera

$$\bigwedge_{i=1}^{n} \bigwedge_{j=i+1}^{n} \sqrt{2}\, \mathrm{d}\omega_{ij} \frac{\mathbf{a}_i \otimes \mathbf{a}_j - \mathbf{a}_j \otimes \mathbf{a}_i}{\sqrt{2}} = \prod_{i=1}^{n} \prod_{j=i+1}^{n} 2^{\frac{n(n-1)}{4}}\, \mathrm{d}\omega_{ij}.$$

Finalement l'élément de volume de $\{\mathbf{H}_n\}$ soit $E_n \underset{s}{\otimes} E_n$ est donc:

$$\prod_{i=1}^{n} \prod_{j=i+1}^{n} (\lambda_i - \lambda_j) \, d\{\mathbf{D}_n\} \, d\{\mathbf{U}_n\}.$$

Le jacobien de l'application sera donc $\prod_{i=1}^{n} \prod_{j=i+1}^{n} (\lambda_i - \lambda_j)$ auquel l'enseignement français du second degré donne le nom de déterminant de Van der Mond

$$V(\lambda, \lambda_2, ..., \lambda_n) = \begin{vmatrix} 1 & \lambda_n & \dots & \lambda_n^{n-1} \\ \vdots & \vdots & \vdots & \vdots \\ 1 & \lambda_2 & \dots & \lambda_2^{n-1} \\ 1 & \lambda_1 & \dots & \lambda_1^{n-1} \end{vmatrix}.$$

Ce calcul particulièrement simple dans l'écriture tensorielle permet de donner une loi généralisant celle de Behrens-Fisher pour plusieurs dimensions; le calcul qui y conduit utilisant la fonction caractéristique multidimensionnelle a été exposé dans le travail [11]. Soient $\mathbf{R}_1$ et $\mathbf{R}_2$ deux variables de Wilks construites à partir de n_1 et n_2 résultats k-dimensionnels ayant tous la même matrice de covariance et une valeur moyenne nulle. Appelons $\Lambda_1, \Lambda_2, ...$ ..., Λ_k les valeurs propres de $\mathbf{R}_1 \mathbf{R}_2^{-1}$, la loi de probabilité de l'ensemble $(\Lambda_1, \Lambda_2, ..., \Lambda_k)$ est la loi non paramétrique

$$C_1 \frac{(\lambda_1 \lambda_2 \dots \lambda_k)^{(n_1-k-1)/2}}{[(1+\lambda_1)(1+\lambda_2)\dots(1+\lambda_k)]^{(n_1+n_2)/2}} V(\lambda_1, \lambda_2, ..., \lambda_k) \, d\lambda_1 \, d\lambda_2 \dots d\lambda_k.$$

Dans le même ordre d'idées, on peut obtenir une loi non paramétrique généralisant la loi de K. Pearson. Appelons $P_1, P_2, ..., P_k$ les valeurs propres de la matrice $\Sigma^{-1}\,\mathbf{R}$, $\mathbf{R}$ étant une matrice de Wilks construite sur n variables k-dimensionnelles ayant Σ pour matrice de covariance et une valeur moyenne nulle. La loi de probabilité élémentaire sera

$$C_2\{\exp[-(\rho_1+\rho_2+\dots+\rho_k)/2]\}(\rho_1\rho_2\dots\rho_k)^{(n-k-1)/2} V(\rho_1, \rho_2, \dots, \rho_k) \, d\rho_1 \, d\rho_2 \dots d\rho_k.$$

Ces méthodes donnent un rôle symétrique aux différentes valeurs propres.

On peut utiliser pour comparer une matrice de Wilks à une matrice de covariance théorique, ou deux matrices de Wilks entre elles, une autre voie que j'ai décrite dans les deux notes [8] et [9].

Introduisons les notations suivantes. Soit une matrice de covariance $k \times k$. Posons $\Phi = \Sigma^{-1}$ la matrice de distribution. Appelons Σ_{lc}, $c = l, ..., k$; $l = 1, ..., k$, les matrices $l \times l$ (σ_{ij}) avec: $i = 1, ..., l$ et $j = 1, ..., l-1, c$. De la même façon $\mathbf{R}$ étant une matrice de Wilks construite sur n vecteurs indépendants de moyenne nulle et de matrice de covariance Σ on définira $\mathbf{R}_{lc}$ de la même façon que Σ_{lc}. On posera $\Sigma_{00} = \mathbf{R}_{00} = \mathbf{I}$ et

$$R_l = \frac{\det \mathbf{R}_{ll}}{\det \mathbf{R}_{(l-1)\,(l-1)}} : \frac{\det \Sigma_{ll}}{\det \Sigma_{(l-1)\,(l-1)}}.$$

On a:

$$\mathsf{P}(R_l < x) = \frac{1}{\Gamma\left(\dfrac{n-l+1}{2}\right) 2^{(n-l+1)/2}} \int_0^x [\exp(-u/2)] u^{(n-l+1)/2} \, du.$$

Si l'on considère maintenant les vecteurs $\mathbf{Y}^l$ $(k - l)$-dimensionnels dont les composantes sont

$$Y_c^l = \sqrt{\frac{\det \mathbf{R}_{ll}}{\det \mathbf{R}_{(l-1)\,(l-1)}}} \left[\frac{\det \mathbf{R}_{lc}}{\det \mathbf{R}_{ll}} - \frac{\det \Sigma_{lc}}{\det \Sigma_{ll}}\right], \quad c = l + 1, \dots, k,$$

on montre que ces vecteurs sont gaussiens de valeur moyenne nulle et de matrice de distribution Φ^{ll} obtenue de la façon suivante. Dans la matrice Φ on prendra la sous matrice $(\varphi_{ij})_{i>l,\, j>l}$. Les vecteurs $\mathbf{Y}^l$ et les quotients R_l sont indépendants dans leur ensemble.

Je ne décrirai pas dans le détail la méthode qui figure dans la note [9]. Elle a pour but de comparer entre elles deux matrices de Wilks. Elle utilise les résultats que je viens d'exposer ainsi que le théorème suivant: Y' et Y'' étant deux variables vectorielles normales indépendantes de moyenne nulle et de même matrice de covariance, (X', X'') étant un couple de variables aléatoires scalaires indépendantes de Y' et Y''

$$\frac{X'Y' + X''Y''}{\sqrt{X'^2 + X''^2}}$$

est une variable normale de moyenne nulle, de matrice de covariance Σ, indépendante de (X', X''). Cette manière de mettre à l'épreuve la qualité de l'ajustement dans le domaine multidimensionnel et de comparer des covariances de Wilks est certainement plus précise que celle que j'ai décrite auparavant. Elle offre toutefois un inconvénient: celui de privilégier l'ordre dans lequel se font les comparaisons des variables aléatoires dégagées. C'est toute la question du pivot comme disent les spécialistes de l'analyse des données qui est posée.

Puis je demande à la sagacité de l'auditoire de bien vouloir retenir l'énoncé des problèmes suivants:

1° Comparaison entre la méthode de [11] et la méthode de [9], un pivot déterminé étant choisi dans cette méthode de [9].

2° Comparaison entre des résultats obtenus avec la méthode de [9] par deux pivots différents.

Dans [11] également je généralise les méthodes d'analyse de variance au cas où la variable est multidimensionnelle. L'analyse de variance ordinaire (je veux dire multifactorielle par opposition au plan d'expériences qui utilise la confusion des interactions, carrés latins et autres méthodes) est fondée sur la décomposition générale de

$$\mathbf{I}_{n_1} \otimes \mathbf{I}_{n_2} \otimes \dots \otimes \mathbf{I}_{n_p} = \left(\mathbf{I}_{n_1} - \frac{1}{n_1}\mathbf{M}_{n_1} + \frac{1}{n_1}\mathbf{M}_{n_1}\right) \otimes \left(\mathbf{I}_{n_2} - \frac{1}{n_2}\mathbf{M}_{n_2} + \right.$$

$$\left. + \frac{1}{n_2}\mathbf{M}_{n_2}\right) \otimes \dots \otimes \left(\mathbf{I}_{n_p} - \frac{1}{n_p}\mathbf{M}_{n_p} + \frac{1}{n_p}\mathbf{M}_{n_p}\right)$$

dans les sommes suivantes:

$$\left(\mathbf{I}_{n_1} - \frac{1}{n_1}\mathbf{M}_{n_1}\right) \otimes \left(\mathbf{I}_{n_2} - \frac{1}{n_2}\mathbf{M}_{n_2}\right) \otimes \ldots \otimes \left(\mathbf{I}_{n_p} - \frac{1}{n_p}\mathbf{M}_{n_p}\right) +$$

$$+\sum_{(1)}\left(\mathbf{I}_{n_1} - \frac{1}{n_1}\mathbf{M}_{n_1}\right) \otimes \ldots \otimes \frac{1}{n_i}\mathbf{M}_{n_i} \otimes \ldots \otimes \left(\mathbf{I}_{n_p} - \frac{1}{n_p}\mathbf{M}_{n_p}\right) +$$

$$+\sum_{(2)}\left(\mathbf{I}_{n_1} - \frac{1}{n_1}\mathbf{M}_{n_1}\right) \otimes \ldots \otimes \frac{1}{n_i}\mathbf{M}_{n_i} \otimes \ldots \otimes \frac{1}{n_j}\mathbf{M}_{n_j} \otimes \ldots \otimes \left(\mathbf{I}_{n_p} - \frac{1}{n_p}\mathbf{M}_{n_p}\right) + \cdots,$$

les différentes sommes étant prises en remplaçant dans la première un facteur $\mathbf{I}_{n_i} - (1/n_i)\,\mathbf{M}_{n_i}$ par $(1/n_i)\,\mathbf{M}_{n_i}$, dans la seconde deux facteurs $\mathbf{I}_{n_i} - (1/n_i)\,\mathbf{M}_{n_i}$ et $\mathbf{I}_{n_j} - (1/n_j)\,\mathbf{M}_{n_j}$ par deux facteurs $(1/n_i)\,\mathbf{M}_{n_i}$ et $(1/n_j)\,\mathbf{M}_{n_j}$. On obtient ainsi 2^p matrices.

En multipliant les variables multifactorielles par des matrices unitaires orthogonalisant les matrices $\mathbf{M}_{n_i}$, comme une des formes diagonales de $(1/n_i)\mathbf{M}_{n_i}$ est une matrice ayant un seul terme différent de 0 le premier terme diagonal égal à 1, on obtient 2^p matrices diagonales sur les diagonales desquelles se répartissent des 0 et des 1. Tous ces 0 et ces 1 s'emboîtent pour reconstituer la matrice $\mathbf{I}_{n_1} \otimes \mathbf{I}_{n_2} \otimes \ldots \otimes \mathbf{I}_{n_p}$. On en déduit l'indépendance de ces différentes formes quadratiques et les rangs de ces matrices donnent ce qu'en analyse de variance on appelle les degrés de liberté. Dans [11], la même écriture tensorielle permet d'étendre cette analyse de variance unidimensionnelle au cas multidimensionnel.

La confusion des interactions c'est-à-dire le regroupement de certaines de ces différentes matrices (et les carrés latins orthogonaux en sont un cas particulier) peut évidemment se généraliser au cas multidimensionnel. Je souhaiterais (un vœu de plus) qu'une théorie matricielle de l'interaction soit élaborée, c'est-à-dire que l'on connaisse les différentes matrices unitaires conduisant à ces regroupements. Dans [7] en utilisant le développement de $\mathbf{I}_{n_1} \otimes \mathbf{I}_{n_2} \otimes \ldots \otimes \mathbf{I}_{n_p}$ indiqué ci-dessus je donne la forme générale pour p dimensions d'un résultat qu'avait établi W. Hoeffding pour deux dimensions. Il s'agit de tests indépendants de la loi. Pour deux dimensions W. Hoeffding avait établi le résultat suivant: H_n étant la répartition empirique de n résultats, si la loi de probabilité est continue et si les deux composantes sont indépendantes la fonction caractéristique de la loi de la variable

$$W_n = n\int_{\mathbf{R}^2} [H_n(x_1, x_2) - H_n(x_1, \infty)\,H_n(\infty, x_2)]^2\, \mathrm{d}H_n(x_1, x_2)$$

tend vers

$$\prod_{k_1=1}^{\infty}\prod_{k_2=2}^{\infty}\left(1 - \frac{2\,iu}{k_1^2 k_2^2\,\pi^4}\right)^{-1/2}.$$

La décomposition matricielle que je viens d'indiquer permet de construire une intégrale dont la fonction caractéristique est

$$\prod_{k_1=1}^{n}\prod_{k_2=1}^{n}\cdots\prod_{k_p=1}^{\infty}\left(1 - \frac{2\,iu}{k_1^2 k_2^2 \ldots k_p^2\,\pi^{2p}}\right)^{-1/2}.$$

Dans [12] j'étends les résultats précédents au cas où l'on a à tester non plus l'indépendance des composantes (unidimensionnelles) d'une variable multidimensionnelle mais l'indépendance de deux matrices, ce qui me permet d'établir un test de normalité multidimensionnelle.

On sait que si deux variables indépendantes X_1 et X_2 sont telles que $X_1 + X_2$ et $X_1 - X_2$ soient indépendantes ce sont des variables normales. Ce théorème qui a d'abord été démontré quand X_1 et X_2 sont des scalaires a été étendu au cas où X_1 et X_2 sont des matrices. Comme la première partie de l'article permet de traiter de l'indépendance de deux matrices, on a ainsi un test de normalité.

J'ai voulu résumer en quelques paragraphes des résultats obtenus à partir des calculs matriciels et tensoriels les plus simples qui soient, ainsi que d'autres résultats que je prépare. J'ai posé quelques problèmes. Je terminerai par un vœu. Celui de voir trouver une structure statistique pour l'enseignement de l'algèbre linéaire et multilinéaire. Dans cet enseignement plusieurs chapitres devraient être consacrés aux matrices «spéciales» (comme on dit fonctions «spéciales»). J'en énumère quelques-unes: matrices symétriques, orthogonales, normales, de permutation, d'incidence (au sens du plan d'expériences), triangulaires (comme la matrice $\mathbf{A}_n$), etc. ...

J'espère que la prochaine réunion de Brașov aura un peu fait avancer ce programme.

BIBLIOGRAPHIE

1. Dugué, D., *Fonctions caractéristiques d'intégrales browniennes.* Rev. Roumaine Math. Pures Appl., **12**, 1207–1215 (1967).
2. Dugué, D., *Fonction caractéristique de ω_n^2 multidimensionnel de von Mises. Fonctions caractéristiques bidimensionnelles liées au mouvement brownien.* C. R. Acad. Sci. Paris, **266**, 834–836 (1968).
3. Dugué, D., *Valeurs propres de matrices ayant un rôle important dans l'étude d'intégrales browniennes.* Rev. Inst. Internat. Statist., **37**, 13–17 (1969).
4. Dugué, D., *Characteristic functions of random variables connected with Brownian motion and of the von Mises multidimensional ω_n^2.* In: Krishnaiah, P. R. (Ed.), *Multivariate Analysis II*, pp. 289–301. New York, Academic Press, 1969.
5. Dugué, D., *Un résultat sur la robustesse de l'analyse de variance. Analyse de variance pour processus laplaciens.* C. R. Acad. Sci. Paris, **270**, 391–394 (1970).
6. Dugué, D., *Sur une extension indépendante de la répartition d'un résultat de M. Paul Lévy et sur la fonction caractéristique d'un couple de variables liées à une courbe brownienne.* C. R. Acad. Sci. Paris, **276**, 395–397 (1973).
7. Dugué, D., *Sur des tests d'indépendance «indépendants de la loi».* C.R. Acad. Sci. Paris, **281**, 1103–1104 (1975).
8. Dugué, D., *Test d'ajustement dans le cas multidimensionnel. Généralisation du test de Student dans le cas multidimensionnel.* C.R. Acad. Sci. Paris, **284**, 623–624 (1977).
9. Dugué, D., *Variables scalaires attachées à deux matrices de Wilks. Comparaison de deux matrices de Wilks en analyse des données.* C.R. Acad. Sci. Paris, **284**, 899–901 (1977).
10. Dugué, D., *Applications of characteristic function in solving some distribution problems.* In: Chakravarti, I. M. (Ed.), *Asymptotic Theory of Statistical Tests and Estimation: In Honor of Wassily Hoeffding*, pp. 65–83. New York, Academic Press, 1980.
11. Dugué, D., *On a multivariate extension of the Behrens-Fisher law.* In: Gani, J., Rohatgi, V. K. (Eds.), *Contribution to Probability: A Collection of Papers Dedicated to Eugene Lukacs*, pp. 247–253. New York, Academic Press, 1981.
12. Dugué, D., *On the way towards a non parametric multidimensional test of normality.* In: Kallianpur, G., Krishnaiah, P.R., Ghosh, J. K. (Eds.), *Statistics and Probability: Essays in Honor of C.R. Rao*, pp. 241–247. Amsterdam, North Holland, 1982.

PAUL EMBRECHTS

Subexponential distribution functions and their applications: A review

1. Introduction

Consider an age dependent branching process $Z(t)$, t positive. Hence particles produce at the end of their life an offspring with particle production generating function $f(s) = \Sigma p_k s^k$. The lifetime of each particle is a random variable with distribution function G. For all t positive, $Z(t)$ denotes the number of particles alive at time t. This general branching process has been studied extensively; for a detailed treatment of its properties, see the book by Athreya and Ney [1]. The behaviour of the process $Z(t)$ essentially depends on whether the mean particle production size $m = \Sigma k p_k$ is larger, smaller or equal to 1. In the supercritical case, i.e. $m > 1$, and $f''(1)$ finite one can prove that

$$e^{-\alpha t} Z(t) \xrightarrow{\mathcal{D}} W \quad (\text{as } t \to \infty)$$

where W is a non-degenerate random variable and α satisfies the equation

$$g(\alpha) = \int_0^\infty e^{-\alpha x} \, dG(x) = m^{-1} \tag{1.1}$$

i.e. $\alpha > 0$.

The parameter α is called the Malthusian parameter of the process $Z(t)$. If $m = 1$ (the critical case), Goldstein proved that whenever $\sigma^2 = f''(1)$ finite, μ the mean of G finite and $1 - G(t) = o(t^{-2})$ as $t \to \infty$, then (see [17])

$$\lim_{t\to\infty} P\left\{\frac{Z(t)}{t} \leqslant x \mid Z(t) > 0\right\} = 1 - \exp(-2\,\mu/\sigma^2)x.$$

The remaining subcritical case ($m < 1$) is much more subtle to solve and evolves around the existence (non-existence) of the Malthusian parameter α as a solution of $g(\alpha) = m^{-1}$, where g denotes the Laplace-Stieltjes transform of the lifesize distribution function G. The existence of α, in this case indicates a faster than exponential tail decay of G and indeed if α exists, it is proved in Athreya and Ney [1] that

$$\lim_{t\to\infty} P\{Z(t) = k \mid Z(t) > 0\} = b_k \tag{1.2}$$

for all k positive. The probability measure $(b_k)_k$ is nondegenerate. A gamma distributed lifetime for instance always yields a solution of (1.1) in this case and hence (1.2) holds. If the Malthusian parameter α does not exist, by virtue of slower than exponential tail decay of G then there still exists an important class of lifetime distributions G for which a result similar to (1.2) holds.

Definition 1.1. A distribution function F with $F(0-)=0$ is called *subexponential*, written $F \in \mathcal{S}$, iff

$$\lim_{x\to\infty} (1-F^{(2)}(x))/(1-F(x)) = 2, \tag{1.3}$$

where $F^{(2)}$ denotes the second convolution of F with itself.

In the next paragraph, we shall discuss the main properties of $\mathcal{S}$ and show that for $G \in \mathcal{S}$ there is no solution to the Malthusian equation (1.1), however

Theorem 1.1. (Chistyakov, Chover-Ney-Wainger) *If $Z(t)$ is an age-dependent branching process with $m<1$ and $G \in \mathcal{S}$, then*

$$\lim_{t\to\infty} P\{Z(t)=1 \mid Z(t)>0\} = 1 \tag{1.4}$$

i.e. the limit in (1.2) *exists and is degenerate in* 1.

This result is proved by Chistyakov [3] for m'small' and $f''(1)$ finite. The general case is due to Chover, Ney and Wainger [4]. The study of distributions satisfying (1.3) constitutes the main aim of this paper.

2. The class $\mathcal{S}$

The reason $\mathcal{S}$ plays a fundamental rôle in the branching process problem comes from the following relation for the mean particle size at time t

$$E(Z(t)) = \sum_{k=0}^{\infty} m^k (G^{(k)}(t) - G^{(k+1)}(t)),$$

where again $G^{(n)}$ denotes the n-th convolution of G with itself. Hence

$$\frac{E(Z)(t))}{1-G(t)} = \sum_{k=0}^{\infty} m^k \left(\frac{(1-G^{(k+1)}(t)}{1-G(t)} - \frac{1-G^{(k)}(t)}{1-G(t)} \right) \tag{2.1}$$

and an asymptotic convolution property of G would imply that

$$E(Z(t)) \sim c(1-G(t)) \text{ as } t\to\infty.$$

This is indeed the basic idea behind the use of $\mathcal{S}$ in this and various other applications. We shall return to this point later.

The following theorem formulates the main known properties of $\mathcal{S}$.

Theorem 2.1. i) *If $F \in \mathcal{S}$ then for all $k \geq 2$,* $\lim_{x\to\infty}(1-F^{(k)}(x))/(1-F(x)) = k$.

ii) *If $F \in \mathcal{S}$ and $\varepsilon > 0$ then* $e^{\varepsilon x}(1-F(x)) \to \infty$ *as* $x\to\infty$.

iii) *If $F \in \mathcal{S}$, then for all real y,* $\lim_{x\to\infty} (1-F(x+y))/(1-F(x)) = 1$ *uniformly over compact y-sets.*

iv) *If $F \in \mathcal{S}$ then for every $\varepsilon > 0$ there exists a finite constant K (independent of n) such that for all x positive and $n = 1, 2, \ldots$*

$$\frac{1 - F^{(n)}(x)}{1 - F(x)} \leqslant K(1+\varepsilon)^n.$$

v) *If $F \in \mathcal{S}$ and* $\lim_{x\to\infty} (1 - F(x))/(1 - G(x)) = c$ *where* $0 < x < \infty$, *then* $G \in \mathcal{S}$.

The first three results were proved by Chistyakov [3]; iv) is due to Kesten and is proved in Athreya and Ney [1]. The last result was first proved by Teugels [26] and various other authors independently (see for instance Pakes [20]). Property i) above gives the desired asymptotic convolution relations whereas iv) allows one to use Lebesgue's dominated convergence theorem in expressions like (2.1). In this latter case it follows that for $G \in \mathcal{S}$

$$E(Z(t)) \sim (1-m)^{-1}(1 - G(t)) \text{ as } t \to \infty.$$

(From now on we shall always write $h_1(t) \sim h_2(t)$ as $t \to \infty$ if $\lim_{t\to\infty} h_1(t)/h_2(t) = 1$.)

The name 'subexponentiality' is justified by property ii), although it holds for the strictly larger class

$$\mathcal{L} = \{F \mid F \text{ is a distribution function on } [0, \infty[\text{ satisfying } \lim_{x\to\infty} (1 - F(x+y))/(1 - F(x)) = 1 \text{ for all real } y\}.$$

The fact that $\mathcal{L} \neq \mathcal{S}$ is not obvious and known counterexamples are rather involved, see Embrechts and Goldie [8] and Pitman [21]. To get a better feeling for the relationship between $\mathcal{S}$ and $\mathcal{L}$ consider

$$(1 - F^2(x))/(1 - F(x)) = 2\int_0^{x/2} (1 - F(x-y))/(1 - F(x))\,\mathrm{d}F(y)$$

$$+ (1 - F(x/2))^2/(1 - F(x)). \tag{2.2}$$

Define now the class of dominated varying distribution functions $\mathcal{D}$ as

$$\mathcal{D} = \{F \mid F \text{ is a distribution function on } [0, \infty[\text{ satisfying } \limsup_{x\to\infty} (1 - F(x/2))/(1 - F(x)) \text{ is finite}\}.$$

An immediate application of Lebesgue's theorem in (2.2) yields $\mathcal{L} \cap \mathcal{D} \subset \mathcal{S}$. For a thorough discussion of the relationship between $\mathcal{S}$ and $\mathcal{D}$, see Goldie [16]. I would like to stress that $\mathcal{D} \not\subset \mathcal{S}$ though some authors claim the converse. An easy counterexample is the Peter-and-Paul distribution: F has mass 2^{-k} at the points 2^k, $k = 1, 2, \ldots$. One easily verifies that $(1 - F(2^k - 1))/(1 - F(2^k)) = 2$ hence $F \notin \mathcal{L}$ (and a fortiori $F \notin \mathcal{S}$) but for all x greater than 2, $(1 - F(x/2))/(1 - F(x)) = 2$ so $F \in \mathcal{D}$. (This example is due to Goldie [16].) Another way for discriminating between $\mathcal{S}$ and $\mathcal{L}$ is that for $\mathcal{L}$-functions, $(1 - F(x-y))/(1 - F(x))$ converges to 1 pointwise (as $x \to \infty$) whereas functions in $\mathcal{S}$ have the property that convergence of $(1 - F(x - .))/(1 - F(x))$ to 1 is in $L^1(F)$, indeed

$$(1 - F^{(2)}(x))/(1 - F(x)) = 1 + \int_0^x (1 - F(x-y))/(1 - F(x))\,\mathrm{d}F(y).$$

In order to find examples of subexponential distribution functions, one seldom uses the definition of $\mathcal{S}$. The theorem below lists the main known sufficient conditions for subexponentiality.

Theorem 2.2. i) $\mathcal{L} \cap \mathcal{D} \subset \mathcal{S}$.

ii) *Suppose F satisfies the following three conditions*

a) $H(x) = -\log(1 - F(x))$ *is concave on some interval* $[x_0, \infty[$;

b) *there exists a function g with* $0 < g(x) \to \infty$, $x - g(x) \to \infty$ *as* $x \to \infty$ *such that*

$$\lim_{x\to\infty} (1 - F(x - g(x)))/(1 - F(x)) = 1;$$

c) $H(x) \exp(-H(g(x))) \to 0$ *as* $x \to \infty$;

then $F \in \mathcal{S}$.

iii) *Denote again* $H(x) = -\log(1 - F(x))$ *and suppose H is differentiable so that* $H'(x)$ *decreases to 0 as* $x \to \infty$. *A necessary and sufficient condition for* $F \in \mathcal{S}$ *in this case is*

$$\lim_{x\to\infty} \int_0^x \exp(yH'(x) - H(y))\, H'(y)\, dy = 1.$$

A sufficient condition is $\exp(yH'(y) - H(y))\, H'(y)$ *is integrable on* $[0, \infty[$.

We already discussed property i) above, ii) is proved in Teugels [26] and iii) is due to Pitman [21]. The above theorem is used to find particular examples of subexponential distribution functions.

Examples of $\mathcal{S}$-functions. i) Suppose $1 - F(x) \sim x^{-r}L(x)$ where r is positive and L a slowly varying function, i.e. for all t positive, $L(tx) \sim L(x)$ as $x \to \infty$ and L is positive, Lebesgue measurable. Then $F \in \mathcal{S}$ as is proved in Feller [15, p. 278] using the so-called Feller-Zolotarev inequalities. It follows immediately from Theorem 2.2.i). Hence all Pareto type distributions are subexponential.

ii) Every lognormal distribution is subexponential. A proof based on Theorem 2.2.ii) ca be found in Embrechts, Goldie and Veraverbeke [11].

iii) Consider

$$1 - F(x) \sim \exp(-x(\log x)^{-b}),\ b > 0,$$

then F is subexponential, as proved by Pitman using Theorem 2.2.iii) ([21]).

iv) If $1 - F(x) \sim \exp(-cx^{-a})$ where the constant c is positive and $0<a<1$, then it follows from Theorem 2.2.ii) that $F \in \mathcal{S}$. It is however *not* true that $F \in \mathcal{S}$ whenever $1 - F(x) \sim \exp(-x^{-a}L(x))$ where $0<a<1$ and L is an arbitrarily slowly varying function. A counterexample was given by Goldie (personal communication).

Define the slowly varying function L as

$$L(x) = \prod_{k=1}^{n} (1 + 1/k),\quad e^n < x \leqslant e^{n+1},$$

thus $L(x) \to \infty$ as $x \to \infty$. But

$$(e^n + 1)^a L(e^n + 1) - (e^n)^a L(e^n) > e^{na}(L(e^n + 1) - L(e^n)) =$$

$$= e^{na} n^{-1} \prod_{k=1}^{n-1} (1 + 1/k) \to \infty$$

as $x \to \infty$. So $H(x+1) - H(x) \not\to 0$, contradicting the $\mathcal{L}$-property of subexponential distributions. This example shows that Corollary 2 ii) in [26] is wrong.

The main disadvantage of $\mathcal{S}$ is it is rather difficult to check definition. There does not exist a nice necessary and sufficient condition for $F \in \mathcal{S}$ in terms of the original tail $1 - F$ itself. Nor does this exist in terms of the Laplace-Stieltjes transform of F. If f denotes the Laplace-Stieltjes transform of F, and if $F \in \mathcal{S}$ then we know that $f(s)$ is *not* analytic at $s = 0$, i.e. 0 is an essential singularity of f. This shows how subexponential distributions give nice alternatives when there is no solution to the Malthusian equation in Paragraph 1. Finally, Teugels [26] remarks that $F \in \mathcal{S}$ iff $\lim_{x\to\infty}(1 - F^{(2)}(x))/(1 - F(x)) = b$ strictly positive and finite and $f(s)$ is not analytic at $s = 0$. However, as discussed in [11, Remark 5.2] there is a gap in the proof leaving this result still unproved.

The main power of $\mathcal{S}$ lies in its many applications; the next paragraph is devoted to a very important one.

3. $\mathcal{S}$ as a characterization class

Denote by $\mathcal{R}$ the class of regularly varying functions, i.e. $F \in \mathcal{R}$ means that there exist a slowly varying function L and an index r positive so that $1 - F(x) \sim x^{-r}L(x)$ as $x \to \infty$. (Remark: without explicit statement, we always assume that $F(0-) = 0$.) Suppose in some probability model there is an 'input' F and an 'output' G and one knows that

$$F \in \mathcal{R} \Leftrightarrow G \in \mathcal{R} \Rightarrow (1 - F(x))/(1 - G(x)) \to c \text{ as } x \to \infty$$

where c is a constant. How far can one enlarge the class $\mathcal{R}$, keeping the implications in force? In several fairly general situations the answer is that one can replace $\mathcal{R}$ by $\mathcal{S}$, and that this is *best possible*, in that the right-hand implication becomes two-way! This is what we mean by '$\mathcal{S}$ as a characterization class'. An important example is the following: suppose G is a proper distribution function on $[0, \infty[$, k strictly positive, define the compound Poisson distribution function

$$F(x) = e^{-k} \sum_{n=0}^{\infty} (k^n/n!)\, G^{(n)}(x).$$

Cohen proved

Lemma 3.1. ([6]). *Suppose r strictly positive and L slowly varying. Equivalent are*

i) $1 - F(x) \sim kx^{-r}L(x)$, $x \to \infty$;
ii) $1 - G(x) \sim x^{-r}L(x)$, $x \to \infty$.
Either of them implies
iii) $\lim_{x\to\infty} (1 - F(x))/(1 - G(x)) = k$.

This result was used in [6] to prove that for the distribution functions of the stationary actual waiting time and of the stationary virtual waiting time of the $GI/G/1$ queueing system, the tails vary regularly iff the tail of the service time distribution is regularly varying. The following more general result is proved in Embrechts, Goldie and Veraverbeke [11, Theorem 3].

Theorem 3.2. *For F and G as above, the following statements are equivalent*

i) $F \in \mathcal{S}$;

ii) $G \in \mathcal{S}$;

(iii) $\lim_{x\to\infty} (1 - F(x))/(1 - G(x)) = k$.

The proof is rather involved, especially that i) implies ii). The main new property of $\mathcal{S}$ needed is the convolution-roots closure of $\mathcal{S}$

Proposition 3.3. ([11]). *If, for some positive integer n, $F^{(n)} \in \mathcal{S}$, then $F \in \mathcal{S}$.*

A special case of Theorem 3.2., i.e. the compound geometric case where for some $0<m<1$

$$F(x) = (1 - m) \sum_{n=0}^{\infty} m^n G^{(n)}(x),$$

was proved by Callaert and Cohen [2] for $\mathcal{R}$ and independently by Teugels [26] and Pakes [20] for $\mathcal{S}$.

In [26] this result was applied to transient renewal theory. Let F be a defective distribution function so that $F(x) < F(\infty) = a$ where $0<a<1$. The renewal function associated with F is given by

$$U(x) = \sum_{n=1}^{\infty} F^{(n)}(x).$$

Since $U(\infty) = a/(1 - a)$ one can compare the common tail behaviour of $F(\infty) - F(x)$ and $U(\infty) - U(x)$.

Corollary 3.4. *The following assertions are equivalent*

i) $F^{-1}(\infty)\, F(x) \in \mathcal{S}$;

ii) $U^{-1}(\infty)\, U(x) \in \mathcal{S}$;

(iii) $\lim_{x\to\infty} (1 - U^{-1}(\infty)\, U(x))/(1 - F^{-1}(\infty)\, F(x)) = (1 - a)^{-1}$.

In Pakes' paper, the above theorem for compound geometric distributions is applied to the $GI/G/1$ queueing system, generalizing the Cohen results.

Theorem 3.5. (Pakes [20]). *Consider the $GI/G/1$ queue with interarrival time distribution A (mean a finite) and service time distribution B (mean b finite) so that $b - a < 0$. If Q denotes the (weak) limit distribution of the successive waiting times and*

$$B_1(x) = b^{-1} \int_0^x (1 - B(t))\, dt$$

then equivalent are

i) $Q \in \mathcal{S}$;

ii) $B_1 \in \mathcal{S}$.

Either of them implies

iii) $1 - Q(x) \sim l(1 - l)^{-1}(1 - B_1(x))$ *as* $x \to \infty$,

where $l = b/a$ is the traffic intensity. For $M/G/1$, iii) *implies* i) *and* ii).

Weaker versions of this result were proved by Smith [23].

4. Some further applications

4.1. *Infinite divisibility.* Let F be an infinite divisible distribution function concentrated on $[0, \infty[$ with Lévy measure N, hence its Laplace-Stieltjes transform f can be represented as (Feller [15, p. 450])

$$f(s) = \exp\left(-as - \int_0^\infty (1 - e^{-sx})\, dN(x)\right), \quad \operatorname{Re} s \geqslant 0,$$

where a is a positive constant and N is a Borel measure on $]0, \infty[$ for which $m = N]1, \infty[$ is finite and $\int_0^1 x\, dN(x)$ is finite. Various known results relate the asymptotic behaviour of $1 - F$ to that of N. See for instance Feller [15, p. 572] for the use of $\mathcal{R}$. The most general result is

Theorem 4.1.1. ([11]). *For F infinitely divisible (as above), the following assertions are equivalent*

i) $F \in \mathcal{S}$;

ii) $m^{-1}N]1, x] \in \mathcal{S}$;

iii) $1 - F(x) \sim N]x, \infty[$ *as* $x \to \infty$.

This result exploits in full the power of $\mathcal{S}$ and is indeed remarkably general. The basic idea of the proof uses the fact that $F = F_1 * F_2$ ($*$ denotes the convolution operator) where

$$\text{i)} \quad 1 - F_1(x) = o(e^{-\varepsilon x}),\ x \to \infty \text{ for every } \varepsilon \text{ positive,}$$

and

$$\text{ii)} \quad F_2(x) = e^{-m} \sum_{n=0}^{\infty} (m^n/n!)\, Q^{(n)}(x)$$

if $Q(x) = m^{-1}N]1, x]$, $x \geqslant 1$; $= 0$ otherwise. The following proposition, interesting in its own right, combines the above information with $\mathcal{S}$-properties.

Proposition 4.1.2. ([11]) *Let $H = F * G$ be the convolution of two proper distribution functions on $[0, \infty[$.*

i) *If $G \in \mathcal{S}$ and $1 - F(x) = o(1 - G(x))$, $x \to \infty$, then $H \in \mathcal{S}$. Moreover $1 - H(x) \sim 1 - G(x)$ as $x \to \infty$.*

ii) *If $H \in \mathcal{S}$ and $1 - F(x) = o(1 - H(x))$, then $G \in \mathcal{S}$ and indeed $1 - G(x) \sim$ $\sim 1 - H(x)$ as $x \to \infty$.*

Rather than comparing the tails of F and N one can ask for similar results if the spectral measure takes one of the alternative forms $dM(x) = x^2\, dN(x)$ or $dP(x)$ where $P[0, x] = a + \int_0^x y\, dN(y)$, x positive. In Embrechts and Goldie [9] we compare the behaviour of $1 - F$ with *all* tail integrals of N weighted by powers of x. It turns out that various other classes besides $\mathcal{S}$ are important. One of novelties of the latter paper is the use of Korenblum's ratio Tauberian theorem (Stadtmüller and Trautner [24]).

4.2. *Random walk theory.* The results listed below were derived by Veraverbeke [28] and complemented by Embrechts and Goldie [10]. Suppose F is a distribution function (general). If $X_1, X_2, \ldots$ are i.i.d. r.v.'s with distribution F define the ladder variables

$$N = \min\left\{n > 0 \mid S_n = \sum_{j=1}^{n} X_j > 0\right\}$$

and

$$\bar{N} = \min \{n > 0 \mid S_n \leqslant 0\}.$$

If f, f_+ and f_- are the Fourier-Stieltjes transforms of the random variables X_1, S_N and $S_{\bar{N}}$ then the famous Wiener-Hopf decomposition holds (Feller [15]):

$$1 - f(t) = (1 - f_+(t)).(1 - f_-(t)), \; t \text{ real}.$$

Denote the distribution functions of the Wiener-Hopf factors by F_+ and F_-. In general it is very difficult to obtain explicit expressions of these factors, hence the need for asymptotic formulae as provided for the right Wiener-Hopf factor by the following theorem of Veraverbeke [28]. We denote by $F_0(x)$ the distribution function $F(x)$ truncated at 0. Moreover denote $A = \Sigma n^{-1} P(S_n \leqslant 0)$.

Theorem 4.2.1. i) *If A is finite, then $F_0 \in \mathcal{S}$ iff $F_+ \in \mathcal{S}$. Each of them implies*

$$\lim_{x \to \infty} (1 - F(x))/(1 - F_+(x)) = e^{-A}.$$

ii) *If A is infinite and $-\infty < E(S_{\bar{N}}) < 0$ then $F_1 \in \mathcal{S}$ iff $F_+ \in \mathcal{S}$ and each of the above implies*

$$\lim_{x \to \infty} \int_x^{\infty} (1 - F(t))/(1 - F_+(x))\, dt = \int_{-\infty}^{0+} |t|\, dF_-(t).$$

The reader is referred to [28] for proofs and further references. Also consult [10] for complementary remarks and some corrections on results presented in [28].

4.3. *Ruin theory*. Applications of $\mathcal{S}$ to the calculation of Cramér-type bounds for the probability of ruin function in the collective theory of risk goes back to the paper by Teugels and Veraverbeke [27]. An up-to-date version is given in Embrechts and Veraverbeke [14]. Some related remarks can be found in Embrechts [7].

An insurance company, starting with initial capital x, receives claims of size $(A_n)_n$ at interclaim times $(T_n)_n$ where both processes consist of i.i.d. r.v.'s and are independent of each other. Denote $K(u) = P(T_1 \leqslant u)$, $L(u) =$

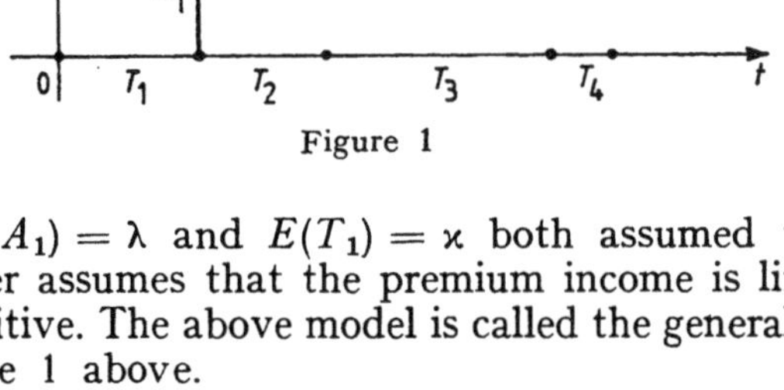

Figure 1

$= P(A_1 \leqslant u)$, $E(A_1) = \lambda$ and $E(T_1) = \varkappa$ both assumed throughout to be finite. One further assumes that the premium income is linear in time with rate c strictly positive. The above model is called the general Sparre Andersen model, see Figure 1 above.

If $N(t)$ denotes the number of claims before t, then ruin occurs iff for some t positive, $\sum_{i=1}^{N(t)} A_i > x + ct$. If $R(x)$ stands for the probability that no ruin ever occurs, then a Cramér-type estimate is an estimate for $1 - R(x)$ as $x \to \infty$. We shall (for brevity) restrict ourselves to the important Poisson case, i.e. $K(u) = 1 - e^{-u/\varkappa}$ and $L(0+) = 0$, hence claims arrive according to a Poisson process and claims are always positive. If the tail of L decays fast enough, so that a strictly posive solution ν to the Lundberg equation

$$(4.2.1) \qquad \int_0^\infty e^{\nu t}(1 - L(t))\,dt = c\varkappa$$

exists, then Cramér's famous theorem states that $1 - R(x) \sim Ce^{-\nu x}$ as $x \to \infty$. The constant C can be calculated explicitely. It is always assumed that $c\varkappa/\lambda > 1$. This theorem is however *not* applicable if L is Pareto or lognormal say, though both distributions are often considered in practice to model large claims. The basic result in the latter case is the following theorem.

Theorem 4.3.1. ([14]). *Suppose in the Poisson model that* $m = \lambda - c\varkappa < 0$ *and denote* $L_1(t) = \lambda^{-1}\int_0^t (1 - L(u))\,du$. *The following statements are equivalent*

i) $L_1 \in \mathcal{S}$;
ii) $R \in \mathcal{S}$;
iii) $1 - R(x) \sim -m^{-1}\int_x^\infty (1 - L(t))\,dt$ *as* $x \to \infty$.

Statement iii) above replaces the Cramér estimate if the claim size distribution is heavy tailed (i.e. $L_1 \in \mathcal{S}$). This result carries over to the general Andersen model though the proofs heavily depend upon the results of the previous paragraph. The same estimate as iii) holds if the integrated right tail of L is subexponential (see [14, Theorem 4.6.]).

4.4. *Renewal theory.* In Corollary 3.4 we already discussed an application of $\mathcal{S}$ to transient renewal theory. Ney ([18]) recently studied rates of convergence in Blackwell's renewal theorem. We shall only state his results in the non-singular case. Let $(X_n)_{n=0}^\infty$ be independent non-negative r.v.'s with the same distribution function F for $n \geq 1$. The initial random variable X_0 may have a different distribution. Let U be the renewal function associated with $(X_n)_n$. Blackwell's renewal theorem states that $U(t + x) - U(t) \to x/m$ where m is the mean of X_1. Introduce the notation $X \stackrel{\mathcal{D}}{=} O(Y)$ iff $P(X > t) \leqslant cP(Y > t)$ for some finite constant c and for all t positive. If X is any r.v. with finite mean m we denote by $\hat{X}$ the random variable with distribution

$$P(\hat{X} \leqslant x) = m^{-1}\int_0^x P(X > t)\,dt.$$

Finally, a r.v. N is of geometric order if $P(N > k) \leqslant ca^k$ for some $a < 1$. Using coupling arguments, Ney proved the following result.

Theorem 4.4.1. ([18]). *If m is finite then there exist r.v.'s $Z_i \stackrel{\mathcal{D}}{=} O(\hat{X}_i)$ $i \geqslant 1$, and a r.v. N of geometric order, which are independent of each other and of X_0, and are such that, if F is a non-singular distribution, then*

$$| U(t + x) - U(t) - x/m | \leqslant cP\left(X_0 + \sum_{i=1}^N Z_i > t\right).$$

This result is very general, though the bound on the right is in general very difficult to calculate more explicitely. Hence

Corollary 4.4.2. *If $X_0 \underline{\mathcal{D}}\ O(\hat{X}_1)$ and the distribution function F_1 of $\hat{X}_1$ is subexponential, then*

$$| U(t+x) - U(t) - x/m| = O\left(\int_t^\infty (1-F(s))\,ds\right) \text{ as } t \to \infty.$$

More results of this kind are given in Embrechts and Omey [13] and Omey [19]. See also references in these papers.

5. Some related classes of distribution functions

5.1. *The classes $\mathcal{S}(\gamma)$.* There is no immediate need to assume that the limit in Definition 1.1. of $\mathcal{S}$ has to be 2. Indeed any limit greater than 2 is possible. This leads to the following definition.

Definition 5.1.1. ([4], [5]). A distribution function on $[0, \infty[$ belongs to $\mathcal{S}(\gamma)$ with $\gamma \geqslant 0$ iff

i) $\lim_{x\to\infty} (1-F^{(2)}(x)) \,/\, (1-F(x)) = c$, finite;

ii) $\lim_{x\to\infty} (1-F(x-y)) \,/\, (1-F(x)) = e^{\gamma y}$, for all y real;

iii) $f(-\gamma) = \int_0^\infty e^{\gamma y}\, dF(y)$ is finite.

If $F \in \mathcal{S}(\gamma)$, then necessarily $c = 2f(-\gamma)$, i.e. $\mathcal{S} = \mathcal{S}(0)$. In a recent paper Embrechts and Goldie [10] study various properties of $\mathcal{S}(\gamma)$. We shall not review them in detail but show via the ruin problem in the Poisson case how these classes enter very naturally. The Lundberg equation (4.2.1) can be re-written as $l_1(-\gamma) = \int_0^\infty e^{\gamma t}\, dL_1(t) = c\varkappa/\lambda > 1$, where L_1 is the integrated tail distribution. $\mathcal{S}$ essentially dealt with Cramér's theorem if the left abscissa of convergence of l_1 equals 0. The other case, where l_1 has a strictly negative l.a.c. $-\gamma$ but $l_1(-\gamma)$ is too small to yield a solution to the Lundberg equation, can be complemented by the classes $\mathcal{S}(\gamma)$. Indeed, Theorem 6.3 in [14] proves that if $L \in \mathcal{S}(\gamma)$, and $l_1(-\gamma) < c\varkappa/\lambda$ then $1 - R(x) \sim C(1 - L(x))$ as $x \to \infty$. Examples of $\mathcal{S}(\gamma)$ functions and further remarks on the ruin problem are given in [7].

5.2. *Discrete subexponentiality.* If $(p_n)_n$ is a discrete infinitely divisible probability measure on N, then there exists a probability measure $(q_n)_n$ and a positive constant k such that

$$P(z) = \sum_{n=0}^{\infty} p_n z^n = \exp\left[-k\left(1 - \sum_{j=1}^{\infty} q_j z^j\right)\right], \quad |z| \leqslant 1.$$

Theorem 4.1.1 can be used to compare the behaviour of $\sum_{k>[x]} p_k$ with that of $\sum_{j>[x]} q_j$. In this case one is however more interested in comparing p_n with q_n and indeed, most applications rely on this kind of results. In [5] the following class of probability measures is defined.

Definition 5.2.1. A probability measure $(p_n)_n$ belongs to $\mathcal{SD}(R)$ with $R \geqslant 1$ iff

i) $\lim_{n\to\infty} p_n^{(2)}/p_n = 2P(R)$ finite;

ii) $\lim_{n\to\infty} p_n/p_{n+1} = R^{-1}$.

This is the discrete analogue of subexponentiality. In Embrechts and Hawkes [12] the following theorem is proved

Theorem 5.2.2. *Suppose* $(p_n)_n$ *is infinitely divisible, as above. Then for* $R \geqslant 1$, *equivalent are*

i) $(p_n)_n \in \mathcal{SD}(R)$;

ii) $(q_n)_n \in \mathcal{SD}(R)$;

iii) $\lim_{n\to\infty} p_n/q_n = kP(R)$ *and* $q_n \sim Rq_{n+1}$ *as* $n \to \infty$.

This result has many applications, one of which is to fluctuation theory for random walks proving a Spitzer-Rogozin type of result when the first upgoing ladder variable $L^+ = \min\{k|S_k > 0\}$ has a finite first moment thus complementing the classical arc-sine law. (See [12], Theorem 3]).

6. Conclusion

I have tried to bring together some of the main properties and applications of the class $\mathcal{S}$ of subexponential distributions. By lack of space I had to leave out many more considerations, however the reader should have an up-to-date picture of $\mathcal{S}$ and its potential use in probability theory. Paragraph 5 contains just a first encounter with the various extensions of $\mathcal{S}$; the interested reader is referred to the literature for further reading on this subject.

REFERENCES

1. Athreya, K. B., Ney, P. E., *Branching Processes.* Berlin—Heidelberg—New York, Springer 1972.
2. Callaert, H., Cohen, J.W., *A lemma on regular variation of a transient renewal function.* Z. Wahrsch. Verw. Gebiete, 24, 275—278 (1972).
3. Chistyakov, V. P., *A theorem on sums of independent, positive random variables and its applications to branching processes.* Theory Probab. Appl., **9**, 640—648 (1964).
4. Chover, J., Ney, P., Wainger, S., *Degeneracy properties of subcritical branching processes.* Ann. Probab., 1, 663—673 (1973).
5. Chover, J., Ney, P., Wainger, S., *Functions of probability measures.* J. Analyse Math., **26**, 255—302 (1973).
6. Cohen, J. W., *Some results on regular variation for distributions in queueing and fluctuation theory.* J. Appl. Probab., 10, 343—353 (1973).
7. Embrechts, P., *A property of the generalized inverse Gaussian distribution, with some applications.* J. Appl. Probab., 20, 537—544 (1983).
8. Embrechts, P., Goldie, C.M., *On closure and factorization properties of subexponential and related distributions.* J. Austral. Math. Soc. Ser. A, **29**, 243—256 (1980).
9. Embrechts, P., Goldie, C. M., *Comparing the tail of an infinitely divisible distribution with integrals of its Lévy measure.* Ann. Probab., **9**, 468—481 (1981).
10. Embrechts, P., Goldie, C. M., *On convolution tails.* Stochastic Process. Appl., **13**, 263—278 (1982).
11. Embrechts, P., Goldie, C.M., Veraverbeke, N., *Subexponentiality and infinite divisibility* Z. Wahrsch. Verw. Gebiete, **49**, 335—347 (1979).
12. Embrechts, P., Hawkes, J., *A limit theorem for the tails of discrete infinitely divisible laws with applications to fluctuation theory.* J. Austral. Math. Soc. Ser. A, **32**, 412—422 (1982).

13. Embrechts, P., Omey, E., *Functions of power series.* Yokohama Math. J., 32 (1984).
14. Embrechts, P., Veraverbeke, N., *Estimates fcr the probability of ruin with special emphasis on the possibility of large claims.* Insurance: Math. and Econom., **1**, 55–72(1982).
15. Feller, W., *An introduction to Probability Theory and Its Applications. Vol. 2, 2nd Ed.* New York, Wiley, 1971.
16. Goldie, C. M., *Subexponential distributions and dominated-variation tails.* J. Appl. Probab., **15**, 440–442 (1978).
17. Goldstein, M. I., *Critical age-dependent branching processes: single and multitype.* Z. Wahrsch. Verw. Gebiete, **17**, 74–88 (1971).
18. Ney, P., *A refinement of the coupling method in renewal theory.* Stochastic Process. Appl., **11**, 11–26 (1981).
19. Omey, E., *Multivariate reguliere variatie en toepassingen in kanstheorie.* Ph. D. Thesis, Uni versity of Leuven, 1982.
20. Pakes, A. G., *On the tails of waiting-time distributions.* J. Appl. Probab. **12**, 555–564(1975).
21. Pitman, E. J.G., *Subexponential distribution functions.* J. Austral. Math. Soc. Ser. A, **29**, 337–347 (1980).
22. Seneta, E., *Regularly Varying Functions.* Lecture Notes in Math., Vol. **508**. Berlin, Springer, 1976.
23. Smith, W., *On the tails of queueing-time distributions.* Dept. of Statistics, Mimeo Ser., No 830. Univ. of North Carolina at Chapel Hill, 1972.
24. Stadtmüller, U., Trautner, R., *Tauberian theorems for Laplace transforms.* J. Reine Angew. Math., **311/312**, 283–290 (1979).
25. Teugels, J. L., *The subexponential class of probability distributions.* Theory Probab. Appl., **19**, 821–822 (1974).
26. Teugels, J. L., *The class of subexponential distributions.* Ann. Probab. **3**, 1000–1011 (1975).
27. Teugels, J.L., Veraverbeke, N., *Cramér type estimates for the probability of ruin.* Discussion Paper, No. 7316. Centre for Operations Research and Econometrics, Heverlee (Belgium), 1973.
28. Veraverbeke, N., *Asymptotic behaviour of Wiener-Hopf factors of a random walk.* Stochastic Process. Appl., **5**, 27–37 (1977).

CARL-GUSTAV ESSEEN

On the application of the theory of probability to two combinatorial problems involving permutations

1. Introduction

The Eulerian numbers $A(n, k)$ were introduced by Leonhard Euler in his *Institutiones Calculi Differentialis* (1755) as the coefficients in the power series development

$$\frac{1-\lambda}{1-\lambda e^{u(1-\lambda)}} = \sum_{n=0}^{\infty}\left(\sum_{k=1}^{n} A(n, k)\lambda^k\right)\frac{u^n}{n!}. \tag{1.1}$$

Only about 200 years later it was observed that these numbers have the following combinatorial property. Consider a permutation σ of $\{1, 2, \ldots, n\}$. By definition σ has a *fall* at j if $\sigma(j) > \sigma(j+1)$. By convention there is always a fall at n. It can be shown that the number of permutations of $\{1, 2, \ldots, n\}$ with exactly k falls, $1 \leqslant k \leqslant n$, is equal to $A(n, k)$. Thus $\sum_{k=1}^{n} \frac{A(n, k)}{n!} = 1$ and we can regard $\left\{\frac{A(n, k)}{n!}\right\}_1^n$ as a probability distribution. For a more detailed account of the Eulerian numbers and their properties, cf. Knuth (1973).

Consider again a permutation σ of $\{1, 2, \ldots, n\}$. By definition there is a *peak* at j, $2 \leqslant j \leqslant n-1$, if $\sigma(j) > \sigma(j+1)$, $\sigma(j) > \sigma(j-1)$. As to the boundaries 1 and n there are different possibilities. In accordance with Carlitz (1974) we say that there is a peak at 1 if $\sigma(1) > \sigma(2)$, a peak at n if $\sigma(n) > \sigma(n-1)$. With this convention we define the peak number $M(n, k)$ as the number of permutations of $\{1, 2, \ldots, n\}$ with exactly k peaks. It is not difficult to see that $M(n, 0) = 0$, $M(n, k) > 0$ for $1 \leqslant k \leqslant \left[\frac{n+1}{2}\right] = k_n$. Thus $\sum_{k=1}^{k_n} \frac{M(n, k)}{n!} = 1$ and once again we can regard $\left\{\frac{M(n, k)}{n!}\right\}_1^{k_n}$ as a probability distribution. In the following k_n is always defined as above.

Another possibility consists in considering only "inner" peaks, possibly situated at j, $2 \leqslant j \leqslant n-1$, there being no peaks at the boundaries 1 and n.

Thus, the permutation {6, 1, 5, 2, 3, 4} has 3 peaks according to the first definition but only 1 inner peak. By definition, let $\widetilde{M}(n, k)$ be the number of permutations of $\{1, 2, ..., n\}$ with exactly k inner peaks. It can be shown that

$$\widetilde{M}(n, k) = M(n, k + 1). \tag{1.2}$$

In David, Barton (1962) or David, Kendall, Barton (1966) the peak numbers are defined as $\widetilde{M}(n, k)$.

Quite a lot of "numbers" occurring in classical combinatorial theory turn out to be approximately normally distributed, for instance the binomial coefficients, the Stirling numbers of the second kind, the signless Stirling numbers of the first kind. It is known that the Eulerian numbers and the peak numbers have this property. In the following we shall give a simple proof of this, together with some asymptotic expressions. We also discuss a process of runs and, sketchily, a process of peaks in close connection with the numbers $A(n, k)$ and $M(n, k)$ respectively. The theory of the Eulerian numbers being well-known, we start by giving a summary of the properties of the peak numbers.

2. On the peak numbers $M(n, k)$

The numbers $M(n, k)$ satisfy the recurrence

$$M(n + 1, k) = (n - 2k + 3)M(n, k - 1) + 2kM(n, k), \tag{2.1}$$

cf. Carlitz (1974). By a minor modification of the proof of an inequality due to Kurtz (1972) it follows that the sequence $M(n, k)$ has the strong log concavity property, i.e.

$$M(n, k)^2 > M(n, k - 1)M(n, k + 1). \tag{2.2}$$

Let the probability generating polynomial of the peak number distribution be

$$Q_n(\lambda) = \sum_{k=1}^{k_n} \frac{M(n, k)}{n!} \lambda^k \tag{2.3}$$

and define the generating functions $K(\lambda, u)$ and $K_1(\lambda, u)$ by

$$K(\lambda, u) = \sum_{n=1}^{\infty} \left(\sum_{k=1}^{k_n} M(n, k)\lambda^{k-1} \right) \frac{u^{n-1}}{(n - 1)!} \tag{2.4}$$

$$K_1(\lambda, u) = \sum_{n=1}^{\infty} \left(\sum_{k=1}^{k_n} M(n, k)\lambda^{k-1} \right) \frac{u^n}{n!}. \tag{2.5}$$

From (2.1) it follows that $Q_n(\lambda)$ satisfies the recurrence

$$Q_{n+1}(\lambda) = \lambda Q_n(\lambda) + \frac{2\lambda(1 - \lambda)}{n + 1} \frac{dQ_n(\lambda)}{d\lambda} \tag{2.6}$$

and from (2.6) and (2.4) that $K(\lambda, u)$ satisfies the partial differential equation

$$2\lambda(1 - \lambda) \frac{\partial K}{\partial \lambda} + (\lambda u - 1) \frac{\partial K}{\partial u} + 2K = 0 \tag{2.7}$$

with the boundary condition $K(\lambda, 0) = 1$. Solving (2.7) we find that

$$K(\lambda, u) = \left(\cosh(u\sqrt{1-\lambda}) - \frac{\sinh(u\sqrt{1-\lambda})}{\sqrt{1-\lambda}}\right)^{-2} \tag{2.8}$$

an expression obtained by Carlitz (1974) in another way. Integrating (2.8) with respect to u we get

$$K_1(\lambda, u) = \left(1 - \frac{\tanh(u\sqrt{1-\lambda})}{\sqrt{1-\lambda}}\right)^{-1} - 1. \tag{2.9}$$

From the recurrence (2.6) it is not hard to prove by induction that the zeros of $Q_n(\lambda)$ are real, simple and non-negative.

Let the probability distribution $\left\{\frac{M(n, k)}{n!}\right\}$ have the mean value μ_n, the variance σ_n^2, the third central moment $\mu_{3,n}$ and let $\varkappa_{k,n}$ be the cumulant of order k. By repeated differentiation of (2.6) it follows that

$$\begin{cases} \mu_n = \varkappa_{1,n} = \frac{1}{3}(n+1) \text{ for } n \geqslant 2 \\ \sigma_n^2 = \varkappa_{2,n} = \frac{2}{45}(n+1) \text{ for } n \geqslant 4 \\ \mu_{3,n} = \varkappa_{3,n} = -\frac{2}{945}(n+1) \text{ for } n \geqslant 6. \end{cases} \tag{2.10}$$

A more general result can be proved by using another method. Let the function

$$\varphi(z) = \frac{\sqrt{1-e^z}}{\tanh^{-1}(\sqrt{1-e^z})}, \varphi(0) = 1, \tag{2.11}$$

where z is complex and $\tanh^{-1}(\cdot)$ is the suitably defined inverse of $\tanh(\cdot)$, be defined as regular analytic in the neighbourhood of $z = 0$. The singularities of $\varphi(z)$ are the branch points $2\pi ik$, $k = \pm 1, \pm 2, \ldots$, and there are no other finite singularities. By cutting the imaginary axis from $2\pi i$ to $+i\infty$ and from $-2\pi i$ to $-i\infty$, $\varphi(z)$ is made one-valued. Consider the power series development

$$\log \varphi(z) = \sum_{k=1}^{\infty} \frac{\varkappa_k}{k!} z^k$$

valid for small values of z. It can be shown that

$$\varkappa_{k,n} = (n+1)\varkappa_k, \quad k = 1, 2, \ldots; \; n \geqslant 2k. \tag{2.12}$$

By the calculus of residues applied to (2.5) and (2.9), the following explicit expression can be established. For every $t > 0$ and for $k = 1, 2, \ldots, k_n$

$$\frac{M(n, k)}{n!} = \frac{1}{2\pi i}\int_{-t-i\infty}^{-t+i\infty} (\varphi(z))^{n+1} e^{-kz} dz = \frac{1}{2\pi}\int_{-\infty}^{\infty} (\varphi(iy))^{n+1} e^{-iky} dy, \tag{2.13}$$

where $\varphi(iy) = \lim_{t \downarrow 0} \varphi(-t + iy)$. It is, however, disappointing that $\varphi(iy)$ is not a characteristic function of a probability distribution.

3. Convergence to normality. Asymptotic expressions

Different proofs of the convergence to normality of the Eulerian number distribution $\left\{\frac{A(n,k)}{n!}\right\}$ as $n \to \infty$ have been given by David, Barton (1962), Carlitz, Kurtz, Scoville, Stackelberg (1972), Tanny (1973). Perhaps the simplest way to prove a global central limit theorem is the following. Let $\{X_j\}_0^\infty$ be a sequence of i.i.d. r.v.'s with a continuous distribution. Set

$$I_j = \begin{cases} 1 \text{ if } X_j > X_{j+1} \\ 0 \text{ otherwise.} \end{cases}$$

It is easy to see that

$$\left\{\frac{A(n,k)}{n!}\right\} \stackrel{d}{=} 1 + \sum_{i=1}^{n-1} I_j = T_n.$$

The indicator variables I_j are 1-dependent and stationary. We get by a simple calculation

$$ET_n = \frac{n+1}{2}, \quad \text{Var } T_n = \frac{n+1}{12}.$$

From the well-known central limit theorem for a sum of m-dependent, stationary r.v.'s it follows that

$$\frac{T_n - \dfrac{n+1}{2}}{\sqrt{\dfrac{n+1}{12}}} \xrightarrow{d} N(0, 1), \; n \to \infty.$$

Thus, for large n, the Eulerian number distribution is approximately $N\left(\frac{n+1}{2}, \frac{n+1}{12}\right)$.

In an analogous way we treat the peak number distribution. Set

$$J_j = \begin{cases} 1 \text{ if } X_j > X_{j-1}, X_j > X_{j+1} \\ 0 \text{ otherwise.} \end{cases}$$

The indicator variables J_j are 2-dependent and stationary. Now

$$\left\{\frac{\widetilde{M}(n,k)}{n!}\right\} \stackrel{d}{=} \sum_{j=2}^{n-1} J_j$$

$$\left\{\frac{M(n,k)}{n!}\right\} \stackrel{d}{=} 1 + \sum_{j=2}^{n-1} J_j = U_n$$

and $EU_n = \frac{1}{3}(n+1)$, Var $U_n = \frac{2}{45}(n+1)$ in accordance with (2.10).

Thus $\left\{\frac{M(n,k)}{n!}\right\}$ is approximately distributed as $N\left(\frac{1}{3}(n+1), \frac{2}{45}(n+1)\right)$ for large n.

It is perhaps more interesting that also a local central limit theorem holds for these distributions so that asymptotic expressions for $A(n, k)$ and $M(n, k)$ can be found for large n. Concerning the Eulerian numbers, see Bender (1973), Carlitz et al. (1972), Tanny (1973), Siraždinov (1980). The peak numbers can be treated analogously. Combining the strong log concavity property (2.2) with the global central limit theorem and using only elementary inequalities (cf. Carlitz et al. (1972) for the corresponding Eulerian case) it is found that

$$\frac{M(n,k)}{n!} = \sqrt{\frac{45}{4\pi(n+1)}}\, e^{-x^2/2} + o\left(\frac{1}{\sqrt{n}}\right) \tag{3.1}$$

uniformly for all x where $k = \frac{1}{3}(n+1) + x\sqrt{\frac{2}{45}(n+1)}$.

To obtain additional terms in the asymptotic expression and a better error term we have to use more complicated methods. For the peak numbers one possibility is to start from the representation (2.13). Another way is the following. Let $f_n(t)$ be the characteristic function of $\left\{\frac{M(n,k)}{n!}\right\}$. Then

$$\frac{M(n,k)}{n!} = \frac{1}{2\pi}\int_{-\pi}^{\pi} f_n(t)\, e^{-ikt}\, dt.$$

It was observed in §2 that the zeros of the probability generating polynomial $Q_n(\lambda)$ are real, simple and non-negative. Thus, cf. Harper (1967), the peak number distribution is identical with the distribution of a sum of independent, 0,1-valued r.v.'s ξ_j and

$$f_n(t) = \prod_{j=1}^{k_n} \varphi_j(t), \tag{3.2}$$

where $\varphi_j(t)$ is the characteristic function of ξ_j. One can use (3.2) and (2.10) in the usual way to obtain an asymptotic development of $f_n(t)$ for $|t| \leqslant c$ where c is a suitably chosen constant. For $c \leqslant |t| \leqslant \pi$ we apply the following

Lemma 3.1. *If the r.v. X only takes integer values 0, 1, ..., N and if the zeros of the corresponding probability generating polynomial are real and non-negative, then the characteristic function $f(t)$ of X satisfies the inequalities*

$$|f(t)| \leqslant |f(c)| \leqslant |\exp\{-(1-\cos c)\operatorname{Var} X\} \quad \textit{for } 0 < c < \pi,\ c \leqslant |t| \leqslant \pi.$$

In this way the following improvement of (3.1) can be established.

Theorem 3.1. *Let* $k = \frac{1}{3}(n+1) + x\sqrt{\frac{2}{45}(n+1)}$. *Then*

$$\frac{M(n,k)}{n!} = \sqrt{\frac{45}{4\pi(n+1)}}\, e^{-x^2/2}\left(1 - \frac{1}{126}\sqrt{\frac{45}{2(n+1)}}\,(x^3-3x)\right) + o\left(\frac{1}{n\sqrt{n}}\right) \tag{3.3}$$

uniformly for all x.

Remark 1. It is of course possible to obtain still more terms in the asymptotic expression.

Remark 2. For the Eulerian numbers Siraždinov (1980) has shown that

$$(3.4)\quad \frac{A(n,k)}{n!} = \sqrt{\frac{6}{\pi(n+1)}}\, e^{-x^2/2}\left(1 - \frac{1}{20(n+1)}(x^4 - 6x^2 + 3)\right) + O\left(\frac{1}{n^{5/2}}\right)$$

where $k = \frac{n+1}{2} + x\sqrt{\frac{n+1}{12}}$.

The asymptotic expressions (3.3) and (3.4) disregarding the error terms, are good approximations only for central values of k, i.e. for k in the vicinity of the respective mean value. To obtain good approximations for a much larger range of k, special representations of the Eulerian numbers and of the peak numbers should be used.

From the generating function (1.1) of the Eulerian numbers it can be deduced that

$$(3.5)\qquad A(n,k) = \sum_{j=0}^{k}(-1)^j\binom{n+1}{j}(k-j)^n,\quad 1 \leqslant k \leqslant n.$$

Let $\{Y_j\}_1^\infty$ be a sequence of i.i.d. r.v.'s uniformly distributed on (0,1) and let $\sum_1^n Y_j$ have the distribution function $G_n(x)$ and the density $g_n(x)$. From (3.5) and the well-known explicit expression of $G_n(x)$ it follows, cf. Tanny (1973), that

$$\frac{A(n,k)}{n!} = G_n(k) - G_n(k-1).$$

For our purposes it is, however, simpler to use the identity

$$(3.6)\qquad \frac{A(n,k)}{n!} = g_{n+1}(k),\quad 1 \leqslant k \leqslant n.$$

Applying a local central limit theorem specialized to the case $\sum_1^{n+1} Y_j$, cf. Petrov (1975), it is easy to prove Siraždinov's expression (3.4), starting from (3.6).

Since Y_j has the moment generating function

$$(3.7)\qquad \varphi(t) = \frac{e^t - 1}{t},\quad -\infty < t < \infty,$$

and the characteristic function $\varphi(it)$ we have from (3.6) with $k = (n+1)a$

$$(3.8)\qquad \frac{A(n,k)}{n!} = \frac{1}{2\pi}\int_{-\infty}^{\infty}(\varphi(it))^{n+1}\, e^{-it(n+1)a} = \frac{1}{2\pi i}\int_{-i\infty}^{+i\infty}(\varphi(z)\, e^{-az})^{n+1}\, dz.$$

Because of symmetry, let $\frac{1}{2} \leqslant a < 1$ without loss of generality. We now apply the saddle point method to (3.8), changing the path of integration in a suitable way. Let

$$m(a) = \min_{0 \leqslant t < \infty} e^{-at}\varphi(t). \tag{3.9}$$

Minimum is assumed for a unique $t = t^*$ and $0 < m(a) < 1$ for $\frac{1}{2} < a < 1$. Further, set

$$\sigma^2(a) = \begin{cases} (t^*)^{-2} - (e^{t^*/2} - e^{-t^*/2})^{-2} \text{ for } \frac{1}{2} < a < 1 \\ 1/12 \text{ for } a = 1/2. \end{cases} \tag{3.10}$$

The following theorem can be proved.

Theorem 3.2. *Let* $k = (n+1)\,a$, $1/2 \leqslant a < 1$, *and let* $m(a)$ *and* $\sigma(a)$ *be defined by* (3.9) *and* (3.10) *respectively. Then*

$$\frac{A(n,k)}{n!} = \frac{(m(a))^{n+1}}{\sigma(a)\sqrt{2\pi(n+1)}}\left(1 + O\left(\frac{1}{n}\right)\right), \tag{3.11}$$

where the error term is uniformly of order $1/n$ *for* $\frac{1}{2}(n+1) \leqslant k \leqslant n$.

Remark 1. Additional terms can be determined in order to obtain still better approximations, for instance

$$\frac{A(n,k)}{n!} = \frac{(m(a))^{n+1}}{\sigma(a)\sqrt{2\pi(n+1)}}\left(1 + \frac{q_2(0)}{n+1} + O\left(\frac{1}{n^2}\right)\right),$$

where $q_2(0)$ is explicitly expressed by means of t^* and the error term is uniformly of order $1/n^2$ for $\frac{1}{2}(n+1) \leqslant k \leqslant n$.

Remark 2. The major part of the proof of Theorem 3.2 consists in showing that the error term is uniformly of order $1/n$. The asymptotic expression (3.11) has earlier been given by Bender (1973), using another method, in the form

$$\frac{A(n,k)}{n!} \sim \frac{(m(a))^{n+1}}{\sigma(a)\sqrt{2\pi(n+1)}}, \quad n \to \infty,$$

uniformly for $\frac{1}{2}(n+1) \leqslant k \leqslant n(1-\varepsilon)$ for every fixed value of $\varepsilon > 0$

Remark 3. The fact that $\varphi(it)$ is a characteristic function considerably facilitates the proof of Theorem 3.2.

For the peak numbers $M(n, k)$ it is also possible to obtain an expression analogous to (3.11), starting from the representation (2.13) and using the saddle point method. The corresponding error term can be shown to decrease uniformly to zero as $n \to \infty$ for $1 \leqslant k \leqslant n\left(\frac{1}{2} - \varepsilon\right)$ for every fixed $\varepsilon > 0$. Since now $\varphi(it)$ is no longer a characteristic function, the proof is more complicated and we refrain from going into details.

4. A process of runs

Let $\{X_j\}_0^\infty$ be a sequence of i.i.d. r.v.'s with a continuous distribution and let the indicator variables I_j be defined as in §3. Set

$$S_n = \sum_1^n I_j, \; N_k = \inf\{n : n \geqslant 1, \; S_n = k\}, \; k = 1, 2, \ldots,$$

$$L_1 = N_1, \; L_k = N_k - N_{k-1} \text{ for } k \geqslant 2.$$

The processes $\{N_k\}$ and $\{L_k\}$ have been studied by several authors, cf. for instance Knuth (1973) and Pittel (1980), (1981). They have a close connection with the Eulerian numbers. Thus EN_k and EL_k can be explicitly expressed by these numbers. Obviously $N_k + 1$ is the number of X-observations necessary to decide whether for the first time k runs up have been obtained and L_k is the length of the kth run up.

We can look at the $\{N_k\}$-process as a renewal process generated by the 1-dependent, stationary r.v.'s I_j. The following relation is easily verified

$$P(N_k \geqslant n) = P(S_{n-1} \leqslant k - 1). \tag{4.1}$$

From the central limit theorem for a sum of m-dependent, stationary r.v.'s it follows that

$$\frac{S_n - n\mu}{\sigma\sqrt{n}} \xrightarrow{d} N(0, 1), \; n \to \infty, \tag{4.2}$$

where $\mu = 1/2$, $\sigma^2 = 1/12$. Using solely elementary methods (cf. Feller (1968) for the classical case with i.i.d. r.v.'s generating the renewal process) we get from (4.1) and (4.2)

$$\frac{N_k - k/\mu}{\sqrt{k\sigma^2/\mu^3}} \xrightarrow{d} N(0,1), \text{ i.e. } \frac{N_k - 2k}{\sqrt{\frac{2}{3}k}} \xrightarrow{d} N(0, 1) \tag{4.3}$$

as $k \to \infty$. By (4.3) it is plausible that

$$EN_k = 2k + o(k), \quad \text{Var } N_k = \frac{2}{3}k + o(k).$$

In fact, it follows from general theorems due to Janson (1982) that

$$EN_k = 2k + O(1), \quad \text{Var } N_k = \frac{2}{3}k + O(k^{1/2}).$$

Still sharper results have been obtained by Knuth (1973), using, in this special case, the representations of EN_k and Var N_k by means of the Eulerian numbers.

It is not hard to get a multi-dimensional version of (4.3). By an immediate multi-dimensional generalization of (4.1) and using the 1-dependence property, one can prove the following result, still applying solely elementary methods

Let $0 < t_1 < t_2 < \ldots < t_r$. Then

$$(4.4)\qquad \left(\frac{N_{[kt_1]}-2kt_1}{\sqrt{\frac{2}{3}k}}, \ldots, \frac{N_{[kt_r]}-2kt_r}{\sqrt{\frac{2}{3}k}}\right) \xrightarrow{d} (W(t_1), \ldots, W(t_r))$$

as $k \to \infty$, where $W(t)$ is a standard Brownian motion. Pittel (1980) has earlier proved (4.4) by quite different methods.

It is also possible to prove a slightly more advanced result. If we apply a theorem of Vervaat (1972) combined with the well-known fact that the functional central limit theorem holds for a sum of m-dependent, stationary r.v.'s with finite variances, cf. Billingsley (1968), we find that the process

$$\frac{N_{[kt]}-2kt}{\sqrt{\frac{2}{3}k}}$$

converges as $k \to \infty$ to a standard Brownian motion in the spaces $D[0, 1]$ or $D[0, +\infty)$ with the usual topologies. Thus, the functional central limit theorem holds, (4.4) being a special case. Recently, Révész (1981) in an unpublished paper has considered the same problem, independently of the present author. Révész has shown that even the strong invariance principle is valid.

We now turn to the $\{L_k\}$-process. From a theorem of Kac (1947) and Ryll-Nardzewski (1961), also proved in a slightly more general setting in Breiman (1968), it follows that $\{L_k\}_1^\infty$ is a stationary sequence, conditionally on $I_0 = 1$, i.e. $X_0 > X_1$. The stationary distribution is easily found to be

$$(4.5)\qquad P(L_k \geqslant n) = 2\left(\frac{1}{n!} - \frac{1}{(n+1)!}\right), \quad n = 1, 2, \ldots$$

with mean value 2.

Let us return to the unconditioned process $\{L_k\}$. In Pittel (1981) this process has been thoroughly investigated and it is shown, among other things, that L_k converges in distribution exponentially fast to the stationary distribution (4.5) as $k \to \infty$. The simplest way to prove this result is, perhaps, to use the theory of Markov chains with a general state space, see for instance Doob (1953). In the following we shall suppose, without loss of generality, that the r.v.'s X_j are uniformly distributed on (0, 1).

We denote by Z_k the first X-variable in the kth run up. Then $\{Z_k\}_1^\infty$ is a Markov chain with state space (0, 1) and stationary transition probabilities. Let $p(u, x)$ be the density of Z_{k+1} conditional on $Z_k = u$. It can be shown that

$$p(u, x) = \begin{cases} e^{1-u} - e^{x-u} & \text{for } x > u \\ e^{1-u} & \text{for } x < u \end{cases}, \quad 0 \leqslant u, \; x \leqslant 1.$$

The function $p(u, x)$ satisfies the conditions in Case b, Doob (1953), p. 197. Let $f(x)$ be the density of X_1 conditional on $I_0 = 1$, i.e. $X_0 > X_1$. Then

$$f(x) = 2 - 2x, \quad 0 \leqslant x \leqslant 1.$$

By an easy calculation it is found that $f(x)$ satisfies the integral equation

$$f(x) = \int_0^1 f(u)\, p(u, x)\, du.$$

Thus $f(x)$ is the density of the unique stationary distribution of the Markov chain $\{Z_k\}_1^\infty$. Let $f_k(x)$ be the density of Z_k. According to Doob (1953) there exist positive constants C and ρ, $0 < \rho < 1$, such that

$$|f_k(x) - f(x)| \leqslant C\rho^k, \quad 0 \leqslant x \leqslant 1. \tag{4.6}$$

It is easily seen that

$$P(L_k \geqslant n) = \int_0^1 f_k(x) \frac{(1-x)^{n-1}}{(n-1)!} dx. \tag{4.7}$$

Let L have the stationary distribution (4.5). Obviously

$$P(L \geqslant n) = \int_0^1 f(x) \frac{(1-x)^{n-1}}{(n-1)!} dx. \tag{4.8}$$

From (4.6), (4.7) and (4.8) it follows that

$$| P(L_k \geqslant n) - P(L \geqslant n) | \leqslant C \frac{\rho^k}{n!}$$

so that $L_k \xrightarrow{d} L$ as $k \to \infty$, exponentially fast.

Several other properties of the L_k-distribution can be deduced by the same method.

Remark 1. The possibility of applying the Markov chain method to the process of runs is hinted at in Knuth (1973) p. 586 and the method is used to solve another problem concerning infinite permutations (Exercise 23, §5.1.3, pp. 47 and 589; see also Weissblum (1967)), however, without proof of convergence.

Remark 2. Starting instead with the indicator variables J_i introduced in § 3 we can define a process of peaks, the wavelengths between successive peaks now corresponding to the r.v.'s L_k. Results analogous to those obtained for the $\{N_k\}$-and $\{L_k\}$-processes can be proved in the same way. Thus, a functional central limit theorem is valid and the transition probability density for the wavelength-process satisfies the above mentioned conditions in Doob (1953).

REFERENCES

1. Bender, E.A., *Central and local limit theorems applied to asymptotic enumeration.* J. Combin. Theory **15**, 91– 111 (1973).
2. Billingsley, P., *Convergence of Probability Measures.* New York, Wiley, 1968.
3. Breiman, L., *Probability.* Reading, Mass., Addison-Wesley, 1968.
4. Carlitz, L., *Permutations and sequences.* Adv. in Math. **14**, 92– 120 (1974).
5. Carlitz, L., Kurtz, D. C., Scoville, R., Stackelberg, O.P., *Asymptotic properties of Eulerian numbers.* Z. Wahrsch. Verw. Gebiete **23**, 47– 54 (1972).
6. David, F.N., Barton, D.E., *Combinatorial Chance.* London, Griffin, 1962.
7. David, F.N., Kendall, M.G., Barton, D.E., *Symmetric Function and Allied Tables.* Cambridge, Cambridge Univ. Press, 1966.
8. Doob, J.L., *Stochastic Processes.* New York, Wiley, 1953.
9. Feller, W., *An Introduction to Probability Theory and Its Applications, Vol.* 1, *3rd Ed.* New York, Wiley, 1968.
10. Harper, L.H., *Stirling behaviour is asymptotically normal.* Ann. Math. Statist., **38**, 410– 414 (1967).

11. Janson, S., *Renewal theory for m-dependent variables.* Report No. 1, Department of Mathematics, Uppsala, Univ., Uppsala, March 1982.
12. Kac, M., *On the notion of recurrence in discrete stochastic processes.* Bull. Amer. Math. Soc., **53**, 1002– 1010 (1947).
13. Knuth, D.E., *The Art of Computer Programming, Vol.* 3: *Sorting and Searching.* Reading, Mass., Addison-Wesley, 1973.
14. Kurtz, D.C., *A note on concavity properties of triangular arrays of numbers.* J. Combin. Theory, **13**, 135– 139 (1972).
15. Petrov, V.V., *Sums of Independent Randcm Variables.* Berlin, Springer, 1975.
16. Pittel, B.G., *A process of runs and its convergence to the Brownian motion.* Stochastic Process. Appl., **10**, 33– 48 (1980).
17. Pittel, B.G., *Limiting behaviour of a process of runs.* Ann. Probab., **9**, 119– 129 (1981).
18. Révész, P., *Three problems on the lengths of increasing runs.* Unpublished paper (1981).
19. Ryll-Nardzewski, C., *Remarks on processes of calls.* In: *Proc. 4th Berkeley Sympos Math. Statist. Probab., Vol. II,* pp. 455– 465. Berkeley, Univ. California Press, 1961.
20. Siraždinov, S.H., *Asymptotic expression fcr Euler numbers.* Izv. Akad. Nauk UzSSR Ser. Fiz.-Mat. Nauk, *no.* 6, 39– 43 (1979). (Russian)
21. Tanny, S., *A probabilistic interpretation of Eulerian numbers.* Duke Math. J., **40**, 717– 722 (1973).
22. Vervaat, W., *Functional central limit theorems for processes with positive drift and their inverses.* Z. Wahrsch. Verw. Geb. **23**, 245– 253 (1972).
23. Weissblum, W., *Solution of problem* 60– 12, *a sorting problem.* SIAM Review, **9**, 121– 122 (1967).

I. V. EVSTIGNEEV, P. K. KATYSHEV

Stochastic models of economic dynamics and equilibrium: Asymptotic behaviour of equilibrium paths

In the present paper, a stochastic modification of V.M. Polterovich's model [11] is proposed. The main attention is devoted to the stability properties of equilibrium paths: certain types of stochastic turnpike theorems are proved.

1. We start with a description of the model. Let s_t $(t = 1, 2, \ldots)$ be a stochastic process taking values in measurable spaces S_t. (The values of s_t can be regarded as states of the environment or of economic conjuncture etc.). It is assumed that, at each time t, the "history" $s^t = (s_0, \ldots, s_t)$ of the process s_t is known.

Non-negative n-vectors are treated as bundles of goods available in the economic system.

The model describes the production sphere and the consumption sphere of an economy. The description of production is based on the concept of technological process. Technological processes are defined as pairs of non-negative measurable* functions

$$(x(s^{t-1}), y(s^t))$$
$$x\colon S^{t-1} \equiv S_0 \times S_1 \times \ldots \times S_{t-1} \to R^n, \; y\colon S^t \to R^n;$$

x is the input vector, y is the output vector. The input is made at time $t-1$ and depends on the history s^{t-1}, while the output is produced at the next moment t and depends on s^t.

For each t, a set Q_t (technological set) is distinguished in the class of all technological processes $(x(s^{t-1}), y(s^t))$; pairs $z(.) = (x(.), y(.)) \in Q_t$ are interpreted as processes technologically feasible at the time interval $(t-1, t)$. It is convenient to accept that $z \in L^{2n}_{\infty}(S^t)$ for any $z \in Q_t$ and t. By $L^m_{\infty}(S^t)$ (resp., $L^m_1(S^t)$) we denote the space of essentially bounded (resp., integrable) functions taking values in R^m.

A sequence $\xi = \{x_t, y_t\}_a^b$ (t ranges over an interval $[a, b]$ in the set of non-negative integers) is called a *production program* if $(x_{t-1}, y_t) \in Q_t$ $(t \geq 1)$, $y_a, x_b \geq 0$. We say that the vector function $y_0 = y_0(s^0)$ is the initial vector (i.v.) of ξ.

---

* In what follows all functions of s_t and s^t $(t > 0)$ are assumed to be measurable.

A sequence $\{p_t\}$ is said to be a *system of prices* if $p_t \in L_1^n(S^t)$, $p_t \geqslant 0$.

Let $\{p_t\}$ be a system of prices. A production program $\{x_t, y_t\}_0^T$ is said to be *optimal* (with respect to the system $\{p_t\}$) if

$$E(p_t y_t - p_{t-1} x_{t-1}) \geqslant E(p_t y'_t - p_{t-1} x'_{t-1}) \tag{1.1}$$

for any $(x'_{t-1}, y'_t) \in Q_t$, $t = 1, \ldots, T$. The notation px means the scalar product of vectors p and x; E is the symbol of mathematical expectation. The economic sense of the above definition is clear: the program ξ maximizes at each time t the mean value of the profit.

Now we turn to the consumption sphere. Assume that, for each $t \geqslant 0$, and each $p(.) \in L_1^n(S^t)$, there corresponds a set $C_t(p(.))$ consisting of non-negative functions $c(.) \in L_\infty^n(S^t)$. The multivalued mapping $p(.) \to C_t(p_t(.))$ is called the demand function.

Any sequence of non-negative functions $c_t \in L_\infty^n(S^t)$ will be called a *consumption program*. Given a system of prices $\{p_t\}$, a consumption program $\{c_t\}$ is said to be *optimal* if $c_t \in C_t(p_t)$ for each t.

A sequence $\Xi = \{p_t, c_t, x_t, y_t\}_0^T$ is called an *equilibrium trajectory* if: 1) $\{x_t, y_t\}$ is an optimal production program (with respect to the prices $\{p_t\}$); 2) $\{c_t\}$ is an optimal consumption program (with respect to $\{p_t\}$); 3) $c_t \leqslant y_t - x_t$; 4) $p_t(y_t - x_t - c_t) = 0$; 5) $p_T x_T = 0$, $x_T \geqslant 0$. (Unless otherwise stated, all equalities and inequalities between random variables hold almost surely (a.s.).) We call y_0 the initial vector (i.v.) of the trajectory Ξ. The vector function y_0 $(0 \leqslant y_0 \in L_\infty^n(S^0))$ will be fixed in the sequel.

Condition 3) means that the consumption at each time t does not exceed the difference between the output and the input. Conditions 4) and 5) express the fact that the cost of the resources which are used neither in production nor in consumption equals 0.

The main assumption. For each T there exists an equilibrium trajectory of length T with the i.v. y_0.

The central results of the paper (Theorems 2 and 3) state that any two equilibrium production programs of different length are close to each other (in a stochastic sense) most of the time.

Remarks. 1. It can be proved that, in the case when S_t are finite, the model can be reduced to the original deterministic model of V. M. Polterovich [11]. Thus, in the finite case, conditions that guarantee the existence of equilibrium can be easily derived from [11]. The general case is considered in another paper, which is in preparation.

2. It is easily seen (cf. [11], Sect. 8) that the above scheme generalizes various models in which the aim of consumption is described in terms of utility functions (optimization models). The demand functions corresponding to the optimization models have the form

$$C_t(p(.)) = \{c(.) : u_t(s^t, c(s^t)) - p(s^t)\, c(s^t) = \tag{1.2}$$
$$= \max_c\, [u_t(s^t, c) - p(s^t)c]\},$$

where $u_t(s^t, c)$ are utility functions ($t = 1, 2, \ldots$). The problem of existence of equilibrium (= competitive) trajectories in such models has been studied by various authors (see, e.g., [1—6, 13] and the bibliography cited in these papers).

3. It should be stressed that, unlike [1—6, 13], we do not impose any stationarity conditions on the model (particularly, the process $\{s_t\}$ may be non-stationary). This fact is essential for the way of formulating and proving the results.

2. This section contains a key technical result which will be used in the sequel. It will be proved that the equilibrium prices are uniformly bounded in L_1^n.

Assume that the following conditions are satisfied.

P.1. For each t, the set Q_t is convex and closed with respect to the convergence a.s.

P.2. $(0, 0) \in Q_t$, $t = 1, 2, \ldots$.

P.3. There exists an H such that, for each t, $z \in Q_t$, we have $|z| \leqslant H$.*

P.4. Let $z_t = (x_{t-1}, y_t) \in Q_t$, $z'_t = (x'_{t-1}, y'_t) \in Q_t$ and $\Gamma \in \mathcal{F}^{t-1}$ (the σ-algebra generated by s^{t-1}). Then $\chi_\Gamma z_t + (1 - \chi_\Gamma) z'_t \in Q_t$, where $\chi_\Gamma = 1$ on Γ and 0 on the complement of Γ.

P.5 (Slater condition). There exists a production program $\{\mathring{x}_t, \mathring{y}_t\}_0^\infty$ with $\mathring{y}_0 = y_0$, $\mathring{y}_t \geqslant \mathring{x}_t + \delta e$ where $\delta > 0$ does not depend on t and s^t, $e = (1, \ldots, 1)$.

C.1 (Cf. [11]). There is a number B such that

$$c(.) \in C_t(p(.)) \Rightarrow E\, pc \leqslant B.$$

Theorem 1. *There exists a constant K possessing the following properties. If $\Xi = \{p_t, c_t, x_t, y_t\}_0^T$ is an equilibrium path with i.v. y_0, then $E\,|p_t| \leqslant K$ (K is independent of T, t, Ξ).*

Proof. First we note that the optimality of a production program is equivalent to the validity of the inequality

$$E \sum_0^T p_t(y_t - x_t) - E \sum_0^T p_t(\tilde{y}_t - \tilde{x}_t) > 0 \tag{2.1}$$

for every production program $\{\tilde{x}_t, \tilde{y}_t\}$ with $\tilde{y}_0 = y_0$.

In order to prove the theorem, we shall use the method of "programs rebuilding" which was outlined in [7] and applied to proving the boundedness of prices (in the deterministic case) in [8]. This method generalizes Gale's construction of "good" programs suggested in [9].

We reproduce here (in a slightly modified form) the consideration in [7] (Lemmas 4 and 5). By virtue of P.5, we can find a $\gamma \in (0, 1)$ such that

$$\mu_t(v) \equiv \mathring{y}_t - \gamma \mathring{x}_t + (\gamma - 1)\, v \geqslant 0$$

for all v's with $|v| \leqslant H$ (H is described in P.1). Let $\eta = \{v_t, w_t\}_a^b$ be some production program ($0 \leqslant a < b$ — arbitrary integers). The program $\eta' = \{v'_t, w'_t\}_a^b$ where

$$v'_t = \gamma^{t-a+1} \mathring{x}_t + (1 - \gamma^{t-a+1})\, v_t,$$

$$w'_t = \gamma^{t-a} \mathring{y}_t + (1 - \gamma^{t-a})\, w_t$$

will be called the direct rebuilding of the program η.

---

* Remember that all inequalities between measurable functions hold a.s.

Because of P.5, one can choose a $\beta \in (0, 1)$ so that the inequality

$$\nu_t(v) \equiv \mathring{y}_t - \frac{1}{\beta}\mathring{x}_t + \left(\frac{1}{\beta} - 1\right) v \geqslant 0$$

will be satisfied for any v with $|v| \leqslant H$. If $\eta = \{v_t, w_t\}_0^b$ is a program with i.v. y_0, then we define $\eta'' = \{v_t'', w_t''\}_0^b$ where

$$v_t'' = \beta^{b-t}\mathring{x}_t + (1 - \beta^{b-t})\, v_t,$$

$$w_t'' = \beta^{b-t+1} y_t + (1 - \beta^{b-t+1})\, w_t$$

(the inverse rebuilding of η).

Let $\xi = \{x_t, y_t\}_0^T$ be an equilibrium program described in the theorem and $0 < \tau < T$. Consider the production programs

$$\{x_t'', y_t''\} = \{x_0, y_0, \ldots, x_{\tau-1}, y_{\tau-1}\}''$$

$$\{x_t', y_t'\} = \{x_{\tau+1}, y_{\tau+1}, \ldots, x_T, y_T\}',$$

and substitute in (2.1) $\tilde{\xi} = \{\tilde{x}_t, \tilde{y}_t\}$, where

$$\tilde{x}_t = \begin{cases} x_t'', & t < \tau \\ \mathring{x}_\tau, & t = \tau \\ x_t', & t > \tau \end{cases}; \quad \tilde{y}_t = \begin{cases} y_t'', & t < \tau \\ \mathring{y}_\tau, & t = \tau \\ y_t', & t > \tau. \end{cases}$$

The sequence $\tilde{\xi} = \{\tilde{x}_t, \tilde{y}_t\}_0^T$ is a program since $x_{\tau-1}'' = \mathring{x}_{\tau-1}$ and $y_{\tau+1}'' = \mathring{y}_{\tau+1}$. (In order to construct $\tilde{\xi}$, we have used the direct rebuilding of the final part of ξ and the inverse rebuilding of the initial part of ξ.)

By substituting $\tilde{\xi}$ in (2.1) and transforming the left-hand side of this inequality, we get

$$E\sum_0^{\tau-1} \beta^{\tau-t} p_t(y_t - x_t - \nu_t(x_t)) + \sum_{\tau+1}^{T} \gamma^{t-\tau-1} p_t \times \tag{2.2}$$

$$\times\, (y_t - x_t - \mu_t(x_t)) + Ep_\tau(y_\sigma - x_\tau) - Ep_\tau(\mathring{y}_\tau - \mathring{x}_\tau) \geqslant 0.$$

Since $Ep_t(y_t - x_t) = Ep_t c_t \leqslant B$, $\mathring{y}_\tau - \mathring{x}_\tau \geqslant \delta e$, $\mu_t \geqslant 0$ and $\nu_t \geqslant 0$, it follows from (2.2) that

$$Ep_\tau \delta e \leqslant \sum_0^{\tau} \beta^{\tau-t} B + \sum_{\tau+1}^{T} \gamma^{t-\tau+1} B \leqslant \text{const}$$

by virtue of C.1 and $\sum_1^{\infty} \beta^t, \sum_1^{\infty} \gamma^t < \infty$.

We have considered the case $0 < \tau < T$. If $\tau = 0$ or $\tau = T$, the argument becomes simpler (only direct or only inverse rebuilding should be used).

The proof is complete.

3. The aim of this section is to prove that any two equilibrium production programs are close to each other (in the sense of L_1) most of the time. This result can be regarded as a stochastic modification of Radner's [12] turnpike theorem in the framework of a non-stationary equilibrium model.

Stochastic turnpike theorems of various types were obtained for stationary optimization models in [1, 2, 4, 5, 7]. The form of the result of this section is close to the first statement of Theorem 1 in [4].

Suppose that the following conditions are fulfilled.

P.6. For each $\varepsilon > 0$, there exists $\gamma_1(\varepsilon) > 0$ such that, for every t, the relations

$$(x, y) \in Q_t, \ (x', y') \in Q_t, \tag{3.1}$$

$\Gamma \in \mathcal{F}^t$, $|y - y'| \geqslant \varepsilon$ (a.s. on Γ) ($\mathcal{F}^t$ is the σ-algebra generated by s^t) imply that there is a non-negative function v with

$$\left(\frac{x + x'}{2}, \ \frac{y + y'}{2} + v\right) \in Q_t, \tag{3.2}$$

$v \geqslant \gamma_1(\varepsilon)e$ (a.s. on Γ) (uniform strict convexity of technological sets with respect to output).

P.7. Given $\varepsilon > 0$, there exists $\gamma_2(\varepsilon) > 0$ (independent of t) such that the conditions (3.1), $\Gamma \in \mathcal{F}^{t-1}$, $|x - x'| \geqslant \varepsilon$ (a.s. on Γ) imply that

$$\left(\frac{x + x'}{2} - w, \ \frac{y + y'}{2}\right) \in Q_t, \tag{3.3}$$

for some non-negative function w with $w \geqslant \gamma_2(\varepsilon)e$ (a.s. on Γ) (uniform strict convexity of the technological sets with respect to input).

Conditions P.6′ and P.7′ formulated below are simple consequences of P.6 and P.7.

P.6′. For every $\varepsilon > 0$, there exists a number $\alpha_1(\varepsilon) > 0$ such that the conditions (3.1) and $E|y - y'| \geqslant \varepsilon$ imply that there is $v \geqslant 0$ with the properties (3.2) and $E \min_i v^i > \alpha_1(\varepsilon)$.

P.7′. For any $\varepsilon > 0$, there exists $\alpha_2(\varepsilon) > 0$ such that the conditions (3.1) and $E|x - x'| \geqslant \varepsilon$ imply that there exists $w \geqslant 0$ with properties (3.3) and $E \min_i w^i \geqslant \alpha_2(\varepsilon)$.

We need also the following assumptions concerning the demand functions.

C.2. There exists $\pi > 0$ such that if $c \in C_t(p)$ and $|p| < \pi$ then $|c| > H$, where H is described in P.3 (uniform insatiability of the demand).

C.3. If $c_p \in C_t(p)$, $c_q \in C_t(q)$ then

$$E(p - q)(c_p - c_q) \leqslant 0$$

(monotonicity of the correspondence $p \to C_t(p)$).

An important class of demand functions possessing the monotonicity property is given by formula (1.2).

Let $\{c_t, p_t, x_t, y_t\}_0^T$ and $\{c_t^*, p_t^*, x_t^*, y_t^*\}_0^{T^*}$ be two equilibrium trajectories, $T \leqslant T^*$, $z = (x, y) \in Q_{t+1}$. Define

$$\Delta_t(z_{t+1}^*, z) = Ep_{t+1}^*(y_{t+1}^* - y) - Ep_t^*(x_t^* - x). \tag{3.4}$$

We have $\Delta_t \geqslant 0$ since the program $\{x_t^*, y_t^*\}$ is optimal.

Lemma 1. (Cf. [11], Lemma 6.1). *For any* $0 \leqslant j \leqslant r \leqslant T - 1$

$$\sum_j^r \Delta_t(z_{t+1}^*, z_{t+1}) \leqslant E(p_{r+1}^* - p_{r+1})(y_{r+1}^* - y_{r+1}) - \tag{3.5}$$
$$- E(p_j^* - p_j)(y_j^* - y_j).$$

Proof. The argument is very similar to that used in Lemma 6.1, [11].

Since $\{x_t, y_t\}_0^T$ is optimal with respect to $\{p_t\}$,

$$0 \leqslant Ep_{t+1}(y_{t+1} - y^*_{t+1}) - Ep_t(x_t - x^*_t) \tag{3.6}$$

and, in virtue of C.3 and the definition of equilibrium,

$$0 \leqslant - E(p_t - p^*_t)(c_t - c^*_t) \leqslant - Ep_t(y_t - x_t) + \tag{3.7}$$
$$+ Ep_t(y^*_t - x^*_t) + Ep^*_t(y_t - x_t) - Ep^*_t(y^*_t - x^*_t).$$

By summing (3.4) (with $x = x_t$, $y = y_{t+1}$), (3.6), (3.7), we get

$$\Delta_t \leqslant E(p^*_{t+1} - p_{t+1})(y^*_{t+1} - y_{t+1}) - E(p^*_t - p_t)(y^*_t - y_t)$$

which obviously implies (3.5).

Lemma 2. *If* $\{p_t, c_t, x_t, y_t\}_0^T$ *is an equilibrium trajectory, then* $|p_t| \geqslant \pi$.

Proof. It follows from $0 \leqslant c_t \leqslant y_t - x_t$ that $|c_t| < H$ and thus the lemma is a simple consequence of C.2.

Lemma 3. (Cf. [11], Lemma 5.1). *For each* $\varepsilon > 0$, *there is* $\delta_1(\varepsilon) > 0$ *such that the relations*

$$\Delta_t(z^*_{t+1}, z) \leqslant \delta_1(\varepsilon), \quad z = (x, y) \in Q_{t+1}$$

imply $E|y^*_{t+1} - y| \leqslant \varepsilon$.

Proof. If $E|y^*_{t+1} - y| > \varepsilon$, then by virtue of P.6′, there exists $v \geqslant 0$ with $Ev \geqslant \alpha_1(\varepsilon)\, e$ and

$$\left(\frac{x + x^*_t}{2}, \frac{y + y^*_{t+1}}{2} + v\right) \in Q_{t+1}.$$

Because of the optimality of $\{p^*_t, x^*_t, y^*_t\}$, we have

$$Ep^*_{t+1}y^*_{t+1} - Ep^*_t x^*_t \geqslant Ep^*_{t+1}\left(\frac{y + y^*_{t+1}}{2} + v\right) - Ep^*_t \frac{x + x^*_t}{2}.$$

Therefore,

$$\Delta_t(z^*_{t+1}, z) \geqslant 2Ep^*_{t+1}v.$$

But

$$p^*_{t+1}v \geqslant |p^*_{t+1}| \min_{1 \leqslant i \leqslant n} v^i \geqslant \pi \min_{1 \leqslant i \leqslant n} v^i$$

by Lemma 2. Thus, $\Delta_t(z^*_{t+1}, z) \geqslant \pi\alpha_1(\varepsilon)$ and the lemma is proved.

Lemma 4 (Cf. [11], Lemma 6.3). *Whatever be* $\varepsilon > 0$, *there exists* $\delta_2(\varepsilon) > 0$ *with the following property. If* $z = (x, y) \in Q_{t+1}$ *and* $\Delta_t(z^*_{t+1}, z) \leqslant \delta_2(\varepsilon)$, *then* $E|x^*_t - x| \leqslant \varepsilon$.

The proof is similar to that of Lemma 3.

Given two equilibrium trajectories $\Xi = \{p_t, c_t, x_t, y_t\}_0^T$ and $\Xi^* = \{p^*_t, c^*_t, x^*_t, y^*_t\}_0^{T^*}$ with $T \leqslant T^*$, we denote by $t(\varepsilon)$ the quantity of moments t for which $E|z^*_t - z_t| \geqslant \varepsilon$ $(\varepsilon > 0)$.

Theorem 2. *For every* $\varepsilon > 0$, *there is a number* $L(\varepsilon)$ *(independent of* Ξ *and* Ξ^**) such that* $t(\varepsilon) \leqslant L(\varepsilon)$.

Proof. Let $\varkappa > 0$. Denote by $\tau(\varkappa)$ the quantity of t's such that $\Delta_t(z^*_{t+1}, z_{t+1}) \geqslant \varkappa$. It follows from Theorem 1, condition P.3 and formula (3.5) that, whatever be $0 \leqslant j \leqslant r \leqslant T - 1$,

$$\sum_j^r \Delta_t(z^*_{t+1}, z_{t+1}) \leqslant 8KH,$$

where K is the constant described in Theorem 1. Clearly, $\sum_0^{T-1} \Delta_t(z^*_{t+1}, z_{t+1}) \geqslant$ $\geqslant \varkappa\tau(\varkappa)$, and hence $\tau(\varkappa) \leqslant \frac{8KH}{\varkappa}$. Now observe that, by virtue of Lemmas 3, 4, the inequality $t(\varepsilon) \leqslant \tau(\varkappa)$ holds with $\varkappa = \min\left(\delta_1\left(\frac{\varepsilon}{2}\right), \delta_2\left(\frac{\varepsilon}{2}\right)\right)$. Consequently

$$t(\varepsilon) \leqslant L(\varepsilon) \equiv \frac{8KH}{\varkappa}, \quad \varkappa = \min\left(\delta_1\left(\frac{\varepsilon}{2}\right), \delta_2\left(\frac{\varepsilon}{2}\right)\right),$$

and the theorem is proved.

4. The main result of the paper (Theorem 3) is established in this section. It is proved that any two equilibrium production programs (of length T and T^*, $T \leqslant T^*$) with the initial vector y_0 are close in the sense of L_1 all the time except for a bounded number of moments at the end of the interval $[1, T]$ (Cf. [10], [7]).

From now on, we assume that

C.1'. There exist functions $\{\beta_t(s^t)\}_0^\infty$ which have the following properties: 1) if $p(s^t) \in L_1^n$, $p \geqslant 0$ and if $c(.) \in C_t(p(.))$, then $c(s_t)\, p(s^t) \leqslant \beta_t(s^t)$; 2) the sequence $\{\beta_t\}_0^\infty$ is uniformly integrable. (A stronger version of C.1.)

Let $\Xi = \{p_t, c_t, x_t, y_t\}_0^T$ be an equilibrium trajectory. Fix a moment τ and suppose that $y_\tau(s^\tau) \geqslant \varphi e$ for all $s^\tau \in \Delta$ where $\Delta \in \mathcal{F}^\tau$ and $\varphi > 0$.

Lemma 5. *There is a function $W(\varphi, \beta)$ (independent of Ξ and τ) possessing the following properties*

1) *whatever be $\varphi > 0$, $W(\varphi, \beta) \to 0$ as $\beta \to 0$.*

2) *$E \mid p_\tau \mid \chi_\Gamma \chi_\Delta \leqslant W(\varphi, \beta)$ for any $\Gamma \in \mathcal{F}^\tau$ with $P(\Gamma) \leqslant \beta$.*

Proof. Without loss of generality, we assume that $\mathring{x}_\tau \leqslant \frac{\varphi}{2} e$; this can be always obtained by multiplying $\{\mathring{x}_t, \mathring{y}\}_0^\infty$ by some $\lambda > 0$ small enough.

Consider the direct rebuilding $\{x'_t, y'_t\}_{\tau+1}^T$ of $\{x_t, y_t\}_{\tau+1}^T$ and define $\Gamma_1 = \Gamma \cap \Delta$,

$$\tilde{x}_t = \begin{cases} x_t, & t < \tau, \\ \chi_{\Gamma_1} \mathring{x}_\tau + (1 - \chi_{\Gamma_1})\, x_\tau, & t = \tau, \\ \chi_{\Gamma_1} x'_t + (1 - \chi_{\Gamma_1})\, x_t, & t > \tau; \end{cases}$$

$$\tilde{y}_t = \begin{cases} y_t, & t \leqslant \tau, \\ y'_t \chi_{\Gamma_1} + (1 - \chi_{\Gamma_1})\, y_t, & t > \tau. \end{cases}$$

The sequence $\{\tilde{x}_t, \tilde{y}_t\}_0^T$ is a production program by virtue of P.4 and the fact that $y'_{\tau+1} = \mathring{y}_{\tau+1}$. By substituting this program in (2.1), we have after some simple transformations (cf. (2.2))

$$(4.1) \qquad E\sum_{\tau+1}^{T} \gamma^{t-\tau-1}\chi_{\Gamma_1}p_t(y_t - x_t - \mu_t) +$$

$$E\chi_{\Gamma_1}p_\tau(y_\tau - x_\tau) - E\chi_{\Gamma_1}p_\tau(y_\tau - \mathring{x}_\tau) \geqslant 0.$$

Furthermore, by using the relations $p_t(y_t - x_t) = p_t c_t \leqslant \beta_t$, $\mu_t \geqslant 0$ and $\chi_{\Gamma_1}(y_\tau - \mathring{x}_\tau) \geqslant \chi_{\Gamma_1}\left(\varphi e - \frac{\varphi}{2} e\right) = \chi_{\Gamma_1} \frac{\varphi}{2} e$, we get from (4.1)

$$Ep_\tau\chi_{\Gamma_1}\frac{\varphi}{2} e \leqslant E\chi_{\Gamma_1}\sum_{\tau+1}^{\infty} \gamma^{t-\tau-1}\beta_t + E\chi_{\Gamma_1}\beta_\tau$$

and the existence of $W(\varphi, \beta)$ follows easily from the uniform integrability of $\{\beta_t\}$. The lemma is proved.

Suppose that the following condition is satisfied (cf. C.2).

C.4. If $c \in C_t(p)$ and $|c| \leqslant H$, then $p > 0$ (remember that we omit "a.s." everywhere).

Theorem 3. *Let $\Xi^* = \{p_t^*, c_t^*, x_t^*, y_t^*\}_0^{T^*}$ and $\Xi = \{p_t, c_t, x_t, y_t\}_0^T$ $(T \leqslant T^* < \infty)$ be equilibrium trajectories with the initial vectors $y_0^* = y_0$. Then, for each $\varepsilon > 0$, there exists a number $M(\varepsilon)$ (independent of Ξ and Ξ^*) such that the inequality*

$$E|z_t^* - z_t| \leqslant \varepsilon \quad [z_t = (x_{t-1}, y_t),\ z_t^* = (x_{t-1}^*, y_t^*)]$$

is valid for any $t \in [1, T - M(\varepsilon)]$.

Proof. Fix a $\delta > 0$. By Theorem 2,

$$(4.2) \qquad E\,|z_\tau^* - z_\tau| \leqslant \delta$$

for some $\tau \in [T - L(\delta), T]$. By using (4.5) with $j = 0$ and $r = \tau - 1$, we have

$$\sum_{0}^{\tau-1} \Delta_t(z_{t+1}^*, z_{t+1}) \leqslant E(p_\tau^* - p_\tau)(y_\tau^* - y_\tau) =$$

$$= Ep_\tau(y_\tau - y_\tau^*) + Ep_\tau^*(y_\tau^* - y_\tau).$$

Let us estimate the first term in the right-hand side (the second one can be estimated similarly). Consider the event

$$D_1 = \left\{\text{there exists } i \text{ such that } y_\tau^i \geqslant \varphi,\ c_\tau^i < \frac{\varphi}{2}\right\}$$

and its complement

$$D_2 = \left\{\text{for every } i,\ y_\tau^i < \varphi \text{ or } c^i \geqslant \frac{\varphi}{2}\right\}.$$

It follows from condition C.4 that $c_\tau = y_\tau - x_\tau$. Hence, if $s^\tau \in D_1$, then $x_\tau^i > \frac{\varphi}{2}$ for some i. By P.2 and P.7, this implies $x_\tau \geqslant \gamma_2\left(\frac{\varphi}{2}\right)e$, and thus $y_\tau \geqslant$

$\geqslant \gamma_2\left(\frac{\varphi}{2}\right) e$ because $c_\tau \geqslant 0$. Now Lemma 5 yields $E\,|\,p_\tau\,|\,\chi_\Gamma \chi_{D_1} \leqslant W\left(\gamma_2\left(\frac{\varphi}{2}\right), \beta\right)$ for each $\Gamma \in \mathcal{F}^\tau$ with $P(\Gamma) \leqslant \beta$. Therefore, for any $\gamma > 0$,

$$E p_\tau(y_\tau - y_\tau^*)\,\chi_{D_1} \leqslant E\,|\,p_\tau\,|\,|\,y_\tau - y_\tau^*\,|\,\chi_{D_1} =$$

$$= E\,|\,p_\tau\,|\,|y_\tau - y_\tau^*\,|\,\chi_{\{|y_\tau - y_\tau^*| \leqslant \gamma\}}\chi_{D_1} +$$

$$+ E\,|\,p_\tau\,|\,|\,y_\tau - y_\tau^*\,|\,\chi_{\{|y_\tau - y_\tau^*| > \gamma\}}\chi_{D_1} \leqslant$$

$$\leqslant \gamma E\,|\,p_\tau\,|\,\chi_{D_1} + 2HE\,|\,p_\tau\,|\,\chi_{\{|y_\tau - y_\tau^*| > \gamma\}}\chi_{D_1}.$$

Furthermore, by Theorem 1, $E|p_\tau| \leqslant K$, and, by Tchebyshev's inequality, $P\{\,|\,y_\tau - y_\tau^*\,| > \gamma\} \leqslant \frac{E|\,y_\tau - y_\tau^*\,|}{\gamma} \leqslant \frac{\delta}{\gamma}$ by virtue of (4.1). Thus,

$$E p_\tau(y_\tau - y_\tau^*)\,\chi_{D_1} \leqslant \gamma K + 2HW\left(\gamma_2\left(\frac{\varphi}{2}\right), \frac{\delta}{\gamma}\right). \tag{4.3}$$

Now let $s^\tau \in D_2$. We split the set $I = \{1, 2, \ldots, n\}$ into two parts: $I_1 = \{i : y_\tau^i < \varphi\}$ and $I_2 = I - I_1$. Observe that $c_\tau^i \geqslant \frac{\varphi}{2}$ for $i \in I_2$ by definition of D_2. Then

$$E p_\tau(y_\tau - y_\tau^*)\,\chi_{D_2} = E\chi_{D_2} \sum_{i \in I_1} p_\tau^i(y_\tau^i - y_\tau^{i*}) + E\chi_{D_2} \sum_{i \in I_2} p_\tau^i(y_\tau^i - y_\tau^{i*}) \leqslant$$

$$\leqslant \varphi E\,|\,p_\tau\,| + E\chi_{D_2} \frac{2\beta_\tau}{\varphi} \sum_{i \in I_2} (y_\tau^i - y_\tau^{i*}),$$

since $c_\tau^i \geqslant \frac{\varphi}{2}$ for $i \in I_2$ and $p_\tau^i c_\tau^i \leqslant p_\tau c_\tau \leqslant \beta_\tau$ which implies $p_\tau^i \leqslant \frac{2\beta_\tau}{\varphi}$ (remember that $y_\tau^i \geqslant 0$ and $c_\tau^i \geqslant 0$, $i = 1, 2, \ldots, n$). Therefore,

$$E p_\tau(y_\tau - y_\tau^*)\chi_{D_2} \leqslant \varphi K + \frac{2}{\varphi} E\beta_\tau\,|\,y_\tau - y_\tau^*|.$$

The uniform integrability of $\{\beta_t\}$ means that: 1) $K_1 \equiv \sup E\,\beta_t < \infty$; 2) there exists a function $W_1(\beta)$ such that $\lim\limits_{\beta \to 0} W_1(\beta) = 0$, and $E\beta_\tau \chi_\Gamma \leqslant W_1(\beta)$ provided $P(\Gamma) \leqslant \beta$. By using these properties of $\{\beta_t\}$ and Tchebyshev's inequality, we obtain

$$E p_\tau(y_\tau - y_\tau^*)\,\chi_{D_2} \leqslant \varphi K + \frac{2}{\varphi}\left[\gamma K_1 + 2HW_1\left(\frac{\delta}{\gamma}\right)\right] \tag{4.4}$$

for each $\gamma > 0$. Finally, it follows from (4.3) and (4.4) that

$$\sum_0^{\tau-1} \Delta_t(z_{t+1}^*, z_{t+1}) \leqslant 2\left[\varphi K + \gamma + \frac{2\gamma K_1}{\varphi} + \right. \tag{4.5}$$

$$\left. + 2HW\left(\gamma_2\left(\frac{\varphi}{2}\right), \frac{\delta}{\gamma}\right) + 2HW_1\left(\frac{\delta}{\gamma}\right)\right].$$

Now it is clear that, whatever be $\varepsilon > 0$, we can choose $\varphi > 0$, $\gamma > 0$ and $\delta > 0$ small enough so that the left-hand side of (4.5) will be less than $m \equiv$

$\equiv \min\left(\delta_1\left(\frac{\varepsilon}{2}\right), \delta_2\left(\frac{\varepsilon}{2}\right)\right)$, where δ_1 and δ_2 are described in Lemmas 3 and 4. In order to conclude the proof, we set $M(\varepsilon) = L(\delta) + 1$. Then

$$\sum_{0}^{T-M(\varepsilon)} \Delta_t \leqslant \sum_{0}^{\tau-1} \Delta_t \leqslant m$$

and thus $\Delta_t \leqslant m$ for every $t \in [1, T - M(\varepsilon)]$, which implies by Lemmas 3 and 4 that $E \mid z_t^* - z_t \mid \leqslant \varepsilon$ for $t \in [1, T - M(\varepsilon)]$.

The proof is complete.

REFERENCES

1. Brock, W.A., Mirman, L.J., *Optimal economic growth and uncertainty: The discounted case.* J. Econom. Theory, **4**, 479–513 (1972).
2. Brock, W.A., Mirman, L.J., *Optimal economic growth and uncertainty: The no discounting case.* Internat. Econom. Rev., **14**, 560–573 (1973).
3. Evstigneev, I.V., *Optimal economic planning taking account of stationary random factors.* Dokl. Akad. Nauk SSSR, **206**, 1040–1042 (1972). (Russian)
4. Evstigneev, I.V., *Asymptotic behaviour of optimal programs in stochastic models of economic development.* Internat. Conf. Probab. Theory Math. Statist., Vilnius: Abstracts of Communications, **I**, pp. 219–222. Vilnius, 1973.
5. Evstigneev, I.V., *Optimal stochastic programs and their stimulating prices.* In: Łós, J., Łós, M.W (Eds.), *Mathematical Models in Economics*, pp. 219–252. Amsterdam–New York–Warszawa, North-Holland, American Elsevier, PWN, 1974.
6. Evstigneev, I.V., *Models of economic dynamics taking account of indeterminacy in the operation of an industrial process.* Dokl. Akad. Nauk SSSR, **223**, 537–540 (1975). (Russian)
7. Evstigneev, I.V., *Turnpike theorems in stochastic models of economic dynamics.* Mat. Zametki, **19**, 279–290 (1976). (Russian)
8. Evstigneev, I.V., *Limits of price modification in developing economy models.* Ekonom. i Mat. Metody, **16**, 592–593 (1980). (Russian)
9. Gale, D., *On optimal development in a multi-sector economy.* Rev. Econom. Stud., **34**, 1–18 (1967).
10. Nikaido, H., *Persistence of optimal growth near the von Neumann ray: A strong version of the Radner turnpike theorem.* Econometrica, **32**, 151–163 (1964).
11. Polterovich, V.M., *Equilibrium trajectories of economic growth.* In: Efimov, B.A. (Ed.), *Methods of Functional Analysis in Mathematical Economics*, pp. 56–97. Moscow, Nauka, 1978. (Russian)
12. Radner, R., *Paths of economic growth that are optimal with regard only to final states: A turnpike theorem.* Rev. Econom. Stud., **28**, 98–104(1961).
13. Radner, R., *Optimal stationary consumption with stochastic production and resources.* J. Econom. Theory, **6**, 68–90 (1973).

K. FLEISCHMANN, A. LIEMANT, K. MATTHES

Critical branching processes with general phase space

1. Introduction

Each particle δ_x at position x in a complete separable metric space A generates a random cluster χ_x of particles situated in A. Different particles produce independent clusters. Intuitively, we get *branching processes* $(\Phi_t)_{t=0,1,\ldots}$ where the states are infinite particle systems.

This model generalizes spatially homogeneous branching processes (cf. for instance [10], Chapter 12) as well as substochastic translation processes (cf. e.g. [1], the references given there and [9, 11]).

Our interests are *convergence* theorems, *stability* behaviour, *structure* assertions for cluster-invariant systems and *mixing* properties in time and space.

The purpose of the paper is to review some main results. Strong formulations and proofs can be found in the references [2—4, 6—8].

2. Model assumptions

Let A be a complete separable metric space, the phase space of our considerations. A particle situated at position x in A is described by the Dirac measure δ_x. Let each particle δ_x generate a *cluster* χ_x of particles. Here χ_x is a stochastic locally finite system of particles, more precisely a random counting measure defined on A with some (probability) distribution $\varkappa_{(x)}$. The mapping $x \curvearrowright \varkappa_{(x)}$ is assumed to be weakly continuous. Looking only at the intensity measure $\Lambda_{\varkappa_{(x)}} = E\chi_x = : J(x, \cdot)$ we get a *kernel* J. We suppose that for all composition powers $J^{[t]}$ of J the measures

$$J^{[t]}(x, \cdot), \; x \in A, \; t \in Z_+ : = \{0, 1, \ldots\},$$

belong to the set N of all locally finite measures defined on A.

Special cases are *substochastic translations* where $\chi_x(A) \leqslant 1$ for all x in A, that is

$$\varkappa_{(x)} = \int J(x, \mathrm{d}y)\, \delta_{\delta_y} + (1 - J(x_1 A))\, \delta_0, \qquad x \in A.$$

Another particular cases are *critical spatially homogeneous clusters* in Euclidean space $R^d = A$. Here χ_x relatively to x coincides in distribution with the cluster χ_0 generated by a particle in the origin O of R^d. Moreover, $J(x, \cdot) = \lambda(\cdot - x)$ for some (probability) distribution λ defined on R^d. Then $J^{[t]}(O, \cdot) = \lambda^t$, the convolution power of λ.

Let φ belong to the set M of all counting measures defined on A. All particles in φ will be clustered independently according to $\varkappa$. If the superposition of all clusters is a.s. locally finite, then its distribution is denoted by $\varkappa_{(\varphi)}$. In this way we get a transition probability $\varkappa_{(\varphi)}(d\psi)$ from M into M and we can consider homogeneous Markov chains $(\Phi_t)_{t \in Z_+}$ with transition probabilities $\varkappa_{(\varphi)}(d\psi)$.

Let P be a (probability) distribution defined on M such that the clustered distribution

$$P_\varkappa := P(d\varphi)\, \varkappa_{(\varphi)}$$

exists. P is called cluster-invariant (with respect to the given $\varkappa$) if $P_\varkappa = P$. The intensity measure Λ_P of a cluster-invariant distribution P fulfils

$$J * \Lambda_P = \Lambda_P.$$

Let ν belong to the set N_J of all measures in N which are invariant with respect to J:

$$J * \nu = J.$$

In order to get a possible *initial distribution* of our branching process (Φ_t), that is the distribution of Φ_0, we introduce the distribution Π_ν of a Poisson point process with intensity measure ν.

Let Q be a distribution on N and

$$\mathfrak{J}(Q) := \int Q(d\gamma)\, \Pi_\gamma$$

the Cox distribution directed by Q. Such Q is called *J-invariant* if $J * \gamma$ belongs to N almost surely with respect to Q and has distribution Q again. If Q is J-invariant then, of course, its intensity measure Λ_Q is also invariant with respect to J. Consequently, if in addition Q is of first order then Λ_Q belongs to N_J. In this case $(\mathfrak{J}(Q))_\varkappa$ exists and $\mathfrak{J}(Q)$ is a possible initial distribution of (Φ_t), too.

3. Convergent branching processes

We start with the following simple convergence theorem.

Proposition 3.1. *Let Q be a J-invariant distribution on N of first order and $(\Phi_t)_{t \in Z_+}$ a branching process (with $\varkappa$) where Φ_0 is distributed according to the Cox distribution $\mathfrak{J}(Q)$. Then Φ_t converges in distribution to some limit population Φ_∞ with distribution $V_{(Q)}$, say, which is cluster-invariant and fulfils* $\mathbf{E}\Phi_\infty < \Lambda_Q$.

The proof is simply based on an elementary monotony property of the Laplace functional of Poisson distributions in connection with clustering.

In order to get the same limits but starting with more general initial states we have to impose some regularity condition. (For instance, for the degenerate cluster field $\varkappa_{(x)} = \delta_{\delta_{x+1}}$ in R each translation invariant distribution would be cluster-invariant.) We shall use the following very general concept.

Let N_J^{reg} denote the set of all measures ν in N_J satisfying

$$\int \nu(\mathrm{d}x)\, J^{[t]}(x, B) \min(1, J^{[t]}(x, B)) \xrightarrow[t\to\infty]{} 0$$

for all B in the ring $\mathfrak{B}$ of all bounded Borel subsets of A.

In the substochastic translation case this just means the well-known deconcentration property

$$\int \nu(\mathrm{d}x)\, (J^{[t]}(x, B))^2 \xrightarrow[t\to\infty]{} 0, \qquad B \in \mathfrak{B},$$

of the transition subprobability J. On the other hand, for a critical spatially homogeneous clustering in R^d all multiples of the Lebesgue measure belong to N_J^{reg} except for the degenerate case $\mathrm{E}\chi_0 = \lambda = \delta_x$ for some x in R^d.

Let L be a distribution on M such that

$$L(J^{[t]} * \gamma \in N) = 1, \qquad t \in Z_+.$$

These L are possible initial distributions for our branching process. If in addition $L(J^{[t]} * \gamma \in \cdot)$ converges weakly to some distribution Q on N then we set $Q = : \mathfrak{W}_L$ and say that $\mathfrak{W}_L$ exists, $\mathfrak{W}_L$ is J-invariant provided that $\Lambda_L \leqslant \nu$ for some ν in N_J.

For instance, in the spatially homogeneous case $\mathfrak{W}_L$ exists for all stationary distributions L with finite intensity.

Now we shall formulate a general convergence theorem.

Theorem 3.2. *Let L be a probability distribution on M such that $\mathfrak{W}_L$ exists and that the intensity measure Λ_L fulfils $\Lambda_L \leqslant \nu$ for some ν in N_J^{reg}. If then Φ_0 is distributed according to L then Φ_t converges in distribution to some limit Φ_∞ with distribution $V_{(\mathfrak{W}_L)}$ and* $\mathrm{E}\Phi_\infty < \nu$.

Remember, the cluster-invariant distributions $V_{(Q)}$ were introduced in Proposition 3.1.

For the proof one can first show that, for an approximation, the initial distribution L can be replaced by the Cox distribution $\mathfrak{J}(L)$. Using some backward technique then one can prove that in the first $t' \leqslant t$ steps in the branching process the clustering can be replaced by "transitions according to J". Therefore we can use $\mathfrak{J}(\mathfrak{W}_L)$ instead of $\mathfrak{J}(L)$ as initial distribution. Finally, Proposition 3.1. will be applied.

To end this section we add here

Lemma 3.3. *For each cluster-invariant distribution P of first order, $\mathfrak{W}_P$ exists and the intensity measure fulfils $\Lambda_{\mathfrak{W}P} = \Lambda_P$.*

4. Stability

Let ν belong to N_J. If Φ_0 has distribution Π_ν then by Proposition 3.1 the limit population Φ_∞ exists, has distribution $V_{(\delta_\nu)} = : V_{(\nu)}$ and fulfils $\mathrm{E}\Phi_\infty \leqslant \nu$. We call ν *stable* if even $\mathrm{E}\Phi_\infty = \nu$ holds. One of the main problems in spatial branching theory is to characterize stability. We write $N_\varkappa$ for the set of all stable ν in N_J and $V_{(\nu)} = :]\varkappa, \nu[$ if ν belongs to $N_\varkappa$.

Stability can be defined also in another way.

Proposifion 4.1. *A measure* ν *in* N_J *is stable if and only if there is a cluster-invariant distribution* P *(of first order) with intensity measure* $\Lambda_P = \nu$.

Let Φ_0 be distributed according to Π_ν with ν in $N_\varkappa$. Then the stability yields the uniform integrability of the sequence

$$E(\Phi_t(B) \mid \Phi_0) = J^{[t]} * \Phi_0(B), \qquad t \in Z_+,$$

with respect to Π_ν, for each B in $\mathfrak{B}$. Now

$$\int\limits_{\{J^{[t]}(x,B) > k\}} \nu(dx)\, J^{[t]}(x, B) \leqslant \int\limits_{J^{[t]} * \varphi(B) > k} \Pi_\nu(d\varphi)\, \varphi(dx)\, J^{[t]}(x, B)$$

so that we obtain the following *necessary first order stability criterion.*

Lemma 4.2. *If* ν *is stable then, for each* B *in* $\mathfrak{B}$, *the sequence* $(J^{[t]}(\cdot, B))_{t \in Z_+}$ *is uniformly integrable with respect to* ν.

Remember that already in the spatially homogeneous case stability cannot be expressed alone in first order terms, cf. [5].

If H is a measure defined on M then the second factorial moment measure $\Gamma_H^{(2)}$ of H is defined by

$$\Gamma_H^{(2)} = \int H \left(\sum_i \delta_{x_i}\right) \sum_{i \neq j} \delta_{[x_i, x_j]} \cdot$$

H is of second order if $\Gamma_H^{(2)}$ is locally finite. Very useful is the following *second order stability criterion.*

Theorem 4.3. *Let* ν *belong to* N_J. *Then* ν *is stable and* $]\varkappa, \nu[$ *is of second order if and only if the measure*

$$\int \nu(dx)\, \Gamma_{\varkappa(x)}^{(2)}(d[y, z]) \sum_{t=0}^{\infty} J^{[t]}(y, \cdot) \times J^{[t]}(z, \cdot)$$

defined on $A \times A$ *is locally finite.*

Let us refer here to the fact that the mentioned measure on $A \times A$ is just the second factorial moment measure of the canonical measure of the infinitely divisible distribution $]\varkappa, \nu[$.

In the spatially homogeneous case, if $\chi_O(A)$ has a positive (finite) variance the local finiteness in the last theorem is just the transience of the symmetrized distribution $\lambda(-\cdot) * \lambda$(if $\nu \neq 0$).

Stability criterions of higher orders can be found in [6, part II].

Stability has some consequences for the cluster-invariant distributions $V_{(Q)}$.

Proposition 4.4. *If in Theorem* 3.1 *the intensity measure* Λ_Q *is additionally stable then we have*

$$\mathfrak{W}^r{}_{(Q)} = \mathfrak{W}_{J(Q)}$$

and $E\Phi_\infty = \Lambda_Q$.

In fact, the second assertion follows from the first one together with Lemma 3.3 and the uniform integrability of the sequence

$$J^{[t]} * \Phi_O(B), \; t \in Z_+,$$

with respect to $\mathfrak{J}(Q)$ (for each B in $\mathfrak{J}$).

5. Structure of cluster-invariant systems

Our aim in this section is to characterize all cluster-invariant distributions with intensity measure ν in $N_{\varkappa}^{\text{reg}} := N_J^{\text{reg}} \cap N_{\varkappa}$. First we shall formulate the following lemma.

Lemma 5.1. *Let Q be a J-invariant distribution on N such that the intensity measure $\Lambda_Q =: \nu$ belongs to $N_{\varkappa}^{\text{reg}}$. Then for the cluster-invariant distributions $V_{(Q)}$ we have*

$$\mathfrak{W}_{V_{(Q)}} = Q.$$

If we now consider a distribution Q as in the preceding lemma, then by Proposition 4.4, we conclude that the intensity measure of $V_{(Q)}$ coincides with ν. On the other hand, if P is any cluster-invariant distribution with intensity measure ν in $N_{\varkappa}^{\text{reg}}$, then by Lemma 3.3 it can be used as an initial distribution in Theorem 3.2. Consequently, this convergence theorem yields $P = V_{(\mathfrak{W}_P)}$. Applying Lemma 5.1 we altogether get the following *structure theorem.*

Theorem 5.2. *Let ν be a measure in $N_{\varkappa}^{\text{reg}}$. Then $Q \curvearrowright V_{(Q)}$ provides a one-to-one mapping of the set of all J-invariant distributions Q on N with intensity measure ν onto the set of all cluster-invariant distributions P with intensity measure ν. The inverse mapping is just given by $P \curvearrowright \mathfrak{W}_P$.*

6. Mixing properties

Let us consider a *stationary* Markov process $(\Psi_t)_{t\in Z}$ with transition probabilities $\varkappa_{(\varphi)}(d\varphi)$ and absolute law $]\varkappa, \nu[$ for a stable ν. (Of course, Z is the set of all integers.) For s in Z the σ-field generated by $(\Psi_t)_{t\leqslant s}$ is denoted by $\mathfrak{F}_s$ and we set $\mathfrak{F}_{-\infty} = \bigcap_{s\in Z} \mathfrak{F}_s$. Our branching process (Ψ_t) is called *past-tail-trivial* if all events in $\mathfrak{F}_{-\infty}$ have probability zero or one.

Theorem 6.1. *The stationary branching process (Ψ_t) with absolute law $]\varkappa, \nu[$ is past-tail-trivial if and only if ν belongs to $N_{\varkappa}^{\text{reg}}$.*

Thus the regularity condition introduced in Section 3 has a nice probabilistic meaning.

From this mixing property in time we can derive the following mixing property in space.

Theorem 6.2. *For all ν in $N_{\varkappa}^{\text{reg}}$ the cluster-invariant distribution $]\varkappa, \nu[$ is tail-trivial (in space), provided that*

$$\varkappa_{(x)}(\chi(A) < \infty) = 1$$

for almost all x with respect to ν.

In the spatially homogeneous case the finiteness of clusters is trivially fulfilled. Thus we get here the tail-triviallity of *all* cluster-invariant distributions $]\varkappa, \nu[$. Combining this result with Theorem 6.1 we obtain that, for non-degenerate λ, the stationary spatially homogeneous branching process (Ψ_t) with absolute law $]\varkappa, \nu[$ is *space-time-mixing of all orders.*

REFERENCES

1. Fichtner, K.H., Kerstan, J., *Asymptotische Invarianz bei Halbgruppen stochastischer Kerne, I.* Serdica, **5**, 229–251 (1979).
2. Fleischmann, K., Matthes, K., *Kritische Verzweigungsprozesse mit allgemeinem Phasenraum, III.* Math. Nachr., **102**, 223–233 (1981).
3. Fleischmann, K., Liemant, A., Matthes, K., *Kritische Verzweigungsprozesse mit allgemeinem Phasenraum, VI.* Math. Nachr., **105**, 307–316 (1982).
4. Fleischmann, K., Hermann, K., Matthes, K., *Kritische Verzweigungsprozesse mit allgemeinem Phasenraum, VIII.* Math. Nachr., **109**, 239–247 (1982).
5. Kallenberg, O., *Stability of critical cluster fields.* Mach. Nachr., **77**, 7–43 (1977).
6. Liemant, A., *Kritische Verzweigungsprozesse mit allgemeinem Phasenraum I.* Math. Nachr., **96**, 119–125 (1980); *II.* Math. Nachr., **99**, 115–120 (1980); *IV.* Math. Nachr., **102**, 235–254 (1981); *VII., IX.* Math. Nachr. (to appear).
7. Liemant, A., *Structure and convergence theorems for critical branching processes with general phase space, I.* Math. Nachr. (to appear).
8. Liemant, A., Matthes, K., *Kritische Verzweigungsprozesse mit allgemeinem Phasenraum, V.*, Math., Nachr., **106**, 235–259 (1982).
9. Liggett, T.M., *Random invariant measures for Markov chains and independent particle systems.* Z. Wahrsch. Verw. Gebiete, **45**, 297–313 (1978).
10. Matthes, K., Kerstan, J., Mecke, J., *Infinitely Divisible Point Processes.* Chichester, Wiley, 1978.
11. Shiga, T., Takahashi, Y., *Ergodic properties of the equilibrium process associated with infinitely many Markovian particles.* Publ. Res. Inst. Math. Sci., **9**, 505–516 (1974).

A. GUT

A central limit theorem for certain stopped sums and some applications

In this paper some joint work, [5] and [6], is summarized. The first of these articles, [5], deals with the theory of chromatography in a probabilistic framework. This work generated [6], where a two-dimensional i.i.d. summation process is considered, with a central limit theorem for the second component evaluated at the first passage times of the first component as a main result. By using a special choice of summation processes, the results from [5] can be rediscovered and with other choices a variety of other applications can be covered.

1. Introduction

Let $\{X_n\}_{n=1}^{\infty}$ be a sequence of independent, identically distributed (i.i.d.) random variables and let $\{S_n\}_{n=1}^{\infty}$ denote their partial sums ($S_0 = 0$). In classical renewal theory one considers nonnegative summands and the main attention has been paid to "the number of renewals in $(0, t]$", i.e. $\max\{n: S_n \leq t\}$, $t \geq 0$. For an introduction to these matters, see e.g. Feller [2], Chapter XI and Prabhu [9], Chapter 5.

Most results have later been generalized to the case of general summands with positive expectation, but instead of studying the number of renewals it is more convenient to study "the first passage time beyond the barrier t", i.e.

$$N(t) = \min\{n: S_n > t\},\ t \geq 0,$$

see e.g. [2], Chapter XI, [9], Chapter 6 and [4].

Motivated by a problem in the theory of chromatography, see [5] and Section 2 below, we were led to another generalization of the above problems.

In [6] a two-dimensional summation process is considered and we investigate the asymptotic behaviour of one of the components stopped at the first passage times of the other one. More specifically, let $\{(W_n, Z_n)\}_{n=1}^{\infty}$ be a sequence of i.i.d. two-dimensional random variables, $U_n = \Sigma_{k=1}^{n} W_k$, $V_n = \Sigma_{k=1}^{n} Z_k$, $n = 1, 2, \ldots$ and set $\tau(t) = \min\{n; U_n > t\}$, $t \geq 0$.

The aim of [6] was to determine the asymptotics of $V_{\tau(t)}$ as $t \to \infty$ under suitable conditions and to present some applications of rather different kinds, which, however, all fit into the same theoretical framework. These matters are discussed in Sections 3 and 4.

In [7] these problems are discussed in a more general setting. For the cases when $W_n = X_n + Y_n$, $Z_n = X_n$ or $W_n = X_n$, $Z_n = X_n + Y_n$, where $\{X_n\}_{n=1}^{\infty}$ and $\{Y_n\}_{n=1}^{\infty}$ are independent sequences of i.i.d. positive random variables, see [11], [10], and [3] and for further references, see [6].

2. The chromatographic problem [5]

The basis for chromatographic separation is the distribution of a sample of molecules between two phases; one which is stationary and has a large surface and one which is mobile (liquid or gas) and percolates through the stationary bed. The sample is injected onto the column and is transported along the column by the moving phase. Part of the time the sample spends sorbed in (attached to) the stationary phase, i.e. there is a reversible process transferring the molecules between the phases. The separation of compounds is caused by their different times spent in the phases. The molecules spread during their travel along the column and the compounds appear as zones (or bands). This zone-spreading originates from multiple paths taken by the molecules during their passage through the column packing, molecular diffusion and mass transfer (reversible transfer of molecules) between the phases.

Chromatographic methods are of great importance in chemistry, biology, medicine, biotechnology etc., where they are employed for separation of mixtures of compounds either for analysis or for purification.

Let us consider one kind of molecules. It has been observed empirically that the concentration profile of the molecules when the peak is eluted (i.e. when the peak has travelled the length of the column) is Gaussian. Also, in order to measure the efficiency of the column one is interested in obtaining the *width* of the concentration profile. This width is defined as "4σ", where σ^2 equals the variance of the Gaussian profile mentioned above (and where 4 really should be $3.92 = 2 \cdot 1.96$).

The specific questions we were faced with were:

1. How far has a molecule travelled at time t?

2. How large is the width of the concentration profile when the peak is eluted?

Since one actually rather measures time than distance in the experiments, one can also wish to consider the question

2′. How large is the width of the *time*-profile when the peak is eluted?

We shall now indicate how the answers to these questions can be obtained. However, here we shall only consider the zone-spreading due to mass transfer. (For further details, see [5], Section 5.)

The basis for the stochastic model is that the chromatographic process is viewed as a series of sorptions and desorptions as the particle travels along the column, such that the probability of a particle changing phase during a short time interval is essentially proportional to the time interval, the proportionality constant being the transition rate constant, and that the motions of a molecule during different time intervals are independent. We are thus led to define a two-state Markov process (or a birth and death process with two states).

Let $\{X_n\}_{n=1}^{\infty}$ ($\{Y_n\}_{n=1}^{\infty}$) denote the successive times a molecule spends in the mobile (stationary) phase. It follows from the model assumptions that these times have exponential distributions (with mean λ^{-1} (and μ^{-1}), say). Further,

$$T_n = X_1 + Y_1 + X_2 + Y_2 + \dots + X_n + Y_n$$

denotes the time that has elapsed when a molecule has been n times in each phase and, by assuming that the longitudinal velocity in the mobile phase is constant, v, (for further details regarding this assumption, see [5]), it follows that

$$v \cdot S_n = v(X_1 + \dots + X_n)$$

denotes the longitudinal distance the molecule has travelled at time T_n.

We now define the first passage time processes

$$M(t) = \min\{n: \ T_n > t\}, \ t > 0$$

and

$$N(t) = \min\{n: \ vS_n > t\}, \ t > 0.$$

This means that $M(t)$ becomes the number of visits into each phase, that a molecule makes in order to pass beyond time t and that $S_{M(t)}$ equals, the time spent in the mobile phase when the molecule has been $M(t)$ times in each phase.

Let $\alpha(t)$ be the amount of time spent in the mobile phase during the time interval $(0, t]$. One can show that $S_{M(t)}$ and $\alpha(t)$ are "close" for large values of t, and so

$S_{M(t)}$ can be viewed as the time spent in the mobile phase during the time interval $(0, t]$.

Consequently, $v \cdot S_{M(t)}$ is the longitudinal distance a molecule has travelled at time t, i.e. this quantity is the relevant one for Question 1. By invoking Theorem 2 of [11] or [10] (cf. Theorem 1' of [6]) we obtain for example that

(I) $v \cdot S_{M(t)}$ is asymptotically normal with mean $\dfrac{v\lambda t}{\lambda + \mu}$ and variance $\dfrac{2\lambda\mu v^2 t}{(\lambda+\mu)^3}$ as $t \to \infty$.

Similarly, $N(L) = \min\{n: v \cdot S_n > L\}$ equals the number of visits a molecule makes into the mobile phase in order to travel the fixed distance, L, e.g. the length of the column. The quantity $T_{N(L)}$ can thus be viewed as the elution time and it follows from Wald's lemma and renewal theory that $ET_{N(L)} \approx \dfrac{\lambda+\mu}{\lambda v} L$ for large L, i.e. $\dfrac{\lambda+\mu}{\lambda v} L$ is the mean elution time for large L.

Thus $v \cdot S_{M(t)}$ with $t = \dfrac{\lambda+\mu}{\lambda v} L$ describes the position of a molecule when the peak is eluted, in particular, the width of the profile at that time point equals

$$\text{(II)} \qquad 4\sqrt{\operatorname{Var}\left(v \cdot S_{M\left(\frac{\lambda+\mu}{\lambda v}L\right)}\right)} \approx 4\sqrt{\frac{2v\mu}{(\lambda+\mu)^2}}\, L \text{ for large } L.$$

This answers Question 2. (Note also that the Gaussian shape has been verified.)

The same reasoning together with [3], pages 114—117 (cf. Theorem 2′ of [6]) yields that

(II′)
$$T_{N(L)} \text{ is asymptotically normal with mean } \frac{\lambda+\mu}{\lambda v} L \text{ and variance } \frac{2\mu L}{\lambda^2 v} \text{ as } L \to \infty.$$

In particular, this answers Question 2′.

3. A more general result [6]

First we observe that the results of the previous section do not depend on the exponential distribution. We require, in fact, only finiteness of the variances. A closer look at the proofs reveals that the theoretical results remain valid also if the "times" are not necessarily positive so long as $E(X_1+Y_1) > 0$ in (I) and (II) and $EX_1 > 0$ in (II′). However, the results can be further generalized.

If we consider the theoretical problems only we find that the common feature for the above questions is that one summation process is stopped and that another summation process is observed at the instant of this stopping.

In this section we present a theorem describing this more general situation and which, with special choices of the summation processes, yields the results of the previous section.

To this end, let $\{(W_n, Z_n)\}_{n=1}^{\infty}$ be i.i.d. two-dimensional random variables, such that $0 < m_w = EW_1 < \infty$, $0 < \sigma_w^2 = \text{Var } W_1 < \infty$ and $0 < \sigma_z^2 = \text{Var } Z_1 < \infty$. Nothing is assumed about $m_z = EZ_1$. Further, let $U_n = \Sigma_{k=1}^{n} W_k$, $V_n = \Sigma_{k=1}^{n} Z_k$ and let $\mathcal{F}_n = \sigma\{(W_k, Z_k);\, k \leqslant n\}$ for $n = 1, 2, \ldots$, with $\mathcal{F}_0 =$ the trivial σ-algebra. Throughout, all random variables with index 0 equal 0.

Define the stopping time (relative to $\{\mathcal{F}_n\}_{n=1}^{\infty}$)

$$\tau(t) = \min\{n\colon U_n > t\},\ t \geqslant 0.$$

Since m_w is assumed to be positive, it follows from the strong law of large numbers that $\tau(t) < \infty$ a.s.

Theorem. *With the above assumptions we have*

$$\mathcal{L}\left(\frac{V_{\tau(t)} - \dfrac{m_z}{m_w} t}{\sqrt{\dfrac{\gamma^2 t}{m_w^3}}}\right) \Rightarrow N(0, 1) \text{ as } t \to \infty,$$

where $\gamma^2 = \text{Var }(m_z W_1 - m_w Z_1) = m_z^2 \sigma_w^2 + m_w^2 \sigma_z^2 - 2 m_w m_z \text{ Cov } (W_1, Z_1)$ *is assumed to be positive.*

The proof of the theorem follows closely the ideas of [10], (cf. also [4], Theorem 2.5) with Anscombe's theorem as an important ingredient (cf. [10], Theorem 1).

By using the Cramér-Wold device the following corollary is obtained

Corollary 1. *If Z_n above is replaced by a vector-valued random variable $(Z_n^{(1)}, Z_n^{(2)}, \ldots, Z_n^{(d)})$, $n = 1, 2, \ldots$, the vector $(V_{\tau(t)}^{(1)}, \ldots, V_{\tau(t)}^{(d)})$ is asymptotically normal as $t \to \infty$.*

By choosing $(Z_n^{(1)}, Z_n^{(2)}) = (Z_n, 1)$ in Corollary 1 we obtain

Corollary 2. *With the assumptions of the Theorem, $(V_{\tau(t)}, \tau(t))$ is asymptotically normal as $t \to \infty$.*

We conclude this section by noting that, if $\{X_n\}_{n=1}^{\infty}$ and $\{Y_n\}_{n=1}^{\infty}$ are defined as in Section 2, (I) follows from the theorem with $\{(W_n, Z_n)\}_{n=1}^{\infty}$ defined by

$$(W_n, Z_n) = (X_n + Y_n, v \cdot X_n) \text{ for } n = 1, 2, \ldots$$

and (II′) follows if we choose

$$(W_n, Z_n) = (v \cdot X_n, X_n + Y_n) \text{ for } n = 1, 2, \ldots .$$

4. Some further applications

4.1 The first application deals with water motion in a river and is taken from [8].

The river is assumed to consist of two layers, such that the longitudinal velocity is constant in the upper layer and 0 in the lower layer. A particle moves back and forth between the two layers in a random way, the randomness being governed by Poisson (exponential) laws.

By interpreting the upper layer as a "mobile phase" and the lower layer as a "stationary phase" it is easy to see how quantities such as "the longitudinal distance a particle travels during the time interval $[0, t]$" and "the time it takes a particle to travel a prescribed distance" (e.g. the distance between two lakes) can be defined.

4.2 A cryptomachine contains a number of (electronic) wheels. Before the coding begins one rotates the wheels into a random startposition, which is a procedure one would like to automize. Thus, the first wheel is rotated with constant velocity during a random period of time (which is determined by a noise generator). When the wheel stops, the second wheel starts etc. When the last wheel stops, the first starts again etc. The procedure is terminated after a fixed time, t.

In order to determine "how randomly" the wheels have been placed one needs the expected value and the variance for the time a particular wheel has rotated during the time interval $[0, t]$.

It is by now clear how one can obtain the desired result. Moreover, let $Z_n^{(i)}$ be the duration of the nth rotation of the ith wheel, and let $W_n = Z_n^{(1)} + \ldots + Z_n^{(d)}$, where d is the number of wheels. Corollary 1 yields a jointly asymptotically normal distribution for the times the different wheels have rotated.

4.3 This example deals with "replacement based on age" which is discussed in Barlow and Proschan, [1], Chapter 3. An age replacement policy means that units are replaced at failure or after a fixed time, a, say, after its installation. The problem is to determine, for example, the number of replacements due to failure during the time period $[0, t]$. For some estimates of the *expected* number of replacements due to failure, see e.g. [1], Chapter 3.3.

By defining $\{X_n\}_{n=1}^{\infty}$ to be the life-times of the components and by putting

$$W_n = \min\{X_n, a\} \text{ and } Z_n = I\{X_n \leqslant a\},$$

$\tau(t)$ becomes the number of replacements in $[0, t]$, $V_{\tau(t)}$ becomes the number of replacements due to failure in $[0, t]$ and the Theorem describes the asymptotic behaviour of $V_{\tau(t)}$ as $t \to \infty$. If F is the distribution function of the life-times we obtain $EW_k = \int_0^a x\, dF(x) + a(1-F(a))$, $\operatorname{Var} W_k = \int_0^a x^2\, dF(x) + a^2(1-F(a)) - (EW_k)^2$, $EZ_k = F(a)$, $\operatorname{Var} Z_k = F(a)(1-F(a))$, $\operatorname{Cov}(W_k, Z_k) = (1-F(a))\left(\int_0^a x\, dF(x) - aF(a)\right)$ and $\gamma^2 = F(a)^2 \cdot \int_0^a x^2\, dF - F(a)\left(\int_0^a x\, dF\right)^2 + a^2F(a)(1-F(a))$, from which everything is obtainable.

If, in particular, $F(x) = 1 - e^{-x/\theta}$, $x > 0$, it follows that

$$\mathscr{L}\left(\frac{V_{\tau(t)} - \frac{t}{\theta}}{\sqrt{\frac{t}{\theta}}}\right) \Rightarrow N(0, 1) \text{ as } t \to \infty$$

which is independent of a. This is, of course, due to the fact that components with exponential life-times only should be replaced upon failure.

4.4 In [1], Chapter 4.2, replacement policies are discussed in terms of the *average* cost per unit time.

The Theorem above can, however, be used to study the asymptotics for the cost itself. Namely, suppose a cost c_1 is suffered for each replacement due to failure and a cost c_2 is suffered for each nonfailed component which is changed $(c_1 > c_2)$. In the notation of the previous example we find that $W_n = X_n \wedge a$ is the time interval between the $(n-1)$st and the nth replacement, $Z_n = c_1 I\{X_n \leq a\} + c_2 I\{X_n > a\}$ is the cost of the nth replacement and $V_{\tau(t)} = \Sigma_{k=1}^{\tau(t)} Z_k$ is the total cost for the replacements made in $[0, t]$.

4.5 Type I counter. Since the interregistration times are i.i.d. random variables, the (asymptotic) behaviour of the number of registrations in $(0, t]$ is given by results from renewal theory. Furthermore, if the dead times are i.i.d. random variables, asymptotics for the total dead time in $(0, t]$ can be obtained by the Theorem. Finally, if the interarrival times are exponentially distributed and independent of the dead times, explicit formulas can be obtained.

4.6 The $G/G/1$-queue. Let $\{X_n\}_{n=1}^{\infty}$ be the busy periods and $\{Y_n\}_{n=0}^{\infty}$ be the idle periods, where Y_n follows X_n. If the interarrival times are exponentially distributed, then Y_n is independent of X_n and results like those of Section 2 are obtainable. In the general case X_n and Y_n may be dependent, but the results remain true (with slightly different expressions for the variances).

4.7 Let $\{X_n\}_{n=1}^{\infty}$ be independent and $R[0, 1]$-distributed. How many X_n are smaller than $1/2$, say, when the sum of all of them reaches t?

Solution. Let $W_n = X_n$ and $Z_n = I\left\{X_n < \frac{1}{2}\right\}$. It follows from the Theorem that

$$\mathscr{L}\left(\frac{\sum_{k=1}^{\tau(t)} I\left\{X_k < \frac{1}{2}\right\} - t}{\sqrt{\frac{7t}{6}}}\right) \Rightarrow N(0, 1) \text{ as } t \to \infty.$$

By using the same arguments applied to the number of the X_n that are larger than $1/2$ one obtains

$$\mathscr{L}\left(\frac{\sum_{k=1}^{\tau(t)} I\left\{X_k > \frac{1}{2}\right\} - t}{\sqrt{\frac{t}{6}}}\right) \Rightarrow N(0, 1) \text{ as } t \to \infty.$$

Note that the normalizing variances above are not the same.

5. Some further results

5.1 The convergence of the moments of $\tau(t)$ under appropriate conditions follows from the theory of first passage times, see [4]. In [6] we prove that the moments of $V_{\tau(t)}$ converge under suitable conditions on the moments of W_1 and Z_1. For example, under the conditions of the theorem we have

$$\mathrm{Cov}(V_{\tau(t)}, \tau(t)) = \frac{m_z\sigma_w^2 - m_w \,\mathrm{Cov}\,(W_1, Z_1)}{m_w^3} t + o(t) \text{ as } t \to \infty.$$

5.2 It is e.g. also possible to prove strong laws of large numbers for $V_{\tau(t)}$ when the variances are not finite (see [6], Remark 7.2).

5.3 The theorem can also be extended to a result on weak convergence to Brownian motion (see [6], Remark 7.5). Furthermore, Corollary 2 can be extended to convergence to a two-dimensional Wiener process (see [6], Theorem 6).

REFERENCES

1. Barlow, R.E., Proschan, F., *Mathematical Theory of Reliability*. New York, Wiley, 1965.
2. Feller, W., *An Introduction to Probability Theory and Its Applications, Vol.* 2. New York, Wiley, 1966.
3. Gnedenko, B.V., Belyayev, Yu.K., Solovyev, A.D., *Mathematical Methods of Reliability Theory*. New York, Academic Press, 1969.
4. Gut, A., *On the moments and limit distributions of some first passage times*. Ann. Probab., 2, 277–308 (1974).

5. Gut, A., Ahlberg, P., *On the theory of chromatography based upon renewal theory and a central limit theorem for randomly indexed partial sums of random variables.* Chemica Scripta, **18**, 248–255 (1981).

6. Gut, A., Janson, S., *The limiting behaviour of certain stopped sums and some applications.* Tech. Report U.U.D.M. 1981: 14, Uppsala.

7. Högfeldt, P., *On the asymptotic behaviour of first passage time processes and certain stopped stochastic processes.* Second Vilnius Conference Probab. Theory Math. Statist.: Abstracts of Communications, **III** pp. 86–87. Vilnius, 1977.

8. Kaijser, T., *A stochastic model describing the water motion in a river.* Nordic Hydrology, **II**, 243–265 (1971).

9. Prabhu, N.U., *Stochastic Processes.* New York, Macmillan, 1965.

10. Rényi, A., *On the asymptotic distribution of the sum of a random number of independent random variables.* Acta Math. Acad. Sci. Hungar., **8**, 193–199 (1957).

11. Takács, L., *On certain sojourn time problems in the theory of stochastic processes.* Acta Math. Acad. Sci. Hungar., **8**, 169–191 (1957).

K. F. HINDERER

On the structure of solutions of stochastic dynamic programs

The present paper gives a survey on general principles for deriving structural properties of the value function and of optimal policies. The emphasis is on monotonicity, concavity and convexity. A number of results extend and unify the existing theory.

1. Introduction

It is well-known that only a few dynamic programs yields "explicit" solutions. Therefore one is interested in structure properties of the solution, in particular when these simplify numerical approaches. Such investigations have been done for special models from the beginning and it would be impossible to survey all the pertinent work. Therefore we concentrate on results for general models, apply these to a relatively general splitting model and fprovide a number of references. Section 2 contains the dynamic programming model (in the discrete time control setting), whereas the splitting model is introduced in Section 3. Then we turn to monotonicity of the value functions, concavity and convexity of the value functions and monotonicity of maximizers. The final section gives some references for additional structure problems.

2. The dynamic programming model

There are many models of differing complexity in use. We restrict ourselves to the discrete-time control model formulation as in Bertsekas [5]. Other models, e.g. non-stationary ones or Semi-Markovian models can be treated in a similar way.

The components of our *control model* (*CM* for short) are the following ones

(i) the *state space* S, the *action space* A and the *disturbance space* M, each endowed with some σ-algebra;

(ii) the non-empty sets $D(s) \subset A$ of actions which are admissible at state s; the so-called *restriction set*

$$D: = \{(s, a) \in S \times A : a \in D(s)\}$$

must be measurable (with respect to the product σ-algebra) and must contain the graph of a measurable map from S into A;
(iii) the *transition law* P, which is a transition probability from D into M. $P(s, \mathrm{a}, B)$ is the probability that the system, being in state s at time n and under the influence of the action a undergoes a disturbance from the set B;
(iv) the *transition function* T, a map from $D \times M$ into S; $T(s, \mathrm{a}, x)$ is the state at time $n+1$ if at time n state s and action a occured and disturbance $x \in M$ was selected by $P(s, \mathrm{a}, \cdot)$;
(v) the one-stage reward function r from D into $\mathbb{P}$; $r(s, \mathrm{a})$ is the reward earned at time n when the system is in state s and action a is taken;
(vi) the terminal reward function V_0 from S into $\mathbb{R}$; $V_0(s)$ is the reward earned at the end of the process when the final state is s;
(vii) a discount factor $\beta \in \mathbb{R}^+ := (0, \infty)$.

Let us point out that the essential components of a CM are restriction set D, transition law P, transition function T, reward function r and — in a lesser degree — the terminal reward function V_0. The properties of these components are crucial for the properties of the solution of the CM.

A *decision rule* is a measurable function from S into A that selects only admissible actions, i.e. for which $f(x) \in D(s)$ for all s. A sequence $\pi = (f_N, f_{N-1}, \ldots$ $\ldots, f_1)$ of decision rules is called an N-stage *policy*. The number N of decision epochs is called the horizon. Given any N-stage policy and any initial state s, there can be defined in a canonical way random variables ζ_n, $0 \leqslant n \leqslant N$, on some probability space $(\Omega, \mathfrak{I}, P_{\pi s})$ with values in S, such that the sequence $(\zeta_n)_0^N$ describes the Markovian (non-stationary) evolution of the states under policy π. We assume that $V_{N\pi}(s)$, the expectation of the total N-stage reward when the process starts in s and when π is used exists for all $n \in \mathbb{N}$ and for all π and s. Then there is also defined

$$V_N(s) = \sup_{\pi} V_{N\pi}(s),$$

the N-stage maximal expected total reward for initial state s. The so-called *value function* $V_N \colon S \to \mathbb{R}$ plays a basic role. A policy π^* is called *optimal* if $V_{N\pi^*}(s) = V_N(s)$ for all s.

It seems to be useful to separate questions of measurability, of existence of optimal policies and of the validity of the so-called value iteration from questions about the structure of the solution. Therefore we assume throughout: There holds the *value iteration*

$$V_n(s) = \sup\{W_n(s, \mathrm{a}) : \mathrm{a} \in D(s)\} \quad \text{for } s \in S,\ n \in \mathbb{N}, \tag{1}$$

where

$$W_n(s, \mathrm{a}) := r(s, \mathrm{a}) + \beta \int P(s, \mathrm{a}, \mathrm{d}x)\, V_{n-1}(T(s, \mathrm{a}, x)). \tag{2}$$

We call a decision rule f_n a *maximizer at stage n* (the stages being counted backwards) if f_n only selects "optimal actions", i.e. if $f_n(s)$ is a maximum point of $W_n(s, \cdot)$ for all s. It is well-known and easy to see that $\pi^* = (f_n)_1^N$ is optimal for the N-stage CM whenever f_n is a maximizer at stage n for $1 \leqslant n \leqslant N$

Now we can make precise what we understand by a *solution* of a CM, namely the sequence $(V_n)_1^\infty$ of value functions and — if it exists — a sequence of maximizers $(f_n)_1^\infty$. By a "structural result" we understand any special dependence of $V_n(s)$ and $f_n(s)$ on stage number n and/or initial state s.

There are important special cases of a CM

(a) If $P(s, \text{a}, \cdot)$ does not depend on (s, a), then we have a model with *independent disturbances.*

(b) If there is no disturbance (i.e. if $|M| = 1$), then we have a *deterministic* DP and P can be dropped.

(c) If $M = S$ and $T(s, \text{a}, x) = x$ for all (s, a, x), then we have an *ordinary* DP. Each CM can be reduced to such a DP, by replacing P by the so-called reduced transition law

$$Q(s, \text{a}, B) := P(s, \text{a}, \{x \in M: T(s, \text{a}, x) \in B\}) \tag{3}$$

3. A general model for the sequential splitting of a renewable resource

We introduce a model that generalizes a classical allocation model (cf. e.g. Bellman [3], p. 3) and which will be used to demonstrate the applicability of the general results given in later sections.

At time n there is available a resource of amount $s_n \in \mathbb{R}_+$ which must be split into two parts for the use in two different activities. Let $s_n - a_n$ and a_n denote that part of s_n that is used for the first and the second activity, respectively. We assume that there is an upper bound $B > 0$ of the resource. It follows that $S = [0, B] = A$ and $D(s) = [0, s]$. There are several possible interpretations which have been used in the literature:

(a) managing a water reservoir,

(b) harvesting of a renewable resource,

(c) pest control,

(d) splitting of capital,

(e) lost sales inventories (after transforming states and actions).

We use a transition function T of the special form

$$T(s, \text{a}, x) = t_1(s - \text{a}, x) + t_2(\text{a}, x) \tag{4}$$

for two measurable functions t_1 and t_2 from $S \times M$ into $\mathbb{R}_+$. In the water reservoir e.g. we would have

$$T(s, \text{a}, x) = \min(B, \text{a} + x) = t_2(\text{a}, x). \tag{5}$$

For the reward function we assume the same structure, except that we do not allow a dependence on the disturbance. Thus we assume

$$r(s, \text{a}) = u_1(s - \text{a}) + u_2(\text{a}) \tag{6}$$

for some measurable functions u_1 and u_2 from S into $\mathbb{R}$. It follows that the value iteration has the form

$$V_n(s) = \sup_{0 \leqslant \text{a} \leqslant s} W_n(s, \text{a}) \tag{7}$$

where

$$W_n(s, \text{a}) = u_1(s - \text{a}) + u_2(\text{a}) + \\ + \beta \int P(s, \text{a}, \text{d}x)\, V_{n-1}(t_1(s - \text{a}, x) + t_2(\text{a}, x)). \tag{8}$$

4. Monotone dependence of the value functions on the initial state

In the sequel a reflexive and transitive relation $\leqslant$ on a set Y will be called a (weak) *order*. It is a partial order if it is also antisymmetric. The relation $y < y'$ means $(y \leqslant y') \wedge (y \neq y')$. By $D(\cdot)$ we denote the correspondence $s \to D(s)$.

The following theorem is one of the simplest structure results.

Theorem 1. *Let the state space S be ordered and assume*
(i) *the correspondence $D(\cdot)$ is increasing, i.e. $D(s) \subset D(s')$ whenever $s \leqslant s'$,*
(ii) *the reduced transition law Q is stochastically increasing, i.e. for each* $a \in A$ *and each measurable increasing function v on S the function $s \to \int Q(s, a, dt)\, v(t)$ is increasing (on its domain),*
(iii) $r(\cdot, a)$ *is increasing for all* a,
(iv) V_0 *is increasing.*

Then $s \to V_n(s)$ is increasing for all n.

The *proof* is not difficult. It runs along the following scheme of six steps for verifying by induction on n that all value functions share a certain property (E)
(a) Assume that V_0 has property (E).
(b) Show that (E) is preserved under integration. This means that in a certain sense the transition law must have property (E).
(c) Show that (E) is invariant under multiplication with positive constants.
(d) Show that (E) is preserved under addition.
(e) Show (or assume!) that r has property (E).
(f) Show that (E) is preserved under $\sup_{a \in D(s)} (\ldots)$. This implies conditions on D.

Then, if V_{n-1} has property (E), it follows from (b) through (e) that also W_n has property (E), and finally (f) ensures that V_n has property (E).

Steps (c) and (d) in the scheme are often simple, but (b) and (f) may be difficult.

Remarks on Theorem 1. 1. Theorem 1 is more or less "folklore". For deterministic CM's it can be found in Beckmann [1], p. 91 and for more general models (including independent disturbances) in Sebastian/Sieber [18], p. 169.

2. By reversing the order on S one easily obtains sufficient conditions for V_n to be decreasing. One only has to replace "increasing" by "decreasing" in assumptions (i), (iii) and (iv).

3. Assumption (i) should often be easy to check. If e.g. $A = \mathbb{R}_+$ and $D(s) = [\underline{d}(s), \bar{d}(s)]$ for some finite functions $\underline{d}$ and $\bar{d}$, with $\underline{d}$ decreasing and $\bar{d}$ increasing, (which is fulfilled in the splitting model) then $D(\cdot)$ is increasing.

4. Useful sufficient conditions for (ii) are well-known for $S = \mathbb{R}$ and also for $S = \mathbb{R}^d$, cf. e.g. Stoyan [23].

5. Condition (ii) is fulfilled in the following special cases:
(a) $P(s, a, \cdot)$ is independent of s and $T(\cdot, a, x)$ is increasing.
(b) The model is an ordinary DP and P is stochastically increasing.

Applying Theorem 1 to the splitting model one easily sees (cf. Remark 5 above) that V_n is increasing for all n if the following holds
(a) $t_1(\cdot, x)$, u_1 and V_0 are increasing,
(b) $P(s, a, \cdot)$ does not depend on s.

In particular, in the reservoir problem with transition function (5) and independent inflows all value functions are increasing if u_1 (or u_2) and V_0 are increasing.

The power of Theorem 1 becomes apparent only when one applies it for various orders, such as the following ones (in round brackets we indicate by cues what it means that a function $g: S \to \mathbb{R}$ is increasing with respect to the corresponding order):

1. The *inverse* $\geqslant$ *of an order* $\leqslant$ (g is decreasing). This order allows to transform any statement about increasing functions into statements about decreasing functions. Thus many of the subsequent theorems have "decreasing versions".
2. The *diagonal order*: $s \leqslant_{\text{diag}} t$ if $s = t$ (g is arbitrary). This order serves for unifying theorems about concavity of value functions, cf. below.
3. The *trivial order*: $s \leqslant_{\text{tr}} t$ for all s, t (g is constant).
4. The *product order* $\underset{i \in I}{\times} (\leqslant_i)$ on $S = \underset{i \in I}{\times} S_i$ where $\leqslant_i$ is an order on S_i (g is componentwise increasing). This order (combined with the inverse order) is useful e.g. for a concise description of monotonicity properties of k-armed bandits, cf. e.g. Benzing/Hinderer/Kolonko [4].
5. The *direct-sum order* on the disjoint union $S = \sum_{i \in I} S_i$, where $\leqslant_i$ is an order on S_i (g is "piecewise" increasing). The special case $S = S_1 + S_2$, with $\geqslant_1$ on S_1 and $\leqslant_2$ on S_2 was extensively studied by Kalin [12], [13]. He calls a function g which is increasing with respect to this direct-sum order (i.e. which is decreasing on S_1 and increasing on S_2) *quasi-concave*. The special case $S = S_1 + \emptyset$ shows that this special order is equivalent to the general concept of an order.
6. The *order induced by a function* $h: S \to R: s \leqslant t$ if $h(s) \leqslant h(t)$ (g is an increasing function of h).
7. Combination of the product orders and direct-sum orders with the other orders gives a wealth of other examples.

Theorem 1 may be useful for numerical computations in the following ways: (a) If V_n is monotone then numerical solutions (necessarily for a finite set of states s) can be monotonely interpolated. (b) If V_{n-1} is monotone and if for some s the function $r(s, \cdot)$ is increasing and if $a \to Q(s, a, \cdot)$ is stochastically increasing then also $W_n(s, \cdot)$ is increasing. It follows that $W_n(s, \cdot)$ assumes its supremum at a largest element of $D(s)$ if such an element exists.

5. Concavity of the value functions

There is a well-known theorem about concavity of V_n in case that $T(\cdot, \cdot, x)$ is affine; cf. Bellman [3], p. 20 and Neumann [16], p. 61 for deterministic models and under continuity assumptions and Wittig [32] (cf. Sebastian/Sieber [18], p. 167) for a more general model; a similar result where D need not be convex is given by Sebastian/Sieber [18], p. 116. Affinity of $T(\cdot, \cdot, x)$ is not fulfilled in the reservoir problem. It is not difficult to prove a version where affinity of $T(\cdot, \cdot, x)$ is weakened to concavity in case isotonaicity of V_n is ensured (cp. the remark in Sebastian/Sieber [18], p. 167). Both cases are contained in the next theorem by choosing for $\leqslant_S$ the diagonal order or the usual product order, respectively. Other choices of $\leqslant_S$ yield more conditions for concavity of V_n.

Let h be a map from the convex set $Y \subset \mathbb{R}^d$ into a convex subset S in $\mathbb{R}^n$ and endow S with an order $\leqslant_S$. We call h *convex with respect to* $\leqslant_S$, if

$$h(\alpha y + (1-\alpha)\, y') \leqslant_S \alpha h(y) + (1-\alpha)\, h(y') \tag{9}$$

whenever $y, y' \in Y$ and $\alpha \in (0, 1)$. (Note that we do not assume an order on Y.)

In the sequel "concave" and "convex" means concave and convex in the usual meaning, i.e. with respect to the usual order on $\mathbb{R}$.

Theorem 2. *Consider the model with independent disturbances. Let D be a convex set in Euclidean space and let $\leqslant_S$ be an order on S. Assume*
(i) *$D(\cdot)$ is increasing and $D(s)$ is bounded for all s,*
(ii) *$T(\cdot, \mathrm{a}, x)$ is increasing and $T(\cdot, \cdot, x)$ is concave with respect to $\leqslant_S$,*
(iii) *$r(\cdot, \mathrm{a})$ is increasing and r is concave,*
(iv) *V_0 is increasing and concave.*
Then V_n and $W_n(s, \cdot)$ are finite and concave (and V_n is increasing).

The proof uses that an increasing and concave function $v: S \to \mathbb{R}$ of a function $h: D \to S$ which is convex with respect to $\leqslant_S$ is concave.

Remarks. 1. Theorem 2 may be useful for computations as follows: (a) If V_n is concave, then numerical solutions may be interpolated in a concave manner. (b) If $W_n(s, \cdot)$ is concave then special methods of maximization (e.g. Fibonacci search or golden section rule) may be used. 2. Inverting the order $\leqslant_S$ one obtains from Theorem 2 another set of sufficient conditions for concavity (not for convexity!) of W_n and $W_n(s, \cdot)$.

Applying Theorem 2 to the splitting model with independent disturbances and $\leqslant_S$ as the usual order on $\mathbb{R}$, one easily checks that V_n and $W_n(s, \cdot)$ are finite and concave if $t_1(\cdot, x)$, $t_2(\cdot, x)$, u_1, u_2 and V_0 are concave and if $t_1(\cdot, x)$, u_1 and V_0 are increasing. Using for $\leqslant_S$ the diagonal order on $\mathbb{R}$ one obtains a similar result with an affinity assumption.

6. Convexity of the value functions

This problem is not "symmetric" to the problem of concavity of the value functions! At first there is a simple result for the special case that $D(s)$ is independent of s, cp. the theorem in D. J. White [31].

Theorem 3. *Let S be a convex subset of a Euclidean space, endowed with an order $\leqslant_S$. Assume that the following holds*
(i) *$D(s) = A$ for all s,*
(ii) *$T(\cdot, \mathrm{a}, x)$ is increasing and convex with respect to $\leqslant_S$,*
(iii) *$P(s, \mathrm{a}, \cdot)$ does not depend on s,*
(iv) *$r(\cdot, \mathrm{a})$ and V_0 are increasing and convex,*
(v) V_n is *finite for all n.*
Then V_n is convex (and increasing) for all n.

The simple proof uses that the supremum of convex functions (if finite) is again convex. For this reasoning it is essential that $D(s)$ does not depend on s; cp. the correction made in D. J. White [31], p. 44 on a statement in Serfozo [19], where $D(s)$ was allowed to depend on s.

Remarks. 1. If V_n is convex, then numerical solutions may be interpolated in a convex manner. 2. For the maximization of $W_n(s, \cdot)$ convexity of that function would be useful, but Theorem 3 does not contain such a statement (even if A would be assumed convex) in contrast to the next theorem. 3. For applications the most important special cases are obtained when $\leqslant_S$ is either the usual order or the diagonal order (the latter implying affinity conditions). 4. If $S = \mathbb{R}$ then condition (iii) can be weakened, cf. the order $\overset{(3)}{\leqslant}$ in Stoyan [23].

Theorem 3 is not applicable to the splitting model because there $D(s)$ depends on s. But the next theorem applies; cf. Bellman [3], p. 19 for a special case.

Theorem 4. *Consider the control model with independent disturbances Let D be a convex set in a Euclidean space. Let $\leqslant_S$ and $\leqslant_A$ be orders on S and A, respectively, and use on D the product order $(\leqslant_S) \times (\leqslant_A)$. Assume*

(i) *$D(\cdot)$ is increasing,*
(ii) *$D(s)$ is the convex hull of the finite family of points $d_i(s) \in A$, $1 \leqslant i \leqslant n$,*
(iii) *the functions $d_i: S \to A$ are convex with respect to $\leqslant_A$,*
(iv) *$T(\cdot, \cdot, x)$ is increasing and convex with respect to $\leqslant_S$.*
(v) *r and V_0 are increasing and convex.*

Then V_n and $W_n(s, \cdot)$ are finite and convex, V_n is increasing and there exists a bang-bang maximizer f_n at stage n (i.e. $f_n(s)$ is an extreme point of $D(s)$ for all s).

Again the most important cases are when $\leqslant_S$ and/or $\leqslant_A$ are either the usual order or the diagonal order. Applying Theorem 4 to the splitting model with independent disturbances and with $\leqslant_S$ and $\leqslant_A$ as the usual orders, we see that the conclusions of Theorem 4 hold (in particular that there are maximizers f_n with $f_n(s) \in \{0, s\}$ for all s) if the following is fulfilled

(i) $t_1 \equiv 0$, $u_1 \equiv 0$,
(ii) $t_2(\cdot, x)$, u_2 and V_0 are increasing and convex.

7. Monotonicity of maximizers

In the last few years there has been intensive research about monotonicity of $s \to f_n(s)$ for appropriately chosen maximizers, provided these exist. If e.g. f_n is increasing then the maximization of $W_n(t, \cdot)$ for $t > s$ can be restricted to those values $a \in D(t)$ for which $a \geqslant f_n(s)$. One of the first papers on a concrete model where increasing maximizers exist is Gessford/Karlin [8]. The question was considered for a special splitting model by Beckmann [2]. A general approach (within the theory of super-modular functions on lattices) is given in the dissertation by Topkis from 1968, part of which is published as Topkis [25]. Our subsequent Theorem 5 is a slight extension of a result of Serfozo [19]. Other pertinent papers include Kalin [11]—[13], Serfozo [20], M. Sobel [22], C.C. White [27]—[29], D.J. White [31]. Numerical consequences for the infinite-stage model are investigated in Tijms [24] and D.J. White [30].

We use the following notation: Two elements a and b of A are *comparable* if either $a \leqslant b$ or $b \leqslant a$. A map f from S into A is called *weakly increasing* ("negatively isotone" in D.J. White [31]) if $s \leqslant t$ and comparability of $f(s)$ and $f(t)$ imply $f(s) \leqslant f(t)$. Let $D_n^*(s)$ denote the (possibly empty) set of maximum points of $W_n(s, \cdot)$. An element $a^* \in D_n^*(s)$ is called *weakly minimal* if $a^* \leqslant a$ for all $a \in D_n^*(s)$ which are comparable with a^*. If a^* is either minimal or smallest element in $D_n^*(s)$, then it is also weakly minimal, but the converse does not hold.

Theorem 5 (cf. Serfozo [19]). *Assume S and A are ordered and that for some n there holds*

(i) *D is "monotonely complete" in the sense that $s \leqslant s'$, $a \leqslant a'$ and $(s, a') \in D$, $(s', a) \in D$ imply that (s, a) and (s', a') belong to D.*

(ii) *W_n is finite and has increasing differences, i.e. if $s < s'$ then $a \to W_n(s', a) - W_n(s, a)$ is increasing on $D(s) \cap D(s')$.*

(iii) *There is a maximizer f_n at stage n such that $f_n(s)$ is weakly minimal in $D_n^*(s)$, $s \in S$.*

Then f_n is weakly increasing.

Remarks. 1. Serfozo assumes that $D(\cdot)$ is increasing and that each set $D(s)$ is decreasing. Then D is monotonely complete, but the converse does not hold. 2. For the application of Theorem 5 one needs criteria for function w on D to have increasing differences. One well-known sufficient condition (cf. Topkis [25] and Sobel [22] in the case $D = \mathbb{R}^2$) is that w is continuous and has a second partial derivative $\frac{\partial}{\partial s}\frac{\partial}{\partial a} w$ which is non-negative. Another criterion (whose special case $g_1(s) = g_2(s) = s$ is known) is given in Lemma 6 below. 3. Reversing the order on S one obtains the criterion given by Kalin [11] for f_n to be decreasing. Reversing the order on A [on S and A] yields another criteria for f_n to be decreasing [increasing].

Lemma 6. *Let $S = A$ be either $\mathbb{R}$ or $\mathbb{R}_+$, let v be a convex [concave] function on S anh let g_1 and g_2 be maps from S into S which are monotone in the same [in the opposite] sense. Then $(s, a) \to v(g_1(s) + g_2(a))$ has increasing differences.*

The proof uses the well-known fact that for a convex function v on an interval in $\mathbb{R}$ and for $h > 0$ the function $a \to v(a + h) - v(a)$ is increasing. (For convex functions in $\mathbb{R}^d$ with $d > 1$ this does not remain true, as can be seen from the example $v(a_1, a_2) := |a_1 - a_2|$. Therefore the proof for Theorem 5.2 in Serfozo [19] and for the Theorem in D.J. White [31], hold only for the one-dimensional case.)

Applications of Theorem 5. a. Under appropriate continuity assumptions there exists a smallest maximizer f_n at each stage of the splitting model with independent disturbances. Assume that $t_1 = 0$, that $t_2(\cdot, x)$, u_1, u_2 and V_0 are concave and u_1 and V_0 are increasing. Then f_n and also $s \to s - f_n(s)$ are increasing. As a consequence, the derivative f_n' of f, which exists a.e. according to Lebesgue satisfies $0 \leqslant f_n' \leqslant 1$. (A result of such a type was already obtained by Gessford/ Karlin [8].)

b. Serfozo [19] considers a controlled random walk with $D(s) \neq A$ and convex V_n. This model can be treated in the case $S = \mathbb{R}$ with the aid of Theorems 4 and 5.

c. Assume that A consists of the two actions $a = 0$ and $a = 1$. (Examples for this are replacement problems, cf. Derman [6], admittance problems in queues, cf. Lippman [14].) Then a smallest maximizer f_n always exists and it is increasing if $W_n(\cdot, 1) - W_n(\cdot, 0)$ is increasing. In the case $S \subset \mathbb{R}$ a policy consisting of increasing maximizers is called a *control-limit policy.*

d. Many applications to inventory models, including a survey on previous work, is given in Kalin [13].

8. Further structural problems

A. Value functions

A1) (Semi-)*continuity* of V_n is a well developped topic, cf. e.g. Chapter 2 of Dynkin/Yushkevich [7].

A2) Papers on *piecewise linear value functions* include Smallwood/Sondik [21], Wittig [32] (cf. Sebastian/Sieber [18]), and Müller/Vogel [15].

A3) At first sight, one will hardly consider *measurability* of V_n with respect to a σ-algebra $\mathcal{S}$ in S as a structural result. However, this seems to be justified only if $\mathcal{S}$ is a "large" σ-algebra (such as the system of Borel sets in $S = \mathbb{R}$). Otherwise, e.g. if $\mathcal{S}$ is the system of symmetric Borel sets in $S = \mathbb{R}$, one obtains indeed structural results. But it seems that these could be better described by means of sufficient statistics (cf. e.g. Hinderer [9]) using a function that generates $\mathcal{S}$.

A4) *Monotonicity* of $n \to V_n(s)$: It is well-known that $n \to V_n(s)$ is increasing for all n if $V_1 \geqslant V_0$ (e.g. if $r \geqslant 0 \equiv V_0$).

B. Maximizers

B1) *Monotonicity* of $n \to f_n(s)$: Assume that A is totally ordered and that f_n and f_{n+1} are smallest maximizers. Then $f_n \leqslant f_{n+1}$ if for some $d > 0$ the function $d\ W_n(s, \cdot) - W_{n+1}(s, \cdot)$ is decreasing for all s.

B2) In rare cases (e.g. in the inventory model treated in Ross [17], p. 175) there exist maximizers f_n independent of n, or even *myopic maximizers* (C.C. White [26]). More generally there may be maximizers f_n with $f_n = f_{N^*}$ for all $n \geqslant N^*$ (existence of *turnpike horizons*, cf. e.g. Hinderer/Hübner [10]).

REFERENCES

1. Beckmann, M.J., *Dynamic Programming of Economic Decisions.* Berlin—Heidelberg—New York, Springer, 1968.
2. Beckmann, M.J., *Resource allocation over time.* In: Łos, J., Łos, M.W. (Eds.), *Mathematical Models in Economics,* pp. 171—178. Amsterdam, North-Holland, 1974.
3. Bellman, R., *Dynamic Programming.* Princeton, Princeton Univ. Press, 1957.
4. Benzing, H., Hinderer K., Kolonko, M., *On the k-armed Bernoulli bandit: Monotonicity of the total reward under an arbitrary prior distribution.* To be published 1982.
5. Bertsekas, D.P., *Dynamic Programming and Stochastic Control.* New York, Academic Press, 1976.
6. Derman, C., *On optimal replacement rules when changes of state are Markovian.* In: Bellman, R. (Ed.), *Mathematical Optimization Techniques,* pp. 201—212. Berkeley and Los Angeles, 1963.
7. Dynkin, E.B., Jushkevich, A.A. *Controlled Markov Processes.* Berlin—Heidelberg—New York, Springer, 1979.
8. Gessford, J., Karlin S., *Optimal policy for hydroelectric operations.* In: Arrow, K.J., Karlin, S., Scarf, H. (Eds.), *Studies in the Mathematical Theory of Inventory and Production,* pp. 179—200. Stanford, Stanford University Press, 1958.
9. Hinderer, K., *Foundations of Non-Stationary Dynamic Programming with Discrete Time Parameter.* Lecture Notes in Operations Research and Mathematical Systems, Vol. 33. Berlin, Springer, 1970.
10. Hinderer, K., Hübner, G., *An improvement of J.F. Shapiro's turnpike-theorem for the horizon of finite stage, discrete dynamic programs.* In: *Trans. 7th Prague Conf. Information*

Theory, Statist. Decision Functions, Random Processes, Vol. A, pp. 245—255. Dordrecht, Reidel, 1977.

11. Kalin, D., *A note on "Monotone optimal policies for Markov decision processes"*. Math. Programming, **15**, 220—222 (1978).
12. Kalin, D., *Über Markoffsche Entscheidungsmodelle mit halbgeordnetem Zustandsraum.* OR-Verfahren, **33**, 233—245 (1979).
13. Kalin, D., *Beiträge zu strukturierten Markoffschen Entscheidungsmodellen.* Habilitationsschrift, Universität Bonn, 1981.
14. Lippman, S.A., *Applying a new device in the optimization of exponential queueing systems.* Oper. Res., **23**, 687—710(1975).
15. Müller, P.H., Vogel, S., *Zur Bestimmung optimaler Strategien für unvollständig beobachtbare stationäre Stopp-Steuer-Probleme.* Math. Operationsforsch. Statist. Ser. Optim., **12**, 91—106 (1981).
16. Neumann, K., *Operations Research Verfahren, Band II: Dynamische Optimierung, Simulation, Lagerhaltung, Warteschlangen.* München—Wien, Hanser, 1977.
17. Ross, S., *Applied Probability Models with Optimization Applications.* San Francisco, Holden-Day, 1970.
18. Sebastian, H.-J., Sieber, N., *Diskrete Dynamische Optimierung.* Leipzig, Akademische Verlagsgesellschaft Geest und Portig, 1981.
19. Serfozo, R.F., *Monotone optimal policies for Markov decision processes.* In: Wets, R.J.-B.(Ed.), *Stochastic Systems II: Optimization.* Mathematical Programming Studies, Vol. 6, pp. 202—215. Amsterdam—New York, North-Holland, 1976.
20. Serfozo, R., *Optimal control of random walks. birth and death processes, and queues.* Adv. in Appl. Probab., **13**, 61—83(1981).
21. Smallwood, R.D., Sondik, E.J, *The optimal control of partially observable Markov processes over a finite horizon.* Oper. Res., **21**, 1071—1088 (1973).
22. Sobel, M. J., *Reservoir management models.* Water Resources Research, **11**, 767—776(1975).
23. Stoyan, D., *Qualitative Eigenschaften und Abschätzungen stochastischer Modelle.* Berlin, Akademie-Verlag, 1977.
24. Tijms, H., *An algorithm for average costs denumerable state semi-Markov decision problems with applications to controlled production and queueing systems.* In: Hartley, R., Thomas, L. C., White, D. J. (Eds.), *Recent Developments in Markov Decision Processes*, pp. 143—179. London, Academic Press, 1980.
25. Topkis, D. M., *Minimizing a submodular function on a lattice.* Oper. Res., **26**, 305—321(1978).
26. White, C. C., *Optimal myopic strategies in combined decision and state estimation problems.* IEEE Trans. Systems Man Cybernet., SMC **8**, 702—705 (1978).
27. White, C.C., *Optimal control-limit strategies for a partially observed replacement problem.* Internat. J. Systems Sci., **10**, 321—331 (1979).
28. White, C. C., *The optimality of isotone strategies for Markov decision processes with utility criterion.* In: Hartley, R., Thomas, L. C., White, D. J., (Eds.), *Recent Developments in Markov Decision Processes*, pp. 261—275. London, Academic Press, 1980.
29. White, C. C., *Monotone control laws for noisy, countable-state Markov chains.* European J. Oper. Res., **5**, 124—132 (1978).
30. White, D. J., *Isotone optimal policies for structured Markov decision processes.* European J. Oper. Res., **7**, 392—402 (1981).
31. White, D. J., *Negatively isotone optimal policies for random walk type Markov decision processes.* OR-Spektrum, **4**, 41—45 (1982).
32. Wittig, W. S., *Numerische Lösungsmethoden in der parametrischen dynamischen Optimierung.* Dissertation, Technische Hochschule Leipzig, 1977.

STURE HOLM

Multiple test unbiasedness

1. Basic concepts

Suppose that a random variable X has a distribution in the class $\mathcal{P} = \{P_\theta: \theta \in \Omega\}$ indexed by a parameter θ ranging over a parameter set Ω. A test of a single hypothesis $H: \theta \in \Omega_H \subseteq \Omega$ against an alternative $K: \theta \in \Omega \setminus \Omega_H$ with test function $\Phi_H(x)$, $0 \leqslant \Phi_H(x) \leqslant 1$, has power function $\beta_H(\theta) = E_\theta[\Phi_H(X)]$ and size $\sup_{\theta \in \Omega_H} \beta_H(\theta)$ which is kept smaller than or equal to a predetermined (small) level of significance α_H.

In a multiple test we have a family of hypotheses to be tested simultaneously. This family can be finite, infinite countable or even infinite non-countable. In order to get satisfactory protection against errors of the first kind, rejecting true hypothesis, the multiple tests are required to have a small multiple level of significance according to the following definition.

Definition 1. A multiple test of the hypotheses in a family $\mathcal{H}$ is said to have a *multiple level of significance* α if the probability of rejecting *any* true hypothesis in $\mathcal{H}$ is at most α for all parameter points.

This definition imposes restrictions in all parameter points where at least one of the hypotheses in $\mathcal{H}$ is true. In a multiple test with a small multiple level of significance α all obtained rejections are *defensible*, since all rejections made in the test are correct with a probability of at least $1 - \alpha$. A rejection of a hypothesis $H: \theta \in \Omega_H$ is a statement $\theta \notin \Omega_H$ or $\theta \in \Omega \setminus \Omega_H$. When accepting this hypothesis a statement $\theta \in \Omega_H$ is *not* defensible and it should not be done. An acceptance of a hypothesis is rather a lack of statement concerning that hypothesis.

If a hypothesis $H_1: \theta \in \Omega_{H_1}$ is rejected and $H_2: \theta \in \Omega_{H_2}$ is another hypothesis satisfying $\Omega_{H_2} \subseteq \Omega_{H_1}$ then the hypothesis H_2 ought also to be rejected because the statement $\theta \in \Omega \setminus \Omega_{H_2}$ is just a *logical consequence* of the defensible statement $\theta \in \Omega \setminus \Omega_{H_1}$. A multiple test satisfying this requirement for all pairs of hypotheses where one of the hypothesis sets is a subset of the other is said to be *coherent*. See e.g. Gabriel (1969).

If two hypotheses $H_1: \theta \in \Omega_{H_1}$ and $H_2: \theta \in \Omega_{H_2}$ satisfy $\Omega_{H_1} \cup \Omega_{H_2} = \Omega$ they ought not to be rejected at the same time because this would imply the contradictory statements $\theta \notin \Omega_{H_1}$ and $\theta \notin \Omega_{H_2}$ with the logical consequence $\theta \notin \Omega$. A multiple test is called *non-contradictory* if it cannot at the same time reject a set of hypotheses whose hypotheses sets have a union equal to the whole parameter space.

It might be objected that a statement $\theta \notin \Omega$ is an indication of the model being false and should be allowed. But if the possibility of the model being incorrect is to be taken into account, the model ought to be considered to correspond to a subset Ω_0 of the whole parameter set Ω and the test of H_0: $\theta \in \Omega_0$ ought to be included in the multiple test procedure.

In a multiple test of a set $\mathcal{H}$ of hypotheses at a small multiple level of significance, all rejections obtained in the test are defensible and the logical consequences (conjunctions and disjunctions) of these rejections are defensible as well. This means that the hypotheses family $\mathcal{H}$ should be chosen in such a way that conjunctions and disjunctions of hypotheses in $\mathcal{H}$ and those rejections themselves include all interesting statements in the problem.

Suppose for example that the parameter is two-dimensional $\theta = (\theta_1, \theta_2)$ and that we want to make statements about the signs of θ_1 and θ_2 (which might have the meaning of positive or negative effects of two types). Considering hypotheses H_1: $\theta_1 \leqslant 0$, H_2: $\theta_1 \geqslant 0$, H_3: $\theta_2 \leqslant 0$, H_4: $\theta_2 \geqslant 0$ and alternatives K_1: $\theta_1 > 0$, K_2: $\theta_1 < 0$, K_3: $\theta_2 > 0$, K_4: $\theta_2 < 0$ the interesting statements are K_1, K_2, K_3, K_4, $K_1 \wedge K_3$, $K_1 \wedge K_4$, $K_2 \wedge K_3$ and $K_2 \wedge K_4$. The natural choice of $\mathcal{H}$ is $\mathcal{H}_1 = \{H_1, H_2, H_3, H_4\}$ but also for instance $\mathcal{H}_2 = \{H_1, H_2, H_3, H_4 \wedge H_1, H_4 \wedge H_2\}$ is possible. In the latter case the statement K_4: $\theta_2 < 0$ can be made only if both $H_4 \wedge H_1$ and $H_4 \wedge K_1$ are rejected and if for instance only $H_4 \wedge H_1$ is rejected the correct statement is "$\theta_2 < 0$ *or* $\theta_1 > 0$" which is not considered to be an interesting statement. It seems to be most natural to choose the family of hypotheses so that the interesting statements are included in the set of rejection statements and their conjunctions only.

Also the other basic multiple test concepts can be illustrated by this example. In the family $\mathcal{H}_1$ as well as in the family $\mathcal{H}_2$ the rejection of both H_1 and H_2 is self-contradictory and should not be allowed in a non-contradictory test. If the multiple test of the hypotheses in $\mathcal{H}_2$ is coherent it can happen that $H_4 \wedge H_1$ is rejected and H_1 is accepted but not that H_1 is rejected and $H_4 \wedge H_1$ is accepted.

A test of a single hypothesis H: $\theta \in \Omega_H$ against the alternative $K: \theta \in \Omega_K = \Omega \setminus \Omega_H$ with a test function $\Phi_H(x)$ is said to be unbiased if

$$\inf_{\theta \in \Omega_K} E_\theta[(\Phi_H X)] \geqslant \sup_{\theta \in \Omega_H} E_\theta[(\Phi_K X)].$$

If each test in a multiple test satisfies this condition we say that the tests are *individually unbiased*. This is a restriction of different parameter points for the same test.

Another reasonable unbiasedness concept for multiple test, which imposes instead restrictions on the different tests in the same point, is the following.

Definition 2. A multiple test is said to be *rejection unbiased* if for all $\theta \in \Omega$ each hypothesis being false in the point θ has a rejection probability which is at least as great as the rejection probability for each hypothesis being true in that point.

In this paper we will investigate the consequences of these unbiasedness conditions together with the basic multiple test concepts in simple examples, in particular examples related to twosidedness of tests and confidence intervals.

2. Some one-dimensional problems

When testing the simple hypothesis H_0: $\theta = \theta_0$ against the two-sided alternative K_0: $\theta \neq \theta_0$ for a one-dimensional parameter θ, the correct statement in case of rejection is $\theta \neq \theta_0$. But in most statistical applications one rather wants to state either $\theta > \theta_0$ or $\theta < \theta_0$ in case of rejecting $\theta = \theta_0$. This means that one ought to consider the multiple test problem with the two hypotheses H_1: $\theta_1 \leqslant \theta_0$ and H_2: $\theta \geqslant \theta_0$ and their alternatives K_1: $\theta > \theta_0$ and K_2: $\theta < \theta_0$.

Our first very simple result gives a uniformly most powerfulness (UMP) property in this situation. The UMP concept is easily generalized from the single test case.

Definition 3. A multiple test is said to be *uniformly most powerful* in a class of multiple tests if for each tested hypothesis the rejection probability with that test is at least as great as with any other test in the class in all points where the hypothesis is false.

Theorem 1. *Let* $\theta \in \Omega$ *be a one-dimensional parameter and suppose that the random variable* X *has a probability density*

$$p_\theta(x) = C(\theta) \exp\,(Q(\theta)\, T(x))$$

with respect to some measure μ, *where* Q *is increasing. For testing the hypotheses*

$$H_1\colon \theta \leqslant \theta_0 \text{ against } K_1\colon \theta > \theta_0,\quad H_2\colon \theta \geqslant \theta_0 \text{ against } K_2\colon \theta < \theta_0,$$

the multiple test with the test functions

$$\Phi_1(x) = \begin{cases} 1 & \text{when } T(x) > C_1 \\ \gamma_1 & \text{when } T(x) = C_1 \\ 0 & \text{when } T(x) < C_1 \end{cases}$$

and

$$\Phi_2(x) = \begin{cases} 1 & \text{when } T(x) < C_2 \\ \gamma_2 & \text{when } T(x) = C_2 \\ 0 & \text{when } T(x) > C_2, \end{cases}$$

where γ_1, C_1, γ_2 *and* C_2 *are chosen to satisfy*

$$E_{\theta_0}[\Phi_1(X)] = E_{\theta_0}[\Phi_2(X)] = \frac{\alpha}{2},$$

is UMP in the class of rejection unbiased non-contradictory tests.

Proof. Each test has a rejection probability which is a continuous function of θ. See e.g. Lehmann (1959), Theorem 9, p. 52. In order to satisfy the condition of rejection unbiasedness the rejection probabilities for the tests of H_1 and H_2 must have the same value when $\theta = \theta_0$. For a given common value α' of the rejection probabilities when $\theta = \theta_0$, the UMP tests of H_1 and H_2 are of the types given in the theorem with

$$E_{\theta_0}[\Phi(X)] = E_{\theta_0}[\Phi(X)] = \alpha'.$$

For a non-contradictory level α multiple test $\alpha' + \alpha' \leqslant \alpha$ and the tests with $\alpha' = \frac{\alpha}{2}$ are at least as powerful as those with $\alpha' < \frac{\alpha}{2}$. Finally the rejection probabilities for the two tests are increasing and decreasing respectively showing that the obtained multiple test is in fact rejection unbiased and has multiple level of significance α. Q.E.D.

Example 1. Suppose that X consists of n components $X_1, X_2, ..., X_n$ which are independent and normally distributed with parameters θ and 1. Then $T(x) = \bar{x} = \frac{1}{n}\sum_{k=1}^{n} x_k$, and in the UMP rejection unbiased test of H_1: $\theta \leqslant 0$ and H_2: $\theta \geqslant 0$ the constants C_1 and C_2 satisfy $C_2 = -C_1$. This multiple test is equivalent to a UMP unbiased test of $\theta = 0$, against $\theta \neq 0$, which at rejection is completed with one of the statements $\theta > 0$ or $\theta < 0$ depending on which side rejection is made.

Example 2. Suppose that X consists of n components $X_1, X_2, ..., X_n$ which are independent and normally distributed with parameters 0 and θ. Then $T(x) = \frac{1}{n}\sum_{k=1}^{n} x_k^2$, and in a UMP rejection unbiased test of H_1: $\theta \leqslant \theta_0$ and H_2: $\theta \geqslant \theta_0$ the constants C_1 and C_2 are determined as θ_0 times the $\frac{\alpha}{2}$ and $1 - \frac{\alpha}{2}$ fractiles in the χ^2 distribution with n degrees of freedom. In this case the multiple test is *not* equivalent to a UMP unbiased test of $\theta = \theta_0$ against $\theta \neq \theta_0$ completed with directional statements at rejection. In the UMP unbiased test of $\theta = \theta_0$ against $\theta \neq \theta_0$ the hypothesis $\theta = \theta_0$ is rejected if $T(x) < C_1'$ or $T(x) > C_2'$ where C_1' and C_2' are determined as θ_0 times fractiles in the χ^2 distribution with n degrees of freedom which are non-symmetric.

There is a close connection between tests and confidence intervals. Confidence intervals can in fact be considered to be collections of test results. Suppose that θ is a one-dimensional parameter and consider the set of hypotheses consisting of $H_{\lambda+}$: $\theta \geqslant \lambda$ and $H_{\lambda-}$: $\theta \leqslant \lambda$ for all λ, $-\infty < \lambda < \infty$.

If $\lambda_1 < \lambda_2$ the hypothesis H_{λ_2+} is included in H_{λ_1+} and a coherent procedure would always reject H_{λ_2+} when it rejects H_{λ_1+}. Analogously H_{λ_1-} is included in H_{λ_2-} and a coherent procedure would always reject H_{λ_1-} when it rejects H_{λ_2-}. This means that the information about the results of the tests of $H_{\lambda+}$, $-\infty < \lambda < \infty$, can be condensed into the half infinite interval $(\bar{\theta}, \infty)$ or $[\bar{\theta}, \infty)$ of λ's for which $H_{\lambda-}$ is rejected. The confidence interval is the set of λ's for which neither $H_{\lambda+}$ nor $H_{\lambda-}$ are rejected. If the tests are non-contradictory, it is a non-empty interval of one of the types

$$(\underline{\theta}, \bar{\theta}), \quad (\underline{\theta}, \bar{\theta}], \quad [\underline{\theta}, \bar{\theta}) \quad \text{or} \quad [\underline{\theta}, \bar{\theta}]$$

depending on the construction of the tests. The probability that this random interval will cover the true parameter is at least one minus the multiple level of significance in the considered multiple test. When a confidence interval $[\underline{\theta}, \bar{\theta}]$ is obtained the final statement can be formulated $\underline{\theta} \leqslant \theta \leqslant \bar{\theta}$. But taking into consideration the two purposes of the confidence interval to estimate θ below and above it is more logical to state that θ is not smaller than $\underline{\theta}$ and θ

is not greater than θ. Observe that if $[\underline{\theta}, \theta]$ misses the true parameter θ, one of the later statements is still true.

We now apply the rejection unbiasedness concept to the above multiple test problem. Let θ be the true parameter point and let θ_1 and θ_2 be two other parameter points such that $\theta_1 < \theta < \theta_2$. Then the hypotheses H_{θ_1+} and H_{θ_2-} are true while H_{θ_1-} and H_{θ_2+} are false. Suppose that the confidence interval is of the closed type $[\underline{\theta}, \overline{\theta}]$. The rejection unbiasedness implies that $P_\theta[\overline{\theta} < \theta_1]$ and $P_\theta[\underline{\theta} > \theta_2]$ should be smaller than or equal to $P_\theta[\underline{\theta} > \theta_1]$ and $P_\theta[\overline{\theta} < \theta_2]$. This means that if θ_1 is not covered by the confidence interval the probability of missing to the right should be at least as great as the probability of missing to the left. For θ_2 it goes in the opposite direction.

Letting θ_1 and θ_2 both tend to θ and supposing that $P_\theta[\underline{\theta} = \theta] = P_\theta[\overline{\theta} = \theta] = 0$ the rejection unbiasedness leads to the condition

$$P_\theta[\underline{\theta} > \theta] = P_\theta[\overline{\theta} < \theta] = \frac{\alpha}{2},$$

i.e. to symmetric confidence intervals.

For one-sided confidence intervals, i.e. lower or upper confidence bounds, the concept of a uniformly most accurate confidence bound is very simple and intuitive. A lower confidence bound $\underline{\theta}$ of θ is said to be *uniformly most accurate* (UMA) if $P_\theta[\underline{\theta} \leqslant \theta_1]$ is minimized for all $\theta_1 < \theta$. This is equivalent to the tests of $H_{\theta_1-} : \theta \leqslant \theta_1$ being uniformly most powerful for $\theta > \theta_1$.

The generalization of the UMA concept to confidence intervals of the type $[\underline{\theta}, \overline{\theta}]$ is not immediately clear. Because of the twofold aim of such a confidence interval the natural requirement is that $P_\theta[\underline{\theta} \leqslant \theta_1]$ should be minimized and *not* that $P_\theta[\underline{\theta} \leqslant \theta_1 \leqslant \overline{\theta}]$ should be minimized. The reason is that the value θ ought to be excluded as the correct θ because it is too small and not because it is too high.

Theorem 2. *Let $\theta \in \Omega$ be a one-dimensional parameter and suppose that the random variable X has a probability density*

$$p_\theta(x) = C(\theta) \exp(Q(\theta)\, T(x))$$

with respect to some measure μ, where Q is increasing. Suppose further that $T = T(X)$ has continuous cumulative distribution function $F_\theta(t)$ for all $\theta \in \Omega$. Then

(*i*) *for testing the hypotheses*

$$H_{\lambda-}: \theta \leqslant \lambda \text{ and } H_{\lambda+}: \theta \geqslant \lambda \quad \text{for all } \lambda \in \Omega$$

at multiple level of significance α, $0 < \alpha < 1$, there exists a UMP rejection unbiased non-contradictory test, wdose results can be condensed in the confidence interval $[\underline{\theta}, \overline{\theta}]$, where

$$\underline{\theta} = \inf\left\{\lambda : F_\lambda(T) = 1 - \frac{\alpha}{2}\right\}, \quad \overline{\theta} = \sup\left\{\lambda : F_\lambda(T) = \frac{\alpha}{2}\right\}.$$

(ii) *the confidence interval $[\underline{\theta}, \overline{\theta}]$ is symmetric, i.e.*

$$P(\underline{\theta} > \theta) = P_\theta(\overline{\theta} < \theta) = \frac{\alpha}{2}.$$

(iii) *among the symmetric confidence intervals* $[\underline{\lambda}, \overline{\lambda}]$ *for* θ *the interval* $[\underline{\theta}, \overline{\theta}]$ *maximizes* $P_\theta[\underline{\lambda} > \theta_1]$ *for all* $\theta_1 < \theta$ *and* $P_\theta[\overline{\lambda} < \theta_2]$ *for all* $\theta_2 > \theta$.

Proof. (i) For testing $H_{\lambda-}: \theta \leqslant \lambda$ and $H_{\lambda+}: \theta \geqslant \lambda$ for a specific λ the UMP rejection unbiased test is given by Theorem 1. Since T has a continuous cumulative distribution function the tests

$$\varphi_2(x) = \begin{cases} 1 & \text{when } T(x) > C_1(\lambda) \\ 0 & \text{else,} \end{cases}$$

$$\varphi_2(x) = \begin{cases} 1 & \text{when } T(x) < C_2(\lambda) \\ 0 & \text{else,} \end{cases}$$

where

$$F_\lambda(C_1(\lambda)) = 1 - \frac{\alpha}{2} \quad \text{and} \quad F_\lambda(C_2(\lambda)) = \frac{\alpha}{2}\cdot$$

are UMP rejection unbiased. The distributions of T are stochastically increasing and thus the functions $C_1(\lambda)$ and $C_2(\lambda)$ are non-decreasing functions of λ. This implies that the multiple test of $H_{\lambda-}: \theta \leqslant \lambda$ and $H_{\lambda+}: \theta \geqslant \lambda$ for all $\theta \in \Omega$ is rejection unbiased and that the test results can be condensed into the confidence interval $[\underline{\theta}, \overline{\theta}]$ where

$$\underline{\theta} = \inf\left\{\lambda : F_\lambda(T) = \frac{\alpha}{2}\right\} \quad \text{and} \quad \overline{\theta} = \sup\left\{\lambda : F_\lambda(T) = 1 - \frac{\alpha}{2}\right\}$$

(ii) The events $\underline{\theta} > \theta$ and $\overline{\theta} < \theta$ are equivalent to rejections of the hypotheses $H_{\theta-}$ and $H_{\theta+}$ which when θ is the real parameter point shall have probabilities $\frac{\alpha}{2}$ according to Theorem 1.

(iii) The event $\underline{\theta} > \theta_1$ is equivalent to rejection of H_{θ_1}. Thus $P_\theta[\underline{\theta} > \theta_1]$ is maximal for $\theta > \theta_1$ since the test of $H_{\theta_1 -}$ is UMP. Q.E.D.

Example 3. Suppose that X consists of n independent components having density

$$p_\theta(x_k) = \begin{cases} (1/\theta)\exp(x_k/\theta), & x_k \geqslant 0 \\ 0, & x_k < 0 \end{cases}$$

with respect to Lebesgue measure. Then $T(x) = \sum_{k=1}^{n} x_k$ and the distribution of T is determined by $2T/\theta$ having a χ^2 distribution with $2n$ degrees of freedom. Thus if $v_{\alpha/2}$ and $v_{1-\alpha/2}$ denote the $\alpha/2$ and $1-\alpha/2$ fractiles in the χ^2 distribution with $2n$ degrees of freedom we get $C_1(\lambda) = \frac{\lambda}{2} v_{1-\alpha/2}$, $C_2(\lambda) = \frac{\lambda}{2} v_{\alpha/2}$, $\underline{\theta} = 2T/v_{\alpha/2}$ and $\overline{\theta} = 2T/v_{1-\alpha/2}$.

Considering the discussion preceding Theorem 2 we can formalize the basic concepts for confidence sets in the following way. A *two-sided confidence interval* with confidence coefficient $1-\alpha$ for a parameter θ is the condensed result of a coherent level α multiple test of the hypotheses $H_{\lambda+}: \theta \geqslant \lambda$ and $H_{\lambda-}: \theta \leqslant \lambda$ for all λ. A confidence interval $[\underline{\theta}, \overline{\theta}]$ is said to be *rejection unbiased* if $P_\theta[\overline{\theta} < \theta_1] \leqslant P_\theta[\underline{\theta} > \theta_1]\ \forall\ \theta_1 < \theta$ and $P_\theta[\underline{\theta} > \theta_2] \leqslant P_\theta[\overline{\theta} < \theta_2]\ \forall \theta_2 > \theta$.

The confidence interval is said to be *non-contradictory* if $P_\theta(\overline{\theta} < \underline{\theta}] = 0$ for every θ.

Further a confidence interval $[\underline{\theta}, \overline{\theta}]$ is said to be *uniformly most accurate* (UMA) in a class of confidence intervals if it minimizes $P_\theta[\underline{\theta} \leqslant \theta_1]$ for all $\theta_1 < \theta$ and $P_\theta[\overline{\theta} \geqslant \theta_2]$ for all $\theta_2 > \theta$ within that class.

The modification of these concepts to confidence intervals with one or both ends open are obvious. Now we have the following reformulation of Theorem 2.

Corollary 1. *Let $\theta \in \Omega$ be a one-dimensional parameter and suppose that the random variable X has a probability measure with a density*

$$p_\theta(x) = C(\theta) \exp(Q(\theta)\, T(x))$$

with respect to some measure μ, where Q is increasing. Suppose further that $T = T(X)$ has continuous cumulative distribution function $F_\theta(t)$ for all $\theta \in \Omega$. Then the confihence interval $[\underline{\theta}, \overline{\theta}]$, where

$$\underline{\theta} = \inf\{\lambda : F_\lambda(T) = 1 - \alpha/2\},$$

$$\overline{\theta} = \sup\{\lambda : F_\lambda(T) = \alpha/2\},$$

is UMA in the class of non-contradictory rejection unbiased confidence intervals for θ.

3. Some multidimensional problems with one-dimensional main parameter

In this section we will introduce nuisance parameters in the simple problems with one-dimensional main parameter. This includes multiple tests and confidence intervals for one of the parameters in a normal distribution when the other is considered to be a nuisance parameter.

We denote the main parameter by θ, the nuisance parameter by ν and the total parameter thus by (θ, ν). When the nuisance parameter is multidimensional we denote its components by $\nu_1, \nu_2, \ldots, \nu_k$.

Like in the purely one-dimensional problems the natural hypotheses to be tested are of the types

$$H_{\theta_0}^{-} : \theta \leqslant \theta_0$$

$$H_{\theta_0}^{+} : \theta \geqslant \theta_0$$

For multiple tests of such hypotheses for some θ_0 we have the following simple lemma.

Lemma 1. *Suppose that the distributions $P_{\theta,\nu}$ are such that the power function of every test is continuous. Then if a level α multiple test of $H_{\theta_0}^{-}$ and $H_{\theta_0}^{+}$ is non-contradictory, rejection unbiased and individually unbiased the power functions for the two tests must have the value $\alpha/2$ for all ν when $\theta = \theta_0$.*

Proof. Let α' and α'' be the levels of the two tests which must be attained for all ν then $\theta = \theta_0$ by continuity and individual unbiasedness. Rejection, unbiasedness and continuity requires $\alpha' = \alpha''$ and the test being non-contradictory requires rejection of both hypotheses to have probability 0, i.e. $\alpha' + \alpha'' = 2\alpha' = \alpha$. Q.E.D.

If the family of distributions $P_{\theta_0,\nu}$ for ν ranging over the set of possible values is boundedly complete, then the tests obtained in Lemma 1 must have Neyman structure with respect to a sufficient statistic T for ν when $\theta = \theta_0$. See Lehmann (1959), Section 4.3.

When using this to find UMP non-contradictory, rejection and individually unbiased multiple tests, one has to find UMP tests of Neyman structure and level $\frac{\alpha}{2}$ and then check the individual unbiasedness and non-contradictory properties.

For exponential classes with densities of the form

$$c(\theta, \nu)\, e^{\left[\theta U(x) + \sum_{i=1}^{k} \nu_i T_i(x)\right]}$$

the critical functions φ^+ and φ^- for the UMP tests take the forms

$$\varphi^-(u, t) = \begin{cases} 1, & \text{when } u > c^-(t) \\ \gamma^-(t), & \text{when } u = c^-(t) \\ 0, & \text{when } u > c^-(t), \end{cases}$$

and

$$\varphi^+(u, t) = \begin{cases} 1, & \text{when } u < c^+(t) \\ \gamma^+(t), & \text{when } u = c^+(t) \\ 0, & \text{when } u > c^+(t), \end{cases}$$

where $\gamma^-(t)$, $c^-(t)$, $\gamma^+(t)$ and $c^+(t)$ satisfy

$$E_{\theta_0}[\varphi^-(U, T) \mid t] = \frac{\alpha}{2} \text{ for all } t$$

and

$$E_{\theta_0}[\varphi^+(U, T) \mid t] = \frac{\alpha}{2} \text{ for all } t.$$

This result can be applied to a number of specific cases by relating to the ordinary UMP unbiased tests for one-sided hypotheses. See Lehmann (1959), Chapters 4 and 5. For instance in a sample of independent $N(\mu, \sigma^2)$ random variables, the UMP non-contradictory rejection and individually unbiased level α test of H^-: $\mu \leqslant \mu_0$ and H^+: $\mu \geqslant \mu_0$ is an ordinary UMP unbiased test of H_0: $\mu = \mu_0$ augmented with directional statements at rejection. In the same situation the UMP non-contradictory rejection and individually unbiased level α test of $H^-: \sigma^2 \leqslant \sigma_0^2$ and $H^+: \sigma^2 \geqslant \sigma_0^2$ is the ordinary percentage symmetric χ^2 test of $H_0: \sigma^2 = \sigma_2^0$ completed with directional statements. In this case however the UMP unbiased test of $H_0: \sigma^2 = \sigma_0^2$ is not the symmetric χ^2 test but an unsymmetric one. See Lehmann (1959), p. 165.

For two independent samples from an $N(\mu, \sigma^2)$ and an $N(\nu, \tau^2)$ distribution the problem of testing $H^-: \tau^2/\sigma^2 \leqslant \Delta_0$ and $H^+: \tau^2/\sigma^2 \geqslant \Delta_0$ can be reformulated in an exponential class problem with one main parameter and three nuisance parameters. The UMP non-contradictory rejection and individually unbiased level α test is then the ordinary percentage symmetric level α test of $H_0 : \tau^2/\sigma^2 = \Delta_0$ augmented with directional statements at rejection. Again the UMP unbiased test of $H_0 : \tau^2/\sigma^2 = \Delta_0$ is not percentage symmetric. In the two last examples the UMP unbiased tests of H_0 are rarely performed in practice while the percentage symmetric test is common. When the hypothesis is rejected directional statements are also usually made, indicating that a multiple test is usually wanted in practice even if it is not explicitly stated.

It might be suspected that the condition of individual unbiasedness is superfluous, the same test being UMP in the larger class of non-contradictory rejection unbiased tests. This is not generally true, which is shown by the following counterexample.

Let X_1 and X_2 be independent $N(\mu, \sigma^2)$. The UMP non-contradictory rejection and individually unbiased level α test of $H^- : \sigma^2 \leqslant \sigma_0^2$ and $H^+ : \sigma^2 \geqslant \sigma_0^2$ consists in a percentage symmetric χ^2 test of $H_0 : \sigma^2 = \sigma_0^2$ based on the ordinary

$$S^2 = \frac{1}{2}(X_2 - X_1)^2$$

with one degree of freedom. Now let F be the cumulative distribution function for the χ^2 distribution with one degree of freedom and introduce the random variables $V_1 = (\bar{X}^2 + S^2)/\sigma_0^2$ and $V_2 = F^{-1}(1 - F(S^2/\sigma_0^2)) + \bar{X}^2/\sigma_0^2$. Then if w is the $\alpha/2$ fractile in the χ^2 distribution with two degrees of freedom, the test consisting in rejecting H^+ when $V_1 \leqslant w$ and rejecting H^- when $V_2 \leqslant w$ has level α, and it is non-contradictory for all reasonably small α. The power functions are continuous, having the same values for both tests on the line $\sigma = \sigma_0$, $-\infty < \mu < \infty$. It is easily seen that the multiple test is rejection unbiased (but not individually unbiased). On the line $\mu = 0$, $\sigma^2 \leqslant \sigma_0^2$ its power for testing H^+ exceeds the corresponding power of the UMP non-contradictory rejection and individually unbiased level α test.

Next we consider confidence intervals in the multidimensional case with one-dimensional main parameter. It means that we are interested in tests of the hypotheses $H_{\lambda+}$: $\theta \geqslant \lambda$ and $H_{\lambda-}$: $\theta \leqslant \lambda$ for all λ, and the confidence interval is the condensed result of a coherent multiple test of those hypotheses. If the confidence interval for θ is $[\underline{\theta}, \overline{\theta}]$, then the rejection unbiasedness of the tests implies $P_{\theta,\nu}[\overline{\theta} < \theta_1] \leqslant P_{\theta,\nu}[\underline{\theta} > \theta_1]$ for all $\theta_1 < \theta$ and all ν as well as $P_{\theta,\nu}[\overline{\theta} > \theta_2] \geqslant P_{\theta,\nu}[\underline{\theta} > \theta_2]$ for all $\theta_2 > \theta$ and all ν. The individual unbiasedness of the tests implies $P_{\theta,\nu}[\overline{\theta} > \theta_1] \geqslant P_{\theta,\nu}[\underline{\theta} > \theta_2]$ and $P_{\theta,\nu}[\overline{\theta} < \theta_1] \leqslant P_{\theta,\nu}[\overline{\theta} < \theta_2]$ for all ν and all $\theta_1 < \theta < \theta_2$. Now suppose that the distributions $P_{\theta,\nu}$ are such that the power function of every test is continuous. Then letting both θ_1 and θ_2 tend to θ in the above formulas we get $P_{\theta,\nu}[\underline{\theta} > \theta] = P_{\theta,\nu}[\overline{\theta} < \theta] = \frac{\alpha}{2}$ for all θ and ν, i.e. symmetric confidence intervals. Like in problems without nuisance parameter we can transform UMP properties of tests to UMA properties of confidence intervals.

For exponential classes there exist UMP results for rejection and individually unbiased tests which give UMA results for the corresponding confidence intervals. This includes the ordinary symmetric confidence intervals for one of the parameters in a normal sample, when the other is considered to be a nuisance parameter and many other important examples.

4. Multidimensional main parameter

Denoting as before the main parameter by θ and the nuisance parameter by ν we suppose now that the main parameter dimension is $m > 1$ and that the main parameter has components $\theta_1, \theta_2, \ldots, \theta_m$. The nuisance parameter ν can be missing, but in general it can have dimension $k \geqslant 1$. The types of interesting statements vary very much with the application when the main parameter is multidimensional. But of course the statements are all related to the main parameter. The characteristic of the nuisance parameter is just that we are not interested in making statements about it.

In the test situation a simple example of a class of interesting statements are conjunctions in the family $\{K_i^+ : i = 1, 2, \ldots, m\} \cup \{K_i^- : i = 1, 2, \ldots, m\}$ where $K_i^+ : \theta_i > 0$ and $K_i^- : \theta_i < 0$. Another richer family of statements in the test situation is $\{K_\lambda : \lambda \in R^m\}$, where K_λ is the statement $\lambda^T\theta > 0$. In the confidence interval situation there are analogous examples. An important special case is when the main parameter consists of the means of m different normal distributions and the nuisance parameter is a common variance in these distributions, i.e. the usual analysis of variance model. We are going to treat here only an even simpler special case where we have independent sufficient statistics for the different parameters.

Suppose that the random variable X has a density of the exponential class form with natural parametrization, i.e. $p_\theta(x) = C(\theta) \exp\left(\sum_{j=1}^{m} \theta_j U_j(x)\right)$ with respect to some measure μ. Then the test with test functions

$$\varphi_j^-(u_j) = \begin{cases} 1, & \text{when } u_j > c_j^- \\ \lambda_j^-, & \text{when } u_j = c_j^- \\ 0, & \text{when } u_j < c_j^-, \end{cases}$$

and

$$\varphi_j^+(u_j) = \begin{cases} 1, & \text{when } u_j < c_j^+ \\ \gamma_j^+ & \text{when } u_j = c_j^+ \\ 0, & \text{when } u_j > c_j^+, \end{cases}$$

where c_j^-, c_j^+, γ_j^-, γ_j^+ are chosen to satisfy

$$E_{\theta_0}(\varphi_j^-(U_j)) = E_{\theta_0}(\varphi_j^+(U_j)) = \frac{1}{2}(1 - (1 - \alpha)^{1/m})$$

is UMP in the class of non-contradictory, individually and rejection unbiased level α tests of the hypotheses $H_j^- : \theta_j \leqslant \theta_{j,0}$ for $j = 1, 2, \ldots, m$ and $H_j^+ : \theta_j \geqslant \theta_{j,0}$ for $j = 1, 2, \ldots, m$.

Since all power functions are continuous in this case (see Lehmann (1959), p. 52) the individual and rejection unbiasedness conditions require all tests to have the same constant power on the boundary ($\theta_j = \theta_{j,0}$) of their hypotheses sets. Let us denote this power by α'. Further $(U_1, U_2, \ldots, U_{j-1}, U_{j+1}, \ldots, U_m)$ is sufficient and boundedly complete on the parameter set $\theta_j = \theta_{j,0}$, implying that the tests of H_j^- and H_j^+ must have Neyman structure with respect to this variable, i.e. they must be based on U_j only. When the tests of H_j^- and H_j^+ are to be non-contradictory, the probability of rejecting at least one of those hypotheses must be $2\alpha'$. Multiple level of significance α is now obtained for $(1 - 2\alpha')^m = 1 - \alpha$, since the test statistics $U_1, \ldots, U_m$ are independent. And finally the most powerful tests at individual level $\alpha' = \frac{1}{2}(1 - (1 - \alpha)^{1/m})$ are of the given form, and those tests are rejection unbiased.

This example shows that requiring both rejection and individual unbiasedness in a multiple test imposes strong restrictions on the test. In a sense the different hypotheses become equalized so that it is not possible to spend more of the level of significance on some hypotheses than on others. And further it is not possible to use the level of significance in the extensive way it is done in a stagewise rejective test. See Holm (1979), Shaffer (1980) and Holm (1981). It seems that the condition of individual unbiasedness is most responsible for excluding those from the class of tests. On the other hand our counterexample shows that even in simple examples with nuisance parameters the condition of individual unbiasedness is almost unavoidable.

It might happen that more suitable (but also more complex) unbiasedness concepts than those we have discussed here include individual unbiasedness with respect to nuisance parameters and ordering of probabilities of combined rejections according to partial orderings of combinations of hypotheses obtained by truth and falseness of hypotheses.

Many of the problems discussed in this paper are studied from a decision theoretic point of view in Lehmann (1957).

REFERENCES

1. Gabriel, K. R., *Simultaneous test procedures — some theory of multiple ccmpariscns.* Ann. Math. Statist., **40**, 224—250 (1969).
2. Holm, S., *A simple sequentially rejective multiple test procedure.* Scand. J. Statist., **6**, 65—70 (1979).
3. Holm, S., *A stagewise directional test for the normal regression situation.* In: *Proc. Sixth Conference Probab. Theory (Brașov, 1979)*, pp. 103—106. București, Ed. Acad. R. S. România, 1981.
4. Lehmann, E. L., *A theory of some multiple decisicn problems, I, II.* Ann. Math. Statist., **28**, 1—25, 547—572 (1957).
5. Lehman, E. L., *Testing Statistical Hypotheses.* New York, Wiley, 1959.
6. Shaffer, J. P., *Control of directional errors with stagewise multiple test procedures.* Ann. Statist., **8**, 1342—1348 (1980).

LARS HOLST

On sequential urn problems

Throw balls at random into r boxes. Continue until k unspecified boxes contain at least m balls each. Let $Y_1, \ldots, Y_r$ be the number of balls in the boxes. Instead of throwing balls we can consider an urn containing r balls of different colours. Balls are drawn at random with replacement until k different colours have been drawn at least m times each. Then $Y_1, \ldots, Y_r$ are the number of times the different colours have been drawn.

For a given function $g(\cdot)$ consider the random variable

$$Z = \sum_{j=1}^{r} g(Y_j).$$

Example 1. Take $k = r$, $g(x) = x$, $m = 1$. Then $Z =$ the number of balls thrown when each box contains at least one ball for the first time.

Example 2. Take $k = r$, $g(x) = I(x = 1)$, $m = 1$. Then $Z =$ the number of boxes containing exactly one ball when all the boxes contain at least one ball.

We call a problem involving such a random variable a sequential urn problem. Such problems, especially waiting times, have been considered by different authors, see e.g. Johnson and Kotz (1977). We will use a general approach given in Holst (1981) to study such problems.

Consider r independent Poisson processes all with intensity one. Each time an event occurs in a process a ball is put in the corresponding box. Stop when k of the processes have at least m events. From the well-known properties of the Poisson process it follows that this scheme is equivalent to the above procedures. Let T denote the stopping time. At this time exactly one process gets its mth event, $k-1$ processes have already at least m events, and $r-k$ processes have at most $m-1$ events. The time until a single process gets its mth event is $\Gamma(m,1)$-distributed. Hence $T = T_{k:r}$ where $T_{k:r}$ is the kth order statistic in a sample $T_1, \ldots, T_r$ from a $\Gamma(m,1)$-distribution.

Introduce a r.v. ξ_t which conditional on $T = t$ is Poisson (t)-distributed. The number of events in the processes with at most $m-1$ events are $U_{t1}, \ldots, U_{t,r-k}$ i.i.d. r.v.'s with the distribution

$$L(U_t) = L(\xi_t \mid \xi_t \leqslant m-1).$$

Analogously, for the $k-1$ processes with at least m events we define $V_{t1}, \ldots, V_{t,k-1}$ i.i.d. r.v.'s with

$$L_t(V) = L(\xi_t \mid \xi_t \geqslant m).$$

Using the construction and notations above it follows that the characteristic function of Z can be written as

$$E(e^{i\theta Z}) = E\left(\exp\left(i\theta\sum_{1}^{r} g(Y_j)\right)\right) =$$

$$\int_0^\infty f_T(t)\, e^{i\theta g(m)}\, (E(e^{i\theta g(V_t)}))^{k-1}\, (E(e^{i\theta g(U_t)}))^{r-k}\, dt,$$

w here

$$f_T(t) = f_{T_{k:r}}(t) = r\binom{r-1}{k-1}(P(\xi_t \geqslant m))^{k-1}(P(\xi_t \leqslant m-1))^{r-k}\, P(\xi_t = m-1).$$

This can be used to get both exact and asymptotic results as the following examples may indicate. For further examples and details see Holst (1981).

Example 3. Take $k = r$, $g(x) = I(x = 1)$, $m = 1$, i.e. consider $Z = Z_r =$ = the number of different colours observed just once when all colours have been observed. After some simple calculation we find

$$E(e^{i\theta Z_r}) = re^{i\theta}\int_0^\infty e^{-t}(e^{i\theta}\, t\, e^{-t} + 1 - e^{-t} - t\, e^{-t})^{r-t}\, dt.$$

The moment-generating function can be written as

$$E(e^{\theta Z}) = e^{\theta}\int_0^\infty r\, e^{-t}(1 + e^{-t}(-1 + (e^{\theta} - 1)\, t))^{r-1}\, dt$$

and by differentiation

$$E(Z_r) = \sum_{j=1}^{r} 1/j \sim \log r.$$

By the change $re^{-t} = v$ we obtain for $\theta < 1$

$$E(e^{\theta Z_r/\log r}) = e^{\theta/\log r}\int_0^r \left[1 - \frac{v}{r}\,(1 + (e^{\theta/\log r} - 1)\,\log(v/r))\right]^{r-1} dv \to$$

$$\to \int_0^\infty e^{-v}\cdot e^{\theta v}\, dv,\ r\to\infty.$$

Hence,

$$Z_r/E(Z_r) \xrightarrow{D} \mathrm{Exp}\,(1),\ r \to \infty$$

Example 4. Consider $Z = W_r =$ the number of balls until k unspecified boxes contain at least m balls, i.e. take $g(x) = x$. Some calculations in the formulas above give the generating function

$$E(z^{W_r}) = E(\exp(r\, T_{k:r}\,(1 - 1/z))).$$

By differentiation it follows that

$$E(W_r) = rE(T_{k:r}).$$

For the case $m = 1$ the simple formula

$$E(W_r) = rE(T_{k:r}) = r\sum_{j=0}^{k-1} 1/(r-j)$$

follows from the representation

$$T_{k:r} = \sum_{j=0}^{k-1} X_j/(r-j)$$

where the X's are i.i.d. Exp (1). For the case $m = 1$ it is particularly simple to get the asymptotics. Consider e.g. $k = r$. For $\theta < 1$

$$E(e^{\theta(T_{r:r} - \log r)}) \to \Gamma(1-\theta), \quad r \to \infty.$$

Hence

$$E(e^{(W_r/r - \log r)}) \to \Gamma(1-\theta), \quad \theta < 1,$$

which e.g. implies

$$W_r/r - \log r \xrightarrow{D} \exp(-e^{-x}), \quad r \to \infty.$$

REFERENCES

1. Holst, L., *On sequential occupancy problems.* J. Appl. Probab., **18**, 435— 442 (1981).
2. Johnson, N. L., Kotz, S., *Urn Models and Their Applicaticns.* New York, Wiley, 1977.

To the memory of my wife,
Sondra Farber Isaac

RICHARD ISAAC

Probabilities related to the random bombardment of a square

1. Introduction

Let S be the unit square. By a semi-square (SS) we mean a square with dimensions $\frac{1}{2} \times \frac{1}{2}$ lying within S and with sides parallel to those of S. Let n darts (i.e., points) be thrown independently and according to the uniform dis tribution onto S. If there exists at least one SS which is intact, that is, which contains none of the darts in its interior, say that an n-success has occurred; otherwise there is an n-failure. It is intuitively clear and easy to show that

$$P(n\text{-success}) = p_n \to 0 \quad \text{as} \quad n \to \infty.$$

In this paper we will study the rate of convergence to 0 of the sequence $\{p_n\}$. The main result is:

Theorem 1. *There exist constants $A > 0$ and $B < \infty$ such that for all* $n \geqslant 1$

$$A < \frac{p_n}{n^2\left(\frac{3}{4}\right)^n} < B. \tag{1.1}$$

Some motivation for interest in this problem is discussed in Section 3. I want to thank Dr. Leopold Flatto for telling me about this problem. The upper bound B was obtained by him (Proposition 2.1).

A summary of definitions and notations used throughout the paper follows.

S: the unit square, taken as target for a sequence of n independent and uniformly distributed points (referred to as "darts"). It will be convenient to speak in terms of the pattern of darts "at time n", that is, the pattern after the nth dart is thrown.

semi-square (SS): a square with dimensions $\frac{1}{2} \times \frac{1}{2}$ lying within S and with sides parallel to those of S.

N: an even positive integer $\geqslant 4$, used to define a grid on S. The sides of S are subdivided into N equal parts with lattice points denoted by the

Cartesian pairs (i, j), $0 \leqslant i, j \leqslant N$. This is the N-grid on S. (The origin (0, 0) is the lower left vertex of S.)

tile: each of the squares of dimension $\frac{1}{N} \times \frac{1}{N}$ composing an N-grid.

interior (canonical) square (ICS): given an N-grid, any square with vertices on lattice points and length of edge $= \frac{1}{2} - \frac{1}{N}$. An ICS contains exactly $\left(\frac{N}{2} - 1\right)^2$ tiles.

exterior (canonical) square (ECS): analogous to ICS, except that length of edge $= 1/2 + 1/N$. Contains $(N/2+1)^2$ tiles.

intact: used in referring to an area within S that has not been hit by any darts (at time n).

c: this notation is used throughout for the ratio $\frac{n}{N}$.

p_n: the probability that there is an n-success, i.e., that there exists at least one intact SS at time n.

$r_{N,n}$: the probability that there exists at least one intact ICS formed from an N-grid at time n.

$q_{N,n}$: the probability that there exists at least one intact ECS formed from an N-grid at time n.

$N_I(N_E)$: the total number of ICS's (respectively, ECS's) $= \left(\frac{N}{2}+2\right)^2$ $\left(\text{respectively, } \left(\frac{N}{2}\right)^2\right)$ formed from an N-grid.

llv: abbreviation for the lower left vertex of an SS, ICS, or ECS.

2. Bounds for p_n

The event that there is an n-success is the union of an uncountable number of events of the form "S_1 is intact" where S_1 is a specific SS. It is not hard, however, to see that the event "an n-success occurs" is measurable so that it makes sense to discuss p_n. The proof of measurability will be omitted.

(1.1) will be proved by finding bounds on $r_{N,n}$ and $q_{N,n}$. The reason this can be done is given in the following lemma.

Lemma 2.1. *Let S_1 be any SS. Then for any N-grid there always exists an ICS I with $I \subset S_1$. If E is any ECS determined from an N-grid, there always exists an SS S_2 with $E \supset S_2$. For all n and $N \geqslant 4$ we have*

$$q_{N,n} < p_n < r_{N,n}. \tag{2.1}$$

Proof. Let the N-grid be given and let (a, b) be the coordinates of the llv of the SS S_1 according to the Cartesian system induced by the grid (see Section 1). There are two numbers k and l such that $0 \leqslant k, l < 1$ and $(a+k, b+l)$ is a lattice point. Consider the ICS I with llv at $(a+k, b+l)$. The x-coordinate of the vertices of the right side of this ICS is

$$a+k+\frac{N}{2}-1 < a+\frac{N}{2}$$

and the y-coordinate of the top is

$$b+l+\frac{N}{2}-1<b+\frac{N}{2}.$$

Since the corresponding coordinates for S_1 are $a+\frac{N}{2}$ and $b+\frac{N}{2}$ respectively, $I\subset S_1$. Now let E be an N-ECS (that is, an ECS determined from an N-grid) and let E have llv (a, b). Then the SS S_2 with llv also at (a, b) is certainly contained within E. Thus, for any time n, if there exists an intact SS there exists *a fortiori* an intact ICS, and if there exists an intact ECS there exists an intact SS. This implies

$$q_{N,n}\leqslant p_n\leqslant r_{N,n}.$$

To obtain the strict inequality (2.1) observe that for any fixed n and fixed N-grid it is possible to have an intact SS but no intact ECS or an intact ICS but no intact SS, and each of these events can happen with positive probability.

The next result giving half of (1.1) is due to Flatto.

Proposition 2.1. *There exists a constant $B<\infty$ such that*

$$p_n<Bn^2\left(\frac{3}{4}\right)^n,\quad n\geqslant 1 \tag{2.2}$$

Proof. For a given N-grid let $\{I_i,\ 1\leqslant i\leqslant N_I\}$ be the set of all ICS's. Summing over all ICS's yields

$$p_n<r_{N,n}\leqslant\sum_i P(I_i \text{ is intact at time } n)\leqslant\sum_i\left(1-\left(\frac{1}{2}-\frac{1}{N}\right)^2\right)^n=$$

$$=N_I\left(\frac{3}{4}\right)^n\left(1+\frac{4}{3N}-\frac{4}{3N^2}\right)^n\leqslant\left(\frac{N}{2}+2\right)^2\left(\frac{3}{4}\right)^n\exp\left(\frac{4n}{3N}\right).$$

Now suppose N has been chosen so that $cN=n$, $c_1\leqslant c\leqslant c_0$ for c_0, c_1 fixed constants. Then there exists a constant K satisfying: $\left(\frac{N}{2}+2\right)^2\leqslant Kn^2$, $n\geqslant 1$ and so there is a constant B satisfying

$$\left(\frac{N}{2}+2\right)^2\exp\left(\frac{4c_0}{3}\right)\leqslant Bn^2\quad n\geqslant 1.$$

2.2) follows from the above inequ alities.

Proposition 2.2. *There exists a constant $A>0$ such that*

$$p_n>An^2\left(\frac{3}{4}\right)^n,\quad n\geqslant 1. \tag{2.3}$$

Proof. References to "intact" below always refer to a fixed number of darts n. The notation E_i refers to an ECS relative to a fixed N-grid.

Let i_0 be a fixed integer, $1\leqslant i_0\leqslant N_E$, and consider the ECS E_{i_0}. Let u be another integer and put

$$p_{i_0u}=P(E_{i_0} \text{ and } E_u \text{ are both intact})=(1-\text{area of } E_{i_0}\cup E_u)^n.$$

First we find an upper bound for

$$p_{i_0} = \sum_{u \neq i_0} p_{i_0,u}.$$

The computation is broken up into two cases, an "essential" part in which the two ECS's are close together, and an "inessential" part in which the two are far apart. Let us suppose that the llv of E_{i_0} is the origin (0,0) and the llv of the fixed ECS E_{u_0} has coordinates (j,k).

Estimate for Case 1: *Both* j *and* k *are* $\leqslant \sqrt{N}/2$. The part of E_{u_0} not overlapping with E_{i_0} consists of two rectangles with areas $\frac{j}{2N}$ and $\frac{k}{2N}$ and these rectangles overlap in a rectangle with area $\frac{j\cdot k}{N^2}$. It follows that

$$p_{i_0,u_0} = \left\{1 - \left(\frac{1}{4} + \frac{1}{N} + \frac{1}{N^2} + \frac{j+k}{2N} - \frac{j\cdot k}{N^2}\right)\right\}^n =$$

$$= \left(\frac{3}{4}\right)^n \left\{1 - \left(\frac{4}{3N} + \frac{4}{3N^2} + \frac{2(j+k)}{3N} - \frac{4j\cdot k}{3N^2}\right)\right\}^n \leqslant$$

$$\leqslant \left(\frac{3}{4}\right)^n \exp\left(-\left\{\left[\frac{4}{3} + \frac{4}{3N} + \frac{2}{3}(j+k) - \frac{4j\cdot k}{3N}\right]\frac{n}{N}\right\}\right),$$

where the inequality $(1-q)^n \leqslant (e^{-q})^n$ valid for $0 \leqslant q \leqslant 1$ has been used, assuming N is large enough. In terms of $c = \frac{n}{N}$ we have

$$p_{i_0,u_0} \leqslant \left(\frac{3}{4}\right)^n e^{-\frac{4c}{3}} e^{\frac{4j\cdot k\cdot c}{3N}} e^{\frac{-2(j+k)c}{3}}. \tag{2.4}$$

For Case 1, $jk \leqslant \frac{N}{4}$ so that (2.4) becomes

$$p_{i_0,u_0} \leqslant \left(\frac{3}{4}\right)^n e^{-\frac{4c}{3}} e^{\frac{c}{3}} e^{\frac{-2(j+k)c}{3}}. \tag{2.5}$$

To find an upper bound for

$$\sum_{u\in U} p_{i_0,u}$$

where U is the set of indices u such that the pair E_{i_0}, E_u fall into Case 1 $u \neq i_0$), we sum the terms $e^{\frac{-2(j+k)c}{3}}$ over all possible pairs (j,k). Although $0 \leqslant j,\ k \leqslant \frac{N-2}{2}$ the sum will be taken for $0 \leqslant j,\ k < \infty$. The matrix of (j,k) pairs that arises is $m_{ij} = (i,\ j)$, $i \geqslant 0$ and $j \geqslant 0$, with the single term (0,0) missing. Adding over these terms for each row yields the sums: for rows 0 and 1

$$e^{-\frac{2}{3}c} + e^{-\frac{2}{3}(2c)} + \ldots = e^{-\frac{2}{3}c}\left(1 - e^{-\frac{2}{3}c}\right)^{-1},$$

and for row k, $k \geqslant 2$,

$$e^{-\frac{2}{3}(kc)} + e^{-\frac{2}{3}(k+1)c} + \ldots = e^{-\frac{2}{3}(kc)}\left(1 - e^{-\frac{2}{3}c}\right)^{-1}.$$

Now since the llv of E_{i_0} is (0, 0), E_u is necessarily positioned "northeast" of E_{i_0}. For E_{i_0} in a general position, however, E_u could have one of four directions relative to E_{i_0}. Therefore we throw in the factor 4 and add up over each row sum above to obtain from (2.5)

$$\sum_{u \in U} p_{i_0,u} \leqslant \tag{2.6}$$

$$\left(\frac{3}{4}\right)^n e^{-\frac{4c}{3}}\left[4\left\{2e^{-\frac{c}{3}}\left(1 - e^{-\frac{2c}{3}}\right)^{-1} + e^{-c}\left(1 - e^{-\frac{2c}{3}}\right)^{-2}\right\}\right].$$

Estimate for Case 2: At least one of j, k $> \sqrt{N}/2$. In this case the two ECS's are far apart: the area of

$$E_{i_0} \cup E_{u_0} \geqslant \left(\frac{1}{2} + \frac{1}{N}\right)^2 + \frac{1}{2}\,\mathrm{Max}\left(\frac{j}{N}, \frac{k}{N}\right) \geqslant \left(\frac{1}{2} + \frac{1}{N}\right)^2 + \frac{1}{2} \cdot \frac{\sqrt{N}}{2N}.$$

Putting $c = \dfrac{n}{N}$ and again using the elementary exponential inequality used in the treatment of Case 1, we get for the pair E_{i_0}, E_{u_0} falling into Case 2

$$p_{i_0,u_0} \leqslant \left\{1 - \left(\frac{1}{4} + \frac{1}{N} + \frac{1}{N^2} + \frac{1}{4\sqrt{N}}\right)\right\}^n$$

$$= \left(\frac{3}{4}\right)^n \left\{1 - \left(\frac{4}{3N} + \frac{4}{3N^2} + \frac{1}{3\sqrt{N}}\right)\right\}^n \leqslant \left(\frac{3}{4}\right)^n e^{-\frac{4c}{3}} e^{-\frac{c\sqrt{N}}{3}},$$

and so, if V is the set of indices u such that the pair E_{i_0}, E_u falls into Case 2, we have

$$\sum_{u \in V} p_{i_0,u} \leqslant N_E \left(\frac{3}{4}\right)^n e^{-\frac{4c}{3}} e^{-\frac{c\sqrt{N}}{3}} = \frac{N^2}{2}\left(\frac{3}{4}\right)^n e^{-\frac{4c}{3}} e^{-\frac{c\sqrt{N}}{3}}. \tag{2.7}$$

If we let n and N increase so that $c = \dfrac{n}{N}$ remains constant and c is chosen large enough, the term in brackets on the right side of (2.6) can be made smaller than $\dfrac{\varepsilon}{4} > 0$. Moreover, as $N \to \infty$ the right side of (2.7) tends to zero. Thus (2.6) and (2.7) imply that we can specify fixed $c_1 = \dfrac{n}{N}$ such that for all n sufficiently large

$$p_{i_0} = \sum_{u \neq i_0} p_{i_0,u} = \sum_{u \in U} p_{i_0,u} + \sum_{u \in V} p_{i_0,u} < \frac{\varepsilon}{2}\left(\frac{3}{4}\right)^n e^{-\frac{4c_1}{3}}. \tag{2.8}$$

Moreover,

$$P(E_{i_0} \text{ is intact at time } n) = \left\{1 - \left(\frac{1}{4} + \frac{1}{N} + \frac{1}{N^2}\right)\right\}^n$$

$$= \left(\frac{3}{4}\right)^n \left\{1 - \left(\frac{4}{3N} + \frac{4}{3N^2}\right)\right\}^n \geqslant \left(\frac{3}{4}\right)^n e^{-\frac{4c_1}{3}} (1 - d_n)$$

where $d_n \to 0$ as $n \to \infty$. By a well-known inequality of Bonferroni (see [1], p. 100)

(2.9) $$p_n > q_{N,n} \geqslant \sum_{i \leqslant N_E} P(E_i) \text{ is intact at time } n) - \sum_{i \leqslant N_E} p_i$$

$$> N_E \left(\frac{3}{4}\right)^n e^{-\frac{4c_1}{3}} \left(1 - d_n - \frac{\varepsilon}{2}\right) > (1 - \varepsilon) e^{\frac{-4c_1}{3}} \frac{N^2}{4} \left(\frac{3}{4}\right)^n$$

$$= (1 - \varepsilon)\, e^{-\frac{4c_1}{3}} \frac{1}{4c_1^2} n^2 \left(\frac{3}{4}\right)^n = A_1 n^2 \left(\frac{3}{4}\right)^n .$$

for all n large enough. Since $p_n \neq 0$ for each n, the constant A_1 in (2.9) can be replaced by a constant A so that the inequality holds for all $n \geqslant 1$, and this is (2.3).

Theorem 1 follows immediately by combining (2.2) and (2.3).

3. Some applications

The problem is related to a variety of geometrical covering type questions (see, e.g. [2]). It also provides a simple model for certain phenomena. Let the small square represent a virus or cell that can be deactivated when a single dart hits it. We suppose the virus can wander within the region S. The estimates on p_n give bounds on the certainty of deactivation at time n. To consider another example, suppose that diseased cells are distributed at random throughout a region of tissue where the cells play the role of the darts and the region is S. Suppose a biopsy is performed, that is, a small sample of tissue is taken from S for examination. The biopsy may be negative because the sample (or semi-square) does not contain any diseased cells. p_n is then a measure of the sensitivity of the biopsy as a test for detecting disease.

4. Final remarks

Flatto has conjectured a significant improvement over Theorem 1: that, in fact,

$$\lim_{n \to \infty} \frac{p_n}{n^2 \left(\frac{3}{4}\right)^n}$$

exists. The present author has recently made considerable progress in proving this conjecture using Theorem 1. Further results will appear elsewhere.

REFERENCES

1. Feller, W., *An Introduction to Probability Theory and Its Applications, Vol. 1, 2nd Ed.*, New York, Wiley, 1963.
2. Kendall, M. G., Moran, P. A. P., *Geometrical Probability*. New York, Hafner, 1963.

PETER JAGERS

The stable asymptotic structure of growing branching populations

Branching processes relate properties of individuals and their reproduction to the growth and composition of collectives. This is a two-way relation: in some contexts (often demographic) the individual behaviour is well-known and the purpose may be to predict the growth of a population or to calculate probability of a random individual being first born; in other situations only certain properties of populations happen to be observable and one would like to infer the underlying, more basic, properties of individual life. This is the case in cell kinetics, where various measures of the population composition like the mitotic index, i.e. the proportion of cells in mitosis, can be obtained experimentally, whereas individual cells can hardly be followed.

This lecture phocuses upon the composition of (old) supercritical branching populations and the purpose is to exhibit a stable population composition, one margin of which will be the celebrated stable age distributions of demography, known in principle already by Euler. But it should also answer questions like: what is the probability of being first born, or what is the sibship size distribution in the population, or to mention a more technical problem, what is the probability in a cell population of having an ancestor who was synthesizing DNA t time units ago?

The work reported is joint with O. Nerman [3]. Though only individuals of one type are considered here, a many type version of the results is under preparation.

Individual life is described by an abstract probability space, $(\Omega, \mathcal{A}, P)$, the *life space*, which could have many forms but should always be rich enough to carry the entitities under consideration. In any case, successive ages at *child bearing* $\tau(1) \leqslant \tau(2) \leqslant \ldots, \tau(j) : \Omega \to [0, \infty]$ should be defined, $\tau(j) < \infty$, $\tau(j+1) = \infty$ being interpreted as the individual obtaining exactly j children. Bearing is also described by the *reproduction* point *process*

$$\xi(A) = \sum_{j=1}^{\infty} 1_A(\tau(j)),$$

where we often write $\xi(a)$ for $\xi([0, a])$, the number of children begotten up to the age a. The *Malthusian parameter* α is defined by

$$E[\hat{\xi}(\alpha)] = E\left[\int_0^{\infty} e^{-\alpha u}\, \xi(du)\right] = 1$$

and we only consider processes such that this α exists and also

$$\beta = \int_0^\infty e^{-\alpha u}\, uE[\xi(du)] < \infty.$$

Such processes are called Malthusian. The number β is known as the average age at child bearing; this interpretation is obtainable from the stable population law able to be discussed. All our results concern the supercritical case, $\alpha > 0$, or equivalently $E[\xi(\infty)] > 1$. The critical case $E[\xi(\infty)] = 1$ has recently been treated in [5]. A technical assumption made throughout is that reproduction is non-lattice: The smallest closed group containing the support of $E[\xi]$ is the whole non-negative line.

If the life space is the first ingredient in the construction of branching populations, then the pedigree structure is the second. In the traditional Ulam-Harris formulation the pedigree starts from an ancestor, 0, goes on to the ancestor's children 1, 2, 3, ..., grand children 11, 12, ..., 21, 22, ... and so on; the space of individuals is

$$I = \bigcup_{n=0}^{\infty} N^n$$

where N is the positive integers and $N^0 = \{0\}$. In our case we want to talk about the ancestry as well as the progeny and sisters of an arbitrary individual, whom we shall term ego in concordance with some anthropological literature. Ego will be denoted by 0 and ego's children and all descendants as the corresponding descendants of the ancestor of the Ulam-Harris space. Ego's mother will be called —1, her children, besides ego, will be denoted by —11, —12, etc. In other words, besides ego and her descendants which are named by the elements of I, ego's mother's progeny are the elements of $\{-1\} \times I$. Ego's grand-mother will be called —2 and so the construction continues. The space of individuals considered here can thus be written

$$J = \bigcup_{k=0}^{\infty} \{-k\} \times I$$

with the convention $(-k, 0) = 0$ and $(0, i_1, ..., i_n) = (i_1, ..., i_n)$. For simplicity we write just $i_1, ..., i_n$ for the latter vector.

To describe the population we need to anchor ego in physical time and in the pedigree. This is done by saying that ego is sampled now at time zero having an age $a \in [0, \infty)$ and an ancestry $(i_k)_{k=1}^{\infty}$, meaning that ego was the i_1th child of her mother, who was the i_2th child of her mother, ... In other words, ego's mother begets the children —11, —12, ..., —1, i_1—1, 0, —1, $i_1 +$ $+ 1$, ... in that order. The basic assumption made in traditional branching processes is then that all the individual lives are independent, besides times of birth of individuals, of course. This leads to the probability space $(\Omega^I, \mathcal{A}^I, P^I)$. In our case similarly the outcome space can be defined as $\Omega^J \times$ $\times N^\infty \times R_+$, where the two latter components anchor ego. The sigma-algebra will be the corresponding products $\Omega^J \times 2^{N^\infty} \times \mathcal{B}_+$, the latter being the trivial and Borel algebras respectively.

The question then arises, what (if any) is the probability measure $\tilde{P}$ corresponding to a randomly sampled ego in an old traditional branching process? To make it more precise we must define this process in time. This is made by saying that the ancestor was born at time zero. Then the bearing ages of an individual $x \in I$ are defined by

$$\tau_x(j)\,(\{\omega_x\}_{x \in I}) = \tau(j)\,(\omega_x)$$

and the succesive birth times by

$$\sigma_0 = 0, \quad \sigma_{xj} = \sigma_x + \tau_x(j).$$

The number of individuals born at time t is thus

$$y_t = \#\{x \in I;\ \sigma_x \leqslant t\}$$

and it is among these we sample and pose questions about their own life and their relatives, old meaning that $t \to \infty$.

The first task is then to describe $\tilde{P}$ from P, the second to show convergence of the sampling measures toward $\tilde{P}$, and finally to investigate what happens if we sample only among some subset of all those born, like those alive (not yet defined) or firstborn.

The probability measure $\tilde{P}$ is characterized by the following properties: Ego's age a now at sampling is exponentially distributed with the parameter being the Malthusian α. Her birth order is independent of her age and

$$\tilde{P}(i_1 = j) = E[e^{-\alpha\tau(j)}], \quad j \in N,$$

where the expectation refers to the life space probability law P. Ego's own life and progeny constitute a traditional branching process on the space $(\Omega^I, \mathcal{A}^I, P^I)$, started at the birth time $-a$, but otherwise independent of everything else in the process. The life of ego's mother is in its turn described by the relation

$$\tilde{P}\,(\omega_{-1} \in A, \quad i_1 = j) = E[e^{-\alpha\tau(j)}; A], \quad A \in \mathcal{A},\ j \in N,$$

and ego's sisters again initiate independent branching processes all with the same underlying life space $(\Omega, \mathcal{A}, P)$ but of course initiated at the birth times of these sisters. Ego's grandmother relates to ego's mother as does the mother to ego herself and so on backwards in the family tree.

Let us have a look at some simple illustrations.

The mentioned *probability of being firstborn* (among all those born, alive or not) is

$$\tilde{P}(i_1 = 1) = E[e^{-\alpha\tau(1)}].$$

The mother's age at her first birth has the distribution

$$P(\tau_{-1}(1) \leqslant u) = E[e^{-\alpha\tau(1)};\ \tau(1) \leqslant u]$$

and generally *age at child bearing* follows the distribution

$$\tilde{P}(\tau_{-1}(i_1) \leqslant u) = \sum_{j=1}^{\infty} \tilde{P}(\tau_{-1}(j) \leqslant u,\ i_1 = j)$$

$$= \sum_{j=1}^{\infty} E\,[e^{-\alpha\tau(j)};\ \tau(j) \leqslant u] = \int_0^u e^{-\alpha v} E[\xi(dv)].$$

Similarly for any life event $A \in \mathcal{A}$ and $u \geqslant 0$

$$\tilde{P}(\omega_{-1} \in A,\ \tau_{-1}(i_1) \leqslant u) = \int_0^u e^{-\alpha v} E[\xi(dv);\ A],$$

which yields

$$\tilde{P}(\omega_{-1} \in A \mid \tau_{-1}(i_1) = u) = \frac{E[\xi(du);\ dA)}{E[\xi(du)]},$$

the ordinary Palm probability of the reproduction point process ξ.

As a last example we mention the expected sibship size:

$$E[\xi_{-1}(\infty)] = \sum_{j=1}^{\infty} [\xi_{-1}(\infty);\ i_1 = j] =$$

$$= \sum_{j=1}^{\infty} E[e^{-\alpha\tau} \xi(\infty)] = E[\hat{\xi}(\alpha)\, \xi(\infty)].$$

More intricate examples can be found in [2] and [3].

As indicated by the exponentially distributed age of ego, which mirrors the exponential growth of branching population, the measure $\tilde{P}$ corresponds to sampling among all those born. For applications it is important to be able to restrict the sampling, e.g. to those alive. To this end we must assume as given some non-negative random variable λ on Ω, the life span, and then we say that an individual x is alive between her birth time σ_x and $\sigma_x + \lambda_x$, where $\lambda_x(\{\omega_x\}) = \lambda(\omega_x)$ is x's life span. The new stable measure simply turns out to be (recall that a is ego's age at sampling)

$$\tilde{P}(\cdot \mid a < \lambda_0),$$

rendering ego's life and age dependent. Indeed, for any $A \in \mathcal{A}$ and $u \geqslant 0$

$$\tilde{P}(\omega_0 \in A,\ a \leqslant u \mid a < \lambda_0) = \frac{\tilde{P}(\omega_0 \in A,\ a \leqslant u,\ \lambda_0 > a)}{\tilde{P}(\lambda_0 > a)} =$$

$$= \frac{\int_0^u P(A \cap \{\lambda > v\})\, \alpha e^{-\alpha v}\, dv}{\int_0^\infty P(\lambda > v)\, \alpha e^{-\alpha v}\, dv}$$

by the definition of $\tilde{P}$. Here $A = \Omega$ yields the stable age distribution. By the independence other marginals remain unchanged, e.g. the probability of being firstborn for sampling among all now alive is

$$\tilde{P}(i_1 = 1 \mid \lambda_0 > a) = \tilde{P}(i_1 = 1) = E[e^{-\alpha\tau(1)}].$$

Further illustrations of these matters can be found in [2]. Generally the stable population measure for sampling among a subset of those born, where this subset is defined by an individual property $A \in \mathcal{A}$ or possibly a property A of an individual and her progeny will be $\tilde{P}(\cdot \mid A_0)$, the index 0 meaning that the property should be attached to ego (and her descendants).

The proofs of these facts follow from the convergence of empirical sampling measures in branching processes, which will now be discussed briefly. A more complete treatment can be found in [3]. Still the remaining part has to be somewhat more technical than the preceding.

For the traditional branching space $(\Omega^I, \mathcal{A}^I, P^I)$ we consider the ancestral shifts, $x \in I$,

$$S_x : \Omega^{\{x\} \times I} \to \Omega$$

mapping $\{\omega_{xy}\}_{y \in I}$ on $\{\omega_y\}_{y \in I}$, thus rendering the individual x the ancestor. If $\{\chi(u)\}_{u \in R}$ is a non-negative jointly measurable stochastic process on $(\Omega^I, \mathcal{A}^I)$, which vanishes for negative u, i.e. a random characteristic [1], we define

$$\chi_x = \chi \circ S_x, \quad x \in I,$$

$\chi_x(u)$ being the characteristic value, or score, of x at her age u, negative age interpreted as time to birth.

The branching process $\{z_t^\chi\}$ counted with characteristic χ is defined to be the sum

$$z_t^\chi = \sum_{x \in I} \chi_x(t - \sigma_x).$$

Several examples of such processes can be found in [1].

In the supercritical and non-lattice case various results hold for $t \to \infty$. In particular, if there is a $\beta < \alpha$ such that $E[\hat{\xi}(\beta)] < \infty$, $E[\sup_u e^{-\alpha u}\chi(u)] < \infty$, and χ has a.s. right and left limits everywhere, then a.s. on the set $\{y_t \to \infty\}$

$$z_t^\chi / y_t \to E[\hat{\chi}(\alpha)] = E\left[\alpha \int_0^\infty e^{-u\alpha}\chi(u)\right] du$$

as $t \to \infty$ [4].

This avails itself to the study of empirical averages of bounded functions φ

$$\varphi : \Omega^J \times N^\infty \times R_+ \to R_+,$$

provided they are determined by a finite number of generations back in the sense that

$$\varphi = \sum_{n=0}^{\infty} \varphi_n,$$

where the value of each φ_n is determined by the projection of its argument onto

$$\Omega^{\bigcup_{k=0}^{n} \{-k\} \times I} \times N^n \times R.$$

We let $\pi_x : \Omega^I \to \Omega^J \times N^\infty$ permute the coordinates so that x becomes ego (0) while keeping all relations (so that if $x = j_1 \ldots j_{n-1} j_n$, then x's mother turns ego's mother i.e. $\omega_{j_1 \ldots j_{n-1}} \curvearrowright \omega_{-1}$) and the part in N^∞ reports x's ancestry i.e. starts by $j_n j_{n-1} \ldots j_1$. Observe that the value of π_x may be arbitrary as concerns the continuation of the ancestry and also of the lives of individuals not in $\bigcup_{k=0}^{n} \{-k\} \times I$. If $x \in N^n$ the value of $\varphi_n(\pi_x, t - \sigma_x)$ is still well determined.

It can be proved [3] that as $t \to \infty$ the importance of undetermined coordinates disappears and that, as $t \to \infty$,

$$\sum_{x \in I} \varphi(\pi_x, t - \sigma_x)/y_t \to \tilde{E}[\varphi] = \int \varphi \, d\tilde{P}$$

a.s. on the set $\{y_t \to \infty\}$.

REFERENCES

1. Jagers, P., *Branching Processes with Biological Applications.* London, Wiley, 1975.
2. Jagers, P., *How probable is it to be firstborn? and other branching process applications to kinship problems.* Math. Biosci., **59**, 1– 15 (1982).
3. Nerman, O., Jagers, P., *The stable doubly infinite pedigree process of supercritical branching populations.* Dept. Math. Chalmers U. Tech. and U. Göteborg, 1981.
4. Nerman, O., *On the convergence of supercritical general (C-M-J) branching process.* Z. Wahrsch. Verw. Gebiete, **57**, 365– 395 (1981).
5. Nerman, O. *The stable pedigrees of critical branching populations.* Dept. Math. Chalmers U. Tech. and U. Göteborg, 1981.

HANS G. KELLERER

Duality theorems and probability metrics

1. Duality theorems for marginal problems

An extension of the well-known transportation problem in linear programming to an infinite set of origins resp. destinations has been considered for the first time by Monge as early as 1781 (see [11]). Combined with a generalization from $n = 2$ to $n \in \mathbf{N}$, it leads to the following measure theoretic "marginal problem":

For fixed distributions μ_i on spaces $X_i (1 \leqslant i \leqslant n)$ and a real function g on their product $X := X_1 \times \ldots \times X_n$

(1) minimize $\int_X g \, d\mu$,

where the distribution μ on X has to satisfy the side condition

$$\mu(\{x \in X : x_i \in B_i\}) = \mu_i(B_i) \text{ for } 1 \leqslant i \leqslant n;$$

(2) maximize $\sum_{1 \leqslant i \leqslant n} \int_{X_i} f_i \, d\mu_i$,

where the real functions f_i on X_i have to satisfy the side condition

$$\sum_{1 \leqslant i \leqslant n} f_i(x_i) \leqslant g(x_1, \ldots, x_n) \text{ for all } x_i.$$

Then the following main questions arise:

(3) is the minimum in (1) actually attained,

(4) is the maximum in (2) actually attained,

(5) are these extreme values identical.

In the discrete case (3) and (4) are obviously true, while the validity of (5) is a consequence of the classical duality theorem.

For a rigorous account of the general case the underlying assumptions from topology and measure theory have to be stated precisely. Concerning the spaces X_i, it suffices to assume them to be Hausdorff spaces henceforth. If $\mathfrak{B}(Y)$ denotes the Borel σ-algebra of a topological space Y, all measures occurring in this section are assumed to belong to the corresponding class

$$M_t(Y) := \{\nu \mid \mathfrak{B}(Y) : \nu \text{ normed and tight}\},$$

where tightness means inner compact-regularity. If $\mathcal{L}(\nu)$ denotes the Lebesgue space of ν-integrable functions, the functions f_i are restricted to $\mathcal{L}(\mu_i)$, and the abbreviation $\mu_i(f_i) := \int_{X_i} f \, d\mu_i$ is then used. Concerning the function g it is advisable to dispense with any condition and to replace the range $\mathbf{R}$ by $\widetilde{\mathbf{R}} := [-\infty, +\infty]$; then the corresponding lower integral $\mu_*(g) := \int_{X*} g \, d\mu$ has to be used.

Now, using the projections $\pi_i : X \to X_i$, the fundamental functionals associated with (1) and (2) can be introduced:

Definition. Let $\mu_i \in M_t(X_i)$ be fixed; then:

$$I(g) := \inf\{\mu_*(g) : \mu \in M_t(X) \text{ and } \pi_i(\mu) = \mu_i \text{ for } 1 \leqslant i \leqslant n\},$$

$$S(g) := \sup\Big\{\sum_{1 \leqslant i \leqslant n} \mu_i(f_i) : f \in \mathcal{L}(\mu_i) \text{ and } \sum_{1 \leqslant i \leqslant n} f_i \circ \pi_i \leqslant g\Big\}$$

for arbitrary functions $g : X \to \widetilde{\mathbf{R}}$ (with the convention $\sup \emptyset := -\infty$).

Then, assuming a sufficiently regular situation (e.g. compact metrizable spaces X and a bounded continuous function g) and applying the theorems of Hahn-Banach and Riesz, it is not hard to prove the "duality theorem" $I(g) = S(g)$ (for a slightly different approach see [5] and [13]). To obtain more general results, some additional notation is needed. Let $\mathcal{C}_*(X)$ denote the class of all lower semicontinuous functions $f : X \to \widetilde{\mathbf{R}}$ and let $\mathcal{A}_*(X)$ consist of all co-Suslin functions with respect to $\mathcal{C}_*(X)$, i.e.

$$f = \inf\{\sup_{k \in \mathbf{N}} f_{n_1, \dots, n_k} : (n_1, n_2, \dots) \in \mathbf{N}^{\mathbf{N}}\},$$

where $f_{n_1, \dots, n_k} \in \mathcal{C}_*(X)$ for $k \in \mathbf{N}$ and $(n_1, \dots, n_k) \in \mathbf{N}^k$.

Then, by a thorough investigation of the functionals I and S and an application of Choquet's capacity theorem as the crucial tool, the following main result can be established:

Theorem. *Without further topological assumptions on the spaces X_i, the equation $I(g) = S(g)$ holds true in each of the following cases*

(a) $g \in \mathcal{C}_*(X)$ *arbitrary,*

(b) $g \in \mathcal{A}_*(X)$ *with* $\sup_{x \in X} g(x) < \infty$.

Moreover, the infimum $I(g)$ is attained in case (a) *and the supremum $S(g)$ is attained in case* (b).

As can be shown by a counterexample, the boundedness condition in case (b) is essential.

Using the notation $\mathfrak{L}(X)$ for the Baire σ-algebra of X, the following result can be deduced from the theorem.

Corollary. *Without further topological assumptions on the spaces X_i, the equation $I(g) = S(g)$ holds true in each of the following cases:*

(c) g *$\mathfrak{L}(X)$-measurable with* $\sup_{x \in X} g(x) < \infty$,

(d) g *$\mathfrak{B}(X_1) \otimes \dots \otimes \mathfrak{L}(X_n)$-measurable with* $\sup_{x \in X} g(x) < \infty$.

Thus, if all spaces X_i are metrizable or second countable, the equation $I(g) = S(g)$ is valid in particular for all $\mathfrak{B}(X)$-measurable functions g with $\sup_{x \in X} g(x) < \infty$.

For the proof of these and related results the reader is referred to [9], where further applications can be found as well.

2. Application to probability metrics

This section is devoted to one of the most important applications of duality theorems for marginal problems. It concerns two metrics in the set of probability measures on a fixed metric space, which are essentially due to Kantorovich-Rubinštein (see [6]) and Wasserštein (see[15]). Above all, their coincidence is shown in Theorem 2 under rather general assumptions and without any moment condition, thus filling a gap in the lecture notes [3] of Dudley (see also the papers [1], [4], [10]).

Throughout this section a space X with a metric d is fixed, where d is not restricted to finite values. Then an essential role is played by the following class of "contractions"

Definition 1. $\mathcal{F}_d(X)$ consists of all functions $f\colon X \to \mathbf{R}$ that are bounded and satisfy the inequality

$$|f(x_1) - f(x_2)| \leqslant d(x_1, x_2) \text{ for all } x_i \in X.$$

An extension theorem for this class will be used repeatedly.

Lemma 1. *For* $\emptyset \neq Y \subset X$ *and* $g \in \mathcal{F}_d(Y)$ *there exists* $f \in \mathcal{F}_d(X)$ *with* $f \mid Y = g$ *and* (*) $\inf_{x \in X} f(x) = \inf_{y \in Y} g(y)$, $\sup_{x \in X} f(x) = \sup_{y \in Y} g(y)$.

Proof. With the abbreviation $\gamma := \sup_{y \in Y} g(y)$ define

$$g_y(x) := (g(y) + d(y, x)) \wedge \gamma \text{ and } f := \inf_{y \in Y} g_y.$$

Then $g_y \in \mathcal{F}_d(X)$ and thus $f \in \mathcal{F}_d(X)$; moreover

$$g_y(y) = g(y) \leqslant g_z(y) \text{ for } y, z \in Y.$$

This implies $f \mid Y = g$, while condition (*) is obviously satisfied.

In addition two simple compactness properties will be needed:

Lemma 2. *The set* $\mathcal{F} := \{f \in \mathcal{F}_d(X)\colon \underline{f} \leqq f \leqq \bar{f}\}$ *with* $\underline{f}, \bar{f} \in \mathcal{F}_d(X)$ *is*

(1) *sequentially compact with respect to pointwise convergence, if X is separable;*

(2) *(sequentially) compact with respect to uniform convergence, if X is totally bounded.*

Proof. (1) If D is a countable dense subset of X, each sequence in $\mathcal{F}$ contains a subsequence converging on D, which in view of the equicontinuity converges everywhere, the limit obviously being contained in $\mathcal{F}$.

(2) Since X is in particular separable, each sequence in $\mathcal{F}$ contains by (1) a subsequence converging everywhere, which in view of the total boundedness must converge uniformly.

For the purpose of this section the set $M_t(X)$ can be enlarged to $M_\tau(X) := \{\mu \mid \mathfrak{B}(X)\colon \mu$ normed and τ-continuous$\}$, where τ-continuity means continuity with respect to increasing nets of open sets (or decreasing nets of closed sets). These measures are known to be characterized by the following criteria.

Lemma 3. *For a normed measure* $\mu \mid \mathfrak{B}(X)$ *the following statements are equivalent*

(0) $\mu \in M_\tau(X)$,

(1) $\sup\ \{\mu(B)\colon B \in \mathfrak{B}(X)$ *totally bounded*$\} = 1$,

(2) $\mu(B) = 1$ *for some separable set* $B \in \mathfrak{B}(X)$.

Proof. The implication (0) ⇒ (1) is the main step in proving that τ-continuous finite measures on complete metric spaces are tight, the implication (1) ⇒ (2) is evident, and the implication (2) ⇒ (0) follows from the fact that finite measures on second countable spaces are always τ-continuous.

It should be mentioned that the question, whether there are in fact finite measures not being τ-continuous, belongs to set theory (see [2], Appendix 3).

A first metric on $M_\tau(X)$, which basically can be traced back to [6], makes essential use of the class $\mathcal{F}_d(X)$.

Definition 2. The distance $\bar{d}$ of two measures $\mu_i \in M_\tau(X)$ is given by $\bar{d}(\mu_1, \mu_2) := \sup\{|\mu_1(f) - \mu_2(f)| : f \in \mathcal{F}_d(X)\}$.

In contrast to the related definition in [8], this distance may be infinite even for a finite metric d. However, it satisfies

Proposition 1. *$\bar{d}$ defines a metric on $M_\tau(X)$.*

Proof. The only point to be checked is the equation $\mu_1 = \mu_2$ for two measures $\mu_i \in M_\tau(X)$ with $\mu_1(f) = \mu_2(f)$ for all $f \in \mathcal{F}_d(X)$. But if $G \subset X$ is open with $F := X \setminus G \neq \emptyset$, the functions $f_k(x) := d(x, F) \wedge \frac{1}{k}$ belong to $\mathcal{F}_d(X)$ and the functions kf_k increase to 1_G, hence $\mu_1(G) = \mu_2(G)$.

This metric is fairly convenient, as shown by the following simple examples:

(1) If d is the discrete metric with $d(x_1, x_2) = 1$ for $x_1 \neq x_2$, modulo additive constants $\mathcal{F}_d(x)$ consist of all functions $0 \leqslant f \leqslant 1$, which implies

$$\bar{d}(\mu_1, \mu_2) = \sup\{|\mu_1(B) - \mu_2(B)| : B \in \mathfrak{B}(X);\}$$

(2) if d is any metric and ε_x is the point measure in x,

$$\bar{d}(\varepsilon_x, \mu) = \int_X d(x, y)\,\mu(dy) \text{ for all } \mu \in M_\tau(X),$$

as follows easily from the equation

$$\varepsilon_x(f) - \mu(f) = \int_X (f(x) - f(y))\,\mu(dy).$$

Generally, the supremum in the definition of $\bar{d}$ is not attained; this happens, however, in an important case:

Proposition 2. *If the metric d is bounded, for each pair of measures $\mu_i \in M_\tau(X)$ there exists $f \in \mathcal{F}_d(X)$ with $\bar{d}(\mu_1, \mu_2) = |\mu_1(f) - \mu_2(f)|$.*

Proof. According to Lemma 3, (2) the space X may be assumed to be separable, and thus Lemma 2, (1) may be applied with $\underline{f} = 0$ and $\bar{f} = \gamma$, where γ is a finite bound for d.

The most important fact about $\bar{d}$ is provided by the following essentially known comparison (see also [3], Section 8):

Proposition 3. *Denote by $\mathfrak{T}$ the usual topology on $M_\tau(Y)$, i.e. the initial topology with respect to the evalutation maps $\mu \mapsto \mu(f)$ for all bounded continuous functions $f: X \to \mathbf{R}$, and by $\bar{\mathfrak{T}}$ the topology induced by the metric $\bar{d}$. Then*

(1) *$\mathfrak{T}$ is coarser than $\bar{\mathfrak{T}}$,*

(2) *$\mathfrak{T}$ and $\bar{\mathfrak{T}}$ coincide, if d is bounded.*

Proof. (1) Choosing an open set G and defining the functions f_k as in the proof of Proposition 1 yield

$$\mu(G) = \sup_{k \in \mathbf{N}} k\mu(f_k) \text{ for all } \mu \in M_\tau(X).$$

Hence, with respect to the topology $\mathfrak{T}$, the maps $\mu \mapsto \mu(G)$ are lower semicontinuous for all open sets G; but $\mathfrak{T}$ is well known to be the coarsest topology on $M_\tau(X)$ meeting this requirement.

(2) Fix a finite bound γ for d and choose arbitrarily $\mu_0 \in M_\tau(X)$ and $\varepsilon > 0$. Then, an application of Lemma 3, (1) yields a totally bounded set $Y \in \mathfrak{B}(X)$ with $\mu_0 (X \setminus Y) < \varepsilon$. Next, using Lemma 2, (2) with $f = 0, \bar{f} = \gamma$ and Y replacing X yield a finite set of functions $g_1, \ldots, g_k \in \mathcal{F}_d(Y)$ such that

$$(*) \quad \begin{cases} 0 \leqslant g_j \leqslant \gamma \text{ and } \min\limits_{1 \leqslant j \leqslant n} \sup\limits_{x \in Y} |g(x) - g_j(x)| \leqslant \varepsilon \\ \text{for all } g \in \mathcal{F}_d(Y) \text{ with } 0 \leqslant g \leqslant \gamma. \end{cases}$$

Finally, extend g_j to $f_j \in \mathcal{F}_d(X)$ as in Lemma 1, denote by G the open ε-neighborhood of Y and define

$$M_0 := \{\mu \in M_\tau(X) : \mu(G \setminus X) < \varepsilon \text{ and } | \mu(f_j) - \mu_0(f_j) | < \varepsilon \text{ for } 1 \leqslant j \leqslant k\}.$$

Obviously this set contains μ_0 and is open with respect to $\mathfrak{T}$. Now consider an arbitrary $\mu \in M_0$ and a function $f \in \mathcal{F}_d(X)$ with $0 \leqslant f \leqslant \gamma$. According to (*) there is an index j such that

$$|f(x) - f_j(x)| 3\varepsilon \leqslant \text{ resp.} \leqslant \gamma \text{ for } x \in G \text{ resp. } x \notin G,$$

and this implies

$$| \mu(f) - \mu_0(f) | \leqslant | \mu(f) - \mu(f_j) | + | \mu(f_j) - \mu_0(f) | + |\mu_0(f_j) - \mu_0(f) |$$
$$\leqslant (3\,\varepsilon + \gamma\varepsilon) + \varepsilon + (3\,\varepsilon + \gamma\varepsilon).$$

Thus $\bar{d}(\mu, \mu_0) \leqslant \varepsilon' := (7 + 2\gamma)\varepsilon$ and the assertion follows.

It should be mentioned that the implication (2) can be reversed, if the metric d is finite. In fact, if d is unbounded, there exist points x and x_k with $d(x, x_k) \geqslant k^2$. For the measures

$$\mu := \varepsilon_x \text{ and } \mu_k := \left(1 - \frac{1}{k}\right)\varepsilon_x + \frac{1}{k}\varepsilon_{x_k}$$

this implies $\mu_k \to \mu$ with respect to $\mathfrak{T}$ and at the same time $\bar{d}(\mu_k, \mu) \to \infty$.

After these basic facts about the first metric a second one will be introduced now. It goes back to [15] and is closely related to the problem treated in Section 1.

Definition 3. The distance $\bar{\bar{d}}$ of two measures $\mu_i \in M_\tau(X)$ is given by $\bar{\bar{d}}(\mu_1, \mu_2) := \inf \{\mu(d) : \mu \in M_\tau(X \times X) \text{ and } \pi_i(\mu) = \mu_i \text{ for } i = 1,2\}$.

Now and henceforth the product measure $\mu_1 \otimes \mu_2$ will always be identified with its extension to a measure in $M_\tau(X \times X)$, which exists uniquely according to Lemma 3, (2). Then there is the simple bound

$$\bar{\bar{d}}(\mu_1, \mu_2) \leqslant (\mu_1 \otimes \mu_2)(d).$$

Proving that $\bar{\bar{d}}$ is in fact a metric is superfluous, since this is a consequence of the main result of this section, the identity $\bar{d} = \bar{\bar{d}}$, which will be proved in two steps in Theorems 1 and 2. The original result of Kantorovich-Rubinštein concerned the case of a compact metric space (see [6]) and was extended to the case of a separable metric space by Dudley under an additional moment condition (see [3], Section 20). As already mentioned, his proof is incomplete; it was completed independently in [1] and [4] for Polish spaces and in [10] for tight measures on arbitrary spaces with a bounded metric. The boundedness condition there being unnecessary, the main result of Section 1 yields

Theorem 1. *Let the measures* $\mu_j \in M_\tau(X)$ *be tight; then*

$$\bar{d}(\mu_1, \mu_2) = \bar{\bar{d}}(\mu_1, \mu_2)$$

and the infimum in the definition of $\bar{\bar{d}}$ *is attained.*

Proof. 1. The inequality $\bar{d} \leqslant \bar{\bar{d}}$ is an immediate consequence of

$$|f \circ \pi_1 - f \circ \pi_2| \leqslant d \text{ for } f \in \mathcal{F}_d(X).$$

Hence, in view of the theorem in Section 1, it remains to show

$$S(d) \leqslant \bar{d}(\mu_1, \mu_2).$$

Now in the definition of $S(d)$ the functions f_i can be replaced by $f_i \wedge k$, $k \in \mathbf{N}$, and may therefore be supposed to be bounded above; but if $f_1, f_2 \leqslant k$, both functions can be replaced by $f_i \vee (-k)$ because of $d \geqslant 0$, thus yielding bounded functions. Hence it suffices to prove

$$(*) \quad \begin{cases} \text{for bounded functions } f_i \in \mathcal{L}(\mu_i) \text{ with } \sum\limits_{i=1,2} f_i \circ \pi_i \leqslant d \\ \text{there exists } f \in \mathcal{F}_d(X) \text{ with } f_1 \leqslant +f \text{ and } f_2 \leqslant -f. \end{cases}$$

2. Without violating the main inequality in (*), the functions f_i may be successively increased to

$$f_1'(x_1) := \inf_{x_2 \in X_2} (d(x_1, x_2) - f_2(x_2)),$$

$$f_2'(x_2) := \inf_{x_1 \in X_1} (d(x_1, x_2) - f_1'(x_1)).$$

These functions are easily seen to be bounded, too, and moreover to belong to $\mathcal{F}_d(X)$, due to the triangle inequality for d. This holds again for $f := \frac{1}{2}(f_1' - f_2')$, where $f_1'(x) + f_2'(x) \leqslant d(x, x) = 0$. Hence $f_1' \leqslant +f$ and $f_2' \leqslant -f$, as was to be proved.

Before extending the first assertion of the preceding theorem to arbitrary measures $\mu_i \in M_\tau(X)$, their tightness will be shown to be essential for the second assertion. To construct a counterexample, let λ denote the Lebesgue measure and $[0,1] = A_1 \cup A_2$ be a nonmeasurable partition with $\lambda^*(A_1) = 1 = \lambda^*(A_2)$. Next define

$$X := (A_1 \times \{1\}) \cup (A_2 \times \{2\}) \subset \mathbf{R}^2$$

and let this space be provided with the Euclidean metric d. Then the sets $B \in \mathfrak{B}(X)$ are of the type

$$B = \bigcup_{i=1,2} (A_i \cap B_i) \times \{i\}) \text{ with } B_. \in \mathfrak{B}([0, 1]).$$

Since $\lambda^*(A_i) = 1$ and X is second countable, for each i the equation

$$\mu_i(B) := \lambda(B_i) \text{ for } B \in \mathfrak{B}(X) \text{ as above}$$

uniquely defines a measure $\mu_i \in M_\tau(X)$ with support $A_i \times \{i\}$. Since $d(x_1, x_2) > 1$ for $x_i \in A_i \times \{i\}$, moreover

$$\mu(d) > 1 \text{ for } \mu \in M_\tau(X \times X) \text{ with } \pi_i(\mu) = \mu_i \text{ for } i = 1, 2.$$

On the other hand, if $k \in \mathbf{N}$ and the sets $C_j \subset X \times X$ are defined by

$$C_j := \left(\left] \frac{j-1}{k}, \frac{j}{k} \right] \times \{1\} \right) \times \left(\left] \frac{j-1}{k}, \frac{j}{k} \right] \times \{2\} \right),$$

it is not difficult to check that

$$\mu(B) := k(\mu_1 \otimes \mu_2)(B \cap \bigcup_{1 \leqslant j \leqslant k} C_j) \text{ for } B \in \mathfrak{B}_\tau(X \times X)$$

yields a measure $\mu \in M_\tau(X \times X)$ with $\pi_i(\mu) = \mu_i$ for $i = 1, 2$ such that $\mu(d) \leqslant 1 + \frac{1}{k}$. Thus $\bar{\bar{d}}(\mu_1, \mu_2) = 1$ and the infimum is not attained.

For the identity $\bar{d} = \bar{\bar{d}}$, however, the tightness of μ_i is dispensable.

Theorem 2. *The equation* $\bar{d}(\mu_1, \mu_2) = \bar{\bar{d}}(\mu_1, \mu_2)$ *holds for all* $\mu_i \in M_\tau(X)$.

Proof. The inequality $\bar{d} \leqslant \bar{\bar{d}}$ follows as in the proof of Theorem 1. For the converse inequality choose $\varepsilon > 0$ and assume X to be separable, which is possible according to Lemma 3, (2). Hence there exists a countable partition $X = \bigcup_{k \in T} B_k$ of $\mathfrak{B}(X)$ such that $B_k \neq \emptyset$ and

$$\delta(B_k) := \sup \{d(x_1, x_2) : x_i \in B_k\} \leqslant \varepsilon \text{ for all } k \in T.$$

Let Y be a subspace of X containing exactly one point y_k from each set B_k and let the measures $\nu_i \in M_\tau(Y)$ be defined by

$$\nu_i(\{y_k\}) := \mu_i(B_k) \text{ for } k \in T \text{ and } i = 1, 2.$$

Then extending $g \in \mathcal{F}_d(Y)$ to $f \in \mathcal{F}_d(X)$ as in Lemma 1 yields the inequality $|f(x) - g(y_k)| \leqslant \varepsilon$ for all $x \in B_k$ and thus

(1) $$\bar{d}(\nu_1, \nu_2) \leqslant \bar{d}(\mu_1, \mu_2) + 2\varepsilon.$$

Now to each measure $\nu \in M_\tau(Y \times Y)$ with $\pi_i(\nu) = \nu_i$ for $i = 1, 2$ a measure $\mu \in M_\tau(X \times X)$ with $\pi_i(\mu) = \mu_i$ for $i = 1, 2$ will be assigned in the following way: there exists a function $h: X \times X \to \mathbf{R}$, being constant on each set $B_{k_1} \times B_{k_2}$, $k_i \in T$, such that

$$\nu(\{y_{k_1}\} \times \{y_{k_2}\}) = h(y_{k_1}, y_{k_2}) \prod_{i=1,2} \nu_i(\{y_{k_i}\}) \text{ for } k_i \in T;$$

the equation

$$\mu(B) := \int_B h d(\mu_1 \otimes \mu_2) \text{ for } B \in \mathfrak{B}(X \times X)$$

then defines a measure $\mu \in M_\tau(X \times X)$, which is easily checked to satisfy the condition $\pi_i(\mu) = \mu_i$ for $i = 1, 2$. Since $d(x_1, x_2) \leqslant d(y_{k_1}, y_{k_2}) + 2\varepsilon$ for $x_i \in B_{k_i}$, obviously $\mu(d) \leqslant \nu(d) + 2\varepsilon$ and thus

$$\bar{\bar{d}}(\mu_1, \mu_2) \leqslant \bar{\bar{d}}(\nu_1, \nu_2) + 2\varepsilon. \tag{2}$$

But Theorem 1 applies to the countable space Y, hence the inequalities (1) and (2) combine to $\bar{\bar{d}}(\mu_1, \mu_2) \leqslant \bar{d}(\mu_1, \mu_2) + 4\varepsilon$ and letting $\varepsilon \downarrow 0$ this implies the missing inequality.

The preceding proof shows in addition that the infimum in the definition of $\bar{\bar{d}}$ can be restricted to those measures μ that are absolutely continuous with respect to the product measure $\mu_1 \otimes \mu_2$.

Returning to the situation of Theorem 1, a more probabilistic version of this result can be stated

Theorem 3. *If the measures $\mu_i \in M_\tau(X)$ are tight and the probability space $(\Omega, \mathfrak{A}, \mathsf{P})$ is given by $\Omega = [0, 1]$, $\mathfrak{A} = \mathfrak{L}(\Omega)$, $\mathsf{P} = \lambda$, then there exist random variables $\xi_i \colon \Omega \to X$ with distributions μ_i such that*

$$\mathsf{E}(d(\xi_1, \xi_2)) = \int_0^1 d(\xi_1(\omega), \xi_2(\omega))\, d\omega = \bar{d}(\mu_1, \mu_2).$$

Proof. In view of the tightness the space X can obviously be supposed to be the union of a countable number of disjoint compact subsets. Since this property carries over to $X \times X$, by standard isomorphism theorems (see for instance [12], Theorem 2.12) there exist a set $B \in \mathfrak{B}(\mathbf{R})$ and a Borel isomorphism $\varphi \colon B \to X \times X$. Let $\mu \in M_\tau(X \times X)$ be chosen according to Theorem 1 and ν its image with respect to φ^{-1}. If $\bar{\nu}$ denotes the trivial extension of ν to $\mathbf{R}$, there exists a real random variable η on $(\Omega, \mathfrak{A}, \mathsf{P})$ with this distribution. Redefining it, if necessary, on a null set, the range of η can be restricted to B. Then the random variables $\xi_i := \pi_i \circ \varphi \circ \eta$ have the distributions μ_i and fulfil

$$\mathsf{E}(d(\xi_1, \xi_2)) = \mu(d) = \bar{d}(\mu_1, \mu_2).$$

An explicit solution can be given in the case $X = \mathbf{R}$: if F_i is the distribution function corresponding to μ_i and

$$\xi_i(\omega) := \sup \{x \in \mathbf{R} \colon F_i(x) \leqslant \omega\} \quad \text{for } i = 1, 2,$$

then it can be shown (see [14]) that

$$\mathsf{E}(d(\xi_1, \xi_2)) = \bar{d}(\mu_1, \mu_2) = \int_{-\infty}^{+\infty} | F_1(x) - F_2(x) |\, dx.$$

Concluding, as another argument for preferring $\bar{d}$ to the more commonly used Prochorov metric, a fairly simple proof of the following fact will be given:

Proposition 4. *If X is complete with respect to the metric d, the space $M_\tau(X)$ is complete with respect to $\bar{d}$.*

Proof. 1. Let $(\mu_k)_{k \in \mathbf{N}}$ be a Cauchy sequence with respect to $\bar{d}$, which — after passage to a subsequence, if necessary — may be assumed to satisfy the condition $\sum_{k\in\mathbf{N}} \bar{d}(\mu_k, \mu_{k+1}) < \infty$. Moreover, in view of Lemma 3, (2) the space X may be assumed to be Polish; hence all measures μ_k are tight and Theorem 1 applies. It yields measures $\mu_{k\,k+1} \in M_\tau(X \times X)$ such that

$$\mu_{k\,k+1}(d) = \bar{d}(\mu_k, \mu_{k+1}) \quad \text{for } k \in \mathbf{N}.$$

Since the space X is Polish, there is a disintegration

$$\mu_{k\,k+1} = \mu \otimes P_{k\,k+1} \text{ for } k \in \mathbf{N}$$

for appropriate Markov kernels $P_{k\,k+1}$ from X to X. Putting them together according to the theorem of Ionescu Tulcea results in

$$\nu := \mu_1 \otimes P_{12} \otimes P_{23} \otimes \ldots,$$

which defines a (Markov) measure in $M_\tau(Y)$ with $Y := X^{\mathbf{N}}$ and which has two-dimensional marginals $\mu_{k\,k+1}$ for $k \in \mathbf{N}$.

2. The canonical random variables $\xi_k := \pi_k$, $k \in \mathbf{N}$, satisfy

$$\int_Y \sum_{k\in\mathbf{N}} d(\xi_k, \xi_{k+1})\, d\nu = \sum_{k\in\mathbf{N}} \int_{X\times X} d(x_k, x_{k+1})\, d\mu_{k\,k+1} = \sum_{k\in\mathbf{N}} \bar{d}(\mu_k, \mu_{k+1}) < \infty.$$

Therefore the set

$$C := \left\{ y \in Y : \sum_{k\in\mathbf{N}} d(\xi_k(y), \xi_1(y)) < \infty \right\} \in \mathfrak{B}(Y)$$

fulfils $\nu(C) = 1$ and, due to the completeness of the space X, the limit $\xi = \lim_{k\to\infty} \xi_k$ exists a.e. and is a random variable, since X is a metric space. Denoting its distribution by μ, this yields

$$\bar{d}(\mu_k, \mu) \leqslant \int_Y d(\xi_k, \xi)\, d\nu \leqslant \int_Y \sum_{l\geqslant k} \bar{d}(\xi_l, \xi_{l+1})\, d\nu = \sum_{l\geqslant k} \bar{d}(\mu_l, \mu_{l+1}) \to 0 \text{ for } k \to \infty.$$

The preceding implication can readily be reversed; if the space $M_\tau(X)$ is complete with respect to $\bar{d}$ and $(x_k)_{k\in\mathbf{N}}$ is a Cauchy sequence in X, due to the isometry $x \to \varepsilon_x$ there is $\mu \in M_\tau(X)$ such that

$$\int_Y d(x_k, x)\, d\mu = \bar{d}(\varepsilon_{x_k}, \mu) \to 0.$$

Hence there is a subsequence $(y_k)_{k\in\mathbf{N}}$ of $(x_k)_{k\in\mathbf{N}}$ satisfying $d(y_k, x) \to 0$ a.e. with respect to μ and thus $y_k \to x$ for a suitable $x \in X$.

REFERENCES

1. De Acosta, A., *Invariance principles in probability for triangular arrays of B-valued random vectors and some applications.* Ann. Probab., 10, 346–373 (1982).
2. Billingsley, P., *Convergence of Probability Measures.* New York, Wiley, 1968.
3. Dudley, R. M., *Probabilities and Metrics.* Lecture Notes Series, No. 45. Aarhus, Matematisk Institut, 1976.
4. Fernique, X., *Sur le théorème de Kantorovitch-Rubinstein dans les espaces polonais.* In: *Séminaire de Probabilités XV.* Lecture Notes in Math., Vol. 850, pp. 6–10. Berlin–Heidelberg–New York, Springer, 1981.

5. Gaffke, N., Rüschendorf, L., *On a class of extremal problems in statistics.* Math. Operationsforschung Statist. Ser. Optim., **12**, 123–135 (1981).
6. Kantorovich, L. V., Rubinstein, G. S., *On a space of completely additive functions.* Vestnik Leningrad. Univ., **13**, 7, 52–59 (1958) (Russian).
7. Kellerer, H. G., *Verteilungsfunktionen mit gegebenen Marginalverteilungen.* Z. Wahrsch. Verw. Gebiete, **3**, 247–270 (1964).
8. Kellerer, H. G., *Markov-Komposition und eine Anwendung auf Martingale.* Math. Ann., **198**, 99–122 (1972).
9. Kellerer, H. G., *Duality theorems for marginal problems.* To be published.
10. Lott, M., *Eine Ordnungsrelation für Radon-Masse.* Diplomarbeit, Universität München, 1981.
11. Monge, G., *Mémoire sur la théorie des déblais et des remblais.* Mém. Math. Phys. Acad. Roy. Sci. Paris, 666–704 (1781).
12. Parthasarathy, K.R., *Probability Measures on Metric Spaces.* New York–London, Academic Press, 1967.
13. Rüschendorf, L., *Sharpness of Fréchet-bounds.* Z. Wahrsch. Verw. Gebiete, **57**, 293–302 (1981).
14. Vallender, S.S., *Calculation of the Wasseršte in distance between probability distributions on the line.* Theory Probab. Appl., **18**, 784–786 (1973).
15. Wasseršte in, L.N., *Markov processes with countable state space describing large systems of automata.* Problems Inform. Transmission, **5**, *3*, 47–52 (1969).

WILLEM K. KLEIN HANEVELD

Distributions with known marginals and duality of mathematical programming with application to PERT

1. Introduction

Suppose that a multivariate probability distribution is unknown, except for the fact that all marginal distributions have been given. In general, there are many joint distributions compatible with the given marginals. In a worst case analysis the following *distribution problem* may be appropriate:

Find the joint distribution which maximizes the expectation of a given criterion function.

This optimization problem has a nice structure: it is a *linear programming* problem of a special type. In fact, it is a generalization of the well-known *transportation* problem. For details we refer to Section 2. In the same section we give sufficient conditions for optimality in terms of a dual linear programming problem. The dual problem involves the upper approximation of the criterion function by a separable function.

In Section 3 we investigate under which conditions on the criterion function the so-called NW Corner Rule generates an optimal solution of the distribution problem. The name refers to the well-known device to get a starting basic feasible solution of a transportation problem. The results are related with [8] and [2].

In Section 4 we apply the results of Section 2 to the project planning model. The objective is the determination of the probability of meeting a specified deadline for a project, whereas the durations of the activities of the project are random. In practical situations there is only very partial information on the joint distribution. PERT is a heuristic for this problem which is often used. Following [5] and [6] we will show that a worst case analysis of the problem is also possible, where the criterion function represents the delay of the completion time of the project. Due to the special form of this criterion it is possible to use a simplified version of the dual problem, which appears to be a finite dimensional convex minimization problem with linear constraints. In fact, it happens to be a stochastic programming problem of the *simple recourse* type which can be easily solved. By means of the dual solution the primal solution, i.e. the most unfavourable distribution, can be characterized. If furthermore it is assumed that even the marginal distributions are not known completely, but that only the ranges and the means are known, the results are still useful since the most unfavourable marginals compatible with the partial information are easily determined ([3], [9]).

2. The distribution problem and its dual

Let r be a natural number, $r \geqslant 2$, and for $i = 1, \dots, r$ let $X_i \subset \mathbf{R}$ be a nonempty Borel set, $e_i: X_i \to \mathbf{R}$ a finite Borel function, $e_i(x_i) \geqslant 1 \ \forall x_i \in X_i$ and μ_i a probability measure on the Borel sets of X_i such that $\int_{X_i} e_i \, d\mu_i < \infty$. One might choose $X_i := \text{support } (\mu_i)$ but this is not necessary. Define $X := \prod_{i=1}^{r} X_i \subset \mathbf{R}^r$, and let $f: X \to \mathbf{R}$ be a Borel function satisfying $|f(x)| \leqslant \sum_{i=1}^{r} M_i e_i(x_i) \ \forall x \in X$ for certain positive constants M_i. Then the distribution problem on X with criterion function f and prescribed marginals $\mu_i \ i = 1, \dots, r$ can be formulated as

$$\text{P} \qquad \text{maximize} \left\{ \int_X f d\mu : \text{proj}_i \mu = \mu_i \ i = 1, \dots, r, \ \mu \geqslant 0 \right\}$$

over the linear space $\mathbf{M}$ of all finite signed measures on the Borel sets of X, and $\text{proj}_i \mu$ denotes the projection of μ on the coordinate space X_i: for a Borel set $B_i \subset X_i$: $(\text{proj}_i \mu)(B_i) := \mu\left(\prod_{j=1}^{r} E_j\right)$ where $E_i := B_i$ and $E_j := X_j$ for $j \neq i$. Notice that the objective as well as the constraints are linear in μ, so that P is a linear programming (LP) problem. It has feasible solutions $\left(\text{e.g. the independent solution } \mu := \hat{\mu} \text{ with } \hat{\mu}\left(\prod_{i=1}^{r} B_i\right) := \prod_{i=1}^{r} \mu_i(B_i)\right)$, and the objective function is bounded on the feasible set $\left(\text{for each feasible } \mu \text{ it holds that } \left|\int f d\mu\right| \leqslant \sum_{i=1}^{r} M_i \int e_i \, d\mu_i < \infty\right)$, so that the optimal value supP is finite.

The functions e_i have been introduced to circumvent finiteness questions. The choice may be adjusted to the special application at hand. For example: if $e_i(x_i) = 1$ then arbitrary marginal probability measures μ_i are allowed, but only bounded criterion functions f are accepted in problem P. If $e_i(x_i) = 1 + |x_i|$ (or $1 + x_i^2$) then μ_i must have a finite first (or second) moment, but the boundedness condition on f can be relaxed.

Sometimes it is pleasant to formulate P by means of r stochastic variables ξ_i each with a given distribution function F_i. Then the objective function gets the form $E_H f(\xi_1, \dots, \xi_r)$ where the expectation is taken with respect to any r-dimensional distribution function H with marginals $F_1, \dots, F_r$: $: H \in \mathbf{H}(F_1, \dots, F_r)$. We shall switch the notations freely.

Problem P is a generalization of the famous transportation problem. Indeed, for $r := 2$, $X_1 := \{s_1, .., s_m\}$, $X_2 := \{t_1, ..., t_n\}$, $p_i := \mu_1(\{s_i\})$, $q_j := \mu_2(\{t_j\})$, $x_{ij} := \mu(\{(s_i, t_j)\})$, $c_{ij} := f(s_i, t_j)$, P is reduced to

$$\text{P}_\circ \qquad \text{maximize}_{x \in \mathbf{R}^{mn}} \left\{ \sum_{i=1}^{m} \sum_{j=1}^{n} c_{ij} x_{ij} : \begin{array}{l} \sum_{j=1}^{n} x_{ij} = p_i, \, i = 1, ..., m \\ \sum_{i=1}^{m} x_{ij} = q_j, \, j = 1, ..., n \end{array} \quad x_{ij} \geqslant 0 \ \forall i, j \right\}$$

The dual problem of $\text{P}_\circ$ can be written as

$$\text{D}_\circ \qquad \text{minimize}_{y_1 \in \mathbf{R}^m, \, y_2 \in \mathbf{R}^n} \left\{ \sum_{i=1}^{m} p_i y_{1i} + \sum_{j=1}^{n} q_j y_{2j} : y_{1i} + y_{2j} \geqslant c_{ij} \ \forall i, j \right\}.$$

Since both the primal problem $\text{P}_\circ$ and the dual problem $\text{D}_\circ$ have feasible solutions they have optimal solutions too, which characterize each other by the well-known LP optimality conditions.

Theorem 1. $x^* \in \mathbf{R}^{mn}$ *is optimal for* $\text{P}_\circ$ *and* $(y_1^*, y_2^*) \in \mathbf{R}^{m+n}$ *is optimal for* $\text{D}_\circ$ *iff the following conditions are true:*

a. x^* *is feasible for* $\text{P}_\circ$: $\Sigma_j x_{ij}^* = p_i$, $\Sigma_i x_{ij}^* = q_j$, $x_{ij}^* \geqslant 0 \ \forall i, j$.

b. (y_1^*, y_2^*) *is feasible for* $\text{D}_\circ$: $y_{1i}^* + y_{2j}^* \geqslant c_{ij}$, $\forall i, j$.

c. *Complementary slackness holds:* $\Sigma_i \Sigma_i x_{ij}^* (y_{1i}^* + y_{2j}^* - c_{ij}) = 0$.

Remark. Consider for a moment $\hat{x} \in \mathbf{R}^{mn}$, $\hat{x}_{ij} := p_i q_j$, which is a feasible solution for $\text{P}_\circ$. It corresponds to the bivariate distribution with independent marginals. Assuming (without loss of generality) that $p_i q_j > 0 \ \forall i, j$, we conclude that $\hat{x}$ is optimal for $\text{P}_\circ$ iff $c_{ij} = y_{1i}^* + y_{2j}^*$ for certain $(y_1^*, y_2^*) \in \mathbf{R}^{m+n}$. In that case the problem $\text{P}_\circ$ is rather meaningless since every feasible solution is optimal. Analogously, we might not expect that the independent distribution $\hat{\mu}$ solves P for any specification of the marginals, except in the case that the criterion function f is *separable*: $f(x_1, ..., x_r) \equiv \sum_{i=1}^{r} f_i(x_i)$. This case is pathological since a separable function f does not discriminate between the feasible distributions of P and does not deserve the name criterion function. On the contrary, one must expect that in general the distributions μ^* which are optimal for P show heavy dependence between the marginals.

We propose the following generalization of $\text{D}_\circ$

$$\text{D} \qquad \text{minimize} \left\{ \sum_{i=1}^{r} \int y_i \, d\mu_i : \sum_{i=1}^{r} y_i \geqslant f \right\}$$

over $\mathbf{Y} := \prod_{i=1}^{r} \mathbf{Y}_i$, where $\mathbf{Y}_i$ is the linear space of all Borel functions y_i on X_i with $\| y_i \| := \sup_{x_i \in X_i} |y_i(x_i)| / e_i(x_i) < \infty$, and where the inequality has the pointwise interpretation. (It would be sufficient to interprete the inequality *a.s.* with respect to $\prod_{i=1}^{r} \mu_i$, but we can do without this refinement.) Problem D is well-defined and it has a finite optimal value inf D because of the assumptions on μ_i and f. We call D a dual problem because of the following theorem.

Theorem 2. *Both* P *and* D *are feasible, and weak duality holds*

$$(1) \qquad -\infty < \sup P \leqslant \inf D < +\infty.$$

If μ^* *is feasible for* P *and* y^* *is feasible for* D *then both are optimal if complementary slackness is true*

$$(2) \qquad \mu^*\left\{x \in X : \sum_{i=1}^{r} y_i^*(x_i) = f(x_1, .., x_r)\right\} = 1.$$

In that case

$$\max P = \min D.$$

Proof. For $\mu \in \mathbf{M}$, $\mu \geqslant 0$, $\text{proj}_i \mu = \mu_i$ $i = 1, .., r$ and $y \in \mathbf{Y}$, $\sum_{i=1}^{r} y_i \geqslant f$ we have that

$$-\infty < \int f \, d\mu \leqslant \int \left(\sum_{i=1}^{r} y_i\right) d\mu = \sum_{1=1}^{r} \int y_i \, d\mu_i < \infty.$$

Under complementary slackness $\leqslant$ can be replaced by $=$. □

Remark. Notice that Theorem 2 gives only sufficient conditions for optimality instead of necessary and sufficient conditions as Theorem 1. Of course it is possible to use modern duality theory (e.g. [7]) in order to formulate regularity conditions which are sufficient for normality and dual solvability (hence sup P = min D) together with a Kuhn-Tucker condition being necessary and sufficient for primal optimality within this framework. However, the results are in terms of weak topologies on $\mathbf{M}_i$ and $\mathbf{Y}_i$ based on the bilinear form $< y_i, \mu_i > := \int y_i d\mu_i$ and do not seem to be very tractable from a practical point of view. Therefore we restrict ourselves to Theorem 2.

3. The NW Corner Rule Solution

Consider the transportation problem P_0. Feasible solutions can be represented by nonnegative $m \times n$-matrices X with prescribed row totals p_i and column totals q_i. The NW Corner Rule Solution is the following extremal feasible solution, uniquely determined by p_i, $i = 1, ..., m$ and q_j, $j = 1, ..., n$

$$(3) \qquad x_{kl}^{NW} := \left[\min\left(\sum_{i=1}^{k} p_i, \sum_{j=1}^{l} q_j\right) - \max\left(\sum_{i=1}^{k-1} p_i, \sum_{i=1}^{l-1} q_j\right)\right]^+,$$

where $[z]^+ := \max(0, z)$. An efficient algorithm to calculate all strictly positive x_{ij}^{NW} is the well-known so-called North West Corner Rule:

a. Start with the initial NW Corner of X, x_{11}, and give it its maximal possible value $x_{11} := \min(p_1, q_1)$; $k := 1, l := 1$.

If $x_{11} = p_1$ set $x_{1j} := 0, j \geqslant 2$ and $k := 2$.

If $x_{11} = q_1$ set $x_{i1} := 0$, $i \geqslant 2$ and $l := 2$.

b. Give the NW Corner of the remaining part of X, x_{kl}, its maximal possible value $x_{kl} := \min\left(p_k - \sum_{j=1}^{l-1} x_{kj},\ q_l - \sum_{i=1}^{k-1} x_{il}\right)$.

If $x_{kl} = p_k - \sum_{j=1}^{l-1} x_{kj}$ then $x_{kj} := 0$, $j > l$ and $k := k+1$.

If $x_{kl} = q_l - \sum_{i=1}^{k-1} x_{il}$ then $x_{il} := 0$, $i > k$ and $l := l+1$.

Procedure b is repeated until the matrix X is filled up completely (i.e. if $k = m+1$ or $l = n+1$).

Notice that the NW Corner Rule Solution is based on a special order of the rows and the columns of X. In fact, every extremal point of the feasible region of P_0 can be found as a NW Corner Rule Solution if a permutation of the rows and a permutation of the columns is applied first. Notice further that the NW Corner Rule Solution does not depend on the objective function, so it might not be expected to be optimal in general. However, it is possible to give conditions on c which guarantee that x^{NW} is optimal for all specifications of p_i and q_j.

Theorem 3. *The vector $x^{NW} = x^{NW}(p, q)$ defined by* (3) *is optimal for P_0 for all values of $p_i \geqslant 0, q_j \geqslant 0$ with $\sum_{i=1}^{m} p_i = \sum_{j=1}^{n} q_j (= 1)$ if and only if for each $k = 1, \ldots, m-1$ and $l = 1, \ldots, n-1$*

$$c_{kl} + c_{k+1,l+1} \geqslant c_{k+1,l} + c_{k,l+1}. \tag{4}$$

Proof. If (4) does not hold for $(\bar{k}, \bar{l})$ then it is easy to show that x^{NW} cannot be optimal if for example $p_{\bar{k}} = p_{\bar{k}+1} = q_{\bar{l}} = q_{\bar{l}+1} = \frac{1}{2}$. On the other hand, if (4) is valid for all (k, l) then the optimality of x^{NW} can be proved by means of Theorem 1. Indeed, x^{NW} is feasible for P_0; and it is possible to define

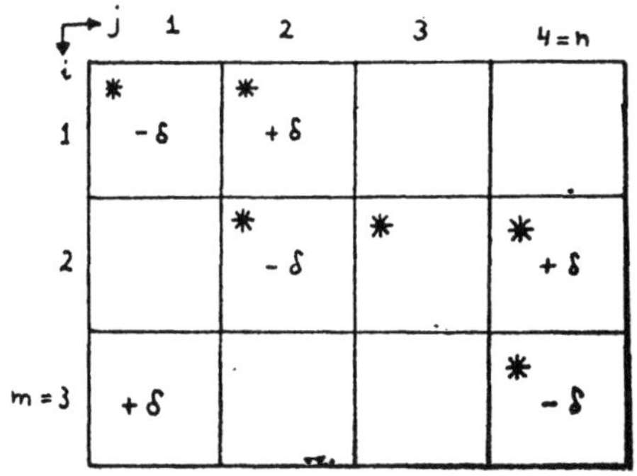

Fig. 1

y_1^* and y_2^* such that the complementary slackness conditions hold, and the feasibility of (y_1^*, y_2^*) for D_0 follows then from (4). An essentially equivalent proof is based on the optimality criterion of the simplex method. Instead of providing a formal proof we give a characteristic example in Figure 1 above.

In the $m \times n$ transportation matrix the positions of the basic variables of the NW Solution are denoted by *. If we allow the nonbasic variable x_{31} to get a positive value δ then the objective function changes with $\delta \cdot c'_{31} := = \delta \cdot (c_{31} - c_{11} + c_{12} - c_{22} + c_{24} - c_{34})$. In order to verify that the reduced cost c'_{31} is nonpositive one only has to add the relations (4) over $(k, l) = = (1, 1), (2, 1), (2, 2)$ and $(2, 3)$. □

As a typical application of Theorem 3 we interpret the transportation model P_0 as a *dynamic production* model. Then i = time period of production, p_i = production capacity in period i, j = time period of demand, q_j = demand in period j, and x_{ij} = production in period i destinated for demand in period j, with corresponding unit cost c_{ij}. Notice that both indices in x_{ij} have a natural order, namely corresponding to time. The NW rule is in that case to be interpreted as a FIFO type of strategy: the oldest demands have highest priority. According to Theorem 3 it gives the minimal cost solution for all values of the capacity parameters p_i and demand parameters q_j (provided the problem is balanced: $\Sigma_i p_i = \Sigma_j q_j$) if the cost vector c satisfies for all (k, l)

(5) $$c_{kl} + c_{k+1, l+1} \leqslant c_{k+1, l} - c_{k, l+1}.$$

This is true for example if $c_{ij} = d + h(i-j)$, h convex. On the other hand, under the same conditions for c the total costs are maximized under the LIFO strategy. The corresponding solution can be called the NE rule solution since high values of j are coupled with low values of i and vice versa.

We shall now try to generalize the results for the transportation problem P_0 to the general distribution problem P. The first question to be answered is then: *what is the definition of the NW Rule Solution of* P? From (3) it follows that

$$\sum_{i=1}^{k} \sum_{j=1}^{l} x_{ij}^{NW} = \min\left(\sum_{i=1}^{k} p_i, \sum_{j=1}^{l} q_j\right)$$

so it is obvious that distribution functions are useful. Let us denote the left-continuous distribution function of the given probability measure μ_i on X_i by F_i, $i = 1, \ldots, r$, then we call μ° the NW Rule Solution of P if its distribution function H° is determined by

(6) $$H^\circ(z_1, \ldots, z_r) = \min_{i=1,\ldots,r} F_i(z_i), \quad z \in \mathbf{R}^r.$$

It is easy to verify that H^0 indeed is a distribution function.

It is illuminating to describe H° by means of inverses of F_i. Let F_i^{-1} be any real function defined on (0, 1) satisfying

$$F_i^{-1}(t) = x_i \rightarrow F_i(x_i) \leqslant t \leqslant F_i(x+0).$$

Then F_i^{-1} is nondecreasing. Let η be a stochastic variable with uniform distribution on (0, 1). Then it is easy to show that $F_i^{-1}(\eta)$ is a stochastic variable with distribution function F_i. For continuous strictly increasing F_i this statement is trivial, but it holds generally, since

(7) $$(0, F_i(x_i)) \subset \{t \in (0, 1) : F_i^{-1}(t) < x_i\} \subset (0, F_i(x_i)]$$

so that $P(F_i^{-1}(\eta) < x_i) = F_i(x_i)$ for all $x_i \in \mathbf{R}$. Hence, the distribution of the stochastic vector

(8) $$\theta := (F_1^{-1}(\eta), F_2^{-1}(\eta), \ldots, F_r^{-1}(\eta))$$

has a distribution which is feasible for P. Using (7) again one shows easily that this distribution is actually the NW distribution H°! The representation of H° by (8) indicates the strong dependence of its marginals: roughly speaking we may say that the probability mass of H° is concentrated on an 'increasing' line in $\mathbf{R}^r$.

Remark. It is possible to use transformations analogous to (8), $\xi_i := F_i^{-1}(\eta_i)$ with η_i uniform on (0, 1), to get a problem P′ equivalent to P with $F_i =$ uniform distribution on (0, 1). The NW Rule Solution is then the main diagonal in the unit cube, and the new criterion function is $f(F_1^{-1}(\cdot), \ldots, F_r^{-1}(\cdot))$.

From (8) useful formulas for μ° and H° can be derived. For example, let $-\infty \leqslant a_i < b_i \leqslant \infty$, $-\infty \leqslant z_i \leqslant \infty$, $I, I_j \subset \{1, \ldots, r\}$ then

$$(9)\quad \begin{cases} P_{H^\circ}(a_i \leqslant \xi_i < b_i,\ i \in I) = [\min_I F_i(b_i) - \max_I F_i(a_i)]^+ \\ P_{H^\circ}(a_i < \xi_i \leqslant b_i,\ i \in I) = [\min_I F_i(b_i+0) - \max_I F_i(a_i+0)]^+, \\ P_{H^\circ}(\xi_i < z_i,\ i \in I_1;\ \xi_i \leqslant z_i,\ i \in I_2;\ \xi_i \geqslant z_i,\ i \in I_3;\ \xi_i > z_i,\ i \in I_4) = \\ = [\min(\min_{I_1} F_i(z_i), \min_{I_2} F_i(z_i+0)) - \max(\max_{I_3} F_i(z_i), \max_{I_4} F_i(z_i+0))]^+. \end{cases}$$

Hence, for any $z \in \mathbf{R}^r$ and any $h, k \in \{1, \ldots, r\}$ we have

$$(10)\quad \begin{cases} P_{H^\circ}(\xi_h < z_h,\ \xi_k \geqslant z_k) \cdot P_{H^\circ}(\xi_h \geqslant z_h,\ \xi_k < z_k) = 0, \\ P_{H^\circ}(\xi_h < z_h,\ \xi_k > z_k) \cdot P_{H^\circ}(\xi_k \geqslant z_h,\ \xi_k \leqslant z_k) = 0, \\ P_{H^\circ}(\xi_h \leqslant z_h,\ \xi_k > z_k) \cdot P_{H^\circ}(\xi_h > z_h,\ \xi_k \leqslant z_k) = 0. \end{cases}$$

This is a characteristic property of the NW distribution. It holds also for disjunct index sets I_1 and I_2 instead of $\{h\}$ and $\{k\}$. Hence, H° has positive probability mass in at most one of each pair of opposite translated orthants in $\mathbf{R}^r$, except when the positive and negative orthants are concerned. For that case we have the characteristic property:

Theorem 4. *Let A be a subset of X of the type $A = A_1 = \{t : t_i < z_i, i \in I_1$ and $t_i \leqslant z_i, i \in I_2\}$ or $A = A_2 = \{t : t_i \geqslant z_i, i \in I_1$ and $t_i > z_i, i \in I_2\}$ for arbitrary $z \in \mathbf{R}^r$ and $I_1, I_2 \subset \{1, \ldots, r\}$. Then*

$$\max_{\mu \in \mathbf{M}} \{\mu(A) : \mathrm{proj}_i \mu = \mu_i, \forall i,\ \mu \geqslant 0\} = \mu^\circ(A).$$

Proof. Without loss of generality we assume that $I_1 \neq \emptyset$, $I_2 \neq \emptyset$ and $I_1 \cap I_2 = \emptyset$. For any $\mu \in \mathbf{M}$ and any $i \in I_1$, $j \in I_2$ it holds that $\mu(A_1) \leqslant P_\mu(\xi_i < z_i) = F_i(z_i)$ and $\mu(A_1) \leqslant P_\mu(\xi_j \leqslant z_j) = F_j(z_j+0)$, so that $\mu(A_1) \leqslant \min(\min_{I_1} F_i(z_i), \min_{I_2} F_i(z_i)) = \mu^\circ(A_1)$, the last equality being true because of (9).

An analogous proof is valid for A_2. □

Let us now consider condition (4). The obvious generalization is quasi-monotonicity (as it is called in [2]; in [8] it would be named 'subadditivity', whereas [1] calls it 'nonnegativity'). A function $f : \mathbf{R}^r \to \mathbf{R}$ is called *quasi-monotone* (QM) if for any $a \in \mathbf{R}^r$, $b \in \mathbf{R}^r$ with $a_i \leqslant b_i$, $i = 1, \ldots, r$, the iterated difference $D_a^b f$ (to be defined) is nonnegative. For $r = 2$ the definition of $D_a^b f$ is simply $f(b_1, b_2) - f(a_1, b_2) - f(b_1, a_2) + f(a_1, a_2)$ and this shows that QM is indeed a generalization of (4). For general r the definition is by iteration: for fixed $a, b \in \mathbf{R}^r$ and arbitrary $x \in \mathbf{R}^r$, $k \in \{1, \ldots, r\}$ define

$$\Delta_k f(x) := f(x_1, \ldots, x_{k-1}, b_k, x_{k+1}, \ldots, x_r) - f(x_1, \ldots, x_{k-1}, a_k, x_{k+1}, \ldots, x_r)$$

and $D_a^b f := \Delta_1 \Delta_2 \ldots \Delta_r f(x)$; clearly the result does not depend on the choice of x and the order in which the difference operators Δ_k have been applied.

For sufficiently differentiable functions the QM condition is equivalent to $\frac{\partial}{\partial x_1} \cdots \frac{\partial}{\partial x_r} f(x) \geqslant 0$, $x \in \mathbf{R}^r$. Hence $f(x) = \prod_{i=1}^{r} x_i$ is QM, but also any function f which is constant in one of its variables. QM is preserved under monotone transformations of the argument x. There is a close relation with measures ν on the Borel sets of $\mathbf{R}^r$. If ν is finite, its distribution function is QM (and left-continuous). On the other hand, if f is QM and left-continuous then it determines a σ-finite measure ν by

$$\nu\left(\prod_{i=1}^{r} [a_i, b_i)\right) := D_a^b f, \qquad a, b \in \mathbf{R}^r,\ a < b.$$

In that case we have by Fubini's theorem for any $\mu \in \mathbf{M}$, $\mu \geqslant 0$,

$$\int (D_a^x f)\, \mu(\mathrm{d}x) = \int_{\geqslant a} P_\mu(\xi > t)\, \nu(\mathrm{d}t)$$

$$\int (D_x^b f)_\mu(\mathrm{d}x) = \int_{<b} P_\mu(\xi \leqslant t)\, \nu(\mathrm{d}t)$$

where $D_a^x f := 0$ if $x \ngeqslant a$, $D_x^b f := 0$ if $x \nleqslant b$. Slightly different results hold for right-continuous QM function f

$$\int (D_a^x f)\, \mu(\mathrm{d}x) = \int_{>a} P_\mu(\xi \geqslant t)\, \nu(\mathrm{d}t),$$

$$\int (D_x^b f)\, \mu(\mathrm{d}x) = \int_{\leqslant b} P_\mu(\xi < t)\, \nu(\mathrm{d}t),$$

$$\text{with } \nu\left(\prod_{i=1}^{r} (a_i, b_i]\right) := D_a^b f \qquad a,\ b \in \mathbf{R},\ a < b.$$

From Theorem 4 we conclude therefore

Corollary 5. *Let $f : \mathbf{R}^r \to \mathbf{R}$ be QM and left-continuous in some variables, and right-continuous in the remaining variables. Then, for any $a \in \mathbf{R}^r$, $b \in \mathbf{R}^r$*

$$\max_{\mu \in \mathbf{M}} \left\{ \int (D_a^x f)\, \mu(\mathrm{d}x) : \mathrm{proj}_i \mu = \mu_i\ \forall i,\ \ \mu \geqslant 0 \right\} = \int (D_a^x f)\, \mu^\circ(\mathrm{d}x),$$

$$\max_{\mu \in \mathbf{M}} \left\{ \int (D_x^b f)\, \mu(\mathrm{d}x) : \mathrm{proj}_i \mu = \mu_i\ \forall i,\ \ \mu \geqslant 0 \right\} = \int (D_x^b f)\, \mu^\circ(\mathrm{d}x).$$

Consequently, the NW distribution is optimal for problem P *if the criterion function f satisfies one of the following conditions:*

1. $f(x) := \nu(\{t \in \mathbf{R}^r : t_i \leqslant x_i, \ i = 1, ..., r\}) < \infty, \ x \in \mathbf{R}^r,$
2. $f(x) := \nu(\{t \in \mathbf{R}^r : t_i \geqslant x_i, \ i = 1, ..., r\}) < \infty, \ x \in \mathbf{R}^r,$

where ν *is any measure on the Borel sets of* $\mathbf{R}^r$.

Proof. The second half of the theorem follows from the first part with $a_i := -\infty \ \forall i$ (hence $D_a^x f \equiv f(x)$), or with $b_i := +\infty \ \forall i$ (hence $D_x^b f \equiv f(x)$).

In the bivariate case it is possible to derive also results for the case in which the measure ν determined by f does not satisfy one of the finiteness conditions of Corollary 5, see [2]. Generalizations to more dimensions seem difficult, probably because of the following reason. For $r = 2$, $D_a^x f - f(x)$ is separable for $x \geqslant a$, so that the difference of $\int_{\geqslant a} (D_a^x f)\mu(dx)$ and $\int_{\geqslant a} f(x)\,\mu(dx)$ is the same constant for all $\mu \in \mathbf{M}$ which are feasible for P. For $r > 2$ this property does not hold.

4. Worst case analysis for PERT

In this section we choose

$$f(x) := \Big[\max_{j=1,...,n} \sum_{i \in B_j} x_i - T\Big]^+, \quad x \in \mathbf{R}^r, \tag{11}$$

as criterion function for P. $B_1, ..., B_n$ are given different nonempty index sets adding up to $\{1, ..., r\}$, and T is an arbitrary scalar. The appropriate choice for $e_i(x_i)$ is $1 + |x_i|$, hence we assume that the given probability measures μ_i on X_i with distribution functions F_i have finite means m_i. The special problem we have in mind is project planning. Suppose that r activities have to be completed in order to finish a project. The precedence relations between the activities can be represented by a directed network with one source and one sink. If B_j, $j = 1, ..., n$, is the index set of the activities of path j from source to sink, and if ξ_i is the duration of activity i and if T is the scheduled project time, then $f(\xi)$ is the necessary delay. ξ_i is supposed to be a stochastic variable. In this application one has $X_i \subset [0, \infty)$, $T \geqslant 0$ and some additional structure in $\{B_1, ..., B_n\}$ but we shall not use this extra information. For the time being we assume complete knowledge on the marginals F_i, but nothing more on the simultaneous distribution function H is supposed to be known. In a worst case analysis one is led to problem P with f defined as in (11). This problem is also studied in [5] and [6] but by the use of the duality Theorem 2 the results become more transparent and more general. Moreover, afterwards we shall relax the assumption that there is complete knowledge on the marginals.

Consider the dual problem D. Since for fixed x_j, $j \neq k$, $f(x) = w + [x_k - z]^+$ for fixed $w, z \in \mathbf{R}$ it looks reasonable to restrict the attention to dual functions $y_i(x_i)$ of the form $y_i(x_i) := w_i + [x_i - z_i]^+$ for arbitrary parameters w_i and z_i. Substitution leads to the *restricted dual* problem $\tilde{\mathrm{D}}$:

$$\text{minimize}_{(z_0, z) \in \mathbf{R}^{r+1}} \Big\{ z_0 + \sum_{i=1}^{r} G_i(z_i) : z_0 \geqslant 0, \ -z_0 + \sum_{i \in B_j} z_i \leqslant T \ \forall j \Big\}, \tag{12}$$

where $z_0 \equiv \sum_{i=1}^{r} w_i$ and $G_i(z_i) := \int [x_i - z_i]^+ \, dF_i(x_i) = \int_{z_i}^{\infty} (1 - F_i(x_i)) \, dx_i$. Elimination of z_0 gives a more concise formulation of $\tilde{D}$:

$$\text{minimize}_{z \in \mathbf{R}^r} C(z) := f(z) + \sum_{i=1}^{r} G_i(z_i).$$

$\tilde{D}$ can be interpreted as the problem to find reference values z_i for the durations of the activities, in such a way that the delay of the project based on z is balanced with the sum of the expected delay of the stochastic activities. Since C is finite and convex (hence continuous) on $\mathbf{R}^r$, bounded from below and increasing to $+\infty$ if $|z_k| \to \infty$, $\tilde{D}$ must have (an) optimal solution(s) z^*. In fact, (12) can be seen as a *simple recourse* model of stochastic programming, for which efficient algorithms are available [4]. Notice that we may choose $z_i^* \in [a_i, b_i]$ if support $(\mu_i) \subset [a_i, b_i]$, $i = 1, \ldots, r$.

Since $\tilde{D}$ is a restriction of D, the duality Theorem 2 also holds for $\tilde{D}$. In fact, by construction one shows that there is no real restriction, hence min D = = min $\tilde{D}$. From Theorem 2 we get

Theorem 6. *Let f be defined in* (11). *In order that a probability measure μ^* is optimal for* P *it is sufficient that μ^* has the correct marginals and that* $\mu^*(X^*(z^*)) = 1$, *where z^* is any optimal solution of $\tilde{D}$ and*

$$X^*(z) := \left\{ x \in \mathbf{R}^r : f(x) = f(z) + \sum_{i=1}^{r} [x_i - z_i]^+ \right\}. \tag{13}$$

Proof. (13) follows from (2) via $\sum_{i=1}^{r} y_i^*(x_i) = f(z^*) + \sum_{i=1}^{r} [x_i - z_i^*]^+$. □

It is not difficult to characterize $X^*(z)$ more explicitly. It appears to be a (union of) translated orthant(s) with apex z. In order to see this we consider first the case $T \approx -\infty$. Since $\forall j \; \forall x \; \forall z$

$$\sum_{i \in B_j} x_i \leqslant \sum_{i \in B_j} z_i + \sum_{i \in B_j} [x_i - z_i]^+ \leqslant \sum_{i \in B_j} z_i + \sum_{i=1}^{r} [x_i - z_i]^+ \tag{14}$$

we have for $f_0(x) := \max_{j=1,\ldots,n} \sum_{i \in B_j} x_i$ the inequality

$$f_0(x) \leqslant f_0(z) + \sum_{i=1}^{r} [x_i - z_i]^+ \quad \forall x \in \mathbf{R}^r, \; \forall z \in \mathbf{R}^r. \tag{15}$$

Let $J(z)$ denote the index set of maximizing j in the definition of $f_0(z)$ and define for each path B_j the orthant

$$K_j = \{x \in \mathbf{R}^r : x_i \geqslant 0 \text{ if } i \in B_j, \; x_i \leqslant 0 \text{ if } i \notin B_j\}.$$

Lemma 7. *In* (15) *the equality holds iff $x - z \in K_l$ for an $l \in J(z)$. In that case, actually $l \in J(x)$ too.*

Proof. Necessity. Suppose that the equality holds for x and z. For any $l \in J(x)$ we have then

$$\sum_{i \in B_l} x_i = f_0(z) + \sum_{i=1}^{r} [x_i - z_i]^+ \geqslant \sum_{i \in B_l} z_i + \sum_{i=1}^{r} [x_i - z_i]^+ \tag{16}$$

so that, since $x_i - z_i = [x_i - z_i]^+ - [x_i - z_i]^-$,

$$\sum_{i \in B_l} [x_i - z_i]^- + \sum_{i \notin B_l} [x_i - z_i]^+ \leqslant 0,$$

or equivalently $x - z \in K_l$. Substitution in (16) gives $l \in J(z)$.

Sufficiency. Suppose $x - z \in K_l$ and let $l \in J(z)$. Then for any j

$$\sum_{i \in B_j \setminus B_l} x_i \leqslant \sum_{i \in B_j \setminus B_l} z_i \leqslant \sum_{i \in B_l \setminus B_j} z_i \leqslant \sum_{i \in B_l \setminus B_j} x_i.$$

Consequently, $\sum_{i \in B_j} x_i \leqslant \sum_{i \in B_l} x_i \; \forall j$, so that $l \in J(x)$. Hence

$$f_0(x) = \sum_{i \in B_l} x_i = \sum_{i \in B_l} z_i + \sum_{i \in B_l} [x_i - z_i]^+ = f_0(z) + \sum_{i=1}^{r} [x_i - z_i]^+. \qquad \square$$

The interpretation of Lemma 7 is clear. If z_i is the duration of activity i, then the set $J(z)$ is precisely the index set of all *critical* (= maximal) paths in the network, and $\cup \{i : i \in B_j \text{ for a } j \in J(z)\}$ is precisely the set of activities without any slack.

Since $f(x) = [f_0(x) - T]^+$, it follows from (15) that

$$f(x) \leqslant f(z) + \sum_{i=1}^{r} [x_i - z_i]^+, \quad \forall x \in \mathbf{R}^r, \quad \forall z \in \mathbf{R}^r. \tag{17}$$

Using Lemma 7 we can characterize equality in (17) explicitly. Define $K_0 : = \{x \in \mathbf{R}^r : x \leqslant 0\}$ and the index sets $J_T(z)$, $z \in \mathbf{R}^r$, by

$$J_T(z) : = \{0\} \qquad \text{if } f_0(z) < T,$$

$$J_T(z) : = \{0\} \cup J(z) \quad \text{if } f_0(z) = T,$$

$$J_T(z) : = J(z) \qquad \text{if } f_0(z) > T.$$

Theorem 8. $X^*(z) = z + \cup \{K_j : j \in J_T(z)\}$.

Proof. First we shall show that the translated orthants are included in $X^*(z)$. If $f_0(z) \leqslant T$ and $x - z \in K_0$ this is trivial since $x \leqslant z$ implies $f_0(x) \leqslant f_0(z) \leqslant T$ so that each term in (17) is zero. If $f_0(z) \geqslant T$ and if equality holds in (15) then $f_0(x) - T = f_0(z) - T + \sum_{i=1}^{r} [x_i - z_i]^+ = f(z) + \sum_{i=1}^{r} [x_i - z_i]^+ \geqslant 0$ so that $f(x) = [f_0(x) - T]^+ = f(z) + \sum_{i=1}^{r} [x_i - z_i]^+$.

In order to prove the reverse inclusions, let $x \in X^*(z)$. Suppose first that $f_0(z) < T$. Then $[f_0(x) - T]^+ = f(x) = f(z) + \sum_{i=1}^{r} [x_i - z_i]^+ = \sum_{i=1}^{r} [x_i - z_i]^+$. From (15) we know, however, that $f_0(x) \leqslant f_0(z) + \sum_{i=1}^{r} [x_i - z_i]^+ < T + \sum_{i=1}^{r} [x_i - z_i]^+$. Consequently, $f_0(x) - T < 0$ and $\sum_{i=1}^{r} [x_i - z_i]^+ = 0$ so that $x - z \in K_0$. Suppose now that $f_0(z) = T$. If $x - z \in K_0$ we have nothing to prove, so assume $\sum_{i=1}^{r} [x_i - z_i]^+ > 0$. Since $x \in X^*(z)$ it holds that $[f_0(x) - T]^+ = f(x) =$

$= f(z) + \sum_{i=1}^{r} [x_i - z_i]^+ > 0$. Hence $f(x) = f_0(x) - T$ and $f(z) = f_0(z) - T$ so that $x \in X^*(z)$ implies that equality must hold in (15). Finally suppose that $f_0(z) > T$. Then equality in (17) implies also equality in (15), since now $f(x) = f_0(x) - T$ because of $f(x) \geqslant f(z) > 0$. □

Having a complete description of $X^*(z)$, we shall now characterize the optimal solution(s) z^* of $\tilde{D}$ by means of the Kuhn-Tucker theorem (in fact: a dualization of $\tilde{D}$ based on perturbation of the right hand sides of (12)). Since the convex program (12) is stable (the optimal value stays finite under perturbations of the right hand sides), z^*, or more precisely: (z_0^*, z^*) with $z_0^* = f(z^*)$, is optimal for $\tilde{D}$ iff there is a vector $\lambda^* \in \mathbf{R}_+^{n+1}$ of multipliers such that (z_0^*, z^*, λ^*) is a saddle point on $\mathbf{R}^{r+1} \times \mathbf{R}_+^{n+1}$ of the Lagrangian

$$
\begin{aligned}
L(z_0, z, \lambda) &:= z_0 + \sum_{i=1}^{r} G_i(z_i) + \lambda_0(-z_0) + \sum_{j=1}^{n} \lambda_j \Big(-z_0 + \sum_{i \in B_j} z_i - T\Big) \\
&= (-T) \sum_{j=1}^{n} \lambda_j + z_0 \Big(1 - \sum_{j=0}^{n} \lambda_j\Big) + \sum_{i=1}^{r} \Big(G_i(z_i) + z_i \Big(\sum_{j \in B_i^{-1}} \lambda_j\Big)\Big).
\end{aligned}
\tag{18}
$$

Here we defined $B_i^{-1} = \{j : i \in B_j\}$ with $B_0 := \emptyset$. Then $B^{-1} \neq \emptyset$ for $i = 1, \ldots, r$ since $\bigcup_{j=1}^{n} B_j = \{1, \ldots, r\}$. From this we get the following form for the dual of $\tilde{D}$:

$$
\max_{\lambda \in \mathbf{R}^{n+1}} \Big\{ -T(1 - \lambda_0) + \sum_{i=1}^{r} \int_0^{\pi_i} F_i^{-1}(1 - t_i)\, \mathrm{d}t_i : \pi_i = \sum_{j \in B_i^{-1}} \lambda_j,\ \lambda_j \geqslant 0,\ \sum_{j=0}^{n} \lambda_j = 1 \Big\}.
$$

We have

Theorem 9. *z^* is optimal for $\tilde{D}$ iff there is a $\lambda^* \in \mathbf{R}^{n+1}$ such that the following conditions hold:*

a. $\lambda_j^* \geqslant 0$, $j = 0, 1, \ldots, n$ *and* $\sum_{j=0}^{n} \lambda_j^* = 1$.

b. $z_i^* \in F_i^{-1}\Big(1 - \sum_{j \in B_i^{-1}} \lambda_j^*\Big)$, *i.e.* $F_i(z_i^*) \leqslant 1 - \sum_{j \in B_i^{-1}} \lambda_j^* \leqslant F_i(z_i^* + 0)$, $i = 1, \ldots, r$.

c. $\lambda_j^* > 0$ *only if* $j \in J_T(z^*)$, $j = 0, 1, \ldots, n$.

Proof. (a) If $\Sigma_j \lambda_j^* \neq 1$ then there can be no saddle point (z_0^*, z^*, λ^*) since $\inf_{z_0} L(z_0, z^*, \lambda^*) = -\infty$. Condition (b) guarantees that $L(z_0^*, z, \lambda^*)$ is minimal in $z = z^*$, since the subgradient in z_i of the convex function $G_i(x_i) + \lambda x_i$ is $[F_i(z_i) - 1 + \lambda, F_i(z_i + 0) - 1 + \lambda]$. Finally, the complementary slackness conditions are $\lambda_0^* \cdot z_0^* = 0$, $\lambda_j^* \cdot \Big(-z_0^* + \sum_{i \in B_j} z_i^* - T\Big) = 0$ for $j = 1, \ldots, n$, whereas the feasibility of (z_0^*, z^*) is equivalent to $z_0^* \geqslant f(z^*)$, but of course only $z_0^* = f(z^*)$ can be optimal. After substitution of $z_0^* := f(z^*)$ it follows from $\lambda_0^* > 0$ that $f_0(z^*) \leqslant T$ whereas $\lambda_j^* > 0$ for some $j \geqslant 1$ implies that $j \in J(z^*)$ and $f_0(z^*) \geqslant T$. □

Theorems 6 and 8 indicate that one might construct an optimal distribution μ^* by *conditioning* upon the translated orthants $z^* + K_j$, $j \in J_T(z^*)$.

That is: define probabilities $\lambda_j \geqslant 0$, $\sum_{j \in J_T(z^*)} \lambda_j = 1$ and define probability measures μ^j on $z^* + K_j$, $j \in J_T(z^*)$, and try $\mu^* := \sum_{j \in J_T(z^*)} \lambda_j \cdot \mu^j$. As we shall show this approach works fine. In order to circumvent problems on the (partly common) boundaries of the translated orthants, we assume for the moment that the given marginal distribution function F_i is continuous in z_i^* for each $i = 1, ..., r$. The question is: how to define λ_j and μ^j? The only condition in Theorem 6 which is not satisfied automatically, is that the marginals of μ^* must be F_i. That is we must have, with $A_i(z_i) := \{x \in \mathbf{R}^r: x_i < z_i\}$,

$$\sum_{j \in J_T(z^*)} \lambda_j \cdot \mu^j(A_i(z_i)) = F_i(z_i) \ \forall z_i \in \mathbf{R} \ \forall i = 1, ..., r. \tag{19}$$

Let us first work out (19) for $z_i = z_i^*$. Since $\mu^j(A_i(z_i^*)) = 1$ if $i \notin B_j$ and $\mu^j(A_i(z_i^*)) = 0$ if $i \in B_j$ for any probability measure μ^j restricted to $z^* + K_j$ (and with probability zero on its boundary) we get conditions only for λ_j, namely

$$\Sigma_j\{\lambda_j : j \in J_T(z^*),\ j \notin B_i^{-1}\} = F_i(z_i^*), \quad i = 1, ..., r.$$

Because of Theorem 9 this condition is true if we put (as we shall do) $\lambda := \lambda^*$, λ^* being any Kuhn-Tucker vector for $\tilde{\mathrm{D}}$. Notice that for $\lambda = \lambda^*$ the condition $j \in J_T(z^*)$ can be dispensed with, since $j \notin J_T(z^*)$ implies $\lambda_j^* = 0$ (Theorem 9c). Let us now investigate the consequences of (19) for μ^j. We have left over the conditions

$$\sum_{j \notin B_i^{-1}} \lambda_j^* \cdot \mu^j(A_i(z_i)) + \sum_{j \in B_i^{-1}} \lambda_j^* \cdot 0 = F_i(z_i), \ z_i < z_i^*,$$

$$\sum_{j \notin B_i^{-1}} \lambda_j^* \cdot 1 + \sum_{j \in B_i^{-1}} \lambda_j^* \cdot \mu^j\,(A_i(z_i)) = F_i(z_i), \ z_i > z_i^*.$$

It is easily verified that these conditions are satisfied by e.g.

$$\begin{aligned} &\mu^j(A_i(z_i)) := \min\left(\frac{F_i(z_i)}{F_i(z_i^*)},\ 1\right) \qquad z_i \in \mathbf{R},\ i \notin B_j, \\ &\mu^j(A_i(z_i)) := \max\left(0,\ \frac{F_i(z_i) - F_i(z_i^*)}{1 - F_i(z_i^*)}\right) \qquad z_i \in \mathbf{R},\ i \in B_j. \end{aligned} \tag{20}$$

The right hand side distributions are precisely the given marginal distributions conditioned upon $(-\infty, z_i^*)$ if $i \notin B_j$ and upon (z_i^*, ∞) if $i \in B_j$, respectively. Denoting the corresponding probability measures by μ_{ij}, which depend of course on z_i^*, then (20) says nothing more than

$$\mathrm{proj}_i\, \mu^j = \mu_{ij}, \quad i = 1, ..., r, \ j = 1, ..., n.$$

These are weak conditions on μ^j which can be satisfied by many probability distributions. In fact, we get for each j a separate distribution problem P_j on $z^* + K_j$ comparable with P on X, with the difference that each feasible solution for P_j is good. We can choose e.g. the independent solution, or the NW Corner Solution etc. on $z^* + K_j$. So we proved the basic result

Theorem 10. *Consider the distribution problem* P *with f defined in* (11). *Let z^* be an optimal solution of $\tilde{\mathrm{D}}$ with corresponding Kuhn-Tucker vector λ^* as in Theorem* 9. *Suppose that each F_i is continuous in z_i^*. Then* max P =

min $\tilde{D}$. *The optimal value* max P *is attained by the probability measure* μ^* *defined by*

$$\mu^* := \sum_{j=0}^{n} \lambda_j^* \cdot \mu^j$$

where μ^j *is any probability measure on* $z^* + K_j$ *with marginals* μ_{ij}. *For any such* μ^* *we have*

$$P_{\mu^*}(\textit{path } B_j \textit{ is critical with length} \geqslant T) = \lambda_j^*, \quad j \geqslant 1,$$

$$P_{\mu^*}(\textit{each path has length} \leqslant T) = \lambda_0^*,$$

$$P_{\mu^*}(\textit{activity } i \textit{ is on critical path with length } \geqslant T) =$$

$$= \sum_{j \in B_i^{-1}} \lambda_j^* = 1 - F_i(z_i^*), \; i = 1, \ldots, r.$$

The only incompleteness in Theorem 10 is the continuity assumption on F_i. Of course, Theorems 6, 8 and 9 do not need this assumption, and even Theorem 10 may be valid for discrete marginal distributions, at least if it is possible to choose z_i^* such that $F_i(\{z_i^*\}) = 0 \; \forall i$, but this is not possible in general. For discrete F_i an approach to construct a μ^* satisfying the conditions of Theorem 6 would be to start with $\bar{K}_j \subset K_j$ such that $\{\bar{K}_j, j \in J_T^*(z^*)\}$ be mutually disjunct and have the same union as the corresponding K_j. But it is not clear how to define $\bar{K}_j$ in general. By the way, max P = min D is true for finite discrete distributions F_i, since then we have finite dimensional LP duality. In that case $\tilde{D}$ can be also formulated as a LP problem, and min D= min $\tilde{D}$. $\tilde{D}$ is to be preferred to D since it has much less variables. A practical approach can be to solve D by LP, and then to try to find a discrete distribution μ^* satisfying Theorem 10, also by LP (first phase of the simplex method). We intend to work out the discrete case in a subsequent paper.

As an example, let $r := 5$, $n := 3$, $B_1 := \{1, 2\}$, $B_2 := \{1, 3, 5\}$, $B_3 = \{4, 5\}$, $F_3(\{0\}) = 1$, $F_i(\{0\}) = F_i(\{1\}) = \frac{1}{2}$, $i = 1, 2, 4, 5$ and $T = 0$. Then a solution of $\tilde{D}$ is $\lambda_0^* = \lambda_2^* = 0$, $\lambda_1^* = \lambda_3^* = \frac{1}{2}$, $z_0^* = z_2^* = z_4^* = 1$, $z_1^* = z_3^* = z_5^* = 0$, and indeed there is, even in this discrete case, a distribution μ^* which satisfies all conditions in Theorem 10:

$$P_{\mu^*}(\xi = (1, 1, 0, 0, 0)) = P_{\mu^*}(\xi = (0, 0, 0, 1, 1)) = \frac{1}{2}.$$

Notice that the critical path has length 2 with μ^*-probability 1.

Finally we shall show that the results on the most unfavourable joint distribution for the project planning problem can be used even if marginal distributions F_i are only known partially, which often is the case. Suppose that the F_i are unknown, except for their ranges $X_i := [a_i, b_i]$, bounded intervals in **R**, and the mean values m_i. This partial information is comparable with the information used by the well-known heuristic PERT. Let $\mathbf{F}_i$ be the class of all distribution functions F_i compatible with the partial information, and $\mathbf{H}_0$ the class of all r-dimensional distributions with marginals in $\mathbf{F}_i$, $i =$

$= 1, \ldots, r$. So $\mathbf{H}_0 = \cup \{\mathbf{H}(F_1, \ldots, F_r): F_i \in \mathbf{F}_i \ \forall i\}$. As a 'worst-case-alternative' for PERT we propose the problem:

$$\text{P}(T) \qquad \text{minimize}_T \ (c(T) + \sup_{H \in \mathbf{H}_0} E_H f_T(\xi))$$

where $f_T(x) \equiv f(x) := [f_0(x) - T]^+$, $f_0(x) := \max_j \sum_{i \in B_j} x_i$. The expected delay with respect to the most unfavourable distribution $H^* \in \mathbf{H}_0$ is balanced with the cost $c(T)$ of having an agreement upon the deadline T for the project. A reasonable choice for $c(T)$ is $(1 - \alpha)\, T$ for $\alpha \in (0, 1)$: then α corresponds to the required reliability of the optimal solution T^*, since the optimal T^* must satisfy

$$P_{H^*}(f_0(\xi) < T^*) \leqslant \alpha \leqslant P_{H^*}(f_0(\xi) \leqslant T^*).$$

We shall use the results on the problem P in order to show that problem P(T) can be handled computationally.

Because of the duality theorem we have

$$\sup_{H \in \mathbf{H}_0} E_H f_T = \sup_{F_i \in \mathbf{F}_i} \max_{H \in \mathbf{H}(F_1, \ldots, F_r)} E_H f_T$$

$$= \sup_{F_i \in \mathbf{F}_i} \min_{z \in \mathbf{R}^r} \left(f_T(z) + \sum_{i=1}^{r} E_{F_i}[\xi_i - z_i]^+ \right).$$

It is well-known (see e.g. [3], [9]) that for any convex function φ

$$\max_{F_i \in \mathbf{F}_i} \int \varphi(x_i)\, \mathrm{d}F_i(x_i) = \int \varphi(x_i)\, \mathrm{d}F_i^*(x_i)$$

where F_i^* represents the discrete distribution

$$F_i^*(\{a_i\}) := \frac{b_i - m_i}{b_i - a_i}, \quad F_i^*(\{b_i\}) := \frac{m_i - a_i}{b_i - a_i}.$$

Notice that F_i^* does not depend on the special function φ! Therefore, we may interchange $\max_{F_i}$ and $\min_z$, and get

$$\sup_{H \in \mathbf{H}_0} E_H f_T = \min_{z \in \mathbf{R}^r} (f_T(z) + \sum_{i=1}^{r} E_{F_i^*}[\xi_i - z_i]^+).$$

Since F_i^* is a discrete distribution, the function to be minimized is convex and piecewisely linear, so that P(T) in fact can be formulated and solved as a *linear programming* problem (the linear representation of f_T is given in (12), and $E_{F_i^*}[\xi_i - z_i]^+$ is linear on $[a_i, b_i]$, the only interval which has to be considered.

By the way, an optimistic bound for the PERT problem is also available, since $\min_{H \in \mathbf{H}_0} E_H f_T$ is attained by the distribution $\tilde{H}$ with $P_{\tilde{H}}(\xi_i = m_i, i = 1, \ldots, r) = 1$ (Jensen's inequality); only $f_0(m)$, the length of the critical path if all stochastic variables are replaced by their means, has to be calculated.

We conclude with another observation. Analogous results can be derived when $\mathbf{F}_i$ is replaced by the class of all distributions on $[a_i, b_i]$ with symmetric unimodular densities. Then F_i^* has to be replaced by the uniform distribution (see [3], [9]). The problem $P(T)$ is then also a simple recourse problem of stochastic programming, which can be solved as a quadratic programming problem (see [4]).

REFERENCES

1. Breiman, L., *Probability*. Reading, Mass., Addison-Wesley, 1968.
2. Cambanis, S., Simons, G., Stout, W., *Inequalities for E k(X,Y) when the marginals are fixed.* Z. Wahrsch. Verw. Gebiete, **36**, 285–294 (1976).
3. Dupačová, J., *Experience in stochastic programming models.* In: Prékopa, A. (Ed.), *Survey of Mathematical Programming (Proc. 9th Internat. Math. Programming Sympos., Budapest,* 1976), *Vol.* 2, pp. 99–105. Amsterdam, North-Holland, 1979.
4. Kall, P., *Stochastic Linear Programming.* Berlin–Heidelberg–New York, Springer, 1976.
5. Meileijson, I., Nádas, A., *Convex majorization with an application to the length of critical paths.* J. Appl. Probab., **16**, 671–677 (1979).
6. Nádas, A., *Probabilistic PERT.* IBM J. Res. Develop., **23**, 339–347 (1979).
7. Rockafellar, R. T., *Conjugate Duality and Optimization.* Philadelphia, SIAM, 1974.
8. Whitt, W., *Bivariate distributions with given marginals.* Ann. Statist., **4**, 1280–1289 (1976).
9. Žačková, J., *On minimax solutions of stochastic linear programming problems.* Časopis Pěst. Mat. **91**, 423–430 (1966).

VIKTOR G. KUROTSCHKA, WERNER WIERICH

Ansätze und Ergebnisse der optimalen Planung von Experimenten mit sowohl qualitativen als auch quantitativen Einflussfaktoren

This paper presents and discusses the present state of the theory of optimal planning of experiments simultaneously influenced by qualitative and quantitative factors.

1. Einführung und Übersicht

Kann ein Einflußfaktor eines Experiments nur an einer fest vorgegebenen endlichen Anzahl von Stufen eingesetzt werden, dann nennt man ihn qualitativ, ist dagegen ein ganzes Intervall von Wirkungsniveaux möglich, dann nennt man ihn quantitativ.

Statistische Modelle spezieller Beispiele von Experimenten mit sowohl qualitativen als auch quantitativen Einflußfaktoren sind in der statistischen Literatur unter dem Namen „Beobachtungsmodelle der Kovarianzanalyse" bekannt. Diese speziellen Modelle beziehen sich auf Versuchsanordnungen, bei denen die qualitativen mit den quantitativen Faktoren keine Wechselwirkungen eingehen, d.h. beide Gruppen einen additiven Beitrag auf das Versuchsergebnis liefern. Außerdem ist bei diesen Modellen lediglich die Wirkung der qualitativen Einflußfaktoren von Interesse, während die Einflüsse der quantitativen Faktoren als Störeffekte erscheinen und elimiert werden, was wegen des Fehlens der Wechselwirkungen besonders einfach gelingt.

Für solche speziellen Modelle publizierte Harville (1975) erste Überlegungen zur optimalen Versuchsplanung. Doch blieben auch weiterhin Lösungen der Planungsprobleme bei diesen mathematisch kamplexeren und für die Anwendungen bedeutungsvollen Beobachtungsmodellen eine immer wieder sowohl von praktisch wie auch theoretisch orientierten Statistikern formulierte Herausforderung, an der wir neben anderen Planungsproblemen in den letzten Jahren an unserem Institut gearbeitet haben. Ein Überblick über die dabei erzielten Ergebnisse wird in dieser Arbeit gegeben.

Im einzelnen enthält der Abschnitt 2 dieser Arbeit eine Beschreibung des allgemeinen Planungsproblems und eine bedingte Darstellung von Versuchsplänen und der Modellgleichungen die sich als einfache aber für die Lösung der einzelenen Planungsprobleme sehr fruchtbare Idee erweist. Mit dieser bedingten Darstellung gelingt nämlich verhältnismäßig einfach die vollständige approximative Lösung — d.h. in Form der Charakterisierung D- und A-optimaler verallgemeinerter Veruschspläne — des Planungsproblems für Experimente, bei denen die qualitativen und quantitativen Einflußfaktoren Wechselwirkungen eingehen (vgl. Abschnitt 3). Aber auch für die vollständige Charakterisierung konkreter, d.h. ganzzahliger Versuchspläne für das einfachste grundlegende Modell solcher Experimente, das sogenannte Modell der linearen Intra-class Regression ist diese Idee, wie am Ende des 3. Abschnitts ausgeführt, hilfreich. Für Experimente, bei denen die qualitativen Einflußfaktoren mit den quantitativen keine Wechselwirkungen eingehen, gelang bisher eine uneingeschränkte Charakterisierung D- und A-optimaler Versuchspläne nur für Experimente, bei denen die qualitativen Einflußfaktoren vollständig in Wechselwirkung stehen, oder was sich als theoretisch äquivalent erweist, für Experimente mit nur einem qualitativen Einflußfaktor (vgl. Abschnitt 4). Für andere Situationen fehlen bisher entsprechende Aussagen über die Vollständigkeit der Klasse der im Abschnitt 4 definierten Produktpläne und stellen damit einen ersten Bereich offener Probleme dar, deren Lösung die Ergebnisse für Modelle, wie sie in Kurotschka (1982) auf der Basis der Resultate von (1971) und (1978) diskutiert werden, für die Anwendung erst voll erschließen würde. Eine zweite Klasse offener Probleme, deren Lösung durch die konditionelle Darstellung von Versuchsplänen hoffnungsvoller erscheint, eröffnet sich in der Weiterführung der exakten Theorie, d.h. in Charakterisierungen optimaler konkreter Versuchspläne, wie sie für einfachere Modelle auch am Ende des Abschnitts 4 vorgestellt sind.

In diesen Zusammenhang gehört auch die Arbeit von Josefa Lopes Troya-Schlottmann, die sie bei uns angefangen hat und über ein Stipendium als Dissertation an der Universität in Berkeley bei Jack Kiefer abgeschlossen hat. Die Ergebnisse — hinreichende Bedingungen für D-optimale Pläne für ein kovarianzanalytischer Modell über die Abschätzung der Determinanten zugehöriger Informationsmatrizen und die Konstruktion exakter D-optimaler Versuchspläne mit Hilfe von Hadamard-und Konferenz-Matrizen konnten hier aus Platzgründen nicht eingehender diskutiert werden, befinden sich in die Arbeit [8].

2. Das allgemeine Planungsproblem

Die Menge T der in Betracht gezogenen Versuchsbedingungen, der sogenannte *Versuchsbereich*, eines Experiments mit K_1 qualitativen und K_2 quantitativen Einflußfaktoren läßt sich formal beschreiben durch

$$T = T_{(1)} \times T_{(2)}, T_{(1)} = \times_{k=1}^{K_1} \{1, \ldots, I_k\}, \mathbb{R}_{K_2} \supset T_{(2)} \supset \times_{k=1}^{K_2} (a_k, b_k) \neq \emptyset$$

und eine *Versuchsbedingung eines Einzelexperiments* durch

$$t = (t_{(1)}, t_{(2)}), t_{(1)} := (i_1, .., i_{K_1}), \; t_{(2)} = (s_1, \ldots, s_{K_2}),$$

wobei $i_k \in \{1, ,,.., I_k\}$ eine mögliche Wirkungsstufe des k'ten qualitativen Faktors, s_k mit $(s_1, \ldots, s_k, \ldots, s_{K_2}) \in T_{(2)}$ ein mögliches Wirkungsniveau des k'ten quantitativen Faktors beschreibt.

Die bei gleichbleibenden Versuchsbedingungen $t \in T$ auftretenden Schwankungen in den resultierenden Beobachtungsergebnissen werden (im Fall reellwertiger Meßwerte und nur dieser sogenannte unvariate Fall wird hier betrachtet) interpretiert, indem man die Beobachtungsergebnisse als Realisierungen von Zufallsgrößen $X(t)$, $t \in T$ und die *„wahre" unbekannte Wirkung* $\mu(t)$ der Versuchsbedingungen $t \in T$ auf den Ausgang des Experiments als Lokationsparameter der Verteilung von $X(t)$, $t \in T$ ansetzt, und zwar in der parametrischen Theorie als Erwartungswert, d.h.

$$\mu : T \ni t \to \mu(t) := EX(t) \in \mathbb{R}$$

in der nichtparametrischen als Verschiebungsparameter, d.h. definiert durch:

$$F_{X(t)}(x - \mu(t)) = F(x), \quad x \in \mathbb{R},$$

wobei $F_{X(t)}$ die Verteilungsfunktion von $X(t)$, $t \in T$ ist.

In beiden Ansätzen werden die Schwankungen

(2.1) $$X(t) - \mu(t) = : Z(t), \quad t \in T$$

als *Zufällige Fehler* interpretiert, die durch geeignete Schätzungen oder Tests über interessierende Aspekte der *Wirkungsfunktion* $\mu : T \to \mathbb{R}$ zu eliminieren sind.

Läßt sich aus praktischen Gründen das Experiment mit K_1 qualitativen und K_2 quantitativen Einflußfaktoren nur an endlich vielen Beobachtungsbedingungen durchführen (nicht durch Beobachten eines ganzen Funktionsverlaufs), handelt es sich also um ein Experiment mit endlichem *Gesamtstichprobenumfang* N, dann wird eine•Vorschrift, daß $N(t_i)$ mal die Versuchsbedingung t_i, $i = 1, \ldots, I$ eingesetzt werden soll als *konkreter Versuchsplan* bezeichnet und beschrieben durch

(2.2) $$d = \begin{pmatrix} t_1 & \ldots t_I \\ N(t_1) & \ldots N(t_I) \end{pmatrix}, \quad N = \Sigma_{i=1}^{I} N(t_i).$$

Die allgemeine Annahme der Unabhängigkeit und der im Mittel gleichen Genauigkeit der daraus resultierenden Einzelbeobachtungen, die durch die klassischen Planungsprinzipien „Randomisierung" (hier zufällige Zuordnung der Versuchsbedingungen an die einzelnen Versuchseinheiten) un „Blockbildung" (hier allgemeiner zu fassen als adäquate Festlegung der möglichen Einflußafktoren und deren Wirkungsniveaux) praktisch zu realisieren ist (und manchmal erst durch zusätzliche Transformation der Beobachtungsgrößen gerechtfertigt ist) reduziert das Planungsproblem auf die Verallgemeinerung, Präzisierung und Durchführung des klassischen Planungsprinzips „Replications" in Richtung der Auszeichnung und Konstruktion der Versuchspläne d so, daß Aussagen über den jeweils interessierenden Aspekt ψ der Wirkungsfunktion μ auf Grund der Beobachtungen X_d möglichst informativ getroffen werden können.

In der Theorie der linearen Beobachtungsmodelle in der die Existenz einer endlichen *linearen Parametrisierung* $\beta = (\beta_1, \ldots, \beta_r)^T$ für die Wirkungsfunktion vorausgesetzt wird:

(2.3) $$\mu = T \ni t \to \mu(t) = a^T(t)\, \beta \in \mathbb{R},$$

wobei $a = (a_1, \ldots, a_r)^T$ ein System bekannter sogenannter *Regressionsfunktionen* darstellt, ist diese Präzisierung durch ein universales Beurteilungsmaß eines Versuchsplans d, nämlich dei *Informationsmatrix* $I_\beta(d)$ eines Versuchsplans d für β getroffen, die definiert ist durch

(2.4) $$I_\beta(d) = \Sigma_{i=1}^{I} N(t_i)\, a(t_i)\, a^T(t_i)$$

und die auch durch bekannte Formeln in Form reduzierter, geeignet transformierter Informationsmatrizen $I_\psi(d)$ Beurteilungsmaße für interessierende (lineare) Teilaspekte $\psi(\beta)$ bestimmt, die im Beobachtungsmodell under d identifizierbar sind, d.h. für die aus $a^T(t_i)\,\beta = a^T\,(t_i)\beta'$ für alle $i = 1, \ldots, I$ $\psi(\beta) = \psi(\beta')$ folgt.

(2.5) Ist T endlich, d.h. sind alle Einflußfaktoren des Experiments qualitativ, dann existiert wegen der Endlichkeit der Menge $\mu(t)$ stets eine lineare Parametrisierung der Wirkungsfunktion μ. Offensichtlich ist eine lineare Parametrisierung der Wirkungsfunktion nicht eindeutig, sondern kann im Hinblick auf anwendungsorientierte oder mathematisch technische Gesichtspunkte gemäß bestehendem a priori Wissen über die Versuchsanordnung (z.B. Fehlen von Wechselwirkungen zwischen gewissen Faktoren, Einbeziehen von Kontrollexperimenten usw.) und a priori Interesse (isolierte Wirkungen einzelner Faktoren oder Faktorgruppen) geeignet festgelegt werden (vgl. z.B. Kurotschka (1978)). Solche Parametrisierungen füren zu den verschiedenen Klassen von Modellen der Varianzanalyse.

(2.6) Ist T der Versuchsbereich eines Experiments mit nur quantitativen Einflußfaktoren, dann werden lineare Parametrisierungen der Wirkungsfunktion zum Teil durch stochastische Modellansätze begründet (Regression der Beobachtungsgrößen bezüglich der kontrollierten Variablen des Experiments), zum Teil durch formale Transformation der Beobachtungsgrößen (bei endlich parametrisierten Wirkungsfunktionen) und zum Teil durch formale Approximation der Wirkungsfunktion erreicht (endliche Entwicklung der Wirkungsfunktion nach einem den theoretischen Grundlagen der Experimentieranordnung angepaßten Funktionssystem a_π, $\pi = 1, \ldots, p$). Solche Parametrisierungen führen zu Modellen der allgemeinen Regressionsanalyse.

Ein natürlicher Zugang für eine endliche lineare Parametrisierung der hier betrachteten Experimente mit $T = T_{(1)} \times T_{(2)}$ wird durch die folgende *konditionelle Darstellung der durch* (2.1) *definierten Modellgleichungen* nahegelegt: Für jedes $t_{(1)} \in T_{(1)}$ sind

$$X_{t_{(1)}}(t_{(2)}) = \mu_{t_{(1)}}(t_{(2)}) + Z_{t_{(1)}}\,(t_{(2)}),\quad t_{(2)} \in T_{(2)}, \tag{2.7}$$

mit $X_{t_{(1)}}\,(t_{(2)}) := X(t_{(1)}, t_{(2)})$ und die anderen entsprechend, Beobachtungsgleichungen auf $T_{(2)}$ für die in (2.6) lineare Parametrisierungen der Wirkungsfunktionen

$$\mu_{t\,(1)} = a^T_{t_{(1)}}\,\beta_{t_{(1)}},\quad t_{(1)} \in T_{(1)} \tag{2.8}$$

begründet wurden, die durch Aggregation gemäß weiteren Annahmen über die Modellgleichungen (vgl. Abschnitte 3 und 4) eine lineare Parametrisierung der Gesamtwirkungsfunktion festlegen.

Diese konditionelle Darstellung der Modellgleichung und die Beschäftigung mit Planungsproblemen bei Versuchsbereichen $T = T_{(1)} \times T_{(2)}$ legen es nahe, auch eine konditionelle Darstellung eines Versuchsplanes einzuführen, und zwar erweist sich folgende Form als fruchtbar.

$$d = (N_{12\ldots K_1},\ S(N_{12\ldots K_1}))$$

$$N_{12\ldots K_1} := (N(t_{(1)}))_{t_{(1)} \in T_{(1)}} = \left((N(i_1, \ldots, i_{K_1}))\, \begin{matrix} i_k = 1, \ldots, I_k \\ k = 1, \ldots, K_1 \end{matrix} \right) \tag{2.9}$$

$$N(t)_{(1)} \in \{0, 1, \ldots, N\},\quad \Sigma_{t_{(1)} \in T_{(1)}} N(t_{(1)}) = N$$

und

$$S(N_{12\ldots K_1}) = \left((s_n(t^{(1)}))\begin{matrix} n = 1, \ldots, N(t_{(1)}) \\ t_{(1)} \in T^{(1)} \end{matrix}\right)$$

mit

$$s_n(t_{(1)}) = (s_{1n}(t_{(1)}), \ldots, s_{K_2 n}(t_{(1)})).$$

Dabei ist $N(t_{(1)}) = N(i_1, \ldots, i_{K_1})$ die Anzahl der Einzelexperimente mit der Stufenkombination $t_{(1)} = (i_1, \ldots, i_K)$ als qualitative Versuchsbedingung und $s_{k_n}(t_{(1)}) = s_{k_n}(i_1, \ldots, i_{K_1})$ das Wirkungsniveau des k'ten quantitativen Faktors bei der n'ten Wiederholung des Einzelexperiments unter der qualitativen Versuchsbedingung $t_{(1)} = (i_1, \ldots, i_{K_1})$

Für einen verallgemeinerten Versuchsplan δ, definiert als Wahrscheinlichkeitsmaß auf einer geeigneten σ-Algebra über T, induzierst diese Darstellung (2.9) folgende *konditionelle Darstellung von* δ

$$\delta = \nu \otimes \sigma, \tag{2.10}$$

wobei ν ein (diskretes) Wahrscheinlichkeitsmaß auf $\mathcal{B}(T_{(1)})$ und σ ein Markov-Kern von $(T_{(1)}, \mathcal{B}(T_{(1)}))$ nach $(T_{(2)}, \mathcal{B}(T_{(2)}))$ ist, mit $\mathcal{B}(T_{(e)}$ als Spur der Borel-σ-Algebra von $\mathbb{R}_{K_e}$ auf $T_{(e)}$, $e = 1, 2$, und der Interpretation, daß $\sigma = \nu \otimes \sigma$ festlegt, die N Beobachtungen gemäß σ auf die durch $t_{(1)} \in T_{(1)}$ induzierten Teilexperimente aufzuteilen und gemäß $\sigma(\cdot, t_{(1)})$ definiert durch σ, die Verteilung der $N \cdot \nu(t_{(1)})$ Beobachtungen innerhalb des durch $t_{(1)}$ indizierten Teilexperiments auf die quantitativen Wirkungsniveaux vorzunehmen.

Mit dieser konditionellen Darstellung von δ vereinfacht sich die Berechnung der Informationsmatrizen verallgemeinerter Versuchspläne

$$I_\beta(\delta) := \int a(t)\, a^T(t)\, \delta(\mathrm{d}t) = \tag{2.11}$$

$$= \sum_{t_{(1)} \in T_{(1)}} \nu(t_{(1)}) \int a(t_{(1)}, t_{(2)})\, a^T(t_{(1)}, t_{(2)})\, \sigma\, (\mathrm{d}t_{(2)}, t_{(1)}).$$

Für die hier behandelten Modelle entfällt das Identifizierungsproblem, denn man kann sich hier ohne Einschränkung der Anwendbarkeit von vornherein auf Parametrisierungen mit linear unabhängigen Funktionen des Systems a auf $T = T_{(1)} \times T_{(2)}$ beschränken, was bedeutet, daß nur Versuchspläne mit nichtsingulären Informationsmatrizen für ganz β in Betracht zu ziehen sind, d.h. solche, die das Beobachtungsmodell voll erhalten.

Da für diese komplizierten Modelle nur für Spezialfälle (vgl. Kurotschka und Wierich (1984) und Wierich (1984)) gleichmäßig optimale Versuchspläne existieren, reduzieren die Ausführungen in diesem Abschnitt das Planungsproblem auf die Charakterisierung D- und A-optimaler Versuchspläne, d.h solcher, welche die Determinante bzw. die Spur der Inversen der Informationsmatrix minimieren, wenn man von anderen nicht so universell zu interpretierenden Optimalitätskriterien absieht.

3. Experimente mit qualitativen und quantitativen Einflußfaktoren in Wechselwirkung

In der konditionellen Darstellung der linear parametrisierten Modellgleichungen bei gegebenem $t_{(1)} \in T_{(1)}$

$$X_{t_{(1)}}(t_{(2)}) = a^T_{t_{(1)}}(t_{(2)})\beta_{t_{(1)}} + Z_{t_{(1)}}(t_{(2)}), \quad t_{(2)} \in T_{(2)} \tag{3.1}$$

bedeutet die Annahme, daß beide Klassen von Faktoren in Wechselwirkung stehen, daß die einzelnen $r(t_{(1)})$ dimensionalen Parametervektoren $\beta_{t_{(1)}}$ von $t_{(1)} \in T_{(1)}$ abhängen, wie man sich an praktischen Beispielen veranschaulichen kann. Ob die Dimension $r(t_{(1)})$ oder die Regressionsfunktionen $a_{t_{(1)}}$ von $t_{(1)} \in T_{(1)}$ abhängen oder nicht, ist nur eine Frage der jeweiligen Wahl der linearen Parametrisierungen für die verschiedenen $t_{(1)} \in T_{(1)}$, etwa der den praktischen Bedürfnissen angepaßten Approximationen der Wirkungsfunktionen $\mu_{t_{(1)}}$ durch Funktionensysteme $a_{t_{(1)}}$ für jedes $t_{(1)} \in T_{(1)}$.

Eine wesentliche Erleichterung für die weitere Darstellung bietet hier eine Durchnumerierung aller qualitativen Wirkungsstufenkombinationen $t_{(1)} \in T_{(1)}$ etwa durch lexikographische Identifikation

$$T_{(1)} \ni t_{(1)} := (i_1, \ldots, i_{K_1}) \leftrightarrow i \in \{1, 2, \ldots, I := I_1 \cdot I_2 \cdots I_{K_1}\}.$$

Mit dieser Numerierung ergibt sich aus der konditionellen Darstellung der Modellgleichungen (3.1) für $i = 1, \ldots, I$

$$X_i(t_{(2)}) = a_i^T(t_{(2)})\beta_i + Z_i(t_{(2)}), \; t_{(2)} \in T_{(2)}$$

mit

$$a_i = (a_{i1}, \ldots, t_{ir(i)})^T \text{ und } t_i = (t_{i1}, \ldots, t_{ir(i)})^T$$

die Gesamtwirkungsfunktion des Experiments

$$\mu : \{1, \ldots, I\} \times T_{(2)} \ni (i, t_{(2)}) \rightarrow \mu(i, t_{(2)}) = a^T(i, t_{(2)})\, \beta \in \mathbf{R}$$

mit $a := (1_{\{1\}}\, a_1^T, \ldots, 1_{\{I\}}\, a_I^T)^T$ und $t := (t_1^T, \ldots, t_I^T)^T$ und nach (2.11) die Informationsmatrix eines verallgemeinerten Versuchsplans $\delta = \nu \otimes \sigma$ für den gesamten Parametervektor β

$$I_\beta(\nu \otimes \sigma) = \Sigma_{i=1}^I \nu(i) \int a(i, t_{(2)})\, a^T(i, t_{(2)})\, \sigma(dt_{(2)}, i) \tag{3.2}$$

$$= \begin{pmatrix} \nu(1) I_{\beta_1}(\sigma(\cdot, 1)) & 0 & \cdot & 0 & \cdot & \cdot \\ 0 \, . & \cdot & & \vdots & \cdot \\ \vdots & \cdot & \cdot & \cdot 0 & \\ 0 \, . & \cdot & 0 & \nu(I)\, I_{\beta_I}(\sigma(\cdot, I)) \end{pmatrix}$$

wobei hier $I_{\beta_i}(\sigma(\cdot, i))$ für jedes $i = 1, \ldots, I$ die Informationsmatrix eines verallgemeinerten Versuchsplans (W-Maß auf $T_{(2)}$) $\sigma(\cdot, i)$ für den Parametervektor σ_i innerhalb des durch $i = 1, \ldots, I$ indizierten reinen Regressionsmodells ist.

Wenn man von einigen speziellen Fragestellungen, bedingt durch spezielle Parametrisierungen absieht, wie sie etwa im Beispiel (3.5) bei linearen intra-class Regressions-Experimenten auftreten können (vgl. dazu Kurotschka und Wierich (1984)) konzentriert sich das Interesse bei den hier in diesem Abschnitt behandelten Modellen entweder auf den Gesamtparametervektor σ oder einer Teilklasse der Parametervektoren $\sigma_{t_{(1)}}$, $t_{(1)} \in T_{(1)}$, wie etwa bei multiplen Vergleichen der Wirkungen einzelner qualitativer Stufenkombinationen, was aber nur eine Reduzierung von $T_{(1)}$ zur Folge hat, so daß im wesentlichen nur Informationsmatrizen der Form (3.2) zu betrachten sind. Da gleichmäßig optimale Versuchspläne bei so strukturierten Informationsmatrizen nicht existieren, wird das Interesse auf die einfach aus (3.2) zu berechnenden

(3.3) $$\det I_\beta(\nu \otimes \sigma) = \pi_{i=1}^{I} \nu\,(i)^{r(i)} \det I_{\beta_i}(\sigma(\cdot, i))$$

und

$$\operatorname{spur} I_\beta^{-1}(\nu\otimes\sigma) = \Sigma_{i=1}^{I}\{1/\nu(i)\} \operatorname{spur} I_{\beta i}^{-1}(\sigma\cdot, 1))$$

gelenkt, deren Maximierung bzw. Minimierung in $\delta = \nu\otimes\sigma$ partiell nach ν und σ relativ einfach gelingt mit dem Ergebnis (vgl. dazu Kurotschka (1982)):

(3.4). **Satz.** *Unter allen verallgemeinerten Versuchsplänen ist* $\delta^* = \nu^*\otimes\sigma^*$ *für den Gesamtparametervektor* β *des Experiments genau dann*

a) *D-optimal ist, wenn*

(i) $$\nu^* = \Sigma_{t_{(1)} \in T_{(1)}} \varepsilon_{t_{(1)}} r(t)_{(1)} / \Sigma_{t'_{(1)} \in T_{(1)}} r(t'_{(1)})$$

und

(ii) σ^* *punktweise definiert ist durch* $\sigma^*(\cdot, t_{(1)})$, *die* $\det I_{\beta t_{(1)}}(\sigma^*(\cdot, (t_{(1)}))$ *maximieren für* $t_{(1)} \in T_{(1)}$

d.h. wenn δ^* *die Gesamtanzahl N der Beobachtungen and die durch* $t_{(1)} \in T_{(1)}$ *indizierten (bedingten) Regressionsmodelle proportional der Anzahl* $r(t_{(1)})$ *ihrer unbekannten Parameter verteilt und in jedem dieser Modelle die quantitativen Stufenkambinationen gemäß einem für das jeweilige Regressionsmodell D-optimalen Versuchsplan* $\sigma^*(\cdot, t_{(1)})$ *auswählt.*

b) *A-optimal, wenn*

(i) $$\nu^* = \Sigma_{t_{(1)} \in T_{(1)}} \varepsilon_{t_{(1)}} c(t_{(1)}) \,/\, \Sigma_{t_{(1)} \in T_{(1)}} c(t'_{(1)}) \quad mit$$
$$c(t_{(1)}) := \{\operatorname{spur} I^{-1}_{\beta t_{(1)}}(\sigma^*(\cdot, t))\}^{1/2}$$

und

(ii) σ^* *punktweise definiert ist durch* $\sigma^*(\cdot\,, t_{(1)})$, *die* $\operatorname{spur} I^{-1}_{\beta_{t_{(1)}}}(\sigma^*(\cdot\,, t_{(1)}))$ *minimieren für* $t_{(1)} \in T_{(1)}$,
d.h. wenn δ^* *in jedem durch* $t_{(1)} \in T_{(1)}$ *indizierten Regressionsmodell die quantitativen Stufenkombinationen gemäß einem A-optimalen Versuchsplan* $\sigma^*(\cdot, t_{(1)})$ *auswählt und dabei die Gesamtanzahl N der Beobachtungen auf die einzelnen durch* $t_{(1)}$ *indizierten Regressionsmodelle proportional der Wurzel der Spur der inversen Informationsmatrix eines A-optimalen Versuchsplans* $\sigma^*(,\cdot\, t_{(1)})$ *verteilt.*

Mit diesem Satz ist das Planungsproblem für die hier behandelten Experimente auf Planungsprobleme reiner Regressionsexperimente zurückgeführt, für die zahlreiche Methoden und Ergebnisse existieren und die hier — anders etwa als für Modelle des nächsten Abschnitts 4 — direkt anwendbar sind.

(3.5) *Beispiel.* Das einfachste Beispiel, das aber für die Anwendung von Bedeutung ist, das sogenannte Intraklassenregressionsmodell, ist als Experiment mit je einem qualitativen und quantitativen Einflußfaktor definiert durch die folgende Wirkungsfunktion

$$\mu : \{1, ..., I\} \times [-1, +1] \ni (i, s) \to \mu(i, s) = \beta_{i1} + \beta_{i2}\, s \in \mathbb{R},$$

wobei hier im Einklang mit der allgemeinen Darstellung $a_i := (1, s)^T$ und $\beta_i := (\beta_{i1}, \beta_{i2})^T$, $i = 1, ..., I$ ist. Für jedes $i \in \{1, 2, ..., I\}$ ist bekanntlich die gleichförmige Verteilung σ^* auf den Endpunkten -1 und $+1$ ein D-optimaler Versuchsplan für β_i im i-ten (konditionellen) Regressionsmodell und nach Satz (3.4) $\delta^* = \nu^* \otimes \sigma^*$ genau dann ein D-optimaler Versuchsplan für $\beta = (\beta_1, ..., \beta_I)^T$ im Intraklassenregressionsmodell, wenn ν^* eine gleichförmige Verteilung auf $T_{(1)} = \{1,, ..., I\}$ ist.

Weitere Beispiele ergeben sich aus der umfangreichen Literatur über optimale Planung von Regressionsexperimenten (vgl. z.B. Fedorov(1972) und Bandemer etc. (1977)). Das einfache Beispiel (3.5) wurde hier ausgewählt, weil es das einzige Modell darstellt von interagierrenden qualitativen und quantitativen Einflußfaktoren, für das uns bisher eine vollständige Charakterisierung konkreter optimaler Versuchspläne für verschiedene Teilklassen der Parameter als auch unter verschiedenen Beobachtungsbeschränkungen gelungen ist (vgl. Kurotschka und Wierich (1984)), und zwar unabhängig von der Ergebnissen des Satzes (3.4), jedoch gestützt auf die konditionelle Darstellung eines Versuchsplans. Zur Illustration der Tragweite und der Grenzen der approximativen Theorie der optimalen Versuchsplanung genügt hieraus der folgende Satz:

3.6. **Satz.** *In der Klasse aller konkreten Versuchspläne ist* $d^* = (N_1^*, S^*(N_1^*))$ *genau dann D-optimal für den gesamten Parametervektor der Intraklassenregression (vgl.* (2.9) *und Beispiel* (3.5)*), wenn*

(i) $N^*(i) \in \{n - 1, n, n + 1\}$,

wobei n die größte gerade ***Zahl*** *kleiner oder gleich* N/I *ist und*

(ii) $s_n^*(t) \in \{-1, +1\}$ *mit* $| \sum_{n=1}^{N^*(i)} s_n^*(i) | \leqslant 1$ *für alle* i *mit* $N^*(i) \geqslant 2$.

Abgesehen von Spezialfällen, ausgedrückt durch verschiedene Teilbarkeitsbedingungen and die Experimentierparameter N, I, $K_{(2)}$ usw — wie etwa N/I eine gerade Zahl im Beispiel (3.5) — und trotz vorhandener Methoden und Algorithmen für Regressionsmodelle, die hier übernommen werden können, aus optimalen verallgemeinerten Versuchsplänen konkrete optimale zu erzeugen, zeigt Satz (3.6), daß die Aussagen des Satzes (3.4) trotz ihrer Allgemeinheit und ihrer Anwendungsbezogenheit nur eine wenn auch wichtige Vorstufe darstellten und nicht alle weiteren Anstrengungen ersparen die exakte Theorie voranzutreiben.

4. Experimente mit qualitativen und quantitativen Einflußfaktoren nicht in Wechselwirkung

Die allgemeine linear parametrisierte Wirkungsfunktion solcher Experimente, die durch die Additivität der Wirkungen der beiden Klassen von Einflußfaktoren definiert ist, lautet

$$\mu : T_{(1)} \times T_{(2)} \ni (t_{(1)}, t_{(2)}) \to \mu(t_{(1)}, t_{(2)}) := a_{(1)}^T(t_{(1)})\beta_{(1)} + a_{(2)}^T(t_{(2)})\,\beta_{(2)} \in \mathbf{R}$$

und ergibt mit $a = (a_{(1)}^T, a_{(2)}^T)^T$ und $\beta = (\beta_{(1)}^T, \beta_{(2)}^T)^T$ für einen allgemeinen Versuchsplan $\delta = \nu \otimes \sigma$ die Informationsmatrix

$$(4.1) \qquad I_\beta(\nu \otimes \sigma) = \begin{pmatrix} I_{11} & I_{12} \\ I_{21} & I_{22} \end{pmatrix}$$

mit

$$I_{11} := \sum_{t_{(1)} \in T_{(1)}} a_{(1)}(t_{(1)})\, a^T_{(1)}(t_{(1)})\, \nu(t_{(1)}) = I_{\beta_{(1)}}(\nu)$$

(4.2) $$I_{12} : I^T_{21} = \sum_{t_{(1)} \in T_{(1)}} a_{(1)}(t_{(1)}) \int a^T_{(2)}(t_{(2)})\, \sigma(dt_{(2)}, t_{(1)})\nu(t_{(1)})$$

$$I_{22} := \sum_{t \in T_{(1)}} \int a_{(2)}(t_{(2)})\, a^T_{(2)}(t_{(2)})\, \sigma\,(dt_{(2)}, t_{(1)})\, \nu(t_{(1)}),$$

die sich für *Produktpläne* $\delta = \nu \otimes \sigma^*$, definiert durch σ^* konstant auf $T_{(1)}$, d.h σ^* ein W-Maß auf $T_{(2)}$, vereinfachen zu

$$I^*_{11} = I_{11} = I_{\beta_{(1)}}(\nu)$$

(4.3) $$I^*_{12} := I^{*T}_{21} = \sum_{t_{(1)} \in T_{(1)}} a_{(1)}(t_{(1)})\, \nu(t_{(1)}) \int a^T_{(2)}(t)_{(2)})\, \sigma^*\,(dt_{(2)})$$

$$I^*_{22} = \int a_{(2)}\,(t_{(2)})\, a^T_{(2)}\,(t_{(2)})\, \sigma^*\,(dt_{(2)}) = I_{\beta_{(2)}}\,(\sigma^*)$$

so daß man hier für die interessierende Informationsmatrix $I_{\beta_{(1)}}\,(\nu\otimes\sigma^*)$ erhält

(4.4) $$I_{\beta_{(1)}}\,(\nu\otimes\sigma^*) = \sum_{t_{(1)} \in T_{(1)}} a_{(1)}(t_{(1)})\, a^T_{(1)}(t_{(1)})\, \nu(t_{(1)}) -$$
$$- c(\sigma^*) \sum_{t_{(1)} \in T_{(1)}} a_{(1)}(t_{(1)})\, \nu(t_{(1)}) \cdot \sum_{t_{(1)} \in T_{(1)}} a^T_{(1)}(t_{(1)})\, \nu(t_{(1)})$$

mit

(4.5) $$c(\sigma^*) = \int a^T_{(2)}(t_{(2)})\, \sigma^*(dt_{(2)}) \left(\int a_{(2)}(t_{(2)})\, a^T_{(2)}(t_{(2)})\, \sigma^*(dt_{(2)})\right)^{-1} \int a_{(2)}(t_{(2)})\, \sigma^*(dt_{(2)})$$

wobei $I_{\beta_{(1)}}(\nu)$ und $I_{\beta_{(2)}}(\sigma)$ die Informationsmatrizen für $\beta_{(1)}$ bzw. $\beta_{(2)}$ in entsprechenden Experimenten, jedoch ohne quantitative bzw. ohne qualitative Einflußfaktoren, sind.

Die auf der konditionellen Darstellung (2.10) basierenden Argumente für die Möglichkeit der Einschränkung auf Porduktexperimente konnten bisher nur für Experimente formal nachgewiesen werden, deren qualitative Einflußfaktoren vollständige Wechselwirkungen eingehen (vgl. Satz (4.7) und Satz (4.8)). Solche Experimente sind dadurch ausgezeichnet, daß die Anzahl der unbekannten Komponenten von $\beta_{(1)}$ gerade die Anzahl aller verfügbaren qualitativen Stufenkombinationen ist, nämlich

$$I := I_1 \cdot I_2 \cdot \ldots \cdot I_{K_1}$$

Eine lexikographische Identifikation

$$T_{(1)} = \times^{K_1}_{k=1} \{1, \ldots, I_k\} \ni (i_1, \ldots, i_{K_1}) \leftrightarrow i \in \{1, \ldots, I\}$$

zeigt hier die Äquivalenz zu Experimenten mit nur einem qualitativen Faktor mit I Wirkungsstufen und ermöglicht eine besonders einfache Darstellung der interessierenden Informationsmatrizen, deren Determinanten und deren Spuren ihrer Inversen (vgl. Kurotschka (1982)) für Produktpläne folgende Form annehmen

(4.6) $$I_\beta(\nu\otimes\sigma) = \begin{pmatrix} I_{11} & I_{12} \\ I_{21} & I_{22} \end{pmatrix}$$

mit

$$I_{11} = \left((\nu(i)\delta_{ii'}) \begin{matrix} i = 1, \ldots, I \\ i' = 1, \ldots, I \end{matrix} \right) = I_{\beta_{(1)}}(\nu),$$

$$I_{12} = I_{21}^T = (\nu(1), \ldots, \nu(I))^T \int a_{(2)}^T (t_{(2)})\, \sigma\, (\mathrm{d}t_{)2)}),$$

$$I_{22} = \int a_{(2)}(t_{(2)})\, a_{(2)}^T (t_{(2)})\, \sigma(\mathrm{d}t_{(2)}) = I_{\beta_{(2)}}(\sigma),$$

$$I_{\beta_{(1)}}(\nu \otimes \sigma) = ((\nu(i)\ \delta_{ii'})) - c(\sigma)\ ((\nu(i) \cdot \nu(i')))$$

mit $c(\sigma)$ definiert in (4.5)

$$\det I_{\beta_{(1)}} (\nu \otimes \sigma) = (1 - c(\sigma)) \prod_{i=1}^{I} \nu(i)$$

$$\operatorname{spur}\ I_{\beta_1}^{-1} (\nu \otimes \sigma) = I c(\sigma)/(1 - c(\sigma)) + \sum_{i=1}^{I} 1/\nu(i)$$

$$\det I_\beta (\nu \otimes \sigma) = \det I_{\beta_{(1)}} (\nu \otimes \sigma) \cdot \det I_{\beta_{(2))}} (\sigma)$$

$$\operatorname{spur}\ I_\beta^{-1} (\nu \otimes \sigma) = \operatorname{spur}\ I_{\beta_{(1)}} (\nu \otimes \sigma) + \operatorname{spur}\ I_{\beta_{(2)}} (\sigma).$$

Inwieweit man diese durch die Produktpläne vereinfachten Beurteilungsgrößen dem Planungsproblem zugrunde legen kann, zeigen die folgenden beiden Sätze (vgl. Wierich (1983)).

(4.7). **Satz.** *Die Klasse Δ^* der Produktpläne ist wesentlich vollständig für $\beta_{(2)}$ aber nicht für $\beta_{(1)}$.*

Hier steht „wesentlich vollständig" — im Sinne der Entscheidungstheorie — für die Existenz eines mindestens so guten Versuchsplans in Δ^* zu jedem außerhalb von Δ^*, wobei sich „gut" hier auf gleichmäßige Optimalität bezieht. Die negative Teilaussage des Satzes (4.7) läßt sich milden und die Produkpläne wegen der relativ seltenen Existenz der gleichmäßig optimalen Versuchspläne ohne Einbuße für die Anwendungen retten durch Einführung der analogen Begriffe der *D*- und *A*-Vollständigkeit bezogen auf die *D*- bzw. *A*-Optimalität der Versuchspläne. Für diese gilt:

(4.8) **Satz.** *Die Klasse der Produktpläne ist wesentlich D- und A-vollständig sowohl für $\beta_{(1)}$ alls auch für ganz β.*

Mit diesem Satz folgt aus den Ausdrücken unter (4.6) eine Verschärfung des Satzes (3.11) aus Kurotschka (1982), der bereits das Ergebnis von Harville (1975) enthält.

(4.9) **Satz.** *In der Klasse aller verallgemeinerten Versuchspläne ist die Teilklasse der Produktpläne $\{\nu^* \otimes \sigma;\ \ \nu^* = \sum \varepsilon_i/I\}_{i=1}^{I}$ bei Experimenten mit qualitativen Einflußfaktoren in vollsdtäniger Wechselwirkung sowohl D- als auch A-vollständig sowohl für $\beta_{(1)}$ als auch für ganz β.*

Dieser Satz bedeutet für die Anwendung, daß stets auch bei diesen komplexeren Experimenten die vollständige Ausgewogenheit des Teilplans β für die qualitativen Einflußfaktoren die bestmögliche Wahl ist und σ unabhängig als *W*-Maß optimal zu bestimmen ist, nämlich gemäß des nun aus (4.6), folgenden Satzes:

(4.10) **Satz.** *In der Klasse aller verallgemeinerten Versuchspläne ist der Produktplan $\delta^* = \nu^* \otimes \sigma^*$ mit $\nu^* = \sum_{i=1}^{I} \varepsilon_i/I$ genau dann*

a) *D-optimal für*
   (i) $\beta_{(1)}$, *wenn* σ^* *minimiert* $c(\sigma)$ *definiert durch* (4.5)
   (ii) β, *wenn* σ^* *maximiert* $(1 - c(\sigma)) \det I_{\beta_{(2)}}(\sigma)$

b) *A-optimal für*
   (i) β, *wenn* σ^* *minimiert* $c(\sigma)/(1 - c(\sigma))$
   (ii) β, *wenn* σ^* *minimiert* $I \cdot c(\sigma)/(1 - c(\sigma)) + \text{spur } I^{-1}_{\beta_{(2)}}(\sigma)$.

Dieser Satz führt das Planungsproblem bei diesen Beobachtungsmodellen wieder auf Planungsprobleme bei reinen Regressionsmodellen zurück für die ein weites Feld von Methoden und Lösungen existiert (vgl. c.B. Fedorov (1972) und Bandemer, etc. (1977)). Dabei werden hier anders als bei Modellen des Abschnittes 3 — und wie es im folgenden Beispiel für die Klasse von Grundmodellen der Kovarianzanalyse möglich ist — die Lösungen wegen der zusätzlichen Minimierung von $c(\sigma)$ nicht immer direkt ablesbar sein.

(4.11) *Beispiel.* Ist die Wirkungsfunktion der quantitativen Einflußfaktoren linear in den quantitativen Wirkungsstufen $s = (s_1, \ldots, s_{K_2})$ (multiple lineare Regression), d.h.

$$a^T_{(2)}(t_{(2)})\beta_{(2)} \equiv s_1\beta_{(2)1} + s_2\beta_{(2)2} + \ldots + s_{K_2}\beta_{(2)K_2}$$

und der Versuchsbereich $T_{(2)} = [-1, +1]^{K_2}$ oder $T'_{(2)} = \{t_{(2)};\ ||t_{(2)}|| \leq 1\}$,

dann ist bekanntlich die gleichförmige Verteilung σ^* der Beobachtungspunkte auf den Eckpunkten des Quadrats $T_{(2)}$, beziehungsweise des von $T'_{(2)}$ umschriebenen achsenparallelen Quadrats ein D-optimaler Versuchsplan für das reine Regressionsmodell. Für diese σ^* ist offensichtlich $c(\sigma^*) = 0$, so daß nach Satz (4.11) a) der Versuchsplan $\delta = (\nu^*, \sigma^*)$ genau dann D-optimal für $\beta_{(1)}$ aber auch für ganz β, und zwar für beide Standardversuchsbereiche, wenn ν^* vollständig ausgewogener Plan über allen Wirkungsstufen der qualitativen Einflußfaktoren ist. Wegen der Äquivarianz D-optimaler Versuchspläne lassen sich bekanntlich die Bereichte $T_{(2)}$ und $T'_{(2)}$ entsprechend praktischen Gegebenheiten transformieren.

Auch hier gilt allgemein, was am Ende des vorherigen Abschnittes 3 über die Konstruktion konkreter Pläne gesagt wurde und wie das zu (3.5) analoge Beispiel (4.11) demonstriert. Auch für diese relativ einfache aber für die Anwendungen wichtige Beispielsklasse liegt unabhängig von den Ergebnissen, hier jedoch unter Zugrundelegung der konditionellen Darstellung eines-Versuchsplans, eine volltsändige Charakterisierung der optimalen konkreten Versuchspläne unter verschiedenen alternativen Beobachtungsbeschränkungen und für alle interessierenden Aspekte der Wirkungsfunktion vor, allerdings nur für die Fälle $K_2 = 1$ (vgl. Kurotschka und Wierich (1984)) und $K_2 = 2$ (vgl. Wierich (1984)). Für allgemeines K_2 sind hinreichende Bedingungen für D-optimale Versuchspläne von Lopes Troya-Schlottmann (1982) für verschiedene spezielle Einschränkungen an die Beobachtungsbedingungen bereits die Vorfront der Forschung, die gerade auf diesem Gebiet einem für die Anwendungen besonders wichtigen Feld von mathematisch interessanten ungelösten Problemen gegenübersteht.

LITERATUR

1. Bandermer, H., u.a., *Theorie und Anwendung der optimalen Versuchsplanung, I, II.* Leipzig Akademie Verlag, 1977.
2. Fedorov, V. V., *Theory of Optimal Experiments.* New York— London, Academic Press, 1972.

3. Harville, D. A., *Computing optimum designs for covariance models.* In: Srivastava, J. N. (Ed,.) *A Survey of Statistical Design and Linear Models,* pp. 209–228. Amsterdam–New York, North-Holland, 1975.
4. Kurotschka, V. G., *Optimale Versuchspläne bei zweifach klassifizierten Beobachtungsmodellen.* Metrika, **18**, 215–232 (1971).
5. Kurotschka, V. G., *Optimal design of complex experiments with qualitative factors of influence.* Comm. Statist. A– Theory Methods, **7**, 1363–1378 (1978).
6. Kurotschka, V. G., *A general approach to optimal design of experiments with qualitative and quantitative factors.* In: *Proc. Golden Jubilee Conf. Indian Statistical Institute.* Calcutta, 1982.
7. Kurotschka, V. G., Wierich, W., *Zur optimalen Plannung von Experimenten mit sowohl qualitativen als auch quantitativen Einflussfaktoren.* Metrika, **31**, (1984).
8. Lopes Troya, Josefa, *Optimal design for covariates' models.* J. Statist. Plann. Inference, **6**, 373–419 (1982).
9. Wierich, W., *Konkrete optimale Versuchspläne für ein lineares Modell mit einem qualitativen und zwei quantitativen Einflussfaktoren.* Metrika, **31**, (1984).
10. Wierich, W., *Complete class theorems for product designs.* Zur Publikation eingereicht (1983)

BO LINDQVIST

On the memory of a Markov chain. A decision theoretical approach

Let $\{X_n\}$ be a non-homogeneous Markov chain and let $E_n^{(m)}$ denote the statistical experiment consisting of observing X_{m+n} when $\theta = X_m$ is the "unknown" parameter. The (asymptotic) memory of the chain is defined to be the sequence $\{E_\infty^{(m)}\}$, where $E_\infty^{(m)} = \lim_{n\to\infty} E_n^{(m)}$ (limit with respect to an experiment topology introduced by LeCam). A characterization of $E_\infty^{(m)}$ in terms of the tail σ-algebra of the chain is given. A simpler representation with certain minimality properties is introduced. Finally, an extension of the model is considered.

1. Introduction

Let $\{X_n\}_{n=0}^{\infty}$ be a non-homogeneous Markov chain with finite or countable state space S. Suppose one observes X_{m+n} and wants information about some past state $X_m (m \geqslant 0, n \geqslant 1)$. In practice, such problems occur e.g. in genetics and reliability analysis. The problem of getting information about hyperparameters in hierarchical models (see [9]) is also of this type.

If the chain is (weakly) ergodic, then intuitively X_{m+n} will contain almost no information about X_m if n is large. In this case one may say that the chain has asymptotically *no* memory. In general, however, there may still be information about X_m present in X_{m+n} even when $n \to \infty$. As an example, if S is finite and the transition matrices $\pi(n) \equiv \{P(X_{n+1} = j \mid X_n = i)\}$ are all permutation matrices, then X_{m+n} provides complete information about X_m. In order to obtain a more precise meaning of "information about X_m contained in X_{m+n}", we shall make a decision theoretical approach. Let $\mathcal{E}_n^{(m)}$ denote the statistical experiment consisting of observing X_{m+n} when $\theta = X_m$ is the unknown parameter and the transition matrices $\pi(n)$ are assumed known. The parameter set of this experiment is S. The problem is that of describing the amount of information contained in the experiments $\mathcal{E}_n^{(m)}$. At our disposal we have then a rather extensive literature on the relationship between information theory and statistical decision theory. In the present paper we shall see how the theory of *comparison of statistical experiments*, mainly developed by D. Blackwell and L. LeCam, can be used in a study of the asymptotic memory of a Markov chain. As it turns out, it is possible to identify limits $E_\infty^{(m)} = \lim_{n\to\infty} E_n^{(m)}$ with respect to a certain reasonable topology on the set of experiments. The sequence $\{E_\infty^{(m)}\}$ will be called the *memory* of the Markov chain. The question of information about X_m contained in X_{m+n} as $n \to \infty$ can thus be regarded as a question of the statistical properties of $\mathcal{E}_{\infty}^{(m)}$.

The limit experiments $\mathcal{E}_{\infty}^{(m)}$ can be characterized in terms of the tail σ-algebra of the Markov chain. Thus, when studying the memory of a Markov chain (as we have defined it), we can draw on the existing literature on the tail σ-algebra (for references, see e.g. Cohn [6]). The present author hopes, on the other hand, that his results will provide additional insight into the nature of the tail σ-algebra of a Markov chain.

The paper extracts some main ideas and results from the author's doctor dissertation [17] which includes [14—16]. Some results and remarks are, however, new.

2. Markov chains. Notations and definitions

A set which is either finite or countably infinite will be called *discrete*. Throughout the paper we shall let $N = \{0, 1, 2, \ldots\}$ and $N^+ = \{1, 2, \ldots\}$.

Let S be a discrete set and let $\Omega = S^N$. For $\omega = (\omega_0, \omega_1, \ldots) \in \Omega$, $n \in N$, let $X_n \colon \Omega \to S$ be defined by $X_n(\omega_n) = \omega_n$. Let for $0 \leqslant m \leqslant n$, $\mathcal{F}_m^n(\mathcal{F}_m^\infty)$ denote the σ-algebra generated by $X_m, X_{m+1}, \ldots, X_n (X_m, X_{m+1}, \ldots)$. For simplicity, set $\mathcal{F} \equiv \mathcal{F}_0^\infty$ and $\mathcal{F}_n \equiv \mathcal{F}_n^n$. The tail σ-algebra $\mathcal{T}$ is defined as $\bigcap_{n \in N} \mathcal{F}_n^\infty$. We will define a shift function T on Ω by $T(\omega_0, \omega_1, \ldots) = (\omega_1, \omega_2, \ldots)$ and call a set $\Lambda \in F$ *invariant* if $T^{-1}\Lambda = \Lambda$. The class $\mathcal{J}$ of invariant sets is a σ-algebra, called the *invariant* σ-algebra.

A *stochastic matrix* is a real matrix $C = (C_{ij})$ with $C_{ij} \geqslant 0$ for all i, j and $\sum_j C_{ij} = 1$ for all i. A *transition sequence* $\pi = \pi\{(n)\}_{n \in N}$ on S is a sequence of stochastic matrices $\pi(n) = (\pi_{ij}(n) \colon i, j \in S)$. Let π be given and let $\tau = (\tau_i \colon i \in S)$ be a probability distribution on S. Then there is a unique measure $P_{\tau,\pi}$ on $(\Omega, \mathcal{F})$ such that $P_{\tau,\pi}(X_0 = i) = \tau_i$ and for $i, j \in S$, $P_{\tau,\pi}(X_{n+1} = j \mid \mathcal{F}_0^n) = \pi_{ij}(n)$ on $\{X_n = i\}$. The stochastic process $\{X_n\}_{n \in N}$ so obtained is a *non-homogeneous Markov chain* (NMC) with initial distribution τ and transition sequence π. If $\pi(n)$ is independent of n, then the chain is called homogeneous (HMC) and the matrix $\pi(n)$ will be denoted simply by π. If $\tau = \delta(i)$, where δ is the Dirac measure, then we will write $P_{i,\pi}$ instead of $P_{\tau,\pi}$. Throughout the paper we shall let ν be a fixed probability distribution on S with $\nu_i > 0$ for all $i \in S$. If π is a transition sequence, then we define $\pi(m, m+n) = \prod_{i=0}^{n-1} \pi(m+i)$. For HMC's we have $\pi(m, m+n) = \pi^n$. Let finally $\pi_{(m)}$ denote the transition sequence $\{\pi_{(m)}(n)\}$ where $\pi_{(m)}(n) = \pi(m+n)$.

We shall say that a transition sequence π is *weakly ergodic* if

$$\lim_{n \to \infty} \sum_k \mid \pi_{ik}(m, m+n) - \pi_{jk}(m, m+n) \mid = 0 \tag{2.1}$$

for all $i, j \in S$, $m \in N$ and *S-uniformly weakly ergodic* if the limit (2.1) holds uniformly in $i, j \in S$. It follows from Iosifescu [10] that π is weakly ergodic if and only if $\mathcal{T}$ is $P_{\nu,\pi_{(m)}}$-trivial for any $m \in N$, i.e. $P_{\nu,\pi_{(m)}}(A) = 0$ or 1 for all $A \in \mathcal{T}$.

3. Comparison of statistical experiments. Basic notations and results

The basic references here are Blackwell [2—3], LeCam [12—13] and Ferguson [8]. A review of the theory of comparison of experiments is given by Torgersen [19].

A statistical experiment is a pair $\mathcal{E} = ((\mathcal{X}, \mathcal{A}), (A_\theta: \theta \in \Theta))$, where $(\mathcal{X}, \mathcal{A})$ is a measurable space and $(A_\theta: \theta \in \Theta)$ is a family of probability measures on $(\mathcal{X}, \mathcal{A})$. We call $(\mathcal{X}, \mathcal{A})$ the *sample space* and Θ the *parameter set* of $\mathcal{E}$.

Suppose a statistician is to select a *decision* from some set D, where the choice $d \in D$ leads to a loss $L(\theta, d)$ when $\theta \in \Theta$ is the "true" value of the parameter. If the statistician bases his decision on the experiment $\mathcal{E}$ through a decision rule ρ, then we can compute the expected loss $R(\theta, \rho)$ when θ is the "true" parameter value. $R(\cdot, \rho)$ is called the *risk function* of ρ. Now suppose that the statistician may instead base his decision on an experiment $\mathcal{G} = ((\mathcal{Y}, \mathcal{B}), (B_\theta: \theta \in \Theta))$ (where Θ is as before). Let then $R(\theta, \sigma)$ denote the risk function of a decision rule σ based on $\mathcal{G}$. The *deficiency of $\mathcal{E}$ with respect to $\mathcal{G}$* (denoted $\delta(\mathcal{E}, \mathcal{G})$) is the *infimum* of all numbers $\alpha \geqslant 0$ such that for all sets D and all loss functions L (with $0 \leqslant L \leqslant 1$), to each decision rule σ available in $\mathcal{G}$ there is a rule ρ available in $\mathcal{E}$ with $R(\theta, \rho) \leqslant R(\theta, \sigma) + \alpha$ for *all* $\theta \in \Theta$.

Put $\Delta(\mathcal{E}, \mathcal{G}) = \max\{\delta(\mathcal{E}, \mathcal{G}), \delta(\mathcal{G}, \mathcal{E})\}$. If $\delta(\mathcal{E}, \mathcal{G}) = 0$, then we say that $\mathcal{E}$ is *more informative than* $\mathcal{G}$ (written $\mathcal{E} \geqslant \mathcal{G}$). If $\Delta(\mathcal{E}, \mathcal{G}) = 0$, then $\mathcal{E}$ and $\mathcal{G}$ are said to be *equivalent* ($\mathcal{E} \sim \mathcal{G}$). It can be shown that Δ is a semi-metric and that the set $E(\Theta)$ of equivalence classes ("experiment types") associated with $\sim$ is a complete metric space. We shall let $\hat{\mathcal{E}}$ denote the equivalence class containing $\mathcal{E}$. If $\mathcal{E} \sim \mathcal{G}$, then $\mathcal{E}$ and $\mathcal{G}$ are called *representations* of $\hat{\mathcal{E}}$. The topology on $E(\Theta)$ induced by Δ is called the *strong experiment topology*.

If $\mathcal{E}$ is as before and F is a subset of Θ then we define the experiment $\mathcal{E}_F = ((\mathcal{X}, \mathcal{A}), (A_\theta: \theta \in F))$. The *weak experiment topology* $w\Delta$ is the topology on $E(\Theta)$ induced by the projections $\mathcal{E} \to (\mathcal{E}_F, \Delta)$ for all *finite* subsets F of Θ. Thus $\mathcal{E}_n \overset{w\Delta}{\to} \mathcal{E}$ as $n \to \infty$ if and only if $\mathcal{E}_{n,F} \overset{w\Delta}{\to} \mathcal{E}_F$ for each finite $F \subseteq \Theta$. The topology $w\Delta$ is compact Hausdorff.

If Θ is finite, then clearly Δ and $w\Delta$ coincide. If Θ is countably infinite, then it can be shown that $(E(\Theta), w\Delta)$ is metrizable.

If both $\mathcal{X}$ and Θ in $\mathcal{E}$ are discreet and $\mathcal{A} = \mathcal{P}(\mathcal{X})$ (the collection of all subsets of $\mathcal{X}$), then $\mathcal{E}$ is completely determined by the stochastic matrix

$$A = (A_{\theta x}: \theta \in \Theta, x \in \mathcal{X})$$

where $A_{\theta x} = A_\theta(\{x\})$. We shall write $\mathcal{E} = \mathcal{E}(A)$ and call it a *discrete* experiment.

If $B = (B_{\theta y}: \theta \in \Theta, y \in \mathcal{Y})$ is another stochastic matrix, then it follows from LeCam [12] that

$$\delta(\mathcal{E}(A), \mathcal{E}(B)) = \inf_M \| AM - B \| \tag{3.1}$$

where M ranges over all stochastic matrices M of appropriate dimension, AM is the usual matrix product and $\| C \| = \sup_i \sum_j | C_{ij} |$ is the matrix norm.

We have $\mathcal{E}(A) \geqslant \mathcal{E}(B)$ if and only if there is a stochastic matrix M with $AM = B$. This is called the *randomization criterion.*

An experiment $\mathcal{L}$ is called totally non-informative whenever $\mathcal{E} \geqslant \mathcal{L}$ for any other experiment $\mathcal{E}$ (with the same parameter set). In this case $\mathcal{L}$ is representable by any experiment with probability distributions independent of the parameter.

4. The memory of a Markov chain

Let π be a transition sequence on a discrete set S. In the introduction we introduced the statistical experiments $\mathcal{E}_n^{(m)}$ consisting of observing X_{m+n} when $\theta = X_m$ was the "unknown parameter". More precisely, we shall define

$$\mathcal{E}_n^{(m)} = \mathcal{E}(\pi(m, m+n))\,; \quad m \in N,\, n \in N^+$$

(see Sections 2 and 3 for notation). An alternative representation of $\hat{\mathcal{E}}_n^{(m)}$ is clearly

$$\mathcal{G}_n^{(m)} = ((\Omega, \mathcal{F}_{\cdot n}),\ (P_{\theta,\pi(m)}\colon\ \theta \in S)). \tag{4.1}$$

If the Markov chain is homogeneous, then $\mathcal{E}_n^{(m)}$does not depend on m, and we will write just $\mathcal{E}_n$.

The limiting information about X_m contained in X_{m+n} as $n \to \infty$ can now be studied by considering the limiting behaviour of $\mathcal{E}_n^{(m)}$ as $n \to \infty$. It is intuitively reasonable that the amount of information contained in X_{m+n} is non-increasing in n. This feeling becomes even more clear from the following.

Let us define the experiment

$$\mathcal{H}_n^{(m)} = ((\Omega, \mathcal{F}_{\cdot n}^{\infty}),\ (P_{\theta,\pi(m)}\colon\ \theta \in S)), \tag{4.2}$$

which can be thought of as the experiment of observing the (infinite) tai $(X_{m+n}, X_{m+n+1}, \ldots)$ when $\theta = X_m$ is the unknown parameter. Since $\mathcal{F}_{\cdot n} \subset \mathcal{F}_{\cdot n}^{\infty}$ it is clear that $\mathcal{H}_n^{(m)} \geqslant \mathcal{G}_n^{(m)}$. However, the conditional probability distribution of $(X_{m+n+1}, X_{m+n+2}, \ldots)$ given X_{m+n} is, by the Markov property, independent of $\theta(= X_m)$, so X_{m+n} is statistically sufficient in the vector $(X_{m+n}, X_{m+n+1}, \ldots)$. Thus $\mathcal{H}_n^{(m)} \sim \mathcal{G}_n^{(m)}$ for all $m \in N$, $n \in N^+$(see Torgersen [19] p. 196). Since the sequence $\{\mathcal{F}_{\cdot n}^{\infty}\}$ of σ-fields is non-increasing, it is clear that for each fixed m we have $\mathcal{H}_1^{(m)} \geqslant \mathcal{H}_2^{(m)} \geqslant \ldots$. Note that this is also a simple consequence of the randomization criterion (Section 3), since $\pi(m, m+n+1) = \pi(m, m+n) \times \times \pi(m+n)$. From this monotonicity and from compactness of the space $(E(S), ,w\Delta)$ it follows that there exists for each m a limit $\hat{\mathcal{E}}_\infty^{(m)}$ with $\hat{\mathcal{E}}_n^{(m)} \xrightarrow{w\Delta} \hat{\mathcal{E}}_\infty^{(m)}$. By the representation (4.2) of $\hat{\mathcal{E}}_n^{(m)}$ it is not surprising that we have

Theorem 4.1 [17]. *The limit $\hat{\mathcal{E}}_\infty^{(m)}$ $(m \in N)$ can be represented by*

$$\mathcal{H}_\infty^{(m)} = ((\Omega, \mathcal{T}),\ (P_{\theta,\pi(m)}\colon\ \theta \in S)).$$

The proof uses a simple backwards martingale argument.

The following definition has already been motivated.

Definition. The sequence $\{\hat{\mathcal{E}}_\infty^{(m)}\}_{m=0}^{\infty}$ is called the *memory* of the transition sequence π.

The next result fits well with our intuition.

Theorem 4.2. [17]. *The transition sequence π is weakly ergodic if and only if $\hat{\mathcal{E}}_\infty^{(m)}$ is totally non-informative for each $m \in N$.*

Proof. "Only if" follows directly from Theorem 4.1 and the remark at the end of Section 2. The "if"-part follows from Theorem 5.2 and the fact that $\hat{\mathcal{L}}$ is representable by $\mathcal{E}(1)$, where 1 is a column of all 1's.

A "strong" version of Theorem 4.2 was proved in [16]: π is S-uniformly weakly ergodic if and only if $\hat{\mathcal{E}}_n^{(m)} \xrightarrow{\Delta} \hat{\mathcal{L}}$ as $n \to \infty$ for each $m \in N$. Thus it should be clear that the sequence $\{\mathcal{E}_n^{(m)}\}_{n \in N^+}$ need not converge in the strong experiment topology Δ.

However, the strong topology is still important, being induced by a metric with a simple interpretation in terms of risks of certain decision problems. We refer to [17] for conditions for strong convergence and on the rate of convergence in terms of Δ.

As the space $(\Omega, \mathcal{T})$ is difficult to work with, it is natural to ask whether "simpler" representations for the limits $\hat{\mathcal{E}}_\infty^{(m)}$ than the $\mathcal{H}_\infty^{(m)}$ are available. Unfortunately, the fact that the $\mathcal{E}_n^{(m)}$ are all discrete experiments does not in general imply that the $\hat{\mathcal{E}}_\infty^{(m)}$ are representable by discrete experiments. However, we shall in the following describe an approach that frequently will lead us to discrete limit representations.

Let us first introduce some notation. If $(\mathcal{Z}, \mathcal{C}, \mu)$ is a probability space, then $C \in \mathcal{C}$ is called *μ-atomic* if $\mu(C) > 0$ and $D \subset C$ implies $\mu(D) = 0$ or $\mu(C)$. An event $W \in \mathcal{C}$ is called *μ-completely non-atomic* if $\mu(W) > 0$ and W contains no μ-atomic subset. It is well known (see e.g. Loeve [18] p. 100) that $\mathcal{Z}$ can be μ-uniquely partitioned as

$$\mathcal{Z} = W \cup (\bigcup_{j \in J} C_j)$$

where W is μ-completely non-atomic and C_j; $j \in J$ are μ-atomic events. The index set J is discrete and may be empty. The event W may be absent, in which case $\mathcal{C}$ is called *μ-atomic.* If $\mathcal{C}$ is μ-atomic, then $\mathcal{C}$ is called, respectively, μ-finite or μ-trivial if J is finite or J contains exactly one index.

Let now $\Omega = W \cup \bigcup_{j \in J} A_j$ be the atomic decomposition corresponding to $(\Omega, \mathcal{T}, P_{\nu,\pi(m)})$. It is not difficult to see that for each $\theta \in S$, $j \in J$ we have either $P_{\theta,\pi(m)}(A_j) = 0$ or A_j is $P_{\theta,\pi(m)}$-atomic. Define an experiment

$$H_\infty^{(m)*} = ((\Omega^*, \mathcal{T}^*), (P^*_{\theta,\pi(m)}\colon \theta \in S))$$

where $\Omega^* = W + J$, $\mathcal{T}^* = (\mathcal{T} \cap W) \oplus \mathcal{P}(J)$ and $P^*_{\theta,\pi(m)}$ is (uniquely) defined by $P^*_{\theta,\pi(m)}(A) = P_{\theta,\pi(m)}(A)$ if $A \subset W$, $A \in \mathcal{T}$ and $P^*_{\theta,\pi(m)}(j) = P_{\theta,\pi(m)}(A_j)$ if $j \in J$. Here $+$ denotes disjoint union and $\mathcal{A} \oplus \mathcal{B}$ is the σ-algebra generated by all sets of the form $A + B$; $A \in \mathcal{A}$, $B \in \mathcal{B}$. It is not difficult to verify that $\mathcal{H}_m^{(m)*} \sim \mathcal{H}_\infty^{(m)}$

Example. Consider a recurrent HMC and let $\{S_c\colon c \in C\}$ denote the partition of S into cyclically moving subclasses. Corollary 2 of Blackwell and Freedman [4] states that $\mathcal{T}$ is equivalent to the σ-algebra generated by the sets $\{X_0 \in S_c\}$ for $c \in C$. Thus the limit $\mathcal{E}_\infty$ is representable by the discrete experiment $\mathcal{E}(Q)$, where $Q = \{Q_{\theta c}\colon \theta \in S, c \in C\}$ and $Q_{\theta c} = 1$ if $\theta \in S_c$ and 0 otherwise.

5. Minimal representations for the memory

In the previous section we introduced the representations $\mathcal{H}_\infty^{(m)*}$ obtained through an atomic decomposition of $(\Omega, \mathcal{T})$. As a consequence, a sufficient condition for $\hat{\mathcal{E}}_\infty^{(m)}$ to be representable by a discrete experiment is that $\mathcal{T}$ is $P_{\nu,\pi(m)}$-atomic. This condition is not necessary, however, as follows from Example 2 in Blackwell and Freedman [4]. They define a HMC with $\mathcal{T}$ completly non-atomic, whereas X_n contains complete information about X_0 for any $n \in N^+$. Thus $\mathcal{E}_\infty$ is completely informative and may be represented by the discrete experiment $\mathcal{E}(I)$, where I denotes the identity matrix.

In Theorem 5.2 we shall see that "necessity" holds if we impose certain conditions of finiteness.

First, we state a Lemma which is esentially a consequence of Kingman [11].

Lemma 5.1. *Let π be a transition sequence. Let $A_1, \dots, A_r$ be a partition of Ω, where $r \geqslant 2$ and $A_j \in \mathcal{T}$ for $1 \leqslant j \leqslant r$. Suppose that for some $k \in N$, $P_{\nu,\pi(k)}(A_j) > 0$ for $1 \leqslant j \leqslant r$. Then there is an $m_0 \in N$ such that for all $m \geqslant m_0$ the r vectors*

$$(P_{\theta,\pi(m)}(A_j);\ \theta \in S);\ 1 \leqslant j \leqslant r$$

are linearly independent.

Theorem 5.2. *Assume that $\mathcal{E}_\infty^{(m)}$ for each $m \in N$ is representable by a discrete experiment $\mathcal{E}(Q_m)$, where Q_m has a finite number r_m of columns. If $\sup_m r_m \equiv r < \infty$, then $\mathcal{T}$ is $P_{\nu,\pi(m)}$-finite for any $m \in N$, where the number of atoms does not exceed r.*

Proof. Suppose the result does not hold. Then there is a partition $A_1, \dots, A_{r+1}$ of Ω with each $A_j \in \mathcal{T}$ and $P_{\nu,\pi(k)}(A_j) > 0$ for all $1 \leqslant j \leqslant r+1$ for some k. Let R_m denote the stochastic matrix

$$R_m = \{P_{\theta,\pi(m)}(A_j) \colon \theta \in S,\ 1 \leqslant j \leqslant r+1\}.$$

It is clear that $\mathcal{H}_\infty^{(m)} \geqslant \mathcal{E}(R_m)$ since $\{A_j\}$ is a partition of $(\Omega, \mathcal{T})$. Thus there is a stochastic matrix M with $R_m = Q_m M$. Here Q_m has rank at most r so this must also be the case of R_m. However, the $r+1$ columns of R_m are linearly independent for m large enough by Lemma 5.1. This contradiction proves the theorem.

In the rest of the section we shall be concerned with the case when $\mathcal{T}$ is atomic and present conditions under which $\mathcal{H}_\infty^{(m)*}$ is minimal in a precise statistical sense.

If $\mathcal{E} = ((\mathcal{X}, \mathcal{A}), (A_\theta \colon \theta \in \Theta))$ is a statistical experiment, then $\mathcal{E}$ is called *minimal sufficient* if, roughly, any experiment obtained from $\mathcal{E}$ by replacing $\mathcal{A}$ by a strict sub-σ-algebra $\mathcal{B} \subset \mathcal{A}$ is *strictly* less informative than $\mathcal{E}$. A condition insuring the minimal sufficiency of $\mathcal{E}$ is that $\mathcal{E}$ is *boundedly complete*, i.e. that for any bounded $\mathcal{A}$-measurable function g such that $\int g \, dA_\theta = 0$ for all $\theta \in \Theta$ one has $g = 0$ a.e. A_θ for all θ. Note that if $\mathcal{E} = \mathcal{E}(A)$ is a discrete experiment where A has a finite number of columns, then $\mathcal{E}$ is boundedly complete if and only if the columns of A are linearly independent.

An easy consequence of Lemma 5.1 is

Theorem 5.3 *Suppose* $P_{\nu,\pi(0)}(X_n = i) > 0$ *for all* $n \in N^+$, $i \in S$. *Then if* $\mathcal{T}$ *is* $P_{\nu,\pi(0)}$ *-finite, there is an* $m_0 \in N$ *such that* $\mathcal{H}_\infty^{(m)*}$ *is boundedly complete for all* $m \geq m_0$.

The conclusion of Theorem 5.3 in the case of HMC's is simply that $\hat{\mathcal{E}}_\infty$ can be represented by a boundedly complete experiment. However, it has been shown in [17] that this conclusion holds under weaker assumptions than needed in Theorem 5.3.

Consider a HMC with transition matrix π. If $A \in \mathcal{T}$ is $P_{\nu,\pi}$-atomic, then following Abrahamse [1] we shall define the asymptotic period of A by

$$N(A) = \inf \{m \geq 1 : T^m A = A \text{ a.s. } P_{\nu,\pi}\}$$

where by convention $\inf \emptyset = \infty$.

As shown in [17] we have

Theorem 5.4. *If* $\mathcal{T}$ *is* $P_{\nu,\pi}$*-atomic with* $N(A_j) < \infty$ *for any atom* A_j, *then* $\mathcal{H}_\infty^{(0)*}$ *is boundedly complete.*

An example in [17] shows that the conclusion of the theorem may not hold if $\mathcal{T}$ is $P_{\nu,\pi}$-atomic and $N(A_j) = \infty$ for some A_j.

Let us finally reconsider the example of Section 4. The asymptotic period $N(A)$ of the atom A equivalent to $\{X_0 \in S_c\}$ coincides with the commonly defined period of S_c, which is always finite. Thus the assumptions of Theorem 5.4 are fulfilled. We conclude that $\mathcal{E}(Q)$ is boundedly complete, which in fact is trivial by the very definition of Q.

6. An extension of the model

The experiment $\mathcal{E}_n^{(m)}$ considered so far consists of observing the chain at a known time instant $m+n$. However, in practice one may want to draw conclusions about X_m also from an observation X_{m+K}, where $K \geq 1$ is an unobservable integer valued random variable. In the present section we consider briefly the case when K is stochastically independent of $\{X_n\}$ with a known probability distribution.

Let, more precisely, $\{K(n)\}_{n \in N}$ be a sequence of such K and put $Y_n^{(m)} = X_{m+K(n)}$. The conditional distribution of $Y_n^{(m)}$ given X_m is determined by the matrix transformation

$$\sum_{k=1}^{\infty} a_{nk}\pi(m, m+k); \quad m \in N,\ n \in N^+$$

where $a_{nk} = \Pr(K(n) = k)$.

Let A denote the stochastic matrix (a_{nk}) and define the discrete experiment

$$(A \cdot \mathcal{E})_n^{(m)} = \mathcal{E}\left(\sum_{k=1}^{\infty} a_{nk}\pi(m, m+k)\right).$$

This can be thought of as the experiment of observing $Y_n^{(m)}$ when $\theta = X_m$ is the unknown parameter.

We shall be concerned with the limiting behaviour of $(A \cdot \mathcal{E})_n^{(m)}$ as $n \to \infty$, under the assumption that $\lim_{n\to\infty} \Pr(K(n) = k) = 0$ for each $k \in N^+$, or equivalently that A is *regular*.

If $a_{nk} = \delta_{nk}$, where δ is the Kronecker delta, then $(A \cdot \mathcal{E})_n^{(m)} = \mathcal{E}_n^{(m)}$. It can be shown that this is asymptotically the most favorable distribution of the $K(n)$ as regards information about X_m. More precisely, we have

Theorem 6.1 [17]. *If A is regular, then for any limit point $\mathcal{Q}$ of $\{(A \times \times \mathcal{E})_n^{(m)}\}_{n \in N^+}$ we have $\mathcal{H}_n^{(m)} \geqslant \mathcal{Q}$.*

Another special case is $a_{nk} = n^{-1}$ for $1 \leqslant k \leqslant n$. The corresponding matrix transformations $\sum_{k=1}^{u} n^{-1} \pi(m, m+n)$, called Cesaro-averages, have been studied in some cases by Bowerman et al. [5]. Consider now a HMC with transition matrix π and let $\mathcal{C}_n$ denote the experiment $(A \cdot \mathcal{E})_n^{(m)}$ resulting from this special choice of A. Recall from Section 4 that the existence of $\lim_{n\to\infty} \hat{\mathcal{E}}_n^{(m)}$ is implied by compactness and the monotonicity property $\mathcal{E}_1^{(m)} \geqslant \mathcal{E}_2^{(m)} \geqslant \ldots$. Unfortunately, we do *not* in general have a corresponding property for the $\mathcal{C}_n$. An elementary proof of the existence of a limit experiment $\mathcal{C}$ can, however, still be given.

Theorem 6.2. *In the case of HMC's there exists a limit experiment $\mathcal{C}$ with $\hat{\mathcal{C}}_n \xrightarrow{w\Delta} \hat{\mathcal{C}}$. Moreover, $\mathcal{C}_n \geqslant \mathcal{C}$ for all $n \in N^+$.*

Proof. By compactness there is a subsequence $\{\mathcal{C}_{n'}\}$ and an experiment $\hat{\mathcal{C}}'$ with $\hat{\mathcal{C}}_{n'} \xrightarrow{w\Delta} \hat{\mathcal{C}}$. Fix $n \in N^+$. Then for all $n' \geqslant n$ we have by (3.1) that

$$\delta(\mathcal{C}_n, \mathcal{C}_{n'}) \leqslant \left\| \left(\frac{1}{n} \sum_{j=1}^{n} \pi^j\right)\left(\frac{1}{n'} \sum_{j=1}^{n'} \pi^j\right) - \left(\frac{1}{n'} \sum_{j=1}^{n'} \pi^j\right) \right\| \leqslant \frac{3n}{n'}$$

Letting $n' \to \infty$ we obtain $\delta(\mathcal{C}_n, \mathcal{C}') = 0$. Thus if $\mathcal{C}''$ is another limit point of $\{\mathcal{C}_n\}$, then $\delta(\mathcal{C}'', \mathcal{C}') = 0$. By symmetry $\mathcal{C}' \sim \mathcal{C}''$ and we are done.

It follows from a result of Derriennic [7] that $\hat{\mathcal{C}}$ may in fact be represented by the experiment $((\Omega, \mathcal{J}), (P_\theta : \theta \in S))$ (see [17]).

Theorem 6.1 provided an asymptotic upper bound for the information about X_m that can be obtained by observing $X_{m+K(n)}$(under our conditions on $K(n)$). The following result gives a lower bound for HMC's.

Theorem 6.3 [17]. *In the case of* HMC's, *if A is regular and $\mathcal{Q}$ is a limit point of $(A \cdot \mathcal{E})_{n'}$, then $\mathcal{Q} \geqslant \mathcal{C}$.*

REFERENCES

1. Abrahamse, A., *The tail field of a Markov chain.* Ann. Math. Statist., **40**, 127–136 (1969).
2. Blackwell, D., *Comparison of experiments.* In: *Proc. Second Berkeley Sympos. Math. Statist. Probab.*, pp. 93–102. Berkeley, Univ. California Press, 1951.
3. Blackwell, D., *Equivalent comparisons of experiments.* Ann. Math. Statist., **24**, 265–272 (1953).
4. Blackwell, D., Freedman, D., *The tail σ-field of a Markov chain and a theorem of Orey.* Ann. Math. Statist., **35**, 1291–1295 (1964).
5. Bowerman, B., David, H. T., Isaacson, D., *The convergence of Cesaro-averages for certain nonstationary Markov chains.* Stochastic Process. Appl., **5**, 221–230 (1977).
6. Cohn, H., *On the asymptotic events of a Markov chain.* Internat. J. Math. Math. Sci., **2**, 537–587 (1979).
7. Derriennic, Y., *Lois «zero ou deux» pour les processus de Markov. Applications aux marches aléatoires.* Ann. Inst. H. Poincaré Sect. B (N.S.), **XII**, 2, 111–129 (1976).
8. Ferguson, T. S., *Mathematical Statistics. A Decision Theoretic Approach.* New York, Academic Press, 1976.

9. Goel, P. K., De Groot, M. H., *Information about hyperparameters in hierarchical models.* J. Amer. Statist. Assoc., **76**, 140–147 (1981).

10. Iosifescu, M., *On finite tail σ-algebras.* Z. Wahrsch. Verw. Gebiete, **24**, 159–166 (1972).

11. Kingman, J. F. C., *Geometrical aspects of the theory of non-homogeneous Markov chains.* Math. Proc. Cambridge Philos. Soc., **77**, 171–183 (1975).

12. LeCam, L., *Sufficiency and approximate sufficiency.* Ann. Math. Statist., **35**, 1419–1455 (1964).

13. LeCam, L., *Limits of experiments.* In: *Proc. Sixth Berkeley Sympos. Math. Statist. Probab., Vol.* 1, pp. 245–261. Berkeley, Univ. California Press, 1972.

14. Lindqvist, B., *How fast does a Markov chain forget the initial state? A decision theoretical approach.* Scand. J. Statist., **4**, 145–152 (1977).

15. Lindqvist, B., *On the loss of information incurred by lumping states of a Markov chain.* Scand. J. Statist., **5**, 92–98 (1978).

16. Lindqvist, B., *A decision theoretical characterization of weak ergodicity.* Z. Wahrsch. Verw. Gebiete, **44**, 155–158 (1978).

17. Lindqvist, B.. *On the memory of a Markov chain. A decision theoretical approach.* Doctoral dissertation, Univ. of Oslo, 1981. (Available from the author).

18. Loève, M., *Probability Theory. 3rd. Ed.* Princeton, Van Nostrand, 1963.

19. Torgersen, E. N., *Comparison of statistical experiments.* Scand. J. Statist., **3**, 186–208 (1976).

K. J. MIESCKE*

Recent results on multi-stage selection procedures

1. Introduction

The problem of how to find the best (in terms of a distributional parameter) of $k \geqslant 2$ populations, by means of observations drawn from them, has been studied by many research workers in the past. A thorough introduction into the field of Ranking and Selection as well as a complete overview over all relevant developments up to 1979 is provided by Gupta and Panchapakesan [16]. The bibliography contained therein can be complemented with the help of Dudewicz and Koo [7].

The purpose of the present paper is to survey recent results in the subfield of sequential or, respectively, multi-stage selection procedures which are not already discussed in [16]. Moreover, several open problems will be pointed out to encourage further work in this direction. The most remarkable publication in this respect, without doubt, has been Bechhofer, Kiefer and Sobel [1], which still serves as an inspiring source of results and ideas.

Let the populations, as usual, be denoted by $\pi_1, \ldots, \pi_k$. From every π_i, a sequence of observations $X_{i1}, X_{i2}, \ldots$ is available to the experimenter. The observations altogether are assumed to be independent. For every i, let the X_{ij}'s have a density f_{θ_i} w.r.t. a sigma-finite measure μ on R, which is the Lebesgue- or a counting measure, respectively. The family of densities $\mathcal{F} = \{f_\theta\}_{\theta \in \Omega}$, $\Omega \subseteq$ R, is assumed to be known. In many papers, $\mathcal{F}$ is a one-parameter exponential family. In the continuous case, the most prominent example consists of k normal populations with unknown means $\theta_1, \ldots, \theta_k \in \Omega =$ R and a common known variance $\sigma^2 > 0$ *(Normal Case)*. In the discrete case, it consists of k Bernoulli populations with unknown success probabilities $\theta_1, \ldots, \theta_k \in \Omega = [0, 1]$ *(Bernoulli Case)*.

A multi-stage selection procedure for finding that π_i with $\theta_i = \max\{\theta_1, \ldots, \theta_k\}$ consists of four different types of decisions which have to be made anew at each subsequent stage $m = 1, 2, \ldots$. More precisely, at stage m,

* Research supported by ONR Contract N00014-75-C-0455 at Purdue University.
Present address: Department of Mathematics, University of Illinois at Chicago Circle, Box 4348, Chicago, Illinois 60680, U.S.A.

based on the observations drawn up to that point, the experimenter has to decide

(a) whether or not he would like to stop *(stopping rule);*

(b) (in case of not stopping) which populations to eliminate from further considerations *(elimination rule),* and what kind of observations to take at the next following stage $m + 1$ *(sampling rule);*

(c) (in case of stopping) which population(s) finally to select *(terminal decision rule)*

Procedures can be categorized in different ways according to their characteristics in (a), (b) or (c), respectively. Some of the terms which are used frequently are the following.

Closed (open) sequential procedure: the number of observations which can be drawn from $\pi_1, \ldots, \pi_k$ is a bounded (unbounded) random variable.

Truncated procedure: the number of stages is a bounded random variable.

q-stage procedure: the number of stages is a fixed constant $q \geqslant 1$.

Procedure with elimination: at every stage, populations (which appear to be inferior) can be eliminated from terminal decisions. The remaining populations constitute a *subset selection.* Typically, eliminated populations are excluded also from further sampling. A justification for this will be given later (cf. Fact 5).

Vector at a time sampling: at every stage m, exactly n_m observations are taken from every non-eliminated population. The sample sizes $n_1, n_2, \ldots$ are determined before the experiment starts.

Adaptive sampling: at every stage, the decision which observations to be taken next depends on the data collected up to that stage. A well known example for the *Bernoulli Case* is the "play the winner sampling rule" which is due to H. Robbins (cf. [16], p. 64).

Subset selection procedure: the final decisions are subsets of $\{\pi_1, \ldots, \pi_k\}$ of random size. A *correct selection* (CS) occurs if the best population is included.

Fixed size t subset selection procedure: the terminal decisions are subsets of $\{\pi_1, \ldots, \pi_k\}$ of fixed size t. In the case of $t = 1$, one calls such a procedure simply a *selection procedure.* A correct selection has the same meaning as before.

All of the terms above are used quite consistently in the literature, except for the term *adaptive.* Some authors call their procedures, though employing vector at a time sampling, adaptive because of certain other reasons. See for example Büringer, Martin and Schriever [5] (the nonparametric part), Hsu and Edwards [19] and Tong [48].

The classical approach to find reasonable procedures is the following. Let $\Omega^k(\delta^*) = \{\Theta \in \Omega^k \mid D(\theta_{[k-1]}, \theta_{[k]}) > \delta^*\}$ be the so-called preference zone and $\Omega^k \setminus \Omega^k(\delta^*)$ be the indifference zone, where D is a distance measure (usually to be $\theta_{[k]} - \theta_{[k-1]}$ or $\theta_{[k]}/\theta_{[k-1]}$, resp.), $\delta^* > 0$ is fixed and $\theta_{[1]} \leqslant \theta_{[2]} \leqslant \ldots \leqslant \theta_{[k]}$ denote the ordered values of $\theta_1, \ldots, \theta_k$. In the *indifference zone approach,* due to R.E. Bechhofer (cf. [16], p. 8), only those procedures are considered which satisfy

$$(1) \qquad \inf\{P_\Theta(\mathrm{CS}) \mid \Theta \in \Omega^k(\delta^*)\} \geqslant P^*,$$

where $P^* > k^{-1}$ is prespecified. A $\Theta \in \Omega^k(\delta^*)$ is called a *least favorable configuration* (LFC) if the infimum in (1) occurs at this Θ. A first step towards establishment of (1) for a suitable type of procedure is thus usually to find its LFC. Among those procedures satisfying (1) one then naturally tries to find a candidate with a small expected terminal subset size and, moreoever, with a small average sample number (ASN).

From a decision point of view, this (minimax type) approach means that the risk w.r.t. a 0—1 terminal decision loss should be less than $1 - P^*$ on $\Omega^k(\delta^*)$ and that, subject to this condition, objective functions (expected terminal subset size and ASN) are tried to get small. One major objection to the indifference zone approach is, however, that nothing is actually controlled if $\Theta \notin \Omega^k(\delta^*)$. Procedures based on this approach will be the main topic of Sections 3-5.

An alternative way of treating the problem is the one prescribed by the *decision theory*. Hereby one has to incorporate all losses due to inappropriate decisions and costs of sampling into one (stage-dependent) loss function, and then to consider the risk function, i.e. the expected loss, as the objective function which has to be minimized. Within the class of permutation invariant procedures, i.e. among those which give no apriori preference to any of the k populations, several optimality results can be derived, especially in the Bayesian approach. This will be described in Section 2 below.

2. The decision theoretic approach

In this section we assume that $\mathcal{F}$ is a one-parameter exponential family. More precisely, let

(2) $\mathcal{F} = \{c(\theta) \exp(\theta x)\, d(x),\ x \in \mathrm{R}\}_{\theta \in \Omega}$, where $\Omega \subseteq \mathrm{R}$ is an interval.

We consider the class $\mathcal{X}_I$, say, of permutation invariant sequential procedures with (or without) elimination, which are based on vector at a time sampling. Let $W_{im} = X_{i1} + X_{i2} + \ldots + X_{i(n_1+\ldots+n_m)}$ denote the sufficient statistic for θ_i, based on all observation which are available from π_i up to stage m, $i = 1, \ldots, k$, and let $W_m = (W_{1m}, \ldots, W_{km})$, $m = 1, 2, \ldots$.

Let $L_m(\Theta, (t_1, \ldots, t_{m+1}))$ be the loss which occurs at $\Theta \in \Omega^k$ if a procedure stops at stage m and finally selects population $t_{m+1} \subseteq \{\pi_1, \ldots, \pi_k\}$, after it has eliminated at stage j population $t_j \subseteq \{\pi_1, \ldots, \pi_k\}$, $j = 1, \ldots, m$, where $t_1, \ldots, t_{m+1}$ is a disjoint decomposition of $\{\pi_1, \ldots, \pi_k\}$. Assume that for every m, L_m has the following properties

(3a) $$L_m(\Theta, (\sigma(t_1), \ldots, \sigma(t_{m+1}))) = L_m((\theta_{\sigma(1)}, \ldots, \theta_{\sigma(k)}), (t_1, \ldots, t_{m+1})),$$

where $\sigma(t_j) = \{\sigma(i) \mid i \in t_j\}$, $j = 1, \ldots, m+1$, for every permutation σ of $(1, \ldots, k)$ and

(3b) $L_m(\Theta, (\tilde{t}_1, \ldots, \tilde{t}_{m+1})) \leqslant L_m(\Theta, (t_1, \ldots, t_{m+1}))$, if for a certain pair (i, j) with $\theta_i < \theta_j$ there exist integers $\alpha < \beta \leqslant m+1$ such that $i \in t_\beta$, $j \in t_\alpha$, $\tilde{t}_\alpha = (t_\alpha \setminus \{j\}) \cup \{i\}$, $\tilde{t}_\beta = (t_\beta \setminus \{i\}) \cup \{j\}$ and $\tilde{t}_\gamma = t_\gamma$ for $\gamma \neq \alpha, \beta$.

Condition (3a) states that L_m is permutation invariant, and (3b) states that a better population should be eliminated later than an inferior one. It is not difficult to see that $-L_m$, if L_m has these two properties, can be represented by a function of Θ and a permutation $(\sigma(1), \ldots, \sigma(k))$ of $(1, \ldots, k)$ which is decreasing in transposition (DT). Functions with this property have been studied by Hollander, Proschan and Sethuraman [18], and their results can be used to derive several optimality results in the present context.

Since the risk function $R(\Theta, \mathcal{D})$, $\Theta, \in \Omega^k$, of a procedure $\mathcal{D}$ from $\mathcal{X}_I$ is permutation symmetric in Θ, uniformly (in Θ) best results in terms of the risk function can be derived more easily in a Bayesian approach under a permutation symmetric prior τ. Thus, let τ be such a prior for the now random parameter vector Θ. Then one can prove, one after another, the following facts (cf. Gupta and Miescke [12, 14, 15]). Let $m \geqslant 1$ be fixed in the sequel.

Fact 1. *The density of W_m, defined on* $\mathrm{R}^k \times \Omega^k$, *is* (DT).

Fact 2. *The posterior density of Θ is* (DT).

Fact 3. — $E\{L_m(\Theta, (t_1, \ldots, t_{m+1})) \mid W_m = W\}$, *as a function of* $(\sigma(1), \ldots, \ldots, \sigma(k))$ *and* W, *is* (DT). *Hereby,* $\{\sigma(1), \ldots, \sigma(q_1)\} = t_1$, $\{\sigma(q_1+1), \ldots, \sigma\{(q_2)\} = t_2$, *and so forth, for certain numbers* $q_1, \ldots, q_{m+1}$ *with* $q_1 \leqslant \ldots \leqslant q_{m+1} = k$.

A *natural terminal decision,* at stage m, selects only those populations among the non-eliminated ones which yield the largest W_{im}-values. In the discrete case, ties have to be split at random to get a procedure within $\mathcal{X}_I$. For one-stage procedures it is well known that this type of decision is optimum in several senses (cf. [16], p. 42 and Miescke [30]). The next result can be considered as a generalization of the so-called "Bahadur-Goodman Theorem" (cf. [16], p. 46).

Fact 4. *For any $\mathcal{D} \in \mathcal{X}_I$, let $\mathcal{D}^*$ be the same procedure as $\mathcal{D}$ except for the terminal decisions where $\mathcal{D}^*$ uses the natural ones. Then*

$$R(\Theta, \mathcal{D}^*) \leqslant R(\Theta, \mathcal{D}), \text{ uniformly in } \Theta \in \Omega^k. \tag{4}$$

Actually, Fact 4 remains unchanged if one assumes that, at every stage m, the complete vector W_m has been observed. With other words, one can state

Fact 5. *Observations from eliminated populations are irrelevant for terminal decisions.*

The next following, rather negative, statement may be considered, in certain situations, as an argument against the use of adaptive sampling techniques.

Fact 6. *For all situations where terminal decisions of a procedure with adaptive sampling are based on unequal numbers of observations from the non-eliminated populations, there is no, uniformly in Θ, optimal terminal decision.*

One might now expect that, within stages where a procedure with elimination does not stop, *natural subset selections* (i.e. subset selections associated with largest W_{im}-values) have similar strong optimality properties as the natural terminal decisions. It turns out, however, that this is only the case under certain circumstances. First of all, results analogous to Fact 4 can be proved only for strongly unimodal exponential families $\mathcal{F}$, i.e. where $f_\theta(x)$ or $d(x)$, respectively, is logconcave. The following is a key-lemma.

Fact 7. *If $\mathcal{F}$ is strongly unimodal, then the conditional density of W_{m+1}, given $W_m = W$, which is derived from the joint distribution of Θ, W_m and W_{m+1}, is* (DT).

If one assumes that at every stage m, all observations W_m are known (just to simplify the proofs), then with backward induction it is not possible to overcome the points where decisions have to be made on *how many* populations to eliminate. Optimal decision would utilize here *all* observations, including those from already eliminated populations. The following three results, however, can at least be proved.

Fact 8. *Let the number of stages q, say, as well as all subset sizes of selections* $r_1 \geq r_2 \geq \ldots \geq r_q$, *say at stages* $1, 2, \ldots, q$ *be fixed. Then the procedure which uses the natural subset selections and natural terminal decision is the unique, uniformly in* Θ, *best procedure in the sub-class of procedures in* $\mathcal{X}_I$ *with these properties.*

Fact 9. *Within the sub-class of fixed size t two-stage procedures in* $\mathcal{X}_I$, *the procedures which use natural subset selections at stage* 1 *and the natural terminal decision at stage* 2 *constitute an essentially complete class.*

Fact 10. *Assume that* L_m *depends on* Θ *only through those* θ_i*'s which are associated with the, at stage m, not eliminated populations,* $m = 1, 2, \ldots$. *If apriori,* $\theta_1, \ldots, \theta_k$ *are independently identically distributed, then every truncated Bayes procedure in* $\mathcal{X}_I$ *uses natural subset selections at all stages, and the natural terminal decision.*

After these considerations under rather mild assumptions upon the loss functions, it is now natural to look for specific procedures which are optimal in more concrete situations. In the control case, where one wishes to select the best population, provided that it is better than the control (i.e. $\theta_{[k]} > \theta_0$), two-stage procedures with elimination have been derived by Miescke [31] and Gupta and Miescke [14]. A Γ-minimax approach, like the one discussed by Miescke [32] for one-stage procedures, appears to be appropriate for such problems, but no work has been done here up to now.

The difficulties arising in concrete problems with the backward induction in sequential selection problems are considerable (cf. Edwards [8]). It seems to be more realistic for future work to look for simple structured procedures which are approximately optimal in a reasonable sense. Ad hoc procedures like the ones proposed by Washburn [52] may have good performance characteristics and deserve to be studied in more detail. Washburn's procedures are open and closed procedures without elimination. They are based on prior knowledge and use adaptive sampling which, as well as the terminal decision, is based on the posterior expectations of $\theta_1, \ldots, \theta_k$. These procedures are not Bayes solutions in a decision theoretic sense since no overall loss is actually considered here. However, as it is shown by Washburn [52], they perform well compared with other procedures given in the literature.

Two papers which give more detailed Bayes solutions in concrete problems, but which do not completely fit into the distributional framework considered so far, are due to Ramey and Alam [41] and Gulati [9]. Ramey and Alam [41] derive a Bayes truncated procedure for the most probable of k cells in a multinomial model under a Dirichlet prior. Gulati [9] considers the problem of finding that one of k uniform distributions which has slipped to the left (and has a shorter support). He finds the Bayes solution with respect to any prior specifying the slipped population, within the class of closed sequential procedures based on adaptive sampling.

In the sections to come, procedures will be discussed which are not derived from the decision theoretic approach. Most of them are fixed size t (especially $t = 1$) subset selection procedures. All of them use the natural subset selections within stages and the natural terminal decisions.

3. The Bernoulli case and related models

The case of k Bernoulli populations with unknown success probabilities $\theta_1, \ldots, \theta_k \in \Omega = [0, 1]$ will be discussed only briefly, since two detailed publications on recent developments in this area are readily available. The first one is Büringer, Martin and Schriever [5]. In their book, various sequential selection procedures with vector at a time as well as play the winner sampling and different stopping rules are studied under the indifference zone approach in great detail. The second one is the paper by Bechhofer and Kulkarni [2] which provides an excellent overview not only over the k-population *selection* problem but also over related areas like clinical trials and multi-armed bandit problems, where the objective functions differ from those in the selection problem. These related areas are also covered by Dudewicz and Koo [7], where further references can be found.

The main topic of Bechhofer and Kulkarni [2], however, is to propose a closed non-eliminating adaptive fixed size t subset selection procedure, which has several optimal properties in terms of the P(CS) and the expected number of observations taken from certain populations. Some results, which are proved for $k = 2$ only, are conjectured to hold also for $k \geqslant 3$. In a subsequent paper, Bechhofer and Kulkarni [3] provide additional results on the performance of their procedure.

Levin and Robbins [27] consider an open non-eliminating procedure with vector at a time sampling which stops if one population has r more "successes" than all the remaining ones. Among others, a conjecture is stated for a procedure with elimination, which is proved to hold for the non-eliminating version.

The related *multinomial case* (note that here independence of the cell-frequencies is clearly not given, but that the joint distribution is (DT)), where the goal is to find the cell with the largest probability, is also treated by Levin and Robbins [27]. A closed sequential (inverse sampling) version of their procedure has been studied already by Ramey and Alam [40], where a conjecture concerning the LFC for $k \geqslant 3$ is stated. A Bayes procedure by Ramey and Alam [41] has been mentioned already in Section 2. Further work has been done by Hwang [24] and Hsuan, Hwang and Parnes [22]. The latter is actually a one-stage result, but the question of whether more sampling is more informative is certainly of relevance for sequential selection problems, too.

4. The normal case and related models

The problem of finding that one of k normal populations $N(\theta_i, \sigma_i^2)$, $i = 1, \ldots, k$, which has the largest mean in more than one stage has been studied under several aspects (but mainly under the indifference zone approach) by many research workers. In the following we shall distinguish between three different situations depending on the status of knowledge about $\sigma_1^2, \ldots, \sigma_k^2$. The *first* one is the simplest one: Here, $\sigma_1^2 = \ldots = \sigma_k^2 = \sigma^2$ and $\sigma^2 > 0$ is known. For this model, Bechhofer, Kiefer and Sobel [1] proposed and studied their, meanwhile classical, open sequential procedure without elimination which is based on vector at a time sampling with $1 = n_1 = n_2 = \ldots$. The terminal decision is the natural one (which is optimum, cf. Fact 4), and the stopping rule is

$$N_{BKS} = \inf \left\{ m \,\middle|\, \sum_{i=1}^{k-1} \exp(-\delta^*(Y_{[k]m} - Y_{[i]m})) \leqslant (1 - P^*)/P^* \right\} \tag{4}$$

$$\text{where } Y_{im} = (X_{i1} + \ldots + X_{im})/\sigma^2, \ i = 1, \ldots, k,$$

which establishes the procedure in the indifference zone approach (i.e. (1)). An upper bound for the first moment of N_{BKS} is given in Huang [23], and asymptotic properties of this stopping rule, i.e. the behavior of the ASN under $P^* \to 1$ and/or $\delta^* \to 0$ have been studied by Bechhofer, Kiefer and Sobel [1], Tong [49] and Jennison, Johnstone and Turnbull [25].

As with the BKS-procedure, many others can be viewed, in one way or another, as being generalizations of Wald's SPRT (cf. [16], p. 127). This is also the case with that one in Mukhopadhyay [37]. It differs from the BKS-procedure only through its stopping rule, N_M, say.

$$N_M = \{\inf m \mid (k-1) \max_{i \neq k} \exp(-\delta^*(Y_{[k]m} - Y_{[i]m})) \leqslant 1 - P^*\}. \tag{5}$$

Apparently, $N_{BKS} \leqslant N_M$ and $E_\Theta(N_{BKS}) < E_\Theta(N_M)$ for all $\Theta \in \mathrm{R}^k$, which implies that the BKS-procedure is more efficient. Usually it is more difficult to compare procedures directly in this manner and asymptotic techniques are then the only way to do this.

Under the indifference zone approach, open sequential procedures with elimination and vector at a time sampling, which are based on paired comparisons of the sample means, are discussed by Swanepoel and Geertsema [44], Kao and Lai [26], Hsu and Edwards [20], Turnbull, Kaspi and Smith [50] and Jennison, Johnstone and Turnbull [25]. In the latter two papers, however, procedures with adaptive sampling are the main topic.

An open sequential procedure without elimination based on vector at a time sampling is proposed by Tong [48] in a more general setting including the normal case. To avoid the indifference zone approach, the single stage P(CS) is hereby estimated stage by stage (by replacing the unknown parameters by estimators), and the procedure stops as soon as this estimate is greater or equal to P^*. It would be interesting to derive a probability guarantee for the P(CS) in $\Omega^k(\delta^*)$. In $\Omega^k \setminus \Omega^k(\delta^*)$, however, one has to encounter (similar as with the BKS-procedure) a large ASN (cf. McCulloch [28]).

Even if the ASN is finite, there remains the uncertainty of how long it actually takes until an open procedure eventually stops (see also Bechhofer and Kulkarni [2], 2.2). From a practical point of view, truncated and q-stage procedures with elimination seem to meet more likely the needs of practitioners. A two-stage procedure whit elimination and vector at a time sampling is proposed by Tamhane and Bechhofer [47]. The elimination hereby is made by means of Gupta's maximum means rule (cf. [16], p. 232):

$$\text{Select } \pi_i \text{ if } Y_{in_1} \geqslant Y_{[k]n_1} - d, \ i = 1, \ldots, k, \tag{6}$$

where $d = d(\delta^*, P^*, n_1, n_2)$ in the present cotext. A multi-stage procedure which is direct generalization of Tamhane and Bechhofer's procedure is proposed by Tamhane [46].

Several optimality properites of Gupta's subset selection rule have been pointed out recently by Gupta and Miescke [11], Miescke [29], Gupta and Kim [10] and Bickel and Yahav [4]. Thus the use of this rule at the first stage is intuitively justified. No theoretical results, however, which support this idea are known at present.

In a conservative approach, Tamhane and Bechhofer [47] use a lower bound for the $P(\mathrm{CS})$ to find a most economical pair (n_1, n_2) in the indifference zone approach. Their conjecture that the slippage configuration $\theta_{[1]} = \ldots = \theta_{[k-1]} = \theta_{[k]} - \delta^*$ is the LFC has been proved to be correct for $k = 3$ by Miescke and Sehr [33]. The case of $k > 3$ is still unproved. Techniques for finding LFC's as well as results for other procedures are given in Gupta and Miescke [13].

For the control problem, several two-stage procedures are considered and discussed in Gupta and Miescke [14] and Miescke [31].

The *second* situation, where still $\sigma_1^2 = \ldots = \sigma_k^2 = \sigma^2$, but $\sigma^2 > 0$ is now unknown, is also considered by Kao and Lai [26] and Jennison, Johnstone and Turnbull [25].

Open sequential procedures without elimination based on vector at a time sampling are studied by Wackerley [51] (in a more general approach) and by Mukhopadhyay and Chou [38]. The latter perform a similar analysis as before Mukhopadhyay [37] has done before. Mukhopadhyay [36] deals with the case of $k = 2$.

The *third* situation, where $\sigma_1^2, \ldots, \sigma_k^2$ are unknown and possibly unequal, is the most difficult one. It is questionable, however, whether it is still reasonable to look for a population with the largest mean which perhaps might also have the largest variance.

Two-stage procedures without elimination, based on the classical Stein-technique (cf. [16], p. 23) to determine the common sample size at stage 2 in dependence of the estimated variances at stage 1, are considered by Rinott [42] and Mukhopadhyay [34] under the indifference zone approach. A three-stage procedure employing the approach, followed by elimination via Gupta's rule at stage 2, is proposed by Hochberg and Marcus [17] in the indifference zone approach. For the case of $k = 2$ populations, other procedures have been considered by Mukhopadhyay [35]. Procedures are also given by Swanepoel and Geertsema [44].

For the control case, open and closed sequential procedures with elimination and vector at a time sampling are proposed by Hsu and Edwards [19].

5. Other results

For the problem of finding the normal population with the smallest variance, Mukhopadhyay and Chou [39] give an open sequential procedure without elimination based on vector at a time sampling. The basic construction is the same as before in Mukhopadhyay [37]. For the linear regression model, another type of sequential procedure is proposed by Hsu and Huang [21].

Finally, several papers dealing with nonparametric sequential procedures are presented by Swanepoel and Venter [45], Swanepoel [43], Carroll [6], Büringer, Martin and Schriever [5] and Hsu and Edwards [20].

Acknowledgement. The author wishes to express his sincere thanks to Professor S. S. Gupta for his encouragement and support. Thanks are also due to Professor S. Panchapakesan for several helpful comments.

REFERENCES

1. Bechhofer, R. E., Kiefer, J., Sobel, M., *Sequential Identification and Ranking Procedures.* Chicago—London, The University of Chicago Press, 1968.
2. Bechhofer, R. E., Kulkarni, R. V., *Closed adaptive sequential procedures for selecting the best of $k \geqslant 2$ Bernoulli populations.* In: Gupta, S. S., Berger, J. O. (Eds.), *Statistical Decision Theory and Related Topics III, Vol.* 1, pp. 61—108. New York, Academic Press, 1982.
3. Bechhofer, R. E., Kulkarni, R. V., *On the performance characteristics of a closed adaptive sequential procedure for selecting the best Bernoulli population.* Tech. Rep. # 532, School of Operations Research and Industrial Engineering, College of Enginering, Cornell University, Ithaca, 1982.
4. Bickel, P. J., Yahav, J. A., *Asymptotic theory of selection procedures and optimality of Gupta's rules.* In: Kallianpur, G., Krishnaiah, P. R., Ghosh, J. K. (Eds.), *Statistics and Probability: Essays in Honor of C. R. Rao,* pp. 109—124. Amsterdam—New York—Oxford, North-Holland, 1982.
5. Büringer, H., Martin, H., Schriever, K.-H., *Nonparametric Sequential Selection Procedures.* Boston—Basel—Stuttgart, Birkhäuser, 1980.
6. Carroll, R. J., *Distributions of stopping times for sequential eliminaticn ranking procedures.* South African Statist. J., 12, 33—45 (1978).
7. Dudewicz, E. J., Koo, J. O., *The Complete Categorized Guide to Statistical Selection and Ranking Procedures.* Columbus, Ohio, American Sciences Press, 1982.
8. Edwards, H. P., *Bayes sequential design procedures: Theory and applications* (Abstract). Bull. Austral. Math. Soc., **21**, 157—158 (1980).
9. Gulati, C. M., *Optimal sequential procedure for finding the slipped population.* Comm. Statist. C, 1,253—261 (1982/83).
10. Gupta, S. S., Kim, W. C., *On the problem of selecting good populations.* Comm. Statist. A, **10**, 1043—1077 (1981).
11. Gupta, S. S., Miescke, K. J., *Optimality of subset selection procedures for ranking means of three normal populations.* Sankhyā Ser. B., **43**, 1—17 (1981).
12. Gupta, S. S., Miescke, K. J., *An essentially complete class of two-stage procedures with screening at the first stage.* Mimeo. Ser. # 81—53, Dept. of Statist., Purdue University, W. Lafayette, 1981.
13. Gupta, S. S., Miescke, K. J., *On the least favorable configurations in certain two-stage selection procedures.* In: Kallianpur, G., Krishnaiah, P. R., Ghosh, J. K. (Eds.), *Statistics and Probability: Essays in Honor of C. R. Rao,* pp. 295—305. Amsterdam—New York—Oxford, North-Holland 1982.
14. Gupta, S. S., Miescke, K. J., *On the problem of finding a best population with respect to a control in two stages.* In: Gupta, S. S., Berger, J. O., (Eds.), *Statistical Decision Theory, and Related Topics III, Vol.* 1, pp. 473—496. New York, Academic Press, 1982.
15. Gupta, S. S., Miescke, K. J., *Sequential selection procedures: A decision theoretic approach* Mimeo. Ser. #82—6, Dept. of Statist., Purdue University, W. Lafayette, 1982
16. Gupta, S. S., Panchapakesan, S., *Multiple Decision Procedures: Theory and Methodology of Selecting and Ranking Populations.* New York—Chichester—Brisbane—Toronto, Wiley, 1979.
17. Hochberg, Y., Marcus, R., *Three stage elimination type procedures fcr selecting the best normal population when variances are unknown.* Comm. Statist. A, 10, 597—612 (1981)
18. Hollander, M., Proschan, F., Sethuraman, J., *Functions decreasing in transposition and their applications in ranking problems.* Ann. Statist., 5, 722—733 (1977).
19. Hsu, J., Edwards, D. G., *Adaptive sequential procedures for comparing new treatments with a standard.* Tech. Rep. #235, Dept. of Statist., Ohio State University, Columbus, 1981.
20. Hsu, J., Edwards, D. G., *Sequential multiple comparison with the best.* Tech. Rep. #260, Dept. of Statist., Ohio State University, Columbus, 1982.
21. Hsu, T. A., Huang, D. Y., *Some sequential selection procedures for good regression models.* Mimeo. Ser. # 81—21, Dept. of Statist., Purdue University, W. Lafayette, 1981.
22. Hsuan, F., Hwang, F. K., Parnes, M., *A class of selection problems for which more sampling is more informative.* Biometrika, **69**, 280—283 (1982).
23. Huang, D. Y., *A note on an efficiency of the Bechhofer-Kiefer-Sobel sequential selection procedure.* Bull. Inst. Math. Acad. Sinica, **4**, 249—252 (1976).
24. Hwang, F. K., *A multi-stage selection scheme for the most probable event.* Unpublished manuscript, 1980.

25. Jennison, C., Johnstone, I. M., Turnbull, B. W., *Asymptotically optimal procedures for sequential adaptive selection of the best of several normal means.* In: Gupta, S. S., Berger, J. O. (Eds.), *Statistical Decision Theory and Related Topics III, Vol. 2,* pp. 55—86. New York, Academic Press, 1982.
26. Kao, S. C., Lai, T. L., *Sequential selection procedures based on confidence sequences for normal populations.* Comm. Statist. A, **9**, 1657—1676 (1980).
27. Levin, B., Robbins, H., *Selecting the highest probability in binomial or multinomial trials.* Proc. Nat. Acad. Sci. U.S.A., **78**, 4663—4666 (1981).
28. McCulloch, C. E., *Conditions under which* $E\{N_1\} = \infty$ *for Tong's adaptive solution to ranking and selection problems.* Comm. Statist. A, **11**, 815—819 (1982).
29. Miescke, K. J., *Bayesian subset selection procedures for additive and linear loss functions.* Comm. Statist. A, **8**, 1205—1226 (1979).
30. Miescke, K. J., *Identification and selection procedures based on tests.* Ann. Statist., **7**, 207—219 (1979).
31. Miescke, K. J., *On two-stage procedures for finding a population better than a control.* Mimeo. Ser. #80—28, Dept. of Statist., Purdue University, W. Lafayette, 1980.
32. Miescke, K. J., *Γ-Minimax selection procedures in simultaneous testing problems.* Ann. Statist., **9**, 215—220 (1981).
33. Miescke, K. J., Sehr, J., *On a conjecture concerning least favorable configurations in certain two-stage selection procedures.* Comm. Statist. A, **9**, 1609—1617 (1980).
34. Mukhopadhyay, N., *Some comments on two-stage selection procedures.* Comm. Statist. A, **8**, 671—683 (1979).
35. Mukhopadhyay, N., *A note on selecting the better treatment.* Tech. Rep. #21, Dept. of Statist., Oklahoma State University, Stillwater, 1981.
36. Mukhopadhyay, N., *Theoretical investigations on some sequential and two-stage procedures to select the larger mean.* Tech. Rep. #22, Dept. of Statist., Oklahoma State University, Stillwater, 1981.
37. Mukhopadhyay, N., *Selecting the largest normal mean through likelihoods.* Statistics and Decisions, **1**, (1982/83).
38. Mukhopadhyay, N., Chou, W. S., *Sequential procedures for selecting the largest normal mean when the common variance is unknown.* Tech. Rep. #30, Dept. of Statist., Oklahoma State University, Stillwater, 1982.
39. Mukhopadhyay, N., Chou, W. S., *Selecting the smallest normal variance through the comparisons of several likelihoods.* Comm. Statist. C, **1**, 121—144 (1982).
40. Ramey, J. T., Alam, K., *A sequential procedure for selecting the most probable multinomial event.* Biometrika, **66**, 171—173 (1979).
41. Ramey, J. T., Alam, K., *A Bayes sequential procedure for selecting the most probable multinomial event.* Comm. Statist. A, **9**, 265—276 (1980).
42. Rinott, Y., *On two-stage selection procedures and related probability-inequalities.* Comm. Statist. A., **7**, 799—811 (1978).
43. Swanepoel, J. W. H., *Nonparametric elimination selection procedures based on robust estimators.* South African Statist. J., **11**, 27—41 (1977).
44. Swanepoel, J.W.H., Geertsema, J. C., *Sequential procedures with elimination for selecting the best of k normal populations.* South African Statist. J., **10**, 9—36 (1976).
45. Swanepoel, J.W.H., Venter, J. H., *A class of elimination selection procedures based on ranks.* South African Statist. J., **9**, 119—128 (1975).
46. Tamhane, A. C., *On a class of multi-stage selection procedures with screening for the normal means problem.* Sankhyā Ser. B, **42**, 197—216 (1980).
47. Tamhane, A. C., Bechhofer, R. E., *A two-stage minimax procedure with screening for selecting the largest normal mean (II): An improved PCS lower bound and associated tables.* Comm. Statist. A, **8**, 337—358 (1979).
48. Tong, Y. L., *An adaptive solution to ranking and selection problems.* Ann. Statist., **6**, 658—672 (1978).
49. Tong, Y. L., *Some new properties of the Bechhofer-Kiefer-Sobel stopping rule.* Ann. Inst. Statist. Math. **31**, 423—433 (1979).
50. Turnbull, B. W., Kaspi, H., Smith, R. L., *Adaptive sequential procedures for selecting the best of several normal populations.* J. Statist. Comput. Simul., **7**, 133—150 (1978).
51. Wackerley, D. D., *An alternative sequential approach to the problem of selecting the best of k populations.* J. Amer. Statist. Assoc., **72**, 464—468 (1977).
52. Washburn, A. R., *NOMBAS—A Bayesian procedure for selecting the greatest mean.* Tech. Rep. # NPS55-82017, Naval Postgraduate School, Monterey, California, 1982.

ARUNAVA MUKHERJEA*

Non-homogeneous Markov chains: Tail idempotents, tail sigma-fields and basis

Let (P_n) be a sequence of stochastic matrices with a finite number of states and (X_n) be the corresponding non-homogeneous Markov chain. For $k < n$, we write: $P_{k,n} = P_{k+1} \dots P_n$. Given any subsequence (n_t) of positive integers, there exists a subsequence $(m_t) \subset (n_t)$ such that for each $k \geq 1$, $P_{k,m_t} \to Q_k$, $Q_{m_t} \to Q = Q^2$ as $t \to \infty$. This Q is called a tail idempotent of the chain (P_n) and was first introduced by I. Csiszàr. There is a natural connection between tail idempotents of a Markov chain and its tail sigma-field. We describe this connection here, and then use it to find the exact number of distinct atoms in the tail sigma-field as well as to study the basis of a general non-homogeneous Markov chain. The concept of basis was first introduced and studied by Maksimov. Finally, we also discuss difficulties and differences in the various definitions of the basis in the case of chains with infinitely many states.

1. Introduction and statements of main results

Let (P_n) be a sequence of s by s stochastic matrices. For $k < n$, we write: $P_{k,n} = P_{k+1} \dots P_n$. Let (X_n) be the associated non-homogeneous Markov chain so that $(P_{k,n}) = P(X_n = j \mid X_k = i)$, whenever $P(X_k = i) > 0$. Let $T_\infty = \bigcap_{n=1}^{\infty} F_n^\infty$, F_n^∞ being the sigma-algebra generated by $\{X_m : m \geq n\}$. Let $\pi = (\pi_1, \dots, \pi_s)$ be the initial distribution of the Markov chain, which is assumed to have all positive entries. T_∞ is called the tail sigma-field of the chain and has been studied in details by Blackwell, Cohn, Freedman, Iosifescu, Kingman, Orey and others. It was proved by Cohn that T_∞ consists of a finite number of atomic sets, and Senchenko [13] shows that the number of these atomic sets does not exceed the number of states. A very interesting proof of Cohn's result was given by Kingman [6], under the assumption that $P(X_n = i) > 0$ for each state i and all n. One of the aims of our paper is to give an elementary and complete proof of a more complete result on the atomic nature of T_∞ (without Kingman's assumption). Another aim of this paper is to present a probabilistic description of the basis of a stochastic chain. The interesting concept of the basis of a Markov chain was first introduced in [7] in the bistochastic case, and then studied in details in different contexts in our earlier papers, specially in [11, 12], in the stochastic case with finite and

* University of South Florida, Tampa, Fla 33620, USA.

countable states. Recently in [4], a probabilistic definition of the basis has been given. Our main aim in this paper is to unify these apparently different concepts of the basis of a Markov chain and to provide a complete picture of the matricial as well as probabilistic properties of the basis. In [4], various martingale convergence theorems have been used. Our methods in this paper are completely different and very elementary. We use a simple idea, that of a tail idempotent, which, as we will show, enables us to present quite general results. In the last section, we will make comments on the suitability of a probabilistic definition of the basis in the infinite dimensional case; a non-trivial example will be discussed in details to point out differences and difficulties.

Before we state our main results, let us present the definitions of a tail idempotent of a stochastic chain, the basis of a tail idempotent, and the basis of a Markov chain.

The definition of a tail idempotent is implicit in the following very simple, but fundamental result (essentially due to Csiszàr, see his Theorem 2.1, Z. Wahrsch. Verw. Gebiete, **5**(1966), 279—295).

Given any sequence (u_t) *of positive integers, there is a subsequence* $(n_t) \subset \subset (u_t)$ *such that for each positive integer* k,

$$P_{k,n_t} \to Q_k, \; Q_{n_t} \to Q = Q^2, \; Q_k Q = Q_k. \tag{1}$$

The idempotent stochastic matrix Q is called a *tail idempotent* of the chain (P_n) corresponding to the subsequence (n_t). Each such Q, being an idempotent, has a uniquc partition (called its *basis*)

$$\{T, \; C_1, \; C_2, \dots, C_p\}$$

whose union is the state space $E = \{1, 2, \dots, s\}$, such that

$$Q_{ij} = 0 \text{ if } j \in T \text{ or } (i \in C_{k_1}, \; j \in C_{k_2} \text{ with } k_1 \neq k_2);$$

$$= Q_{kj} (> 0) \text{ if } i, \; j, \; k \text{ belong to the same } C\text{-class}.$$

Here p is the rank of Q.

We remark that a stochastic chain can have infinitely many different tail idempotents with the same basis. For example, let $s = 4$, and A_n be such that its 1st and 3rd rows are (0 0 1 0) and 2nd and 4th rows are $\left(0 \frac{1}{n} 0 \frac{n-1}{n}\right)$. Then $A_n A_m = A_m$. Define the chain (P_n) as follows: $P_n = A_i$ whenever n is of the form p_i^k, $= I$ otherwise, where p_i's are the primes. Then each A_i is a tail idempotent with the basis $\{T = 1, \; C_1 = \{2, 4\}, \; C_2 = 3\}$. It is interesting to point out that $\inf \{(A_i)_{22}\colon 1 \leqslant i < \infty\} = 0$.

An important property of the tail idempotents is that they all have the same rank. This we will establish in the next section. Now we give several definitions of the basis of a stochastic chain.

Definition (I). A partition $\{T, \; C_1, \; C_2, \dots, C_p\}$ of the state space E is called the *basis* of the chain (P_n) if the following three conditions hold:

(i) for each $k \geqslant 1$,

$$\lim_{n \to \infty} (P_{k,n})_{ij} = 0 \text{ for each } i, \text{ whenever } j \in T;$$

(ii) for any i, j in C_m, $1 \leqslant m \leqslant p$, there is a positive integer k_0 and a $\beta > 0$ such that for each $k > k_0$, there is a positive integer $n(k)$ such that $n \geqslant n(k) \Rightarrow (P_{k,n})_{ij} > \beta$;

(iii) if i and j are in different C-classes, then given $\varepsilon > 0$, there exists k_0 such that for each $k > k_0$, there exists $n(k)$ with the property: $n \geqslant n(k) \Rightarrow (P_{k,n})_{ij} < \varepsilon$.

Let us remark that the above definition of basis was implicit in our earlier papers. For a convergent chain (P_n), i.e., when $\lim_{n \to \infty} P_{k,n} = Q_k$ exists for each k, then all tail idempotents (which are simply limit points of the Q_k's) have the same basis, which turns out to be the basis of the chain according to the above definition. (See [9].)

Definition (II). A partition $\{T, C_1, C_2, \ldots, C_p\}$ of E is called the *basis* of the chain (P_n) if in Definition (I), only the condition (iii) is replaced by the following weaker condition:

(iii)′ for i and j in different C-classes, the condition (ii) in (I) does not hold.

Definition (III). A partition $\{T, C_1, C_2, \ldots, C_p)$ of E is called the *basis* of the chain (P_n) if the following three conditions hold:

(i) for each $k \geqslant 1$,

$$\lim_{n \to \infty} (P_{k,n})_{ij} = 0 \text{ for each state } i, \text{ whenever } j \in T;$$

(ii) $\liminf_{n \to \infty} P(X_n = i) > 0$, whenever $i \notin T$;

(iii) each atom in T_∞ is equivalent to an event $\{X_n = i \text{ i.o.}\}$ for some $i \notin T$; moreover, the events $\{X_n = i \text{ i.o.}\}$ and $\{X_n = j \text{ i.o.}\}$, when $i \notin T$ and $j \notin T$, are equivalent atoms in T_∞ iff they are in the same C-class.

We remark that a definition similar to (but slightly different in the statement of condition (i)) the above definition has been considered only recently in [4]. But in [4], implications of such a definition in terms of properties such as described in Definition (I) have not been studied. Our main theorem will unify all these. Here are our results.

Theorem 1. *Let (X_n) be a non-homogeneous Markov chain and (P_n) be the corresponding (s by s) transition matrices. Let p be the rank of a tail idempotent of the chain (P_n). Let Q be any fixed tail idempotent corresponding to a subsequence (n_t). Let the basis of Q be $\{T, C_1, C_2, \ldots, C_p\}$. Let*

$$p_0 = \text{cardinality of } \{m\colon \lim_{t \to \infty} P(X_{n_t} = i) > 0 \text{ whenever } i \in C_m, 1 \leqslant m \leqslant p\}.$$

Then, T_∞ has exactly p_0 $(< p)$ atoms which generate T_∞.

Theorem 2 (The Main Theorem). *Let $\{T, C_1, \ldots, C_p\}$ be a partition of $E = \{1, 2, \ldots, s\}$. Let (P_n) be a stochastic chain with (X_n) as the corresponding Markov chain with state space E. Then the following three assertions are equivalent*

(a) $\{T, C_1, \ldots, C_p\}$ is a basis of the chain according to (I);

(b) $\{T, C_1, \ldots, C_p\}$ is a basis of the chain according to (III), *if we start the chain at a sufficiently late time (i.e. if we assign the initial distribution to X_k for sufficiently large k);*

(c) all tail idempotents have the same basis, which coincides with $\{T, C_1, \ldots, C_p\}$; moreover, there exists a positive β such that for any tail idempotent Q, we have

(i) $j \notin T \Rightarrow Q_{ij} > \beta$;

(ii) *the corresponding* Q_k*'s converge to* Q *on* T^c, *i.e.*, $\lim_{k \to \infty} (Q_k)_{ij} = Q_{ij}$, *whenever* $i \notin T$.

Theorem 1 will be proven in Section 2 and Theorem 2 in Section 3. Now we state another two theorems that reveal other properties of the basis of a chain. We will not prove these theorems here, since they are implicit in our earlier papers.

Theorem 3. *Let* $\{T, C_1, \ldots, C_p\}$ *be the basis for a chain* (P_n) *according to the weaker Definition* (II). *Suppose that for* $i \in T$, $j \in T$, $\sum_{n=1}^{\infty} (P_n)_{ij} < \infty$. *(This condition means that the states in* T *are transient, that is,* $P(X_n = i \text{ i.o.}) = 0$ *for* $i \in T$*). Then all tail idempotents have the basis* $\{T, C_1, \ldots, C_p\}$; *moreover, for each* m, $1 \leqslant m \leqslant p$, *the restricted chain* $(P_n|_{C_m})$ *is weakly ergodic.* (*Note that here the matrix* $M_n = P_n|_{C_m}$ *is defined as follows: for* $i \in C_m$, $j \in C_m$, $(M_n)_{ij} = (P_n)_{ij} / \left[\sum_{k \in C_m} (P_n)_{ik}\right]$.)

Theorem 4. *Let* (P_n) *be a sequence of infinite-dimensional nonnegative matrices such that for some two states* i *and* j *the following conditions hold*

(i) *for* $k \geqslant n$, *the entries* $(P_{k,n})_{ii}$, $(P_{k,n})_{ji}$ *and* $(P_{k,n})_{jj}$ *are bounded above by some positive number* M *(independent of* n *and* k*)*;

(ii) *there exists* $\beta > 0$ *such that for each* k_0, *there is a* $k > k_0$ *and* $n(k)$ *such that* $n \geqslant n(k) \Rightarrow (P_{k,n})_{ii} > \beta$;

(iii) *there exists a subsequence* (n_t), *a* $u > 0$ *such that*

(*a*) *for each* $k \geqslant$ *some* k_0, *there is a* $t(k)$ *such that* $t \geqslant t(k) \Rightarrow (P_{k,n_t})_{ii} > u$;

(*b*) *given* $\varepsilon > 0$, *there is a* k_0 *such that for each* $k > k_0$, *there is a* $t(k)$ *such that* $t \geqslant t(k) \Rightarrow (P_{k,n_t})_{ij} < \varepsilon$.

Then, the series $\sum_{n=1}^{\infty} (P_n)_{ij} < \infty$.

2. Some useful results on tail idempotents: Probabilistic interpretations and proof of Theorem [1]

Proposition 2.1. *Let* Q *and* R *be any two tail idempotents for the stochastic* (s *by* s) *chain* (P_n). *Then they have the same rank.*

Proof. By definition, there are subsequences (n_t) and (m_t) such that for each $k \geqslant 1$,

$$P_{k,n_t} \to Q_k, \qquad P_{k,m_t} \to R_k,$$

$$Q_k Q = Q_k,\ Q_{n_t} \to Q, \qquad R_k R = R_k,\ R_{m_t} \to R.$$

Let R_0 be a limit point of (R_{n_t}) and Q_0 be a limit point of (Q_{m_t}). Notice that for $k < q$, we have

$$P_{k,q} Q_k = Q_k, \qquad P_{k,q} R_q = R_k. \tag{2}$$

Consequently, we also have

$$(3)\ RQ_0 = Q_0 \qquad (4)\ QR_0 = R_0 \qquad (5)\ Q_0R_0 = R$$

$$(6)\ R_0Q_0 = Q \qquad (7)\ R_0R = R_0 \qquad (8)\ Q_0Q = Q_0.$$

It is immediate from these equations that

$$\text{rank } R = \text{rank } R_0 = \text{rank } Q_0 = \text{rank } Q.$$

We remark that in the above proposition, in general the basis of Q can be quite different from that of R; in fact, the "T" states of Q and R together can be s in number. For example, let $s = 4$ and the matrices A and B be defined by

$$A = \begin{pmatrix} 0 & 0 & 1 & 0 \\ 0 & 0 & 0 & 1 \\ 0 & 0 & 1 & 0 \\ 0 & 0 & 0 & 1 \end{pmatrix}, \qquad B = \begin{pmatrix} 1 & 0 & 0 & 0 \\ 0 & 1 & 0 & 0 \\ 1 & 0 & 0 & 0 \\ 0 & 1 & 0 & 0 \end{pmatrix}.$$

Then $A = A^2$, $B = B^2$, $AB = B$, $BA = A$. Consequently, A and B can be tail idempotents of (P_n) where the P_n's are alternately A and B.

Proposition 2.2. *Let (n_t) be any subsequence of the positive integers. Write: $a_m = P(B|X_{n_m} = j)$, where $j \in E$, $B \in T_\infty$ and $B \subset \{X_{n_t} = j \text{ i.o. in } t\}$. If the sequence (a_m) is well-defined, then it can have only 0 or 1 as limit points.*

Proof. It is sufficient to prove the result when $\lim_{m\to\infty} a_m = a$ exists. We will prove that $a = a^2$. Let $\varepsilon > 0$. Let n_0 be such that

$$m \geqslant n_0 \Rightarrow |a_m - a| < \varepsilon.$$

Let $N > n_0$. By elementary measure theory considerations, we can choose $N' > N$ such that

$$|a_N - P(B;\ X_{n_t} = j \text{ for some } t > N';\ B_{N'}|X_{n_N} = j)| < \varepsilon,$$

where $P(B \Delta B_{N'}) < \varepsilon/2$ and $B_{N'}$ depends only on $\{X_m : m \leqslant N'\}$. Then we also have for this N':

$$|P(B;\ X_{n_t} = j \text{ for some } t > N';\ B_{N'}|X_{n_N} = j) - a_{N'} \cdot a_N|$$

$$= \left|\left\{\sum_{m=N'+1}^{\infty} P(B;\ X_{n_m} = j;\ B_{N'};\ X_{n_u} \neq j \text{ for } N < u < m|X_{n_N} = j)\right\} - a_{N'} \cdot a_N\right|$$

$$= \left|\left\{\sum_{m=N'+1}^{\infty} P(B|X_{n_m} = j) \cdot P(X_{n_m} = j;\ X_{n_u} \neq j \text{ for } N < u < m;\ B_{N'}|X_{n_N} = j)\right\} - a_N a_N\right|$$

$$\leqslant \left|\sum_{m=N'+1}^{\infty} (a_m - a_N)\, P(X_{n_m} = j;\ B_{N'};\ X_{n_u} \neq j \text{ for } N < u < m|X_{n_N} = j)\right|$$

$$+ |a_N P(B_{N'};\ X_{n_t} = j \text{ for some } t > N'|X_{n_N} = j) - a_N a_N|.$$

The proposition now follows easily.

Proposition 2.3. *Let Q be a tail idempotent of the chain (P_n) such that for each $k \geqslant 1$, $P_{k,n_t} \to Q_k$ and $Q_{n_t} \to Q$ as $t \to \infty$. Let $\{T, C_1, \ldots, C_p\}$ be the basis of Q. Let $j \notin T$ and $i \notin E$ such that $P(X_{n_k} = i) > 0$ for large k. Then we have*

(i) $\lim_{k\to\infty} P(X_{n_t} = j \text{ for some } t > k \mid X_{n_k} = i) = [Q_{ij}/Q_{jj}]$;

(ii) *if* $P(X_k = i) > 0$, *then* $P(X_{n_t} = j \text{ i.o.} \mid X_k = i) = [(Q_k)_{ij}/Q_{ij}]$.

Proof. Write: $F_{ij}(n_k, n_r) = P(X_{n_r} = j;\ X_{n_u} \neq j \text{ for } k < u < r \mid X_{n_k} = i)$. Then, for $k < t$ we have

$$(P_{n_k, n_t})_{ij} = \sum_{r=k+1}^{t} F_{ij}(n_k, n_r)(P_{n_r, n_t})_{jj}.$$

Then (i) follows easily by letting first $t \to \infty$ and then $k \to \infty$.

To prove (ii), let $j \in C_m$, $1 \leqslant m \leqslant p$. Then

$$P(X_{n_t} = j \text{ i.o.} \mid X_k = i) = \lim_{N\to\infty} P(X_{n_t} = j \text{ for some } t > N \mid X_k = i)$$

$$= \sum_{u \in E} \Big[\lim_{N\to\infty} P(X_{n_t} = j \text{ for some } t > N \mid X_{n_N} = u)\,(P_{k,n_N})_{iu}\Big]$$

$$= \sum_{u \in C_m} (Q_k)_{iu} \text{ (by using (i) and the fact that } (Q_k)_{iu} = 0 \text{ for } u \in T)$$

$$= (Q_k)_{ij}/Q_{jj} \text{ since } Q_k Q = Q_k.$$

We remark that a result similar to (ii) above was first given in [4] for convergent chains using martingale convergence theorems. However, tail idempotents make it possible to find the natural domain of validity of such results, using only elementary methods.

Our next result can be obtained from our previous results. However, we present it separately because of an interesting method of proof (that can be used in even more general situations).

Proposition 2.4. *Let Q be a tail idempotent corresponding to (n_t) with basis $\{T, C_1, \ldots, C_p\}$. Suppose that $Q_{ij} > 0$. (Here i may be a T state). Then, $\{X_{n_t} = i \text{ i.o.}\} \subset \{X_{n_t} = j \text{ i.o.}\}$ a.s. Therefore, if i and j belong to the same C-class, then these events are equivalent.*

Proof. Let $i \neq j$ and $Q_{ij} > 0$. Notice that

$$P(X_{n_t} = i \text{ i.o.}) = \lim_{m\to\infty} P(X_{n_t} = i \text{ for some } t \geqslant m).$$

Let $N > N'$. Then we have

$$P(X_{n_t} = i \text{ for some } t \geqslant N;\ X_{n_r} \neq j \text{ for each } r \geqslant N')$$

$$\leqslant \sum_{m=N}^{\infty} [P(X_{n_r} \neq j \text{ for each } r > m \mid X_{n_m} = i) \times$$

$$\times P(X_{n_t} \neq i \text{ for } N \leqslant t < m;\ X_{n_m} = i;\ X_{n_r} \neq j \text{ for } N' < r \leqslant N)].$$

Let $Q_{ij} = 2\beta > 0$. Then for m sufficiently large, $(Q_{n_m})_{ij} > \beta$ and therefore, whenever $P(X_{n_m} = i) > 0$,

$$P(X_{n_r} = j \text{ for some } r > m | X_{n_m} = i) \geqslant \lim_{r \to \infty} (P_{n_m, n_r})_{ij} = (Q_{n_m})_{ij} > \beta.$$

It follows that if $N > N'$ with N sufficiently large, then

$$P(X_{n_t} = i \text{ for some } t \geqslant N;\ X_{n_r} \neq j \text{ for each } r \geqslant N') \leqslant$$
$$\leqslant (1 - \beta)\, P(X_{n_t} = i \text{ for some } t \geqslant N;\ X_{n_r} \neq j \text{ for } N' \leqslant r \leqslant N).$$

Now let $N \to \infty$. Then we have

$$P(X_{n_t} = i \text{ i.o.};\ X_{n_r} \neq j \text{ for each } r \geqslant N') \leqslant$$
$$\leqslant (1 - \beta)\, P(X_{n_t} = i \text{ i.o.};\ X_{n_r} = j \text{ for each } r \geqslant N').$$

Letting N' to ∞, it follows easily that

$$P(X_{n_t} = i \text{ i.o.}) = P(X_{n_t} = i \text{ i.o.};\ X_{n_t} = j \text{ i.o.}).$$

The proposition is now clear.

Proposition 2.5. *Suppose that for some i and j in E, there is a positive β such that for sufficiently large m,*

$$P(X_n = j \text{ for some } n > m | X_m = i) \geqslant \beta,$$

whenever $P(X_m = i) > 0$. Then we have

$$\{X_n = i \text{ i.o.}\} \subset \{X_n = j \text{ i.o.}\} \ a.s.$$

Proof. The proof is exactly similar to that of Proposition 2.4.

Proposition 2.6. *Let Q be an idempotent stochastic matrix and Q_k be any stochastic matrix such that $Q_k Q = Q_k$. Let $\{T, C_1, ..., C_p\}$ be the basis of Q. If $j \in C_m$, $1 \leqslant m \leqslant p$, then for any scalars (b_i),*

$$(10) \qquad \sum_i b_i (Q_k)_{ij} = Q_{jj} \sum_{t \in C_m} \sum_i b_i (Q_k)_{it}.$$

Thus, if Q is a tail idempotent with basis $\{T, C_1, ..., C_p\}$ corresponding to the subsequence (n_t), and if $j \in C_m$ $(1 < m < p)$ with $f(j) = \lim_{t \to \infty} P(X_{n_t} = j)$, then

$$(11) \qquad f(j) = Q_{jj} \sum_{u \in C_m} f(u).$$

Proof. The proof is straightforward and omitted.

Proposition 2.7. *Let Q be a tail idempotent corresponding to (n_t) with basis $\{T, C_1, ..., C_p\}$. Then for $i \in T$, $\{X_{n_t} = i \text{ i.o.}\}$, whenever non-null, is an atom in T_∞; this event is non-null iff $f(i) > 0$, where f is defined in Proposition 2.6. In any case, for $i \in C_m$ $(1 \leqslant m \leqslant p)$,*

$$(12) \qquad P(X_{n_t} = i \text{ i.o.}\} = f(j)/Q_{jj} = \sum_{u \in C_m} f(u) \quad (\forall j \in C_m).$$

If $i \in T$, then the event $\{X_{n_t} = i \text{ i.o.}\}$ is also an atom in T_∞, whenever it is non-null, and in this case, this event is equivalent to $\{X_{n_t} = j \text{ i.o.}\}$ for some

$j \in T$. Finally, for $i \notin T$, $j \notin T$, the events $\{X_{n_t} = i \text{ i.o.}\}$ and $\{X_{n_t} = j \text{ i.o.}\}$, when both are atoms, are equivalent iff i and j belong to the same C-class.

Proof. Let $B \in T_\infty$ and $B \subset \{X_{n_t} = i \text{ i.o.}\}$, where $i \in C_m$. Notice that $\lim_{m \to \infty} P(X_{n_t} = j) = 0$ if $j \in T$ and by Proposition 2.3,

$$\lim_{m \to \infty} P(B, X_{n_k} = j) = 0$$

if $j \notin T \cup C_m$. Hence,

$$P(B) = \lim_{k \to \infty} \left[\sum_{j \in C_m} P(B | X_{n_k} = j) \; P(X_{n_k} = j) \right] \tag{13}$$

It is now clear that $P(B) > 0$ iff $\lim_{k \to \infty} P(X_{n_k} = j) > 0$ for some $j \in C_m$ (and by (11), it is positive for each $j \in C_m$). Now let (n'_k) be a subsequence of (n_k) such that for each $j \in C_m$, $\lim_{n \to \infty} P(B | X_{n'_k} = j)$ exists. Hence, if $P(B) > 0$, this limit is 1 by Proposition 2.2 for some j, say j_0, in C_m. But then for any other j in C_m, by taking $n'_k > n_k$ we have

$$P(B | X_{n_k} = j) \geqslant P(B | X_{n'_k} = j_0) \, (P_{n_k, n'_k})_{jj_0}$$

and therefore,

$$\liminf_{k \to \infty} P(B | X_{n_k} = j) \geqslant Q_{jj_0} > 0.$$

By Proposition 2.2, it follows that

$$P(B) > 0 \Rightarrow \lim_{k \to \infty} P(B | X_{n_k} = j) = 1 \tag{14}$$

for each $j \in C_m$. Now (12) follows from (11) and (13).

Now let $i \in T$ and $P(X_{n_t} = i \text{ i.o.}) > 0$. Since for each $t \in T$, the tth column of Q is a zerocolumn, $Q_{ij} > 0$ for some $j \notin T$. By Proposition 2.4, $\{X_{n_t} = i \text{ i.o.}\} \subset \{X_{n_t} = j \text{ i.o.}\}$ a.s. The rest is clear.

Proof of Theorem 1. The theorem follows immediately from Proposition 2.7 since for any $A \in T_\infty$ with $P(A) > 0$, the state space E being finite,

$$A \subset \bigcup_{i \in E} \{X_{n_t} = i \text{ i.o.}\},$$

where (n_t) is the subsequence corresponding to Q in Proposition 2.7.

Remarks. (i) The number p_0 in Theorem 1 depends on the starting time of the chain. Here is an example. Consider the chain (P_n) defined by

$$P_1 = \begin{pmatrix} 1 & 0 & 0 \\ 0 & 1 & 0 \\ 1 & 0 & 0 \end{pmatrix}, \quad \text{and for } n \geqslant 2, \; P_n = \begin{pmatrix} 0 & 1 & 0 \\ 0 & 1 & 0 \\ 0 & 0 & 1 \end{pmatrix}.$$

Then this chain is convergent with basis $\{T = 1, \; C_1 = 2, \; C_2 = 3\}$. Notice that for $n > 2$,

$$P_1 P_2 \dots P_n = \begin{pmatrix} 0 & 1 & 0 \\ 0 & 1 & 0 \\ 0 & 1 & 0 \end{pmatrix}$$

so that if the initial distribution is given to X_0, then state 3 is transient since $P(X_n = 3) = 0$ for each $n \geqslant 1$. But if P_1 is dropped and the chain starts at a later time $k (\geqslant 1)$, then 3 is recurrent.

(ii) In the special case when $P(X_n = i) > 0$ for each $i \in E$ and each n, Theorem 1 could also be proved using Kingman's proof (see page 172, [6]) in the following way: Let Q be a tail idempotent such that $P_{k,n_t} \to Q_k$, $Q_k Q = Q_k$, $Q_{n_t} \to Q$. Let. $\{T, C_1, ..., C_p\}$ be the basis of Q. Notice that for $k < m$, $P_{k,m} Q_m = Q_k$ and thus,

$$\text{rank}(Q_k) \leqslant \text{rank}(Q_m) \leqslant \text{rank}\,(Q) = p.$$

Also, there is a k_0 such that $k \geqslant k_0 \Rightarrow \text{rank}\, Q_k = p$. (To see this, define the p by p matrices M_k and M as follows: Choose $s_i \in C_i$ for each i. Then $(M_k)_{ij} = (Q_k)_{s_i s_j}$, $M_{ij} = Q_{s_i s_j}$. Notice that $M_{n_t} \to M$ so that $\det(M_{n_t}) \to \det(M)$. Since $\det(M) > 0$ for M is a diagonal matrix with positive diagonal, $\det(M_{n_t}) > 0$ for sufficiently large t. Hence, rank $(Q_{n_t}) \geqslant$ rank $(M_{n_t}) = p$ (for large t). Now following Kingman's notation, it is clear that $\Phi_k = Q_k \Phi$. For $m > k \geqslant k_0$, the mapping $P_{k,m}\colon \Phi_m \to \Phi_k$ is bijective. Then it follows easily (following Kingman's argument almost word for word) that p is the number of distinct atoms in T_∞. Notice that Kingman's condition $P(X_n{=}i){>}0$ for each i and each n implies that for $i \notin T$, $\lim_{n\to\infty} P(X_{n_t}{=}i) > 0$.

(iii) Let Q be a tail idempotent corresponding to the subsequence (n_t) with basis $\{T, C_1, ..., C_p)\}$, $p \geqslant 2$. Suppose that $i \in T$, $j \in C_{k_1}$ and $k \in C_{k_2}$ $(k_1 \neq k_2)$ such that $Q_{ij} > 0$ and $Q_{ik} > 0$. Then by Propositions 2.4 and 2.7, $P(X_{n_t} = \text{i i.o.}) = 0$. Thus, for a convergent chain, *such a state i is transient.*

3. The basis of a Markov chain and atoms in T_∞: Proof of the main theorem

Proposition 3.1. Let Q *and* R *be any two tail idempotents as in Proposition* 2.1. *Suppose that the zero columns of* Q *and* R *are the same. Let* $\{T, C_1, ..., C_p\}$ *be the basis of* Q. *Then* R *also has the same basis if* $\lim_{k\to\infty} (Q_k)_{ij} = 0$ *for* i, j *in two different C-classes, and in this case, for* $i \notin T$, $Q_{ij} = \lim_{k\to\infty} (Q_k)_{ij}$ *for each* $j \in E$.

Proof. Let $\{T', C'_1, ..., C'_p\}$ be the basis of R. By hypothesis, $T = T'$. It is sufficient to show that each $C'_i \subset$ some C_j. To prove this, let

(14) $$a \in C'_i \cap C_{k_1} \text{ and } b \in C'_i \cap C_{k_2} \quad (k_1 \neq k_2).$$

By (3), $R_{ab}(Q_0)_{bb} \leqslant (Q_0)_{ab} = 0$ (since we are assuming that $(Q_k)_{ab}$ goes to zero as $k \to \infty$). Since $R_{ab} > 0$, $(Q_0)_{bb} = 0$. By (8), for $c \in E$,

$$(Q_0)_{bc} Q_{cb} \leqslant (Q_0)_{bb} = 0$$

and therefore, $(Q_0)_{bc} = 0$ for each $c \in C_{k_2}$. Also, by (8), $(Q_0)_{bc} = 0$ whenever $c \in T$. Thus, if $\lim_{k\to\infty} (Q_k)_{ij} = 0$ for i and j in different C-classes in the basis of Q, then the bth row of Q_0 is a zero row. This is impossible and therefore (14) is impossible. Thus, Q and R have the same basis. The last assertion follows from (8).

Proposition 3.2. *Let* Q *be a tail idempotent of the chain* (P_n) *corresponding to the subsequence* (n_t) *such that for each* $k \leqslant 1$, $P_{k,n_t} \to Q_k$, $Q_{n_t} \to Q = Q^2$ *as* $t \to \infty$. *Suppose that the basis of* Q *is* $\{T, C_1\, C_2, ..., C_p\}$. *Suppose that for each* $j \notin T$,

(15) $$\liminf_{n\to\infty} P(X_n = j) > 0.$$

Then the following results hold

(a) *The event* $\{X_n = j \text{ i.o.}\}$ *is an atom in* T_∞ *for each* $j \in T$ *iff*

(16) $$\lim_{n\to\infty} (Q_n)_{ij} = 0$$

whenever $i \in C_{k_1}$, $j \in C_{k_2} (k_1 \neq k_2)$;

(b) *if* (16) *holds, then the condition*

17) $$\lim_{n\to\infty} (P_{k,n})_{ji} = 0 \ (\forall k,\ \forall j \in E, \text{ whenever } i \in T)$$

holds iff all tail idempotents have the same basis.

Proof. Part (b) follows easily if we observe that for any tail idempotent with basis $\{T', C_1', \ldots, C_p'\}$, (15) and (17) imply that $T = T'$ and therefore, Proposition 3.1 applies.

Now we prove part (a). First, suppose that (16) holds. Then it is immediate from equation (8) that

(18) $$\lim_{n\to\infty} (Q_n)_{ij} = Q_{ij} \ (\forall j \in E),$$

whenever $i \notin T$. Using (18) and the identity

$$(P_{k,n_t})_{ij} = \sum_{m=k+1}^{n_t} P(X_m = j;\ X_r \neq j \text{ for } k < r < m \mid X_k = i)\,(P_{m,n_t})_{jj}$$

it follows easily (as in Proposition 2.3) that when $i \notin T$ and $j \notin T$,

(19) $\lim_{k\to\infty} P(X_n = j \text{ for some } n > k \mid X_k = i)$ exists and equals 1 or 0 according as i and j belong to the same or different C-classes of Q.

Let $j \in C_m$, $1 \leqslant m \leqslant p$. By Proposition 2.7, the event $\{X_n = j \text{ i.o.}\}$ is a.s. the union of atoms of T_∞. If it is not an atom in T_∞, then there exists $u \notin T \cup C_m$ such that

(20) $$\{X_{n_t} = u \text{ i.o.}\} \cup \{X_{n_t} = j \text{ i.o.}\} \subset \{X_n = j \text{ i.o.}\} \quad \text{a.s.}$$

By (14),

(21) $$\lim_{k\to\infty} P(X_{n_t} = u \text{ i.o.} \mid X_{n_k} = u) = 1.$$

By (20) and (21),

(22) $$\lim_{k\to\infty} P(X_n = j \text{ i.o.} \mid X_{n_k} = u) = 1.$$

But this contradicts (19). Hence, $\{X_n = j \text{ i.o.}\}$ is an atom in T_∞. To prove the converse part, suppose that $\{X_n = j \text{ i.o.}\}$ is an atom for $j \notin T$. Let (r_t) be any subsequence of positive integers. Then there is a tail idempotent R corresponding to a subsequence $(m_t) \subset (r_t)$ with basis $\{T', C_1', \ldots, C_p'\}$. By (15),

$$T' \subset T,\ C_1 \cup \ldots \cup C_p \subset C_1' \cup \ldots \cup C_p'.$$

Now let $i \in C_{k_1}$ and $j \in C_{k_2}$ $(k_1 \neq k_2)$. By Proposition 2.7, the events $\{X_{n_t} = i \text{ i.o.}\} = \{X_n = i \text{ i.o.}\} = \{X_{m_t} = i \text{ i.o.}\}$ (a.s.) and $\{X_{n_t} = j \text{ i.o.}\} = \{X_n = j \text{ i.o.}\} = \{X_{m_t} = j \text{ i.o.}\}$ (a.s.) are two distinct atoms in T_∞, and therefore, $i \in C_{m_1}'$ and $j \in C_{m_2}'$ $(m_1 \neq m_2)$.

This means that

$$\lim_{k\to\infty} (Q_{m_k})_{ij} \leqslant \lim_{k\to\infty} \lim_{t\to\infty} (P_{m_k, n_t})_{ij}$$

$$\leqslant \lim_{k\to\infty} P(X_{n_t} = j \text{ i.o.} | X_{m_k} = i)$$

$$= \lim_{k\to\infty} P(X_{m_t} = j \text{ i.o.} | X_{m_k} = i) = 0,$$

by Proposition 2.3. This establishes (16).

Remarks. (i) Note that for a chain, condition (15) may hold and all the tail idempotents may have the same basis, and yet, (16) may not hold. Perhaps the simplest example to illustrate this is when each $P_n = \begin{pmatrix} 0 & 1 \\ 1 & 0 \end{pmatrix}$. This chain has only one tail idempotent $Q = \begin{pmatrix} 1 & 0 \\ 0 & 1 \end{pmatrix}$ and the Q_k 's which are alternately $\begin{pmatrix} 0 & 1 \\ 1 & 0 \end{pmatrix}$ and $\begin{pmatrix} 1 & 0 \\ 0 & 1 \end{pmatrix}$ do not converge to Q.

(ii) We may also point out that when all tail idempotents have the same basis, then if for some tail idempotent Q (corresponding to (n_t)), the corresponding Q_k 's converge to Q on T^c, then for any other tail idempotent R, the corresponding R_{n_t} 's converge on T^c to R. This follows from equations (6) and (7).

Proposition 3.3. *Suppose that $\{T, C_1, ...; C_p\}$ is a partition of E, the (finite) state space of a non-homogeneous Markov chain (X_n), such that the following conditions hold*

(i) *if $i \in T$, then $\lim_{n\to\infty} (P_{k,n})_{ji} = 0 \; \forall j$ and every positive integer k;*

(ii) *if $i \notin T$, then $\liminf_{n\to\infty} P(X_n = i) > 0$;*

(iii) *for each $i \notin T, \{X_n = i \text{ i.o.}\}$ is an atom in T_∞; but any two such atoms $\{X_n = i \text{ i.o.}\}$ and $\{X_n = j \text{ i.o.}\}$ are distinct iff i and j belong to different C-classes in the partition.*

Then the following results are true

(a) *all tail idempotents have the same basis;*

(b) *if $i \notin T$, then $\{R_{ii}$: R is a tail idempotent$\}$ has a positive infimum;*

(c) *for any C_m, $1 \leqslant m \leqslant p$, if i and j are in C_m, then there exists $\beta > 0$ and a positive integer k_0 such that for each $k > k_0$, there exists a positive integer $n(k)$ with the property $n \geqslant n(k) \Rightarrow (P_{k,n})_{ij} > \beta$;*

(d) *if $i \in C_{m_1}$, $j \in C_{m_2}$ with $m_1 \neq m_2$, then given $\varepsilon > 0$ there exists a positive integer k_0 such that for each $k > k_0$, there exists a positive integer $n(k)$ with the property $n \geqslant n(k) \Rightarrow (P_{k,n})_{ij} < \varepsilon$.*

Proof. (a) Let Q be any tail idempotent with basis $\{T', C'_1, ..., C'_q\}$. By (i) and (ii), $T = T'$. By (ii) and Theorem 1, q is the number of distinct atoms of T_∞. Thus, $q = p$. Since $\bigcup_{m=1}^{p} C_m = \bigcup_{m=1}^{p} C'_m$, it follows from (iii), Propositions 2.3 and 2.4 that each C_m is some C'_m. Thus, $\{T, C_1, ..., C_p\}$ is the basis of Q.

(b) Let Q be a fixed tail idempotent and R be any other tail idempotent. $\geqslant$or $i \notin T$, $\{X_n = i \text{ i.o.}\}$ is an atom and therefore, by (12) we have: $R_{ii} \geqslant FQ_{ii} \liminf_{n\to\infty} P(X_n = i)$ (> 0, by (ii)).

(c) Let Q be a tail idempotent corresponding to (n_t) such that $P_{k,n_t} \to Q_k$, $Q_{n_t} \to Q$ as $t \to \infty$. Let us write

$\inf\{R_{aa}: R$ is a tail idempotent and $a \notin T\} = 2v > 0$

Now let $i \in C_m$, $j \in C_m$ $(1 \leqslant m \leqslant p)$. Then $Q_{ij} = Q_{jj} \geqslant 2v$. Suppose that the assertion (c) in the proposition is false. Then there is a positive integer k_1 and a subsequence (h_t) of positive integers such that

$$(Q_{k_1})_{ij} > v \tag{23}$$

$$(P_{k_1,h_t})_{ij} < v^2 \text{ (for each } t) \tag{24}$$

and for each positive integer k,

$$P_{k,h_t} \to H_k,\ H_{h_t} \to H = H^2. \tag{25}$$

(The existence of the tail idempotent H is immediate from the definition of a tail idempotent, see Section 1.) By Proposition 3.2 (see also (16) and (18)), $\lim_{k\to\infty} (H_k)_{jj} = H_{jj} \leqslant 2v$. Thus, there exists $k_2 > k_1$ such that $k \geqslant k_2 \Rightarrow$

$$(H_k)_{jj} \geqslant \frac{4}{3} v,\ (H_{k_1})_{ij} \leqslant v^2 \text{ (by (24))}. \tag{26}$$

Since for $k > k_1$, $P_{k_1,k} H_k = H_{k_1}$, we have for $k > k_2$,

$$(P_{k_1,k})_{ij} \leqslant (H_{k_1})_{ij}/(H_k)_{jj} \leqslant \frac{3}{4} v,$$

which means that $(Q_{k_1})_{ij} < v$. This contradicts (23). This proves (c).

(d) Let i and j be in different C-classes. It follows from Proposition 2.3 that under assumptions in this proposition,

$$\lim_{k\to\infty} P(X_n = j \text{ i.o.} \mid X_k = i) = 0.$$

Notice that $\limsup_{n\to\infty} (P_{k,n})_{ij} \leqslant P(X_n = j \text{ i.o.} \mid X_k = i)$. The assertion (d) now follows easily.

Proof of the main theorem. Note that if $n \geqslant n(k) \Rightarrow (P_{k,n})_{jj} > \beta > 0$ for some k and if we start the Markov chain at time k (i.e. if the initial distribution is given to X_k), then $\liminf_{n\to\infty} P(X_n = j) > 0$. With this observation, the main theorem follows easily from Propositions 3.2 and 3.3.

4. Comments on the infinite dimensional case: An example

Consider the following stochastic chain (P_n), with state space $\{1, 2, 3, \ldots\}$, defined by

$$(P_n)_{11} = (P_n)_{22} = 1 - \frac{1}{n}; \quad (P_n)_{1,n+2} = (P_n)_{2,n+2} = \frac{1}{n};$$

$$(P_n)_{ii} = 1 \text{ for } 2 < i \leqslant n; \quad (P_n)_{i1} = (P_n)_{i2} = \frac{1}{2} \text{ for } i \geqslant n+1.$$

Then it can be verified that for each positive integer k, $\lim_{n\to\infty} P_{k,n} = Q_k$ exists and $\lim_{k\to\infty} Q_k = Q$ exists where the matrices Q_k and Q are given by

(i) $$(Q_k)_{11} = (Q_k)_{12} = (Q_k)_{21} = (Q_k)_{22} = \frac{1}{2};$$

$$(Q_k)_{i1} = (Q_k)_{i2} = \frac{1}{2} \text{ for } i > k+1;$$

$$(Q_k)_{ii} = 1 \text{ for } 2 < i \leqslant k+1.$$

(ii) $$Q_{11} = Q_{12} = Q_{21} = Q_{22} = \frac{1}{2}; \quad Q_{ii} = 1 \text{ for } i > 1.$$

(This example was considered in another context in [12]. See [12] for proofs of the above assertions.) Note that $(Q_k)_{ij} = 0$ for each i, whenever $j > k+1$ so that if we assign the initial distribution to X_{k_0}, then $\lim_{n\to\infty} P(X_n = j) = 0$, whenever $j > k_0 + 1$. It is clear that if we define a basis for this chain according to Definition (I), then the basis of Q will be the basis of the chain, which is given by

(27) $$(b_1): \{C_1 = \{1, 2\},\ C_2 = \{3\},\ C_3 = \{4\}, \dots\}.$$

But for this chain, a basis according to Definition (III) does not exist; but if this definition is modified by replacing the condition (i) in this definition by the condition that $\lim_{n\to\infty} P(X_n = i) = 0$ iff $i \in T$, which is the definition used in [4], then the basis in this sense for this chain is given by

(28) $$(b_2): \{T = \{i: i > k_0 + 1\},\ C_1 = \{1, 2\}, C_2 = \{3\}, \dots, C_{k_0} = \{k_0 + 1\}\}.$$

Note that Propositions 2.2, 2.3 and 2.4 hold for this example with (n_t) replaced by the entire sequence of positive integers. In this example, for the basis (b_2), if $j \in C_m$, then obviously by Proposition 2.3, $\lim_{m\to\infty} P(X_n = j \text{ i.o.} | X_m = i) = 0$ (uniformly in $i \notin T \cup C_m$). However, this uniformity condition does not hold for the basis (b_1), since $(Q_m)_{i2}$ does not go to zero as $m \to \infty$ uniformly in $i (> 2)$. The above uniformity condition has been used in a result in [4].

Finally, we conclude this paper by pointing out the following important difference for an infinite convergent chain.

Suppose that $\{T, C_1, C_2, \dots\}$ is the basis according to Definition (I) and $\{T', C_1', C_2', \dots\}$ is the basis according to [4]. Then it is verified easily by Proposition 2.3 that each C_i' equals some C_j; however, there may be many C-classes that may be contained in T', as evidenced by our example. Thus, it is not clear how one can obtain a result involving infinitely many C-classes in the first basis by only considering the second basis.

REFERENCES

1. Blackwell, D., *Finite non-homogeneous Markov chains.* Ann. of Math., **46**, 594–599 (1945).
2. Cohn, H., *Finite non-homogeneous Markov chains: Asymptotic behaviour.* Adv. in Appl. Probab., **8**, 502–516 (1976).

3. Cohn, H., *Countable non-homogeneous Markov chains: Asymptotic behaviour.* Adv. in Appl. Probab., **9**, 542–552 (1977).
4. Cohn, H., *On a class of non-homogeneous Markov chains.* Preprint, Dec. 1981.
5. Iosifescu, M., *On finite tail sigma-algebras.* Z. Wahrsch. Verw. Gebiete, **24**, 159–166 (1972).
6. Kingman, J. F. C., *Geometrical aspects of the theory of non-homogeneous Markov chains.* Math. Proc. Cambridge Philos. Soc., **77**, 171–185 (1975).
7. Maksimov, V. M., *Convergence of non-homogeneous bistochastic Markov chains.* Theory Probab. Appl., **15**, 604–618 (1970).
8. Mukherjea, A., *Limit theorems: Stochastic matrices, ergodic Markov chains, and measures on semigroups.* In: Bharucha-Reid, A. T. (Ed.), *Probabilistic Analysis and Related Topics, Vol.* 2, pp. 143–203. New York, Academic Press, 1979.
9. Mukherjea, A., *A new result on the convergence of non-homogeneous stochastic chains.* Trans. Amer. Math. Soc., **262**, 505–520 (1980).
10. Mukherjea, A., Chaudhuri, R., *Convergence of non-homogeneous stochastic chains. II.* Math. Proc. Cambridge Philos. Soc., **90**, 167–182 (1981).
11. Mukherjea, A., Nakassis, A., *Convergence of non-homogeneous stochastic chains. III.* To appear.
12. Mukherjea, A., Nakassis, A., *Convergence of non-homogeneous stochastic chains with countable states: An application to measures on semigroups.* In: Heyer, H. (Ed.), *Probability Measures on Groups.* Lecture Notes in Math., Vol. 928, pp. 315–336. Berlin, Springer, 1982.
13. Senchenko, D. V., *The final sigma-algebra of an inhomogeneous Markov chain with a finite number of states.* Math. Notes, **12**, 610–613 (1972).

INGEMAR NÅSELL

The role of the breakpoint in schistosomiasis eradication*

Schistosomiasis (bilharzia, snail fever) is a parasitic infection that is widespread in the tropical areas of the world. Mathematical models for the transmission of schistosomiasis have shown that eradication of schistosomiasis from a community can be achieved in two conceptually different ways, namely either by permanent reduction of parameters to certain threshold values or by momentary reduction of the infection level to a so-called breakpoint. An eradication strategy based on the breakpoint has for a long time been considered very attractive, since the associated costs might be considerably lower than the costs for eradication based on the threshold. The present paper investigates the usefulness of the breakpoint for eradication. This is done by studying the relation between meaningful quantifications of distances to threshold and breakpoint. We conclude that the breakpoint does not present a feasible mode of eradication for schistosomiasis.

1. Introduction

Human schistosomiasis is a parasitic infection that is widespread in tropical areas of the world. It is caused by parasitic flatworms, schistosomes, with complex life cycles. Sexual reproduction takes place in the definitive host (man) and asexual reproduction in an intermediate host (freshwater snail). Infection of snails is caused by poor hygienic conditions that allow parasite eggs in human body wastes to reach the water areas where the snails are found. Infection of human beings can only take place by exposure to water containing cercariae emitted by infective snails.

The transmission of schistosomiasis in a community is a complex and variable phenomenon that is influenced by a multitude of biological, environmental, behavioral and hygienic factors. Many of these factors show large variabilities in both space and time. In spite of the complexity and variability mathematical models have been established and used to study transmission and control and eradication of schistosomiasis. Among the results of the models we note especially the important qualitative conclusion that eradication of a schistosome infection from a community can be achieved in two conceptually different ways, namely either by permanent reduction of parameters to certain threshold values or by momentary reduction of the infection level to a so-called breakpoint. Thresholds appear in other parasitic infections that have been studied mathematically, such as malaria (Mac-

* This work has been supported by the Swedish Natural Science Research Council.

donald (1957), (1973)), and hermaphroditic helminthiasis (Nåsell and Hirsch (1972)). The importance of the breakpoint for schistosomiasis eradication was first pointed out by Macdonald (1965). It has since been further studied by Nåsell and Hirsch (1973), Nåsell (1976b), (1977 a) and May (1977 a), (1977 b). The breakpoint in schistosomiasis models is caused by the fact that the sexually mature parasites are dioecious and that mating between male and female schistosomes is necessary for the completion of the parasite life cycle.

The breakpoint offers a mode of eradication of schistosomiasis that is not available for malaria or for helminthiasis caused by hermaphroditic worms. The models predict that a mass chemotherapy program that reduce the worm load to below the breakpoint leads to eradication. The purpose of the present paper is to study the practical usefulness of this mode of eradication. Our finding is that the efficiency required of a mass chemotherapy program in order for the worm load to be reduced to below the breakpoint is unrealistically high for practically all communities. The only exceptions are given by communities that lie very close to the threshold and where therefore the effort required to achieve eradication (via the threshold) is minor.

A number of features that influence the epidemiologiy of schistosomiasis are known, e.g. snail latency, modes of mating of sexually mature schistosomes, immunity among human hosts, aggregation of parasites in the human host. These features have generally been studied by appending them one at a time to a basic transmission model and evaluating their consequences. We follow the same approach, and begin by describing the basic transmission model in the next session.

2. The model

The analysis is based on the schistosomiasis model of Nåsell and Hirsch (1973), with hypotheses essentially equivalent to those introduced by Macdonald (1965). Let w denote the expected number of worms per human host ("the worm load"), and let y denote the expected proportion of infected snails. These two state variables satisfy the following system of differential equations

(1) $$w' = T_1\mu_1 y - \mu_1 w,$$

(2) $$y' = \frac{1}{2} T_2\mu_2\Psi(w)(1 - y) - \mu y.$$

The mating between mature male and female worms is assumed to be monogamous. The number of paired parasites per host is $\Psi(w)$, where the function Ψ is given by

(3) $$\Psi(w) = w(1 - \exp(-w)(I_0(w) + I_1(w))),$$

and I_0 and I_1 are modified Bessel functions of the first kind of orders 0 and 1, respectively. Furthermore, the transmission factors T_1 and T_2 are defined in terms of basic parameters through the relations

(4) $$T_1 = p_1\lambda_2 N_2/\mu_1$$

and

(5) $$T_2 = p_2\lambda_1 N_1/\mu_2.$$

Here, N_1 is the number of human beings, N_2 is the number of snails, λ_1 is the rate of egg-laying per paired female schistosome, λ_2 is the rate of shedding of cercariae per infective snail, p_1 is the probability for a cercaria to infect a given human being, p_2 is the probability for an egg to infect a given snail, μ_1 is the death rate per mature schistosome, and μ_2 is the death rate per snail (or, equivalently, the recovery rate per infected snail).

The critical points of the system of equations (1), (2) correspond to stationarity (endemicity) of the infection in the community. The coordinates of any critical point $(\bar{w}, \bar{y})$ of (1), (2) are related by

$$\bar{y} = \bar{w}/T_1, \tag{6}$$

while the stationary worm load $\bar{w}$ satisfies the relation

$$T_2(T_1 - \bar{w})\Psi(\bar{w}) = 2\bar{w}. \tag{7}$$

This relation determines the possible values of the steady state worm load $\bar{w}$ as a function of the two transmission factors T_1 and T_2. A threshold function in the T_1, T_2-plane separates two domains of the parameter space where the solutions of (1), (2) show qualitatively different behavior. The origin, i.e. the point $\bar{w} = \bar{y} = 0$, is a locally stable stationary point for all positive values of the transmission factors T_1 and T_2. Clearly, it corresponds to the absence from the community of any infection. The origin is the only critical point if T_1 and T_2 are sufficiently small. For values of the transmission factors T_1 and T_2 above the threshold function, i.e. $T_2 > \overline{\overline{T}}_2(T_1)$, or, equivalently, $T_1 > \overline{\overline{T}}_1(T_2)$, where $\overline{\overline{T}}_1$ is the inverse of $\overline{\overline{T}}_2$, there exist two additional critical points. One of these is unstable and the other one is stable. The unstable point is the so-called breakpoint while the stable point corresponds to an endemic infection level. The steady state worm load (i.e. the value of $\bar{w}$) is denoted by $B_1 = B_1(T_1, T_2)$ at the breakpoint and by $B_2 = B_2(T_1, T_2)$ at the associated stable critical point.

We proceed to consider the roles of the threshold and the breakpoint for eradication purposes.

3. The threshold

Reduction of either or both of the two transmission factors T_1 and T_2 to new values below the threshold function will lead to eradication of the infection. It is required that action is taken to ensure that the transmission factors are permanently kept below the threshold function. The eradication is not achieved instantaneously. Rather, the worm load will approach its new steady state value 0 at a rate determined by the smaller of μ_1 and μ_2, see (1) and (2). The average life length $1/\mu_1$ of nature schistosomes is of the order of a few years, while the average life length $1/\mu_2$ of snails is of the order of months. Thus, μ_1 is less than μ_2, and the rate at which the worm load approaches 0 is determined by the death rate μ_1 of mature schistosomes. We note that the time for the worm load to approach 0 is further prolonged if the values of the transmission factors after the reduction lie only slightly below the threshold.

The amount of parameter modification required to achieve eradication is meaningfully measured by the ratio $T_1T_2/T_{10}T_{20}$, where T_1, T_2 are the pre-action values, and T_{10}, T_{20} are the post-action values. The minimum value of this ratio is achieved by reducing T_1 only and leaving T_2 unchanged, see Nåsell (1977a). This minimum value is called the minimum eradication effort E. It is determined as a function of the pre-action values T_1 and T_2 through the relation

$$E(T_1, T_2) = T_1/\overline{\overline{T}}_1(T_2). \tag{8}$$

Note that the transmission factor T_2 can be expressed as a function of the transmission factor T_1 and the minimum eradication effort E as follows

$$T_2 = \overline{\overline{T}}_2(T_1/E). \tag{9}$$

Variations in both space and time occur naturally of many of the basic parameters of the model. These variations can easily amount to an order of magnitude (a factor of 10) or more. A community with a high value of the (average) minimum eradication effort will, in spite of this variability, show a rather constant (stable) value of the endemic infection level. The endemicity is said to be high if E exceeds 100 and low if E is less than 10.

4. The breakpoint

Eradication follows from a momentary reduction to sufficiently low values of both of the state variables worm load $w(t)$ and snail prevalence $y(t)$. As noted above, this mode of eradication was first pointed out by Macdonald (1965). The Macdonald model contains one differential equation for w, in contrast to the two equations for w and y given in (1), (2). The Macdonald equation is an approximation to the system (1), (2), derived by setting $y' = 0$. This approximation is to some extent justified if the y-variable is "faster" than the w-variable. This is indeed the case, since, as noted above, the death rate μ_2 of snails is larger than the death rate μ_1 of mature schistosomes. The approximation has the consequence that the snail prevalence $y(t)$ immediately takes on a value that is determined as a function of the worm load $w(t)$ by setting the derivative y' equal to 0 (see equation (2)). In particular, the Macdonald model predicts that eradication follows as a result of a lowering of the worm load to just below the worm load B_1 at the breakpoint.

For the model treated here, it is shown in Nåsell (1976b) that a momentary reduction of the worm load all the way down to 0 will lead to eradication if $\mu_2/\mu_1 > B_2/B_1 - 1$, but that the worm load $w(t)$ will even then again approach the endemic worm load B_2 if $\mu_2/\mu_1 < (B_2/B_1 - 1)/3$. Note that these results follow from the unrealistic hypothesis that the chemotherapy campaign that serves to reduce the worm load is 100% efficient. There is a qualitative agreement between the conclusions of Macdonald's model and that of ours: eradication follows from the prescribed momentary lowering of the worm load if the ratio μ_2/μ_1 of snail death rate to schistosome death rate is sufficiently large.

We make the observation that if the worm load is reduced to a value between 0 and B_1, and if actions are taken to ensure that the worm load is kept at this low level until the snail prevalence has become reduced to a value sufficiently close to its corresponding steady state value, then our model predicts that eradication will follow.

Eradication by use of the breakpoint requires that the worm load be reduced from the endemic level B_2 to a value below B_1. We introduce $R =$

$= (B_2 - B_1)/B_2$ as a measure ot the proportionate reduction of the worm load required to achieve eradication.

Reductions of the worm load can in practice be achieved through mass administration of antischistosomal drugs. If such a chemotherapy campaign is to lead to eradication, then its efficiency, as measured by the proportionate reduction of the worm load in the community, must clearly exceed the ratio $R = (B_2 - B_1)/B_2$ defined above. The efficiency of a chemotherapy campaign is always less than 100%, and even 90% may be hard to reach. Limitations are caused by the difficulty of identifying and getting cooperation from all individuals in the community, and also by the inefficiency of available drugs.

Eradication using the breakpoint is possible because of the existence of two stable critical points of the system of equations (1), (2). Such eradication requires that the state vector (w, y) of equations (1), (2) be brought from the domain of attraction of the endemic infection (the critical point with worm load B_2) to the domain of attraction of the origin. The required efficiency R of a chemotherapy campaign designed to accomplish this is a measure of the size of the domain of attraction of the endemic infection level.

We note that both the minimum eradication effort E and the required chemotherapy campaign efficiency R are measures of what Holling (1973) terms "resilience". Both are measures of the maximum amount of perturbation that an endemic infection can suffer without the qualitative change that consists in the disappearance of the parasite population. Equivalently, they measure the minimum amount of perturbation that is necessary for eradication. We proceed in the next section to analyze the relation between E and R. The fact that the minimum eradication effort E measures modifications in parameter space, while the required chemotherapy campaign efficiency R measures perturbations in state space is of conceptual importance. Webster, Waide and Patten (1975) and Harrison (1979) have suggested that the former type of measure be termed "resistance" and that the term "resilience" be preserved for measures of perturbations in state space. We support this distinction in terminology.

5. The required efficiency of a chemotherapy campaign

Successful eradication using the breakpoint requires that the chemotherapy campaign have an efficiency exceeding $(B_2 - B_1)/B_2$. For what communities can this required efficiency be achieved? To answer this question, we study the relation between the required efficiency and the minimum eradication effort E. By using relation (9) to eliminate T_2, we find that the required efficiency R can be written as a function of the transmission factor T_1 and the minimum eradication effort E as follows

$$R(T_1, E) = 1 - B_1(T_1, \overline{\overline{T}}_2(T_1/E))/B_2(T_1, \overline{\overline{T}}_1/E). \tag{10}$$

This function cannot be written in explicit form, since both the worm load functions B_1 and B_2 and the threshold function $\overline{\overline{T}}_2$ are only defined implicitly.

The conclusions below concerning the feasibility of eradication using the breakpoint makes use of two properties of the function R, i.e. that it is increasing in each of its two arguments. It is readily proved that $\partial R/\partial E > 0$, but an analytic proof that $\partial R/\partial T_1 > 0$ has not been found. For the latter property, we refer to numerical evaluations that have been made of R for selected values of T_1 and E. They all support our contention; some of them are illustrated by solid curves in Fig. 1. Here, the required proportionate worm load reduction R is shown as a function of the transmission factor T_1 for three values of the minimum eradication effort E. A logarithmic scale along

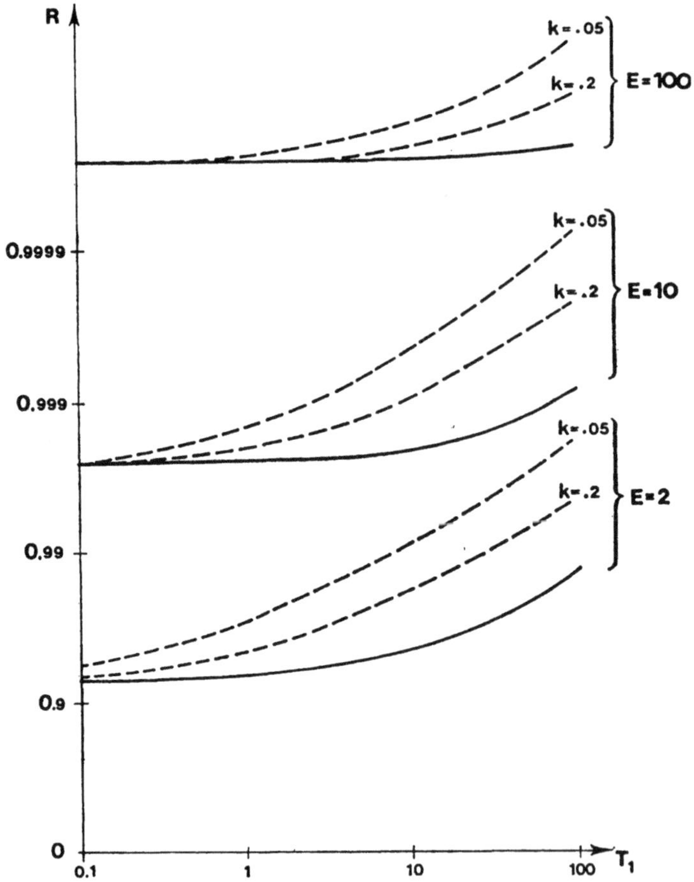

Fig. 1. The required chemotherapy campaign efficiency R is shown as a function of the transmission factor T_1 for three values of the minimum eradication effort E. Solid curves refer to the homogeneous case. Dashed curves refer to aggregation with sexes distributed *together*. The clumping parameter k has the values 0.05 and 0.2, as indicated.

the abscissa is used to cover a range of values to which T_1 is confined in most actual epidemiological situations. The logarithmic scale along the ordinate allows an expansion of the values near one to which R appears to belong for all communities except those that are very close to the threshold.

Based on the numerical results illustrated in Fig. 1, we conclude that R is an increasing function of T_1 for each $E > 1$. It follows therefore that R is larger than its asymptotic value as $T_1 \to 0$. The latter value can be evaluated explicitly and is found to be

$$\lim_{T_1 \to 0} R(T_1, E) = R_0(E) = 2\sqrt{E^2 - 1}/(E + \sqrt{E^2 - 1}), \quad E > 1. \tag{11}$$

We conclude therefore that the following inequality holds between the required chemotherapy program efficiency R and the minimum eradication effort E

$$R > R_0(E) = 2\sqrt{E^2 - 1}/(E + \sqrt{E^2 - 1}). \tag{12}$$

Values of the lower bound $R_0(E)$ are given in Table I for a few selected values of E.

Conclusions regarding the feasibility of eradicating schistosomiasis through a chemotherapy campaign can be drawn on the basis of the entries in Table I. We note that if the minimum eradication effort E exceeds 10 (a moderately large value), then eradication by use of the breakpoint requires that the efficiency of the chemotherapy campaign be larger than 99.75%. This efficiency is in practice impossible to achieve and therefore we conclude that eradication using the breakpoint is not possible in a community with E larger than 10. Indeed, if the chemotherapy campaign efficiency is 90%, then Table I shows that eradication by use of the breakpoint is not possible unless the minimum eradication effort E is less than 1.74, i.e. unless the endemicity is very low.

Table I

The lower bound $R_0(E)$ of the required chemotherapy campaign efficiency is listed as a function of the minimum eradication effort E

E	$R_0(E)$
1.74	0.9
2	0.928
5.05	0.99
10	0.9975
100	0.999975

The lower bound $R_0(E)$ of the required chemotherapy campaign efficiency is listed as a function of the minimum eradication effort E

6. The influence of aggregation

The transmission model that we have used above is based on assumptions of homogeneity in exposure and in susceptibility of human beings. One of its consequences is that the number of parasites per host has a Poisson distribution.

Heterogeneities will influence both the parasite distribution, the threshold, and the breakpoint. Studies of these influences have been made by Barbour (1978), Dietz (1981) and May (1977a). Barbour has established a model to study heterogeneity under the simplifying assumption that every female lays fertilized eggs, no matter if male parasites are present or not. In this situation, no breakpoint exists. Dietz has applied a modification of the Barbour model to the special case of one group of snails (one pond) only, and has concluded that heterogeneity lowers the threshold if there is positive correlation between exposure rate and contamination rate, but does not affect it if the correlation is zero. This implies that heterogeneity in susceptibility of human beings does not affect the threshold in schistosomiasis when mating between male and female parasites is ignored.

Heterogeneity of either exposure or susceptibility leads to an overdispersed distribution of parasites per human host, i.e. a distribution whose variance is larger than its mean. Such a deviation from the Poisson distribution influences the mating probability of female parasites, and this in turn affects both the threshold and the breakpoint. May (1977a) has derived the influence of aggregation in the parasite distribution on threshold and breakpoint. The above comments on Barbour's model show that May's model is consistent with heterogeneity of human susceptibility but not of human exposure. We proceed to apply May's model to study how the relation between the required efficiency of a chemotherapy campaign R and the minimum eradication effort E is influenced by the aggregation. Our main result is that the inequality (12) that relates R and E under assumptions of homogeneity holds also for May's model with aggregation.

May uses a negative binomial distribution to represent the distribution of the number of parasites per host. This distribution is characterized by two parameters, the mean w and the clumping parameter k. Its variance is $w + w^2/k$. The parameter k takes positive values. The smaller k is, the more overdispersed is the distribution. The negative binomial distribution approaches the Poisson distribution as $k \to \infty$. The model used above and based on assumptions of homogeneity is recovered as a limiting case from May's more general model by letting $k \to \infty$.

The joint distribution of male and female parasites per human host is of importance in determining the mating probability. Two cases are distinguished by May, i.e. the cases when the sexes are aggregated "together" or "separately". The sexes are said to be aggregated together if the total number of worms per host has a negative binomial distribution with mean w and clumping parameter k, and the sex of each worm is male or female with probability 1/2. The sexes are said to be aggregated separately if the number of female worms and the number of male worms are independent radom variables, each with a negative binomial distribution with mean $w/2$ and clumping parameter k. May shows that the mating probability increases as the overdispersion is increased (k is lowered) in the "together" case, but that the mating probability decreases as the overdispersion is increased in the "separate" case. The two cases merge to give the same mating probability when the overdispersion disappears, i.e. as $k \to \infty$.

The May model leads to a system of differential equations for the worm load w and the snail prevalence y that closely resembles equations (1) and (2). The only difference is that $\Psi(w)$, which stands for the expected number of paired parasites per human host, is replaced by the k-dependent quantity $w\varphi(w, k)$, where $\varphi(w, k)$ is the mating probability. The mating probability φ carries the subscript $t(s)$ when the sexes are aggregated "together" ("separately"). Explicit expressions for the mating probabilities $\varphi_t(w, k)$ and $\varphi_s(w, k)$ are given in the paper by May.

The threshold function for the transmission factor T_2 depends on the transmission factor T_1, the clumping parameter k, and on the mode of aggregation of the sexes; we denote it by $\overline{\overline{T}}_{2j}(T_1, k)$, where the subscript j equals $t(s)$ when the sexes are aggregated together (separately). Numerical evaluations indicate that $T_1\overline{\overline{T}}_{2j}(T_1, k)$ is a decreasing function of T_1 for fixed k and for $j = t$ or $j = s$. This implies that the minimum eradication effort can be defined as the ratio between T_1 and the threshold value $\overline{\overline{T}}_{1j}(T_2, k)$, in analogy with expression (8), as follows

$$E_j(T_1, T_2, k) = T_1/\overline{\overline{T}}_{1j}(T_2, k). \tag{13}$$

Our analysis of the May aggregation model is based on numerical evaluations of the required chemotherapy campaign efficiency R as a function of the transmission factor T_1, the minimum eradication effort E, and the clumping parameter k. The defining relation is the analogue of (10), but with the obvious added complication that all of the functions R, B_1, B_2 and $\overline{\overline{T}}_2$ depend on the clumping parameter k and on the mode of aggregation of the sexes ("together" or "separately").

Results of such evaluations for the case when the sexes are aggregated together are shown with dashed curves in Fig. 1. These results indicate that the required chemotherapy campaign efficiency R is increased as the clumping parameter k is lowered, i.e. as the overdispersion is increased. The figure also indicates that the required chemotherapy campaign efficiency R is an increasing function of the transmission factor T_1.

Results for the case when the sexes are aggregated separately are similar: The required chemotherapy campaign efficiency R is a decreasing function of the clumping parameter k and an increasing function of the transmission factor T_1. The results for the two modes of aggregation are compared in Fig. 2. The figure indicates that the required chemotherapy campaign

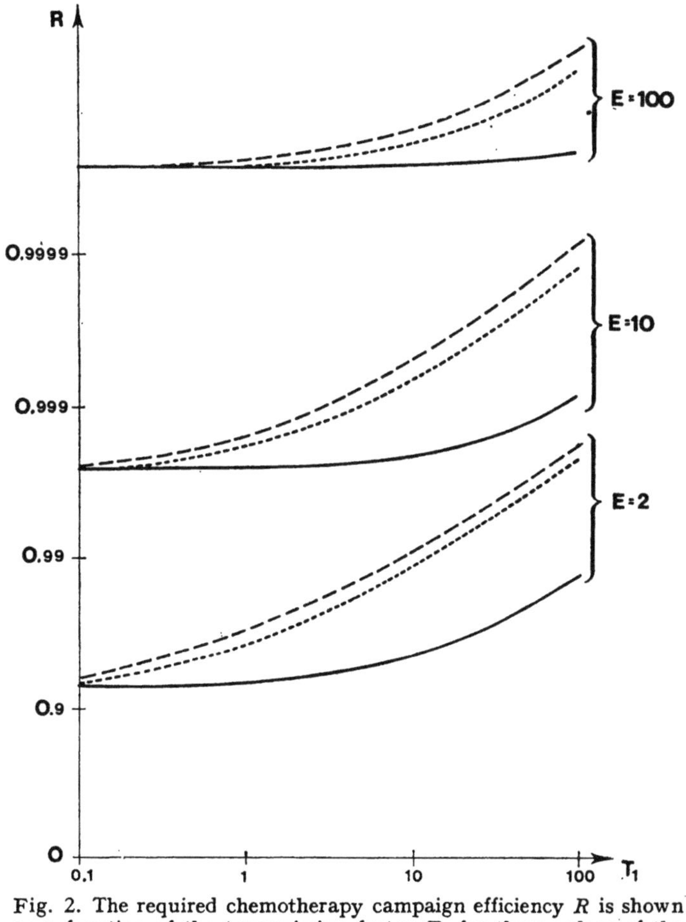

Fig. 2. The required chemotherapy campaign efficiency R is shown as a function of the transmission factor T_1 for three values of the minimum eradication effort E. Solid curves refer to the homogeneous case. Aggregation *together* is shown by dashed curves, and aggregation *separately* by dotted curves. All aggregation cases have the clumping parameter $k = 0.05$.

efficiency is somewhat lower when the sexes are aggregated separately than when they are aggregated together.

These results may appear to contradict a noteworthy finding by May (1977a) and Bradley and May (1978), i.e. that increased overdispersion when the sexes are aggregated separately raises the threshold and enhances the breakpoint phenomenon. This means that both the minimum eradication effort E and the required chemotherapy campaign efficiency R are lowered in the situation considered by May, where the transmission factor T_1 and T_2 are fixed. Our results show that R is raised as a result of increased overdispersion in the separate case when E and T_1 are fixed. Our approach has been used to analyze the influence of the value of k on the relation between threshold and breakpoint.

Our results show that the required chemotherapy campaign efficiency R is an increasing function of the transmission factor T_1 for either mode of aggregation of the sexes. A lower bound of R is therefore given by the asymptotic expression for R as $T_1 \to 0$. We evaluate the latter as follows. The explicit expressions for the mating probabilities $\varphi_t(w, k)$ and $\varphi_s(w, k)$ in May (1977a) give the following asymptotic expressions as the worm load w approaches 0:

(14) $$\varphi_t(w, k) \sim (k + 1)w/(2k), \quad w \to 0,$$

(15) $$\varphi_s(w, k) \sim w/2, \quad w \to 0.$$

Furthermore, the threshold function for the transmission factor T_2 has the following asymptotic behavior as $T_1 \to 0$:

(16) $$\overline{\overline{T}}_{2t}(T_1, k) \sim 16\, k/((k + 1)T_1^2), \quad T_1 \to 0,$$

(17) $$\overline{\overline{T}}_{2s}(T_1, k) \sim 16/T_1^2, \quad T_1 \to 0.$$

The asymptotic expression for R as $T_1 \to 0$ is found to be

(18) $$R(T_1, E, k) \sim 2\sqrt{E^2 - 1}/(E + \sqrt{E^2 - 1})$$

and this expression holds for both of the modes of aggregation of the sexes. We conclude that the lower bound for the required chemotherapy campaign efficiency R given in inequality (12) and listed in Table I of the preceding section holds also for the cases of aggregation treated by May.

7. Concluding comments

The models dealt with in this paper have been extended in the literature in several ways to account for increased realism. Thus, studies have been made of latency in the snail population, of promiscuous mating among sexually mature schistosomes, of immune reactions among human beings, and of external infection. We consider the expected consequences of each of these complicating factors on our results.

Latency in the snail population is expected not to influence our results. This expectation is based on the finding by May (1977a) and Nåsell (1976a) that latency among snails is accounted for by redefining the transmission factors in terms of the basic parameters, and that the relation between the stationary worm load w and the new transmission factors is the same as the corresponding relation (7) in the case without snail latency. Note that both threshold and breakpoint, and hence both minimum eradication effort E and required chemotherapy campaign efficiency R, are derived from this relation.

Promiscuous mating among mature schistosomes has been studied by May (1977a) and Nåsell (1978). This mating behavior is closer to hermaphroditic egglaying than monogamous mating is. Since the breakpoint is absent when egglaying is hermaphroditic, we expect that promiscuous mating will lead to a further suppression of the breakpoint compared to what has been exhibited above for monogamous mating.

The epidemiological consequences of concomitant immunity among human beings are studied by Nåsell (1977b). The model is based on an assumption of promiscuous mating among mature schistosomes. The immune mechanism is hypothesized to take the form that the rate of immigration of new parasites into a human being is proportional to the difference between a saturation level and the number of live parasites. It is shown in the quoted paper that this form of immune reaction has only a minor influence on the minimum eradication effort, while its ability to depress the endemic worm load can be substantial. We conjecture that an accounting for the immune reaction may serve to lower the proportionate worm load reduction required to achieve eradication using the breakpoint and hence to facilitate such eradication.

The influence of external infection has been treated by Nåsell (1975). It is no longer possible to achieve full eradication if infective material is continually impressed on a community from sources outside its boundaries. Even so, the model with external infection exhibits features that are reminiscent of the threshold and the breakpoint. The paper quoted shows that the existence of external sources of infection will reduce the portion of the parameter space where there exists a counterpart to the breakpoint, and that this "breakpoint" is completely wiped out if the amount of external infection is sufficiently large. Thus we expect that external infection will strengthen the results already obtained, namely that (near-)eradication using the breakpoint is available only for communities very close to the threshold.

To summarize these concluding comments, we note that the basic result in this paper is that the required chemotherapy campaign efficiency R is so high that eradication using the breakpoint is available only for communities very close to the threshold. This result is further strengthened if promiscuous mating or external infection are taken into account, it is unaffected by snail latency, but it may be weakened by consideration of immunity or any other density dependent mechanism in the human population.

Our results lead to the conclusion that a mass chemotherapy program conducted only once cannot be expected to lead to eradication if applied alone. On the other hand, the beneficial effects of mass chemotherapy, when conducted in conjunction with other control actions, should not be denied. It serves to lower the endemic infection level and the disease prevalence, and it can play the very important role of speeding up the results of control actions since it serves to reduce the long time constant given by the average life length $1/\mu_1$ of mature schistosomes.

REFERENCES

1. Barbour, A.D., *Macdonald's model and the transmission of bilharzia.* Trans. Roy. Soc. Trop. Med. Hyg., **72**, 6– 15 (1978).
2. Bradley, D.J., May, R. M., *Consequences of helminth aggregation for the dynamics of schistosomiasis.* Trans. Roy. Soc. Trop. Med. Hyg., **72**, 262– 273 (1978).
3. Dietz, K., *Models for vector-borne parasitic diseases.* In: *Vito Volterra Symposium on Mathematical Models in Biology.* Lecture Notes in Biomath., Vol. 39, pp. 264– 277. Berlin– Heidelberg– New York, Springer, 1980.

4. Harrison, G. W., *Stability under environmental stress: resistance, resilience, persistence and variability*. Amer. Natur., **113**, 659–669 (1979).
5. Holling, C. S., *Resilience and stability of ecological systems*. Annu. Rev. Syst., **4**, 1–23 (1973).
6. Macdonald, G., *The Epidemiology and Control of Malaria*. London, Oxford University Press, 1957.
7. Macdonald, G., *The dynamics of helminth infections, with special reference to schistosomes*. Trans. Roy. Soc. Trop. Med. Hyg., **59**, 489–506 (1965).
8. Macdonald, G., *Dynamics of Tropical Disease*. London, Oxford University Press, 1973.
9. May, R. M., *Togetherness among schistosomes: its effects on the dynamics of the infection*. Math. Biosci., **35**, 301–343 (1977 a).
10. May, R. M., *Thresholds and breakpoints in ecosystems with a multiplicity of stable states*. Nature, **269**, 471–477 (1977b).
11. Nåsell, I., *Schistosomiasis in a community with external infection*. In: *Proceedings of the 8th International Biometric Conference*, pp. 123–131. Bucureşti, Ed. Acad. R. S. Romania, 1975.
12. Nåsell, I., *A hybrid model of schistosomiasis with snail latency*. Theoret. Population Biol., **10**, 47–69 (1976a).
13. Nåsell, I., *On eradication of. schistosomiasis* Theoret. Population Biol., **10**, 133–144 (1976b).
14. Nåsell, I., *On transmission and control of schistosomiasis, with comments on Macdonald's model*. Theoret. Population Biol. **12**, 335–365 (1977a).
15. Nåsell, I., *Schistosomiasis with concomitant immunity*. Bull. Internat. Statist. Inst. **47**(2), 3–21 (1977b).
16. Nåsell, I., *Mating models for schistosomes*. J. Math. Biol., **6**, 21–35 (1978).
17. Nåsell, I., Hirsch, W. M., *A mathematical model of some helminthic infections*. Comm. Pure Appl. Math., **25**, 459–477 (1972).
18. Nåsell, I., Hirsch, W. M., *The transmission dynamics of schistosomiasis*. Comm. Pure Appl. Math., **26**, 395–453 (1973).
19. Webster, J. R., Waide, J. B., Patten, B. C., *Nutrient recycling and the stability of ecosystems*. In: *Mineral Cycling in Southeastern Ecosystems*. AEC Sympos. Ser. Conf. 74–0513, 1975.

F. PESARIN

On randomized statistical procedures

1. Introduction

A lot of statistical methods which need auxiliary random elements in an essential manner are known. They are usually concerned with test of hypotheses, estimation and general decision theory. One of the first examples of such methods is the well known Fisher-Tocher test. This is a test on a 2×2 contingency table whose sampling distribution becomes continuous by considering an auxiliary random experiment from a continuous uniform random variable over a proper interval, which takes place under the condition of complete independence of empirical data. By such an artifice it is possible to apply the Neymann-Pearson lemma for every value of first kind risk and to obtain a known optimal test.

Nevertheless many authors consider all these methods unsatisfactory. Their perplexities are mainly in that two different researchers with the same set of data can arrive at two different conclusions. And this due to the effect of the auxiliary random data.

In our opinion this argument is incorrect. In fact it tends to compare a well defined decision procedure with another one, strictly conditioned to observational data. In general, this is not a correct statistical decision procedure because the preference criterion based on risk functions is meaningless for decision procedures conditioned to observational data.

In many cases auxiliary randomization increases the value of the risk function, this happens in almost all regular cases. But there are cases, especially when the loss function is not convex, see Hodges and Lehmann (1950), or when the sets of nature states and of final actions are finite, where for every non randomized decision function there exists a randomized one having a better risk function. On the other hand, as in the opinion of many authors, randomized statistical procedures allow a convenient treatment of statistical theory. This mainly because the treatment becomes stronger, more general and more elegant.

Also, statistical methods with auxiliary randomization are still of practical usefullness, particularly when there are statistical problems otherwise considered as unsolvable.

Literature presents a lot of such methods. We mention, for instance, those suggested by Scheffé (1943), Stevens (1950), Durbin (1961), Warner (1965), Bell and Doksum (1965), Wagle (1968), Cucconi (1968 and 1970) Csörgő and Seshadri (1970), Pesarin (1972 and 1975), Diana and Pesarin,

(1973), Csörgő, Seshadri and Yalowsky (1975), Sarkadi (1977), Bondesson (1982), Diana and Taroni (1980), Landenna, Marasini and Ferrari (1982).

The statistical decision theory with auxiliary randomization is treated from a general point of view in Iannotta and Pesarin (1977), and in Fiorin, Iannotta and Pesarin (1979).

In this paper we present a simplified version of the theorem which states the equivalence among the different kinds of statistical random procedures. We also recall from literature some interesting examples of such randomized statistical procedures and in the last section we shortly present a new one concerning a nonparametric test for the independence of two continuous random variables.

2. Three kinds of randomized statistical decision procedures

We first consider the general structure of a statistical decision problem. As known, a statistical decision problem can be formally defined on the basis of the following elements:

i) a non emply set Θ of nature states θ, $\Theta = \{\theta\}$;

ii) a non emply set A of actions (or final decissions) a, $A = \{a\}$;

iii) a loss function L which maps $\Theta \times A$ into $\mathbf{R}^1$;

iv) an n-dimensional observable random variable X whose cumulative distribution function (c.d.f.) $F_\theta(x)$ depends on θ.

In statistical decision theory these elements are interconnected as they are in a sentence as the following: at first nature selects an element $\theta \in \Theta$, then a statistician gets information about θ by observing the value x of the observable random variable X, and chooses the action $a \in A$, according to a suitable decision function $d\colon \mathbf{R}^n \to A$, the loss associated with this procedure being $L(\theta, d(x))$.

Definition 1. Every function $d\colon \mathbf{R}^n \to A$ such that the risk function $R(\theta, d) = \int_{\mathbf{R}^n} L(\theta, d(x))\,\mathrm{d}F_\theta(x)$ exists for every θ is called a *simple decision function*. The set of all simple decision functions is denoted by D. A simple statistical decision problem is denoted by the triplet (Θ, D, R).

Decision theory, showing how it is possible to study the statistician's behaviour according to pre-established optimality criteria, assumes the loss function L to be measurable whatever θ is. This implies that over A we must define a suitable σ-algebra $\mathcal{A}$. To do this, let us call $\mathcal{B}_1$ the Borel σ-algebra generated by all open intervals of $\mathbf{R}_1$ and let $L_\theta(a)$ be the loss function for a fixed θ. So the set of inverse images of $\mathcal{B}_1$, i.e. $\mathcal{A}_\theta = L_\theta^{-1}(\mathcal{B}_1)$, is a σ-algebra of sets of A. Therefore we can define $\mathcal{A}$ as the least σ-algebra containing all $\mathcal{A}_\theta$, or containing the union $\bigcup_\theta \mathcal{A}_\theta$. Then $\mathcal{A}$ is the least σ-algebra over A such that L is measurable whatever θ is.

From the assumed measurability of the risk function R considered as a map from $\Theta \times D$ into $\mathbf{R}^1$ we can also define a σ-algebra $\mathcal{D}$ over the set D of all simple decision functions, such that R is measurable whatever θ is. In analogy with the previous one we can consider the risk function for a fixed value of θ, that is $R_\theta(d)$, as a map from D into $\mathbf{R}^1$. So $\mathcal{D}_\theta = R_\theta^{-1}(\mathcal{B}_1)$ is a σ-algebra on D. Then we have $\mathcal{D} = \bigcup_\theta \mathcal{D}_\theta$.

As well known, in the Bayesian approach it is assumed that nature chooses randomly its value θ, so that it is necessary to define a σ-algebra $\mathcal{T}$ over Θ and a probability distribution, or at least a σ-finite non-negative measure, over $(\Theta, \mathcal{T})$. In such a situation L and R are assumed to be measurable, in an obvious manner, also with respect to $\mathcal{T}$. In our treatment the Bayesian approach is not involved, so that in what follows we never mention it; we stress only the fact that all our general conclusions remains still valid both in the Bayesian and in classical approaches.

A first kind of random statistical procedures is introduced as follows. Consider the set A^* of all probability distributions over the measurable space $(A, \mathcal{A})$. Then we can assign to the sample value x the task of indicating the probability distribution by which to draw at random an action $a \in A$. So we can consider a function $\delta: \mathbf{R}^n \to A^*$, which for every x determines the probability distribution $\delta_x \in A^*$ that allows us to draw $a \in A$. In such a case we define the average loss function as $\hat{L}(\theta, \delta_x) = \int_A L(\theta, a)\, d\delta_x(a)$ and the risk function as $\hat{R}(\theta, \delta) = \int_{\mathbf{R}^n} \hat{L}(\theta, \delta_x)\, dF_\theta(x)$.

Definition 2. Every function $\delta: \mathbf{R}^n \to A^*$ whose risk function exists for every θ is called (Ferguson, 1967) a *behavioural decision function*. The set of behavioural decision functions is denoted by Δ. A behavioural statistical decision problem is denoted by the triplet $(\Theta, \Delta, \hat{R})$.

In accordance with this first randomized scheme, a statistician solves his decision problem by choosing at first a behavioural decision function $\delta \in \Delta$. In this manner he selects, on the basis of the sample x, drawn from the observable random variable X, the probability distribution $\delta_x \in A^*$ which allow him to draw randomly the action $a \in A$, the risk being $\hat{R}(\theta, \delta)$.

Second, let us consider the statistical decision procedures with auxiliary randomization. This second scheme consists in choosing the action $a \in A$ as a joint function of the sample x, drawn from the observable random variable X, and of a sample z drawn from a suitable auxiliary m-dimensional random variable Z, stochastically independent of X and having c.d.f. G. Formally this second scheme is defined as follows.

Definition 3. i) Let m, the dimensionality of the auxiliary random variable Z, be sufficiently large so that there exists a regular measurable function Φ from $\mathbf{R}^m$ *onto* A (in such a manner that every point of A is image by Φ of at least a point of $\mathbf{R}^m$), where regularity means that for every measure η defined on $(\mathbf{R}^m, \mathcal{S}_\Phi = \Phi^{-1}(\mathcal{A}))$ there exists at least a measure μ on $(\mathbf{R}^m, \mathcal{B}_m)$ such that η is a restriction of μ to $\mathcal{S}_\Phi$: i.e. $\mu(B) = \eta(B)$, $\forall B \in \mathcal{S}_\Phi$.

ii) Let Z be an arbitrary m-dimensional nondegenerate continuous random variable, stochastically independent of X and having c.d.f. G not dependent on θ.

iii) Let Ψ be the set of all function ψ which map $\mathbf{R}^{n+m}$ into A, such that the partial functions ψ_x, considered as functions of the m-dimensional vector z from $\mathbf{R}^m$ into A, are measurable for almost all x, and such that the risk function $\bar{R}(\theta, \varphi) = \int_{\mathbf{R}^{n+m}} L(\theta, \psi(x, z))\, dG(z)\, dF_\theta(x)$ exists finite for every θ.

In this context Z is called an *auxiliary* random variable for the decision problem and the functions ψ are called statistical decision functions *with auxiliary randomization.* These procedures are denoted by $(\Theta, \Psi, \bar{R})$.

According to this second scheme a statistician: 1st selects a function with auxiliary randomization $\psi \in \Psi$; 2nd he draws the outcome x from X; 3rd he draws the outcome z from Z; 4th he gets the action $a = \psi(x, z)$, the risk being $\bar{R}(\theta, \psi)$.

It is easy to see that Ψ contains all almost z-invariant functions; i.e. all functions such that $\psi(x, z) = d(x)$ almost everywhere with respect to the auxiliary random variable Z. In fact, if $\mathcal{A}$ contains all single points a, every z-invariant decision function with auxiliary randomization is partially measurable, then it is a member of Ψ if its risk function exists finite. So it coincides with probability one with a simple decision function. So we can conclude that $D \subset \Psi$.

As known, the third scheme of random procedures is more abstract than the others. It considers over the measurable space $(D, \mathcal{D})$ the set D^* of all probability distributions P, called *randomized* decision functions, such that the risk function $\mathring{R}(\theta, P) = \int_D R(\theta, d)\, \mathrm{d}P(d)$ exists for every θ.

By this third randomized scheme a statisticians chooses a randomized decision function $P \in D^*$, i.e. he draws at random, according to the probability distribution P, the simple decision function d by which he selects the action $a = d(x)$ as a function of the experimental outcome x, the risk being $\mathring{R}(\theta, P)$. These procedures are denoted by the triplet $(\Theta, D^*, \mathring{R})$. This third scheme is practically useful only when D has a finite (and small) number of elements.

3. The equivalence of the three random procedures

Wald and Wolfowitz (1951) stated, in a sequential context, the equivalence between the first scheme $(\Theta, \Delta, \hat{R})$ and the third one $(\Theta, D^*, \mathring{R})$. The equivalence of all the three was stated by Fiorin, Iannotta and Pesarin (1979). Here we present a simplified version of the equivalence theorem between the first $(\Theta, \Delta, \hat{R})$ and the second $(\Theta, \Psi, \bar{R})$.

Theorem. a) *For every fixed* $\delta \in \Delta$ *there exists at least a* $\psi_\delta \in \Psi$ *such that* $\hat{R}(\theta, \delta) = \bar{R}(\theta, \psi_\delta), (\forall\theta)$; *b) For every* $\psi \in \Psi$ *there exists at least a* $\delta_\psi \in \Delta$ *such that* $\bar{R}(\theta, \psi) = \hat{R}(\theta, \delta_\psi)$, $\forall\theta$. *So we can say that* $(\Theta, \Delta, \hat{R})$ *and* $(\Theta, \Psi, \bar{R})$ *are equivalent.*

Proof. a) Let us fix $x \in \mathbf{R}^n$ and $\theta \in \Theta$, and consider a $\delta \in \Delta$. So $\delta(x) = \delta_x \in A^*$ is a probability measure over $(A, \mathcal{A})$. Let us consider over $(R^n, \mathcal{S}_\Phi)$ the probability measure ${}_\delta\eta_x$ induced by Φ^{-1}. By regularity conditions of Φ there exists at least one probability measure ${}_\delta\mu_x$ over $(\mathbf{R}^m, \mathcal{B}_m)$ such that ${}_\delta\eta_x(B) = {}_\delta\mu_x(B)$, $\forall B \in \mathcal{S}_\Phi$. By the extension theorem to ${}_\delta\mu_x$ it uniquely corresponds a cumulative distribution function ${}_\delta H_x(z)$. Now, we can consider a random variable transformation ${}_\delta\xi_x$ from $\mathbf{R}^m$ into $\mathbf{R}^m$ that transforms the c.d.f. G of the auxiliary random variable Z into ${}_\delta H_x$. Then, as x varies, we can consider the composite function $\Psi_\delta = \Phi \circ {}_\delta\xi_x$, which maps $\mathbf{R}^{n+m}$ into A.

Such a function induces over $(A, \mathcal{A})$ for every x the same probability measure as δ does. Moreover its risk function is

$$\bar{R}(\theta, \psi_\delta) = \int_{\mathbf{R}^n} \int_{\mathbf{R}^m} L(\theta, \psi_\delta(x, z)) \, dG(z) \, dF_\theta(x) =$$

$$= \int_{\mathbf{R}^n} \int_A L(\theta, a) \, d\delta_x(a) \, dF_\theta(x) = \hat{R}(\theta, \delta), \ \forall \theta$$

and is finite. So $\psi_\delta \in \Psi$.

We notice, of course, that in practice there are many members of Ψ equivalent to a fixed δ.

b) Let us fix x and δ and consider a $\psi \in \Psi$. The function ψ_x from $\mathbf{R}^m$ into A, being measurable by hypothesis, induces over $(A, \mathcal{A})$ the probability measure ${}_\psi G_x \in A^*$. Then let us consider the function $\delta_\psi(x) = \psi_x$. Such a function maps $\mathbf{R}^n$ into A^* and its risk function, being $\hat{R}(\theta, \delta_\psi) = \bar{R}(\theta, \psi)$, $\forall \theta$, is finite. Then by definition $\delta_\psi \in \Delta$. This completes the proof of the equivalence of $(\Theta, \Delta, \hat{R})$ and $(\Theta, \Psi, \bar{R})$.

4. Some examples of statistical procedures with auxiliary randomization

With the purpose of constructing an exact goodness of fit test for normality, Durbin (1961) and, independently, Cucconi (1970) suggested a very elegant method explicitely based on auxiliary random elements. The Durbin-Cucconi's method is as follows: Let $X_1, X_2, \ldots, X_n$ be independent, normally distributed random variables with unknown mean μ and variance σ^2. Let $\bar{X}$ and S^2 be the sample mean and variance of this random sample. Further let $Y_1, \ldots, Y_n$ be an independent auxiliary random sample from a standard normal distribution $N(0, 1)$. Define $\hat{X}_i$ by the equations $X_i = (\hat{X}_i - \bar{X})S_y/S + \bar{Y}$, $i = 1 \ldots, n$, where $\bar{Y}$ and S_y^2 are the sample mean and variance of $Y_1, \ldots, Y_n$.

Durbin proved that $\hat{X}_i$ so defined are independent $N(0, 1)$ variables and consequently the goodness of fit composite hypothesis concerning normality of X_i can be tested as simple hypothesis concerning standard normality of $\hat{X}_i$.

In the same paper Durbin suggests a general method of replacing composite hypothesis by equivalent simple ones. Suppose in general that $X_1, \ldots, \ldots, X_n$ have multivariate c.d. f. $F(x_1, \ldots, x_n; \theta)$ depending on a set of parameters θ for which a sufficient set of statistics T_1 is available. Suppose that a transformation λ, independent of θ and carrying $X_1, \ldots, X_n$ into (T_1, T_2), exists, T_2 being another set of statistics distributed independently of T_1, such that the inverse transformation λ^{-1} carries (T_1, T_2) into $X_1, \ldots, X_n$ uniquely. Let $G(t_1, \theta)$ be the c.d.f. of T_1 and let T_1^* be an independent auxiliary random vector with c.d.f. $G(t_1^*, \theta_0)$, where θ_0 is a known value of θ. Let $\hat{X}_1, \ldots, \ldots, \hat{X}_n$ be the values obtained by applying λ^{-1} to (T_1^*, T_2). Then $\hat{X}_1, \ldots, \hat{X}_n$ are distributed with c.d.f. $F(\hat{x}_1, \ldots, \hat{x}_n; \theta_0)$. So any composite hypothesis concerning $X_1, \ldots X_n$ is reduced to a simple one concerning $\hat{X}_1, \ldots, \hat{X}_n$, the nuisance parameters being out.

An improved version of the Durbin theorem can be found in Bondesson (1982).

Following Durbin's idea Pesarin (1972) proposed a test for the homogeneus Poisson process. Let $X_1, \ldots, X_n$, $n > 2$, be the arrival times of an homogeneus Poisson process, then the inter-arrival times $Y_1 = X_1$, $Y_2 = X_2 - X_1, \ldots$ $\ldots, Y_n = X_n - X_{n-1}$ are independent exponentially distributed with density function $\theta \exp\{-\theta y\}$, $\theta > 0$ unknown.

Let the auxiliary random variable Z be a gamma random variable with n degrees of freedom and unit scale parameter, then the transforms $\hat{Y}_i = Z \cdot Y_i \Big/ \sum_{j=1}^{n} Y_j$, $i = 1, \ldots, n$, are independent exponentially distributed with unit parameter. So, it is now possible to apply a goodness of fit test for simple hypothesis in an obvious manner.

The same theorem allows to obtain a test for the Erlang process. As known in such a process the interarrivals $Y_i = X_{i+1} - X_i$ are i.i.d. gamma with unknown scale parameter and known r degrees of freedom. So the transforms

$$\hat{Y}_i = \frac{Y_i}{\sum_{j=1}^{n} Y_j} \cdot Z,$$

where Z is an auxiliary gamma random variable with unit scale parameter and $n \cdot r$ degrees of freedom, are i.i.d. gamma with r degrees of freedom and unit scale parameter.

Using the same idea Diana and Pesarin (1973) proposed exact tests for the problem of the homogeneity of k variances in the normal case.

Cucconi (1968) suggested a method to reduce a functional composite hypothesis to a simple one. The problem has a large interest in quality control when the quality of an object is defined as a function in the unit interval (0,1) of, say, k different items. Let $X_{i1}, \ldots, X_{ik}$ be the k items of the i-th object and $Y_i = \Psi(X_{i1}, \ldots, X_{ik})$, $i = 1, \ldots, n$, its relative quality, which is a random variable supposed continuous and having unknown density function f. We wish to test sequentially the hypothesis $H_0 = E(Y) > \mu_0$ against $H_1 = E(Y) < \mu_1$, $0 < \mu_1 < \mu_0 < 1$. As known, this is possible only if the likelihood ratio is known. For this consider the random transforms $\hat{Y}_i = 0$ if $Z_i > Y_i$ and $\hat{Y}_i = 1$ if $Z_i < Y_i$, where the Z_i are independent uniformly distributed random variables $U(0, 1)$. The $\hat{Y}_i$ are independent elementary binomially distributed random variables $Bi(1, \mu = E(Y_i))$. In fact we have $P\{\hat{Y} = 1\} = \int_0^1 P\{Z < Y \mid Y = y\} f(y)\, dy = \int_0^1 y f(y)\, dy = E(Y)$, where f represents the unknown density of the quality random variable.

The solution now is obvious, the unknown function f being out.

Auxiliary random elements may also be used with random subsamples drawn from the given main sample. For the application of these methods in the area of goodness of fit, characterization properties of the involved random variables are generally considered. Csörgő, Seshadri and Yalowsky (1975) consider a lot of such problems.

One of these problems concerns a test for normality. Consider a random sample $X_1, \ldots, X_n$, $n > 6$, and consider the transforms

$$V_k = \left[\sum_{i=1}^{k} X_i - kX_{k+1}\right] \Big/ \sqrt{k(k+1)},$$

$$\hat{X}_k = V_{k+1}\sqrt{k} \Big/ \sqrt{\sum_{i=1}^{k} V_i^2}\,, \quad k = 1, \ldots, n-2.$$

The $\hat{X}_k$, $k = 1, \ldots, n-2$, are independent and distributed as Student's t with k degrees of freedom if and only if the X_i, $i = 1, \ldots, n$, are normally distributed. At this stage it is obvious how to proceed for testing normality. Here auxiliary randomization operates by choosing a random permutation of X_i's which brings to the $\hat{X}_k$'s transforms.

Another problem concerns a test for exponentiality. Consider a random sample $X_1, \ldots, X_n$, $n > 3$ and consider the transforms

$$\hat{X}_k = \sum_{i=1}^{k} X_i \Big/ \sum_{i=1}^{n} X_i, \quad k = 1, \ldots, n-1.$$

The multivariate distribution of $\hat{X}_1, \ldots, \hat{X}_{n-1}$ is the same as that of an ordered sample of size $n-1$ drawn from the uniform random variable $U(0, 1)$ if and only if the X_i, $i = 1, \ldots, n$, are independent exponentially distributed. Then for testing exponentiality it is sufficient to test the ordered uniformity of the $\hat{X}_k$ by means of a simple goodness of fit test.

An interesting randomized procedure to obtain unbiased goodness of fit tests is due to Diana and Taroni (1980). Let X_j, $j = 1, \ldots, n$ be a random sample from a continuous random variable X, whose c.d.f. under H_0 is $F_0(x)$, and let m $(m > 2)$ be an arbitrary integer. Divide the unit interval $(0, 1)$ into m disjoint and equal subintervals $I_1 = [(i-1) / m, i/m)$, $i = 1, \ldots, m$, and call f_i the number of transformations $r_1, \ldots, r_m$ from the $m!$ equally likely permutations of the first m integers and associate with the interval I_i the frequency f_{r_i} originally belonging to I_{r_i}. Consider, then, m independent auxiliary random samples of size f_{r_i}, $i = 1, \ldots, m$, respectively drawn from the auxiliary uniform distribution on I_{r_i}, $i = 1, \ldots, m$. In this manner the original data are replaced by auxiliary data $\hat{X}_j$, $j = 1, \ldots, n$, to which usual goodness of fit tests can be applied. The authors proved that in this way usual goodness of fit tests become unbiased.

5. Testing independence of two continuous variables

In this last section we introduce a nonparametric test for the independence of two continuous random variables which is based on a transformation of the original data by means of auxiliary random normal data having the same ranks. The idea for such a problem is due to Bell and Doksum (1965).

Consider a bivariate random sample (X_i, Y_i), $i = 1, \ldots, n$, drawn from the bivariate continuous random variable (X, Y), whose c.d.f. under H_0 is supposed of the form $F(x, y) = F_1(x) \cdot F_2(y)$, $(x, y) \in \mathbf{R}^2$, where F_1 and F_2 are the c.d.f. of the marginal X and Y respectively. Under H_1 this relation is supposed not true in some sense.

Let $\mathcal{R}(X_i)$ and $\mathcal{R}(Y_i)$ be respectively the ranks of X_i and Y_i, $i = 1, \ldots, n$, that is the place occupied by X and Y in their increasing arrangegements.

Consider, apart, two independent random samples, each of size n, drawn from an auxiliary standard normal distribution; namely $S_1, \ldots, S_n$ $S_1^*, \ldots, S_n^*$ respectively. Then consider the transformations of the original data

$$\hat{S}_i = S(\mathcal{R}(X_i)) = \{S : \mathcal{R}(S) = \mathcal{R}(X_i);\ S \in (S_1, \ldots, S_n)\},$$

$$\hat{S}_i^* = S^*(\mathcal{R}(Y_i)) = \{S^* : \mathcal{R}(S^*) = \mathcal{R}(Y_i);\ S^* \in (S_1^*, \ldots, S_n^*)\},\ i = 1, \ldots, n$$

that is the $S(S^*)$ value which has the same rank as $X_i(Y_i)$.

The Bell and Doksum test (1965) is based on the statistic

$$W = \frac{1}{n} \sum_{i=1}^{n} \hat{S}_i \cdot \hat{S}_i^*,$$

that is the sampling covariance between the normal random ranks. Spurr (1980) obtained its exact distribution under H_0, which is as the difference of two independent chisquare with n degrees of freedom.

It is easy to see that this test is consistent only for monotonic alternatives. Bell and Doksum proved that, if the observable random variable (X, Y) is a bivariate normal one, the correlation coefficient between W and the sampling correlation coefficient $r(x, y)$ is asymptotically one, then W is asymptotically optimum, in a natural sense against linear alternatives. So, asymptotically speaking, no information is lost by the replacement of the original data by means of random normal ranks.

The idea of the random normal ranks can be utiliezd to improve the Bell and Doksum test, that is to obtain a test consistent against a wider set of alternatives.

Consider the two bivariate arrangements of the original data (X_i', Y_i) and (X_i, Y_i'), $i = 1, \ldots, n$, where $X_1' < X_2' < \ldots < X_n'$ and $Y_1' < Y_2' < \ldots < Y_n'$ are the order statistics; that is consider a first bivariate arrangement according to the increasing X-values and a second one according to the increasing Y-values.

Consider, then, the two rank transformations $(i, \mathcal{R}(Y_i))$ and $(\mathcal{R}(X_i), i)$, $i = 1, \ldots, n$, and, like Bell and Doksum, the two replacements of the original data by means of the random normal ranks $(i, \hat{S}_i^*)$ and $(\hat{S}_i, i)$, $i = 1, \ldots, n$.

The normal random ranks replacement variables $\hat{S}$ and $\hat{S}^*$ under H_0 are independent normal random samples. So the two processes $(i, \hat{S}_i^*)$ and $(\hat{S}_i, i)$ are two independent discrete time white noise processes. Then for testing independence we could use tests on white noise without estimation of the underlying model. As an example we can consider the well known statistic

$$\bar{T} = \frac{1}{2(n-1)} \sum_{i=1}^{n-1} (\bar{S}_{i+1} - \bar{S}_i)^2,$$

where $\bar{S}$ stands for $\hat{S}$ or $\hat{S}^*$, so this leads to the two statistics $\hat{T}$ and $\hat{T}^*$ respectively.

This statistic under H_0 has mean value $E(\bar{T}) = 1$ and variance $\sigma^2(\bar{T}) = (3n - 4) / (n - 1)^2$, which vanishes as n increases. The sampling distribution of $2(n-1)^2T/(3n-4)$ is approximately distributed as a χ^2 with $2(n-1)^2/(3n-4)$ degrees of freedom. Its distributional approximation seems good even for moderate values of n, practically n greater than 10 or 15 is adequate.

As a test statistic we could have also chosen the Box-Pierce or Box-Ljung statistics, instead of the von Neumann one. In a series of Monte Carlo experiments we found the von Neumann statistic always more efficient than the others, at least for linear, quadratic, exponential and periodic alternatives for which the three were performed.

Consider now the behaviour of $\bar{T}$ under H_1. For this let us suppose that

a) The regression functions $\tilde{E}(X_i|Y_i')$, $\tilde{E}(Y_i|X_i')$, $E(\hat{S}_i|Y_i') = \mu_i$ and $E(\hat{S}_i^* | X_i') = \mu_i^*$ are regular that is with at most a finite number of discontinuities, where $\tilde{E}$ denotes the median value and E, of course, the mean value.

b) The residuals $\bar{Z}_i = \bar{S}_i - \bar{\mu}_i$ and $\bar{Z}_j = \bar{S}_j - \bar{\mu}_j$ are uncorrelated i.e. $E(\bar{Z}_i \cdot \bar{Z}_j) = 0$, $i \neq j$; where again $(\bar{S}_i, \bar{\mu}_i)$ stands for $(\hat{S}_i, \mu_i)$ or $(\hat{S}_i^*, \mu_i^*)$.

Under these conditions the following theorem holds.

Theorem. *$\bar{T}$ converges in probability, as n goes to infinity, to $E(\sigma^2(\bar{Z})) \leqslant 1$; 1 being the total variance of $\bar{S}$ and the variance of all $\bar{Z}_i$ under H_0.*

In fact, as n tends to infinity, we have that $\Delta\bar{\mu}_i = \bar{\mu}_{i+1} - \bar{\mu}_i \xrightarrow{P} 0$ for almost all i because of the supposed regularity conditions of the regression functions and of the continuity of the involved random variables. Moreover, as $\bar{S}_{i+1} - \bar{S}_i = \Delta\bar{\mu}_i + \bar{Z}_{i+1} - \bar{Z}_i$, we have also $E(\bar{S}_{i+1} - \bar{S}_i)^2 = \Delta\bar{\mu}_i^2 + \sigma^2(\bar{Z}_{i+1}) + \sigma^2(\bar{Z}_i)$ which leads to

$$\bar{T} = \frac{1}{2(n-1)} \left\{ \sum_{i=1}^{n-1} \Delta\bar{\mu}_i^2 + \sum_{i=1}^{n-1} \sigma^2(\bar{Z}_{i+1}) + \sum_{i=1}^{n-1} \sigma^2(\bar{Z}_i) \right\}.$$

Then as n increases $\bar{T} \xrightarrow{P} E(\sigma^2(\bar{Z})) \leqslant 1$.

This implies that small values of $\bar{T}$ are significant and that $\bar{T}$ is consistent against alternatives having non constant regression, that is for first order dependence alternatives.

Since non constant regression of X on Y does not imply non constant regression of Y on X, as a first order dependence test we can consider the statistic $T = \min(\hat{T}, \hat{T}^*)$.

The null distribution of $T \cdot 2(n-1)^2 / (3n-4)$ is approximately as the least of two independent χ^2 with $2(n-1)^2/(3n-4)$ degrees of freedom, this because $\hat{T}$ and $\hat{T}^*$ under H_0 are independent.

Different overall, or combined, statistics could be used instead of the proposed one. One of them for instance could be $Q = \hat{T} + \hat{T}^*$. This, if the χ^2 approximation is acceptable, is distributed as a $\chi^2 \cdot (3n-4)/2(n-1)^2$ with $4(n-1)^2/(3n-4)$ degrees of freedom.

A choice among the possible solutions could be done by means of Monte Carlo experiments.

To achieve a test consistent against first and second order alternatives, that is against non constant conditional variability and/or non constant regression we can consider the two solutions which follow.

Solution A

1st — Draw at random two independent numbers from the discrete uniform distribution $U(1, 2, ..., n)$, namely I_1 and I_2.

2nd — Consider the transformations of the original data

$$(X_i, \mid Y_i - Y_{I_1} \mid = \overline{Y}_i),$$
$$(\mid X_i - X_{I_2} \mid = \overline{X}_i, \ Y_i), \ i = 1, ..., n$$

These randomized transformations have the effect to convert a non constant conditional variability into a non constant regression, because of the absolute values, and to preserve non constant regression.

3rd — Apply, then, the $\hat{T}$ statistic to the n points $(X_i, \overline{Y}_i)$ and the $\hat{T}^*$ statistic to the points $(\overline{X}_i, Y_i)$. It is useful to observe that the $\hat{S}$ random normal rank associated with $\overline{Y}_{I_1} = 0$ simply is the least among $S_1, ..., S_n$ and the one associated with $\overline{X}_{I_2} = 0$ is the least among $S_1^*, ..., S_n^*$.

4th — Consider the test statistic $T^* = \min(\hat{T}, \hat{T}^*)$. Because of the randomized transformation applied, this test statistic is consistent against first and second order alternatives. The approximate null distribution of $T^* \cdot 2(n-1)^2/(3n-4)$ is the same as the least of two independent χ^2 with $2(n-1)^2/(3n-4)$ degrees of freedom.

Solution B

Alternatively we could consider the following two phase solution:

i) Apply to the original data the $\hat{T}$ and $\hat{T}^*$ statistics previously described; let us call them $\hat{T}_1$ and $\hat{T}_1^*$ respectively.

ii) Let $\hat{X}$ and $\hat{Y}$ be the sample medians values and consider the transformations

$$(X_i \mid Y_i - \widetilde{Y}_i \mid = \hat{Y}_i), \ (\mid X_i - \hat{X} \mid = \hat{X}_i, Y_i), \ i = 1, ..., n.$$

Then, apply to the n points $(X_i, \hat{Y}_i)$ the $\hat{T}$ statistic and to the $(X_i, \hat{Y}_i)$ the $\hat{T}^*$ statistic; let us call them $\hat{T}_2$ and $\hat{T}_2^*$ respectively.

All these statistics are approximately distributed as χ^2 random variables with $2(n-1)^2/(3n-4)$ degrees of freedom, apart of the scale coefficient. They are very weakly dependent if in both phases the same auxiliary random normal ranks are used. This dependence quickly vanishes as n increases. They are independent if in the second phase different auxiliary random normal ranks are used.

As a final test statistic we could consider

$$T^{**} = \min(\hat{T}_1, \hat{T}_1^*, \hat{T}_2, \hat{T}_2^*).$$

The null distribution of T^{**} is approximately as the least of four χ^2 random variable with $2(n-1)^2/(3n-4)$ degrees of freedom, apart of the scale coefficient. The approximation is good even for moderate values of n if the two phases are computed with different random normal ranks; otherwise larger sample sizes are needed.

The proposed solutions can be easily extended for testing pairwise independence in the multivariate case. For this, let us suppose that $(X_{1i}, X_{2i}, ..., X_{ki})$, $i = 1, ..., n$, is a random sample from the k-variate continuous random variable $(X_1, ..., X_k)$ whose c.d.f. under H_0 has all bivariate marginals of the form $F_{lh}(x_l, x_h) = F_l(x_l) \cdot F_h(x_h)$ for every $1 \leqslant l < h \leqslant k$. In this situation as a test procedure we sugest: i) apply the T^{**} statistic to each bivariate subsample (X_{li}, X_{hi}), $i = 1, ..., n$ and $1 \leqslant l < h \leqslant k$, and let us call it T^{**}_{lh}; ii) as an overall statistic consider then $T^{(k)} = \min_{l,h} (T^{**}_{lh})$. This test statistic is approximately distributed, apart of the scale coefficient, as the least of $2 \cdot k(k-1)$ independent χ^2 with $2(n-1)^2/(3n-4)$ degrees of freedom, provided that n is sufficiently large and that in each replacement of original data by means of random normal ranks different random standard normal deviates are used.

REFERENCES

1. Ash, R. B., *Real Analysis and Probability*. New York, Academic Press, 1972.
2. Bell, C. B., Doksum, K. A., *Some new distribution-free statistics*. Ann. Math. Statist. **36**, 203–214 (1965).
3. Blackwell, D., Girshick, M. A., *Theory of Games and Statistical Decisions*, New York, Wiley, 1954.
4. Bondesson, L., *To reduce a composite hypothesis to a simple one by sampling from the structural or a conditional distribution*. Scand. J. Statist., **9**, 129–132 (1982).
5. Checcucci, V., Tognoli, A., Vesentini, E., *Lezioni di Topologia Generale*. Milano, Feltrinelli, 1968.
6. Csörgő, M., Seshadri, V., *On the problem of replacing composite hypothesis by equivalent simple ones*. Rev. Internat. Statist. Inst., **38**, 351–368 (1970).
7. Csörgő, M., Seshadri, V., Yalowsky, M., *Applications of characterizations in the area of goodness, of fit*. In: Patil, G. P., et al (Eds.) *Statistical Distributions in Scientific Work Vol*. 2, pp. 79–90. Dordrecht, Reidel, 1975.
8. Cucconi, O., *Contributi all'analisi sequenziale nel controllo di accettazione della qualità*. Atti A.I.C.Q., **6**, 171–186 (1968).
9. Cucconi, O., *Un Test di Normalità*. Padova, CLEUP, 1970.
10. Diana, G., Pesarin, F., *Sul problema della verifica dell'uguaglianza di k varianze*. Metron, **31**, *3–4*, 336–348 (1973).
11. Diana, G., Taroni, G., *Alcune considerazioni sufficienti per la non distorsione dei tests funzionali*. Giornale degli Economisti e Annali di Economia, *5*, 671–677 (1980).
12. Durbin, V., *Some methods of constructing exact tests*. Biometrika, **48**, 41–55 (1961).
13. Dvoretsky, A., Wald, A., Wolfowitz, J., *Elimination of randomization in certain statistical decision procedures and zero sum two person games*. Ann. Math. Statist., **22**, 1–21 (1951).
14. Ferguson, T. S., *Mathematical Statistics: A Decision Theoretic Approach*. New York, Academic Press, 1967.
15. Folsom, R., Greenberg, B., Horvitz, D.G., *The two alternate questions randomized response model for human surveys*. J. Amer. Statist. Assoc., **68**, 525–530 (1973).
16. Greenberg, B., Abdel Latif, A., Simmons, W., Horvitz, D.G., *The unrelated question randomized response model: Theoretical framework*. J. Amer. Statist. Assoc., **64**, 520–– 539 (1969).
17. Hodges, J. L., Lehmann, E. L., *Some problems in minimax point estimation*. Ann. Math Statist., **21**, 520–539 (1950).
18. Horvitz, D. G., Shah, B. V., Simmons, W., *The unrelated response model*. In: *Social Statist. Section Proc. Amer. Statist. Assoc.*, pp. 65–72. Washington, D. C., 1967.
19. Iannotta, P., Pesarin, F., *Sulla teoria delle decisioni con casualizzazione ausiliaria*. Statistica (Bologna), **37**, 515–524 (1977).
20. Kendall, M. G., *On the reconciliation of theories of probability*. Biometrika, **36**, 101–116 (1949).
21. Lancaster, H. O., *The Chi-squared Distribution*. New York, Wiley, 1969.
22. Landenna, G., Marasini, D., Ferrari, P., *Schemi di campionamento a due stadi con risposte casualizzate*. Atti S.I.S., **31**, *3*, 23–26 (1982).

23. Lanke, J., *On the choice of the unrelated questions in Simmons' version of randomized response.* J. Amer. Statist. Assoc., **70**, 80–83 (1975).
24. Lehmann, E. L., *Testing Statistical Hypothesis.* New York, Wiley, 1959.
25. Milhøj, A., *A test of fit in time series model.* Biometrika, **68**, 177–187 (1981).
26. Moors, J. J. A., *Optimization of the unrelated question randomized response model.* J. Amer. Statist. Assoc., **66**, 627–629 (1971).
27. Naddeo, A., *Contributi alla Teoria Statistica dei Piccoli Campioni.* Milano, Giuffrè, 1960.
28. Neveu, J., *Bases Mathématiques du Calcul des Probabilités.* Paris, Masson, 1970.
29. Pearson, E. S., *On questions raised by the combination of tests based on discontinuous distributions.* Biometrika, **37**, 383–398 (1950).
30. Pesarin, F., *Un test per la verifica delle condizioni caratteristiche del processo di Poisson.* Atti S.I.S., **27**, 193–205 (1972).
31. Pesarin, F., *Su un nuovo test di verifica d'ipotesi funzionale.* L'Elaborazione Automatica, **1**, 45–54 (1973).
32. Pesarin, F., *Un test funzionale corretto.* Atti S.I.S., **28**, 1–8 (1975).
33. Sarkadi, K., *Characterization and testing for normality.* In: Gyires, B. (Ed.), *Analytic Function Methods in Probability Theory.* Colloq. Math. Soc. J. Bolyai, Vol. 21, pp. 317–329. Amsterdam, North-Holland, 1980.
34. Scheffé, H., *On solutions of the Behrens-Fisher problem, based on the t distribution.* Ann. Math. Statist., **14**, 35–44 (1943).
35. Spurr, B.O., *The exact distribution of the Bell-Doksum test of independence.* Austral. J. Statist. **22**, 166–171 (1980).
36. Stevens, W. L., *Fiducial limits of the parameter of a discontinuous distribution.* Biometrika, **37**, 117–129 (1950).
37. Tocher, K. D., *Extension of Neyman-Pearson theory of tests to discontinuous variates.* Biometrika, **37**, 130–144 (1950).
38. Wagle, B., *The multivariate beta distribution and a test for multivariate normality.* J. Roy. Statist. Soc. Ser. B, **30**, 511–515 (1968).
39. Wald, A., Wolfowitz, J., *Two methods of randomization in statistics and the theory of games.* Ann. of Math., **53**, 581–586 (1951).
40. Wald, A., *Statistical Decision Functions.* New York, Wiley, 1950.
41. Warner, S. L., *Randomized response: A survey technique for eliminating evasive answer bias.* J. Amer. Statist. Assoc., **60**, 63–69 (1965).
42. Zacks, S., *The Theory of Statistical Inference.* New York, Wiley, 1971.

AGNIESZKA PLUCIŃSKA

Some stochastic equations for quasi-diffusion processes

Let $(\Omega, \mathcal{F}, P)$ be a probability space, $T_0 = [0, T]$, $\{X_t, t \in T_0\}$ — a real valued stochastic process, $E(X_t) = 0$, $E(X_t^2) < \infty$, $\{W_t, t \in T_0\}$ — a Wiener process.

It is well known that under some regularity assumptions the solution of the stochastic equation

$$X_t - X_{t_1} = \int_{t_1}^{t} \tilde{a}(s, X_s)\, ds + \int_{t_1}^{t} \tilde{b}(s, X_s)\, dW_s \tag{1}$$

is a diffusion process. Here

$$\begin{aligned} \tilde{a}(t, X_t) &= \lim_{\Delta \to 0+} \frac{1}{\Delta} E((X_{t+\Delta} - X_t) \mid X_t) \text{ a.s.} \\ \tilde{b}^2(t, X_t) &= \lim_{\Delta \to 0+} \frac{1}{\Delta} E((X_{t+\Delta} - X_t)^2 \mid X_t) \text{ a.s.} \end{aligned} \tag{2}$$

The definition of integrals and the definition of solution of (1) are understood in the well-known typical sense [3]. There are different definitions of diffusion processes. We adopt the definition given in [3, p. 64].

The aim of this paper is to give stochastic differential equations similar to (1) such that the solutions of these equations are quasi-diffusion processes.

Let $t_1 < \ldots < t_n < t$, $\mathbf{t}_n = (t_1, \ldots, t_n)$, $\mathbf{x}_n = (x_1, \ldots, x_n) \in R^n$, $\mathbf{X}_n = (X_{t_1}, \ldots, X_{t_n})$. We remind [5] that $\{X_t, t \in T_0\}$ is a quasi-diffusion process if for all $n \geqslant 1$, $\mathbf{t}_n$, $\mathbf{x}_n$, $\varepsilon > 0$,

$$\lim_{\Delta_n \to 0} \frac{1}{\Delta_n} \int_{|x_{n+1} - x_n| > \varepsilon} (x_{n+1} - x_n)^2 P^{(n)}(\mathbf{t}_n, \mathbf{x}_n; t_{n+1}, dx_{n+1}) = 0 \tag{3}$$

$$\lim_{\Delta_n \to 0} \frac{1}{\Delta_n} \int_{R^1} (x_{n+1} - x_n) P^{(n)}(\mathbf{t}_n, \mathbf{x}_n; t_{n+1}, dx_{n+1}) = \tilde{a}(\mathbf{t}_n, \mathbf{x}_n) \tag{4}$$

$$\lim_{\Delta_n \to 0} \frac{1}{\Delta_n} \int_{R^1} (x_{n+1} - x_n)^2 P^{(n)}(\mathbf{t}_n, \mathbf{x}_n; t_{n+1}, dx_{n+1}) = \tilde{b}^2(\mathbf{t}_n, \mathbf{x}_n), \tag{5}$$

where $\Delta_n = t_{n+1} - t_n$, and $P^{(n)}(\mathbf{t}_n, \mathbf{x}_n; t_{n+1}, A) = P(X_{t_{n+1}} \in A \mid \mathbf{X}_n)$.

We shall use the following notation for the conditional mathematical expectation and variance

$$\mu(\mathbf{t}_{n+1}, \mathbf{X}_n) = E(X_{t_{n+1}} \mid \mathbf{X}_n)$$

$$\sigma^2(\mathbf{t}_{n+1}, \mathbf{X}_n) = E((X_{t_{n+1}} - \mu(\mathbf{t}_{n+1}, \mathbf{X}_n))^2 \mid \mathbf{X}_n)$$

Throughout the paper we assume that there exist the right hand side derivatives

$$a(\mathbf{t}_{n+1}, \mathbf{x}_n) = \frac{\partial^+}{\partial t_{n+1}} \mu(\mathbf{t}_{n+1}, \mathbf{x}_n), \; b^2(\mathbf{t}_{n+1}, \mathbf{x}_n) = \frac{\partial^+}{\partial t_{n+1}} \sigma^2(\mathbf{t}_{n+1}, \mathbf{x}_n)$$

The relations between a, $\tilde{a}$ and $\tilde{b}^2$, b^2 are evident, namely by virtue of (4), (5) we have

$$\tilde{a}(\mathbf{t}_n, \mathbf{x}_n) = \lim_{\Delta_n \to 0+} \frac{1}{\Delta_n} (\mu(\mathbf{t}_{n+1}, \mathbf{x}_n) - \mu(\mathbf{t}_n, t_n, \mathbf{x}_n)) = a(\mathbf{t}_n, t_n, x_n) \tag{6}$$

$$\tilde{b}^2(\mathbf{t}_n, \mathbf{x}_n) = \lim_{\Delta_n \to 0+} \frac{1}{\Delta_n} E(X_{t_{n+1}} - X_{t_n})^2 \mid \mathbf{X}_n = \mathbf{x}_n) = \tag{7}$$

$$= \lim_{\Delta_n \to 0+} \frac{1}{\Delta_n} E((X_{t_{n+1}} - \mu(\mathbf{t}_{n+1}, \mathbf{x}_n))^2 \mid \mathbf{X}_n = \mathbf{x}_n) =$$

$$= \frac{\partial^+}{\partial t_{n+1}} \sigma^2(t_{n+1}, \mathbf{x}_n) \Big|_{t_{n+1} = t_n} = b^2(\mathbf{t}_n, t_n, \mathbf{x}_n).$$

For fixed n (the role of parameter n will be explained in examples) we shall consider the stochastic equation

$$X_t - X_{t_n} = \int_{t_n}^{t} \alpha(\mathbf{t}_n, s, \mathbf{X}_n)\, ds + \int_{t_n}^{t} \beta(\mathbf{t}_n, s, \mathbf{X}_n)\, dW_s, \; t > t_n, \tag{8}$$

with initial condition $\mathbf{X}_n$, where $\alpha(\mathbf{t}_n, \cdot, \mathbf{x}_n) \in C^0$, $\beta(\mathbf{t}_n, \cdot, \mathbf{x}_n) \in C^0$. For given $\mathbf{t}_n$, $\mathbf{X}_n$ the integrals in (8) are defined in a well-known usual sense, as in Ito's equation [3].

Property 1. *If* (8) *holds, then*

$$\alpha(\mathbf{t}_{n+1}, \mathbf{x}_n) = \frac{\partial^+}{\partial t_{n+1}} \mu(\mathbf{t}_{n+1}, \mathbf{x}_n) = a(\mathbf{t}_{n+1}, \mathbf{x}_n) \tag{9}$$

$$\beta^2(\mathbf{t}_{n+1}, \mathbf{x}_n) = \frac{\partial^+}{\partial t_{n+1}} \sigma^2(\mathbf{t}_{n+1}, \mathbf{x}_n) = b^2(\mathbf{t}_{n+1}, \mathbf{x}_n) \tag{10}$$

Proof. It follows from (8) and the general properties of stochastic integrals that for $\Delta > 0$

$$a(\mathbf{t}_{n+1}, \mathbf{x}_n) = \lim_{\Delta \to 0} \frac{1}{\Delta} (\mu(\mathbf{t}_n, t_{n+1} + \Delta, \mathbf{x}_n) - \mu(\mathbf{t}_{n+1}, \mathbf{x}_n)) =$$

$$= \lim_{\Delta \to 0} \frac{1}{\Delta} E((X_{t_{n+1}+\Delta} - X_{t_{n+1}}) \mid \mathbf{X}_n) = \lim_{\Delta \to 0} \frac{1}{\Delta} \int_{t_{n+1}}^{t_{n+1}+\Delta} \alpha(\mathbf{t}_n, s, \mathbf{x}_n)\, ds = \alpha(\mathbf{t}_{n+1}, \mathbf{x}_n),$$

$$b^2(\mathbf{t}_{n+1}, \mathbf{x}_n) = \lim_{\Delta\to 0} \frac{1}{\Delta} (\sigma^2(\mathbf{t}_n, t_{n+1} + \Delta, \mathbf{x}_n) - \sigma^2(\mathbf{t}_{n+1}, \mathbf{x}_n)) =$$

$$= \lim_{\Delta\to 0} \frac{1}{\Delta} E((X_{t_{n+1}+\Delta} - \mu(\mathbf{t}_n, t_{n+1} + \Delta, \mathbf{x}_n))^2 -$$

$$- (X_{t_{n+1}} - \mu(\mathbf{t}_{n+1}, \mathbf{x}_n))^2) \mid \mathbf{X}_n = \mathbf{x}_n) =$$

$$= \lim_{\Delta\to 0} \frac{1}{\Delta} E((X_{t_{n+1}+\Delta} - X_{t_{n+1}})^2 \mid \mathbf{X}_n = \mathbf{x}_n) =$$

$$= \lim_{\Delta\to 0} \frac{1}{\Delta} E\left(\left(\int_{t_{n+1}}^{t_{n+1}+\Delta} \beta(\mathbf{t}_n, s, \mathbf{X}_n)\, dW_s\right| \mathbf{X}_n = \mathbf{x}_n\right) =$$

$$= \lim_{\Delta\to 0} \frac{1}{\Delta} \int_{t_{n+1}}^{t_{n+1}+\Delta} \beta^2(\mathbf{t}_n, s, \mathbf{x}_n)\, ds = \beta^2(\mathbf{t}_{n+1}, \mathbf{x}_n), \quad \text{q.e.d.}$$

By virtue of Property 1 equation (8) can be written in the following form

$$X_t = \mu(\mathbf{t}_n, t, \mathbf{X}_n) + \int_{t_n}^{t} b(\mathbf{t}_n, s, \mathbf{X}_n)\, dW_s \tag{8'}$$

Equation (8) has a form very similar to (1). The relationship between the coefficients of (1) and (8) are given by (6), (7).

We shall investigate the properties of the solutions of (8').

Property 2. *Equation* (8) *has the unique strong solution.*

Proof. For fixed $\mathbf{t}_n$, $\mathbf{X}_n$ equation (8) can be treated as a special case of (1), namely for

$$\tilde{a}(t, x) = \tilde{a}(t) = a(\mathbf{t}_n, t, \mathbf{x}_n)$$

$$\tilde{b}(t, x) = \tilde{b}(t) = b(\mathbf{t}_n, t, \mathbf{x}_n)$$

where $\mathbf{t}_n$, $\mathbf{X}_n$ are parameters, which have no impact on the proof of the theorem of existence and uniqueness of the solution of (1). In such a situation

$$\tilde{a}(t, x) - \tilde{a}(t, y) = \tilde{b}(t, x) - \tilde{b}(t, y) = 0.$$

Moreover, $a(\mathbf{t}_n, \cdot, \mathbf{x}_n)$, $b(\mathbf{t}_n, \cdot, \mathbf{x}_n)$ being continuous in the closed interval $[0, t]$, they are bounded. Thus all the assumptions of the theorem of existence and uniqueness of the solution of (1) are satisfied. q.e.d.

Now we shall investigate properties of (8) for the case of linear coefficients.

Let us denote $k_{ij} = E(X_{t_i} X_{t_j}) = K(t_i, t_j)$, $K^{(n)} = \det[k_{ij}]_{i, j=1,\dots,n}$, $K_{ij}^{(n)}$ — the cofactor of k_{ij}, $\mathbf{x}_{n(-i)} = (x_j, j = 1, \dots, n, j \neq i)$, $\mathbf{X}_{n(-i)} = (X_{t_j}, j = 1, \dots, n, j \neq i)$.

It is well known that for Gaussian processes

$$\mu_{n(-i)} = E(X_{t_i} \mid \mathbf{X}_{n(-i)}) = -\frac{1}{K_{ii}^{(n)}} \sum_{\substack{j=1\\ j\neq i}}^{n} K_{ij}^{(n)} X_{t_j} \tag{9}$$

(10) $$\sigma^2_{n(-i)} = E((X_{t_i} - \mu_{n(-i)})^2 \mid \mathbf{X}_{n(-i)}) = \frac{K^{(n)}}{K^{(n)}_{ii}}$$

under the natural assumption $K^{(n)} > 0,\ n > 1$.

If moreover the covariance function K of a Gaussian process has one-hand side derivatives the considered process is a quasi-diffusion process [5].

In our considerations we shall use a Lipschitz condition: there exist $L,\ \Delta_0$, such that for $\Delta < \Delta_0$

(11) $$| K(t_i + \Delta, t_j) - K(t_i, t_j) | < L\Delta.$$

Our further reasoning will be based on (see [6])

Property 3. *If $E(X_t^4) < \infty$, (9)—(10) for $n > 1$ and (11) hold, then $\{X_t, t \in T_0\}$ is a Gaussian process. If (9), (10) are satisfied only for $n \leqslant N + 1$ then only $\mathbf{X}_N$ has a Gaussian distribution.*

Now we are going to prove

Property 4. *Let for some n equation (8′) with initial condition $\mathbf{X}_n$ be satisfied and assumptions of Property 3 hold for $n' \leqslant n + 1$. Then the solution of (8′) has a Gaussian distribution and can be written in the following form*

(12) $$X_t = \mu(\mathbf{t}_n, t, \mathbf{X}_n) + Y_t,\ t > t_n,$$

where $\{Y_t,\ t > t_n\}$ is a Gaussian martingale such that $E(Y_t) = 0,\ E(Y_t^2) = \sigma^2(\mathbf{t}_n, t)$.

Proof. By virtue of Property 3, X_n has a Gaussian distribution, thus every linear combination of $X_{t_1}, \ldots, X_{t_n}$ has also a Gaussian distribution.

Let us put for $t > t_n$

$$Y_t = \int_{t_n}^{t} \left(\frac{\partial}{\partial s}\sigma^2(\mathbf{t}_n, s)\right)^{1/2} dW_s$$

Then it follows from the general properties of the Ito integral that $\{Y_t,\ t > t_n\}$ is a Gaussian martingale [7] and it is obvious that $E(Y_t) = 0$. Next

$$E(Y_t^2) = E\left[\int_{t_n}^{t} \left(\frac{\partial}{\partial s}\sigma^2(\mathbf{t}_n, s)\right)^{1/2} dW_s\right]^2 = \int_{t_n}^{t} \frac{\partial}{\partial s}\sigma^2(\mathbf{t}_n, s)\, ds = \sigma^2(\mathbf{t}_n, t).$$

Moreover, Y_t does not depend on X_n.

Thus (12) is satisfied and X_t given by (12) has a Gaussian distribution.

Remark 1. Equations (1) and (8) have a very similar form. There is an insignificant difference between the interpretation of the coefficients of these equations. In equation (8) the coefficients are strictly connected with differentials of μ and σ^2, in equation (1) we take these differentials at the point $t_2 = t_1$. This small difference implies important consequences. It is easier to solve (8) than (1). In the considered case of linear coefficients (8) is equivalent to (8′) and the solution of (8′) is given by (12).

Remark 2. Let $\{X_t, t \in T_0\}$ be an n-Markov process [4] or, in other words an n-composed Markov process [2]. It is known that in this case we can replace the investigation of $\{X_t,\ t \in T_0\}$ by the investigation of some n-dimensional

Markov process. Consequently we get a system of n equations of the form (1). The method proposed in the present paper gives for an n-Markov process only one equation with n initial conditions which appear as parameters in equation (8).

Remark 3. The assumption of Property 4 that conditions (9), (10) are satisfied for $n' \leqslant n+1$ can be replaced by the following assumptions: (9), (10) are satisfied for $n+1$ and $\mu \in C^0$, $\sigma^2 \in C^0$. The assumption $\mu \in C^0$ means that

$$(13) \qquad \lim_{t_{i+1} \to t_i} E(X_{t_n} \mid \mathbf{X}_{n-1}) \mid x_{t_{i+1}} = x_{t_i} =$$

$$= E(X_{t_n} \mid X_{t_1}, \ldots, X_{t_i}, X_{t_{i+2}}, \ldots, X_{t_{n-1}}), \quad i+2 \leqslant n-1$$

Note that the function $\mu(\mathbf{t}_n, \mathbf{x}_{n-1})$ is not defined on the hyperplanes $H_i : t_i = t_{i-1}, x_i \neq x_{i-1}, i = 2, \ldots, n-1$, since by putting $t_i = t_{i-1}, x_i \neq x_{i-1}$ we reach a contradiction. Thus all considerations concern only the values outside H_i.

Now we are going to give some examples. In all the examples we shall assume that $E(X_t^4) < \infty$ and (11) holds.

Example 1. Equation (8) for linear coefficients has a strict connexion with ARMA models [1]. Solution (12) of equation (8) explains this fact in an obvious manner.

Example 2. Let $\{X_t, t \in T_0\}$ be a Markov process and (9), (10) hold for $n = 2$. By virtue or Property 4 the solution of (8) is

$$X_t = \frac{K(t_1, t)}{K(t_1, t_1)} X_{t_1} + Y_t,$$

where $\{Y_t, t > t_1\}$ is a Gaussian martingale such that $E(Y_t) = 0$ and $E(Y_t^2) = (K(t, t)\, K(t_1, t_1) - K^2(t_1, t))\, (K(t_1, t_1))^{-1}$, and X_t has a Gaussian distribution.

Example 3. Let $K(t_1, t_2) = 1 + t_1 t_2$. Then $K^{(2)} = (t_2 - t_1)^2$, $K^{(3)} = 0$. We assume that (8)—(10) are satisfied for $n = 3$.

Then the solution of (8) is

$$X_t = \frac{t - t_1}{t_2 - t_1} X_{t_2} - \frac{t - t_2}{t_2 - t_1} X_{t_1}.$$

It is obvious that the function

$$\mu(\mathbf{t}_3, \mathbf{x}_2) = \frac{t_3 - t_1}{t_2 - t_1} x_2 - \frac{t_3 - t_2}{t_2 - t_1} x_1$$

is continuous outside the hyperplane: $t_1 = t_2$, $x_1 \neq x_2$ and the conditional variance $\sigma^2(\mathbf{t}_3, \mathbf{x}_2) = 0$. The process considered is a 2-Markov process. This process was obtained in another manner in [4], Example 4.8.

Example 4. Let $K(t_1, t_2) = 1 + t_1 + t_1 t_2$ and (9), (10) hold. Then taking into account (9) and subtracting two times nth row from $(n+1)$st row we get

$$K^{(n-1)}E(X_{t_n} \mid \mathbf{X}_{n-1}) = -(t_2-t_1)(t_3-t_2)(t_4-t_3)\dots(t_{n-1}-t_{n-2}) \times$$

$$\times \begin{vmatrix} 1+t_1+t_1^2 & 1+t_1+t_1t_2 & 1+t_1+t_1t_3 & 1+t_1+t_1t_4 & \cdots & 1+t_1+t_1t_n \\ t_1 & 1+t_2 & 1+t_3 & 1+t_4 & \cdots & 1+t_n \\ t_1 & t_2 & 1+t_3 & 1+t_4 & \cdots & 1+t_n \\ t_1 & t_2 & t_3 & 1+t_4 & \cdots & 1+t_n \\ \cdot & \cdot & \cdot & \cdot & \cdots & \cdot \\ \cdot & \cdot & \cdot & \cdot & \cdots & \cdot \\ \cdot & \cdot & \cdot & \cdot & \cdots & \cdot \\ X_{t_1} & X_{t_2} & X_{t_3} & X_{t_4} & \cdots & 0 \end{vmatrix} =$$

$$= -(t_2-t_1)(t_3-t_2)(t_4-t_3)\dots(t_{n-1}-t_{n-2}) \times$$

$$\times \begin{vmatrix} 1+t_1+t_1^2 & 1+t_1+t_1t_2 & 1+t_1+t_1t_3 & 1+t_1+t_1t_4 & \cdots & 1+t_1+t_1t_n \\ t_1 & 1+t_2 & 1+t_3 & 1+t_4 & \cdots & 1+t_n \\ 0 & -1 & 0 & 0 & \cdots & 0 \\ 0 & 0 & -1 & 0 & \cdots & 0 \\ \cdot & \cdot & \cdot & \cdot & \cdots & \cdot \\ \cdot & \cdot & \cdot & \cdot & \cdots & \cdot \\ \cdot & \cdot & \cdot & \cdot & \cdots & \cdot \\ X_{t_1} & X_{t_2} & X_{t_3} & X_{t_4} & \cdots & 0 \end{vmatrix}$$

and

$$K^{(n)} = (t_2-t_1)(t_3-t_2)\dots(t_n-t_{n-1})(1+t_n+t_1t_n).$$

Therefore

$$E(X_{t_n} \mid \mathbf{X}_{n-1}) =$$

$$= \frac{1+t_n+t_1t_n}{1+t_{n-1}+t_1t_{n-1}} X_{t_{n-1}} - \frac{t_n-t_{n-1}}{1+t_{n-1}+t_1t_{n-1}} X_{t_1}.$$

By virtue of (10)

$$\sigma^2(\mathbf{t}_n, \mathbf{x}_{n-1}) = (t_n - t_{n-1}) \frac{1+t_n+t_1t_n}{1+t_{n-1}+t_1t_{n-1}}$$

Thus the conditional moments of $X_{t_n} \mid \mathbf{X}_{n-1}$ depend only on X_{t_1}, $X_{t_{n-1}}$, t_1, t_{n-1}, t_n, i.e. on the preceding state $X_{t_{n-1}}$ and the initial condition (state) X_{t_1}.

The conditional moments depend on two preceding states. Then we take $n = 2$ in (8). Let the assumptions of Property 4 be satisfied. Then the solution of (8) is

$$X_t = \frac{1+t+t_1t}{1+t_2+t_1t_2} X_{t_2} - \frac{t-t_2}{1+t_2+t_1t_2} X_{t_1} + Y_t, \tag{14}$$

where $\{Y_t,\ t > t_2\}$ is a Gaussian martingale such that $E(Y_t) = 0$ and

$E(Y_t^2) = (t - t_2)(1 + t + t_1 t)(1 + t_2 + t_1 t_2)^{-1}$, and X_t has a Gaussian distribution.

It is clear that the conditional mean value μ is continuous. Condition (13) has the following form

$$\lim_{t_2 \to t_1} E(X_{t_3} \mid X_{t_2}, X_{t_1}) \mid_{X_{t_2} = X_{t_1}} = \frac{1 + t_1 + t_1 t_3}{1 + t_1 + t_1^2} X_{t_1} = E(X_{t_3} \mid X_{t_1})$$

and is obviously satisfied.

Example 5. Let $K(t_1, t_2) = 1 + t_1 + t_2$ for $0 < t_1 < t_2 < 2$ and conditions (9), (10) hold. This example is similar to Example 4.

By virtue of (9) after transformations similar to those in Example 4 we get

$$E(X_{t_n} \mid \mathbf{X}_{n-1}) = \frac{1}{2 - t_n + t_1}((2 - t_n + t_1) X_{t_{n-1}} - (t_n - t_{n-1}) X_{t_1}).$$

It follows from (10) that

$$\sigma^2(\mathbf{t}_n, \mathbf{x}_{n-1}) = \frac{2 - t_n + t_1}{2 - t_{n-1} + t_1}(t_n - t_{n-1}).$$

Thus the conditional moments of $X_{t_n} \mid \mathbf{X}_{n-1}$ depend only on X_{t_1}, $X_{t_{n-1}}$, t_1, t_{n-1}, t_n. We take $n = 2$ in (8). Let the assumptions of Property 4 be satisfied. Then the solution of (8) with initial condition $\mathbf{X}_2$ is

$$X_t = \frac{2 - t + t_1}{2 - t_2 + t_1} X_{t_2} - \frac{t - t_2}{2 - t_2 + t_1} X_{t_1} + Y_t,$$

where $\{Y_t, 0 < t < 2\}$ is a Gaussian martingale such that $E(Y_t) = 0$ and $E(Y_t^2) = (t - t_2)(2 - t + t_1)(2 - t_2 + t_1)^{-1}$, and X_t has a Gaussian distribution.

It is clear that the conditional mean value μ is continuous. Condition (13) has the following form

$$\lim_{t_2 \to t_1} E(X_{t_3} \mid X_{t_1}, X_{t_2}) \mid_{X_{t_2} = X_{t_1}} = (1 + t_1 - t_3) X_{t_1} = E(X_{t_3} \mid X_{t_1})$$

and is obviously satisfied.

REFERENCES

1. Anderson, O. D., *Time Series Analysis and Forecasting.* London, Wiley, 1976.
2. Doob, J. L., *Stochastic Processes.* New York, Wiley, 1953.
3. Gihman, I. I., Skorohod, A. V., *Stochastic Differential Equations.* Kiev, Naukova Dumka, 1968. (Russian)
4. Meditch, J.S., *Stochastic Optimal Linear Estimation and Control.* New York, McGraw-Hill 1969.
5. Plucińska, A., *Remarks on prospective equations.* Teor. Verojatnost. i Primenen., **25**, 350—358 (1980).
6. Plucińska, A., *On a stochastic process determined by the conditional mean value and the conditional variance.* (To appear)
7. Stroock, D.W., Varadhan, S.R.S., *Multidimensional Diffusion Processes.* New York, Springer 1979.

N. U. PRABHU

Wiener-Hopf factorization of Markov semigroups — I. The countable state space case

Wiener-Hopf factorization techniques play an important role in the theory of random walks (sums of independent and identically distributed random variables). In earlier extensions of this factorization to continuous time, the results obtained were for the Fourier exponent of processes with stationary independent increments (Lévy processes). In 1972 I obtained this factorization as a property of the infinitesimal generator of the convolution semigroup associated with Lévy processes. In this paper I extend this result to Markov processes with an ordered countable state space having finite transition rates, the birth-death process being a special case.

1. Introduction

Let $\{X(t), t \geqslant 0\}$ be a Markov process whose state space is the real line, and $\{T_t, t \geqslant 0\}$ the associated transition semigroup. Let A be the infinitesimal generator fo T_t. We seek a factorization of the form

$$\theta I - A = (\theta_1 I - A_+)(\theta_2 I - A_-) \quad (\theta > 0) \tag{1}$$

where I is the identity operator, A_+ and A_- are generators of Markov processes with non-decreasing and non-increasing sample functions respectively, and the constants θ_1, θ_2 and the generators A_+, A_- depend on θ. In the special case where X is a Lévy process (with stationary independent increments) I established this result about ten years ago (Prabhu [7]). Recently, Rogers and Williams [8] have obtained a factorization for Markov processes; see also Barlow, Rogers and Williams [2]. However, their technique is very different from the one I used for the Lévy process case, and they do not refer to my earlier result. In this paper I present a factorization for continuous time Markov processes with an ordered countable state space having finite transition rates, the birth-death process being a special case. My approach is similar to the one I used for Lévy processes. In related work, Kaspi [5] established a factorization for continuous time Markov additive processes, extending the discrete time results of Arjas and Speed [1]. For sub-stochastic matrices Dinges [3] obtained a Wiener-Hopf factorization. The fluctuation theory of discrete time Markov chains has recently been investigated by Kulkarni and Prabhu [6]. Stone ([9], 10]) has dealt with the supremum of Markov and semi-Markov chains.

The generators A_+ and A_- in the factorization (1) for Lévy processes are constructed out of the processes Z_+ and Z_- defined as follows. Let

(2) $$M(t) = \sup_{0 \leqslant s \leqslant t} X(s),$$

L_+ the local time of the process $M(t) - X(t)$ at zero (unique up to a multiplicative constant) and T_+ its inverse. Let $Z_+(\tau)$ be the process $\{T_+(\tau), M(T_+(\tau))\}$, and $Z_-(\tau)$ the process obtained similarly, starting with the infimum functional of X. Fristedt [4] proved that Z_+ and Z_- are Lévy processes, and obtained their Laplace-Fourier exponents. Because of their importance in the factorization (1) I called them Fristedt processes (Prabhu [7]). When X is a Markov process, Z_+ and Z_- turn out to be Markov additive processes, but in the factorization (1) only one of their generators occurs. We shall prove these statements in the case where X has an ordered countable state space.

Accordingly, let $\{X(t),\ t \geqslant 0\}$ be a time-homogeneous Markov process with the state space $\{0, 1, 2, \ldots\}$, and let

(3) $$P_{jk}(t) = P\{X(s+t) = k \mid X(s) = j\}$$

denote its transition probabilities, with $P_{jk}(0) = \delta_{jk}$ and $P_{jk}(t) \to \delta_{jk}$ as $t \to 0+$. We assume that $\sum_{k=0}^{\infty} P_{jk}(t) = 1$, the limits

(4) $$q_{jk} = \lim_{t \to 0+} \frac{P_{jk}(t) - \delta_{jk}}{t}$$

exist and $q_{jk} > 0$ $(k \neq j)$ and $q_{jj} > -\infty$. Then $q_{jj} = -\sum_{k \neq j} q_{jk} > 0$. The matrix $Q = (q_{jk})$ is the generator matrix of the process X. Our first main result is the verson of (1) expressed in terms of generator matrices Q_+ and Q_-. We identify the matrix Q_+ in terms of the Fristedt process Z_+ as follows. Let M be the supremum functional (2) and

(5) $$\zeta = \{t \geqslant 0 : M(t) - X(t) = 0\}$$

(6) $$L_+(t) = \text{Lebesgue measure of } \zeta \cap (0, t]$$

(7) $$T_+(\tau) = \inf\{t : L_+(t) > \tau\}.$$

As before $Z_+(\tau) = \{T_+(\tau), M(T_+(\tau))\}$. Our second main result shows the connection between Q_+ and the generator of Z_+. We state these results as follows.

Theorem 1. *Given the Markov process X described above with the generator matrix Q, there exist two other generator matrices Q_+ and Q_-, corresponding to Markov processes with the same state space but having non-decreasing and non-increasing sample functions respectively, and such that for $\theta > 0$*

(8) $$I - \theta^{-1} Q = (I - Q_-)(I - Q_+)$$

where I is the identity matrix.

Theorem 2. *For the Markov process X described above, the process Z_+ is a Markov-additive one with the state space $[0, \infty) \times [0, \infty)$. If A_+ is its generator, then for any bounded sequence $u = \{u(j),\ j = 0, 1, 2, \ldots\}$ and $\theta > 0$ we have*

(9) $$(A_+ e^{-\theta t} u)(j) = -\theta e^{-\theta t} (C_+(I - Q_+) u)(j)$$

where C_+ is a diagonal matrix.

In the case of the birth-death process the matrices Q_+ and Q_- are the generators of a pure birth process and a pure death process respectively. The Markov component of the process Z_+ in this case is a pure birth process.

2. Some preliminary results

For the Markov process X defined by (3)—(4), let

(10) $$T_i = \inf\{t \geq 0 : X(t) \geq i\}$$

be the first entrance time into the set $\{i, i+1, \dots\}$. For $\theta > 0$ let

(11) $$\zeta^i_{jk}(\theta) = E_j[e^{-\theta T_i};\quad X(T_i) = k]\ (k \geq i \geq j)$$

(12) $$\eta^i_{jk}(\theta) = \int_0^\infty \theta e^{-\theta t} P_j\{T_i > t,\ X(t) = k\}\, dt\ (i > \max(j, k))$$

(13) $$\beta_{jk}(\theta) = \frac{1}{\theta} \sum_{l \leq k} q_{jl} \zeta^k_{lk}(\theta)\ (k \leq j).$$

For convenience we shall write these as ζ^i_{jk}, η^i_{jk}, β_{jk} respectively. Note that $\zeta^j_{jk} = \delta_{jk}$. Finally, let

(14) $$R_{jk} \equiv R_{jk}(\theta) = \int_0^\infty e^{-\theta t} P_{jk}(t)\, dt.$$

We have then the following results.

Proposition 1. *We have*

(i) $\zeta^i_{jk} = \sum_{j \leq l < i} \zeta^l_{jl} \zeta^{l+1}_{lk}\ (k \geq i > j)$

(ii) $\eta^i_{jk} = \sum_{\max(j,k) \leq l < i} \zeta^l_{jl} \eta^{l+1}_{lk}\ (i > \max(j, k))$

(iii) $\theta R_{jk} = \sum_{l \geq \max(j,k)} \zeta^l_{jl} \eta^{l+1}_{lk}\ (j, k = 0, 1, \dots).$

Proof. (i) For $k \geq i > j$ let

$$B^i_{jk}(dt) = P_j\{T_i \in (t, t+dt],\ X(T_i) = k\}.$$

Then

$$B^i_{jk}(dt) =$$

$$= \sum_{j \leq l < i} \int_0^t P_j\{T_l \in (\tau, \tau + d\tau], X(T_l) = l, X(s) \leq l\ (T_l < s \leq t), X(t + dt) = k\} =$$

$$= \sum_{j \leq l < i} \int_0^t P_j\{T_l \in (\tau, \tau + d\tau], X(T_l) = l\}\, P_l\{X(s) \leq l\ (0 \leq s \leq t - \tau), X(t + dt - \tau) = k\}.$$

Thus

(15) $$B^i_{jk}(dt) = \sum_{j \leq l < i} \int_0^t B^l_{jl}(d\tau)\, B^{l+1}_{lk}(dt - \tau).$$

Multiplying both sides of (15) by $e^{-\theta t}$ and integrating over $(0, \infty)$ we obtain (i).

(ii) For $i > \max(j, k)$ let

$$v^i_{jk}(t) = P_j\{T_i > t,\ X(t) = k\}.$$

We have

$$v^i_{jk}(t) =$$

$$= \sum_{j \leqslant l < i} \int_0^t P_j\{T_l \in (\tau, \tau + d\tau],\ X(T_l) = l,\ X(s) \leqslant l\,(T_l < s \leqslant t),\ X(t) = k\} =$$

$$= \sum_{\max(j,k) \leqslant l < i} \int_0^t P_j\{T_l \in (\tau, \tau, + d\tau], X(T_l) = l\}\, P_l\{X(s) \leqslant l\,(0 \leqslant s \leqslant t - \tau), X(t - \tau) = k\}.$$

Thus

$$v^i_{jk}(t) = \sum_{\max(j,k) \leqslant l < i} \int_0^t B^l_{jl}(d\tau)\, v^{l+1}_{lk}(t - \tau). \tag{16}$$

Taking transforms of both sides of (16) we obtain (ii).

(iii) Letting $i \to \infty$ in (ii) we obtain (iii) for all j, k.

Proposition 2. *We have*

(i) $\zeta^{j+1}_{jk} = \frac{1}{\theta} q_{jk} + \sum_{m \leqslant j} \beta_{jm} \zeta^{m+1}_{mk} \quad (k > j)$

(ii) $\eta^{j+1}_{jk} = \delta_{jk} + \sum_{k \leqslant m \leqslant j} \beta_{jm} \eta^{m+1}_{mk} \quad (k \leqslant j)$

(iii) $\beta_{jk} = \frac{1}{\theta} q_{jk} + \sum_{m < k} \beta_{jm} \zeta^{m+1}_{mk} \quad (k \leqslant j).$

Proof. (i) Considering all possibilities of the first transition in $(0, t]$ we find that for $k > j$

$$B^{j+1}_{jk}(dt) = e^{q_{jj}t} q_{jk}\, dt + \sum_{l < j} \int_0^t e^{q_{jj}t} q_{jl}\, d\tau\, B^{j+1}_{lk}(dt - \tau) \cdot$$

This gives

$$\zeta^{j+1}_{jk} = \frac{q_{jk}}{\theta - q_{jj}} + \sum_{l < j} \frac{q_{jl}}{\theta - q_{jj}} \zeta^{j+1}_{lk}$$

or

$$\theta \zeta^{j+1}_{jk} = q_{jk} + \sum_{l \leqslant j} q_{jl} \zeta^{j+1}_{lk}.$$

Using Proposition 1 (i) in this we find that

$$\theta \zeta^{j+1}_{jk} = q_{jk} + \sum_{l \leqslant j} q_{jl} \sum_{l \leqslant m \leqslant j} \zeta^m_{lm} \zeta^{m+1}_{mk} = q_{jk} + \sum_{m \leqslant j} \left(\sum_{l \leqslant m} q_{jl} \zeta^m_{lm} \right) \zeta^{m+1}_{mk} =$$

$$= q_{jk} + \theta \sum_{m \leqslant j} \beta_{jm} \zeta^{m+1}_{mk},$$

which is the desired result (i).

(ii) Again considering all possibilities of the first transition in $(0, t]$ we obtain for $k < j$

$$v_{jk}^{j+1}(t) = \sum_{l<j} \int_0^t e^{q_{jj}\tau} q_{jl} \, d\tau \, v_{lk}^{j+1}(t-\tau).$$

This gives

$$\theta \eta_{jk}^{j+1} = \sum_{l \leqslant j} q_{jl} \eta_{lk}^{j+1}.$$

Using Proposition 1 (ii) in this we find that

$$\theta_{jk}^{j+1} = \sum_{l \leqslant j} q_{jl} \sum_{\max(l,k) \leqslant m \leqslant j} \zeta_{lm}^{m} \eta_{mk}^{m+1} = \sum_{k \leqslant m \leqslant j} \left(\sum_{l \leqslant m} q_{jl} \zeta_{lm}^{m} \right) \eta_{mk}^{m+1} = \theta \sum_{k \leqslant m \leqslant j} \beta_{jm} \eta_{mk}^{m+1}.$$

This is (ii) for $k < j$. For $k = j$ we have as before

$$v_{jj}^{j+1}(t) = e^{q_{jj}t} + \sum_{l<j} \int_0^t e^{q_{jj}\tau} q_{jl} \, d\tau \, v_{lj}^{l+1}(t-\tau).$$

This gives

$$\theta \eta_{jj}^{j+1} = \theta + \sum_{l \leqslant j} q_{jl} \eta_{lj}^{j+1}.$$

Proceeding as before we obtain $\eta_{jj}^{j+1} = 1 + \beta_{jj} \eta_{jj}^{j+1}$, which agrees with (ii) for $k = j$.

(iii) In the definition of β_{jk} substituting for ζ_{lk}^{k} from Proposition 1 (i) we obtain

$$\beta_{jk} = \frac{1}{\theta} q_{jk} + \frac{1}{\theta} \sum_{l<k} q_{jl} \sum_{l \leqslant m < k} \zeta_{lm}^{m} \zeta_{mk}^{m+1} =$$

$$= \frac{1}{\theta} q_{jk} + \sum_{m<k} \left(\frac{1}{\theta} \sum_{l \leqslant m} q_{jl} \zeta_{lm}^{m} \zeta_{mk}^{m+1} \right) = \frac{1}{\theta} q_{jk} + \sum_{m<k} \beta_{jm} \zeta_{mk}^{m+1}$$

as was required to be proved.

3. The matrices Q_+ and Q_-

Let $Q_+ = (q_{jk}^+)$, $Q_- = (q_{jk}^-)$, where

$$q_{jk}^+ = \frac{\zeta_{jk}^{j+1}}{1 - \zeta_{j\cdot}^{j+1}} \quad (k > j)$$

(17)
$$= - \frac{\zeta_{j\cdot}^{j+1}}{1 - \zeta_{j\cdot}^{j+1}} \quad (k = j)$$

$$= 0 \qquad (k < j),$$

$$q_{jk}^- = 0 \qquad (k > j)$$

(18)
$$= \zeta_{j\cdot}^{j+1} + \beta_{jj}(1 - \zeta_{j\cdot}^{j+1}) \quad (k = j)$$

$$= \beta_{jk}(1 - \zeta_{k\cdot}^{k+1}) \qquad (k < j)$$

and

(19) $$\zeta_{j\cdot}^{j+1} = \sum_{k>j} \zeta_{jk}^{j+1}.$$

Clearly, $0 < q_{jk}^{+} < \infty$ $(k > j)$ and $0 < q_{jk}^{-} < \infty$ $(k < j)$. Also

(20) $$q_{jj}^{+} = -\sum_{k>j} q_{jk}^{+} < 0.$$

We shall prove a similar result for q_{jj}^{-}.

Proposition 3. *For* q_{jj}^{-} *defined by* (18) *we have*

(21) $$q_{jj}^{-} = -\sum_{k<j} q_{jk}^{-} < 0.$$

Proof. From Proposition 2 (i) we obtain

$$\zeta_{j\cdot}^{j+1} = \sum_{k>j} \zeta_{jk}^{j+1} = \frac{1}{\theta}\sum_{k>j} q_{jk} + \sum_{k>j}\sum_{m\leqslant j} \beta_{jm}\zeta_{mk}^{m+1} = -\frac{1}{\theta}\sum_{k\leqslant j} q_{jk} +$$

$$+\sum_{m\leqslant j} \beta_{jm} \sum_{k>j} \zeta_{mk}^{m+1} = -\frac{1}{\theta}\sum_{k\leqslant j} q_{jk} + \sum_{m\leqslant j} \beta_{jm}\left(\zeta_{m\cdot}^{m+1} - \sum_{m<k\leqslant j} \zeta_{mk}^{m+1}\right) =$$

$$= \sum_{m\leqslant j} \beta_{jm}\zeta_{m\cdot}^{m+1} - \sum_{k\leqslant j}\left(\frac{1}{\theta} q_{jk} + \sum_{m<k} \beta_{jm}\zeta_{mk}^{m+1}\right).$$

Using Proposition 2 (iii) in this we find that

$$\zeta_{j\cdot}^{j+1} = \sum_{m\leqslant j} \beta_{jm}\zeta_{m\cdot}^{m+1} - \sum_{k\leqslant j} \beta_{jk} = -\Sigma\beta_{jm}(1 - \zeta_{m\cdot}^{m+1}).$$

We have thus proved (21).

4. Proofs of the theorems

We are now in a position to prove the main results of the paper.

Proof of Theorem 1. From Proposition 1 (i) we have

$$\zeta_{jk}^{k} = \delta_{jk} + \sum_{j\leqslant l<k} \zeta_{jl}^{l}\zeta_{lk}^{l+1} \quad (k \geqslant j).$$

In matrix notation this can be written as

(22) $$(\zeta_{jk}^{k})\,[I - (\zeta_{jk}^{j+1})] = I.$$

Similarly Proposition 2 (ii) can be written as

(23) $$[I - (\beta_{jk})]\,(\eta_{jk}^{j+1}) = I.$$

From (22)—(23) we thus obtain

(24) $$(\zeta_{jk}^{k})^{-1} = I - (\zeta_{jk}^{j+1}), \quad (\eta_{jk}^{j+1})^{-1} = I - (\beta_{jk}).$$

Proposition 1 (iii) can be written as

(25) $$(\theta R_{jk}) = (\zeta_{jk}^{k})\,(\eta_{jk}^{j+1}).$$

Inverting this identity and using (24) we obtain

$$I - \theta^{-1}Q = [I - (\beta_{jk})]\,[I - (\zeta_{jk}^{j+1})]. \tag{26}$$

Now it is verified that

$$\delta_{jk} - \zeta_{jk}^{j+1} = (1 - \zeta_{j\cdot}^{j+1})\,(\delta_{jk} - q_{jk}^{+}),$$

and so

$$I - (\zeta_{jk}^{j+1}) = [(1 - \zeta_{j\cdot}^{j+1})\,\delta_{jk}]\,(I - Q_{+}). \tag{27}$$

Similarly

$$I - (\beta_{jk}) = (I - Q_{-})\left(\frac{\delta_{jk}}{1 - \zeta_{j\cdot}^{j+1}}\right). \tag{28}$$

Therefore

$$[I - (\beta_{jk})]\,[I - (\zeta_{jk}^{j+1})] = (I - Q_{-})\,(I - Q_{+}). \tag{29}$$

The required result thus follows from (26)—(29). It is clear that Q_+ and Q_- are the generator matrices of Markov processes with non-decreasing and non-increasing sample functions respectively.

Proof of Theorem 2. (i) Let $\mathcal{F}_t$ be the sigma-field generated by $X(s)$ $(0 \leqslant s \leqslant t)$. Then $M(t)$ is $\mathcal{F}_t$-measurable and $\{(X(t), M(t)),\ t \geqslant 0\}$ is a Markov process. Since L is an additive functional of (X,M), its inverse T_+ is a stopping time for (X,M). We have

$$T_+(\tau + h) = T_+(\tau) + \hat{T}_+(h) \tag{30}$$

where $\hat{T}_+(h)$ is computed from $\{(X(T_+(\tau)+t),\ M(T_+(\tau)+t)),\ t \geqslant 0\}$ by the same rule as $T_+(h)$ is from $\{X(t),\ M(t)\}$. Using the strong Markov property it follows that

$$P\{X(T_+(\tau+h)) = l,\ M(T_+(\tau+h)) = k \mid \mathcal{F}_{T_+(\tau)}\} =$$
$$= P\{X(T_+(h)) = l,\ M(T_+(h)) = k \mid X(0) = i,\ M(0) = j\}$$

if $X(T_+(\tau)) = i$, $M(T_+(\tau)) = j$. Since $X(T_+(\tau)) = M(T_+(\tau))$ for all $\tau \geqslant 0$ we can write the last identity as

$$P\{M(T_+(\tau+h)) = k \mid \mathcal{F}_{T_+(\tau)}\} = Q_h(j;k) \tag{31}$$

on $M(T_+(\tau)) = j$. This shows that the process $\{M(T_+(\tau)),\ \tau \geqslant 0\}$ is Markovian. Writing $\hat{M}(\tau) = M(T_+(\tau))$ and $\hat{\mathcal{F}}_\tau = \mathcal{F}_{T_+(\tau)}$ we can prove similarly that

$$P\{T_+(\tau+h) - T_+(\tau) \leqslant t,\ \hat{M}(\tau+h) = k \mid \hat{\mathcal{F}}_\tau\} = Q_h(j;k,t) \tag{32}$$

on $\hat{M}(\tau) = j$. This shows that $\{T_+(\tau),\ M(T_+(\tau))\}$ is a Markov additive process.

(ii) We have

$$Q_h(j;k,t) = (1 + q_{jj}h)\,\delta(t-h) + \sum_{l<j} q_{jl}B_{lj}^{j}(t-h) + o(h) \quad (k = j)$$
$$= q_{jk}h\delta(t-h) + \sum_{l<j} q_{jl}B_{lk}^{j}(t-h) + o(h) \quad (k > j)$$

where $\delta(x) = 0$ for $x < 0$ and $= 1$ for $x \geqslant 0$. For $t \geqslant 0$, $j = 0, 1, 2, \ldots$, let $f(t, j)$ be a bounded function such that for each fixed j, f is continuous and has a bounded continuous derivative $\partial f/\partial t$. The infinitesimal generator A_+ of $\{T_+(\tau),\ M(T_+(\tau))\}$ is given by

$$(A_+f)\,(t, j) = \lim_{h\to 0+} h^{-1} \sum_{k=0}^{\infty} \int_{0-}^{\infty+} [f(t+v, k) - f(t, j)]\, Q_h(j;\, k, \mathrm{d}v).$$

Substituting for $Q_h(j;\, k, \mathrm{d}v)$ in this and simplifying we obtain

$$(33) \qquad (A_+f)\,(t, j) = \frac{\partial f}{\partial t} + \sum_{k\geqslant j} q_{jk} f(t, k) + \sum_{k\geqslant j} \int_{0-}^{\infty+} f(t+v, k) \sum_{l<j} q_{jl} B^j_{lk}(\mathrm{d}v)$$

(iii) Choose $f(t, j) = \mathrm{e}^{-\theta t} u(j)$ where $\theta > 0$ and $u = \{u(j),\ j = 0,1, \ldots\}$ is a bounded sequence. Then (33) yields the result

$$(A_+f)\,(t, j) = -\theta\, \mathrm{e}^{-\theta t}\, u(j) + \mathrm{e}^{-\theta t} \sum_{k\geqslant j} \left(q_{jk} + \sum_{l<j} q_{jl} \zeta^j_{lk} \right) u(k) =$$

$$= -\left(\theta - \sum_{l\leqslant j} q_{jl} \zeta^j_{lj} \right) \mathrm{e}^{-\theta t}\, u(j) + \sum_{l>j} \left(q_{jk} + \sum_{l<j} q_{jl} \zeta^j_{lk} \right) \mathrm{e}^{-\theta t}\, u(k).$$

Now

$$\sum_{l\leqslant j} q_{jl} \zeta^j_{lj} = \theta \beta_{jj}$$

by definition, and

$$q_{jk} + \sum_{l<j} q_{jl} \zeta^j_{lk} = q_{jk} + \sum_{l<j} q_{jl} \sum_{l\leqslant m<j} \zeta^m_{lm} \zeta^{m+1}_{mk} = q_{jk} + \sum_{m<j} \left(\sum_{l\leqslant m} q_{jk} \zeta^m_{lm} \right) \zeta^{m+1}_{mk} =$$

$$= q_{jk} + \sum_{m<j} \theta \beta_{jm} \zeta^{m+1}_{mk} = \theta(1 - \beta_{jj})\, \zeta^{j+1}_{jk}$$

on account of Proposition 1 (i) and Proposition 2 (i). Therefore

$$(A_+f)\,(t, j) = -\theta\, \mathrm{e}^{-\theta t}\, (1 - \beta_{jj}) \left[u(j) - \sum_{k>j} \zeta^{j+1}_{jk} u(k) \right] =$$

$$= -\theta\, \mathrm{e}^{-\theta t}\, c^+_{jj} \left[u(j) - \sum_{k\geqslant j} q^+_{jk} u(k) \right]$$

where

$$(34) \qquad c^+_{jj} = (1 - \beta_{jj})\, (1 - \zeta^{j+1}_{j\cdot}) = 1 - q_{\bar{j}\bar{j}}.$$

Let $C_+ = (c^+_{jj} \delta_{jk})$. Then we can write

$$(35) \qquad (A_+f)\,(t, j) = -\theta\, \mathrm{e}^{-\theta t}\, (C_+(I - Q_+)\, u)\,(j),$$

which is the desired result.

5. The birth-death process

For this process we have

$$(36) \qquad q_{jj+1} = \lambda_j,\ q_{jj-1} = \mu_j,\ q_{jj} = -\lambda_j - \mu_j \quad (j \geqslant 0)$$

with $\mu_0 = 0$), all other q_{jk} being zero. It is clear that $\zeta^i_{jk} = \zeta^j_{ii}\delta_{ik}$. We shall (write ζ_{jk} for $\zeta^k_{jk}(k \geq j)$. It turns out that all relevant quantities can be expressed in terms of ζ_{jj+1}. We introduce the quantities $\pi_j (j \geq 0)$, where

$$\pi_0 = 1, \ \pi_j = \frac{\lambda_0 \lambda_1 \dots \lambda_{j-1}}{\mu_1 \mu_2 \dots \mu_j} \quad (j \geq 1). \tag{37}$$

Proposition 4. *For the birth-death process with* (36) *we have*

(i) $\beta_{jj-1} = \frac{1}{\theta}\mu_j, \ 1 - \beta_{jj} = \frac{\lambda_j}{\theta \zeta_{jj+1}}, \ \beta_{jk} = 0 \quad (k \neq j, \ k \neq j-1)$

(ii) $\zeta_{jk} = \zeta_{jj+1}\zeta_{j+1,j+2} \dots \zeta_{k-1,k} \quad (k > j)$

(iii) $\eta^{j+1}_{jk} = \frac{\theta \pi_k}{\lambda_j \pi_j} \zeta_{k,j+1} \quad (k \leq j)$.

Proof. (i) From the definition of β_{jk} we see that $\beta_{jk} = 0$ for $k < j-1$, and $\theta\beta_{jj-1} = \mu_j$. Proposition 2 (i) yields the relation

$$\zeta_{jj+1} = \frac{1}{\theta}\lambda_j + \beta_{jj}\zeta_{jj+1}.$$

(ii) Proposition 1 (i) gives $\zeta_{jk} = \zeta_{jk-1}\zeta_{k-1,k} \ (k > j)$. By induction we obtain ζ_{jk} for $k > j$ in terms of $\zeta_{lc+1} (j \leq l \leq k-1)$.

(iii) From Proposition 2 (ii) we find that

$$\eta^{l+1}_{lk} = \delta_{lk} + \beta_{ll}\eta^{l+1}_{lk} + \beta_{ll-1}\eta^{l}_{l-1,k}.$$

Using (i) we can write this as

$$\frac{\lambda_l}{\zeta_{ll+1}} \eta^{l+1}_{lk} = \theta\delta_{lk} + \mu_l \eta^l_{l-1,k} \quad (k \leq l) \tag{38}$$

for $l = k$ this gives

$$\eta^{k+1}_{kk} = \frac{\theta}{\lambda_k}\zeta_{kk+1} \tag{39}$$

which agrees with the desired result. Again, for fixed k, putting $l = k$ $k+1, \dots, j \ (> k)$ in (38) and multiplying the resulting identities together we obtain

$$\frac{\lambda_k \lambda_{k+1} \cdots \lambda_j}{\zeta_{kj+1}} \cdot \eta^{j+1}_{jk} = \theta \mu_{k+1}\mu_{k+2} \dots \mu_j$$

or

$$\eta^{j+1}_{jk} = \frac{\theta \mu_{k+1}\mu_{k+2} \cdots \mu_j}{\lambda_k \lambda_{k+1} \lambda_{k+2} \cdots \lambda_j} \cdot \zeta_{k,j+1} = \frac{\theta}{\lambda_j} \cdot \frac{\pi_k}{\pi_j} \zeta_{k,j+1}$$

This proves the desired result for $k > j$.

Theorem 1 in this case takes a simple form, as stated in

Theorem 3. *For the birth-death process with* (36) *we have*

(38)
$$\left(I-\frac{1}{\theta}\lambda_j\Delta-\frac{1}{\theta}\mu_j\Delta^{-1}\right)u(j)=$$
$$=\left[I-\frac{1}{\theta}\mu_j(1-\zeta_{j-1.j})\,\Delta^{-1}\right]\left[I-\frac{\zeta_{jj+1}}{1-\zeta_{jj+1}}\,\Delta\right]u(j)$$

where $\Delta^r u(j)=u(j+r)-u(j)$ $(r=+1,\ -1)$, *and* $Iu(j)=u(j)$.

Proof. The infinitesimal generator of the process is given by

(39)
$$(Au)(j)=(\lambda_j\Delta+\mu_j\Delta^{-1})\,u(j)$$

so that the operator on the left side of (38) equals $I-\theta^{-1}A$. Further,

(40)
$$q^+_{jj+1}=\frac{\zeta_{jj+1}}{1-\zeta_{jj+1}},\quad q^+_{jj}=-q^+_{jj+1}$$
$$q^-_{jj-1}=\theta^{-1}\mu_j(1-\zeta_{j-1.j}),\quad q^-_{jj}=-q^-_{jj-1},$$

all other q^+_{jk} and q^-_{jk} being zero. The desired result now follows from Theorem 1.

REFERENCES

1. Arjas, E., Speed, T.P., *Symmetric Wiener-Hopf factorization in Markov additive processes.* Z. Wahrsch. Verw. Gebiete, **26**, 105–118 (1973).
2. Barlow, M.T., Rogers, L. C., G., Williams, D., *Wiener-Hopf factorization for matrices.* In: *Séminaire de Probabilités XIV.* Lecture Notes in Math., Vol. 784, pp. 324–331. Berlin, Springer, 1980.
3. Dinges, H., *Wiener-Hopf Factorisierung für substochastische Übergangsfunktionen in angeordneten Räumen.* Z. Wahrsch. Verw. Gebiete, **11**, 152–164 (1969).
4. Fristedt, B., *Sample functions of stochastic processes with stationary independent increments.* In: Ney, P., Port, S. (Eds.), *Advances in Probability and Related Topics, Vol.* 3, pp. 241–396. New York, Marcel Dekker, Inc., 1974.
5. Kaspi, H., *On the symmetric Wiener-Hopf factorization for Markov additive processes.* Z. Wahrsch. Verw. Gebiete, **59**, 179–196 (1980).
6. Kulkarni, V. G., Prabhu, N. U., *A fluctuation theory for Markov chains.* Tech. Report No. 539, School of Operations Research and Industrial Engineering, Cornell University, 1982.
7. Prabhu, N. U., *Wiener-Hopf factorization for convolution semigroups.* Z. Wahrsch. Verw. Gebiete, **23**, 103–113 (1972).
8. Rogers, L. C.G., Williams, D., *Time-substitution based on additive functionals (Wiener-Hopf factorization for infinitesimal generators).* In: *Séminaire de Probabilités XIV.* Lecture Notes in Math., Vol. 784, pp. 332–342. Berlin, Springer, 1980.
9. Stone, L. D., *On the distribution of the maximum of a semi-Markov process.* Ann. Math. Statist., **39**, 947–956 (1968).
10. Stone, L. D., *On the distribution of the supremum functional for semi-Markov processes with continuous state-spaces.* Ann. Math. Statist., **40**, 844–853 (1969).

K. SARKADI

On the asymptotic behavior of the Shapiro-Wilk test

1. Introduction

This paper deals with the asymptotic behavior of the Shapiro-Wilk test [14] and of its asymptotic version.

The practical importance of the Shapiro-Wilk test for normality lies in the fact that this test has good power against many alternatives. In the light of Monte Carlo experiments [15, 4, 8] it is superior to most of the classical goodness of fit tests.

Let $X_1, ..., X_n$ be independent, identically distributed random variables with distribution function $F(x)$. Let $X_{1n} \leqslant ... \leqslant X_{nn}$ the rearrangement of $X_1, ..., X_n$ according to the order of magnitude. Let $F_0(x)$ a distribution function whose first two moments are 0 and 1, respectively.

Suppose we wish to test the hypothesis

$$H_0: F(x) = F_0[(x - \theta_1)/\theta_2]$$

$(\theta_2 > 0)$, where θ_1 and θ_2 are unspecified in the null hypothesis.

The Shapiro-Wilk test [14] can be defined by the test statistic

$$W = \left(\frac{\hat{s}}{s}\right)^2$$

(Here and in the sequel unimportant constant factors are neglected.) In this formula

$$s = \sqrt{\frac{\Sigma(X_i - \bar{X})^2}{n-1}}, \quad \bar{X} = \Sigma X_i/n, \quad \hat{s} = \Sigma c_{in} X_{in},$$

and the constants c_{in} are determined in such a way that

$$E(\hat{s} \mid H_0) = \theta_2, \quad D^2(\hat{s} \mid H_0) = \min.$$

Formulae and tables for the coefficients c_{in} are given in [14].

Low values of W indicate significance.

Let us introduce the following notation. We denote by $\mathcal{F}$ the set of all distribution functions $F(x)$ which are three times continuously differentiable, moreover, $F'(x) = f(x)$ is positive for any real x and $\lim_{x\to\pm\infty} x^2 f(x) = 0$. We denote by $\mathcal{F}_0$ the set of all distribution functions $F(x)$ such that $F(x)\in\mathcal{F}$, $F(x)$ has the first two moments 0 and 1, respectively, and its fourth moment exists.

Throughout the paper let $F_1(x)\in\mathcal{F}_0$ denote a distribution function which differs from $F_0(x)$. We use the notation $Q_i(x) = F_i^{-1}(x)$ for the inverse function of $F_i(x)$ $(i = 0, 1)$. The hypothesis $F(x) = F_1[(x-\theta_1)/\theta_2]$ will be denoted by H_1.

Further, if $F_0(x)\in\mathcal{F}_0$, we denote by $\tilde{s}$ the Bennett estimator of the scale parameter θ_2, see Eqs. (3.1) and (3.13) in [1],

$$\tilde{s} = \Sigma J\left(\frac{i}{n+1}\right) X_{in}/n,$$

where

$$J(x) = \frac{J_0(x)}{\int_0^1 J_0(y)\, Q_0(y)\, dy}, \tag{1.1}$$

$$J_0(x) = J_2(x)\int_0^1 J_1(y)\, dy - J_1(x)\int_0^1 J_2(y)\, dy,$$

$$J_1(x) = \frac{Q_0'''(x)}{[Q_0'(x)]^3} - \frac{2[Q_0''(x)]^2}{[Q_0'(x)]^4} \tag{1.2}$$

$$J_2(x) = \frac{Q_0''(x)}{[Q_0'(x)]^2} + \frac{Q_0(x)\, Q_0'''(x)}{[Q_0'(x)]^3} - \frac{2Q_0(x)\,[Q_0''(x)]^2}{[Q_0'(x)]^4}.$$

[1, (3.13)] is an alternative equivalent expression for $J(x) = J_4(x)$.

If H_0 holds, $\tilde{s}$ is an asymptotically efficient estimator of θ_2, [1, Eq. (3.13)] and thus, $\hat{s}$ and $\tilde{s}$ are asymptotically equivalent:

$$\tilde{s} = \hat{s} + o_p(n^{-1/2}) \tag{1.3}$$

We shall denote by μ_i the integral

$$\mu_i = \int_0^1 Q_i(x)\, J(x)\, dx, \quad i = 0, 1.$$

Notice that, by virtue of (1.1),

$$\mu_0 = 1. \tag{1.4}$$

The test based on the following statistic will be called the asymptotic version of the Shapiro-Wilk test [10, p. 323]

$$\tilde{W} = \left(\frac{\tilde{s}}{s}\right)^2$$

We denote by s_i^* the statistic

$$s_i^* = \Sigma Q_i\left(\frac{i}{n+1}\right) X_{in}/n, \quad i = 0, 1.$$

If H_i is valid, s_i^* is an asymptotically unbiased estimator of θ_2. The test statistic of DeWet and Venter [3] is

$$r = \frac{s_0^*}{s}$$

In the following sections we are dealing with the asymptotic behavior of $\widetilde{W}$ and W.

2. Testing for normality

In this section we suppose that

$$F_0(x) = \Phi(x) = (2\pi)^{-1/2}\int_{-\infty}^{x} \exp(-y^2/2)\, dy.$$

It is well known that in this case s is an asymptotically efficient estimator of θ_2, therefore (cf. (1.3))

$$s = \hat{s} + o_p(n^{-1/2}) = \tilde{s} + o_p(n^{-1/2}) \tag{2.1}$$

under H_0.

In this case (1.1) gives $J(x) = Q_0(x)$ (cf. [1], Example 3) and thus

$$\tilde{s} = s_0^*, \quad \widetilde{W} = r^2, \tag{2.2}$$

i.e. $\widetilde{W}$ is the test statistic of DeWet and Venter [3]. (2.1) implies that

$$W = 1 + o_p(n^{-1/2})$$

and

$$\widetilde{W} = r^2 = 1 + o_p(n^{-1/2}) \tag{2.3}$$

under the null hypothesis.

Remark 1. The proper asymptotic null distribution of $\widetilde{W} = r^2$ has been determined by DeWet and Venter [3].

Consider now the case of the alternative hypothesis H_1. From [12, Theorem 1] we obtain the following result.

Theorem 1. *If the assumptions of Theorem 1 of Shorack [16] are fulfilled, by $Q_0(x) = J(x)$ and $Q_1(x) = g(x)$ then, under the alternative hypothesis H_1.*

$$n^{1/2}(\widetilde{W} - \mu_1^2) = n^{1/2}(r^2 - \mu_1^2) \tag{2.4}$$

is asymptotically normally distributed with expectation 0 and positive variance

Proof. Below we show that the assertion on the positivity of the variance is true; the rest of the proof is a straightforward application of [12, Theorem 1].

Suppose the contrary, i.e. that the variance of the asymptotic distribution of (2.4) vanishes. Then $\widetilde{W} = \mu_2^1 + o_p(n^{-1/2})$, consequently, $s = \tilde{s} + o_p(n^{-1/2})$, or, by virtue of (2.2), $s = s_0^* + o_p(n^{-1/2})$ under H_1. Using a theorem of Csörgő and Révész [2, Lemma 5.8.4] we obtain that $s_1^* = s + o_p(n^{-1/2})$ under H_1 which implies that

$$s_1^* - s_0^* = o_p(n^{-1/2}). \tag{2.5}$$

The variance of the asymptotic distribution of $n^{1/2}(s_1^* - s_0^*)$ under H_1 can be written in the form

$$\int_0^1 (1-x)^{-2}\left[\int_x^1 [Q_1(y) - Q_0(y)]\, Q_1'(y)\,(1-y)\,\mathrm{d}y\right]^2 \mathrm{d}x$$

(see [1, Eq. (2.22)], [16, Eq. (12)]). Our assumptions assure that this expression is positive, i.e. (2.5) cannot hold. This proves the theorem.

Remark 2. The variance of the asymptotic distribution of the expression (2.4) can be evaluated according to [12, p. 247].

Remark 3. For an alternative way of proving Theorem 1 of [12] under slightly modified assumptions, we refer to Govindarajulu [6, Section (6)] and [7, Theorem 9.1].

Since $\mu_1 < 1$ by virtue of the Cauchy inequality, (2.3) and Theorem 1 imply that the test $\widetilde{W}$ is consistent. This is in accordance with [11] where it has been proved, under more general assumptions, that the DeWet-Venter test is consistent.

It is likely that the asymptotic behavior of W agrees with that of $\widetilde{W}$ even under the alternative hypothesis, and, accordingly, the W test for normality is consistent. However, it seems difficult to prove that Eq. (2.1) would suggest that the coefficients in the expressions of $\hat{s}$ and $\tilde{s}$ agree asymptotically. In fact, Tusnády [17] has proved the pointwise convergence $\lim_{n\to\infty} nc_{[nx],n} = Q_0(x)$; the proof is tedious. We do not know, however, whether this convergence is uniform or not and thus, it is not known whether the relation $\hat{s} = \tilde{s} + o_p(n^{-1/2})$ can be extended to the case of the alternative hypothesis or not.

Remark 4. The results concerning the DeWet-Venter statistic r^2 above are more generally valid even in the non-normal case. The extension is assured by the mentioned theorem of Csörgő and Révész [2, Lemma 5.8.4].

The same is true for the test of Shapiro and Francia [13] and for some related tests, see [12].

3. The case of non-normal null hypothesis

This section concerns the asymptotic behavior of $\widetilde{W}$ and W in the case when $F_0(x)$ is not normal. We suppose, however, that $F_0(x) \in \mathcal{F}_0$.

We have the folllowing result.

Theorem 2. *If the assumptions of Theorem 1 of Shorack [16] are fulfilled by $J(x)$ and $Q_i(x) = g(x)$ then, under the hypothesis H_i, $i = 0, 1$, $n^{1/2}(\widetilde{W} - \mu_i^2)$ is asymptotically normally distributed with expectation 0. The variance of the asymptotic distribution is positive for $i = 0$ ($\mu_0 = 1$).*

Proof. The first assertion follows again from [12, Theorem 1] in a straightforward way. Below we prove the validity of the assertion about the positivity of the variance.

Assuming that this variance is zero it follows that

$$s = \tilde{s} + o_p(n^{-1/2})$$

under the null hypothesis.

Arguments parallel to those used in the proof of Theorem 1 show that under H_0

$$s_0^* = \tilde{s} + o_p(n^{-1/2})$$

and further on that $J(x) = Q_0(s)$ $(0 < x < 1)$.

Consequently the density function of $F_0(x)$ takes on the form of Eq. (7) in [10], cf. also pp. 326—327 therein. Taking the assumption $F_0(x) \in \mathcal{F}_0$ into account this means that the distribution $F_0(x)$ is normal, in contradiction to our assumption. This proves the theorem.

Using (1.3) and the last theorem we obtain

Theorem 3. *Under H_0 the asymptotic distribution of $n^{1/2}(W - 1)$ is normal with expectation 0 and positive variance.*

Remark 5. The variances of the distributions mentioned in Theorems 2 and 3 can be calculated by the formulae given in [12, p. 247].

Now we prove

Theorem 4. *$\tilde{W}$ is inconsistent.*

Proof. We prove that $F_1(x)$ can be chosen in such a way that $\mu_1 = 1$ but $F_1(x)$ differs from $F_0(x)$.

In the proof of Theorem 2 we have shown that the identity $J(x) = Q_0(x)$ cannot hold. Now, if we take into account the assumptions on the moments of $F_0(x)$ and Eqs. (1.2) and (1.4) we see that

$$\int_0^1 (J(x) - Q_0(x))\,dx = \int_0^1 Q_0(x)\,[J(x) - Q_0(x)]\,dx = 0,$$

therefore both integrands cross the abscissa axis at least once in (0, 1). $Q_0(x)$ crosses it once. Consequently we can find a value x_0 for which $Q_0(x_0) = J(x_0) \neq 0$. For reasons of continuity we can then find four different open intervals I_{ij} $(i, j = 1, 2)$ such that

$$Q_0(x)\,[J(x) - Q_0(x)] > 0 \quad \text{if} \quad x \in I_{1j},$$
$$Q_0(x)\,[J(x) - Q_0(x)] < 0 \quad \text{if} \quad x \in I_{2j},$$
$$\operatorname{sgn} Q_0(x) = \operatorname{sgn} Q_0(x_0) \quad \text{if} \quad x \in I_{ji},\ i, j = 1, 2.$$

We suppose, further, that the endpoints of the four intervals I_{ij} lie in the interior of the interval [0, 1].

Now let us define the functions $Q_{ij}(x) \in \mathcal{F}$ in such a way that

$$Q_0(x)\,[Q_{i1}(x) - Q_0(x)] > 0 \quad \text{if} \quad x \in I_{i1},$$
$$Q_0(x)\,[Q_{i2}(x) - Q_0(x)] < 0 \quad \text{if} \quad x \in I_{i2},$$
$$Q_{ij}(x) = Q_0(x) \quad \text{if} \quad x \notin I_{ij},\ i, j = 1, 2.$$

Evidently, this choice of the functions $Q_{ij}(x)$ is possible.

Since $\int_0^1 Q_0(x)\,dx = 0$, the integrals $\int_0^1 Q_{i1}(x)\,dx$ and $\int_0^1 Q_{j2}(x)\,dx$ are of different sign. If we define

$$Q_{30}(x) = \lambda_1 Q_{12}(x) + (1 - \lambda_1)\,Q_{21}(x),$$
$$Q_{40}(x) = \lambda_2 Q_{11}(x) + (1 - \lambda_2)\,Q_{22}(x),$$

then an appropriate choice of $0 < \lambda_1, \lambda_2 < 1$ can ensure that

$$\int_0^1 Q_{i0}(x)\,\mathrm{d}x = 0, \quad i = 3, 4.$$

Further, an easy calculation shows that if

$$Q_i(x) = \lambda_i Q_{i0}(x) + (1 - \lambda_i)\, Q_0(x), \quad i = 3, 4,$$

then, if $0 < \lambda_3, \lambda_4 \leqslant 1$ are appropriately chos en (cf. (1.4)):

$$\frac{\int_0^1 Q_3(x)\, J(x)\,\mathrm{d}x}{\left[\int_0^1 [Q_3(x)]^2\,\mathrm{d}x\right]^{1/2}} < 1 < \frac{\int_0^1 Q_4(x)\, J(x)\,\mathrm{d}x}{\left[\int_0^1 [Q_4(x)]^2\,\mathrm{d}x\right]^{1/2}}.$$

Let us define

$$Q_5(x) = \lambda_5 Q_3(x) + (1 - \lambda_5)\, Q_4(x),$$

where $0 < \lambda_5 < 1$ is chosen in such a way that

$$\frac{\int_0^1 Q_5(x)\, J(x)\,\mathrm{d}x)}{\left[\int_0^1 [Q_5(x)]^2\,\mathrm{d}x\right]^{1/2}} = 1.$$

This choice is possible by continuity reasons. Then put

$$Q_1(x) = \frac{Q_5(x)}{\left[\int_0^1 [Q_5(x)]^2\,\mathrm{d}x\right]^{1/2}}.$$

It follows that $\mu_1 = 1$. Further, let $x_1 \in I_{11}$, $x_2 \in I_{12}$. Then $Q_0(x)\,[Q_1(x_1)\,Q_0(x_2) - - Q_1(x_2)\,Q_0(x_1)] > 0$, i.e. $Q_1(x)$ differs from $Q_0(x)$.

Therefore, the asymptotic distribution of $n^{1/2}(\widetilde{W} - 1)$ is normal with expectation 0 both under H_0 and H_1 and we know that this distribution is proper normal under H_0. If we fix the size of the test, the probability of the error of second kind tends to a positive limit with increasing n, i.e. the test is inconsistent.

It is again likely that W shares the asymptotic properties of its asymptotic version, $\widetilde{W}$, and, in particular, that W is inconsistent. However, in trying to prove this, we meet difficulties of the same kind as in the preceding section.

Remark 6. A case where the assumptions of the present paper are not fulfilled has been treated in [12, p. 251] where it was proved that W is inconsistent if the null hypothesis is the exponentiality.

4. Conclusion

It has been shown that the $\widetilde{W}$ test, the asymptotic version of the Shapiro-Wilk test, is inconsistent in general; the exception is the case of normal null distribution, where the $\widetilde{W}$ test is consistent. Heuristic arguments support the conjecture that the Shapiro-Wilk test shares the asymptotic properties of its asymptotic version.

Let be reminded, that the tests of Shapiro-Francia, DeWet-Venter and some related tests are consistent, irrespectively of the form of the null distribution [5, 9, 11].

REFERENCES

1. Chernoff, H., Gastwirth, J., Johns, M. V., *Asymptotic distribution of linear combinations of functions of order statistics.* Ann. Math. Statist., **38**, 52–72 (1967).
2. Csörgő, M., Révész, P., *Strong Approximations in Probability and Statistics.* London, Academic Press, 1981.
3. DeWet, T., Venter, J.H., *Asymptotic distribution of certain test criteria of normality.* South African Statist. J., **6**, 135–149 (1972).
4. Dyer, A. R., *Comparison of tests for normality with a cautionary note.* Biometrika, **61**, 185–189 (1974).
5. Gerlach, B., *A consistent correlation-type goodness of fit test.* Math. Operationsforsch. Statist. (Statist.), **10**, 427–452 (1979).
6. Govindarajulu, Z., *Asymptotic normality of linear combinations of functions of order statistics II.* Proc. Nat. Acad. Sci. U.S.A., **59**, 713–719 (1968).
7. Govindarajulu, Z., *Asymptotic normality of linear combinations of functions of order statistics in one and several samples.* In: Gnedenko, B. V., Puri, M. L., Vincze, I. (Eds.), *Nonparametrical Statistical Inference.* Colloq. Math. Soc. J. Bolyai, Vol. 32, pp. 315–349. Amsterdam, North-Holland, 1982.
8. Pearson, E. S., D'Agostino, R. B., Bowman, K. O., *Tests for departure from normality: Comparison of powers.* Biometrika, **64**, 231–246 (1977).
9. Sarkadi, K., *The consistency of the Shapiro-Francia test.* Biometrika, **62**, 445–450 (1975).
10. Sarkadi, K., *Characterization and testing for normality.* In: Gyires, B. (Ed.), *Analytic Function Methods in Probability Theory.* Colloq. Math. Soc. J. Bolyai, Vol. 21, pp. 317–329. Amsterdam, North-Holland, 1980.
11. Sarkadi, K., *On the consistency of some goodness of fit tests.* In: *Proc. Sixth Conference Probab. Theory (Braşov, 1979),* pp. 195–204. Bucureşti, Ed. Acad. R.S. România, 1981.
12. Sarkadi, K., *The asymptotic distribution of certain goodness of fit test statistics.* In: *The First Pannonian Sympos. Math. Statist.* Lecture Notes in Statistics, Vol. 8, pp. 245–253. New York, Springer, 1981.
13. Shapiro, S. S., Francia, R. S., *Approximate analysis of variance test for normality.* J. Amer. Statist. Assoc., **67**, 215–216 (1972).
14. Shapiro, S. S., Wilk, M. B., *An analysis of variance test for normality.* Biometrika, **42**, 591–611 (1965).
15. Shapiro, S. S., Wilk, M. B., Chen, N. J., *A comparative study of various tests for normality.* J. Amer. Statist. Assoc., **63**, 1343–1372 (1968).
16. Shorack, G. R., *Functions of order statistics.* Ann. Math. Statist., **43**, 412–427 (1972).
17. Tusnády, G., *Personal communication (cf. also* [10]*)* (1974).

W. SCHAAFSMA

Me and the anthropologist

For a statistician, physical anthropology is an outstanding area to acquire consultation experiences. The introduction (Section 0) describes the author's cooperation with the anthropologist G. N. Van Vark which has lasted for more than twenty years. The author and his students became motivated to do research in various areas among which (1) the problem of dimensionality in discriminant analysis (Section 1), (2) the accuracy of estimates for posterior probabilities (Section 2), (3) problems with missing data (Section 3). The paper is an attempt to transfer this motivation.

0. Introduction

Consultation experiences are of the utmost importance to the mathematical statistician and his science. They provide insights into the relevance of existing theory and they may stimulate the development of new theory and interest. Asymmetrical consultations where know-how flows almost exclusively from the statistician to his client, should not be avoided because they may pave the way for those where the exchange is better balanced. A beautiful example of the fruitfulness of the cooperation between dissimilar partners was given in the film "Me and the Colonel". The title of this paper is a reminiscence of the peculiar cooperation between a Polish Jew and a Polish warrior shown in those famous movies.

Is there a reason why so many statisticians have been interested in physical anthropology? The following exposition explains why physical anthropology is an outstanding area to acquire consultation experiences. Statistics is about drawing inferences concerning wholes from information concerning parts. Such inferences cannot be drawn with certainty. The ethics of statistics require respect for the data and for the process of drawing these inferences. The involved uncertainties should be made visible unless they are ignorable. Physical anthropology is an outstanding area of application because (1) it is very natural to have respect for the bones of one's ancestors (other reasons follow), (2) the actual information is often very limited, especially when dealing with the evolution of our own kind which is one of the main topics of physical anthropology.

Respect for the data can be obtained by learning about the difficulties involved in defining measurements and the accuracy which is needed to perform these measurements. One should note that anthropometrical observations can in principle be repeated by competing fellow-anthropologists.

The ethics of physical anthropology require that the objects of measurement are well preserved and placed at the colleague's disposal if a request is made. Hence physical anthropologists will bestow great care on their data.

The available data are often very limited because (i) sample sizes can be small and (ii) skeletal remains are frequently incomplete. It is interesting to emphasize that sample sizes for e.g. prehistoric populations like Homo sapiens neanderthaliensis cannot be increased at will. This underlines the necessity to express the uncertainties, and the importance of the choice of procedure. It makes very clear that statistics is not some kind of asymptotics (though asymptotic arguments may be of interest as a last resource). Another interesting phenomenon is that there are no natural limitations to the number p of measurements to be taken e.g. from complete skulls. This raises interest in an unusual kind of asymptotics where p tends to infinity whereas the sample sizes are kept fixed (or are increasing together with p). Finally it is obvious that the incompleteness of many skulls from prehistoric populations raises interest in missing data problems.

The author is an old-fashioned objectivist who regards as the most crucial task of statistics, to express uncertainties whether the client likes it or not. There is a tendency among physical anthropologists to ignore the uncertainties involved in arguing from the sample to the population. They do this by foccussing on the sample. Some of them argue that lack of statistical significance is a drawback of classical statistical methodology and not an intrinsic consequence of lack of anthropometrical information. Their attention is shifted from classical methods towards data-analytical techniques which can be used to provide nice representations of the data. The reader will understand that the author feels unhappy about this tendency.

The author's cooperation with the anthropologist G.N. Van Vark started about 1960. During the years 1960—75 the author was used more or less as a consultant and a thesis advisor. Van Vark's anthropological work was mainly devoted to the description of the differences between various recent and non-recent subpopulations of Homo sapiens sapiens. This entails the problem to describe the sexual dimorphism of various populations. Another aspect of Van Vark's work was that he expressed criticism of standard multivariate statistical methodology. He adapted existing procedures and invented new ones whenever he thought that this was necessary. The author was reluctant to get involved because his research had always been devoted to the construction of "optimal" procedures. It was obvious that many problems of Van Vark were too complicated for an optimality approach while the other ones were more or less standard. Nevertheless, there are many examples in the author's research which arose from consultations with Van Vark. Chapter 3 of Schaafsma (1966) is devoted to decision problems where $\mathcal{A} = \{a_0, a_1, a_2\}$, a_0 standing for "remain undecided", a_1 for "accept the hypothesis", a_2 for "accept the alternative". This reformulation of two-sided testing problems originated from Van Vark's desire to express the uncertainties in sex-diagnoses by using doubtful regions as described in Rao (1952). Later "uniformly most powerful tests among the tests which are both invariant and similar size-α" and "uniformly minimum risk procedures among the invariant ones" were derived to solve univariate classification problems (references can be found in Schaafsma-Van Vark (1977)).

At the end of the mid-seventies the cooperation was intensified after the decision to write a joint paper for Statistica Neerlandica. Long and sometimes even quarrelsome discussions were needed to bridge the gaps between

Van Vark's attitude and interests and those of the author. It was soon clear that the cooperation, though interesting and stimulating, would not be very fruitful unless a technical obstacle was removed. The author needed some insight into the inverse Wishart distribution. At a certain moment (probably in 1975) the following simple but crucial lemma was obtained (later the author learned from M.L. Eaton that exact expectations and covariance matrices can be obtained).

Lemma *If* S *has the* $W_p(f, I_p)$ *distribution, then* $f^{1/2}(fS^{-1}-I_p)$ *has the same limit distribution as* $f^{1/2}(f^{-1}S-I_p)$ *as* $f \to \infty$; *this latter distribution follows from the central limit theorem.*

The proof of the lemma is straightforward. Write $f^{-1}S=I_p+(f^{-1}S-I_p)$ and $fS^{-1} \approx I_p-(f^{-1}S-I_p)$ because $f^{-1}S \approx I_p$. Complete the proof by showing that $f^{1/2}[(f^{-1}S-I_p)+\{I_p-(f^{-1}S)^{-1}\}]$ tends to 0 in probability and that the normal limit law has expectation 0 and is invariant under reflection w.r.t. the origin.

One of Van Vark's most interesting problems was about selecting variables in discriminant analysis. His approach was as follows (the author's approach will be described in Section 1, it is based on the above-mentioned lemma). It is convenient to start from some deliberate ordering of the variables (preferably based on other data). In that case selections are determined by the corresponding number p of variables selected (always the first p ones, of course). It would be naive to expect that p should be chosen as large as possible. There will exist some optimum number p^* which can be obtained by studying the actual performance as a function of p. Different approaches may lead to different concepts of performance. Van Vark dealt with two approaches, a third one will be discussed at the end of Section 2.

Approach 1 dealt with classification aspects (in the sense of allocation). For a given set of variables, the performance should refer to the probability (or probabilities) of making correct classifications if individuals are taken at random from some population and are classified by means of the allocation rule based on the values of the selected variables for the given samples. Such probabilities of correct classification are seriously overestimated if one uses the resubstitution method. Therefore Van Vark used the leaving-one-out method which leads to estimators which are almost unbiased. Plotting these estimates of the actual classification performance as a function of p, Van Vark obtained graphs which displayed a maximum at some value p^*. The literature on pattern recognition provides many other examples of this "peaking phenomenon" and proposes terminology where p^* is called the optimal measurement complexity (see Van Vark (1976), Chandrasekaran-Jain (1975), Van Ness (1977) a.o.).

Approach 2 dealt with discrimination aspects. Such aspects are more relevant to physical anthropology than those of making allocations (the opposite will be true in clinical decision making). Van Vark generalized the $\bar{U}$ method of Lachenbruch-Mickey (1968) to situations with more than 2 populations. This leaving-one-out method gives an indication of the actual performance of a set of discriminant functions (e.g. the canonical variates, see Van Vark (1976)). Plotting the generalized $\bar{U}$ estimates as a function of p, Van Vark observed a similar peaking phenomenon as mentioned before.

It was obvious to the author that Van Vark had entered an interesting problem area. Classical multivariate analysis focusses on cases where the variables are given. It is very essential that the choice of variables is discussed. The author did not like the idea of using leaving-one-out estimates because the underlying procedures (e.g. plug-in classification rules or canonical variates) were essentially based on assumptions of normality and homogeneity of covariances. Moreover, he did not understand what Van Vark's generalized $\bar{U}$ estimates are exactly estimating. So he tried to contribute by restricting the attention to the case of $k = 2$ populations and by making the above-mentioned usual assumptions. Concerning approach 1 he could refer to existing theory based on asymptotic expansions of certain characteristic functions to get approximations for the true misclassification probabilities of classical allocation rules as functions of the unknown parameter (μ_1, μ_2, Σ). In this connection it is of some interest to note that the author had obtained exact expressions for the univariate case (see Schaafsma-Van Vark (1977) for further references). The author wanted to construct confidence intervals for the misclassification probabilities. With respect to approach 2 the author developed the theory to be presented in Section 1.

An interesting related issue appeared when discussing the first draft for the joint paper in Statistica Neerlandica. Correcting certain statements of Van Vark, the author expressed that misclassification probabilities are properties of the underlying classification procedure. These properties depend on the unknown parameters (that is why confidence intervals are needed). They do not depend on the vector of scores of the individual that has to be allocated. Van Vark argued that such misclassification probabilities could not be very relevant because some of the individuals to be allocated are almost beyond doubt whereas other individuals have larger probabilities to be misclassified. The author admitted that this was a very good point. It was clear that Van Vark's misclassification probabilities are determined if one knows the individual's posterior probabilities and classification. This requires a Bayesian setting with prior and posterior probabilities of group membership. The author developed the idea that in many classification situations, especially in clinical decision making, the cooperation between statistician and client will be improved considerably by putting responsibilities where they belong. The statistician should retreat from the area of making diagnoses and focus on the construction of confidence intervals for the posterior probabilities. Asymptotic methods will be indicated in Section 2.

Another interesting issue appeared a few years ago. Physical anthropologists are extremely interested in the evolution of our own kind (the genus Homo) and its forerunners. Can a line of development be perceived from Homo erectus through Homo sapiens praesapiens, and possibly even Homo sapiens neanderthaliensis, to Homo sapiens sapiens, etc.? Statements of this kind should follow from anthropometrical data. It is obvious that such data are scarce and often of low quality e.g. because skulls are only partly available. Two schools of anthropologists can be distinguished: (1) those who rely on visual inspection and subsequent discussion, (2) those who try to use numerical methods in spite of all inherent complications. Van Vark gained the impression that the statements made by the visualists were more to the point than those of the second school. Contemplating about the pheonomenon that the opinions of the visualists show more coherency than the other ones, Van Vark distinguished as reasons that the members of the second school work with very different sets of variables and, more importantly, that the visualists use their understanding of natural variations within the recent world population more or less intuitively, as a common yardstick.

Van Vark realized that the results of the second school will obtain practical significance only if they are based on a reliable estimate for the covariance matrix Σ of the recent world population. He travelled to Harvard to ask Howells' assistance. Howells' famous data bank provided a stratified sample from a large part of the present-day world population. An unbiased estimator for the covariance matrix in this part was easily derived. The corresponding outcome is regarded as the required reliable estimate. It will be used as if it were the true Σ. Data for Older Hominids were obtained from Stringer of the British Museum. The sample sizes are so small that the data cannot be used to reject the hypothesis that all covariance matrices are equal to the obtained estimate. It will be seen in Section 3 that it is very convenient if the above-mentioned hypothesis is postulated.

1. The problem of dimensionality in discriminant analysis, $k = 2$, the discriminatory aspect (jointly with A.G.M. Steerneman)

If the first p variables are considered, then the data consist of two independent random samples $X_{h1}, \ldots, X_{hn_h}$ from $N_p(\mu_h, \Sigma)$ $(h = 1, 2)$, where $\theta = (\mu_1, \mu_2, \Sigma)$ is unknown. We want to construct a linear discriminant function $g(x) = w^T x$ or a linear classification statistic $g(x) = aw^T x + b$ where the constants a and b provide some extra flexibility e.g. because one wants to use 0 as cutting-point. Our problem is to determine the column vector w of weights (and a, b). Our linear statistics can be compared by means of their population discriminatory value

$$\{E_2 g(X) - E_1 g(X)\}/[\mathrm{var}\,\{g(X)\}]^{1/2}$$

$$= w^T(\mu_2 - \mu_1)/(w^T \Sigma w)^{1/2} = d(w) \tag{1.1}$$

which satisfies $-d(\omega) \leqslant d(w) \leqslant d(\omega)$ where $\omega = \Sigma^{-1}(\mu_2 - \mu_1)$ is the vector of weights of Fisher's population linear discriminant function and

$$d(\omega) = \{(\mu_2 - \mu_1)^T \Sigma^{-1}(\mu_2 - \mu_1)\}^{1/2} = \Delta \tag{1.2}$$

is the population Mahalanobis distance. It is a natural consequence of these definitions that Δ is a non-decreasing function of p. In principle results should be improving if p is enlarged because then more information becomes available. In actual practice it may work out the other way because $\theta = (\mu_1, \mu_2, \Sigma)$ is unknown and the involved uncertainty increases if p increases, e.g. because the intrinsic dimensionality of θ is equal to $\frac{1}{2}p(p+5)$. An elaboration for discriminant functions is as follows. Let X_1 and X_2 denote the sample means and $S = \Sigma\Sigma(X_{hi} - X_{h\cdot})(X_{hi} - X_{h\cdot})^T$ the SSP matrix. Fisher's sample linear discriminant function, like some other classification statistics, is based on the sample vector of weights

$$W = \{S/(n_1 + n_2 - 2)\}^{-1}(X_{2\cdot} - X_{1\cdot}). \tag{1.3}$$

The corresponding discriminatory value

$$D = d(W) = W^T(\mu_2 - \mu_1)/(W^T \Sigma W)^{1/2} \tag{1.4}$$

is a random variable whose distribution is concentrated on $[-\Delta, +\Delta]$.

Theorem (Schaafsma (1976)). *If* $n = n_1 + n_2 \to \infty$ *while* $\varkappa = n^2 n_1^{-1} n_2^{-1}$, μ_1, μ_2 *and* Σ *are kept fixed (and* Σ *is nonsingular), then*

$$n(\Delta - D) \xrightarrow{L} \frac{1}{2} (\varkappa\Delta^{-1} + \Delta)\, \chi^2_{p-1}. \tag{1.5}$$

The crux of the proof of this theorem is constituted by the lemma of the introduction. If Σ would have been known like in the case of Older Hominids (see the end of the introduction), then one could have used $\Sigma^{-1}(X_{2.} - X_{1.})$ instead of (1.3) and the corresponding discriminatory value can be studied without using the lemma. The result

$$n[\Delta - d\{\Sigma^{-1}(X_{2.} - X_{1.})\}] \to \frac{1}{2} \varkappa\Delta^{-1} \chi^2_{p-1} \tag{1.6}$$

of Schaafsma-Steerneman (1981), shows that the problem of dimensionality does not disappear if singularity of estimators for Σ is avoided by simply postulating Σ.

Illustrating the problem of dimensionality. The pros and cons of Fisher's sample linear discriminant function are most accurately expressed by the whole distribution of D. In practice one will prefer some summary. It is obvious that the $\bar{U}$ estimates of Lachenbruch-Mickey (1968) are intended to estimate something like $E_\theta D$. Result (1.5) leads to the approximation

$$E_\theta D \approx \Delta - \frac{1}{2} n^{-1} (\varkappa\Delta^{-1} + \Delta)(p - 1). \tag{1.7}$$

Other approximations are studied in Schaafsma (1982). Simulation experiments by Steerneman did not suggest worthwile improvements over (1.7).

If one knew Δ as a function of p then one could plot the right-hand side of (1.7) as a function of p. If the graph of Δ is sufficiently flat and n is sufficiently small, then (1.7) will be decreasing for p sufficiently large and an optimal measurement complexity p^* will appear. Illustrations of this kind were presented in Schaafsma-Van Vark (1979) and discussed in Schaafsma (1982).

In practice one does not know Δ as a function of p. One might use the $\bar{U}$ method though the author does not know whether the bias of this estimator for $E_\theta D$ is sufficiently small. (Asymptotically for $n \to \infty$ there is no problem because everything tends to Δ, more precision is needed because $\Delta - D$ is the quantity of interest.) The author prefers to exploit the assumptions of normality and homogeneity of covariances as follows. The positive bias of the plug-in (resubstitution) estimator

$$\hat{\Delta}^2 = (X_{2.} - X_{1.})^T \{S/(n_1 + n_2 - 2)\}^{-1} (X_{2.} - X_{1.}) \tag{1.8}$$

disappears completely if $(n-2)^{-1}(n - p - 3)\hat{\Delta}^2 - n_1^{-1} n_2^{-1} np$ is considered. The latter best unbiased estimator for Δ^2 is improved if negative estimates are made impossible. Hence let

$$\tilde{\Delta}^2 = \max\{0, (n-2)^{-1}(n-p-3)\hat{\Delta}^2 - n_1^{-1} n_2^{-1} p\}. \tag{1.9}$$

It is interesting to plot $\hat{\Delta}$, $\tilde{\Delta}$ and

$$\widetilde{E_\theta D} = \tilde{\Delta} - \frac{1}{2} n^{-1}(\varkappa\tilde{\Delta}^{-1} + \tilde{\Delta})(p - 1) \tag{1.10}$$

as functions of p. The first two graphs will tend to be increasing, the last graph may show a maximum.

The origin of the problem of dimensionality. Naive intuition suggests that performances cannot decrease if the vector of measurements is extended and if trivialities like losses due to costs, or errors, are ignored. If they do decrease (for sufficiently large dimensionality) than this suggests that something is wrong, e.g. the choice of procedure. One might try to robustify procedures against the lack of performance which results if the dimensionality is increased beyond the optimal measurement complexity p^*. The whole discussion becomes very complicated. Schaafsma-Van Vark (1979), Schaafsma-Steerneman (1981) and Schaafsma (1982) use the Scylla and Charybdis metaphor. Charybdis stands for the degradation of performance which appears if p or some other complexity, like the generality of the probabilistic model, is increased beyond its optimum which depends on the sample sizes and the choice of procedure. Scylla stands for the dangers which appear if the data are used twice, first for selecting the set of variables or the model, next for evaluating the data as if model and variables were predetermined. The origin of Charybdis lies in sampling phenomena which conceal the truth. If one knew the true values of all parameters involved then it cannot be counterproductive if complexities are increased, at least not if optimal procedures are used and trivial losses are ignored. The statistician never knows the truth completely. Increasing the complexity, he increases the intrinsic uncertainties. This may lead to the loss of performance. The author hopes very much that Krishnaiah-Kanal (1982) will contain many revealing contributions with respect to the problem of dimensionality. Interested readers are invited to write to A.G.M. Steerneman, Dept. Mathematics, Postbox 800, Groningen, especially if they have comments or suggestions.

2. Interval estimates for posterior probabilities (jointly with A.W. Ambergen)

The introduction mentioned that the author's attitude towards classification problems changed completely after discussions with Van Vark. He had been interested in classification problems because he wanted to replace subjective arguments in various areas like clinical decision making, judging examination papers etc. by the application of optimal classification procedures. He started to realize that a clear distinction should be drawn between situations where routine assignment of individuals is required and those where the individual under classification is essentially unique. Routine assignment requires a classification procedure together with (interval) estimates for the corresponding misclassification probabilities. On the other hand, if we are interested in classifying a unique individual, as in medical diagnosis or in anthropological sex diagnosis, then the computation of confidence intervals for the posterior probabilities is usually much more relevant than making a crude classification. One might argue that it is difficult to define the prior probabilities. This is often true but it is not to the credit of the application of some classical classification procedure because such procedures are essentially based on the assumption of equal prior probabilities.

If one wants to discuss the group membership of a unique individual like in clinical decision making (Ambergen-Schaafsma (1983) contains elaborations for the anthropological problem to discuss the origin of the Border Cave cranium) then the following stages should appear:

(1) measuring the vector of scores x for the individual under investigation;

(2) a priori discussions concerning the populations to which the individual might belong; it would be nice if these populations were mutually exclusive and if prior probabilities $\rho_1, \ldots, \rho_k$ could be defined;

(3) discussions concerning the typicality of the individual; one needs samples $x_{h1}, \ldots x_{hn_h}$ from the various populations ($h = 1, \ldots, k$) in order to discuss whether they are populations to which the individual might belong; these discussions can be carried out by testing the null-hypothesis H_h that x and $x_{h1}, \ldots, x_{hn_h}$ are outcomes of i.i.d. random vectors ($h = 1, \ldots, k$), it may also be of some interest to construct confidence intervals for typicality probabilities (see Ambergen-Schaafsma (1983));

(4) after removing impossible populations and deciding upon the appropriate number k of populations and their prior probabilities, the main task of the statistician begins: he should estimate the posterior probabilities

$$\rho_{t|x} = \rho_t f_t(x) \Big/ \sum_{h=1}^{k} \rho_h f_h(x) \qquad (t = 1, \ldots, k) \tag{2.1}$$

on the basis of the available data

$$x;\ x_{11}, \ldots, x_{1n_1};\ \ldots;\ x_{k1}, \ldots, x_{kn_k} \tag{2.2}$$

and express the involved uncertainties because these can easily exceed one's expectations (especially if complexities are increased);

(5) using the estimates of the posterior probabilities, the corresponding standard errors and other experiences as input, the actual decision maker chooses from the available options for further action.

Elaborating stage (4). In (2.1) the prior probabilities $\rho_1, \ldots, \rho_k$ are regarded as predetermined. They will usually be provided by the actual decision maker. The data (2.2) is needed to obtain estimates for the values $f_h(x)$ of the unknown densities $f_h (h = 1, \ldots, k)$. The assumption has to be made that (2.2) is the outcome of the random vectors $X, X_{11}, \ldots, X_{kn_k}$ which, conditionally given that the individual is from population t (which event has probability ρ_t), are independent with X having density f_t and $X_{h1}, \ldots, X_{hn_h}$ having density $f_h (h = 1, \ldots, k)$. There can be various reasons why these assumptions are unpleasant, e.g. because patients may get cured before their precise group membership is revealed. These inconveniences should be discussed but they should not be used to ignore the sampling errors in estimating the posterior probabilities (unless the estimates are also ignored, of course).

One needs formulas for the asymptotic variances of estimators $R_{t|x}$ for the posterior probabilities $\rho_{t|x}$. Always regarding $\rho_1, \ldots, \rho_k$ as predetermined, results have successively been derived in Schaafsma-Van Vark (1977) for the case [$p = 1$, $k = 2$, normality, $\sigma_1^2 = \sigma_2^2$ unknown], in Schaafsma (1976) and Schaafsma-Van Vark (1979) for the case [$p \geqslant 1$, $k = 2$, normality, $\Sigma_1 = \Sigma_2$ unknown], in Ambergen (1981) and Ambergen-Schaafsma (1983a) for the cases [$p \geqslant 1$, $k \geqslant 2$, normality, $\Sigma_1 = \ldots = \Sigma_k$ unknown] and [$p \geqslant 1$, $k \geqslant 2$, normality, $\Sigma_1 \ldots, \Sigma_k$ unknown]. Ambergen-Schaafsma (1983b) is a review paper where also obvious nonparametric density estimators for the f_h s are applied in case [$p \geqslant 1$, $k \geqslant 2$] and where attempts are made to discuss how many observations it costs if valid assumptions are dropped and how biased conclusions will be if they are based on invalid assumptions. Interested readers are invited to write to A. W. Ambergen, Mathematisch Centrum, Kruislaan 413, 1098 SJ Amsterdam.

Another approach to the dimensionality problem in discriminant analysis. The theory of Section 1 was restricted to the case $k = 2$, with normality. It was also presupposed more or less that the variables were ordered. The classification approach in the introduction was less restrictive in principle. It led Van Vark to the idea that if he had to make classifications, he would use p^* variables where p^* is the optimal measurement complexity (a term from the literature on statistical pattern recognition). Of course p^* depends on the sample sizes and on the unknown parameters. Hence it has to be estimated. The above-mentioned asymptotic variances can be used to make an alternative proposal where the selected subset $\{j_1(x), ..., j_q(x)\}$ of the set of variables is allowed to depend upon the vector x of scores of the individual under investigation. The idea is as follows. Choose some confidence level $1 - \alpha$ ($1 - \alpha = .50$ if one wants to be daring, $1 - \alpha = .95$ if one likes to be a bit conservative). Given some selected subset $\{j_1, ..., j_q\}$, determine the k approximate confidence intervals for the posterior probabilities (note that these probabilities themselves depend upon the choice of variables). One would like to have the largest lower bound close to 1 and the other upper bounds close to 0. This can be translated into mathematical language. Let $[R_{t|x}, \bar{R}_{t|x}]$ $(t = 1, ..., k)$ denote the confidence intervals for the $\rho_{t|x}$'s (with respect to some selected set of variables). Let t^* denote the population such that

$$R_{t^*|x} = \max_{t=1, ..., n} R_{t|x} \tag{2.3}$$

Select the set of variables for drawing inferences concerning this very particular individual, by making $R_{t^*|x}$ as large or

$$1 - R_{t^*|x} + \sum_{\{t \neq t^*\}} \bar{R}_{t|x} \tag{2.4}$$

as small as possible.

3. Best unbiased estimators for evaluating incomplete observation vectors exist if Σ is known (jointly with W.H.V. de Goede)

Mathematically oriented statisticians tend to ignore "missing data" problems because (1) such problems are usually nasty (if Σ is unknown then incomplete exponential families appear and best unbiased estimators do not exist), (2) the missing is often caused by neglect, inaccuracy or laziness of the experimenter. In this last respect physical anthropologists are not to be blamed. It is not their fault that skulls of Older Hominids are often incomplete. Rao's story of the broken skulls (Rao (1965) 4.g.3) shows that one should do one's utmost to include all available information into the analysis.

Van Vark had used the (unbiased estimates of) mean μ and covariance matrix Σ of the present-day world population to compute estimates or rather predictions, for the missing observations. He simply applied linear regression. It is obvious that this procedure is inappropriate from a theoretical point of view because it is based on the assumption that the incomplete skull is a random drawing from the recent world population. Campbell criticized Van Vark's approach on the basis of such arguments. Van Vark claimed that his method provided reasonable results in many practical situations (Van Vark (1982)). Rao emphasized that a solution should be obtained.

The attention is restricted to the estimation of the vector ν of unknown expectations of e.g. Homo sapiens neanderthaliensis, on the basis of a partly observed independent random sample $X_1, \ldots, X_n$ from $N_p(\nu, \Sigma)$ where Σ is taken equal to the known covariance matrix of the present-day world population. Let $Y_1, \ldots, Y_n$ with $Y_i = A_i X_i$ denote the vectors of actual observations. The j-th row of the $r_i \times p$ matrix A_i consists of $p - 1$ zeros and one unity, namely at the h-th place if X_{ih} is the j-th observed measurement for individual i.

It has to be assumed that there is no dependence between the values of the scores and their possible missing. More precisely, the conditional distribution of $(X_1, \ldots, X_n)$ given $(A_1, \ldots, A_n)$ should always be that of an i.r.s. from $N_p(\nu, \Sigma)$. Moreover, every variable should have been observed at least once. (Otherwise it is omitted, of course.)

In the theory $A_1, \ldots, A_n$ are considered known and predetermined. $Y_1, \ldots, Y_n$ are independent, Y_i having the $N_{r_i}(A_i\nu, A_i\Sigma A_i^T)$ distribution. The corresponding p.d.f. $p_\nu(y_1, \ldots, y_n)$ is equal to

$$c_1 \exp\left\{-\frac{1}{2}\sum_{i=1}^{n}(y_i - A_i\nu)^T(A_i\Sigma A_i^T)^{-1}(y_i - A_i\nu)\right\}$$

(3.1)

$$= c_2(y)\, c_3(\nu) \exp\left\{\nu^T \sum_{i=1}^{n} A_i^T(A_i\Sigma A_i^T)^{-1}y_i\right\}.$$

Hence theory for multiparameter exponential families provides $\sum_{i=1}^{n} A_i^T(A_i\Sigma A_i^T)^{-1}Y_i$ as a complete sufficient statistic. The corresponding expectation is equal to $\sum_{i=1}^{n} A_i^T(A_i\Sigma A_i^T)^{-1}A_i\nu$. Hence

$$\left\{\sum_{i=1}^{n} A_i^T(\Sigma A_i^T)^{-1}A_i\right\}^{-1} \sum_{i=1}^{n} A_i^T(A_i\Sigma A_i^T)Y_i \tag{3.2}$$

is the best unbiased, efficient, m.l. estimator for ν. This result can be used as the basis for a theory to derive unbiased estimators for various Mahalanobis distances. These results can then be used to estimate posterior probabilities and typicality probabilities and to estimate the corresponding accuracy. The applicability of the theory is seriously restricted by the assumption that "each variable should have been observed at least once (for each of the involved populations)". This will imply that different choices of populations will lead to Mahalanobis distances, posterior probabilities and typicality probabilities with respect to different sets of variables. Except for Homo sapiens neanderthaliensis and Homo sapiens sapiens, all hominid samples contain only one or a few elements which are nearly always very incomplete. This explains Van Vark's desire to keep working with his own method because this involves no restrictions at all (only nasty assumptions). The author's reaction is Wittgenstein's "Wovon man nicht sprechen kann, darüber muss mann schweigen". Interested readers are invited to write to W.H.V. De Goede, Verzetsstrydersiaan 66, Groningen.

REFERENCES

1. Ambergen, A. W., *Approximate confidence intervals for posterior probabilities.* Master Thesis, Report TW-224, Dept. Math., Univ. Groningen, 1981.
2. Ambergen, A. W., Schaafsma, W., *Interval estimates for posterior probabilities.* Submitted for publication (1983 a).
3. Ambergen, A. W., Schaafsma, W., *The asymptotic variance of estimators for the posterior probabilities. A useful tool which should not be neglected.* In: Van Vark and Howells (Eds.) (1983 b).
4. Chandrasekaran, B., Jain, A. K., *Independence, measurement complexity, and classification performance.* IEEE Trans. Systems Man Cybernet. SMC-**5**, 240–244 (1975).
5. Krishnaiah, P. R., Kanal, L., (Eds.), *Classification, Pattern Recognition and Reduction of Dimension.* Handbook of Statistics, Vol. 2. Amsterdam, North-Holland, 1982.
6. Lachenbruch, P. A., Mickey, M.R., *Estimation of error rates in discriminant analysis.* Technometrics, **10**, 1–11 (1968).
7. Rao, C. R., *Advanced Statistical Methods in Biometric Research.* New York, Wiley, 1952.
8. Rao, C. R., *Linear Statistical Inference and Its Applications.* Wiley, New York, 1965.
9. Schaafsma, W., *Hypothesis Testing Problems with the Alternative Restricted by a Number of Inequalities.* Groningen, Noordhoff, 1966.
10. Schaafsma, W., *The asymptotic distribution of some statistics from discriminant analysis.* Report TW-176, Dept. Math., Univ. Groningen, 1976.
11. Schaafsma, W., *Selecting variables in discriminant analysis for improving upon classical procedures.* In: Krishnaiah and Kanal (Eds.) (1982).
12. Schaafsma, W., Steerneman, T., *Discriminant analysis when the number of features is unbounded.* IEEE Trans. Systems. Man Cybernet., SMC-**11**, 144–151 (1981).
13. Schaafsma, W., Van Vark, G.N., *Classification and discrimination problems with applications. I.* Statist. Neerlandica, **31**, 25–45 (1977).
14. Schaafsma, W., Van Vark, G.N., *Classification and discrimination problems with applications. II a.* Statist. Neerlandica, **33**, 91–126 (1979).
15. Van Ness, J. W., *Dimensionality and classification performance with independent coordinates.* IEEE Trans. Systems Man Cybernet., SMC-**7**, 560–564 (1977).
16. Van Vark, G.N., *Some statistical procedures for the investigation of prehistoric human skeletal material.* Thesis, Groningen, 1970.
17. Van Vark, G. N., *A critical evaluation of the application of multivariate statistical methods to the study of human populations from their skeletal remains.* HOMO, 94–114 (1976).
18. Van Vark, G. N., *On the reconstruction of human phylogeny with the aid of multivariate statistical methods.* In: *Proc. Golden Jubilee Conf. Indian Statist. Institute.* Calcutta, 1982.
19. Van Vark, G.N., van der Sman, P. G. M., *New discrimination and classification techniques in anthropological practice.* Z. für Morphologie und Anthropologie, **72**, 1 (1982).
20. Van Vark, G. N., Howells, W. W., (Eds.), *Multivariate Statistical Methods in Physical Anthropology.* Dordrecht, Reidel, 1983.

B. SCHNEIDER

Bayesian approach to the analysis of clinical studies

0. Introduction

The topic of this paper is concerned with the application of the mathematical theory of probability and statistics to clinical trials. This application needs an interpretation of the formal mathematical theories in terms of clinical events and situations or — in other words — an interpretative model. Clearly the principles of the construction and identification of these models must be outside the mathematical theory and are to some extent arbitrary. Therefore different models, i.e. different approaches to the interpretation and identification of mathematical theories may be used as adequate for the application.

In this paper the use of the Bayesian approach to clinical research is investigated.

1. Clinical studies

Clinical studies are concerned with the behaviour of the health state — or better illness state — of patients. Their final goal is to get valid and reliable information about drugs and manoeuvres for the improvement of the health state, or more optimistically, for the cure of the various diseases. This information is contained in the observation about various quantitative or qualitative variables (mostly called by clinicians "parameters") obtained from the patients under different conditions (e.g. with and without a specific treatment). The first and fundamental step of an analysis of these data is the acceptance of an adequate model for the interrelation between the data and the therapeutic information to be analyzed. It is usual to accept a model which I shall call the "collective model". The observation x_i obtained from a patient i is considered as randomly drawn from a collective of similar measurements. This means that the observed results $x_1, ..., x_n$ are assumed as realizations of n independently and identically distributed random variables X_i with the distribution function $F(x, \theta)$ depending on a parameter or parameter vector θ. This parameter represents the relevant information about the health state or the patient's reaction to the treatments to be analyzed. We may for example think of Bernoulli variables X with the 2 states "cured" (corresponding to 1) and "not cured" (corresponding to 0) and θ the probability of cure. The crucial point of this model is the assumption that the parameter θ is a constant fixed value for all patients of the random sample, i.e. patients with the same treatment conditions, etc. My primary statement is that this assumption is not appropriate to clinical studies; on the contrary, one must assume that the health state or the patient's reaction to treatments varies from patient to patient and from application to application. Therefore the parameter θ representing this health state or patient's reaction cannot be considered as fixed (but unknown), but must be considered as a random variable. Such models are usually called "Bayesian models". These Bayesian models are more appropriate for the analysis of clinical studies than the usually used random sample models.

This can be seen from the analogy with medical diagnoses. The usually accepted stochastic model for the diagnostic procedure is the Bayes model: The diagnosis d of a patient's disease is considered as a parameter, influencing the probability distribution $F(x, d)$ of the patient's symptoms x. Within the population of patients coming to a physician the diagnoses form a random variable D. It is the final goal of the physician's diagnostic procedure to decide about the "most probable" diagnosis d_i of a certain patient on the basis of the observed symptoms x. I cannot see any essential difference between this diagnostic process and the therapeutic decision of the physician. Also in the therapeutic process we must assume that different patients will react differently to a treatment A. We must in addition assume that the same patient i will react randomly to treatment A, if this treatment will be applied repeatedly to him in the same situation (if this were possible). The reaction of the patient to treatment A can be characterized by the value θ_{Ai} of a parameter θ_A, i.e. the parameter of the probability distribution of the patient's reaction to repeated application of A. For example the reaction may be "cured" or "not cured" and the parameter θ_A the probability p_A of cure. The reaction of a population of patients to treatment A is characterized by the distribution $F(\theta_A)$ of the "reaction parameter" θ_A, e.g. the distribution of the cure probability p_A. This distribution corresponds to the so-called a priori distribution or prior distribution in Bayes theorem.

One difference should be mentioned between the diagnostic and therapeutic process: Whereas in the diagnostic process the physician makes his decision about the most probable diagnosis d_i after observing the symptom x_i, in the therapeutic process the physician must decide about the best treatment, i.e. the treatment with highest cure probability p_{Ai} for the patient i before he can observe the patient's reaction x_i. Here comes the role of clinical studies. They should provide the physician with relevant information about the distribution of the parameters θ_A, θ_B, ... after different treatments A, B, ... within the population of his patients.

2. Bayesian analysis of clinical studies

If the parameters θ are considered as random variables, the basis for statistical analysis is Bayes formula

$$\mathrm{pr}(\theta | x) = \frac{l(x \mid \theta) \cdot q(\theta)}{P(x)} \cdot$$

$\mathrm{pr}(\theta \mid x)$ is the a posteriori probability (or density) for θ after observing x, $l(x \mid \theta)$ the probability (or density) for x if the parameter value θ holds (this corresponds to the "likelihood"), $q(\theta)$ the a priori probability (or density) for θ and $P(x) = \int l(x \mid \theta) q(\theta) d(\theta)$ the unconditional probability for x.

The crucial point in the application of this a posteriori probability for statistical inferences is the assumption of the a priori probabilities (or densities) $q(p)$. This assumption must be made by the experimenter, physician or statistician more or less arbitrarily as expression of his own opinion on the distribution of θ. If comparable information from earlier experiments or observations is available, then the experimenter could and should use this information for his choice of the a priori probability. This was formalized in the method of "empirical prior distribution" (see e.g. Wetherill [7]).

The arbitrariness of the choice of a priori distributions can be much reduced if the experimenter is willing to restrict his choice to a certain class of a priori distributions according to the choice of the class of likelihood functions. These classes are the conjugate prior distributions with the property that the a posteriori distribution belongs to the same class as the a priori distribution

$$k(\theta \mid x; \beta) = \frac{l(x \mid \theta)\, k(\theta; \alpha)}{P(x)},$$

where α, β are parameters of the class of functions k.

For the most important classes of likelihood functions the corresponding classes of conjugate prior distributions exist (see e.g. Barnard [1], Raiffa and Schlaifer [5], Novick and Grizzle [4]. For normal likelihood distributions the classes of conjugate priors are also the normal distributions. For binomial likelihoods with parameter p the class of conjugate priors are the Beta-distributions with density

$$k(p; m, s) = \frac{p^s (1 - p)^{m-s}}{\beta(s + 1, m - s + 1)},$$

where $\beta(s + 1, m - s + 1)$ is the Beta-function.

If the observation on n individuals showed r "successes", the likelihood is the binomial probability

$$l(r; n, p) = \binom{n}{r} p^r (1 - p)^{n-r}$$

and the a posteriori density is a Beta-density with parameters $n + m$ and $r + s$

$$\mathrm{pr}(p \mid r) = \frac{p^{r+s}(1 - p)^{n+m-r-s}}{\beta(r + s + 1,\ m + n - (r + s) + 1)}.$$

This shows an essential feature of conjugate priors: If the observation is based on a sufficiently high number of cases, the a posteriori distribution will be nearly independent of the choice of the parameter in the prior density.

For example the "most probable" value for p according to the a posteriori distribution (i.e. the mode of the a posteriori density) is

$$\hat{p} = \frac{r + s}{n + m}.$$

If s is small compared with r and m small compared with n, these values are approximately r/n, the maximum likelihood estimation in the classical approach. It can be shown that this property holds in general for conjugate priors under certain conditions (see e.g. Raiffa and Schlaifer [5], Edwards, Lindeman and Savage [3]). So, with the use of conjugate priors unrealistic assumptions in the choice of the prior distribution will not influence the a posteriori distribution, if the observations are based on sufficiently many cases.

A special case of the Beta-distribution for $m = 0$ and $s = 0$ is the uniform density

$$k(p; 0, 0) = 1, \qquad \text{for } 0 \leqslant p \leqslant 1$$
$$= 0, \qquad \text{otherwise.}$$

This shows, that the choice of the uniform density is not the worst one, if the experimenter does not have any preference about p or knowledge of earlier experiments.

If m and s are integers, the a priori Beta-density can be interpreted as the probability of s successes in m observations. If former experiments showed such results or if the experimenter expected such a result, the choice of these parameters for the a priori distribution seems a reasonable behaviour. By the new experiment this choice will either be improved or disproved. In either case the a posteriori distribution gives a better estimate of the real distribution of the parameters.

If the likelihood function depends on additional parameters (like the normal distribution on the variance σ^2), these parameters must be estimated. So we get also an estimation of the a posteriori distribution. The accuracy of this estimation can be expressed by confidence intervals, as Schaafsma [6] has pointed out.

3. Bayesian regression

If we go back to the physician, who wants to provide the success of a treatment for concrete patients, and ask what gain will this physician have by knowing the a posteriori probability distribution of the expected success probability in the population of his patients, we must answer: not too much. This distribution gives him some idea about how the success probabilities may be spread within the population of his patients; but he does not know whether the concrete patient he has to treat belongs to the group with low success probability or with high success probability for a treatment. So for the physician — and for the individual patient — the a posteriori distribution of the success probability is rather uninteresting. What he wants are indications or prognostic variables, which tell him with sufficiently high precision whether a concrete patient has a high or low probability of success for a treatment A.

Formally this means that the success probability p is considered as dependent on a number of observable variables $z_1, z_2, \ldots$ (like age, duration of illness, state of illness, etc.), which allow a sufficiently precise prediction of p in an individual case.

This dependence is what the physician wants to know from clinical studies.

For the analysis of this dependence one will usually assume some parametric model, e.g. the logistic model with linear expression for the prognostic variables

$$\log \frac{p}{1-p} = \alpha_0 + \alpha_1 z_1 + \alpha_2 z_2 + \ldots .$$

In the Bayesian approach the parameters $\alpha_0, \alpha_1, \ldots$ are considered as random variables. For a given a priori density (e.g. uniform density in a given interval of the parameter space) and observations (x, z) the a posteriori density for the parameters α_i can be calculated (or estimated) and one can use as estimates for the parameters the most probable values.

Using these estimates the relevance of the different prognostic variables z can be judged and regions of indicator values with high success probabilities for different treatments A, B can be defined. These regions can give the physician some idea about which of the alternative treatments may be more appropriate for specific patients and which should be avoided.

4. Bayesian treatment comparison

The aim of clinical studies is not only to get information about the success or failure of one treatment, but to compare the therapeutic or toxic effects of two or more treatments. So clinical studies are for example concerned with the relation between 2 success probabilities p_A and p_B of 2 treatments A and B within a defined population of patients. The classical way to handle this problem is the use of significance tests, assuming p_A and p_B as fixed but unknown values and deciding between the null-hypothesis

$$H_0: p_A = p_B$$

and the alternative hypothesis

$$H_1: p_A \neq p_B.$$

In the Bayesian approach, assuming p_A and p_B as random values, such hypotheses do not make much sense. There must be at least an additional assumption about the probability with which such hypotheses should hold; and then the problem is entirely undefined.

Pure Bayesians like Lindley, Savage or Cornfield would add to the assumption of random parameters the philosophical concept of personal probabilities and the principle that all decision should be based only on the data (i.e. the likelihood) and not on additional concepts like error probabilities. Then they would not hesitate to consider the truth of the null-hypothesis as a random event (in a personalistic sense) with a given a priori probability. Then for given observations x the a posteriori probability of the truth of the null-hypothesis can be calculated and acceptance or rejection decided on the basis of this probability. So, Cornfield [2] proposed a decision procedure where the null-hypothesis is accepted as long as the a posteriori probability for the truth of H_0 is above some level β (e.g. 10%) and is rejected if it is less than this level. He could show that this is a feasible procedure, which does not lead to probability 1 for rejection of H_0 with increasing number of observations, if for the truth of the null-hypothesis H_0 an a priori probability $q > 0$ is given.

But I do not want to introduce concepts which cannot be accepted by objectivists and therefore will not assume personal probabilities for the truth of hypotheses. So the treatment comparison cannot be based on acceptance or rejection of hypotheses but solely on the a posteriori distribution for the success probabilities p_A and p_B.

As the clinical experiments with the two treatments are normally conducted with different patients, one can assume p_A and p_B as independently distributed.

So we get as a posteriori distribution for combined values of p_A and p_B:

$$\mathrm{pr}(p_A, p_B \mid x_A, x_B) = \mathrm{pr}(p_A \mid x_A) \cdot \mathrm{pr}(p_B \mid x_B)$$

This combined distribution is the Bayesian basis for treatment comparisons.

To get an operational procedure one can decide to prefer treatment A over treatment B, if the a posteriori probability for the event $(p_A > p_B)$ is higher than a fixed level (e.g. 90%). This probability is given by the integral of the bivariate a posteriori distribution over the region $p_A > p_A$, i.e. the area above the diagonal line of the unit square in the first quadrant of the parameter space (p_A, p_B) (see Figure 1):

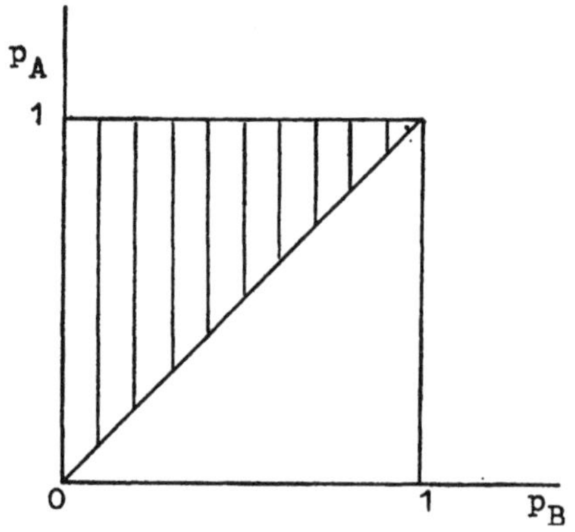

Figure 1

$$P(p_A > p_B \mid x_A, x_B) = \iint_{p_A > p_B} \mathrm{pr}\,(p_A \mid x_A) \cdot \mathrm{pr}\,(p_B \mid x_B)\, \mathrm{d}p_A\, \mathrm{d}p_B$$

This integral can be calculated by using methods of numerical integration.

This procedure which was first proposed by Novick and Grizzle [4] can be extended immediately to the comparison of more than 2 treatments by using the third or more dimensional a posteriori distribution of the parameters $p_A, p_B, p_C, \ldots$ and dividing the parameter space into appropriate disjoint sets. For 3 treatments one may choose the 3 sets preference set $A: p_A > \max(p_B, p_C)$

preference set $B: p_B > \max(p_A, p_C)$

preference set $C: p_C > \max(p_A, p_B)$.

For each of these sets the a posteriori probabilities pr(A), pr(B), pr(C) (given the observations $x_1, ..., x_n$) may be calculated and the "preference" of the 3 treatments can be ordered according to these 3 probabilities. This shows an intrinsic feature of this sort of analysis: The analysis offers to the physician "preferences" and not "decisions" and error probabilities. It is up to the physician to use these preferences for his own responsible decision.

To illustrate this procedure we use an example from Novick and Grizzle [4], where 2 surgical techniques for the operation of duodenal ulcer are compared. With technique A applied to 100 patients there died 7, with technique B applied to 100 patients there died only 1 in a definite time (within 6 months after operation). So we get the following contingency table

Table 1

	survived	dead	
technique A	93	7	100
technique B	99	1	100
	192	8	200

Using Chi²-method with Yates correction we get a value of 3.26 for 1 degree of freedom, which indicates no significant differences at a level α of 0.05.

Using the Bayesian approach we have to calculate the a posteriori probability for $p_A > p_B$, given the observations of the contingency table 1. Here p_A and p_B are considered as probabilities for dying within 6 months after operation A resp. B. Assuming for both probabilities the uniform density as a priori probability the a posteriori probability is given by

$$P(p_A > p_B \mid 100, 7; 100, 1) =$$

$$= \frac{(101\,!)^2}{7\,!\,93\,!\,99\,!} \iint_{p_A > p_B} p_A^7(1-p_A)^{93}\, p_B(1-p_A)^{99}\, dp_A\, dp_B.$$

By calculating the integral with numerical methods we get a value greater than 0.98. This means (in the frame of a frequency interpretation of probability) that one can expect for more than 98% of the patients with duodenal ulcer to have a higher death probability with technique A than with technique B. So, technique B has a very high preference over technique A.

In my opinion such an information is more understandable and valuable for the physician than the usual declaration of "significance". But if somebody is not satisfied with the degree of preference given by the a posteriori probability $P(p_A > p_B \mid x, y)$ and wants a decision on acceptance or rejection of one of the treatments A or B, such a procedure can be easily constructed by fixing two probability levels P_1 and P_0 ($P_1 > P_0$). If $P(p_A > p_B \mid x, y)$

exceeds P_1 treatment B is accepted. If $P(p_A > p_B \mid x, y)$ is exceeded by P_0, treatment A is accepted and if the a posteriori probability lies between both levels, both treatments are considered as equivalent. In the last case experimentation can be continued.

The choice of the levels P_0 and P_1 can be made arbitrarily by the physician. A more sophisticated decision-theoretical approach was proposed by Wetherill [7], who defined loss functions for the possible decision situations and calculated for each decision the a posteriori risk. Then the decision with minimal risk is chosen at any step of experimentation.

In the frame of Neyman-Pearson statistics one would like to base the decision on error probabilities, but in the Bayesian approach this procedure does not make much sense. If the probabilities p_A and p_B are random variables, what meaning can have an error of rejecting A or B? One possible approach could be to consider the likelihood $l(x, y \mid p_A = p_B)$ for the observed results x, y (i.e. the frequencies of deaths under both treatments) on the straight line $p_A = p_B$ in the parameter space. With the aid of this likelihood there can be defined for a given size $\alpha (0 < \alpha < 1)$ a critical region R in the sample space. This critical region contains all points (x, y) in the extreme sites of the sample space (i.e. for $\mid x - y \mid > \delta$ for some δ) with

$$\iint_{|x-y|>\delta} l(x, y \mid p_A = p_B) \, dx \, dy = \alpha.$$

This means that δ has to be chosen in such a way that the above condition holds. As the a posteriori probability $P(p_A > p_B \mid x, y)$ is a continuous function of the points (x, y) of the sample space, the levels P_0 and P_1 can be defined in such a way that the points with

$$P(p_A > p_B \mid x, y) > P_1 \text{ or } P(p_A > p_B \mid x, y) < P_0$$

are exactly the points of the critical region R.

In this case preference of A over B or B over A is equivalent to rejection of the null-hypothesis $p_A = p_B$ in a significance test with error probability α. This shows that preference rules in the Bayesian approach can be defined equivalently to a Neyman-Pearson test. This means that A is preferred to B or B to A for exactly the same sample points, for which the null-hypothesis is rejected at level α. But the Bayesian statistician will consider this equivalence only as a mathematical toy; as to him the null-hypothesis is meaningless. He will restrict his statistical task to give the physician preferences and let the decision how to proceed with these preferences to the physician.

REFERENCES

1. Barnard, G. A., *Sampling inspection and statistical decisions.* J. Roy. Statist. Soc. Ser. B, **16,** 151– 174 (1954).
2. Cornfield, J., *A Bayesian test of some classical hypotheses — with applications to sequential clinical trials.* J. Amer. Statist. Assoc., **61,** 577– 594 (1966).
3. Edwards, W., Lindeman, H., Savage, L. J., *Bayesian statistical inference for psychological research.* Psych. Rev., **70,** 193– 242 (1963).
4. Novick, M. R., Grizzle, J. E., *A Bayesian approach to the analysis of data from clinical trials.* J. Amer. Statist. Assoc., **60,** 81– 96 (1965).
5. Raiffa, H., Schlaifer, R., *Applied Statistical Decision Theory.* Boston, Harvard Business School, 1961.
6. Schaafsma, W., *Me and the anthropologist.* These Proceedings, pp. 333—343.
7. Wetherill, G. B., *Sequential Methods in Statistics.* London, Chapman and Hall, 1980.

JULIET POPPER SHAFFER

Issues arising in multiple comparisons among populations

1. Introduction

Let X_{ij}, $j = 1, ..., n_i$; $i = 1, ..., k$; be random variables with joint distribution P belonging to some family of distributions $\mathcal{P}$, and let P_i be the marginal distribution of $X_{i1}, ..., X_{in_i}$. We are interested in comparing these marginal distributions on the basis of some functions $f_i(P_i)$. If a set of distributions differs on the values of the corresponding functions, the set is called *heterogeneous;* otherwise it is called *homogeneous*. The true situation can be represented as a partition of the distributions

$$(1) \qquad (i_1, ..., i_{v_1}) \, (i_{v_1+1}, ..., i_{v_1+v_2}) \, ...$$

where $i_1, ..., i_k$ is a permutation of the integers $1, ..., k$; the partition (1) corresponds to the situation

$$[f_{i_1}(P_{i_1}) = f_{i_2}(P_{i_2}) = ... = f_{i_{v_1}}(P_{i_{v_1}})] \neq$$

$$[f_{i_{v_1}}(P_{i_{v_1+1}}) = f_{i_{v_1+2}}(P_{i_{v_1+2}}) = ... = f_{i_{v_1+v_2}}(P_{i_{v_1+v_2}})] \neq ... .$$

A set of decisions with respect to the populations is generally based on a series of hypothesis tests. The set of acceptances and rejections resulting from the application of a multiple testing procedure is called a *pattern* or *decision pattern.*

The hypotheses may refer to partitions or to subsets of the distributions. Rejection of a partition means that the partition is declared incorrect, i.e., not the true one given in (1). Rejection of a subset means rejection of the hypothesis that the populations in that subset are homogeneous; such rejection is equivalent to declaring the subset heterogeneous. As usual, acceptance of a hypothesis means only that there is insufficient evidence to reject it. We consider only testing procedures for which a hypothesis is never accepted if any hypothesis implied by it is rejected; such procedures have been termed *coherent* (Gabriel, 1969). Thus, rejection of a subset implies rejection of all subsets containing it and all partitions of which it is a member, while rejection of a partition implies no further rejections.

It is assumed that all the populations are of equal interest and that there is no prior information on the relations among them (or no wish to make use of any prior information there may be). It is further assumed that, if the f_i are real-valued, the magnitudes of the differences among them are not of primary interest. Where the f_i values may be completely ordered, i.e. where the true partition is ordered, it may or may not be of interest to specify the order or direction of the differences among distributions.

As an example, consider the comparison of weight gain or loss under a variety of diets. There is likely to be particular interest in pair comparisons and in ordering diets; in other words, in conclusions such as "Diet A results in relatively greater weight gain (or loss) than Diet B." The conclusion that a subset of three or more diets is heterogeneous, that the three do not produce equal changes in weight, is of less value but might permit concentration on that subgroup for further study.

In general, there is usually more interest in heterogeneity conclusions on pairs than on larger subsets, as in this example, and in order relations among distributions, that is, in drawing directional conclusions. There may also be special interest in the partition as a whole, rather than its separate parts. If, for instance, the diets in the example consists of systematic combinations of different components, knowledge of the true partition would make it possible to draw conclusions on the effectiveness or ineffectiveness of these various components in the diet.

Thus, interest may lie primarily in conclusions with respect to pairs, subsets in general, and/or partitions of the distributions. The division into pairs, subsets and partitions, which can be thought of as three levels of analysis, is useful heuristically in considering a number of different issues.

2. Accuracy of decision pattern

The objective of the multiple testing procedure is a decision pattern as close as possible to the true partition. In a decision-theoretic approach, the difference between the pattern and the true partition is expressed in the form of a loss function, and one seeks procedures minimizing expected loss. An alternative approach sets an upper bound on some function of the probabilities of rejecting true hypotheses and tries to maximize the probabilities of rejecting false hypotheses. In both approaches, the losses or errors in the pattern depend upon whether distributions are to be ordered if found heterogeneous. If ordering is not involved, there are only two kinds of undesirable decisions: rejecting homogeneous subsets or true partitions, and accepting heterogeneous subsets or false partitions. If ordering is involved, there is an additional (third) type of undesirable decision in relation to pairs and partitions: rejecting a heterogeneous pair but ordering it incorrectly, or accepting an ordered partition which is true except for incorrect ordering of the members.

Optimal procedures for the situation in which the distributions are compared on a real-valued parameter θ were developed by Lehmann (1957a, 1957b) and Spjøtvoll (1972). Both took the set of pair hypotheses as the basic set to be tested. Lehmann considered the directional problem by formulating the set of $k(k-1)$ directional pairwise hypotheses H_{ij}: $\theta_i \leq \theta_j$ (rather than the $k(k-1)/2$ pairwise hypotheses $\theta_i = \theta_j$), where θ_i is the value of the parameter for P_i. Rejection of H_{ij} implies the decision $\theta_i > \theta_j$. Lehmann used a decision-theoretic framework, assuming loss a for rejecting H_{ij} if true, loss b for accepting H_{ij} if false, and loss zero otherwise. The loss connected with a decision pattern is the sum of the losses of the decisions on each hypothesis H_{ij}. Then under some continuity and regularity conditions, if each hypothesis is tested at level $b/(a+b)$, and if each of these tests has certain optimal properties as a test of its individual hypothesis, the product procedure (i.e. the joint set of tests) uniformly minimizes the risk among all risk-unbiased multiple test procedures.

Spjøtvoll obtained an optimal result for tests of pairwise hypotheses under both directional and nondirectional formulations. He set an upper bound γ on the expected number of rejections of true hypotheses (i.e. on the sum of the probabilities of rejection over the true hypotheses).[1] Subject to this restriction, he found procedures which maximize minimum average pairwise power and also minimum power of the individual pair tests over certain alternatives. For Spjøtvoll as well as for Lehmann, the optimal procedures for the multiple test reduce to testing each individual hypothesis at a fixed level with a test which has some optimal properties for that particular hypothesis, i.e. the multiple test is the product of the individually optimal tests.

Both Lehmann's and Spjøtvoll's formulations imply that the undesirability of any decision pattern increases additively with the number of false pair rejections. Yet, if the decision or inference to be reached depends substantially on the true partition, a single rejection of a true pair hypothesis may be almost as misleading as a set of false rejections. An alternative measure of loss, then, assigns a constant loss to any decision pattern with one or more incorrect rejections, or distinguishes only between correct patterns (i.e., those having no incorrect rejections) and incorrect patterns (those having one or more incorrect rejections). While an ideal formulation would probably be somewhere between these extremes, most work in multiple comparisons has adopted the second, non-additive, formulation.

For any procedure δ, the probability of obtaining a pattern including one or more rejections of correct hypotheses, given some distribution P, is known as the experimentwise error rate of δ for P. In correspondence with most of the work in this area, the remainder of this paper will restrict attention to procedures with experimentwise error rates $\leqslant \alpha$ for all $P \in \mathcal{D}$, and will compare the procedures on their probabilities of rejection of false hypotheses.

For given k, let $H_1, \ldots, H_q$ be a list of all subset and partition hypotheses in some order. For a given distribution P let $I_P = \{i : H_i \text{ is true given } P\}$. For a given procedure δ and distribution P, let $\pi(P, \delta) = (\pi_1(P, \delta), \ldots, \pi_q(P, \delta)$, where $\pi_i(P, \delta)$ is the probability of rejecting H_i. A procedure δ is *at least as powerful* as δ^* if

$$\pi_i(P, \delta) \geqslant \pi_i(P, \delta^*) \text{ for all } i \notin I_P. \tag{2}$$

A procedure δ is *uniformly more powerful* than δ^* if (2) holds for all $P \in \mathcal{D}$ and at least one inequality is strict for some $P \in \mathcal{D}$.

When the criterion is a specified bound on the experimentwise error rate, there are no optimality results comparable to those of Lehmann and Spjøtvoll, but there are procedures of more complicated structure which are uniformly more powerful than the product procedures optimal under the

[1] Spjøtvoll set an upper bound on the expected number of rejections given that all hypotheses are true, but showed that his procedure was also optimal under this more general bound.

Lehmann and Spjøtvoll additive formulations. The remainder of this paper will describe some procedures of this kind. The nondirectional problem, in which no ordering of pairs or partitions is attempted, will be considered first. Later, the directional problem will be considered briefly.

3. Partition, subset, and pairwise sequentially-rejective multiple test procedures for nondirectional decisions

Three general approaches have been used in constructing tests whichs are more powerful than product tests and which control the experimentwise error rate at some specified level α. These are conveniently labelled the partition approach, the subset approach, and the pairwise sequentially-rejective approach, corresponding to the three levels of analysis described earlier.

The partition approach is related to work by Cox and Spjøtvoll (1982) and is a closed testing procedure as defined by Marcus, Peritz, and Gabriel (1976). In this approach, each possible partition is tested by a single overall test of the set of homogeneity hypotheses implied by the partition, using any test with significance level α. It follows directly that the true partition is included among the accepted set with probability $1 - \alpha$. A subset is rejected if all subsets containing it and all partitions of which it is a member are rejected.

The subset approach has been described by Tukey (1953) and Lehmann and Shaffer (1979). In this approach, the homogeneity of each subset is provisionally tested, using any test depending only on the distributions in that subset, with significance level α_m, where m is the size of (i.e., the number of distributions in) the subset, $2 \leqslant m \leqslant k$. The subset is either provisionally accepted or provisionally rejected, depending upon the outcome of the test. A subset is rejected if and only if all subsets containing it are provisionally rejected. A partition is rejected if and only if at least one member of the partition is rejected. The set of significance levels α_m, $2 \leqslant m \leqslant k$, are selected in any way which keeps the experimentwise error rate less than or equal to α.

The sequentially-rejective approach was formulated explicitly in a general context by Holm (1977, 1979). Given a set of hypotheses of interest, let $\{H_i\}$, $i = 1, \ldots, n$, be the subset of *minimal* hypotheses (Gabriel, 1969), that is, the set of all hypotheses which are not equivalent to intersections of other hypotheses. The approach involves stagewise testing of the minimal hypotheses, followed by rejection of any nonminimal hypothesis if and only if at least one minimal hypothesis implied by it is rejected. It is assumed that the n minimal hypotheses H_i have associated test statistics Y_i for which small values lead to rejection,[2] and which satisfy the two conditions (Holm, 1977):

A. For each α and m, there exists a $\gamma_{\alpha,m}$ such that

$$\text{Prob}\,(\min_{i \in I} Y_i \leqslant \gamma_{\alpha,m}) \leq \alpha \tag{3}$$

whenever the number of elements in the index set I is equal to m and all hypotheses in the set $\{H_i : i \in I\}$ are true.

[2] For convenience in the present context, the description has been slightly modified from Holm (1977), who used a statistic for which large values lead to rejection.

B. The constants $\gamma_{\alpha,1}, ..., \gamma_{\alpha;n}$ satisfy the inequalities

$$\gamma_{\alpha,1} \geqslant \gamma_{\alpha,2} \geqslant ... \geqslant \gamma_{\alpha,n}.$$

Let $Y^{(1)} \leqslant Y^{(2)} \leqslant ... \leqslant Y^{(n)}$ be the order statistics of the Y_i, and let $H^{(1)}, H^{(2)}, ..., H^{(n)}$ be the correspondingly ordered minimal hypotheses. Then at Stage 1, $H^{(1)}$ is rejected if $Y^{(1)} \leqslant \gamma_{\alpha,n}$; otherwise all hypotheses are accepted. If $H^{(1)}$ is rejected, $H^{(2)}$ is rejected if $Y^{(2)} \leqslant \gamma_{\alpha,n-1}$; otherwise $H^{(2)}, ...$ $..., H^{(n)}$ are accepted. Similarly, for $j = 2, ..., n$, Stage j is reached if $H^{(1)}, ..., H^{(j-1)}$ have been rejected, and in that case $H^{(j)}$ is rejected if $Y^{(j)} \leqslant \gamma_{\alpha,n-j+1}$; otherwise all $H^{(l)}$, $l \geqslant j$, are accepted.

In the context of nondirectional comparisons among distributions, the minimal hypotheses are the $n = \binom{k}{2}$ pair homogeneity hypotheses; thus, the procedure in this case is here designated the pairwise sequentially-rejective procedure.

4. Comparison of the three classes of procedures in the nondirectional case

The properties of and relations among pairwise sequentially-rejective, subset, and partition approaches will be illustrated by considering their application under the following two conditions:

a. There is a statistic T_i for testing each pair homogeneity hypothesis H_i, $i = 1, ..., \binom{k}{2}$, such that large values of T_i lead to rejection.

b. Any test required is generated from the T_i through the use of the Bonferroni inequality. In (3), the Y_i are conveniently taken to be the uniformly distributed random variables $1 - \inf_{H_i} F_i(T_i)$, and the $\gamma_{\alpha,m}$ to be α/m for each m. (Each Y_i can be equivalently defined as $\sup_{H_i} P(T_i > t_i)$, where t_i is the observed value of T_i; Y_i is often called the *significance probability*.)

The sequentially-rejective approach, when applied to the set of all pairwise homogeneity hypotheses among a set of k distributions, can be uniformly improved without increasing the maximum experimentwise error rate. If the general sequentially-rejective procedure described above does not terminate before Stage j, then $H^{(j)}$ is rejected if $Y^{(j)}$ is smaller than or equal to $\alpha/[n - (j - 1)]$. Under the modified pairwise procedure, $[n - (j - 1)]$ in this expression is replaced by n^*_{j-1}, where n^*_{j-1} equals the maximum number of homogeneous pairs given $j - 1$ heterogeneous pairs (Shaffer, 1982). As will be illustrated, the modified sequentially-rejective procedure permits the use of larger significance levels at some stages for testing individual hypotheses than does the general sequentially-rejective procedure, thereby increasing power, without increasing the maximum experimentwise error rate.

Table 1 gives the successive significance levels for the unmodified and modified pairwise sequentially-rejective procedures for several values of k. To illustrate how the values in the modified procedure are obtained, consider $k = 5$. Initially, no hypotheses have been rejected, so n^*_0 equals the maximum

Table 1

Comparison of sequentially-rejective procedure (SR) and modified sequentially-rejective procedure (MSR) for $3 \leqslant k \leqslant 6$ and experimentwise error rate = α

(Entries are significance levels)

Stage	$k = 3$		$k = 4$		$k = 5$		$k = 6$	
	SR	MSR	SR	MSR	SR	MSR	SR	MSR
1	$\alpha/3$	$\alpha/3$	$\alpha/6$	$\alpha/6$	$\alpha/10$	$\alpha/10$	$\alpha/15$	$\alpha/15$
2	$\alpha/2$	$\alpha/1$	$\alpha/5$	$\alpha/3$	$\alpha/9$	$\alpha/6$	$\alpha/14$	$\alpha/10$
3	$\alpha/1$	$\alpha/1$	$\alpha/4$	$\alpha/3$	$\alpha/8$	$\alpha/6$	$\alpha/13$	$\alpha/10$
4			$\alpha/3$	$\alpha/3$	$\alpha/7$	$\alpha/6$	$\alpha/12$	$\alpha/10$
5			$\alpha/2$	$\alpha/2$	$\alpha/6$	$\alpha/6$	$\alpha/11$	$\alpha/10$
6			$\alpha/1$	$\alpha/1$	$\alpha/5$	$\alpha/4$	$\alpha/10$	$\alpha/10$
7					$\alpha/4$	$\alpha/4$	$\alpha/9$	$\alpha/7$
8					$\alpha/3$	$\alpha/3$	$\alpha/8$	$\alpha/7$
9					$\alpha/2$	$\alpha/2$	$\alpha/7$	$\alpha/7$
10					$\alpha/1$	$\alpha/1$	$\alpha/6$	$\alpha/6$
11							$\alpha/5$	$\alpha/4$
12							$\alpha/4$	$\alpha/4$
13							$\alpha/3$	$\alpha/3$
14							$\alpha/2$	$\alpha/2$
15							$\alpha/1$	$\alpha/1$

number of possibly homogeneous pairs, ten. Given one heterogeneous pair, the true partition (1) must have at least two members, which implies that at most six pairs are homogeneous (consider the partition (1) (2, 3, 4, 5)), so $n_1^* = 6$. Given two, three, or four heterogeneous pairs, there may still be six homogeneous pairs (e.g. pairs 12, 13, 14 and 15 are heterogeneous given the partition considered above). Given five heterogeneous pairs, however, at most four pairs can be homogeneous (as in the partition (1, 2) (3, 4. 5), for example), etc.

In order to compare the modified pairwise sequentially-rejective procedure with a subset procedure, the set of levels α_m for the latter must be specified. The levels most frequently used were proposed by Tukey (1953) and by Ryan (1960): $\alpha_k = \alpha_{k-1} = \alpha$; $\alpha_m = (m/k)\alpha$, $2 \leqslant m \leqslant k - 2$. The choice of levels was considered by Lehmann and Shaffer (1979), who found the Tukey-Ryan levels to be near-optimal in terms of a minimax criterion they considered.[3]

In Table 2, the modified sequentially-rejective procedure is compared with the subset procedure using the Tukey-Ryan levels, for $k = 4$ and 5 by giving in each row i the level at which a pair hypothesis may be rejected given $i - 1$ previous pair rejections. (For $k = 3$, the two procedures are identical.) Since the subset procedure levels depend on both the particular pairs already rejected and the one being tested when $k > 3$, they are not completely determined by the number of previous pair rejections, as are the

[3] Lehmann and Shaffer adopted a somewhat less general framework than has been adopted here; this statement represents the appropriate generalization of their result.

Table 2

Comparison of modified sequentially-rejective procedure (MSR) and Tukey-Ryan subset procedure (TR Sub) for $k = 4$ and 5 and experimentwise error rate $\leqslant \alpha$. (Entries are significance levels)

Stage	$k = 4$		$k = 5$	
	MSR	TR Sub	MSR	TR Sub
1	$\alpha/6$	$\alpha/6$	$\alpha/10$	$\alpha/10$
2	$\alpha/3$	$\alpha/3$	$\alpha/6$	$\alpha/6$
3	$\alpha/3$	$\alpha/2$ or $\alpha/3$	$\alpha/6$	$\alpha/5$ or $\alpha/6$
4	$\alpha/3$	$\alpha/2$ or $\alpha/3$	$\alpha/6$	$2\alpha/5$ or $\alpha/5$ or $\alpha/6$
5	$\alpha/2$	$\alpha/2$	$\alpha/6$	$2\alpha/5$ or $\alpha/5$ or $\alpha/6$
6	$\alpha/1$	$\alpha/2$	$\alpha/4$	$2\alpha/5$ or $\alpha/5$
7			$\alpha/4$	$2\alpha/5$ or $\alpha/5$
8			$\alpha/3$	$2\alpha/5$ or $\alpha/5$
9			$\alpha/2$	$2\alpha/5$
10			$\alpha/1$	$2\alpha/5$

Note. See Footnote 4.

sequentially-rejective levels.[4] It is easily shown that for $k > 3$, for both procedures, the levels are $2\alpha/[k(k-1)]$ and $2\alpha/[(k-1)(k-2)]$ at Stages 1 and 2, respectively; but that at Stage 3 the level for the modified sequentially-rejective procedure is $2\alpha/[(k-1)(k-2)]$, while that for the subset procedure with the Tukey-Ryan levels is either $2\alpha/[(k-1)(k-2)]$ (if the two pairs already rejected do not have any distribution in common), or $2\alpha/[(k(k-3)]$ for some pairs (if the two pairs already rejected have a distribution in common). This superiority of the Tukey-Ryan subset over the modified sequentially-rejective procedure holds for the first k stages. By a similarly straightforward comparison it can be shown that the modified sequentially-rejective procedure is superior to the Tukey-Ryan subset procedure over the last $[k/2]$ stages, where $[k/2]$ is the largest integer $\leqslant k/2$. It then follows that, in general, neither procedure is uniformly more powerful than the other.

For every subset procedure, there are associated partition procedures which are uniformly more powerful, for $k \geqslant 4$. These can be described as replacing the levels α_m adopted for the particular subset procedure by levels $\alpha_{i,m}$, where i is an index ranging over all partitions. To test any partition i, each member of size m is tested at level $\alpha_{i,m}$, and the partition is rejected if at least one member is rejected. The levels $\alpha_{i,m}$ can be chosen so that $\alpha_{i,m} \geqslant \alpha_m$ for all i and m with the inequality strict for some i and m. In the case of the Tukey-Ryan levels, for example, where $\alpha_m = (m/k)\alpha$, $2 \leqslant m \leqslant k-2$, the levels $\alpha_{i,m}$ can be taken as $[m/(k-u_i)]\alpha$, where $u_i =$ the number of members of the partition i of size 1. For instance, given $k = 6$ and the partition (1, 2) (3) (4) (5) (6) as the only one not already rejected, it (and thus the pair 1, 2) would be rejected if and only if the significance probability

[4] Under the subset procedure, for $k \geqslant 5$, it is possible at some intermediate stages to reject subsets of more than two distributions without rejecting any individual pairs and thus without progressing to the following stage. These additional subset rejections do not require modification of any comparisons in this paper between the subset and modified sequentially-rejective procedures.

for the subset (1, 2) were $\leqslant (2/6)\alpha$ under the Tukey-Ryan subset procedure, while it would be rejected under the associated partition method if the significance probability for (1, 2) were $\leqslant [2/(6-4)]\alpha = \alpha$.

From the definitions of the subset and partition approaches, it follows directly that the partition approach is at least as powerful as the subset approach whenever $\alpha_{i,m} \geqslant \alpha_m$ for all partitions i. Furthermore, as illustrated above, after all except one pair has been rejected, there exists a partition method for which that pair can be rejected at level α, while under the subset method the level must be strictly less than α; thus, there is always an associated partition method which is uniformly more powerful than any subset method. The disadvantage of partition methods is that they are considerably more complicated to apply than subset or pairwise methods if k is large. This is the result of the potentially extremely large number of possible tests required under a partition approach. Table 3, which gives, for $2 \leqslant k \leqslant 8$, the numbers of pair hypotheses, subset hypotheses, and partition hypotheses which might have to be tested in applying the pairwise sequentially-rejective approach, the subset approach, and the partition approach, respectively, illustrates the rapid increase in number of partition hypotheses as k increases.

Table 3

Number of homogeneity or partition hypotheses for possible test

	Function of k	$k=2$	3	4	5	6	7	8
Pair hypotheses	$\frac{k(k-1)}{2}$	1	3	6	10	15	21	28
Subset hypotheses	$2k-k-1$	1	4	11	26	57	120	247
Partition[a] hypotheses	$\sum_{t=1}^{k} S(k,t)-1$	1	4	14	51	202	876	4139

[a] $S(k, t)$ are the Stirling numbers of the second kind:

$$S(k,t) = \frac{1}{t!}\sum_{j=0}^{t} (-1)^j \binom{t}{j} (t-j)^k.$$

There have apparently been no comparisons of subset procedures, or partition procedures, under different choices of the α_m; most work on these procedures has used the Tukey-Ryan choice or slight modifications of it. Given the Tukey-Ryan levels, it has been shown above that there are partition methods which are uniformly more powerful than their associated subset methods, but those partition methods are not uniformly more powerful than the pairwise sequentially-rejective method. There exists a choice of the α_m and $\alpha_{i,m}$, however, for which the partition method is uniformly more powerful than both the associated subset method and the pairwise sequentially-rejective method, namely,

$$\alpha_m = \binom{m}{2} \Big/ \left[\binom{m}{2} + \binom{k-m}{2}\right] \quad \text{and} \quad \alpha_{i,m} = \binom{m}{2} \Big/ \sum_j \binom{s_{ij}}{2},$$

where s_{ij} is the number of distributions in member j of partition i, and the sum is over all members of partition i.

5. Directional decisions

In practice, directional decision patterns are often generated from nondirectional multiple test procedures by considering estimates $\hat{f}_i(P_i)$ [of $f_i(P_i)$] and $\hat{f}_j(P_j)$ [of $f_j(P_j)$], appropriately related to the test statistics, for each rejected pair of populations P_i, P_j, and declaring that the sign of $f_i(P_i) - f_j(P_j)$ is the same as the sign of $\hat{f}_i(P_i) - \hat{f}_j(P_j)$. The question arises as to whether the directional procedures so generated have the same maximum experimentwise error rates as their nondirectional counterparts. While the answer is positive for many product procedures of the kind described in Section 2, it is unknown for the more powerful types of procedures considered in Section 3. Holm (1979) proposed a general sequentially-rejective procedure, applicable to directional comparisons among populations, which controls experimentwise error at the required level and is intermediate in power between a product procedure and a nondirectional sequentially-rejective procedure supplemented by directional decisions as described above. Both empirical evidence and intuition suggest that in many circumstances the more powerful supplemented nondirectional procedures do control experimentwise error at the nominal level. (For results in a simpler context, see Holm (1981) and Shaffer (1980).) Further theoretical work on this issue is very much needed.

6. Comparisons involving nonequivalence relations

The comparisons among distributions treated in this paper do not exhaust the types of comparisons which are possible. For example, for any two continuous distributions P_i and P_j, one may be interested in the hypothesis $P(X_i < X_j) = P(X_j < X_i) = .5$, where X_i and X_j are independent observations from P_i and P_j, respectively. The Wilcoxon test has been generalized to test this hypothesis (Potthoff, 1963: Zaremba, 1962). The general sequentially-rejective method of Holm can be used to test this set of hypotheses for all pairs i and j, $i \neq j = 1, \dots, k$. However, the modified sequentially-rejective, subset, and partition methods are not applicable because the relation $P(X_i < X_j) = P(X_j < X_i)$ does not generate an equivalence relation among distributions: it may be true, for populations i, j and k, that $P(X_i < X_j) = P(X_j < X_i)$, $P(X_j < X_k) = P(X_k < X_j)$, but $P(X_i < X_k) \neq P(X_k < X_i)$. Thus, the true situation cannot be represented as a partition of the distributions. Note that the relation $P(X_i < X_j) = P(X_j < X_i)$ cannot be expressed as $f_i(P_i) = f_j(P_j)$ for any function f; thus, it is outside the class of comparisons considered in this paper.

REFERENCES

1. Cox, D. R., Spjøtvoll, E., *On partitioning means into groups.* Scand. J. Statist., 9, 147–152 (1982).
2. Gabriel, K. R., *Simultaneous test procedures — some theory of multiple comparisons.* Ann. Math. Statist. 40, 224–250 (1969).
3. Holm, S., *Sequentially rejective multiple test procedures.* Research Report 1977– 1, Institute of Mathematics and Statistics, University of Umeå, 1977.

4. Holm, S., *A simple sequentially rejective multiple test procedure.* Scand. J. Statist., **6**, 65—70 (1979).
5. Holm, S., *A stagewise directional test for the normal regression situation.* In: *Proc. Sixth Conference Probab. Theory (Braşov,* 1979), pp. 103— 106. Bucureşti, Ed. Acad. R.S. România, 1981.
6. Lehmann, E. L., *A theory of some multiple decision problems, I.* Ann. Math. Statist., **28**, 1— 25 (1957 a).
7. Lehmann, E. L., *A theory of some multiple decision problems, II.* Ann. Math. Statist., **28**, 547— 572 (1957 b).
8. Lehmann, E. L., Shaffer, J. P., *Optimum significance levels for multistage comparison procedures.* Ann. Statist. **7**, 27— 45 (1979).
9. Marcus, R., Peritz, E., Gabriel, K. R., *On closed testing procedures with special reference to ordered analysis of variance.* Biometrika, **63**, 655—660 (1976).
10. Potthoff, R. F., *Use of the Wilcoxon statistic for a generalized Behrens-Fisher problem.* Ann Math. Statist., **34**, 1596— 1599 (1963).
11. Ryan, T. A., *Significance tests for multiple comparison of proportions, variances, and other statistics.* Psych. Bull., **57**, 318— 328 (1960).
12. Shaffer, J. P., *Control of directional errors with stagewise multiple test procedures.* Ann. Statist., **8**, 1342— 1347 (1980).
13. Shaffer, J. P., *Modified sequentially-rejective multiple test procedures.* (In preparation, 1982).
14. Spjøtvoll, E., *On the optimality of some multiple comparison procedures.* Ann. Math. Statist., **43**, 398— 411 (1972).
15. Tukey, J. W., *The problem of multiple comparisons.* Unpublished manuscript, 1953.
16. Zaremba, S. K., *A generalization of Wilcoxon's test.* Monatsh. Math., **66**, 359 — 370 (1962).

JORDAN M. STOYANOV

Problems of estimation in continuous-discrete stochastic models

Continuous-time stochastic processes depend on unknown parameters which should be estimated on the basis of discrete data. General notions and classification of the estimation problems and of the available data are described. Unbiased and consistent estimators are found for the parameters of linear diffusion process by observations at random times. A short survey of previous results in this field is given. Related problems are discussed.

1. Introduction

This paper is devoted to statistical problems for continuous-time stochastic processes, and especially to problems of estimation of unknown parameters by using discrete data obtained as point observations. Below ξ denotes a stochastic process, θ its unknown parameters.

First, let ξ be a discrete-time stochastic process: $\xi = (\xi_n,\ n = 1, 2, \ldots)$ and assume we make our measurements at times $n = 1, 2, \ldots$. Then for each N the observations are $\xi_1, \ldots, \xi_N$ and any estimator of θ has the form $\theta_N = \varphi_N(\xi_1, \ldots, \xi_N)$ where $\varphi(\cdot)$ is a measurable function of N variables. Evidently this case can be called the *purely discrete case.*

Second, let ξ be a continuous-time stochastic process and assume we are able to observe ξ continuously in time. Let $T > 0$ and $\xi_0^T = \{\xi_s, 0 \leqslant s \leqslant T\}$ be the observations in the interval $[0, T]$. Then any estimator of θ has the form $\tilde{\theta}_T = \psi_T(\xi_0^T)$ where ψ_T is a measurable functional on the space of the realizations of ξ. Since both the process and the observations are continuous in time, this case can be called the *purely continuous case.*

Now we shall concentrate our attention on a case which differs from both given above. Suppose ξ is a continuous-time stochastic process depending on unknown parameters θ. The problem is to estimate θ on the basis of observations on ξ. The main of our assumptions is the following: we are not able to observe ξ continuously in time but all measurements can be made at times $t \in \mathcal{W}$ where $\mathcal{W}$ is some discrete set, $\mathcal{W} \subset R_+ = (0, \infty)$. For example $W = \{t_1, t_2, \ldots, \}$ with $0 < t_1 < t_2 < \ldots$ Thus for each natural N we have at our disposal the observations $\xi_{t_1}, \ldots, \xi_{t_N}$ obtained at times $t \in \mathcal{W}_N = \{t_1, \ldots, t_N\}$ and any estimator $\hat{\theta}_N$ is a measurable function of these N observations ($\hat{\theta}_N$ depends also on $t_1, \ldots, t_N$).

Further on we shall treat namely a scheme of continuous-discrete type. Let us note that there are important practical situations which can be modelled just in this way, see e.g. [17], [30], [31], [42].

The paper is organized as follows: Section 2 contains definitions of the basic notions as well as a classification of the estimation problems and of the available data. The example in Section 3 illustrates the possibility to find solutions of continuous-discrete estimation problems. In Section 4 we present a short survey of previous results and papers. This survey shows that the treatment of the continuous-discrete models is reduced to interesting stochastic problems. Moreover such problems can be useful in the analysis of important practical phenomena. One specific but interesting model is considered in Section 5: a linear diffusion process is observed at random times. Unbiased and consistent estimators are suggested for its coefficients. Other problems and possibilities related to those presented here are discussed in Section 6.

2. Basic notions. Classifications

All random variables and stochastic processes considered in this paper are defined on a complete probability space $(\Omega, \mathfrak{F}, \mathbf{P})$ and for simplicity take values in $R_1 = (-\infty, \infty)$.

Let $\xi = (\xi_t, t > 0)$ be a continuous-time stochastic process. Suppose that the trajectories of ξ are $\mathbf{P}$-a.s. right continuous with left hand limits.

The general problem discussed here is the following. Suppose ξ depends on unknown parameters, say $\theta : \xi = \xi(\theta)$. Let $\mathcal{W} = \{t_1, t_2, ..., t_N, ...\}$ be a discrete set, $\mathcal{W} \subset R_+$ and $t_1 < t_2 < ...$ Our aim is to estimate θ on the basis of observations on ξ only at times $t \in \mathcal{W}$. Thus we have to solve an estimation problem for continuous-time stochastic process by using discrete data.

Definition. The pair $(\xi, \mathcal{W})$ where ξ is a continuous-time stochastic process and $\mathcal{W}$ is a discrete subset of R_+ is said to be a *continuous-discrete stochastic model*. The set $\mathcal{W}$ consisting of all times when we can observe ξ is said to be a ***discrete window***. (Another suitable name for $\mathcal{W}$ is *t-set for sampling* ξ.)

Evidently, these models are mixtures of the purely discrete and purely continuous cases described briefly in the Introduction. If $(\xi, \mathcal{W})$ is a model of this type then for solving problems of estimation of θ we have at our disposal the discrete data $\mathcal{D} = \{\xi_{t_1}, \xi_{t_2}, ..., \xi_{t_N}, ...\}$.

It is essential to note that there are cases when estimation problems cannot be solved by the observations ξ_0^T. Then naturally we look for solutions based on discrete data with the hope that the best "discrete" solution will be near in some sense to the real "continuous" solution.

The elements of the set $\mathcal{W}$ can be of two types: (i) deterministic: t_j are fixed positive numbers and $\mathcal{W}$ is called a *deterministic window*.; (ii) random: $\mathcal{W}$ consists of a sequence of random variables. In this case we shall use Greek letters τ_j, $j = 1, 2, ...$, i.e. $\mathcal{W} = \{\tau_1, \tau_2, ...\}$ and $\mathcal{W}$ is called a *random window*. In both cases, $\mathcal{W} = \{t_1, t_2, ..., \}$ and $\mathcal{W} = \{\tau_1, \tau_2, ...\}$, it is supposed that $0 < t_1 < t_2 < ...$ and $0 < \tau_1 < \tau_2 < ...$ $\mathbf{P}$-a.s.

One particular case of a deterministic window is the following: $\mathcal{W} = \{\Delta, 2\Delta, ..., N\Delta, ...\}$ where $\Delta = \text{const} > 0$. Here the times for sampling are equally spaced and $\mathcal{D} = \{\xi_\Delta, \xi_{2\Delta}, ..., \xi_{N\Delta}, ...\}$ are known as *equally spaced data*. If $t_1, t_2, \ldots$ are arbitrary, then $\mathcal{D} = \{\xi_{t_1}, \xi_{t_2}, \ldots\}$ represent *unequally spaced data*. Of course, if $\mathcal{W} = \{\tau_1, \tau_2, \ldots\}$ is a random window then with probability 1 the data $\mathcal{D}$ are unequally spaced.

The case of equally spaced data is simpler from a theoretical point of view. However there are many situations where this simple and convenient scheme of sampling is not realizable and the only kind of data which we can get and use are discrete unequally spaced data.

Let $\mathcal{W} = \{\tau_1, \tau_2, \ldots\}$ be a random window and $\mathcal{D} = \{\xi_{\tau_1}, \xi_{\tau_2}, \ldots\}$ be the corresponding data. Actually, in this case $\{\tau_n\}$ form a point type process (in particular it can be a counting process) and thus we have to solve some statistical problem on the basis of the point observations $\mathcal{D}$.

If $\mathcal{W} = \{t_1, t_2, \ldots\}$ is a deterministic window and $\tilde{\theta}$ is an estimator of θ based on observations on ξ at $t \in \mathcal{W}$, then $\tilde{\theta}$ as well as the variance of $\tilde{\theta}$ depend explicitly on $t_1, t_2, \ldots$. If $\mathcal{W} = \{\tau_1, \tau_2, \ldots\}$ is a random window, then of course the variance of $\tilde{\theta}$ depends on the probabilistic structure of $\mathcal{W}$, in particular, on those parameters defining $\mathcal{W}$. However these parameters are either given, or they are unknown. So we come to the following estimation problems for the models $(\xi, \mathcal{W})$.

Direct estimation problem. The observable process ξ depends on unknown parameters θ and the window $\mathcal{W}$ is well defined, i.e. $\mathcal{W}$ does not contain unknown parameters and its probabilistic structure is given. The problem to estimate θ using observations on ξ at times $t \in \mathcal{W}$ is called a direct estimation problem.

Inverse estimation problem. Suppose that the process ξ is given completely but $\mathcal{W}$ depends on unknown parameters, say μ. Then the estimation of μ on the basis of observations on ξ at times $t \in \mathcal{W}$ is called an inverse estimation problem.

Combined estimation problem. The classification of the above problems as direct and inverse is conditional, of course. The most general case is the following: both components ξ and $\mathcal{W}$ depend on unknown parameters, θ and μ. The process ξ is observed only for $t \in \mathcal{W}$. Then the joint estimation of θ and μ is called a combined estimation problem.

The following question is quite natural. What sort of data are possible when we consider a continuous-discrete stochastic model? There are three answers, and they are as follows:

Extended data: $\hat{\mathcal{D}} = \{\xi_{\tau_1}, \xi_{\tau_2}, \ldots, \tau_1, \tau_2, \ldots\}$. We know both the observations ξ_{τ_j} and the times τ_j, $j = 1, 2, \ldots$

Ordered data: $\tilde{\mathcal{D}} = \{X_1, X_2, \ldots\}$ where $X_j = \xi_{\tau_j}$, $j = 1, 2, \ldots$. We only know the observations on the process ξ and they are given in their natural order of appearance.

Unordered data: $\bar{\mathcal{D}} = \{Y_{i_1}, Y_{i_2}, \ldots\}$ where $Y_{i_k} = \xi_{\tau_{i_k}}$, $k = 1, 2, \ldots$. We have observations on ξ obtained at times $t \in \mathcal{W}$ but their order of appearance is unknown.

It is clear now that each of the problems formulated above can be treated separately depending on the available data. Of course, in each case the estimators which we construct should be functions of the corresponding data only.

In this section we mentioned only problems of estimation of unknown parameters. There are also other important problems which are included naturally into the framework of the continuous-discrete models. Some possibilities will be discussed in Section 6.

3. Example

Our aim is to illustrate how an unknown parameter of a continuous-time process can be estimated by using discrete observations. A few cases of sampling as well as a few different estimators are suggested.

Let $w = (w_t,\ t \geqslant 0)$ be the standard Wiener process (a continuous process with independent and Gaussian distributed increments, $w_0 = 0$ **P**-a.s., $\mathbf{E}w_t = 0$, $\mathbf{E}w_t^2 = t$). Consider the process $\xi = (\xi_t,\ t \geqslant 0)$ where

$$\xi_t = \sigma w_t \tag{1}$$

with an unknown parameter σ which should be estimated. Take the interval $[0, T]$, $T > 0$ and the subdivision $\{0 = t_0 < t_1 < \ldots < t_N = T\}$. The process ξ is observed at the times of the deterministic window $\mathcal{W}_N = \{t_1, t_2, \ldots, t_N\}$ and thus we get the data $\mathcal{D}_N = \{\xi_{t_1}, \xi_{t_2}, \ldots, \xi_{t_N}\}$. Put

$$b_N^2 = \frac{1}{T}\sum_{j=1}^{N}(\xi_{t_j} - \xi_{t_{j-1}})^2. \tag{2}$$

The quantity b_N^2 from (2) as an estimator of σ^2 is suggested by Doob [9]. It is easy to find the expectation $\mathbf{E}\{b_N^2\}$ and the variance $\mathbf{D}\{b_N^2\}$:

$$\mathbf{E}\{b_N^2\} = \sigma^2, \quad \mathbf{D}\{b_N^2\} = \frac{2\sigma^4}{T^2}\sum_{j=1}^{N}(t_j - t_{j-1})^2.$$

Therefore b_N^2 is an unbiased estimator of σ^2. However $\mathbf{D}\{b_N^2\} \to 0$ if and only if the diameter of the subdivision $d_N = \max_j (t_j - t_{j-1}) \to 0$ as $N \to \infty$, i.e. b_N^2 is a consistent estimator of σ^2 if and only if $d_N \to 0$. Moreover $d_N \to 0$ implies

$$\lim_{N\to\infty} b_N^2 = \sigma^2 \tag{3}$$

in the L_2-sense and **P**-a.s. Equation (3) is a simple consequence of the well-known Lévy's theorem (see e.g. Liptser and Shiryaev [23]).

Evidently, the number d_N is a characteristic of the window $\mathcal{W}_N$ and $d_N \to 0$ means that for large N, $\mathcal{W}_N$ is "almost" the whole interval $[0, T]$, i.e. (2) is a consistent estimator of σ^2 if "practically" we observe ξ continuously during the interval $[0, T]$. However it can happen that $N \to \infty$ but $d_N \not\to 0$. In this case b_N^2 does not satisfy the standard requirement to be consistent. It is natural to try to construct a "good" estimator in this "bad" case. Indeed, if we take

$$\hat{b}_N^2 = \frac{1}{N}\sum_{j=1}^{N}\frac{(\xi_{t_j} - \xi_{t_{j-1}})^2}{t_j - t_{j-1}}, \tag{4}$$

then $\mathbf{E}\{\hat{b}_N^2\} = \sigma^2$ and $\mathbf{D}\{\hat{b}_N^2\} = 2\sigma^4/N$. Therefore for an arbitrary window $\mathcal{W}_N = \{t_1, \ldots, t_N\}$ (without the assumption $d_N \to 0$) and observations on ξ at $t \in \mathcal{W}_N$, $\hat{b}_N^2$ defined by (4) is an unbiased and consistent estimator of σ^2.

Suppose now that the process ξ given by (1) is observed through a random window. Following Žilinskas and Senkienė ([43], [44]) for each N we can consider the random window $\mathcal{W}_N = \{\tau_1, \ldots, \tau_N\}$, where the variables τ_j are strictly increasing in j and τ_j depends on $\tau_1, \ldots, \tau_{j-1}, \xi_{\tau_1}, \ldots, \xi_{\tau_{j-1}}$. Let $\tau_0 = 0$ and ξ be observed at $t \in \mathcal{W}_N$. Put

$$\tilde{b}_N^2 = \frac{1}{N} \sum_{j=1}^{N} \frac{(\xi_{\tau_j} - \xi_{\tau_{j-1}})^2}{\tau_j - \tau_{j-1}}. \tag{5}$$

Evidently, the estimator (5) depends explicitly on both the observations $\xi_{\tau_1}, \ldots, \xi_{\tau_N}$ and the times $\tau_1, \ldots, \tau_N$, i.e. for constructing this estimator we use extended data. One can show that $\tilde{b}_N^2$ is an unbiased and consistent estimator of the parameter σ^2 (for details see [43], [44]).

4. A short survey of previous results

There are many publications on statistical problems for stochastic processes. The reference list given at the end includes papers and books dealing with problems related to those described here. The cited papers are devoted to different processes and problems and contain various results. Nevertheless the following classes of processes are studied in many papers: stationary processes, diffusion type processes, point type processes.

As far as we know, Bartlett [2] was the first author to consider, in our terminology, a continuous-discrete model $(\xi, \mathcal{W})$ where ξ is a continuous-time stationary process (in fact a second order autoregressive process) and $\mathcal{W}$ is the deterministic window $\mathcal{W} = \{\ldots, -2\Delta, -\Delta, 0, \Delta, 2\Delta, \ldots\}$, $\Delta =$ const. This is the case of equally spaced data. In that paper problems of estimation of unknown parameters are studied. Other papers in the same direction are those of Durbin [11], Pandit and Wu [28].

Grenander [14] was the first to suggest to use observations on a continuous-time stationary process at random times which form a point type process.

The papers by Brillinger ([7], [8]) and Robinson [32] are devoted to interesting problems for stationary or point type processes. The approach in these papers is similar to that used in the present paper. The book by Basawa and Prakasa Rao [3] contains some results in this field as well as additional references not included here. Blum and Boyles [5] consider the model $(\xi, \mathcal{W})$ where ξ is a continuous-time stationary process and $\mathcal{W}$ is generated by a point process. They find an estimator of the finite-dimensional distributions of ξ.

The papers mentioned above mainly treat problems of estimation of unknown parameters of stationary type processes. Another important problem is to find estimators for the unknown spectral densities of stationary processes. Beutler [4], Masry [25], [26] and Robinson [32] have found solutions to problems of this kind.

It is not surprising that many authors have investigated statistical problems for diffusion processes, see the books by Liptser and Shiryaev [23], Ibragimov and Hasminskii [16], Kutoyantz [20], Basawa and Prakasa Rao [3] and other references given therein. However, these authors suppose that the process is observed continuously in time and the estimators depend explicitly on stochastic integrals. Of course, from a theoretical point of view the purely continuous approach is attractive enough. Now this field is one of the most beautiful branches in Stochastics. Estimation problems for the parameters of a diffusion process on the basis of discrete observations were recently investigated. Interesting results were obtained by Le Breton [21], [22], Žilinskas and Senkienė ([43], [44]), Robinson [31], Wilks [42], Loges [24].

Dorogovchev [10] and Knopov [18] have studied the important problem to estimate a continuous function by observing its values at discrete points with random errors. This situation can be considered as a functional version of a continuous-discrete model, see also Sonnevend [33].

The present paper is a continuation with further extensions of the previous papers by Stoyanov ([34], [35]), and Stoyanov and Vladeva ([36], [37] and [38]). We shall consider the model $(\xi, \mathcal{W})$ where ξ is associated with a linear stochastic differential equation and $\mathcal{W}$ is a random window generated by the partial sums of a sequence of i.i.d. random variables.

5. Parameter estimation in a linear stochastic differential equation

First let us define the process ξ and the window $\mathcal{W}$. Suppose the process $\xi = (\xi_t, t \geqslant 0)$ has the form

$$\xi = at + \sigma w_t \tag{6}$$

where a and σ are constants, $-\infty < a < \infty$, $\sigma^2 > 0$ and w is the standard Wiener process. Thus ξ is a simple diffusion process with constant drift and diffusion coefficients a and σ^2, respectively.

Let α be a positive random variable and $\alpha_1, \alpha_2, \ldots, \alpha_N, \ldots$ a sequence of independent copies of α. Put $\mathcal{W} = \{\tau_1, \tau_2, \ldots, \tau_N, \ldots\}$, where $\tau_1 = \alpha_1$, $\tau_2 = = \alpha_1 + \alpha_2, \ldots, \tau_N = = \alpha_1 + \ldots + \alpha_N, \ldots$, i.e. $\mathcal{W}$ is a random window of renewal type.

Remark. Let the process $\eta = (\eta_t, t \geqslant 0)$ satisfy the linear stochastic differential equation (in Ito or Stratonovich sense):

$$d\eta_t = a_0 \eta_t \, dt + \sigma_0 \eta_t dw_t, \quad \eta_0 = c. \tag{7}$$

By Ito formula, see e.g. Liptser and Shiryaev [23], we can find η explicitly. Then it is easy to see that the new process $\tilde{\eta} = \ln(\eta/c)$ has just the form (6). Suppose a_0 and σ_0 in (7) are unknown and should be estimated from observations on η at times $t \in \mathcal{W}$. This problem can be reduced to a similar problem for a simpler process of the type (6) with new constant coefficients. This means that if we have a solution of an estimation problem for the process (6), then we can easily solve the same problem for the process (7). Moreover, our aim is to construct "good" estimators. "Good" is each estimator which satisfies the minimal requirements to be unbiased and consistent.

In the sequel we shall suppose that the window $\mathcal{W}$ and the Wiener process (the noise process) w are independent. Introduce for convenience τ_0: $\tau_0 = 0$ implying $\xi_{\tau_0} = X_0 = 0$. Put also $m_k = \mathbf{E}\{\alpha^k\}$, $k = 1, 2, 3, 4$ and suppose that $m_4 < \infty$.

Now we shall formulate a few propositions assuming that ξ and $\mathcal{W}$ satisfy the conditions described above.

Direct estimation problem. Recall that a and σ^2 in (6) are unknown and the window $\mathcal{W}$ does not contain unknown parameters.

Proposition 1. *Suppose that for each N, $N \geq 2$, the extended data $\widehat{\mathcal{D}}_N = \{\xi_{\tau_1}, \ldots, \xi_{\tau_N}; \tau_1, \ldots, \tau_N\}$ are available. Then*

$$\hat{a}_N = \frac{\xi_{\tau_N}}{\tau_N} \text{ and } \hat{\sigma}_N^2 = \frac{1}{N-1}\left[\sum_{j=1}^{N} \frac{(\xi_{\tau_j} - \xi_{\tau_{j-1}})^2}{\tau_j - \tau_{j-1}} - \frac{\xi_{\tau_N}^2}{\tau_N}\right]$$

are unbiased and strong consistent estimators of a and σ^2, respectively.

Proposition 2. *Let for each N, $N \geq 2$, the ordered data $\widetilde{\mathcal{D}}_N = \{X_1, \ldots, X_N\}$ be given and $r = m_2/m_1^2$. Then*

$$\tilde{a}_N = \frac{X_N}{m_1 N} \text{ and } \tilde{\sigma}_N^2 = \frac{N+r-1}{m_1 N(N-1)}\left[\sum_{j=1}^{N} (X_j - X_{j-1})^2 - \frac{r}{N+r-1} X_N^2\right]$$

are unbiased and strong consistent estimators of a and σ^2, respectively.

Proposition 3. *If the unordered data $\overline{\mathcal{D}}_N = \{Y_{i_1}, \ldots, Y_{i_N}\}$ are available, then*

$$\bar{a}_N = \frac{2}{m_1 N(N+1)} \sum_{k=1}^{N} Y_{i_k}$$

is an unbiased and strong consistent estimator of a.

Inverse estimation problem. In this case we know a and σ^2 in (6) but the window $\mathcal{W}$ contains unknown parameters. Suppose that some of the moments m_k, $k = 1, 2, 3, 4$ are unknown. Our aim is to find estimators for these parameters from observations on ξ at $t \in \mathcal{W}$.

Evidently if we have at our disposal the extended data $\widehat{\mathcal{D}}_N$ then the unknown parameters of $\mathcal{W}$ can be estimated by using only a part of the data $\widehat{\mathcal{D}}_N$, namely $\tau_1, \ldots, \tau_N$. But this is classical.

Proposition 4. *Let the ordered data $\widetilde{\mathcal{D}}_N = \{X_1, \ldots, X_N\}$ be given, the drift coefficient $a \neq 0$ and the first moment m_1 be unknown. Then $\widetilde{m}_{1N} = \dfrac{X_N}{aN}$ is an unbiased and strong consistent estimator for m_1.*

If m_2 is unknown and m_1 is given, then $\widetilde{m}_{2N} = \dfrac{X_N^2}{a^2 N} - (N-1)m_1^2 - \dfrac{\sigma^2}{a^2} m_1$ is an unbiased and strong consistent estimator for m_2.

If both parameters m_1 and m_2 are unknown, then the estimators for m_1 and m_2 are the same as above but m_1 in $\widetilde{m}_{2N}$ should be replaced by $\widetilde{m}_{1N}$ in order that $\widetilde{m}_{2N}$ preserve its consistency.

Proposition 5. *Let again the ordered data $\widetilde{\mathcal{D}}_N$ be given, the first moment m_1 be unknown but the drift $a = 0$. Then each of the statistics*

$$\widetilde{m}_{1N}^{*} = \frac{X_N^2}{\sigma^2 N} \text{ and } \widetilde{m}_{1N}^{**} = \frac{1}{\sigma^2 N} \sum_{j=1}^{N} (X_j - X_{j-1})^2$$

is an unbiased and strong consistent estimator of m_1.

If m_2 is unknown and m_1 is given, then $\widetilde{m}_{2N}^{*} = \dfrac{X_N^4}{3\sigma^4 N} - (N-1)m_1$ *is an unbiased and consistent estimator of m_2. If in addition m_1 is also unknown it should be replaced by its estimator, and then $\widetilde{m}_{2N}^{*}$ preserves the consistency property.*

Proposition 6. *Let the unordered $\overline{\mathcal{D}}_N = \{Y_{i_1}, \ldots, Y_{i_N}\}$ be available, the parameter m_1 be unknown and the drift coefficient $a \neq 0$. Then* $\overline{m}_{1N} = \dfrac{2}{aN(N+1)} \sum_{k=1}^{N} Y_{i_k}$ *is an unbiased and strong consistent estimator of m_1.*

Combined estimation problem. Here both components of the model $(\xi, \mathcal{W})$ depend on unknown parameters which should be estimated from observations on ξ at times $t \in \mathcal{W}$. However this problem has not always a solution. In general, this happens if we have ordered or unordered data.

Proposition 7. *Suppose the extended data $\hat{\mathcal{D}}_N = \{\xi_{\tau_1}, \ldots, \xi_{\tau_N}; \tau_1, \ldots, \tau_N\}$ are available and the parameters a, σ^2, m_1 are unknown. Then $\hat{a}_N$ and $\hat{\sigma}_N^2$ from Proposition* 1 *together with* $\hat{m}_{1N} = \dfrac{1}{N}(\tau_1 + \ldots + \tau_N)$ *are unbiased and consistent estimators of a, σ^2, m_1, respectively.*

Remark. We can consider a few other cases when the model $(\xi, \mathcal{W})$ depends on unknown parameters in another way and construct corresponding estimators.

The estimators suggested in Propositions 1 and 7 are modified maximum likelihood estimators. For finding the other ones the method of moments and some closely related arguments are used.

The restricted size of the paper is the reason why the proofs of Propositions 1—7 are omitted. Actually the proofs are not so difficult. They are based on the properties of ξ and $\mathcal{W}$.

In [36], [37] we considered the model $(\xi, \mathcal{W})$ where ξ is the process defined by (6) and $\mathcal{W}$ is the so-called λ-Poisson random window, i.e. $\{\tau_n\}$ is the usual Poisson point process.

The processes (6) and (7) introduced above and the Ornstein-Uhlenbeck process ($dx_t = \beta x_t dt + \gamma dw_t$) have a lot of interesting properties which can be used for solving various problems, for example estimation of unknown parameters. At the same time these processes play an essential rôle in the analysis of phenomena from other fields. Recall that processes of this type appear to be limits of complicated processes arising in queueing theory (see Borovkov [6]) or in branching process theory (see Wang [41] and the references therein to previous papers by W. Feller, M. Jirina, P. Jagers, T. Kurtz). Examples from Physics, Chemistry and Biology can be given.

6. Some remarks and further problems

Let us briefly mention a few other interesting and important problems for continuous-discrete stochastic models. Some aspects or particular cases of these questions are discussed in literature.

1) *Investigation of other asymptotic properties of the estimators* (asymptotic normality, loglog law, etc.). The results and techniques in Basawa and Prakasa Rao [3], Robinson [31], Le Breton [22] and Loges [24] will be very useful. The martingale limit theorems, see Hall and Heyde [15], Feigin [12], are another powerful method.

2) *Testing of hypotheses* for continuous-time processes on the basis of discrete observations (at fixed or random times). The case of equally spaced data is studied by Newman and Stuck [27].

3) *Estimation of parameters of general classes of processes* (with independent increments, Markov, linear and nonlinear models with stationary noises, etc.). See e.g. Stoyanov and Vladeva [37], Vladeva [40].

4) *Estimation problems by indirect observations.* We observe another process $\zeta = \zeta(\xi)$ at times $t \in \mathcal{W}$; see e.g. Loges [24]. Moreover we can take θ to be a stochastic process, too. Then we have to solve the smoothing, filtering and prediction problems for θ given observations on ξ only at times $t \in \mathcal{W}$. Two aspects of this problem in the Markovian case are treated by Jazwinski [17] and Gill [13]. For the Gaussian case see Takeuchi and Akashi [39] and Komornik [19]. The diffusion case in a little different context is studied by Pardoux [29].

5) *Approximation of "continuous"* (in time) *estimators by "discrete" ones.* The paper by Le Breton [21] is a good reference in this respect.

The comparison of different estimators is also an interesting problem, see e.g. Stoyanov and Vladeva [36], [37] and Vladeva [40].

6)' *Two optimization problems.* Let $\mathcal{W} = \{t_1, t_2, ...\}$ be a deterministic window and assume that for solving some problem we are given a certaifn criterion. A natural and important question arises: For which location op the times t_j shall we get the best solution? For some results see Le Breton [22], Wilks [42], Newman and Stuck [27], Sonnevend [33]. If $\mathcal{W}$ is a random window we can ask: How to choose the parameters defining $\mathcal{W}$ in order to get the best solution? A little different optimization problems are considered by Barhatova [1] and Komornik [19]. In both papers the optimal controls depend essentially on a sequence of times (fixed or random).

Acknowledgements. I am grateful to the organizers of the 7th Braşov Conference on Probability Theory, and especially to Prof. Marius Iosifescu, for their interest in my problematics and for the invitation to take part in the Conference as one of the invited speakers.

During the last 2—3 years I had the pleasure to discuss these problems with many stochasticians. Now I use this opportunity to express my thanks for useful comments and suggestions to: Prof. A. Shiryaev, Prof. B. Gnedenko, Prof. D. G. Kendall, Prof. D. Chibisov, Prof. E. Hmaladze, Prof. Yu. Kutoyantz, Prof. J. Zabczyk, Prof. M. Musiela and Prof. E. Platen.

REFERENCES

1. Barhatova, I. V., *On optimal moments of observation of certain linear systems.* In: *Theory of Random Processes, No.* 9, pp. 12—20. Kiev, Naukova Dumka, 1981. (Russian)
2. Bartlett, M. S., *On the theoretical specification of sampling properties of autocorrelated time series.* J. Royal Statist. Soc. Suppl., **8**, 27—41 (1946).
3. Basawa, I. V., Prakasa Rao, B. L. S., *Statistical Inference for Stochastic Processes.* London, Academic Press, 1980.

4. Beutler, F. J., *Alias free randomly timed sampling of stochastic processes.* IEEE Trans. Inform. Theory, IT– **16**, 147– 152 (1970).
5. Blum, J. R., Boyles, R. A., *Random sampling from a continuous parameter process.* In: *Analytical Methods in Probability Theory.* Lecture Notes in Math., Vol. 861, pp. 15– 24. Berlin—Heidelberg—New York, Springer, 1981.
6. Borovkov, A. A., *Asymptotic Methods in Queueing Theory.* Moscow, Nauka, 1980. (Russian)
7. Brillinger, D. R., *The spectral analysis of stationary interval functions.* In: *Proc. Sixth Berkeley Sympos. Math. Statist. Probab., Vol.* 1, pp. 483– 513. Berkeley, Univ. California Press, 1972.
8. Brillinger, D. R., *Statistical inference for stationary point processes.* In: Puri, M. L. (Ed.), *Stochastic Processes and Related Topics,* pp. 55– 99. New York, Academic Press, 1975.
9. Doob, J. L., *Stochastic Processes.* New York, Wiley, 1953.
10. Dorogovchev, A. Ya., *Estimation Theory of the Parameters of Stochastic Processes.* Kiev, Kiev State Univ., 1982. (Russian)
11. Durbin, J., *Efficient fitting of linear models for continuous stationary time series from discrete data.* Bull. Internat. Statist. Inst., **38**, 273– 282 (1961).
12. Feigin, P. D., *Maximum likelihood estimation for continuous-time stochastic processes.* Adv. in Appl. Probab., **8**, 712– 736 (1976).
13. Gill, R.D., *Nonparametric estimation based on censored observations of a Markov renewal process.* Z. Wahrsch. Verw. Gebiete, **53**, 97– 116 (1980).
14. Grenander, U., *Stochastic processes and statistical inference.* Ark. Mat., **1**, 195– 277 (1951).
15. Hall, P., Heyde, C. C., *Martingale Limit Theorem and Its Applications.* New York, Academic Press, 1980.
16. Ibragimov, I. A., Hasminskii, R.Z., *Statistical Estimation. Asymptotic Theory.* New York— Heidelberg— Berlin, Springer, 1981. (Translation of the Russian Edition, Moscow, Nauka, 1979)
17. Jazwinski, A., *Stochastic Processes and Filtering Theory.* New York— San Francisco— London, Academic Press, 1970.
18. Knopov, P. S., *Optimal Estimation of Parameters of Stochastic Systems.* Kiev, Naukova Dumka, 1981. (Russian)
19. Komornik, J., *Optimal control under discrete observation of continuous stochastic systems with time delay.* Kybernetika (Prague), **14**, 102– 109 (1978).
20. Kutoyantz, Yu. A., *Estimation of Parameters of Random Processes.* Erevan, Armenian Acad. Sci., 1980. (Russian; English translation in preparation).
21. Le Breton, A., *On continuous and discrete sampling for parameter estimation in diffusion type processes.* In: Wets, R. J.-B. (Ed.), *Stochastic Systems I: Modeling and Identification.* Mathematical Programming Studies, Vol. 5, pp. 124– 144. Amsterdam— New York, North-Holland, 1976.
22. Le Breton A., *Estimations des paramètres d'un système bilinéaire autonome homogène.* Rapport Rech. No. 166, Univ. Grenoble, 1979.
23. Liptser, R. S., Shiryaev, A. N., *Statistics of Random Processes. I: General Theory; II. Applications.* New York— Heidelberg— Berlin, Springer, 1977; 1978. (Translation of the Russian Edition, Moscow, Nauka, 1974).
24. Loges, W., *Estimation of parameters for Hilbert space valued partially observable stochastic processes.* Preprint, Math. Inst. Ruhr-Univ., Bochum, 1982.
25. Masry, E., *Poisson sampling and spectral estimation of continuous-time processes.* IEEE Trans. Inform. Theory, IT-24, 173– 183 (1978).
26. Masry, E., *Discrete-time spectral estimation of continuous-time processes. The orthogonal series method.* Ann. Statist., **8**, 1100– 1109 (1980).
27. Newman, C. M., Stuck, B.W., *Optimal sampling of independent increment processes.* Stochastics, 2, 213– 225 (1979).
28. Pandit, S. M., Wu, S. M., *Unique estimates of the parameters of a continuous stationary stochastic process.* Biometrika, **62**, 497– 501 (1975).
29. Pardoux, E., *Filtering of a diffusion process with Poisson type observation.* In: *Stochastic Control Theory and Stochastic Differential Equations.* Lecture Notes in Control and Information Sci., Vol. 16, pp. 510– 518. Berlin— Heidelberg—New York, Springer, 1979.
30. Ripley, B.D., Rasson, J. P., *Finding the edge of a Poisson forest.* J. Appl. Probab., **14**, 483– 491 (1977).
31. Robinson, P. M., *Estimation of a time series model from unequally spaced data.* Stochastic Process. Appl., **6**, 9– 24 (1977).
32. Robinson, P. M., *Continuous model fitting from discrete data.* In: Brillinger, D., Tiao, G. (Eds.), *Directions in Time Series.* pp. 263– 278. Hayward, Cal., Inst. Math. Statist., 1980.

33. Sonnevend, G., *On optimal sensor location in stochastic differential systems and in their deterministic analogous.* In: *Stochastic Differential Systems.* Lecture Notes in Control and Information Sci., Vol. 36, pp. 165–174. Berlin–Heidelberg–New York, Springer, 1981.
34. Stoyanov, J., *An approach to estimation problems for continuous-time stochastic processes.* Report at the Conference "Mathematische Stochastik" (L. Arnold and H. Strasser, organizers), Oberwolfach, March 28–April 3, 1982.
35. Stoyanov, J., *Continuous-discrete stochastic models. Basic notions and some results.* (In print)
36. Stoyanov, J., Vladeva, D., *Estimation of unknown parameters of continuous time stochastic processes by observations at random moments.* C.R. Acad. Bulgare Sci., **35**, 153–156 (1982).
37. Stoyanov, J., Vladeva, D., *Estimation of the parameter of a Poisson random window by observations of a diffusion process.* In: *Mathematics and Education in Mathematics* (Proc. 12th Conf. Union Bulgar. Math.), pp. 431–435. Sofia, Bulgar. Acad. Sci., 1982.
38. Stoyanov, J., Vladeva, D., *Parameter estimation in a continuous-discrete stochastic model with a stationary noise.* In: *Trans. 9th Prague Conf. Information Theory, Statist Decision Functions, Random Processes (Prague,* 1982*)*, pp. 205–210. Prague Academia, 1983.
39. Takeuchi, Y., Akashi, H., *Nonlinear filtering formulas for discrete-time observations.* SIAM J. Control Optim., **19**, 244–261 (1981).
40. Vladeva, D., *Estimation of parameters in a diffusion-geometric model.* (In preparation)
41. Wang, F. J. S., *Diffusion approximations of age-and-position dependent branching processes.* Stochastic Process. Appl., **13**, 59–74 (1982).
42. Wilks, A. R., *Optimal Design of Observation Times for Continuous Time Series.* Ph. D. Thesis, Princeton Univ., Dept. Statistics, October 1980.
43. Žilinskas, A., Senkienė, E., *Estimation of the parameter of the Wiener process.* Litovsk. Mat. Sb., **18**, 3, 59–62 (1978). (Russian)
44. Žilinskas, A., Senkienė, E., *Estimation of Wiener stochastic field parameter from observations at random dependent points.* Kybernetika (Kiev), *6*, 107–109 (1979). (Russian)

H. THÖNI

Ein statistisches Modell für Bonituren

A statistical model is proposed which describes scores („Bonituren") as a discrete transformation of the scored random variable. A conditional probability function is used to take into account scoring errors.
Computations using scoring rules found in the agricultural literature show a smooth and monotone dependence between the expected values of the scores and those of the scored variables. The variance of the scores is related to the mean in a typical way: the variance vanishes when the mean approaches the minimum and maximum value of the range of the scores, and may have more than one relative maximum at intermediate values. This irregular dependence between mean and variance makes it difficult to fiind variance-stabilizing transformations.

1. Einleitung

Bonituren werden in vielen Bereichen angewandt, um quantitative Merkmale anhand einer nur wenige (meist zwischen drei und neun) Stufen aufweisenden Skala zu bewerten. Beispiele hierfür sind die Bewertung des Pilzbefalls von Kulturpflanzen oder die Verunkrautung der Parzellen in Pflanzenschutzversuchen, die Bewertung der Ausprägung bestimmter Merkmale bei der Beurteilung von Neuzüchtungen in der Pflanzenzüchtung, die sensorische Bewertung von Nahrungsmitteln, u.a.m.

Die statistische Auswertung derartiger Versuchsergebnisse erfolgt vielfach mit Hilfe varianzanalytischer Methoden. Lageunterschiede der verschiedenen experimentellen Gruppen werden auf Grund der Differenzen der arithmetischen Mittelwerte beschrieben und beurteilt, und Hypothesen über schätzbare Funktionen der Modellparameter werden über t-bzw. F-Verteilung geprüft.

Die Erfahrung zeight jedoch, dass Bonituren oft deutlich heteroskedastisch sind. Als Beispiel seien Ergebnisse aus einer Untersuchung des Cercosporella-Befalls bei Winterweizen angeführt. Der Befall wurde nach einer neunstufigen Boniturskala bewertet. Die Varianzen der (unterschiedlich grossen) Stichproben sind im Mittel in Stichproben mit kleinen Mittelwerten kleiner als in solchen, deren Mittelwerte im mittleren Bereich der Boniturskala liegen (Abb. 1). Diese Heteroskedastizität stellt jedoch eine Verletzung einer der wesentlichsten Voraussetzungen der Varianzanalyse dar. Zudem wird wegen des "ordinalen" Characters der Zufallsvariablen "Bonitur" die Berechnung arithmetischer Mittelwerte und Varianzen als fragwürdig bezeichnet.

Es existieren Vorschläge für varianzstabilisierende (Weiling, 1959) und normalisierende (Burghausen, 1964; Fisher and Yates, 1963) Transformationen, doch ist deren Brauchbarkeit und Wirksamkeit nur spärlich untersucht. Allgemein anwendbare Ansätze zum Aufsuchen geeigneter varianzstabilisierender und/oder normalisierender Transformationen sind nicht bekannt da über die Gesetzmässigkeiten des Zusammenhanges zwischen Erwartungswert und Varianz und über Verteilungsformen Untersuchungen bisher fehlen.

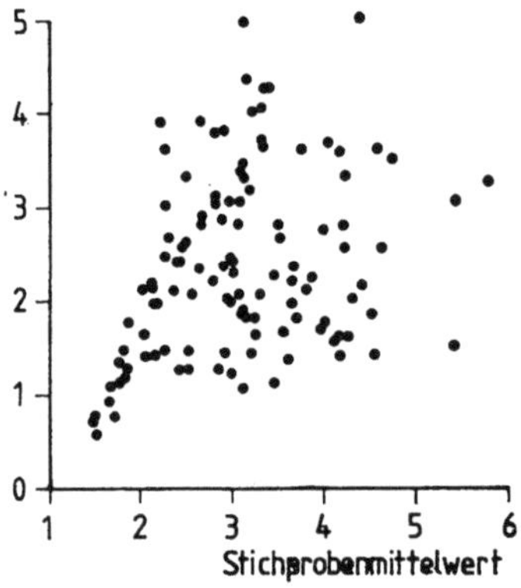

Abbildung 1. Empirische Beziehung zwischen Stichprobenmittelwert und Stichprobenvarianz für ein bonitiertes Merkmal: Bonitierung des Cercosporella-Befalls auf Halmen von Winterweizen mittels einer neunstufigen Boniturskala; 120 Stichproben mit wechselnden Stichprobenumflängen zwischen 17 und 89 (Dr. E. Kübler, Hohenheim, pers. Mitt.).

Im Folgenden soll ein statistisches Modell entwickelt werden, welches gestattet, derartige Gesetzmässigkeiten zu ermitteln, und welches erlaubt, die Zusammenhänge zwischen der Verteilung des bonitierten Merkmals der Boniturvorschrift und der Verteilung der Boniturnoten zu erhellen.

2. Das Modell

Jedes bonitierte Merkmal kann als eine Zufallsvariable X mit einer Verteilungsfunktion $F(x \mid \theta)$ aufgefasst werden, für welche wir voraussetzen, dass Erwartungswert $E(X \mid \theta) = \xi(\theta)$ und Varianz $V(X \mid \theta) = \eta^2(\theta)$ existieren. θ sei der Parameter bzw. -vektor der Verteilungsfunktion, ξ und η^2 sind in der Regel Funktionen dieser Parameter.

Eine Boniturvorschrift ordnet bestimmten Werten der Zufallsvariablen $X = x_i$ bzw. bestimmten Intervallen $X \in (x_{i-1}, x_i]$ feste Boniturwerte $Y = y_i$ zu. Als Boniturwerte werden i.d.R. die Reihe der natürlichen Zahlen von Eins bis K bzw. von Null bis $K-1$ gewählt. Die Boniturvorschrift hat die Form einer Treppenfunktion und kann als Transformation $y = T(x)$ des bonitierten Merkmals X aufgefasst werden. Ist die Verteilungsfunktion bekannt, so kann die Wahrscheinlichkeitsfunktion $g(y)$ für die diskrete Zufallsvariable Y daraus berechnet werden

(2.1) $$g(y) = P\{Y = y_i\} = P\{X \in (x_{i-1}, x_i]\}$$

Mit Hilfe dieser Wahrscheinlichkeitsfunktion können Erwartungswert $E(Y) = \mu$ und Varianz $V(Y) = \sigma^2$ der Bonituren berechnet werden, die wiederum über die Verteilungsfunktion $F(x \mid \theta)$ von den Parametern θ abhängen.

Betrachtet man für ein bestimmtes Merkmal X alle Verteilungsfunktionen $F(x \mid \theta)$ über den zulässigen Wertebereich von θ, so kann daraus die

Beziehung zwischen ξ und μ, sowie zwischen μ und σ^2 und deren Abhängigkeit von der Boniturvorschrift $T(x)$ untersucht werden. Die erste Beziehung kann Aufschluss über die Frage geben, wie sinnvoll die Berechnung von Mittelwerten aus Boniturnoten ist und ob daraus Rückschlüsse auf die Lage der Verteilung des Merkmals X gezogen werden können. Die zweite Beziehung könnte Hinweise auf mögliche varianzstabilisierende Transformationen für Bonituren geben.

3. Bonitierfehler

Bonitierung bedeutet subjektive Beurteilung der Ausprägung des Merkmals X. Dessen Ausprägung wird somit nicht exakt ermittelt, sondern nur geschätzt. Dadurch kann es vorkommen, dass insbesondere in der Umgebung der Intervallgrenzen x_i falsche (d.h. zu hohe bzw. zu niedrige) Boniturwerte vergeben werden. Solche Bonitierfehler können in dem Modell durch die Einführung einer Gewichtsfunktion $w_i(x)$ berücksichting werden. Diese ist die bedingte Wahrscheinlichkeit dafür, dass der Boniteur die Boniturnote $Y = y_i$ vergibt, wenn das Merkmal die Ausprägung $X = x$ aufweist

$$w_i(x) = P\{Y = y_i \mid X = x\} \tag{3.1}$$

Setzen wir $w_i(x) = 1$ wenn $x \in (x_{i-1}, x_i]$ und $w_i(x) = 0$ sonst, so ergibt sich daraus eine „fehlerfreie" Bonitur.

Nehmen wir zur Vereinfachung zunächst an, dass nur Bonitierfehler um eine Note zu hoch bzw. zu tief vorkommen, dann erscheint es sinnvoll, die Gewichte auf den Intervallgrenzen den Wert 0.5 annehmen zu lassen. Wir erhalten die folgenden Bedingungen

$$w_i(x) = \begin{cases} 0 \leqslant w_i(x) \leqslant 0.5 \\ 0.5 \leqslant w_i(x) \leqslant 1 \\ 0.5 \geqslant w_i(x) \geqslant 0 \\ \quad w_i(x) = 0 \end{cases} \quad \text{falls} \begin{cases} x \in (x_{i-2}, \ x_{-1}] \\ x \in (x_{i-1}, \ x_i] \\ x \in (x_i, \ x_{i+1}] \\ x \notin (x_{i-2}, x_{i+1}] \end{cases} \tag{3.2}$$

Die $w_i(x)$ müssen die Bedingung

$$\sum_{i=1}^{k} w_i(x) = 1$$

an jeder Stelle von x erfüllen, und sollen in der Mitte eines Intervalls den Wert 1 erreichen, und in den benachbarten Intervallen auf Null absinken.

Die Wahrscheinlichkeitsfunktion $g(y)$ lässt sich sodann als gewogene Summe bzw. als gewogenes Integral darstellen

$$g(y) = \begin{cases} \sum_{x=0}^{\infty} w_i(x)\, f(x) & x \in \{\mathrm{N}, 0\} \quad (3.3) \\ \int_0^1 w_i(x)\, \mathrm{d}F(x)\, \mathrm{d}x & x \in [0, 1] \quad (3.4) \end{cases}$$

wenn wir Vorschriften zur Bonitierung von abzählbaren Merkmalen ($X \in \in \{\mathrm{N}, 0\}$) bzw. von Relativwerten (Prozentzahlen, Anteilen, $X \in [0, 1]$) als für die Praxis wichtige Fälle betrachten.

4. Modellbeispiele

4.1 *Abzählbare Merkmale.* Für abzählbare Merkmale, wie z.B. die Anzahl der Frassstellen auf einem Blatt, die Anzahl der Unkrautpflanzen auf einer Parzelle, die Anzahl der Rostflecken auf einem Pflanzenteil, eignen sich Poisson-und Negative Binomialverteilung als Wahrscheinlichkeitsfunktion $f(x \mid \theta)$ zur Beschreibung der Verteilung des dikreten Merkmals X. Platter und Krämer (1962) haben für solche Merkmale eine sechsstufige Boniturskala vorgeschlagen, welche die Noten $y_1 = 0$ für $X = 0$, $y_2 = 1$ für $X = 1$, $y_3 = 2$ für $X \in [2, 5]$, $y_4 = 3$ für $X \in [6, 10]$, $y_5 = 4$ für $X \in [11, 20]$ und $y_6 = 5$ für $X \geqslant 21$ vergibt.

Die Ausprägungen $X = 0, 1, 2, 3$ dürften in der Regel fehlerfrei zu bonitieren sein. Bei grösseren Anzahlen dagegen sind Fehler bei der Schätzung der Anzahl möglich. An den Sprungstellen der Treppenfunktion $x_i = 5, 10, 20$ ist mit der Möglichkeit zu rechnen, dass der Boniteur falsche Noten vergibt, z.B. für die Ausprägung $X = 5$ nur mit Wahrscheinlichkeit 0.9 den korrekten Wert $Y = 2$ und mit Wahrscheinlichkeit 0.1 den falschen Wert $Y = 3$; umgekehrt dafür bei $X = 6$ mit Wahrscheinlichkeit 0.1 die Note $Y = 2$ und mit Wahrscheinlichkeit 0.9 die Note $Y = 3$. Ähnliche Verhältnisse mit etwas breiterer Unsicherheitszone dürften für die anderen Sprungstellen realistisch sein. Die sich daraus ergebenden Gewichte $w_i(x)$, die den nachfolgenden Berechnungen zu Grunde liegen, sind in Tabelle 1 zusammengestellt.

Tabelle 1

Bedingte Wahrscheinlichkeiten (Gewichtsfunktionen) $w_i(x)$ für die Vergabe der Boniturnote y_i bei der Bonitierung von zählbaren Merkmalen mittels der Boniturvorschrift nach Platter und Krämer (1962)

Boniturklasse i	absolute Häufigkeit des Merkmals X	Boniturnote y_i: 0	1	2	3	4	5
1	0	1	0	0	0	0	0
2	1	0	1	0	0	0	0
	2	0	0	1	0	0	0
3	3	0	0	1	0	0	0
	4	0	0	1	0	0	0
	5	0	0	0.90	0.10	0	0
	6	0	0	0.10	0.90	0	0
	7	0	0	0.05	0.95	0	0
4	8	·	0	0	1	0	0
	9	·	0	0	0.95	0.05	0
	10	·	0	0	0.90	0.10	0
	11		·	0	0.10	0.90	0
	12		·	0	0.05	0.95	0
	13		·	0	0	1	0
	14			·	0	1	0
5	15			·	0	1	0
	16			·	0	1	0
	17				0	1	0
	18				0	0.95	0.05
	19				0	0.90	0.10
	20				0	0.80	0.20
	21				0	0.20	0.80
	22				0	0.10	0.90
6	23				0	0.05	0.95
	24 und mehr				0	0	1

Unter Verwendung von (3.3) könen die Wahrscheinlichkeitsfunktionen $g(y)$ für die Bonituren und daraus die Erwartungswerte und Varianzen in Abhängigkeit des Erwartungswertes von X für Poisson-verteilte und negativ binomialverteilte Variable berechnet werden. Die sich daraus ergebenden Beziehungen zwischen $E(X)$ und $E(Y)$ sowie $E(Y)$ und $V(Y)$ sind in den Abbildungen 2 und 3 dargestellt.

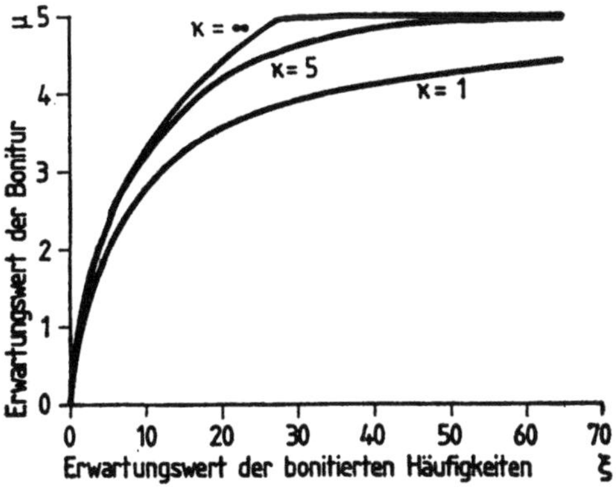

Abbildung 2. Abhängigkeit des Erwartungswertes μ der Bonitur vom Erwartungswert ξ des bonitierten Merkmals: Poisson- ($\varkappa = \infty$) bzw. negativ binomial- ($\varkappa = 5$ bzw. 1) verteilte Häufigkeiten, bonitiert nach der Boniturvorschrift von Platter und Krämer (1962).

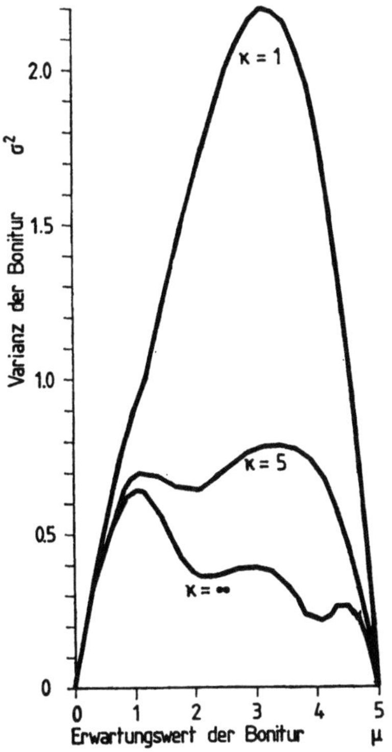

Abbildung 3. Abhängigkeit der Varianz σ^2 der Bonitur von ihrem Erwartungswert μ. Gleiches Modell wie Abb. 2.

Der Erwartungswert der Bonitur ist eine monoton nicht-fallende Funktion des Erwartungswertes von X, welcher asymptotisch gegen den maximal möglichen Grenzwert $\mu_\infty = 5$ strebt. Der Anstieg dieser Beziehung ist umso flacher, je grösser die Varianz von X bei gleichem Erwartungswert ist (bei gegebenem Erwartungswert negativ binomialverteilter Variabler wächst die Varianz mit kleiner werdendem Exponenten $\varkappa$). Die Varianz der Bonitur strebt an den Enden der Boniturskala gegen Null. Je grösser die Varianz von X, umso grösser ist auch die Varianz von Y. Bei Poisson-verteilten Merkmalen zeigt der Verlauf von σ^2 zwei relative Minima auf, welche mit abnehmendem Wert von $\varkappa$ weniger deutlich ausgeprägt ausfallen und bei $\varkappa = 1$ ganz verschwinden.

4.2 *Merkmale, welche in Anteilen beurteilt werden.* Viele Merkmale werden nach Prozenten gemessen (Prozent Verunkrautung, Prozent Rostbefall, usw.). Solche Merkmale können als stetige Zufallsvariable im Intervall [0, 1] beschrieben werden. Ihre Zufallsverteilung kann mit Hilfe der Beta-Verteilung modelliert werden, wobei durch entsprechende Wahl der Parameter α und β Verteilungen mit unterschiedlichen Varianzen erzeugt werden können.

Es existieren zahlreiche Boniturvorschriften, welche das Intervall von 0% bis 100% in eine unterschiedliche Anzahl von Teilintervallen ungleicher Länge einteilen (eine Übersicht findet sich in O'Svath, Peil und Geidel, 1979). Für das dort erwähnte sog. „alte Boniturschema" z.B. mit fünf Stufen können die Intervallgrenzen durch folgende Sequenz gebildet werden: $x_i = (2/3)^j$, $j = 2^{(4-i)}$, $i = 1, 2, 3, 4$. Man erhält sodann folgende Boniturklassen

Bonitur klasse i	untere Grenze x_{i-1}	obere Grenze x_i	Boniturnote y_i
1	0	0.0390	0
2	0.0390	0.1975	1
3	0.1975	0.4444	2
4	0.4444	0.6667	3
5	0.6667	1	4

Die Gewichtsfunktion $w(x)$ soll an den Sprungstellen sinusförmig ansteigen bzw. abfallen und auf den Intervallgrenzen den Wert 0.5 annehmen. Die folgende Funktion erfüllt diese Bedingungen

$$(4.1)\quad w_i(x) = \begin{cases} \frac{1}{2}\left(1 + \sin\left(\frac{\pi}{2}\frac{x - x_{i-1}}{\delta_{i-1}}\right)\right) & \text{für } x_{i-1} - \delta_{i-1} \leqslant x \leqslant x_{i-1} + \delta_{i-1} \\ 1 & \phantom{\text{für }} x_{i-1} + \delta_{i-1} \leqslant x \leqslant x_i - \delta_i \\ \frac{1}{2}\left(1 - \sin\left(\frac{\pi}{2}\frac{x - x_i}{\delta_i}\right)\right) & \phantom{\text{für }} x_i - \delta_i \leqslant x \leqslant x_i + \delta_i \end{cases}$$

mit $\delta_i = x_i(1 - x_i)/3$.

Wählen wir als Verteilung von X Betaverteilungen mit den Parameterwerten $\alpha + \beta = 5$, 10 bzw, 20, so ergeben sich ähnliche Verhältnisse wie bei Zählungen:

1. Der Erwartungswert der Bonitur steigt von $\mu = 0$ bei $\alpha = 0$ $(\xi = 0)$ monoton auf $\mu = 4$ bei $\alpha = 5$ bzw. 10 bzw. 20 $(\xi = 1)$ an.

2. Die Varianz der Bonitur strebt an den Enden der Boniturskala gegen Null. Bei mittleren Erwartungswerten fällt die Varianz der Bonitur umso

grösser aus, je grösser die Varianz der angenommen Betaverteilung ist. Bei $\alpha + \beta = 20$ (d.h. relativ kleiner Varianz von X) treten mehrere relative Minima auf, bei $\alpha + \beta = 5$ (relativ grösserer Varianz von X) fehlen solche (vgl. auch Thöni, 1982).

Manche Boniturvorschriften weisen Noten nicht Intervallen, sondern ausgezeichneten Punkten der Prozentskala zu. Eine Vorschrift des Bundessortenamtes (1977) zur Bonitierung der Krautfäule z.B. ordnet in einer neunstufigen Skala die ungeraden Noten $y_i = 1, 3, 5, 7, 9$ den Ausprägungen „klein Befall", „ca. 5%", „ca. 50%", „ca. 90%" und „vollständig befallen" zu; die geradzahligen Notenwerte sollen für die dazwischenliegenden Befallsgrade verwendet werden. Hier müssen die Gewichte $w_i(x)$ so gewählt werden, dass sie an diesen Stellen von X jeweilen den maximalen Wert 1 annehmen.

Es ergibt sich folgende Boniturvorschrift

Bonitur-klasse i	Befallsgrad X	Bonitur-note y_i
1	0	1
2	$0 < x < 0.05$	2
3	0.05	3
4	$0.05 < x < 0.5$	4
5	0.5	5
6	$0.5 < x < 0.9$	6
7	0.9	7
8	$0.9 < x < 1$	8
9	1	9

Die zugehörigen Gewichtsfunktionen können folgendermassen festgelegt werden

$$w_i(x) = \frac{1}{2}\,(1 + \cos(\pi(x - x_i)/\delta_i)$$

für $x_i - \delta_i \leqslant x \leqslant x_i + \delta_i$, i ungerade und $0 \leqslant x \leqslant 1$,

$$(4.2) \qquad w_i(x) = \begin{cases} 1 - w_{i-1}(x) & \\ 1 & \text{für} \\ 1 - w_{i+1}(x) & \end{cases} \begin{cases} x_{i-1} \leqslant x \leqslant x_{i-1} + \delta_{i-1} \\ x_{i-1} + \delta_{i-1} \leqslant x \leqslant x_{i+1} - \delta_{i+1} \\ x_{i+1} - \delta_{i+1} \leqslant x \leqslant x_{i+1} \end{cases}$$

i gerade, $w_i(x) = 0$ sonst. Für δ_i setzen wir $\delta_i = x_i(1 - x_i)$ für $i = 3, 5, 7$, $\delta_1 = x_3 - \delta_3$ und $\delta_9 = 1 - (x_7 + \delta_7)$.

Auch für dieses Modell ergaben Modellrechnungen einen monoton nichtfallenden Verlauf für die Abhängigkeit des Erwartungswertes der Bonitur vom Erwartungswert von X, und eine ähnliche Abhängigkeit der Varianz der Bonitur von deren Erwartungswert und der Verteilung von X wie im vorangehenden Beispiel (Thöni, 1982).

5. Einfluss des Bonitierfehlers

Um den Einfluss des Bonitierfehlers auf die Verteilung der Boniturnoten zu untersuchen, vergleichen wir Erwartungswerte und Varianzen der Bonituren für Modellansätze mit „fehlerfreien" Bonitierungen mit solchen „mit Fehlern".

In den untersuchten Modellansätzen ist der Einfluss des Bonitierfehlers relativ gering. Die Erwartungswerte der Bonitur werden durch Bonitierfehler gegenüber denjenigen für fehlerfreie Bonitur etwas gegen mittlere Werte verschoben. Die Varianz der Bonitur wird durch Bonitierfehler leicht erhöht.

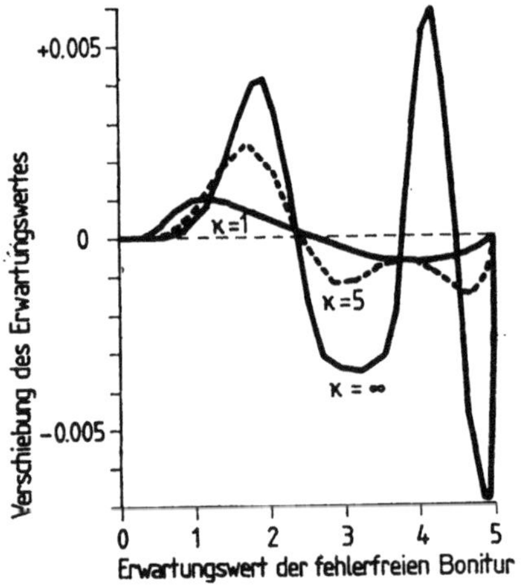

Abbildung 4. Verschiebung des Erwartungswertes der Bonitur durch den Bonitierfehler. Durch den Bonitierfehler wird der Erwartungswert um die aufgetragenen Differenzen gegenüber demjenigen bei fehlerfreier Bonitur erhöht bzw. erniedrigt. Gleiches Modell wie Abb. 2.

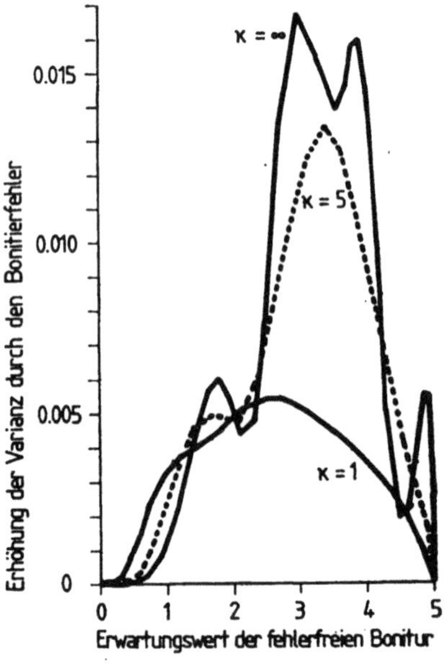

Abbildung 5. Erhöhung der Varianz der Bonitur durch den Bonitierfehler. Durch den Bonitierfehler wird die Varianz gegenüber derjenigen bei fehlerfreier Bonitur durchwegs erhöht. Gleiches Modell wie Abb. 2.

Für das in Abschnitt 4.1 beschriebene Modellbeispiel sind diese Unterschiede in den Abbildungen 4 und 5 graphisch dargestellt. Die Verschiebung des Erwartungswertes nimmt extremere Werte an, wenn wir für X eine Poisson-Verteilung unterstellen, und verflacht zunehmend, wenn wir für X

negative Binomalverteilungen mit kleiner werdenden Werten des Exponenten $\varkappa$ einsetzen. Die Vergrösserung der Varianz ist ebenfalls von der Verteilung von X abhängig und bei Poisson-Verteilung deutlich grösser als bei negativen Binomialverteilungen mit kleinen $\varkappa$-Werten.

Berechnungen anhand des „alten Boniturschemas" für stetige, im Intervall [0, 1] verteilte Anteile ergaben entsprechende Verhältnisse. Auch in diesem Modellansatz fällt die Erhöhung der Varianz der Bonitur umso grösser aus, je kleiner die Varianz von X angenommen wird, und ist die Verschiebung des Erwartungswertes der Bonitur umso grösser, je keiner die Varianz von X.

6. Diskussion

Boniturnoten lassen sich als diskrete Zufallsvariable beschreiben. Sie stellen eine diskretisierende Transformation der bonitierten Merkmale dar. Ihre Wahrscheinlichkeitsverteilung kann aus der Verteilungsfunktion der bonitierten Merkmale hergeleitet werden. Zwischen den Erwartungswerten der bonitierten Merkmale und den Erwartungswerten der Bonituren besteht eine glatte, monotone Beziehung. Diese Feststellung legt den Schluss nahe, dass Stichprobenmittelwerte, welche aus Boniturnoten berechnet werden, Information über die Lage der Verteilungsfunktion des bonitierten Merkmals enthalten, und Bonituren somit mehr als nur ordinalen Charakter aufweisen.

Die Varianz der Boniturnoten hängt über die Verteilung des bonitierten Merkmals mit ihrem Erwartungswert zusammen. Für diese Abhängigkeit konnte noch keine geschlossene funktionale Form ermittelt werden. Wie weit es gelingen wird, evtl. durch Modifikationen der Boniturvorschriften diese Beziehung zwischen Erwartungswert und Varianz zu vereinfachen und so zu gestalten, dass erfolgreiche varianzstabilisierende Transformationen gefunden werden können, muss in weiteren Untersuchungen geklärt werden.

Anhang

Die Wahrscheinlichkeitsfunktion der negativen Binomialverteilung wird in der Form

(A. 1) $$f(x) = \frac{\Gamma(x + \varkappa)}{x!\,\Gamma(\varkappa)} \left(\frac{\varkappa}{\xi + \varkappa}\right)^{\varkappa} \left(\frac{\xi}{\xi + \varkappa}\right)^{x}$$

geschrieben. Es gilt sodann $E(X) = \xi$ und $V(X) = \xi + \frac{\xi^2}{\varkappa}$. Die negative Binomialverteilung geht für $\varkappa \to \infty$ in die Poisson-Verteilung über mit $E(X) = V(X) = \xi$.

Die Beta-Verteilung mit den Parametern α und β hat die Form

f(A. 2) $$F(x) = \frac{\Gamma(\alpha + \beta)}{\Gamma(\alpha)\,\Gamma(\beta)} \int_0^x t^{\alpha-1}(1 - t)^{\beta-1}\, dt$$

ür welche gilt $E(X) = \alpha/(\alpha + \beta)$ und $V(X) = \alpha\beta/(\alpha + \beta)^2\,(\alpha + \beta + 1) = \xi(1 - \xi)/(\alpha + \beta + 1)$.

LITERATUR

1. Bundessortenamt, *Richtlinien für die Durchführung von landwirtschaftlichen Wertprüfungen und Sortenversuchen.* Hannover, 1977.
2. Burghausen, R., *Statistische Auswertung von Vegetationsbeobachtungen in Feldversuchen.* Abh. Deutsch. Akad. Wissensch. Berlin, Mathematische Klasse Nr. 4, 43– 48 (1964).
3. Fisher, R.A., Yates, F., *Statistical Tables for Biological, Agricultural and Medical Research, 6th Edition.* Edinburgh, Oliver and Boyd, 1963.
4. O'Svath, J., Peil, H., Geidel, H., *Über das Messen und die Skalenarten im Pflanzenschutzversuch.* Mitt. Biol. Bundesanstalt f. Land-u. Forstwirtsch., Heft 192. Berlin—Dahlem, 1979.
5. Platter, Th., Krämer, F. W., *Erfahrungen bei der Ausbringung von Pflanzenschutzmitteln durch Beregnungsanlagen.* Der Landwirt, **16**, 247– 250 (1962).
6. Thöni, H., *Ein statistisches Modell für Bonituren.* EDV in Medizin und Biologie, Supplementum, 1982.
7. Weiling, F., *Die Bedeutung der Transformation fur die Versuchsanalyse, insbesondere bei Bonituren mit beliebiger Breite der Boniturskala.* Der Züchter, **19**, 281– 284 (1959).

R. J. TOMKINS *

Rates of convergence for the stability of maxima

1. Introduction

Throughout this article, $X_1, X_2, \ldots$ will be a sequence of independent, identically distributed (i.i.d.) random variables (r.v.) with common distribution function (d.f.) $F(x) = P[X_1 \leqslant x]$. Define $M_n = \max(X_1, \ldots, X_n)$ for $n \geqslant 1$.

If X_1 is unbounded above (i.e. $F(x) < 1$ for all x), the sequence $\{M_n\}$ is called *almost surely relatively stable* if $M_n/a_n \to 1$ almost surely (a.s.) for some real sequence $\{a_n\}$. It is known (see Theorem 4.5 of Barndorff-Nielsen [1] or Theorem 1 of Resnick and Tomkins [6]) that a.s. relative stability obtains if and only if (iff)

$$\int_1^\infty [1 - F(\delta y)]^{-1} \,\mathrm{d}F(y) < \infty \tag{1.1}$$

for every $\delta < 1$, in which case $a_n \sim b_n \equiv F^{-1}(1 - n^{-1})$, where $F^{-1}(y) \equiv \inf\{x : F(x) \geqslant y\}$ and "$a_n \sim b_n$" means that $a_n/b_n \to 1$ as $n \to \infty$.

The contents of this paper stem from a quest for a criterion equivalent to the complete convergence (in the sense of Hsu and Robbins [4]) of M_n/a_n to one $\left(\text{i.e. for } \sum_{n=1}^\infty P[\,|M_n/a_n - 1| > \varepsilon] > \infty \text{ for every } \varepsilon > 0\right)$. Section 2 will present necessary and sufficient conditions for the convergence of $\sum_{n=1}^\infty \Phi(n)\, P[\,|M_n/b_n - 1| > \varepsilon]$ for every $\varepsilon > 0$ for a certain class of functions Φ. The particular case $\Phi(n) = n^\alpha$, $\alpha \geqslant 1$, will be considered in detail.

If $x_1 \equiv F^{-1}(1) < \infty$, then $M_n \to x_1$ completely and, hence, a.s. Section 3 addresses the question: When does a sequence $\{a_n\}$ exist such that $a_n(x_1 - M_n) \to 1$ a.s.? It will be shown that the answer is directly related to a.s. relative stability.

---

* This work was supported by a grant from the Natural Sciences and Engineering Research Council of Canada.

2. The unbounded case

Throughout this section, $M_n = \max(X_1, ..., X_n)$, where $\{X_n\}$ are i.i.d. with common d.f. F obeying $F(x) < 1$ for all x. Before presenting the main results, two lemmas will be established.

Lemma 2.1. *Let $C_n \uparrow \infty$ be a real sequence. Suppose $-\infty < p \leqslant 1$.*

(i) *If* $\sum_{n=1}^{\infty} n^{-p}[1 - F^n(C_n)] < \infty$ *then* $F^n(C_n) \to 1$ *as* $n \to \infty$;

(ii) *If* $\sum_{n=1}^{\infty} n^{-p} F^n(C_n) < \infty$ *then* $F^n(C_n) \to 0$ *as* $n \to \infty$.

Proof. Both parts are easy to prove if $p \leqslant 0$, because the summands of a convergent series converge to zero of necessity.

Assume, therefore, that $0 < p \leqslant 1$. In this case,

$$1 - F^n(C_{2n}) = 2 \sum_{j=n+1}^{2n} (2n)^{-1}(1 - F^n(C_{2n})) \leqslant 2 \sum_{j=n+1}^{2n} j^{-p}(1 - F^j(C_j)).$$

Under the hypothesis of (i), then, $F^n(C_{2n}) \to 1$ and, hence, $F^{2n}(C_{2n}) \to 1$. But $1 \geqslant F^{2n+1}(C_{2n+1}) \sim F^{2n}(C_{2n+1}) \geqslant F^{2n}(C_{2n}) \to 1$, so (i) is proved.

Similarly $F_n^2(C_n) \leqslant 2 \sum_{j=n+1}^{2n} j^{-p} F^j(C_j)$, from which (ii) follows readily.

Lemma 2.2. *Let Φ be a continuous, non-negative function defined on $(0, \infty)$ obeying*

(2.1) $\limsup_{n \to \infty} \Phi(n+1)/\Phi(n) < \infty$ *and* $x^q \Phi(x)$ *is non-decreasing for some* $q \geqslant 0$.

Let $\{C_n\}$ be a real sequence satisfying $C_n \uparrow \infty$ and $\liminf_{n \to \infty} F^n(C_n) > 0$. Then the following statements are equivalent:

(i) $\sum_{n=1}^{\infty} \Phi(n)[1 - F^n(C_n)] < \infty$;

(ii) $\sum_{n=1}^{\infty} n\Phi(n)[1 - F(C_n)] < \infty$;

(iii) $\int_1^{\infty} x\Phi(x)[1 - F(C(x))]\, dx < \infty$

for some (equivalently, every) function $C(x) \uparrow \infty$ such that $C(n) = C_n$.

Proof. It is well known (e.g., see [2], p. 66) that

$$t \leqslant -\log(1 - t) \leqslant t/(1 - t) \text{ for } 0 \leqslant t < 1. \tag{2.2}$$

Letting $t = 1 - F^n(C_n)$ and noting that $F^n(C_n) \geqslant \delta > 0$ for some δ and all large n by hypothesis, it follows that $1 - F^n(C_n) \leqslant -n \log F(C_n) \leqslant \delta^{-1}[1 - F^n(C_n)]$ for large n. But $1 - y \sim -\log y$ as $y \to 1$ so $-n \log F(C_n) \sim n[1 - F(C_n)]$. Clearly (i) and (ii) are equivalent.

(2.1) guarantees that $\Phi(n) \leqslant K\Phi(n-1)$ for some $K > 0$ and all large n. For such values of n, then,

$$n\Phi(n)(1-F(C_n)) \leqslant Kn\,\Phi(n-1)(1-F(C_n)) \sim$$

$$\sim K(n-1)^{q+1}\,\Phi(n-1)(1-F(C_n))/n^q \leqslant K\int_{n-1}^{n} x\Phi(x)(1-F(C(x)))\,dx$$

by (2.1). Therefore (iii) implies (ii). Similarly,

$$\int_{n}^{n+1} x\Phi(x)[1-FC(x))]\,dx \leqslant K(n+1)^{q+1}\,\Phi(n)[1-F(C_n)]/n^q \sim$$

$$\sim Kn\,\Phi(n)[1-F(C_n)],$$

so (ii) implies (iii).

Theorem 2.3. *Let $X_1, X_2, \ldots$ be i.i.d. with common d.f. F, where $F(x) < 1$ for all x. For $n \geqslant 1$, define $M_n = \max(X_1, \ldots, X_n)$ and $b_n = F^{-1}(1-n^{-1})$ Let $\Phi(x)$ be a continuous, non-negative function defined on $(0, \infty)$ and obeying (2.1). Define $H(x) = \int_0^x y\Phi(y)\,dy$ for $x \geqslant 0$.*

For any $\gamma > 1$, the following statements are equivalent

(i) $\sum_{n=1}^{\infty} \Phi(n)\, P[M_n > \gamma b_n] > \infty$;

(ii) $\sum_{n=1}^{\infty} n\,\Phi(n)\, P[X_1 > \gamma b_n] < \infty$;

(iii) $\int_1^{\infty} x\Phi(x)[1-F(\gamma F^{-1}(1-x^{-1}))]\,dx < \infty$;

(iv) $\int_{F^{-1}(0)}^{\infty} H([1-F(y/\gamma)]^{-1})\,dF(y) < \infty.$

If, in addition,

(2.3) $\qquad \Phi(x) = x^m \eta(x)$, *where $m \geqslant -1$ and $\eta(x) \downarrow$ as $X \to \infty$,*

then each of the preceding assertions is equivalent to

(v) $\sum_{n=1}^{\infty} \Phi(n)\, P[\,|M_n/b_n - 1\,| > \gamma - 1] < \infty.$

In particular, these five statements are equivalent if $\Phi(n) = n^\alpha$ for some $\alpha \geqslant -1$.

Proof. By definition of b_n, $F^n(\gamma b_n) \geqslant (1-n^{-1})^n \to e^{-1}$, so the equivalence of (i), (ii) and (iii) is established by Lemma 2.2.

For brevity, let $\delta = \gamma^{-1}$ and $x_0 = F^{-1}(0)$. Generalizing a technique of [6] yields

$$\int_{x_0}^{\infty} H([1 - F(\delta y)]^{-1})\, dF(y) = \int_{y=x_0}^{\infty} \int_{x=0}^{[1-F(\delta y)]^{-1}} x\Phi(x)\, dx\, dF(y) =$$

$$= \int_{x=1}^{\infty} x\Phi(x) \int_{y=\gamma F^{-1}(1-x^{-1})}^{\infty} dF(y)\, dx =$$

$$= \int_{1}^{\infty} x\Phi(x)\, [1 - F(\gamma F^{-1}(1 - x^{-1}))]\, dx,$$

so (iii) and (iv) are equivalent.

Since (v) clearly implies (i), it will suffice to show that (iv) implies (v) under (2.3). Since (i) and (iv) are the same, it need only be shown that $\sum_{n=1}^{\infty} \Phi(n)\, F^n((2 - \gamma)b_n) < \infty$. But $2 - \gamma < \delta$, so it will be enough to show that (iv) implies

(2.4) $$\sum_{n=1}^{\infty} \Phi(n)\, F^n(\delta b_n) < \infty.$$

To this end, note that

$$\Phi(n)F^n(\delta b_n) \sim n^q\Phi(n)\, F^{n+1}(\delta b_n)/(n + 1)^q$$

$$\leqslant \int_{n}^{n+1} x^q\Phi(x)\, F^x(\delta F^{-1}(1 - x^{-1}))x^{-q}dx$$

$$= \int_{n}^{n+1} \Phi(x) \exp\{x \log F(\delta F^{-1}(1 - x^{-1}))\}\, dx.$$

By virtue of (2.2), then, (2.4) holds if

(2.5) $$I \equiv \int_{1}^{\infty} \Phi(x) \exp\{-x[1 - F(\delta F^{-1}(1 - x^{-1}))]\}\, dx < \infty.$$

By (2.3), $H(x) \geqslant \eta(x)x^{m+2}/(m + 2)$, so

(2.6) $$(m + 2)\, H([1 - F(\delta y)]^{-1} \geqslant \eta([1 - F(y)]^{-1})/[1 - F(\gamma y)]^{m+2}$$

since $\eta(x)$ is non-increasing.

Let $A > 0$ satisfy $x^{m+2}\, e^{-x} < A$ for all $x \geqslant 0$. Making the change of variable $y = F^{-1}(1 - x^{-1})$ in (2.5) yields

$$I = \int_{x_0}^{\infty} \Phi([1 - F(y))]^{-1} \exp\{-(1 - F(\delta y))/(1 - F(y))\}\, [1 - F(y)]^{-2}\, dF(y)$$

$$\leqslant A \int_{x_0}^{\infty} \eta([1 - F(y)]^{-1})\, [1 - F(\delta y)]^{-(m+2)}\, dF(y)$$

$$\leqslant (m + 2)\, A \int_{x_0}^{\infty} H([1 - F(\delta y)]^{-1})\, dF(y)$$

by (2.6). Therefore (iv) implies (2.5).

Example. If $F(x) = 1 - e^{-x}$ for $x \geqslant 0$, then, for every $r \geqslant 0$,

$$\sum_{n=1}^{\infty} (\log n)\, n^{-1} P[\,|\, M_n/\log n - 1 \,|\, > \varepsilon] < \infty$$

for every $\varepsilon > 0$. To verify this, let $\Phi(x) = (\log x)^r/x$. Then $\int_1^{\infty} (\log x)^{-\gamma} dx$, the integral in (iii), converges for every $\gamma > 1$.

The next theorem focusses on the special case $\Phi(x) = x^{r-2}$, $r \geqslant 1$.

Theorem 2.4. *Let $\{X_n\}$ be i.i.d. r.v. with common d.f.F. with $F(x) < 1$ for all x. Define $b_n = F^{-1}(1 - n^{-1})$ and $M_n = \max(X_1, \ldots, X_n)$. Let $r \geqslant 1$. Then the following statements are equivalent:*

(i) $\sum_{n=1}^{\infty} n^{r-2} P[\,|\, M_n/a_n - 1 \,|\, > \varepsilon] < \infty$ *for every* $\varepsilon > 0$ *and some sequence* $a_n \uparrow \infty$;

(ii) $\sum_{n=1}^{\infty} n^{r-2} P[\,|M_n/b_n - 1| > \varepsilon] < \infty$ *for every* $\varepsilon > 0$;

(iii) $\sum_{n=1}^{\infty} n^{r-2} P[M_n > (1 + \varepsilon) b_n] < \infty$ *for every* $\varepsilon > 0$;

(iv) $\sum_{n=1}^{\infty} n^{r-1} P[X_1 > (1 + \varepsilon)\, b_n] < \infty$ *for every* $\varepsilon > 0$;

(v) *for every* $\delta < 1$, *the integral*

$$I(r, \delta) \equiv \int_{F^{-1}(0)}^{\infty} [1 - F(\delta\, y)]^{-r}\, dF(y) \tag{2.7}$$

converges.

Proof. Using $\Phi(x) = x^{r-2}$ in Theorem 2.3 yields the equivalence of (ii), (iii), (iv) and (v). Clearly, (ii) implies (i). To prove the converse, assume (i) holds. By Lemma 2.1, $M_n/a_n \to 1$ in probability. But then $a_n \sim b_n$ by Lemma 2 of Resnick [5]. It follows easily that (ii) holds.

The following corollary provides a new criterion tantamount to a.s. relative stability.

Corollary 2.5 *Suppose* $a_n \uparrow \infty$. *Then* $M_n/a_n \to 1$ a.s. *iff*

$$\sum_{n=1}^{\infty} n^{-1} P[\,|M_n/a_n - 1 \,|\, > \varepsilon] < \infty \tag{2.8}$$

for every $\varepsilon > 0$.

Proof. Theorem 2.4 with $r = 1$ shows that (2.8) is equivalent to (1.1) which, in turn, is equivalent to a.s. relative stability (cf. [1], [6]).

Fix $r \geqslant 1$. In light of Theorem 2.4, it is natural to seek an example of a d.f. F such that $I(s, \delta) < \infty$ for all $\delta < 1$ iff $s \leqslant r$. However, the following result demonstrates that such a search is futile if $r > 1$.

Theorem 2.6. *Let F be d.f. such that $F(x) < 1$ for all x. Define $I(r, \delta)$ by (2.7) and $L(x) = -\log(1 - F(x))$. Then the following assertions are equivalent:*

(i) $I(r, \delta) < \infty$ *for every* $\delta < 1$ *and some (equivalently, every)* $r > 1$;

(ii) $\lim_{x\to\infty} L(x)/L(\delta x) = \infty$ *for every* $\delta < 1$;

(iii) *for some* $r > 1$ *and every* $\delta < 1$,

$$\liminf_{x\to\infty} L(x)/L(\delta x) \geqslant r; \tag{2.9}$$

(iv) *for every* $\delta < 1$ *and some (equivalently, every)* $r > 1$,

$$\lim_{x\to\infty} [1 - F(x)]/[1 - F(\delta x)]^r = 0. \tag{2.10}$$

Proof. Notice that (2.10) holds iff

$$L(\delta y)[L(y)/L(\delta y) - r] \to \infty \text{ as } y \to \infty.$$

Inasmuch as $L(\delta y) \geqslant 0$, clearly (iv) implies (iii). Moreover, $L(x) \uparrow \infty$, so (ii) readily implies (iv) for every $r > 1$.

Now, to prove that (iii) implies (ii), fix any $s > 1$ and assume (2.9) holds for some $r > 1$ and all $\delta < 1$. Select an integer $N \geqslant 1$ such that $r^N > s$. For any $\beta > 1$, define $\delta = \beta^{1/N}$. In view of (2.9),

$$\liminf_{x\to\infty} \frac{L(x)}{L(\beta x)} = \liminf_{x\to\infty} \prod_{i=1}^{N} \frac{L(\delta^{i-1} x)}{L(\delta^i x)} \geqslant r^N > s.$$

Since the choices of $\beta < 1$ and $s > 1$ are otherwise unrestricted, (ii) holds. (It also follows now that (iv) holds for some $r > 1$ iff it holds for all $r > 1$.)

If $I(r, \delta) < \infty$ for some $r > 1$, then

$$[1 - F(x)]/[1 - F(\delta x)]^r = \int_x^\infty [1 - F(\delta x)]^{-r} \, dF(y) \leqslant$$

$$\leqslant \int_x^\infty [1 - F(\delta y)]^{-r} \, dF(y) \to 0,$$

so (i) implies (iv) for some $r > 1$ and, therefore, for every $r > 1$.

To complete the proof, it will suffice to show that (ii) implies (i).

Fix $r > 1$, $\delta < 1$ and assume (ii) holds. Then a number x_r exists such that $L(x) > 2\,rL(\delta x)$ if $x > x_r$. Therefore,

$$\int_{x_r}^\infty [1 - F(\delta x)]^{-r} d\,F(x) = \int_{x_r}^\infty e^{rL(\delta x) - L(x)} \, dL(x)$$

$$\leqslant \int_{x_r}^\infty e^{-L(x)/2} \, dL(x) = 2e^{-L(x_r)/2}.$$

Remarks. 1. It follows from Theorem 2.6 that

$$\sum_{n=1}^\infty n^\alpha P[\,|M_n/b_n - 1| > \varepsilon] < \infty \tag{2.11}$$

for every $\varepsilon > 0$ and *some* $\alpha > -1$ iff (2.11) holds for *every* $\alpha > -1$. In particular, $M_n/b_n \to 1$ completely (in the sense of Hsu and Robbins [4]) iff (2.11) holds for every $\alpha > -1$.

2. Every d.f. F such that $F(x) < 1$ for all x possesses exactly one of the following properties: A. $I(1, \delta) = \infty$ for some $\delta < 1$; B. $I(1, \delta) < \infty$ for all $\delta < 1$, but for each $r > 1$, there exists $\delta_r < 1$ such that $I(r, \delta_r) = \infty$; and C. $I(r, \delta) < \infty$ for every $r \geqslant 1$ and every $\delta > 1$. For instance, if $1 - F(x) = x^{-\beta}$ for some $\beta > 0$ and all $x \geqslant 1$, then F has property A. Any exponential distribution has property B, while the d. f. $F(x) = 1 - e^{-e^x}$, $-\infty < x < \infty$, possesses property C.

Let $T(x)$ be a non-increasing function on $(0, \infty)$ such that $T(\gamma x)/T(x) \to t(\gamma) \leqslant \infty$ for every $\gamma > 1$ as $x \to \infty$. Then (cf., for instance, page 268 of [3]), $t(\gamma) = \gamma^{-\rho}$ for some $0 \leqslant \rho \leqslant \infty$. Moreover, $T(x)$ is called rapidly-, regularly- or slowly-varying accordingly as $\rho = \infty$, $0 < \rho < \infty$ or $\rho = 0$, In light of Theorem 2.6, F possesses property C iff $1/L(x)$ is rapidly-varying. Moreover, if $1/L(x)$ is regularly-varying, then F has property B; in this case, for each $\delta < 1$ and some finite $\rho > 0$, $L(x)/(\delta x) \to \delta^{-\rho}$, so if $\varepsilon < 1 - \delta^{\rho}$, $L(\delta x) - L(x) < -\varepsilon L(x)$ for all large x. It follows from an argument in proof of Theorem 2.6 that $I(1, \delta) < \infty$.

3. The bounded case

Let F be a d.f. such that $x_1 \equiv F^{-1}(1) < \infty$. Let $M_n = \max(X_1, \ldots, X_n)$, where $\{X_n\}$ are i.i.d. with d.f.F; then $X_1 \leqslant x_1$ a.s. It is easy to check that $M_n \to x_1$ completely and, hence, a.s. and in probability. To elucidate the rate of convergence, it is reasonable to ask when a sequence $a_n \uparrow \infty$ exists such that $a_n(x_1 - M_n) \to 1$ a.s. To this end, let $X_n^* = (x_1 - X_n)^{-1}$ and $M_n^* = \max(X_1^*, \ldots, X_n^*)$. Then $a_n(x_1 - M_n) = a_n/M_n^*$, so the query is equivalent to seeking criteria for $M_n^*/a_n \to 1$ a.s.

If F^* represents the d.f. of X_1, then $F^*(x) = F(x_1 - x^{-1})$, $0 < x < \infty$, and, hence, $F^*(x) < 1$ for all x. Inasmuch as $\{X_n^*\}$ are i.i.d., it follows from [1] or [6] that $M_n^*/a_n \to 1$ a.s. iff (1.1) holds for every $\delta < 1$ with F^* in lieu of F. Consequently, the following result holds.

Theorem 3,1. Let $M_n = \max(X_1, \ldots, X_n)$, *where* $X_1, X_2, \ldots,$ *are i.i.d. with d.f.F. Assume* $x_1 \equiv F^{-1}(1) < \infty$. *Then* $a_n(x_1 - M_n) \to 1$ *a.s. for some sequence* $\{a_n\}$ *iff* $\int_1^{\infty} [1 - F(x_1 - (\delta y)^{-1})]^{-1} dF(x_1 - y^{-1}) < \infty$, *for every* $\delta < 1$, *in which case* $a_n \sim [x_1 - F^{-1}(1 - n^{-1})]^{-1}$.

Remark. It is evident from the discussion preceding Theorem 3.1 that $\sum_{n=1}^{\infty} \Phi(n) P[\,|a_n(x_1 - M_n) - 1| > \varepsilon] < \infty$ for every $\varepsilon > 0$ iff $\sum_{n=1}^{\infty} \Phi(n) P[M_n^*/a_n - 1| > \varepsilon] < \infty$ for every $\varepsilon > 0$. In light of Theorems 2.3 and 2.4, these two conditions are each equivalent to certain integral conditions if Φ satisfies (2.1) and (2.3) and, in particular, if $\Phi(x) = x^{\alpha}$ for some $\alpha \geqslant -1$. Clearly, then, by exploiting the duality between d.f. with $x_1 < \infty$ and those with $x_1 = \infty$, given by $F^*(x) = F(x_1 - x^{-1})$, rates of convergence for $a_n(x_1 - M_n)$ are immediate from the results of Section 2. Note, too, that every d.f. F with $x_1 < \infty$ lies in exactly one of three classes A, B or C accordingly as F^* has property, A, B or C (cf. end of Section 2). For example, uniform distributions are in class A (i.e. $a_n(x_1 - M_n)$ does not converge a.s. for any $\{a_n\}$). $F(y) =$

$= 1 - e^{-(x_1-y)^{-1}}$ for $-\infty < y < x_1$ is in class B (since F^* is exponential), and $F(y) = 1 - e^{-\{e^{(x_1-y)^{-1}}\}}$, $-\infty < y < x_1$, belongs to class C (because $F^*(x) = 1 - e^{-e^x}$).

REFERENCES

1. Barndorff-Nielsen, O., *On the limit behavior of extreme order statistics.* Ann. Math. Statist., **34**, 992–1002 (1963).
2. Feller, W., *An Introduction to Probability Theory and Its Applications, Vol. 1, 3rd, Ed.* New York, Wiley, 1968.
3. Feller, W., *An Introduction to Probability Theory and Its Applications, Vol. 2.* New York, Wiley, 1966.
4. Hsu, P. L., Robbins, H., *Complete convergence and the law of large numbers.* Proc. Nat. Acad. Sci. U.S.A., **33**, 25–31 (1947).
5. Resnick, S. I., *Stability of maxima of random variables defined on a Markov chain.* Adv. in Appl. Probab., **4**, 285–295 (1972).
6. Resnick, S. I., Tomkins, R. J., *Almost sure stability of maxima.* J. Appl. Probab., **10**, 387–401 (1973).

ISTVÁN VINCZE

Indistinguishability of particles and exchangeable random variables

1. The two concepts occurring in the title suggest by themselves a certain connection between them.

The physical notion came up in connection with the Bose-Einstein statistics due to the ingenious collaboration of S. N. Bose and A. Einstein; this concept induced an interpretation which plays a fundamental role in the modern quantum physics. The significance of the exchangeable random variables was recognized by B. de Finetti, who established their basic properties.

The connection of these two notions appears in the modern quantum theory in the following way. The wave-function of a system which follows the Bose-Einstein statistics is symmetric, it is a mixture of the wave-functions of the single particles; this construction is based on the "principle of the indistinguishability of the particles". Our aim is to investigate this symmetry but by going back to the origin, i.e. to turn to the derivation of the energy distribution in the case of the black body radiation.

For solving this problem Planck applied the maximum probability principle which was established and used successfully in statistical physics by S. Boltzmann. By using this method Planck had to overcome combinatorial difficulties coming from the continuous nature of the energy. For surmounting them he assumed that the total energy consists of a well defined finite number of parts of the same type and introduced the well known quantity h; this hypothesis became a radical turning point in physics. As a result of this ingenious step its consequences were justified in many respects in the past.

To derive the Planck distribution Bose and Einstein looked for a different way and they came to the mentioned principle of indistinguishable particles. It was really surprising that for the derivation of a continuous distribution a hypothesis is needed, which assumes the discreteness of a variable. Although the quantity in question is really discrete — which was the ingenious discovery of Planck — in mathematics and in physics it frequently happened that sums or other quantities or functions of discrete nature were approximated by means of integrals or by other continuous functions with a very high precision. This circumstance might be one reason for physicists to derive repeatedly the Planck distribution. As known, this question aroused Einstein's attention as well.

In the next paragraph we shall outline a derivation of a maximum probability principle valid in the case of discrete and of continuous random variables (even in the case of a mixture of them), furthermore extended beyond independent to dependent random variables. This helps to solve the following type of problem: Let us have a system containing a large number of particles of the same type. We consider the energy (or another quantity) belonging to the single particles and we look for the distribution of this variable assuming that the total energy is given. For the independent case this principle was formulated by S. Kullback [3] and earlier for the discrete case by E. T. Jaynes [1].

According to Boltzmann the essence of this principle consists of the maximization of the "entropy", i.e. of a certain probability. But what is the entropy in the continuous case? Planck's combinatorial difficulty was essentially the same as Shannon's problem as he tried to extend Boltzmann's discrete entropy to continuous random variables. This problem can be settled on the ground of Kolmogorov's foundation of the probability theory.

An interpretation of the I-divergence and its direct derivation from the Boltzmann-Shannon formula was given in [5, 6, 8], later Jaynes [2] came to the same result using a different reasoning. The extension of the I-divergence (or negative entropy) to the dependent case by using a theorem of I. N. Sanov [4] and the formulation of the general form of the maximum probability principle came first in my lecture given at the International Mathematical Meeting in Amsterdam (1968), and was published in [7] and [8]. The procedure is obvious and does not need any principle unrelated to physical reasoning. An interesting feature of its application to the derivation of the Planck distribution is that it does not need in advance any assumption of discreteness: the relation obtained is suitable for discrete or continuous variables as well.

At the same time the procedure makes questionable the principle of "indistinguishability" even if as an interpretation it is justifiable. We obtain this conclusion from the following: Whenever the energies of the particles, viewed as random variables are independent we come in the discrete case to the Boltzmann, while, in the continuous case to the Gibss distribution. In the case of dependent random variables the primary problem is the determination of the joint distribution of the energies viewed as random variables. As far as I know this is not known. As we shall see in our investigations of the next paragraph Bose and Einstein avoided this step and turned immediately to the second step using a formula for the entropy which can be interpreted as a property of indistinguishability of the particles.

Concerning the unknown joint distribution it seems that *the energies of the particles form a set of exchangeable random variables.* The knowledge of the joint distribution may suggest another interpretation, something beyond indistinguishability, but including it.

Our reasoning is supported also by the following heuristic argumentation: When tossing two coins the possible outcomes are HH, HT, TT. Whenever the tosses are independent these outcomes will occur with probabilities 1/4, 1/2, 1/4, regardless of the fact that the coins are distinguishable or not. But whenever the tosses are not independent we may have different probabilities p_{HH}, p_{HT}, p_{TT} depending on the type of dependence. This example shows that dependence and not distinguishability is decisive in the resulting distribution.

Now we turn to sketch the model and the procedure based on our investigations in [8] and [9] adding some remarks to them; at the same time we omit certain mathematical motivations.

2. Let us reconstruct the maximum probability method given by Boltzmann and applied by Planck.

Let $X_1, X_2, \dots X_N, \dots$ be a sequence of identically distributed random variables. We assume that X_i is associated with the ith particle of a gas system, which may be in several states. The set of states will be denoted by $\mathcal{T}$. We assume that the initial state of the system is $T_0 \in \mathcal{T}$, which means that in this state the distribution of a single particle is $P(X_i < x \mid T_0) = F(x; T_0)$, $i = 1, 2, \dots$.

Suppose now that due to a certain external physical effect the state of the system will change into $T \in \mathcal{T}$. The problem is the determination of $F(x; T)$. Following Boltzmann we determine the *probability* of the location of the first N variables in a system of intervals belonging to a partition of the real line; then we choose that distribution for which this probability is largest; the extreme value has to be taken under the condition (state) T.

Let us divide the real axis into r parts

$$-\infty \leqslant a_{1,N} < a_{2,N} < \dots < a_{r,N} \leqslant \infty,$$

with the usual assumption $\max_i (a_{i+1,N} - a_{i,N}) \to 0$, as r and N tend to infinity. Denoting by $\nu_{i,N}$ the random variable giving the number of the X_i among $X_1, X_2, \dots X_N$ which lie in the interval $(a_{i-1,N}, a_{i,N})$, the corresponding probability is

$$(2.1) \qquad P_{r,N} = P\left(\nu_{i,N} = n_{i,N},\ i = 1, 2, \dots r;\ \sum_1^r \nu_{i,N} = \sum_1^r n_{i,N} = N\right).$$

We are looking for that system $(n_{1,N}, \dots, n_{r,N})$ for which (2.1) is maximal. But when N is large this probability approaches zero and for this reason we take the logarithm of its N-th root; this does not change the extreme value. For the evaluation of this quantity the joint distribution of the X_i

$$(2.2) \qquad P(X_1 < x_1, X_2 < x_2, \dots X_N < x_N \mid T_0)$$

is needed.

From now on we shall distinguish two cases.

a) *The independent case.* When the random variables are independent the joint distribution in question is the multinomial one; in this case we have to study the expression

$$\frac{1}{N} \log \frac{N!}{n_{1,N}!\, n_{2,N}! \dots n_{r,N}!}\, p_{1,N}^{n_{1,N}}\, p_{2,N}^{n_{2,N}} \dots p_{r,N}^{n_{r,N}}$$

with $p_{i,N} = F(a_{i,N} \mid T_0) - F(a_{i-1,N} \mid T_0)$, $i = 1, 2, \dots r$. Applying Stirling's formula we have asymptotically

$$\sum_1^r \frac{n_{i,N}}{N} \log \frac{p_{i,N}}{\frac{n_{i,N}}{N}} = -\sum \frac{n_{i,N}}{N} \log \frac{\frac{n_{i,N}}{N}}{p_{i,N}}.$$

Due to the external physical effect the relative frequencies $n_{i,N}/N$ will approach the probabilities $F(X_i \mid T) - F(X_{i-1} \mid T)$ associated with state T. Consequently, for large N and r, we may write $n_{i,N}/N \sim F(X_i|T) - F(X_{i-1}|T) \sim \sim \mathrm{d}F(X_i';T)$; also $p_{iN} \sim \mathrm{d}F(X_i''; T_0)$. We obtain for our expression the form

$$-\sum_{1}^{r} \mathrm{d}F(X_i'; T) \log \frac{\mathrm{d}F(X_i'; T)}{\mathrm{d}F(X_i''; T_0)} \tag{2.3}$$

which is to be maximized under the conditions

$$\int_0^\infty \mathrm{d}F(x, T) = 1 \tag{2.4}$$

and

$$\int_0^\infty x\, \mathrm{d}F(x; T) = m \neq m_0 = \int x\, \mathrm{d}F(x; T_0), \tag{2.5}$$

the latter assuming state T. (I.e., it is assumed that the total energy — in consequence of the external effect — changed from m_0 to m.) Expression (2.3) is the entropy of the system with respect to $F(x; T)$.

Kullback [3] and Vincze [7, 8] showed that this procedure is equivalent to minimizing the relative entropy

$$\int_0^\infty \mathrm{d}F(x; T) \log \frac{\mathrm{d}F(x; T)}{\mathrm{d}F(x; T_0)} \tag{2.3*}$$

which is always nonnegative. Under the conditions (2.4) and (2.5) the extremum problem has the solution

$$\mathrm{d}F(x; T) = C\mathrm{e}^{cx}\, \mathrm{d}F(x; T_0),$$

which corresponds to the Boltzmann distribution in the case of discrete variables, while to the Gibbs distribution for the continuous case. Consequently, we have a unified, mathematically exact reformulation of the maximum probability principle initiated by Boltzmann.

b) *The dependent case.* Let $\Phi(x)$ be the common distribution function of the (independent) X_i and let $F(x)$ be a further distribution function such that the expression $\int_{-\infty}^{\infty} \mathrm{d}F(x) \log \frac{\mathrm{d}F(x)}{\mathrm{d}\Phi(x)}$ exists. The following relation is due to Sanov [4]

$$\lim_{\varepsilon\to 0} \lim_{N\to\infty} \frac{1}{N} \log P(\sup_x |\Phi_N(x) - F(x)| < \varepsilon) = -\int_{-\infty}^{\infty} \mathrm{d}F(x) \log \frac{\mathrm{d}F(x)}{\mathrm{d}\Phi(x)}, \tag{2.6}$$

where $\Phi_N(x)$ denotes the empirical distribution function associated with the X_i, $1 \leqslant i \leqslant N$, and P is taken according to the law $\Phi(x)$. This relation is true for discrete or continuous random variables as well. This means that the expression $\exp\left\{-N \int_{-\infty}^{\infty} \mathrm{d}F(x) \log \frac{\mathrm{d}F(x)}{\mathrm{d}\Phi(x)}\right\}$ has the following interpretation: it is (asymptotically) the probability that although the random variables are distributed according to the law $\Phi(x)$, nevertheless — due to random circumstances (in our case due to a certain effect) — the sample (i.e. the empirical distribution) will be close to $F(x)$. Furthermore this probability can be approximately determined from the probability (2.1) — as we showed under a).

From these considerations it is obvious that we can adopt as entropy for the dependent case too, the expression

$$-k \lim_{\varepsilon \to 0} \lim_{N \to \infty} \frac{1}{N} \log P(\sup_x |\Phi_N(x) - F(x; T)| < \varepsilon | T_0),$$

where our earlier notation is used. To determine this quantity the joint distribution of the random variables $X_1, X_2, \ldots$ given in (2.2) is needed. The ingenious step by Bose and Einstein was that they assumed immediately the knowledge of the probability (2.1) in such a way that it should imply the Planck distribution. The expression for the probability (2.1), given and interpreted by Bose and by Einstein, has the form:

$$P_{r,N} = \frac{\prod_{i=1}^{r} \binom{N_i + z_i - 1}{n_i}}{\binom{N + Z - 1}{N}} \frac{1}{r^r} \tag{2.7}$$

where the n_i particles falling in the interval $(a_{i-1,N}\; a_{iN})$ are divided into z_i cells $\left(\sum_1^r z_i = Z\right)$. It is assumed that each interval has the same probability $1/r$; the form of this formula implies that the particles are indistinguishable.

We mention again our main points: the primary task, i.e. the determination of the joint distribution (2.2) was — in an ingenious way — avoided; it seems to be really obvious that the quantities $X_1, X_2, \ldots, X_N$ form a set of exchangeable random variables, the nature of which may yield a more informative interpretation for physics.

Now we turn to the application of our method for the dependent case — accepting the indistinguishability situation.

We take the logarithm of (2.7) and divide it by N, then let $N \to \infty$.

Using the same combinatorial technique as before, after neglecting a certain term we obtain for the I-divergence the value

$$\int dF(x|T) \log \frac{dF(x|T)}{(1 - \Psi(x))\, dF(x|T_0)},$$

where $\Psi(x) = \dfrac{dF(x|T)}{dF(x|T_0)}$. This is to be minimized under the conditions (2.4) and (2.5). Denoting by λ and μ the corresponding multipliers we obtain

$$dF(x|T) = \frac{c(\lambda, \mu)\, dF(x|T_0)}{e^{\lambda + \mu x} - 1}$$

in agreement with the Planck-Bose-Einstein formula. As mentioned, we neglected a certain term. In our original considerations [8] a more complicated (not explicit) result was obtained, featuring the exponent $1 + \dfrac{1}{\Psi(x)}$

to the term $(1 + \Psi(x))$. Again we emphasize the possibility that the (non-explicit) result corresponding to that exponent in [8] may yield distributions which describe the behaviour of gases related to not quite exactly ideal bosons.

Acknowledgements. The author is very indebted to Professor Bruno de Finetti for his encouragement, which gave the possibility to write again on this topic.

REFERENCES

1. Jaynes, E. T., *Information theory and statistical mechanics. I, II.* Phys. Rev., **106**, 620–630; **108**, 171–190 (1957).
2. Jaynes, E. T., *Information theory and statistical mechanics.* In: *Statistical Physics* (Brandeis Summer Institute, 1962, Vol. 3), pp. 181–218. New York, W. A. Benjamin, 1963.
3. Kullback. S., *Information and Statistics.* New York, Wiley, 1959.
4. Sanov, I. N., *On the probability of large deviations of random variables.* In: *Selected Translations in Mathematical Statistics and Probability, Vol.* 1, pp. 213–244. Providence, R. I., Amer. Math. Soc., 1961.
5. Vincze, I., *An interpretation of the I-divergence of information theory.* In: *Trans. Second Prague Conf. Information Theory, Statist. Decision Functions, Random Processes,* pp. 681–684. Prague, Academia, 1959.
6. Vincze, I., *Some questions concerning the probabilistic problem of information theory.* In: *Selected Translations in Mathematical Statistics and Probability, Vol.* 5, pp. 373–380. Providence, R. I., Amer. Math. Soc., 1965.
7. Vincze, I., *On some distribution laws of statistical physics.* MTA III. Oszt. Közl., **19**, 117–130 (1969). (Hungarian).
8. Vincze, I., *On the maximum probability principle of statistical physics.* In: *Progress in Statistics* (Proc. European Meeting of Statisticians, Budapest, 1972). Colloq. Math. Soc. J. Bolyai, Vol. 9, pp. 869–893. Amsterdam, North-Holland, 1974.
9. Vincze, I., *On some problems in connection with the Bose–Einstein statistics.* Sankhya Ser. B, **37**, 355–362 (1975).

COMMUNICATIONS

I. PROBABILITY

GH. BOCŞAN

On measurability of the inverse of a random mapping with random domain

In this paper a natural class of random mappings with random domain having ranges and inverses which are measurable is introduced and studied. This class contains some classes studied by many authors and thus some known results are generalized and improved.

For the notions relative to multifunctions and Souslin families see [13].

Let $(\Omega, \mathcal{X})$ be a measurable space (the σ-algebra $\mathcal{X}$ being supposed to be a Souslin family), X and Y two complete separable metric spaces, $E: \Omega \to \mathcal{P}(X) = \{A \subseteq X: A \neq \emptyset\}$ a multifunction with $\mathrm{Gr}E = \{(\omega, x) \in \Omega \times X: x \in E(\omega)\}$, $\mathcal{A}_X$ the trace on $\mathrm{Gr}E$ of the product σ-algebra $\mathcal{X} \times \mathcal{B}_X$($\mathcal{B}_X$ is the Borel σ-algebra on X) and let $E^{-}(C) = \{\omega \in \Omega: E(\omega) \cap C \neq \emptyset\}$ $(C \subseteq Y)$.

The range of a map $T: \mathrm{Gr}E \to Y$ is the multifunction $R: \Omega \to \mathcal{P}(Y)$ defined by $R(\omega) = \{T(\omega, x): x \in E(\omega)\}$. Clearly, $R(\omega)$ is the range of the map $T(\omega): X \to Y$ defined by $T(\omega)x = T(\omega, x)$. T is said to be "invertible" if each map $T(\omega): E(\omega) \to R(\omega)$, $\omega \in \Omega$, is invertible. In this case the inverse T^{-1} of T is the map $T^{-1}: \mathrm{Gr}R \to X$ defined by $T^{-1}(\omega, y) = T^{-1}(\omega)y$.

A mapping $T: \mathrm{Gr}E \to Y$ will be called a random mapping with random domain E if E is measurable and for all $x \in X$ and $C \in \mathcal{F}_Y$ (the family of all closed subsets of Y), the set $D = \{\omega \in \Omega: x \in E(\omega)$ and $T(\omega, x) \in C\}$ is in $\mathcal{X}$. This notion was introduced by Engl (see [8]).

Let $\mathcal{S}(\mathcal{T})$ be the Souslin family generated by the family $\mathcal{T}$ of parts of a given set. The Souslin family $\mathcal{S}(\mathcal{T})$ where $\mathcal{T} = \{A \times B: A \in \mathcal{X},\ B \in \mathcal{F}_X\}$ is denoted by $\mathcal{S}_X$. We note that $\mathcal{X} \times \mathcal{B} \subset \mathcal{S}_X$ [6].

We denote by $\mathcal{M}(E, X; Y)$ the set of all $(\mathcal{A}_X, \mathcal{B}_Y)$-measurable mappings $T: \mathrm{Gr}E \to Y$ such that $\mathrm{Gr}E \in \mathcal{S}_X$.

Proposition 1. *If* $T \in \mathcal{M}$ $(E. X; Y)$, *then*

(i) *T is a random mapping with random domain E*

(ii) *for each measurable selector $x: \Omega \to X$ of E, the map $\omega \to T(\omega, x(\omega))$, from Ω into Y, is measurable.*

Proof. Since $\mathrm{Gr}E$, $\Omega \times \{x\}$ and $T^{-1}(C)$ are in $\mathcal{S}_X$ and since $D = \mathrm{pr}_\Omega (T^{-1}(C) \cap (\Omega \times \{x\})) \in \mathcal{X}$ [7, p.28] and E is measurable [6, Corollary 2], (i) is proved.

Now, if y is the map $y(\omega) = (\omega, x(\omega))$ and Σ is the σ-algebra generated by $\mathcal{S}_X$, it follows that y is $(\mathcal{X}, \Sigma)$ -measurable because $y^{-1}(A) \in \mathcal{X}$ for every $A \in \mathcal{S}_X$. Since $\mathrm{Gr}E \in \Sigma$, T is measurable and $T(\omega, x(\omega)) = T(y(\omega))$, (ii) is proved.

We set $Z = \{A \times C : A \in \mathcal{A}_X,\ C \in \mathcal{F}_Y\}$, $\mathcal{T} = \mathcal{X} \times \mathcal{B}_Y$, $\mathcal{S}$ respectively $\mathcal{S}_1$ the Souslin families generated by $\{E \times F : E \in \mathcal{T},\ F \in \mathcal{F}_X\}$ respectively $\{K \times M : K \in \mathcal{X},\ M \in \mathcal{F}_{X \times Y}\}$.

Theorem 1. *The range R of a mapping $T \in \mathcal{M}(E, X; Y)$ is a measurable multifunction. Moreover,* $\mathrm{Gr}R \in \mathcal{S}_Y$ *and R is a multifunction of Souslin type.*

Proof. We have $\mathrm{Gr}T \in \mathcal{A}_X \times \mathcal{B}_Y$, $Z \subset \mathcal{S}$ and $\mathcal{A}_X \times \mathcal{B}_Y \subset \mathcal{S}(Z) \subset \mathcal{S}$, therefore $\mathrm{Gr}T \in \mathcal{S}$. Since $\mathrm{Gr}R = \mathrm{pr}_{\Omega \times Y}(\mathrm{Gr}T) \in \mathcal{S}(\mathcal{T}) \subset \mathcal{S}_Y$ [7, p.28], from [6, Theorem 6 and Corollary 2] the theorem is proved.

Theorem 2. *If $T \in \mathcal{M}(E, X; Y)$, then there exists an (at most) countable set $\mathcal{U} \subset \mathcal{M}(R, Y; X)$ such that for every $S \in \mathcal{U}$ we have $T(\omega, S(\omega, y)) = y$ for all $(\omega, y) \in \mathrm{Gr}R$.*

Moreover, the multifunction F: $\mathrm{Gr}R \to \mathcal{P}(X)$ defined by

$$F(\omega, y) = \{x \in E(\omega) : T(\omega, x) = y\}$$

is of weak Souslin type and $\mathcal{U}$ is a Castaing representation for F (the set $\mathrm{Gr}R$ *is endowed with the σ-algebra $\mathcal{A}_Y$).*

Proof. For the multifunction $H(\omega) = \{(x, T(\omega, x)) : x \in E(\omega)\}$, $\omega \in \Omega$, we have $\mathrm{Gr}H \in \mathcal{S}_1$ because $\mathrm{Gr}H = \mathrm{Gr}T$, and $\mathrm{Gr}T \in \mathcal{A}_X \times \mathcal{B}_Y$, $Z \subset \mathcal{S}_1$ and $\mathcal{A}_X \times \mathcal{B}_Y \subset \mathcal{S}(Z) \subset \mathcal{S}_1$.

Thus, H is a multifunction of Souslin type [6, Theorem 6], therefore $H(\omega) = g(D(\omega))$, $\omega \in \Omega$ where $g: P \to X \times Y$ is a continuous map, $D: \Omega \to \mathcal{P}(P)$ is a closed-valued measurable multifunction and (P, d) is a complete separable metric space.

We now define the closed-valued multifunctions S and Q from $\mathrm{Gr}R$ into $\mathcal{P}(P)$ by $S(\omega, y) = D(\omega)$ and $Q(\omega, y) = \{p \in S(\omega, y) : f((\omega, y), p) = 0\}$, where $f((\omega, y), p) = \mathrm{dist}(\mathrm{pr}_Y g(p), y)$, $(\omega, y) \in \mathrm{Gr}R$, $p \in P$ is a measurable function.

Since $S^-(C) = \mathrm{Gr}R \cap (H^-(C) \times Y)$, the multifunction S is measurable, therefore S has a Castaing representation $\mathcal{U}$ [13, Theorem 4.2 (d) and (e)]. Thus, using $\mathcal{U}$, as in [4, Theorem 1] we show that $Q^-(C) \in \mathcal{A}_Y$ for every compact set $C \subset P$, hence Q is a weakly measurable multifunction [13, Theorem 4.2.(e)].

Clearly, since $F(\omega, y) = \mathrm{pr}_X g(Q(\omega, y))$, $(\omega, y) \in \mathrm{Gr}R$ the theorem is proved, taking as $\mathcal{U}$ a Castaing representation of F[13, Theorem 6.1].

It can be easily seen that if $T \in \mathcal{M}(E, X; Y)$ and the random equation $T(\omega, x) = y(\omega)$ has a solution in wide-sense, then it has a random solution. Moreover, if $\mathcal{U}$ is as in Theorem 2, then the mapping $\omega \to x_S(\omega) = S(\omega, y(\omega))$ is a random solution of this equation for each $S \in \mathcal{U}$, the multifunction $G(\omega) = \{x \in E(\omega) : T(\omega, x) = y(\omega)\}$ is measurable, $\mathrm{Gr}G \in \mathcal{S}_Y$ and $\{x_S : S \in \mathcal{U}\}$ is a Castaing representation of G. Thus, generalizations and improvements of some results due to Nowak [9], Römisch [11] and Ždanok [14] are obtained.

Similar results can be proved for the random equation $T(\omega, x) = S(\omega, x)$, where $T, S \in \mathcal{M}(E, X; Y)$. Clearly, if $X = Y$ and $S(\omega, x) = x$, $(\omega, x) \in \mathrm{Gr}E$, then a (random or wide-sense) solution is a (random or wide-sense) fixed point of T and the generalizations of some results due to Engl [5], Nowak [10] and Bocşan [2], [3], [4] can be obtained.

From Theorems 1 and 2 above we have

Theorem 3. *Suppose* $T\colon \mathrm{Gr}E \to Y$ *is invertible. Then* $T \in \mathcal{M}(E, X; Y)$ *if and only if* $T^{-1} \in \mathcal{M}(R, Y; X)$.

This theorem extends some results due to Špaček, Hanš, Nashed and Salehi, Nashed and Engl (see [8]) and Ždanok [14].

Let $(\Omega, \mathcal{K}, P)$ be a complete and σ-finite measure space, hence $\mathcal{K}$ is a Souslin family [13, p. 864].

Clearly, if $\mathcal{R}(E, X; Y)$ is the set of all mappings $T\colon \mathrm{Gr}E \to Y$ such that $T^{-1}(B) \in \mathcal{K} \times \mathcal{B}_X$; $B \in \mathcal{B}_Y$ [11], then $\mathcal{R}(E, X; Y) \subseteq \mathcal{M}(E, X; Y)$, with equality if E is closed-valued.

Also, denoting by $\mathcal{N}(E, X; Y)$ the set of all the separable and invertible random mappings $T\colon \mathrm{Gr}E \to Y$ such that the range R of T is a separable multifunction and the inverse T^{-1} of T is continuous [8], we have $\mathcal{N}(E, X; Y) \subseteq \subseteq \mathcal{M}(E, Y; X)$. (If $T \in \mathcal{N}$, R is weakly measurable [8, Theorem 1.9] hence $\mathrm{Gr}R \in \mathcal{K} \times \mathcal{B}_Y$ [13, Th. 4.2.(e)]. Since T^{-1} is continuous and R is separable from [11, Lemma 2] we have $T^{-1} \in \mathcal{R}(R, Y; X) \subset \mathcal{M}(R, Y; X)$. Thus, by Theorem 3 it follows that $T \in \mathcal{M}$.)

REFERENCES

1. Bharucha-Reid, A. T., *Fixed point theorems in probabilistic analysis.* Bull. Amer. Math. Soc. **82**, 641–657 (1976).
2. Bocşan, Gh., *The existence of random fixed points for random operators on separable Banach spaces.* Sympos. Funct. Models in Operator Theory Timişoara, 1978); Abstracts, pp. 27–28. Univ. of Timişoara and INCREST Bucharest, 1978. (Romanian)
3. Bocşan, Gh., *A general random fixed point theorem and application to random equations.* Rev. Roumaine Math. Pures Appl., **26**, 375–379 (1981).
4. Bocşan, Gh., *Some properties of continuous random functions on random domains with applications to continuous extensions and random fixed point theorems.* Seminarul de Teoria Probabilităţilor şi Aplicaţii, nr. 45. Univ. Timişoara, 1979.
5. Engl, H. W., *Random fixed point theorems for multivalued mappings.* Pacific J. Math, **76**, 351–360 (1978).
6. Leese, S. J., *Multifunctions of Souslin type.* Bull. Austral. Math. Soc., **11**, 395–411 (1974).
7. Leese, S. J., *Measurable selections and the uniformisation of Souslin sets.* Amer. J. Math., **100**, 19–41 (1978).
8. Nashed, M. Z., Engl, H. W., *Random generalized inverses and approximate solutions of random operator equations.* In: Bharucha-Reid, A. T. (Ed.), *Approximate Solution of Random Equations*, pp. 149–210. New York, Elsevier–North Holland, 1979.
9. Nowak, A., *Random solutions of equations.* In: *Trans. Eighth Prague Conf. Information Theory, Statist. Decision Functions, Random Processes, Vol. B*, pp. 77–82. Dordrecht, Reidel, 1978.
10. Nowak, A., *A note on random fixed point theorems.* Prace Matematyczne, **11**, 33–35 (1981).
11. Römisch, W., *On the approximate solution of random operator equations.* Preprint Nr. 5, Humboldt-Universität zu Berlin, Sektion Mathematik, 1980.
12. Skorokhod, A. V., *Random Linear Operators.* Kiev, Naukova Dumka, 1978. (Russian)
13. Wagner, D. H., *Survey of measurable selections theorems.* SIAM J. Control Optim., **15**, 859–903 (1977).
14. Ždanok, T. A., *Random linear operators with finite-dimensional kernel.* In: *Theory of Random Processes, No. 8*, pp. 50–59. Kiev, Naukova Dumka, 1980. (Russian)

GH. BOCŞAN, GH. CONSTANTIN, V. RADU

Some properties of random operators and applications to existence theorems for random equations

1. The study of classical linear and nonlinear integral equations with random right-hand sides, random kernels, or defined on random domains has been initiated by Špaček and Hanš [9, 11]. Especially the contributions by A. T. Bharucha-Reid, his book [2] and papers [1,3] initiated an essential improvement of the theory of random equations. Also several classes of random equations have been studied extensively by Tsokos and Padgett [13].

It is well known that random fixed point theorems play a very important rôle in the study of existence, uniqueness, and measurability of the random solutions of random operator equations.

The aim of the first part is to give a random version of the extension property of a separable metric space with respect to a given metric space, which has been used to establish the measurability of a mapping associated with a random function and of a solution of a random differential equation [4, 5].

Let $(\Omega, \mathcal{X}, P)$ be a complete probability measure space and X a separable metric space. If $\mathcal{P}(X)$ is the family of all subsets of X and $E: \Omega \to \mathcal{P}(X)$ a multifunction on Ω into X, then E is said to be a *measurable* multifunction if

$$\{\omega: E(\omega) \cap C \neq \emptyset\} \in \mathcal{X}$$

for all closed sets $C \subseteq X$. If $E(\omega)$ is closed for every $\omega \in \Omega$, then E is said to be a *closed* multifunction.

A mapping $u: \Omega \to X$ is said to be a *selector* of E if $u(\omega) \in E(\omega)$, for every $\omega \in \mathrm{Dom}(E)$. If u is an X-valued random variable ($u^{-1}(B) \in \mathcal{X}$ for every $B \in \mathcal{B}(X) =$ the σ-algebra of all Borel subsets of X) it is said to be a *measurable* selector of E.

A *Castaing representation* of a multifunction E is at most a countable set U of measurable selectors of E such that $E(\omega) = \bar{U}(\omega)$, for all $\omega \in \mathrm{Dom}\,(E)$.

Definition 1. Let X and Y be two metric spaces and $E: \Omega \to \mathcal{P}(X)$ a measurable multifunction. A mapping $T: \{(\omega, x) \in \Omega \times X: x \in E(\omega)\} = \mathrm{Gr}\,(E) \to Y$ is said to be a *random mapping* of X into Y with random domain E, if $\{\omega \in \Omega: x \in E(\omega)$ and $T(\omega, x) \in C\} \in \mathcal{X}$, for every closed subset C of Y and $x \in X$.

T is said to be *continuous* if the mappings $x \to T(\omega, x)$ are continuous on $E(\omega)$, for all $\omega \in \mathrm{Dom}(E)$.

If E is a closed measurable multifunction and there exists a Castaing representation U of E such that the mappings $\omega \to T(\omega, u(\omega))$ are measurable, for all $u \in U$, then E is called a *random subset* of X and T.

Definition 2. Let X and Y be two metric space. Y is said to be a metric space with the *random extension property* with respect to X, if for every random closed subset E of X, every random continuous mapping $T\colon \mathrm{Gr}(E) \to Y$ can be extended to a random continuous mapping $T\colon \Omega \times X \to Y$.

Theorem 1. [4] *Every separable metric space Y has the random extension property with respect to a separable metric space X.*

Lemma 1. [4] *Let $f\colon \Omega \times X \to R$ be a random continuous function and E a random closed set of X. If the function*

$$a(\omega) = \inf_{x \in E(\omega)} f(\omega, x) < \infty, \text{ for every } \omega \in \Omega,$$

then $a(.)$ is a random variable.

Proposition 1. [4] *If f is a random continuous function with a random domain E and if*

$$a(\omega) = \inf_{x \in E(\omega)} f(\omega, x) < \infty, \text{ for every } \omega \in \Omega,$$

then $a(.)$ is a random variable.

We recall that, if X is a Banach space and A is a bounded set of X, then the infimum of all $\varepsilon > 0$ such that there exists a decomposition of A into a finite number of subsets of diameter less than ε is called the *Kuratowski number of A*.

A continuous mapping $T\colon X \to X$ is said to be *α-Lipschitz* with modulus h, if for every bounded set $A \subseteq X$

$$\alpha(T(A)) \leqslant h\alpha(A), \quad 0 \leqslant h < \infty.$$

If $T(\omega)$ is a continuous random operator on $\Omega \times X$ into Y, where $(X, \|\,.\,\|)$ is a separable Banach space, and $h(\omega)$ is a nonnegative real-valued random variable such that for any bounded set $A \subseteq X$

$$\alpha(T(\omega)\,(A)) \leqslant h(\omega)\,\alpha(A)$$

for every $\omega \in \Omega$, then $T(\omega)$ is said to be a random *α-Lipschitz* operator with modulus h.

Let the random variables $a\colon \Omega \to (0, \infty)$, $r\colon \Omega \to (0, \infty)$, $t_0\colon \Omega \to R$ and the X-valued random variable $x_0\colon \Omega \to X$ be given; let $\omega \to I_a(\omega) = \{t \in R\colon |t - t_0(\omega)| \leqslant a(\omega)\}$ and $\omega \to S_r(\omega) = \{x \in X\colon \|x - x_0(\omega)\| \leqslant r(\omega)\}$, $\omega \in \Omega$, denote the closed random interval I_a of R, respectively the closed random sphere S_r of X.

Assume that:

(A) The continuous random function f with random domain $\omega \to I_a(\omega) \times S_r(\omega)$, $\omega \in \Omega$, i.e.

$$f\colon \{(t, x, \omega) \in R \times X \times \Omega\colon (t, x) \in I_a(\omega) \times S_r(\omega)\} \to X$$

is such that $(t, x) \to f(t, x, \omega)$ is a uniformly continuous mapping for each $\omega \in \Omega$, $\omega \to f(t, x, \omega)$ is $\mathcal{X}$-measurable on $\Omega_{(t,x)} = \{\omega \in \Omega,\ (t, x) \in I_a(\omega) \times S_r(\omega)\} \in \mathcal{X}$ for every (t, x), and $x \to f(t, x, \omega)$ is a random α-Lipschitz operator with modulus h.

Definition 3. [5] A continuous random function x with domain a random interval $\omega \to I(\omega)$ of R, $x:\{t, x) \in R \times \Omega: t \in I(\omega)\} \to X$, is called a *measurable solution* of the differential equation with random parameters

$$\frac{dx}{dt} = f(t, x, \omega) \tag{1}$$

with random initial condition

$$x(t_0(.),.) = x_0(.) \tag{1'}$$

on the random interval I, where t_0 is a selector of I, if the following conditions are satisfied for all $\omega \in \Omega$,

(a) $x(t, \omega) \in S_r(\omega)$, $t \in I(\omega)$;

(b) $\frac{dx(t, \omega)}{dt} = f(t, x(t, \omega), \omega)$, $t \in I(\omega)$;

(c) $x(t_0(\omega),\omega) = x_0(\omega)$.

Theorem 2. *If* (A) *is satisfied, then there exists at least a measurable solution of equation* (1) *with random initial condition* (1') *on the closed random interval* $\omega \to I_\delta(\omega)$, *where*

$$\delta(\omega) = \min \left\{ \delta'(\omega),\ a(\omega),\ \frac{r(\omega)}{m(\omega)} \right\},$$

$$m(\omega) = \max\{\| f(t, x, \omega) \|,\ (t, x) \in I_a(\omega) \times S_r(\omega)\}$$

and $\delta'(\omega)$ *is a real random variable such that* $\delta'(\omega)\, h(\omega) < 1$ *for all* $\omega \in \Omega$.

2. We shall introduce a class of random operators on Fréchet spaces and give a fixed point theorem which can be used in the theory of random differential equations and of random Mikusiński operators [14]. These theorems have been given for the deterministic case on locally convex vector space by Hadžić and Stanković [7] (see also [6]).

We recall the concept of a τ-stratifiable σ-algebra.

Definition 4. Let (X, τ) be a topological space and $\mathcal{B}$ a σ-algebra in X. We say that $\mathcal{B}$ is *τ-stratifiable* iff there exists $\tau_0 \subseteq \tau$ such that $\mathcal{B}$ is the σ-algebra generated by τ_0 and

$$\forall E \in \tau_0,\ \exists \{E_n\}_1^\infty \subseteq \tau_0 \ni \bigcup_{n=1}^{\infty} \bar{E}_n = E = \bigcup_{n=1}^{\infty} E_n$$

where $\bar{E}_n$ is the τ-closure of E_n.

It is easy to check that the σ-algebra $\mathcal{B}_\tau$ generated by the metric topology τ, the σ-algebra $\mathcal{B}_S$ generated by the neighborhood bases $S(x, r)$, $x \in X$, $r \in R_+$ and the algebra $\mathcal{B}_W$ generated by the neighborhood bases $W(x, f_k, r)$, $x \in X$, $f_k \in X^* - \{0\}$, $k \in K$ (finite), $r \in R_+$, of the weak topology of a Banach space X are τ-stratifiable. Also, $\mathcal{B}_\tau$, the metric topology of a Fréchet space, is stratifiable and therefore $\mathcal{M}(\mathcal{X}, \mathcal{B}_\tau)$ the space of $(\mathcal{X}, \mathcal{B}_\tau)$ measurable mappings is τ-sequentially closed [12].

Definition 5. Let (X, τ) be a Fréchet space. We say that a mapping $v: (\Omega, \mathcal{X}, P) \to (X, \mathcal{B}_\tau)$ is an *X-valued* random variable if v is $(\mathcal{X}, \mathcal{B}_\tau)$-measurable and the range $v(\Omega)$ is separable.

A mapping of the Cartesian product space $\Omega \times X$ into the space X is called a random operator if $T(.,x)$ is an X-valued random variable for every $x \in X$.

In a similar way as in [8] we can prove

Proposition 2. *Let $(\Omega, \mathcal{K}, P)$ be a probability space with a complete probability measure P, and (X, τ) a non-empty Fréchet space. Let T be a continuous random transform of the Cartesian product $\Omega \times X$ into the space X and v an X-valued random variable. Then the mapping u of Ω into X defined for every $\omega \in \Omega$ by the formula*

$$u(\omega) = T(\omega, v(\omega))$$

is an X-valued random variable.

Theorem 3. *Let E be a closed subset of the Fréchet space X and $T: \Omega \times \times E \to E$ a continuous operator satisfying the following conditions*

i) *For every $j \in J$ there exists a nonnegative real-valued random variable $q_j(\omega)$ such that, a.s., $|T(\omega, x) - T(\omega, y)|_j \leqslant q_j(\omega) \, |x - y|_{\varphi(j)}$, $\forall \, x, y \in \mathcal{D}(T)$ a.s., where $|.|_j$, $j \in J$, is a saturated family of seminorms;*

ii) *For every $j \in J$ there exists n_j such that for $n \geqslant n_j$ we have $q_{\varphi^n(j)}(\omega) \leqslant q(\omega, j) < 1$ a.s.;*

iii) *There exists $x_0: \Omega \to E$ such that*

$$|x_0(\omega) - T(\omega, x_0(\omega)|_{\varphi^n(j)} \leqslant p(\omega, j) < \infty, \text{ a.s., } \varphi^0(j) = j.$$

Then there exists one and only one random solution of the equation $T(\omega, x(\omega)) = x(\omega)$ a.s., where $x(\omega)$ is an E-valued random variable which also satisfies the condition

iv) $|x(\omega) - x_0(\omega)|_{\varphi^n(j)} < p(x(\omega), j) < \infty$ a.s., $n \geqslant 0$.

Proof. As in [7] we shall construct the sequence $\{x_n(\omega)\}_1^\infty \subseteq E$, such that $x_n(\omega) = T(\omega, x_{n-1}(\omega))$ a.s. with $x_0: \Omega \to E$. This is a Cauchy sequence a.s. since

$$|x_n(\omega) - x_{n-1}(\omega)|_j = |T(\omega, x_{n-1}(\omega)) - T(\omega, x_{n-2}(\omega))|_j \leqslant$$

$$q_j(\omega) \, |x_{n-1}(\omega) - x_{n-2}(\omega)|_{\varphi(j)} \leqslant q_j(\omega) \, q_{\varphi(j)}(\omega) \, |x_{n-2}(\omega) - x_{n-3}(\omega)|_{\varphi^2(j)} \leqslant \ldots \leqslant$$

$$\prod_{i=0}^{n-2} q_{\varphi^i(j)}(\omega) \, |x_1(\omega) - x_0(\omega)|_{\varphi^{n-1}(j)} \leqslant \left(\prod_{i=0}^{n_j} q_{\varphi^i(j)}(\omega) \right) q^{n-2-n_j}(\omega, j) \, p(\omega, j), \text{ a.s.},$$

for $n \geqslant n_j + 2$, and

$$|x_{n+m}(\omega) - x_n(\omega)|_j \leqslant \sum_{i=1}^{m} |x_{n+i}(\omega) - x_{n+i-1}(\omega)|_j \leqslant$$

$$\leqslant p(\omega, j) \left(\prod_{i=0}^{n_j} q_{\varphi^i(j)}(\omega) \right) q^{n-1-n_j}(\omega, j) \frac{1}{1 - q(\omega, j)} .$$

Then the limit mapping $x(.)$ is unique a.s. since $\mathcal{M}(\mathcal{K}, \mathcal{B}_\tau)$ is τ-sequentially closed and $T(\omega, x(\omega)) = x(\omega)$ a.s. One can also show without difficulty that condition iv) is satisfied.

A random generalization of Theorem 2 and of the generalized Krasnoselski theorem from [7] can be also proved.

REFERENCES

1. Bharucha-Reid, A. T., *On the theory of random equations*. In: *Proc. Sympos. Appl. Math. Vol.* 16, pp. 40–69. Providence, R. I., Amer. Math. Soc., 1964.
2. Bharucha-Reid, A. T., *Random Integral Equations*. New York, Academic Press, 1972.
3. Bharucha-Reid, A. T., *Fixed points theorems in probabilistic analysis*. Bull. Amer. Math. Soc. (N.S.), **82**, 641–657 (1976).
4. Bocşan, Gh., Constantin, Gh., Radu, V., *On the random extension property of a separable metric space*. Seminarul de Teoria Probabilităţilor şi Aplicaţii, nr. 53. Univ. Timişoara, 1980.
5. Bocşan, Gh., Constantin, Gh., Radu, V., *Some applications of the existence of random fixed points*. Seminarul de Teoria Probabilităţilor şi Aplicaţii, nr. 54. Univ. Timişoara, 1980.
6. Deleanu, A., Marinescu, G., *A fixed point theorem and implicit functions in locally convex spaces*. Rev. Roumaine Math. Pures Appl. **8**, 51–59 (1963).
7. Hadžić, O., Stanković, B., *Some theorems on the fixed point in locally convex spaces*. Publ. Inst. Math. (Beograd) (N.S.), **10(24)**, 9–19 (1974).
8. Hanš, O., *Generalized random variables*. In: *Trans. First Prague Conf. Information Theory, Statist. Decision Functions, Random Processes*, pp. 61–103. Prague, Publ. House Czechoslovak Acad. Sci., 1957.
9. Hanš, O., *Random fixed point theorems*. In: *Trans. First Prague Conf. Information Theory, Statist. Decision Functions, Random Processes*, pp. 105–125. Prague, Publ. House Czechoslovak Acad. Sci., 1957.
10. Hanš, O., *Random operator equations*. In: *Proc. 4th Berkeley Sympos. Math. Statist. Probab., Vol.* 2, pp. 185–202. Berkeley, Univ. California Press, 1961.
11. Hanš, O., Špaček, A., *Random fixed point approximation by differentiable trajectories*. In: *Trans. 2nd Prague Conf. Information Theory, Statist. Decision Functions, Random Processes*, pp. 203–213. Prague, Publ. House Czechoslovak Acad. Sci., 1960.
12. Masani, P., *Measurability and Pettis integration in Hilbert spaces*. J. Reine Angew. Math., **297**, 92–135 (1978).
13. Tsokos, C. P., Padgett, W. J., *Random Integral Equations with Applications to Life Sciences and Engineering*. New York, Academic, Press, 1974.
14. Ullrich, M., *Random Mikusiński operators*. In: *Trans. 2nd Prague Conf. Information Theory, Statist. Decision Functions, Random Precesses*, pp. 639–659. Prague, Publ. House Czechoslovak Acad. Sci., 1960.

W. BRYC

Conditional expectation with respect to dependent σ-fields

In this note we will deal with random variables (X_k), which are the conditional expectation of some random variable X with respect to σ-fields $(\mathbf{F}_k)$. In the case where $(\mathbf{F}_k)$ are independent σ-fields such random variables were characterized by Bryc and Kwapień [2] and by a similar technique part of their results was carried over to some dependent σ-fields by Bryc [3]. The asymptotic behaviour of (X_k) for Markov dependent σ-fields $(\mathbf{F}_k)$ was studied by Isaac [4].

Here we will concentrate on the case of two σ-fields $\mathbf{F}_1$, $\mathbf{F}_2$ and we will look for the conditions close to being necessary for all reasonable X_1, X_2 to be of the form $X_k = E^{\mathbf{F}_k}(X)$, $k = 1, 2$; cf. the theorem below.

We assume in the sequel that each $\mathbf{F}_k$, $k = 1, 2$, is a P-complete σ-field. Let $\mathbf{G} = \mathbf{F}_1 \cap \mathbf{F}_2$ and denote by $L_p^0(\mathbf{F}_k)$ the Banach subspace of L_p, $1 \leqslant p \leqslant \infty$, generated by $\mathbf{F}_k$-measurable random variables Y such that $E^{\mathbf{G}}(Y) = 0$, $k = 1, 2$ (L_p denotes the Banach space of all p-integrable random variable with the usual norm). Let us observe that the conjugated space to $L_p^0(\mathbf{F}_k)$ is isometrically isomorphic to $L_{p'}^0(\mathbf{F}_k)$, where $p < \infty$, p and p' are conjugated numbers, i.e. $1/p + 1/p' = 1$ $(\infty' = 1)$.

First, let us observe that if there exists a random variable $X \in L_1$ such that $X_k = E^{\mathbf{F}_k}(X)$, $k = 1, 2$, then $E^{\mathbf{G}}X_1 = E^{\mathbf{G}}X_2$. Moreover it is possible to assume that $E^{\mathbf{G}}X_k = 0$, $k = 1, 2$, since evidently the assertions

(i) $\forall\ X_k \in L_p(\mathbf{F}_k)$, $E^{\mathbf{G}}(X_1) = E^{\mathbf{G}}(X_2) \Rightarrow \exists\ Z \in L_p$ such that $X_k = E^{\mathbf{F}_k}(Z)$, $k = 1, 2$,

and

(ii) $\forall\ X_k \in L_p^0(\mathbf{F}_k)$ $\exists$ $Z \in L_p$ such that $X_k = E^{\mathbf{F}_k}(Z)$, $k = 1, 2$,

are equivalent. This observation motivates the introduction of the spaces L_p^0.

Lemma 1. *Let $1 \leqslant p \leqslant \infty$. Then the following conditions are equivalent*

(A) *For every $X_k \in L_p^0(\mathbf{F}_k)$, $k = 1, 2$, there exists $Z \in L_p$ such that $X_k = E^{\mathbf{F}_k}(Z)$, $k = 1, 2$;*

(B) *There exists $C > 0$ such that for every $Y_k \in L_{p'}^0(\mathbf{F}_k)$*

$$\|Y_1\|_{p'} + \|Y_2\|_{p'} \leqslant C\,\|Y_1 + Y_2\|_{p'}.$$

Proof. We will prove the lemma only for $1 \leqslant p < \infty$. If $p = \infty$ some changes are needed in the reasoning (a conjugate version of the Banach theorem has to be used). Let $T\colon L_p \to L_p^0(\mathbf{F}_1) \times L_p^0(\mathbf{F}_2)$ be defined by $T(X) = ((E^{\mathbf{F}_k} - E^{\mathbf{G}})(X))_{k=1,2}$. Then $T^*\colon L_{p'}^0(\mathbf{F}_1) \times L_{p'}^0(\mathbf{F}_2) \to L_{p'}$ is easily seen to be defined by $T^*((Y_1, Y_2)) = Y_1 + Y_2$. By the Banach theorem (see, e.g., Rudin [6], Th. 4.15) the linear operator T is "onto" if the adjoint operator

T^* is an isomorphic embedding, i.e. there exists a constant C such that $|||(Y_1, Y_2)||| \leqslant C \| T^*((Y_1, Y_2)) \|_{p'}$, where $||| \cdot |||$ is any of the equivalent norms on $L^0_{p'}(\mathbf{F}_1) \times L^0_{p'}(\mathbf{F}_2)$. Choosing $|||(Y_1, Y_2)||| = \| Y_1 \|_{p'} + \| Y_2 \|_{p'}$ we conclude the proof.

Let $\rho = \sup\{E(X_1 X_2) : X_k \in L^0_2(\mathbf{F}_k), \| X_k \|_2 = 1, k = 1, 2\}$. Note that $0 \leqslant \rho \leqslant 1$. In the case where $\mathbf{G}$ is the trivial σ-field ρ is known as the *maximal correlation coefficient*. The rôle of ρ in the context of conditional expectations is partially explained by

Corollary 1. *Let $\mathbf{F}_1, \mathbf{F}_2$ be σ-fields such that for every $X_k \in L^0_2(\mathbf{F}_k)$, $k = 1, 2$, there exists $Z \in L_2$ such that $X_k = E^{\mathbf{F}_k}(Z)$, $k = 1, 2$. Then $\rho < 1$.*

Proof. On account of Lemma 1 the corollary follows from the inequality $E|X_1 + X_2|^2 \leqslant (1 - \rho)(E|X_1|^2 + E|X_2|^2)$.

Note that since the inequality in the proof of Corollary 1 can be arbitrarily close to equality the converse implication holds, too. This will be reproved in more constructive a way in our main result

Theorem. *Let $1 < p < \infty$ and assume $\mathbf{F}_1, \mathbf{F}_2$ are such that $\rho < 1$. Then for every $X_1 \in L^0_p(\mathbf{F}_1)$, $X_2 \in L^0_p(\mathbf{F}_2)$ there exists $Z \in L_p$ such that $X_k = E^{\mathbf{F}_k}(Z)$, $k = 1, 2$.*

As an immediate consequence of the theorem we have

Corollary 2. *If $\mathbf{F}_1, \mathbf{F}_2$ are σ-fields such that for every $X_k \in L^0_2(F_k)$ there exists $Z \in L_2$ such that $X_k = E^{\mathbf{F}_k}(Z)$, $k = 1, 2$, then for every $X_k \in L^0_p(\mathbf{F}_k)$ there exists $Z \in L_p$ such that $X_k = E^{\mathbf{F}_k}(Z)$, $k = 1, 2$, $\forall\ 1 < p < \infty$.*

For the proof of the theorem we will need one more lemma which strengthens the convergence in the "alterniende Verfahren" (Rota [5]).

Lemma 2. *Let $1 < p < \infty$, $\rho < 1$. If $E^{\mathbf{F}_1}E^{\mathbf{F}_2}$ is considered as a linear operator acting on L_p, then $(E^{\mathbf{F}_1}E^{\mathbf{F}_2})^k \to E^{\mathbf{G}}$ as $k \to \infty$ in the norm topology. The same holds for $E^{\mathbf{F}_2}E^{\mathbf{F}_1}$.*

Proof. Let $Q = E^{\mathbf{F}_1}E^{\mathbf{F}_2}$. First, let us observe that if Q is considered as acting on L^0_2, then $\| Q \| \leqslant \rho$. Indeed, $\| Qf \|^2 = E(E^{\mathbf{F}_1}E^{\mathbf{F}_2}(f)\, E^{\mathbf{F}_1}E^{\mathbf{F}_2}(f)) = E(E^{\mathbf{F}_2}(f)\, E^{\mathbf{F}_1}E^{\mathbf{F}_2}(f)) \leqslant \rho \| f \|_2 \| Qf \|_2$. Therefore $Q^k \to 0$ as $k \to \infty$ when Q is considered as an operator acting on L^0_2, and since $\| Q^k - E^{\mathbf{G}} \|_{L_2 \to L_2} = \| Q^k(I - E^{\mathbf{G}}) \|_{L_2 \to L_2} \leqslant 2 \| Q^k \|_{L^0_2 \to L^0_2} \to 0$, the lemma is proved for $p = 2$.

Now, let $1 < p < 2$. Then, since $\| Q^k - E^{\mathbf{G}} \|_{L_1 \to L_1} \leqslant 2$, by the Riesz convexity theorem ([1], Th. 1.1.1.) $\| Q^k - E^{\mathbf{G}} \|_{L_p \to L_p} \leqslant C \| Q^k - E^{\mathbf{G}} \|^{2(p-1)/p}_{L_2 \to L_2} \to 0$ as $k \to \infty$.

A similar argument with L_∞ instead of L_1 proves the lemma in the case $2 < p < \infty$.

By symmetry the same holds for $E^{\mathbf{F}_2}E^{\mathbf{F}_1}$.

Proof of the theorem. By Lemma 2 both series $\sum_{k=0}^{\infty} (E^{\mathbf{F}_1}E^{\mathbf{F}_2})^k$ and $\sum_{k=0}^{\infty} (E^{\mathbf{F}_2}E^{\mathbf{F}_1})^k$ are convergent in the operator norm, when considered as linear operators acting on the Banach space L^0_p. Indeed, $(E^{\mathbf{F}_1}E^{\mathbf{F}_2})^k = (E^{\mathbf{F}_1}E^{\mathbf{F}_2})^k - E^{\mathbf{G}}$ on L^0_p, hence $\| (E^{\mathbf{F}_1}E^{\mathbf{F}_2})^{n_0} \| < 1$ for some n_0 large enough and $\| (E^{\mathbf{F}_1}E^{\mathbf{F}_2})^{kn_0} \| \leqslant \| (E^{\mathbf{F}_1}E^{\mathbf{F}_2})^{n_0} \|^k$.

To complete the proof it suffices now to notice that if $X_k \in L^0_p(\mathbf{F}_k)$, $k = 1, 2$, then $X_1 - E^{\mathbf{F}_1}(X_2)$ and $X_2 - E^{\mathbf{F}_2}(X_1)$ are in L^0_p. Therefore

$$Z = \sum_{k=0}^{\infty} (E^{\mathbf{F}_1}E^{\mathbf{F}_2})^k (X_1 - E^{\mathbf{F}_1}(X_2)) + \sum_{k=0}^{\infty} (E^{\mathbf{F}_2}E^{\mathbf{F}_1})^k (X_2 - E^{\mathbf{F}_2}(X_1))$$

is a well-defined p-integrable random variable. A direct computation shows that $E^{\mathbf{F}_k}(Z) = X_k$, $k = 1, 2$.

Remark. During the Braşov Conference the author was convinced that the theorem holds at least for $p=1$ without any assumptions on the dependence of σ-fields. This appeared not to be true. To see this let us consider the interval $[0, 1]$ with the uniform probability distribution and the following σ-fields: $\mathbf{F}_1$ is generated by a family I of intervals such that $I = \bigcup_{n=0}^{\infty} I_n$ where I_n is any finite partition of $(2^{-n-1}, 2^{-n})$ consisting of intervals of length less than $n^{-1}2^{-n-1}$; $\mathbf{F}_2$ is generated by a family J of intervals such that $J = \bigcup_{n=0}^{\infty} J_n$, where J_n is a finite partition of $(2^{-n-1}, 2^{-n})$ generated by the centers of intervals in I_n (thus the intervals in J_n are of length less than $n^{-1}2^{-n-1}$, too). Then the σ-field $\mathbf{G} = \mathbf{F}_1 \cap \mathbf{F}_2$ is generated by the intervals $(2^{-n-1}, 2^{-n})$. To prove that there are $\mathbf{F}_k$-measurable random variables X_k, $k = 1, 2$, such that $E^{\mathbf{G}}(X_1) = E^{\mathbf{G}}(X_2)$ and which cannot be of the form $X_k = E^{\mathbf{F}_k}(Z)$, $k = 1, 2$, for any integrable Z, we use Lemma 1. Namely, we will show that there is no constant $C > 0$ such that for every $Y_1 \in L^0_\infty(\mathbf{F}_1)$, $Y_2 \in L^0_\infty(\mathbf{F}_2)$

$$\|Y_1\|_\infty + \|Y_2\|_\infty \leqslant C\, \|Y_1 - Y_2\|_\infty \tag{1}$$

To this end let $n \in N$ be fixed and define

$$Y(\omega) = \begin{cases} 2^{n+1}\omega - 1 & \text{if} \quad 2^{-n-1} < \omega < 2^{-n} \\ 0 & \text{otherwise} \end{cases}$$

and

$$Y_k = E^{\mathbf{F}_k}(Y), \ k = 1, 2.$$

Then, since J_n consists of intervals of length less than $n^{-1}2^{-n-1}$ and Y is a Lipschitz function with constant 2^{n+1}, we have $\|Y - Y_2\|_\infty < n^{-1}$ and, similarly, $\|Y - Y_1\|_\infty < n^{-1}$. Thus, $\|Y_1 - Y_2\|_\infty < 2n^{-1}$, which contradicts (1) since n is arbitrary and $\|Y_1\|_\infty, \|Y_2\|_\infty \geqslant (n-1)/n$.

Acknowledgements. The author wishes to thank T. Bojdecki whose critical remark on the author's Ph. D. thesis suggested the topic of this note. Our thanks are also due to L. Blake who acquainted the author with "alterniende Verfahren", to S. Kwapień who suggested the example given in the remark above, and to B. Singer and G. Zbăganu for helpful conversations during the Braşov Conference. Finally, the author thanks G. Zbăganu for further comments and remarks, which in particular improved the presentation of the paper.

REFERENCES

1. Bergh, J., Löfström, J., *Interpolation Spaces.* Berlin, Springer, 1976.
2. Bryc, W., Kwapień, S., *On the conditional expectation with respect to a sequence of independent σ-fields.* Z. Wahrsch. Verw. Gebiete, **46**, 221—225 (1979).
3. Bryc, W., *Sequences of weakly dependent random variables.* Ph. D. Dissertation (1981).
4. Isaac, R., *Markov dependent σ-fields and conditional expectations.* Ann. Probab., **7**, 1088—1091 (1979).
5. Rota, G.C ., *An "alterniende Verfahren" for general positive operators.* Bull. Amer. Math Soc., **68**, 95—102 (1962).
6. Rudin, W., *Functional Analysis.* New York, McGraw-Hill, 1973.

MIOARA BUICULESCU

Changes of time associated with non-homogeneous regular Markov processes

Let X be a non-homogeneous regular Markov process whose phase space is assumed to be Borelian. It is shown that a right process $\hat{X}$ can be obtained from X by means of a three fold transformation: an isomorphism of the phase space, a continuous change of time and a canonical change of the probability spaces. The main underlying fact is the following property: if A is a finite continuous additive functional of X and τ its right continuous inverse, then (τ, X_τ) is a regular Markov process with temporally homogeneous transition function. A one-to-one correspondence between predictable (respectively continuous) additive functionals of X and predictable (respectively, continuous) additive functionals of $\hat{X}$ is also established.

1. Change of time and the associated process

1.1. We shall be concerned with a process X satisfying the regularity conditions introduced in [7]. Let $(E_t, \mathbf{E}_t)$, $t \in R$, denote the Borelian state spaces, $P_{tu}(x, \Gamma)$, $t, u \in R$, $x \in E_t$, $\Gamma \in \mathbf{E}_u$, the transition function, $(E, \mathbf{E})$ the phase space and $[\alpha, \beta)$ the random life time interval of the process. Further, let $\mathbf{F}^\circ$ be the σ-algebra generated by the variables of the process on Ω and $\mathbf{K}$ the family of probabilities on $(\Omega, \mathbf{F}^\circ)$ associated with X; let $\mathbf{F}^P$ denote the completion of $\mathbf{F}^\circ$ relative to $P \in \mathbf{K}$, $\mathbf{N}^P$ the family of all P-null sets in $\mathbf{F}^P$ and $\mathbf{F} = \bigcap_{P \in \mathbf{K}} \mathbf{F}^P$. Similarly, for $T \subseteq R$, let $\mathbf{F}^\circ(T)$ be the σ-algebra generated by $\{X_s, s \in T\}$, $\mathbf{F}^P(T) = \mathbf{F}^\circ(T) \vee \mathbf{N}^P$ and $\mathbf{F}(T) = \bigcap_{P \in \mathbf{K}} \mathbf{F}^P(T)$. In fact T arises as an interval of the form $(-\infty, t]$, $[s, t]$ or $[t, \infty)$, the corresponding σ-algebra being respectively denoted by $\mathbf{F}_t$, $\mathbf{F}_{st}$, $\mathbf{F}^t$.

We assume that the following additional conditions are satisfied: the phase space is Borelian, the transition function separates the states and $P_{tx}(X_t = x) = 1$ for all $(t, x) \in E$.

Following [9] we shall call a process *reconstructable* if it is measurable with respect to the σ-algebra generated on $R \times \Omega$ by the real valued processes $(\eta_t)_{t \in R}$ that are a.s. right continuous and $\eta_t \in \mathbf{F}^t$ for each t. Among other things the class $\mathfrak{R}$ of all reconstructable processes is important because it makes possible to write the strong Markov property in the following very convenient way: let $\eta \in \mathfrak{R}_+$ or $\eta \in b\mathfrak{R}$ and τ be an $(\mathbf{F}_t)_t$-stopping time; then for each $P \in \mathbf{K}$ we have

$$P\{\eta_\tau \mid \mathbf{F}_\tau\} = P_{\tau X_\tau}\{\eta_\alpha\}, \ P\text{-a.s. on } \{\alpha \leqslant t < \beta\}.$$

Finally, let $\mathbf{E}^{(n)}$ denote the σ-algebra of those $\Gamma \subseteq E$ such that for any $P \in \mathbf{K}$ there exist $\Gamma_1, \Gamma_2 \in \mathbf{E}$ satisfying $\Gamma_1 \subseteq \Gamma \subseteq \Gamma_2$ and $P\{T^{\Gamma_2 \setminus \Gamma_1} = \infty\} = 1$, where $T^{\Gamma_2\setminus\Gamma_1} = \inf\{u;\ (u, X_u) \in \Gamma_2 \setminus \Gamma_1\}$.

1.2. By a *finite additive functional* of X (an AF of X) we mean a process $(A_t)_{t \in R}$ such that

(i) the following conditions hold almost surely: its paths are increasing and right continuous, it takes finite values in $(-\infty, \beta)$, $A_\alpha = 0$ and $A_t = A_\beta$ for all $t > \beta$.

(ii) for any $s < t$, $A_t - A_s \in \mathbf{F}_{st}$.

If an AF A of X is such that its paths are a.s. continuous on R then A is said to be a *continuous* additive functional of X(a CAF of X).

We associate with A the a.s. perfect random set C of its points of increase, and its fine support $\Phi = \{(t, x):\ P_{tx}(S = t) = 1\}$, where $S(\omega) = \inf\{u > \alpha(\omega) : (u, \omega) \in C\}$.

Let A be a CAF of X and $(\tau_t)_t$ its right continuous inverse, i.e. $\tau_t = \inf\{u: A_u > t\}$; then $\tau_t = -\infty$ for all $t < 0$, $\tau_0 = S$, its paths are a.s. right continuous, strictly increasing and with finite values on $[0, A_\beta)$; for each t, τ_t is an $(\mathbf{F}_t)$-stopping time.

For given $s, t \in R$, $s \leqslant t$ we define $(\eta_u^{st})_{u \in R}$ to be the process given by

$$\eta_u^{st} = \inf\ \{v: A_{u+v} - A_u > t - s\} = \tau_{A_u + t - s} - u$$

so that $\eta^{st} \in \mathfrak{R}$; we have $\eta_\alpha^{st} = \tau_{t-s} - \alpha$, a.s. and

$$\eta_{\tau_s}^{st} = \tau_t - \tau_s \quad \text{a.s. on} \quad \{\tau_s < \beta\}.$$

1.3. Our main object will be the process $Y = (Y_t, \mathbf{G}_t, \mathbf{K})$, where $Y_t = (\tau_t, X_{\tau_t})$, $\mathbf{G}_t = \mathbf{F}_{\tau_t}$, whose state space is $(E, \mathbf{E})$. First, we are going to show that Y has the Markov property on the random interval $[0, A_\beta)$, with temporally homogeneous transition function given by $Q_t((s, x);\ \Gamma) = P_{sx}\{(\tau_t, X_{\tau_t}) \in \Gamma\}$, for $t > 0$, $(s, x) \in E$, $\Gamma \in \mathbf{E}$. Indeed, for any $P \in \mathbf{K}$, $\Gamma \in \mathbf{E}$, $t \geqslant s$, we have

$$P\{Y_t \in \Gamma \mid \mathbf{G}_s\} = P\{(\tau_s + \eta_{\tau_s}^{st},\ X_{\tau_s + \eta_{\tau_s}^{st}}) \in \Gamma \mid \mathbf{F}_{\tau_s}\} =$$

$$= P_{\tau_s X_{\tau_s}}\{(\alpha + \eta_\alpha^{st},\ X_{\alpha + \eta_\alpha^{st}}) \in \Gamma\} = P_{\tau_s X_{\tau_s}}\{(\tau_{t-s}, X_{\tau_{t-s}}) \in \Gamma\} =$$

$$= Q_{t-s}(Y_s, \Gamma),\ P\text{-a.s. on} \quad \{s < A_\beta\}.$$

1.4. We now turn to showing that for any $f \in b\mathbf{E}$, $u \in R_+$, the process $Z_t^{u,f} = Q_{u-t}(Y_t, f) I_{[0,\, u \wedge A_\beta)}(t)$ is a.s. right continuous on R_+. This holds when f belongs to $\mathcal{C}$, the family of all functions in $b\mathbf{E}$ with the property that $t \to f(t, X_t)$ is a.s. right continuous on $[\alpha, \beta)$, because $t \to Q_t(Y_s; f) I_{[0, A_\beta)}(s)$ is right continuous whenever $f \in \mathcal{C}$ and, on the other hand, X being regular, $t \to Q_s(Y_t; f)$ is a.s. right continuous on $[0, A_\beta)$ for any $f \in b\mathbf{E}$, $s \in R_+$. Thus the process $Z_t^{u,f}$ which is a supermartingale over $\{\Omega, (\mathbf{G}_t^P), P\}$ for any $f \in b\mathbf{E}$, $u \in R_+$, $P \in \mathbf{K}$ is a.s. right continuous whenever $f \in \mathcal{C}$. An application of Lemma 2.4. from [7] leads now to the desired result.

Perhaps it is worth noting that the set of non-branching points for Y is Φ and that the probabilities in $\mathbf{K}$ when restricted to the σ-algebra generated by Y are of the form

$$P(\Gamma) = \int_\Phi P_{sx}(\Gamma)\, \mu(ds\ dx)$$

where μ is the probability on $(\Phi, \mathbf{E}^* \mid_\Phi)$ given by $\mu(B) = P\{(S, X_S) \in B\}$

1.5. Let $Z_t = (Y_t, \tilde{\tau}_t)$, where $\tilde{\tau}_t = \tau_t - \tau_0$, $t \in R_+$. An application of the strong Markov property as in 1.3 implies that $(Z_t, \mathbf{G}_t, P)$ is a Markov additive process (in the sense of [4]), for each $P \in \mathbf{K}$, with temporally homogeneous semimarkovian transition function given by $Q_t((s, x); \Gamma \times C) = P_{sx}\{Y_t \in \Gamma, \tilde{\tau}_t \in C\}$. (Note that, in case all spaces coincide, this provides a description of the random occupation time set for a regular state, as being the image of the additive part of a Markov additive process with homogeneous transition function, shifted by the hitting time of the state.)

2. Transformation of X into a right process

2.1. From now on we consider a very simple change of time, namely the one coming from the CAF $(C_t)_t$ given by $C_t = 0$ for $t < \alpha$, $C_t = t - \alpha$ for $\alpha \leqslant t < \beta$ and $C_t = \beta - \alpha$ for $t \geqslant \beta$, whose right continuous inverse satisfies $\tau_t = t + \alpha$ for $t \in [0, \beta - \alpha)$.

In attempting to transform X into a right process we have to somehow relate it to a right continuous process. With this in mind we consider the map h introduced in [8] from $(E, \mathbf{E})$ to the Hilbertian cube $(H, \mathbf{H})$. If $\tilde{E} = h(E)$ and $\tilde{\mathbf{E}} = \mathbf{H} \mid_{\tilde{E}}$ then, under our assumption that $(E, \mathbf{E})$ is a Borelian space, it follows that h is an isomorphism between the Borelian spaces $(E, \mathbf{E})$ and $(\tilde{E}, \tilde{\mathbf{E}})$. Recall that h is such that after removing a set of probability zero for any $P \in \mathbf{K}$ — the process $(h(t, X_t))_t$ is right continuous on $[\alpha, \beta)$. Therefore the process $\tilde{X}_t = h(\tau_t, X_{\tau_t})$ with state space $(\tilde{E}, \tilde{\mathbf{E}})$ is right continuous on $[0, \beta - \alpha)$. Let $\tilde{\mathbf{F}}$, $(\tilde{\mathbf{F}}_t)_t$ have their usual meaning relative to X which is a Markov process with respect to the family $\tilde{\mathbf{K}} = \{\tilde{P}^{\tilde{\mu}} : \tilde{P}^{\tilde{\mu}} = P_{\tilde{\mu} \circ h}$, with $\tilde{\mu}$ probability measure on $(\tilde{E}, \tilde{\mathbf{E}})\}$, and has temporally homogeneous transition function

$$\tilde{P}_t(\tilde{y}, \Gamma) = P_{h^{-1}(\tilde{y})}\{(\alpha + t, X_{\alpha+t}) \in h^{-1}(\tilde{\Gamma})\}, \quad t \in \tilde{R}_+, \ \tilde{y} \in \tilde{E}, \ \tilde{\Gamma} \in \tilde{\mathbf{E}}.$$

By the results in Section 1 $\tilde{X}$ is a regular Markov process.

The process $\hat{X} = (\hat{\Omega}, \hat{\mathbf{F}}, \hat{\mathbf{F}}_t, \hat{\theta}_t, \hat{P}^{\tilde{y}})$ canonically associated with $\tilde{X}$ is a right process. If φ denotes the canonical map then $\hat{X} \circ \varphi = \tilde{X}$, $\hat{P}^{\cdot} = \tilde{P}^{\cdot} \circ \varphi^{-1}$ and $\varphi^{-1}(\hat{\mathbf{F}}_t) \subseteq \tilde{\mathbf{F}}_t$ for each $t \in R_+$.

A number of elements (defined for X as a non-homogeneous Markov process and for $\tilde{X}$, $\hat{X}$ as processes with temporally homogeneous transition function) are invariant under continuous time change and canonical change of the probability space: excessive functions, nearly Borelian sets, regular points, fine topology, hitting distributions. Consequently, as far as these are concerned it is only the map h that counts. For instance, h induces an isomorphism between E and $\tilde{E}$ when these are respectively endowed with the σ-algebras of nearly Borelian sets and a homeomorphism between E and $\tilde{E}$ when considered as topological spaces endowed with the corresponding fine topology. Therefore it is to be expected that all properties involving only elements invariant up to the ismorphism can be transferred from the right process $\hat{X}$ to X. Technically, for many results it is equivalent to get them directly for nonhomogeneous Markov processes or to transfer them from right processes (it is the case of 2.2 below in which actually we do not need assuming that $(E, \mathbf{E})$ is Borelian). For other properties, however, by transferring them from the corresponding ones for right processes a lot of technical difficulties can be avoided (as illustrated by the characterization of semimartingale functions in 2.3).

2.2. *Essential base of a set.* For any $B \in \mathbf{E}^{(n)}$ we consider $M_B = \{(t, \omega): (t, X_t(\omega)) \in B\}$, the hitting process $T_t^B(\omega) = \inf \{u > t: (u, \omega) \in M_B\}$, $T^B = T_\alpha^B$, the penetration process $\sigma_t^F(\omega) = \inf \{u > t: (-\infty, u] \cap M_B(\omega)$ is uncountable$\}$, $\sigma^B = \sigma_\alpha^B$ and E_B the set of condensation points for M_B. Obviously, $(T_t^B) \in \mathfrak{R}$; also $(\sigma_t^B) \in \mathfrak{R}$ (by [13], Lemma 1.5). Define the *essential base* of B to be the set

$$\rho(B) = \{(t, x): P_{tx}(\sigma^B = t) = 1\},$$

so that $\rho(B) \subseteq B^r$ and $M_{\rho(B)}$ is a.s. right closed.

The essential base was introduced in [11] in the framework of potential theory and the above description was obtained there after proving that $\sigma^B = T^{\rho(B)}$ a.s.

In fact, standard application of the section theorem and of the strong Markov properties implies $\overline{M}_{\rho(B)} \doteq \overline{E}_B$, taking also into account that $\overline{M}_{\rho(B)} = \bigcup_{r \in Q} [\![T_r^{\rho(B)}]\!]$ and $\overline{E}_B = \bigcup_{r \in Q} [\![\sigma_\alpha^B]\!]$.)

Recall now from [8] that a set $C \in \mathbf{E}^{(n)}$ is said to be *scanty* if M_C is a.s. at most countable. Because of the above property we have

(i) $B \setminus \rho(B)$ is scanty,

(ii) B is scanty if and only if $\rho(B) = \emptyset$,

(iii) any $F \in \mathbf{E}^{(n)}$ such that $F^r \subseteq F$ and $B \setminus F$ is scanty satisfies $\rho(B) \subseteq F$.

In view of all these properties the set $B \setminus \rho(B)$ may be called the *scanty part* of B. Note also that in case there exists a reference measure, the term "scanty" may be replaced by "semipolar" (this is actually the situation in [11]).

Assume now that $[\alpha, \beta) \subseteq R_+$ and that the process $(\sigma_t^B)_t$ is quasi-left-continuous. Since $\rho(B)^r = \rho(B)$, by [3] there exists an a.s. continuous local time $(L_t)_t$ for $\rho(B)$. Thus the set of increase points for L is $\overline{E}_B$ (to be related to the general result in [6], Ch. VI, Th. 39) and Supp $L = \rho(B)$.

2.3. *Characterization of semimartingale functions for X.* We now impose the condition $\beta = \infty$.

A function $f: E \to R$ is called a *semimartingale function* for X if the process $f(\alpha, X_\alpha) I_{t<\alpha} + f(t, X_t) I_{t \geq \alpha}$ is a semimartingale over $(\Omega, \mathbf{F}, \mathbf{F}_{t \vee \alpha}, P_{sx})$ for all $(s, x) \in E$.

The following description of semimartingale functions for X restates in fact Theorem 4.6. from [5]; $\mathbf{E}^e$ denotes the σ-algebra generated on E by the excessive functions.

A real function f on E is a semimartingale function for X if and only if the following condition is satisfied: there exists a sequence (E_n) of finely open $\mathbf{E}^e$-measurable sets with $E = \bigcup_n E_n$ such that $\sup_n T^{E_n^c} = \infty$ a.s. and for each n there exist bounded $1-(X, T^{E_n^c})$ excessive functions g_n, h_n on E such that $f I_{E \setminus (E_n^c)^r} = g_n - h_n$.

This is a consequence of [12], (10.12) and Remark (10.40), Theorem 4.6 from [5].

3. Correspondence between additive functionals

Throughout this section we assume that the time parameter for X ranges over R_+. When referring to "additive functionals of X" we mean processes as in 1.2, while the "additive functionals of $\hat{X}$" are considered with their usual meaning for right processes (but finite on the life time interval as well). If A is an AF of X, then we denote by $\tilde{A}$ the process (A_{τ_t}) and if $\hat{A}$ is an AF of $\hat{X}$, then $\bar{A}$ denotes the process $\bar{A}_t = \hat{A}_t \circ \varphi$ When A is $(\mathbf{F}_t)$-predictable, then $\tilde{A}$ is $(\tilde{\mathbf{F}}_t)$-predictable (this follows since $\tilde{A}$ coincides with $\tilde{A}'$, where $\tilde{A}'_t = A_{\sigma_t}$, with $\sigma_t = \alpha + t$ for all $t \in R_+$). Also, when $\hat{A}$ is $(\hat{\mathbf{F}}_t)$-predictable, $\bar{A}$ is $(\tilde{\mathbf{F}}_t)$-predictable.

The main result of this section is the following:

(3.1) *There is a one-to-one correspondence between* $(\mathbf{F}_t)$*-predictable (respectively continuous)* AF's *A of X and* $(\hat{\mathbf{F}}_t)$*-predictable (respectively continuous)* AF's *$\hat{A}$ of $\hat{X}$ such that the processes $\tilde{A}$ and $\bar{A}$ (or equivalently (A_t) and $(\bar{A}_{c_t})$) are P-indistinguishable for each $P \in \mathbf{K}$.*

We will only prove the statement referring to the existence of $\hat{A}$ with the required property when A is given since the proof of the remaining part goes on the same line. We first consider the case when A is such that u_A^0 is bounded on E. Denote by $\hat{A}$ the $(\hat{\mathbf{F}}_t)$-predictable (respectively continuous) AF of X such that $u_A^0 \circ h^{-1} = \hat{u}_{\hat{A}}^0$. Using this we get $P(\bar{A}_\infty - \bar{A}_t | \tilde{\mathbf{F}}) = P(\tilde{A}_\infty - \tilde{A}_t | \tilde{\mathbf{F}}_t)$ for any $P \in \mathbf{K}$, $t \in R_+$ so that $\bar{A}$ and $\tilde{A}$ are P-indistinguishable.

For the general case we use the approximation result in [14] and consider the sequence of AF's of X, $(A^{(n)})_n$ such that $A = \sup_n A^{(n)}$ and $u_{A^{(n)}}^0$ is bounded in E. It follows from [14] that $A^{(n)}$ are continuous in case A is continuous; on the other hand, when A is $(\mathbf{F}_t)$-predictable, replacing each $A^{(n)}$ by its dual predictable projection (which exists and is independent of the considered probability by [2]), one may assume the processes in the approximation sequence as being $(\mathbf{F}_t)$-predictable.

For each n there exists $\hat{A}^{(n)}$, AF of $\hat{X}$, $(\hat{\mathbf{F}}_t)$-predictable (respectively continuous) such that $\bar{A}^{(n)}$ and $\tilde{A}^{(n)}$ are P-indistinguishable for each $P \in \mathbf{K}$ The process $\hat{A} = \sup_n \hat{A}^{(n)}$ is the required AF of $\hat{X}$.

Note that in the above proof two results regarding AF's of X are already involved, namely the one in [2] (both for the existence of the dual predictable projections and for establishing the correspondence from $\hat{A}$ to A) and the approximation theorem in [14]. But other results on nonhomogeneous additive functionals follow at once from the established correspondence. For instance

(3.2) **Motoo's theorem** ([5]). *Let A, B be* CAF's *of X such that for any $f \in b\mathbf{E}$*

$$\int_\alpha^t f(u, X_u)\, dA_u = 0 \ a.s. \Rightarrow \int_\alpha^t f(u, X_u)\, dB_u = 0 \ a.s.$$

Then there exists $g \in \mathbf{E}^e$ such that $B_t = \int_\alpha^t g(u, X_u)\, dA_u \ a.s.$

(3.3) Existence of equilibrium local times: *If* $B \in \mathbf{E}^{(n)}$ *is such that* $B = B^r$, *then for each* $\lambda \in (0, \infty)$ *there exists a unique predictable local time* $(L_t^{(\lambda)})_t$ *for* B *satisfying*

$$u_{L^{(\lambda)}}^{\lambda}(t, x) = P_{tx}\{e^{-\lambda(T^B - t)}\} \text{ for all } (t, x) \in E.$$

The local time $(L_t^{(\lambda)})_t$ *is a.s. continuous if and only if the process* $\{T_t^B\}_t$ *is quasi-left-continuous.*

This actually covers the result in [3], but there the phase space is not assumed to be Borelian.

(3.4) Uniqueness of a continuous local time for a set up to a continuous time change: *Let* A *and* B *be* CAF's *of* X *such that* Supp A = Supp B. *Then there exists a strictly increasing* CAF F *of the process* $\{h(a_t, X_{a_t})\}_t$ *such that* $b = a_f$, *where* a, b, f *denote the right continuous inverses of* A, B, F *respectively.*

This follows from the results in [10] extended to right processes using (3.26) (ii) from [5].

REFERENCES

1. Blumenthal, R. M., Getoor, R. K., *Markov Processes and Potential Theory.* New York, Academic Press, 1968.
2. Buiculescu, M., *Representation of excessive functions as potentials of additive functionals for regular Markov processes.* Teor. Verojatnost. i Primenen., **26**, 392–394 (1981)
3. Buiculescu, M., *On local times for non-homogeneous Markov processes.* Z. Wahrsch. Verw. Gebiete, **53**, 175–182 (1980).
4. Çinlar, E., *Markov additive processes I.* Z. Wahrsch. Verw. Gebiete, **24**, 85–93 (1972).
5. Çinlar, E., Jacod, J., Protter, P., Sharpe, M. J., *Semimartingales and Markov processes.* Z. Wahrsch. Verw. Gebiete, **54**, 161–219 (1982).
6. Dellacherie, C., *Capacités et Processus Stochastiques.* Berlin—Heidelberg—New York, Springer, 1972.
7. Dynkin, E. B., *Regular Markov processes.* Uspehi Mat. Nauk, **28**, *1(170)*, 37–64 (1973). (Russian)
8. Dynkin, E. B., *Additive functionals of Markov processes and stochastic systems.* Ann. Inst. Fourier (Grenoble), **25**, 177–200 (1975).
9. Dynkin, E. B., *Markov systems and their additive functionals.* Ann. Probab., **5**, 653–677 (1977)
10. Getoor, R. K., *Some remarks on continuous additive functionals.* Ann. Math. Statist., **38**, 1655–1660 (1967).
11. Hansen, W., *Semi-polar sets and quasi-balayage.* Preprint.
12. Jacod, J., *Calcul Stochastique et Problèmes de Martingales.* Lecture Notes in Math., Vol. 714. Berlin—Heidelberg—New York, Springer, 1980.
13. Kuznečov, S. E., *Behaviour of excessive functions along the trajectories of a Markov process.* Teor. Verojatnost. i Primenen., **24**, 486–502 (1979). (Russian).
14. Šur, M. G., *On approximation of additive functionals of Markov processes.* Uspehi Mat. Nauk, **29**, *6*, 183–184 (1974). (Russian).

I. GOLEŢ

Quelques remarques sur les métriques probabilistes

0. Soit D^+ l'ensemble des fonctions F de R dans $[0, 1]$ telles que F est croissante, continue à gauche et $F(0) = 0$.

Soit $(S, \mathbf{F})$ une paire ordonnée, où S est un ensemble quelconque et $\mathbf{F}$ une application de $S \times S$ dans D^+. Si

(0.1) $\mathbf{F}(p, q) = H_0$ si et seulement si $p = q$, où H_0 est l'élément maximal dans D^+ par rapport à l'ordre naturel ($F \leqslant G \Leftrightarrow F(x) \leqslant G(x)$ pour tout $x \in R$) avec $H_0(x) = 0$ pour $x \leqslant 0$ et $H_0(x) = 1$ pour $x > 0$;

(0.2) $$\mathbf{F}(p, q) = \mathbf{F}(q, p),$$

alors $(S, \mathbf{F})$ s'appelle *espace semi-métrique probabiliste* (semi-PM-espace). Si

(0.3) $$\mathbf{F}(p, q) \geqslant \tau(\mathbf{F}(p, r), \mathbf{F}(r, q))$$

est aussi satisfaite, où τ est une fonction triangulaire [14], $(S, \mathbf{F}, \tau)$ s'appelle espace métrique probabiliste (PM-espace).

Dans ce qui suit, nous allons analyser la possibilité d'obtenir de nouvelles métriques probabilistes par la composition avec des fonctions $f: D^+ \to D^+$.

En ce qui concerne le même problème pour les métriques habituelles, on connaît certaines conditions nécessaires et suffisantes pour que l'on obtienne, à l'aide de la composition avec des fonctions $f: R_+ \to R_+$, de nouvelles métriques [5], [8], [2].

Dans [14] Sibley définit une métrique naturelle $\mathbf{L}$ sur D^+ appelée la métrique de Lévy *modifiée* et montre que l'espace métrique $(D^+, \mathbf{L})$ est compact.

Dans [15] pour chaque $F \in D^+$ on définit la fonction $F^\varepsilon \in D^+$, $\varepsilon \in [0, 1]$, telle que

$$F^\varepsilon(x) = \begin{cases} 0 & \text{pour } x \leqslant 0 \\ \min\,(F(x+\varepsilon)+\varepsilon,\, 1) & \text{pour } x \in \left(0, \dfrac{1}{\varepsilon}\right] \\ 1 & \text{pour } x > \dfrac{1}{\varepsilon}\cdot \end{cases}$$

1. Puisque $(D^+, \mathbf{L})$ est compact, on a le résultat suivant, utile dans ce qui suit.

Lemme 1.1. *Soit f une fonction continue* $f\colon (D^+, \mathbf{L}) \to (D^+, \mathbf{L})$. *Alors pour tout* $\varepsilon > 0$ *il existe* $\delta_0 > 0$ *tel que* $\mathbf{L}(f(F), f(F^\delta)) < \varepsilon$ *pour tout* $\delta < \delta_0$ *et* $F \in D^+$.

Dans [15] pour chaque semi-*PM*-espace et pour une fonction $\Phi \in D^+$, appelée *fonction profil*, on définit un opérateur de fermeture C_Φ sur S et avec celui-ci une Φ-topologie naturelle sur S.

Théorème 1.2. *Soient* $(S, \mathbf{F}, \tau)$ *un PM-espace par rapport à la fonction triangulaire* τ, $f\colon D^+ \to D^+$ *et* $\Phi \in D^+$ *une fonction profil.*

a) *Si les conditions suivantes sont remplies*

a₁) $f(F) = H_0$ *si et seulement si* $F = H_0$;

a₂) $f(F) \geqslant f(G)$ *si* $F \geqslant G$;

a₃) $f(\tau(F, G)) \geqslant \tau(f(F), f(G))$;

alors $(S, \mathbf{F}^f, \tau)$ *est un PM-espace, où* $\mathbf{F}^f(p, q) = f(\mathbf{F}_{pq})$.

b) *Si, de plus f est continue, alors la* Φ*-topologie définie par le système de voisinages* $\{\mathbf{N}_p^{\mathbf{F}}(\Phi) \mid p \in S\}$ *est plus fine que la* $f(\Phi)$*-topologie définie par le système* $\{\mathbf{N}_p^{\mathbf{F}^f}(f(\Phi)) \mid p \in S\}$.

Démonstration. a) En effet nous avons (en écrivant $\mathbf{F}(p,q) = F_{p,q}$).

1) $F^f_{pq} = H_0$ si et seulement si $p = q$ (puisque $f(F) = H_0$ si et seulement si $F = H_0$);

2) $F^f_{pq} = f(F_{pq}) = f(F_{qp}) = F^f_{qp}$,

3) $F^f_{pr} = f(F_{pr}) \geqslant f(\tau(F_{pq}, F_{qr})) \geqslant \tau(f(F_{pq}, F_{qr})) = \tau(F^f_{pq}, F^f_{qp})$.

Il en résulte que $(S, \mathbf{F}^f, \tau)$ est un *PM*-cspace.

b) Soit $N^{\mathbf{F}^f}(f(\Phi), \varepsilon) \in \mathbf{N}^{\mathbf{F}^f}(f(\Phi))$; du lemme 1.1 il résulte qu'il existe $\delta_0 > 0$ tel que $f(F)(x + \varepsilon) + \varepsilon \geqslant f(F^\delta)(x)$ pour tout $\delta < \delta_0$, $F \in D^+$ et $x \in [0, 1/\varepsilon]$.

Soit $r \in N_p(\Phi, \delta)$, alors selon [15], lemme 2.5, $F^\delta_{pr} \geqslant \Phi$; il en résulte que

$$F^f_{pq}(x + \varepsilon) + \varepsilon = f(F_{pr})(x + \varepsilon) + \varepsilon \geqslant f(F)(x) \geqslant f(\Phi)(x)$$

pour tout $x \in \left[0, \dfrac{1}{\varepsilon}\right)$, ce qui implique $N_p(\Phi, \delta) \subset N_p(f(\Phi), \varepsilon)$ et alors (b) a lieu.

Remarque 1.3. Si $f\colon (D^+, \mathbf{L}) \to (D^+, \mathbf{L})$ est bijective alors, puisque $(D^+, \mathbf{L})$ est compact, f est bicontinue et, donc dans ce cas, les deux topologies sont équivalentes.

Remarque 1.4. Si l'on prend $\Phi = H_0$ alors $f(\Phi) = H_0$ et le système de (ε, λ)-voisinages est équivalent au système de (H_0, ε)-voisinages. Dans ce cas il résulte que les (ε, λ)-topologies des métriques $\mathbf{F}, \mathbf{F}^f$ coïncident.

Notons par $\mathbf{M}_\tau$ l'ensemble des fonctions $f\colon D^+ \to D^+$ telles que pour chaque *PM*-espace $(S, \mathbf{F})$ par rapport à la fonction triangulaire τ, $(S, \mathbf{F}^f)$ est aussi un *PM*-espace par rapport à τ.

Alors on a

Proposition 1.5. a) $\mathbf{M}_\tau$ *est un semi-groupe par rapport à l'opération de composition des fonctions.*

b) *Pour* $f, g \in \mathbf{M}_\tau$ *considérons l'application* $\bar\tau_{fg}\colon D^+ \to D^+$ *définie par la formule* $\bar\tau_{fg}(F) = \tau(f(F), g(F))$, $F \in D^+$. *Alors* $\bar\tau_{fg} \in \mathbf{M}_\tau$. *Définissons* $\bar\tau\colon \mathbf{M}_\tau \times \times \mathbf{M}_\tau \to \mathbf{M}_\tau$ *par* $\bar\tau(f, g) = \bar\tau_{fg}$. *Alors* $(\mathbf{M}_\tau, \bar\tau)$ *est un semi-groupe.*

Soit $f:[0, 1] \to [0, 1]$ et définissons la fonction $\bar{f}$ sur D^+ par $\bar{f}(F) = f \circ F$:

Théorème 1.6. *Si $(S, \mathbf{F}, \tau_T)$ est un PM-espace où T est une t-norme continue à gauche, et $f:[0, 1] \to [0, 1]$ est telle que $f(0) = 0$, $f(1) = 1$, f continue, strictement croissante et $f(T(x, y)) \geqslant T(f(x), f(y))$, alors $(S, \mathbf{F}^{\bar{f}}, \tau_T)$ est un PM-espace. Si Φ est une fonction profil, alors la Φ-topologie définie par $\mathbf{F}$ coïncide avec la $f(\Phi)$-topologie définie par $\mathbf{F}^{\bar{f}}$.*

Le théorème 1.6 admet la généralisation suivante, qui peut être démontrée directement.

Théorème 1.7. *Soient $(S, \mathbf{F}, T)$ un espace Menger avec la t-norme T telle que $\sup_{x<1} T(x, x) = 1$ et f: $[0, 1] \to [0, 1]$ remplissant les conditions du théorème 1.6. Alors*

(a) *$(S, \bar{f} \circ \mathbf{F}, \tilde{T})$ est un espace Menger où*

$$\tilde{T}(x, y) = f(T(f^{-1}(x), f^{-1}(y)) ;$$

(b) *Si $f(T(x, y)) \geqslant T(f(x), f(y))$ alors $(S, \bar{f} \circ \mathbf{F}, T)$ est un espace Menger;*

(c) *Les systèmes des (ε, λ)-uniformités correspondant à $\mathbf{F}$ et respectivement à $\bar{f} \circ \mathbf{F}$ coïncident.*

2. Soit $\mathbf{D}^+ = \{F \in D^+ \mid F(\infty) = 1\}$. On sait que $(\mathbf{D}^+, \mathbf{L})$ n'est pas un espace métrique compact.

Soient $\mathbf{F}: S \times S \to \mathbf{D}^+$ et $\tilde{\tau}^s_T : \mathbf{D}^+ \times \mathbf{D}^+ \to \mathbf{D}^+$

$$\tilde{\tau}^s_T(F, G)(x) = \sup_{t<x} \sup_{s(u, v)=t} T(F(u), G(v))$$

la fonction triangulaire définie dans [1], où T est une t-norme continue à gauche et $s: R_+ \times R_+ \to R_+$ une opération continue, telle que (R_+, s) est un semi-groupe avec élément neutre 0.

Nous appellerons [7] *iséomorphisme* de deux semi-groupes tout isomorphisme algébrique qui est aussi un homéomorphisme.

Alors nous avons les résultats suivants.

Théorème 2.1. *Soient T_1 et T_2 deux t-normes continues à gauche, s une opération continue sur $R_+ = [0, \infty)$ et f: $([0, 1], T_1) \to ([0, 1], T_2)$ un iséomorphisme de semi-groupes.*

Si $(S, \mathbf{F}, \tilde{\tau}^s_{T_1})$ est un PM-espace, alors $\bar{f}(F) = f \circ F$ définit une application de $\mathbf{D}^+$ dans lui-même telle que $(S, \mathbf{F}^{\bar{f}}, \tilde{\tau}^s_{T_2})$ est un PM-espace et les (ε, λ)-topologies des deux espaces coïncident.

Démonstration. Selon la définition de la t-norme il résulte que $f(0) = 0$ et $f(1) = 1$; comme f est aussi bijective il résulte que f est strictement croissante. Il en résulte que $\bar{f}(F) \in \mathbf{D}^+$ pour tout $F \in \mathbf{D}^+$ et les conditions suivantes sont vérifiées:

(1) $F^{\bar{f}}_{pq} = H_0$ si et seulement si $p = q$;

(2) $F^{\bar{f}}_{pq} = F^{\bar{f}}_{qp}$;

(3) $F^{\bar{f}}_{pq}(x) \geqslant \tau\ (F^{\bar{f}}_{pr}, F^{\bar{f}}_{rq})(x)$.

Par conséquent, $(S, \mathbf{F}^{\bar{f}}, \tilde{\tau}^s_{T_2})$ est un PM-espace.

Dans la suite on a à démontrer que $\bar{f}$ est bicontinue par rapport à la topologie de la convergence faible donnée par $\mathbf{L}$. Pour cela on suppose que $F_n \to F \in \mathbf{D}^+$. Alors pour tout point de continuité x de F nous avons $F_n(x) \to F(x)$, d'où $f(F_n(x)) \to f(F(x))$, f étant continue. Puis, f étant strictement croissante, $\bar{f}(F)$ est continue en x si et seulement si F est continue en x. Ainsi on a $\bar{f}(F_n) \to \bar{f}(F)$, d'où il résulte que $\bar{f}$ est continue. On montre de la même manière que $\bar{f}^{-1}$ est continue.

Soit $\{p_n\}_{n \in N}$ une suite de points de S. Alors $F_{p_n p} \to H_0$, si et seulement si $F^{\bar{f}}_{p_n p} \to H_0$. Il en résulte que les deux (ε, λ)-topologies associées aux métriques $\mathbf{F}$, $\mathbf{F}^{\bar{f}}$ coïncident.

Théorème 2.2. *Soient T une t-norme continue à gauche, s_1 et s_2 deux opérations continues sur R_+ et $f\colon (R_+, S_1) \to (R, S_2)$ un iséomorphisme de semigroupes. Si $(S, \mathbf{F}, \tilde{\tau}_T^{s_1})$ est un PM-espace, alors $\bar{f}(F) = F \circ f^{-1}$ définit une application de $\mathbf{D}^+$ en lui-même telle que $(S, \mathbf{F}^{\bar{f}}, \tilde{\tau}_T^{s_2})$ est un PM-espace et les (ε, λ)-topologies correspondant à $\mathbf{F}$ et respectivement à $\mathbf{F}^{\bar{f}}$ coïncident.*

Corollaire 2.3. *Soient T une t-norme archimédienne avec le générateur multiplicatif h, s une opération continue sur R_+ et $(S, \mathbf{F}, \tilde{\tau}_T^s)$ un PM-espace.*

a) Si T est strictement croissante, alors $\bar{h}(F) = h \circ F$ définit une application de $\mathbf{D}^+$ dans lui-même telle que $(S, \mathbf{F}^{\bar{h}}, \tilde{\tau}^s_{\mathrm{Prod}})$ est un PM-espace et les (ε, λ)-topologies des deux espaces coïncident (Prod $(x, y) = x \cdot y$, $x, y \in [0, 1]$).

b) Si T n'est pas strictement croissante, alors $\bar{g}(F) = g \circ F$, où $g(x) = 1 - \log h(x)/\log h(0)$, définit une application de $\mathbf{D}^+$ dans lui-même telle que $(S, \mathbf{F}^{\bar{g}}, \tilde{\tau}^s_{T_m})$ est un PM-espace et les (ε, λ)-topologies des deux espaces coïncident $(T_m(x, y) = \max\{x + y - 1, 0\}$, $x, y \in [0, 1])$.

BIBLIOGRAPHIE

1. Bocșan, Gh., Radu, V., *On the triangle inequality in PM-spaces II*. Seminarul de Teoria Funcțiilor și Matematici Aplicate, A. Spații Metrice Probabiliste, nr. 47, pp. 1–6. Univ. Timișoara, 1979.
2. Borsík, J., Doboš, J., *O functsiah kompozitsia kotoryh s metrikoi iavliaetsia metrikoi*. Math. Slovaca, **31**, 3–12 (1981).
3. Constantin, Gh., Istrățescu, I., *Elemente de analiză probabilistă și aplicații*. București, Ed. Acad. R.S.România, 1981.
4. Goleț, I., *Iseomorphisms between $\tilde{\tau}_T^s$-semigroups*. Seminarul de Teoria Probabilităților și Aplicații, nr. 59, pp. 1–9. Univ. Timișoara, 1981.
5. Kelley, J. L., *Obshchaia topologhia*. Moskva, Nauka, 1968.
6. Moynihan, R., *On the class of τ_T semigroups of probability distribution functions*. Aequationes Math., 12, 249–261 (1975).
7. Moynihan, R., *On τ_T semigroups of probability distribution functions II*. Aequationes Math., **17**, 19–40 (1978).
8. Nagy, I., *Vybrané partie z moderni matematiky*. Praha, SMTL, 1976.
9. Radu V., *On the triangle inequality in PM-spaces*. Seminarul de Teoria Funcțiilor și Matematici Aplicate, A. Spații Metrice Probabiliste, nr. 39, pp. 1–12. Univ. Timișoara, 1978.
10. Schweitzer, B., *Associative functions and statistical triangle inequalities*. Publ. Math., Debrecen, **8**, 169–185 (1961).
11. Schweitzer, B., *Associative functions and abstract semigroups*. Publ. Math. Debrecen, **10**, 69–81(1963).
12. Schweitzer, B., *Multiplications on the spaces of distribution functions*. Aequationes Math., **12**, 156–183 (1975).
13. Serstnev, A. N., *Triangle inequalities for random metric spaces*. Učen. Zap. Kazan. Gos. Univ. **125**, *no.* 6, 90–93 (1965). (Russian)
14. Sibley, D. A., *A metric for weak convergence of distribution functions*. Rocky Mountain J. Math., **1**, 427–430 (1971).
15. Tardiff, R., *Topologies for probabilistic metric spaces*. Pacific J. Math., **65**, 223–251 (1976).

B. P. HARLAMOV

Continuous semi-Markov processes and an extremal property of Markov processes

A semi-Markov process $(P_x)_{x \in X}$ with a metric phase space X is considered, for which $(\forall\, x \in X)\ E_x(\tau_r) \to 0$ as $r \to 0$, where τ_r is the first exit time from a spherical r-neighbourhood of the initial point of the trajectory. The conditional distribution of passage times given a fixed sequence of states is investigated. If the semi-Markov process is a Markov one and $(\forall \lambda > 0)\ E_x(1 - e^{-\lambda\tau_r})/E_x(\tau_r) \to \lambda\ (r \to 0)$ uniformly in $x \in \Delta$ for every compact $\Delta \subset X$ then this conditional distribution is degenerated. The Wiener process is an example.

1. Preliminaries

Let X be a complete separable locally compact metric space, $\mathfrak{D}$ the set of all functions $\xi\colon R_+ \to X$ which are right continuous and without discontinuities of the second kind. A random process with trajectories in $\mathfrak{D}$ will be regarded as a measurable family (P_x) of probability measures on the σ-algebra $\mathcal{F} = \sigma(\pi_t,\ t \in R_+)$, where $\pi_t\xi = \xi(t)$ and $P_x(\pi_0 = x) = 1$. We will denote $E_x(f, A) = \int_A f P_x(d\xi)$, where $A \in \mathcal{F}$, f is a measurable function on $\mathfrak{D}$ and also $E_x(f, \mathfrak{D}) = E_x(f)$. Put $(\forall t \in R_+)\ \mathcal{F}_t = \sigma(\pi_s,\ s \leqslant t)$. Let $\mathfrak{T}$ be the class of all mappings $\tau\colon \mathfrak{D} \to \bar{R}_+ = R_+ \cup \{\infty\}$, and set $\mathfrak{T}_+ = \{\tau \in \mathfrak{T}\colon (\forall t \in R_+)\{\tau < t\} \in \mathcal{F}_t\}$, $\mathfrak{T}_0 = \{\tau \in \mathfrak{T}\colon (\forall t \in R_+)\ \{\tau \leqslant t\} \in \mathcal{F}_t\}$. Then $(\forall \tau \in \mathfrak{T}_0)\ \mathcal{F}_\tau = \{B \in \mathcal{F}\colon (\forall t \in R_+) B \cap \{\tau \leqslant t\} \in \mathcal{F}_t\}$ is the σ-algebra of all events anterior to the stopping time τ; $(\forall \tau \in \mathfrak{T}_+)\ \mathcal{F}_\tau = \sigma(\tau, \alpha_\tau)$ where $(\forall t \in \bar{R}_+)\ \alpha_t\colon \mathfrak{D} \to \mathfrak{D}$ is a stopping mapping: $(\forall s \in R_+)\ (\alpha_t\xi)(s) = \xi(s \wedge t)$ and $(\forall \tau \in \mathfrak{T})\ \alpha_\tau\xi = \alpha_{\tau(\xi)}\xi$. We have $\mathfrak{T}_0 \subset \mathfrak{T}_+$ and $(\forall \tau \in \mathfrak{T}_0)\ \mathcal{F}_\tau = \sigma(\alpha_\tau) = \alpha_\tau^{-1}\mathcal{F}$ [4, p. 22]. For all $\Delta \subset X$ we denote by τ_Δ the first exit time from the set Δ: $\tau_\Delta\xi = \inf\{t \geqslant 0\colon \xi(t) \notin \Delta\}$ ($\inf \varnothing = \infty$). Let $\mathfrak{A}$ ($\tilde{\mathfrak{A}}$) be the class of all open (closed) subsets of the set X, $\mathfrak{K}$ the class of all $\Delta \in \mathfrak{A}$ with compact closure. We have $(\forall\, \Delta \in \mathfrak{A})\ \tau_\Delta \in \mathfrak{T}_0$ and $(\forall \Delta \in \tilde{\mathfrak{A}})\ \tau_\Delta \in \mathfrak{T}_+$ [1, p. 194]. Let $(\forall t \in R_+)\ \theta\colon \mathfrak{D} \to \mathfrak{D}$ be a shift mapping: $(\forall s \in R_+)(\theta_t\xi)(s) = \xi(s + t)$ and $(\forall \tau \in \mathfrak{T})(\forall \xi \in \{\tau < \infty\})\ \theta_\tau\xi = \theta_{\tau(\xi)}\xi$. For all $\tau_1, \tau_2 \in \mathfrak{T}$ let $\tau_1 \dot{+} \tau_2 = \tau_1 + \tau_2 \circ \theta_{\tau_1}$ when $\tau_1 < \infty$ and $\tau_1 \dot{+} \tau_2 = \infty$ when $\tau_1 = \infty$. We denote $\tau_{(\Delta_1, \Delta_2, \ldots, \Delta_n)} = \tau_{\Delta_1} \dot{+} \tau_{\Delta_2} \dot{+} \ldots \dot{+} \tau_{\Delta_n}$ where $n \in N = \{1, 2, \ldots\}$, $\Delta_i \subset X$. If every Δ_i is open then $\tau_{(\Delta_1, \ldots, \Delta_n)} \in \mathfrak{T}_0$ and has the following property: $(\forall t \in R_+)$

$$\{t < \tau_{(\Delta_1, \ldots, \Delta_n)}\} = \{\tau_{(\Delta_1, \ldots, \Delta_n)} \circ \alpha_t = \infty\}.$$

A stopping time $\tau \in \mathcal{T}_+$ is called a *regenerative time* of the process (P_x) ($\tau \in \mathrm{RT}$) if $(\forall x \in X)$ $(\forall B \in \mathcal{F})$ $P_x(\theta_\tau^{-1}B|\mathcal{F}_\tau) = P_{\pi_\tau}(B)$, P_x-a.s. on the set $\{\tau < \infty\}$, where $\pi_\tau\xi = \pi_{\tau(\xi)}\xi$. The family (P_x) is called a *semi-Markov* (SM) process if $(\forall \Delta \in \mathfrak{A})\ \tau_\Delta \in \mathrm{RT}$. We have also $(\forall n \in N)$ $(\forall \Delta_i \in \mathfrak{A})$ $\tau_{(\Delta_1, \dots, \Delta_n)} \in \mathrm{RT}$. The family (P_x) is called a *Markov* process if $(\forall t \in R_+)\ \tau_t \in \mathrm{RT}$, where $\tau_t \equiv t$ and a *strong* Markov process if $(\forall \tau \in \mathcal{T}_0)\ \tau \in \mathrm{RT}$. The SM class contains the class of strong Markov processes and also the class of all regular semi-Markov step processes of Lévy and Smith [3, p. 27] but the SM class is essentially wider than the union of these subclasses. We will be interested in continuous SM processes for which $(\forall x \in X) m_{\tau_r}(x) \to 0$ $(r \to 0)$ where $m_\tau(x) = E_x(\tau, \tau < < \infty)$, $\tau_r\xi = \tau_\Delta\xi$ and $\Delta = S(\xi(0), r)$ is an open spherical r-neighbourhood of the initial point of the trajectory ξ. For such a process we consider the so-called λ-characteristic operator: $(\forall \lambda \geqslant 0)$ $A_\lambda\varphi(x) = \lim_{r \to 0} A_\lambda^r\varphi(x)$, where $A_\lambda^r\varphi(x) = (E_x(e^{-\lambda\tau_r} \cdot \varphi \circ \pi_{\tau_r}, \tau_r < \infty) - \varphi(x))/m_{\tau_r}(x)$, and φ is any $\mathcal{B}(X)$-measurable function. We say the process is *pseudolocal* at the point $x \in X$ if $(\forall \lambda > 0)$ $(\forall R > 0)$ $E_x(1 - e^{-\lambda\tau_r},\ \tau_r = \tau_R)/m_{\tau_r}(x) \to 0$ as $r \to 0$. As an example we can quote any process with P_x-a.s. continuous trajectories. For a pseudolocal at the point x process $A_\lambda\varphi(x) = A_0\varphi(x) + (A_\lambda 1(x) - A_0 1(x))\varphi(x)$, where $1(x) \equiv 1$, $\varphi \in V_0(x) \cap C(x)$, $V_0(x)$ is the domain of the operator A_0 at x, and $C(x)$ is the set of all functions continuous at x. The subclass of Markov processes in our class of SM processes is characterized by the following property. The function $A_\lambda 1(x)$ for a Markov process is a linear function of λ [8] and if $(\exists a > 0)(\forall r > 0)\ E_x((\tau_r)^2, \tau_r < \infty)/E_x(\tau_r, \tau_r < \infty) \leqslant a$ then $A_\lambda 1(x) = -\lambda$.

For a given sequence $\delta = (\Delta_n)_1^\infty$ $(\Delta_n \in \mathfrak{A} \cup \widetilde{\mathfrak{A}})$ the sequence $(\tau_\delta^n)_{n=1}^\infty (\tau_\delta^n = = \tau_{(\Delta_1, \dots, \Delta_n)})$ is non-decreasing: $(\forall \xi \in \mathfrak{D})\ \tau_\delta^n\xi \leqslant \tau_\delta^{n+1}\xi$ $(n \in N)$. The sequence δ is said to be *outleading* ($\delta \in \mathrm{OS}$) if $(\forall \xi \in \mathfrak{D})$ $\lim_{n \to \infty} \tau_\delta^n \xi = \infty$. For every open covering $\mathfrak{A}_1 \subset \mathfrak{A}$ of the set X there exists an outleading sequence whose terms belong to this covering [5]. Denote by $\mathrm{OS}(\mathfrak{A}_1)$ the class of these sequences. For every outleading sequence δ the mapping $L_\delta\colon \mathfrak{D} \to \mathfrak{D}$ is well defined, where $L_\delta\xi(t) = \pi_{\tau_\delta^n}\xi$ for $\tau_\delta^n\xi \leqslant t < \tau_\delta^{n+1}\xi$. Define rank $\delta = \sup\{\mathrm{diam}\Delta : \Delta \in \in \delta\}$. If rank$\delta$ is equal to $r > 0$ then $(\forall t \in R_+)$ $\rho(\xi(t), L_\delta\xi(t)) \leqslant r$ (ρ is the metric in X). Thus a sequence $(\delta_n)_1^\infty$, $\delta_n \in \mathrm{OS}(r_n)$ (rank$\delta_n \leqslant r_n$) with $r_n \downarrow 0$, corresponds to a uniformly converging sequence of step functions $L_{\delta_n}\xi \to \xi$ on R_+. Let $r > 0$ and $\tau_r^k = \tau_r^{k-1} \dot{+} \tau_r$ $(k \in N,\ \tau_r^0 = 0)$ be an iterated first exit time from spherical r-neighbourhoods. We have $(\forall \xi \in \mathfrak{D})\ \tau_r^k\xi \to \infty\ (k \to \infty)$. Define $L_r\xi(t) = \pi_{\tau_r^k}\xi$ for $\tau_r^k\xi \leqslant t < \tau_r^{k+1}\xi$. Evidently $L_r\xi \to \xi (r \to 0)$ uniformly on R_+.

For arbitrary $\Delta \in \mathfrak{A} \cup \widetilde{\mathfrak{A}}$ we have $\tau_{\Delta^{-r}} \uparrow \tau \leqslant \tau_\Delta$, where $\Delta^{-r} = \{x \in \Delta : :\rho(x, \partial\Delta) > r\}$, $\partial\Delta = [\Delta] \setminus \mathrm{int}\,\Delta$. For $\Delta \in \mathfrak{A}$ we have also $\tau_{\Delta^{+r}} \downarrow \tau_{[\Delta]} \geqslant \tau_\Delta$, where $\Delta^{+r} = \{x \in X : \rho(x, \Delta) < r\}$. This holds for $\Delta \in \widetilde{\mathfrak{A}}$ too, but in this case $[\Delta] = \Delta$ ($[\Delta]$ is the closure of Δ). We are interested in the property $\tau\xi = \tau_\Delta\xi = \tau_{[\Delta]}\xi$. Let $\Pi(\Delta)$ be the set of all functions $\xi \in \mathfrak{D}$ such that: 1) $\xi(0) \in \partial\Delta$; 2) $\beta_{\tau_{\Delta^{-r}}}\xi \to \beta_{\tau_\Delta}\xi$; 3) $\beta_{\tau_{\Delta^{+r}}}\xi \to \beta_{\tau_\Delta}\xi (r \downarrow 0)$, where

$$\beta_\tau\xi = \begin{cases} \overline{\infty}, & \text{if } \tau(\xi) = \infty \quad (\overline{\infty} \notin R_+ \times X) \\ (\tau(\xi), \pi_\tau\xi), & \text{otherwise.} \end{cases}$$

We say that $\Pi(\Delta)$ is a *correct exit set* from Δ. For every probability measure P on $\mathcal{F}$ an arbitrary set $\Delta \subset X$ can be "slightly" deformed: $\Delta \to \Delta'$, to have $P(\Pi(\Delta')) = 1$ [7]. For example, $(\forall x \in X)$ there are only denumerably many $r > 0$ such that $P(\Pi(S(x, r))) < 1$. Let $\Pi(\Delta_1, \Delta_2) = (\Pi(\Delta_1) \cap \{\tau_{\Delta_1} = \infty\}) \cup (\Pi(\Delta_1) \cap \theta^{-1}_{\tau_{\Delta_1}} \Pi(\Delta_2))$ $(\Delta_i \in \mathfrak{A} \cup \tilde{\mathfrak{A}})$ and $\Pi(\Delta_1, ..., \Delta_m) = \left(\bigcup_{k=1}^{m-1} \bigcap_{i=1}^{k} \theta^{-1}_{\tau_{i-1}} (\Pi(\Delta_i)) \cap \{\tau_i = \infty\}\right) \cup \bigcap_{i=1}^{m} \theta^{-1}_{\tau_{i-1}} \Pi(\Delta_i)$, where $\tau_0 = 0$, $\tau_i = \tau_{(\Delta_1, ..., \Delta_i)}$. $\Pi(\Delta_1, ..., \Delta_m)$ is a correct exit set from the sequence $(\Delta_1, ..., \Delta_m)$. Similarly, let $\Pi_1(r) = \{\xi : \xi \in \Pi(S(\xi(0), r))\}$ and $\Pi_m(r) = \left(\bigcup_{k=1}^{m-1} \bigcap_{i=1}^{k} \theta^{-1}_{\tau_r^{i-1}} \Pi_1(r) \cap \{\tau_r^k = \infty\}\right) \cup \bigcap_{i=1}^{m} \theta^{-1}_{\tau_r^{i-1}} (\Pi_1(r)$. For any $r > 0$ there exists $\delta \in \mathrm{OS}(r)$ with open or closed elements for which $P(\Pi(\delta)) = 1$, where $\Pi(\Delta_1, \Delta_2, ...) = \bigcap_{m=1}^{\infty} \Pi(\Delta_1, ..., \Delta_m)$, and $(\exists r' \leqslant r)\ P(\Pi(r')) = 1$, where $\Pi(r) = \bigcap_{m=1}^{\infty} \Pi_m(r)$.

2. Conditional distributions given a sequence of states

Let Φ be the class of all continuous non-decreasing functions $\varphi: R_+ \to R_+$ for which $\varphi(0) = 0$, $\varphi(t) \to \infty$ $(t \to \infty)$. The functions $\xi_1, \xi_2, \in \mathfrak{D}$ are said to have *identical* sequences of states (curves) if $(\exists\, \varphi_1, \varphi_2 \in \Phi)\ \xi_1 \circ \varphi_1 = \xi_2 \circ \varphi_2$ [2, p. 109; 6]. Under this relation we mean by a *sequence of states* $L\xi$ of a function ξ the corresponding equivalence class. This definition is a natural one, but it is inconvenient for investigating conditional distributions [6]. For this reason we will consider a rather coarser equivalence class partition of $\mathfrak{D}$. Let

$$\gamma_\tau \xi = \begin{cases} \overline{\infty}, & \text{if } \tau(\xi) = \infty \ (\overline{\infty} \notin X) \\ \pi_\tau \xi, & \text{otherwise.} \end{cases}$$

Let $T(\mathcal{T}_1)$ be a semigroup with respect to the operation $\dot{+}$ generated by the family $\mathcal{T}_1 \subset \mathcal{T}_+$ of stopping times. We say the functions $\xi_1, \xi_2 \in \mathfrak{D}$ have *identical* sequences of essential states if $(\forall \tau \in T(\tau_\Delta, \Delta \in \tilde{\mathfrak{A}}))\ \gamma_\tau \xi_1 = \gamma_\tau \xi_2$. The corresponding equivalence class $l\xi$ of a function ξ is called its *sequence of essential states.* Such a partition has the advantage of being characterized by a denumerable set of values. Let $\tilde{\mathcal{F}} = \sigma(\gamma_\tau, \tau \in T(\tau_\Delta, \Delta \in \tilde{\mathfrak{A}}))$.

Theorem 1 (Countable engendering). *$\tilde{\mathcal{F}}$ is a countably generated σ-algebra.*

Proof. Let $r_n \downarrow 0 (n \to \infty)$, $\delta_n \in \mathrm{OS}(\tilde{\mathfrak{A}})$, $\mathrm{rank}\,\delta_n \leqslant r_n$ and $\tilde{\mathcal{F}}' = \sigma(\pi_0, \gamma(\tau^k_{\delta_n});\ k, n \in N)$. Then $\tilde{\mathcal{F}}' \subset \tilde{\mathcal{F}}$. On the other hand $(\forall k \in N)(\forall \Delta_i \in \tilde{\mathfrak{A}})\ (\forall \xi \in \mathfrak{D})\ \beta_{\tau_{(\Delta_1, ..., \Delta_k)}} \xi = \lim_{n \to \infty} \beta_{\tau_{(\Delta_1, ..., \Delta_k)}} L_{\delta_n} \xi$ and hence $\gamma_{\tau_{(\Delta_1, ..., \Delta_k)}} \xi = \lim_{n \to \infty} \gamma_{\tau_{(\Delta_1, ..., \Delta_k)}} L_{\delta_n} \xi$. Therefore $\tilde{\mathcal{F}} \subset \tilde{\mathcal{F}}'$. From separability of X the countable engendering of $\tilde{\mathcal{F}}'$ follows. The theorem is proved.

We will write $B_1 \overset{P}{=} B_2$ if $P(B_1 \Delta B_2) = 0$, where $B_1 \Delta B_2 = (B_1 \setminus B_2) \cup \cup (B_2 \setminus B_1)$ $(B_1, B_2 \in \mathcal{F})$ and also $\mathcal{F}_1 \overset{P}{=} \mathcal{F}_2$ if $(\forall B_1 \in \mathcal{F}_1)(\exists B_2 \in \mathcal{F}_2)\, B_1 \overset{P}{=} B_2$ and $(\forall B_2 \in \mathcal{F}_2)(\exists B_1 \in \mathcal{F}_1)\, B_1 \overset{P}{=} B_2$ $(\mathcal{F}_1, \mathcal{F}_2 \subset \mathcal{F})$.

Theorem 2 (Equivalence of σ-algebras). *Let* $\tau = \tau_r^k$ *and* $P(\Pi_k(r)) = = 1 (r > 0,\ k \in N)$ *and also* $r_n \downarrow 0$, $\delta_n \in \mathrm{OS}(\mathcal{A})$ $\operatorname{rank}\delta_n \leqslant r_n$ *and* $P\left(\bigcap_{n=1}^{\infty} \Pi(\delta_n)\right) = = 1$. *Denote* $\overline{\mathcal{F}} = \sigma(\pi_0,\ \gamma(\tau_{\delta_n}^k),\ k,\ n \in N)$ *and* $\widetilde{\mathcal{F}}_{(\tau)} = \sigma(\alpha_\tau^{-1}\ \overline{\mathcal{F}},\ \theta_\tau^{-1}\widetilde{\mathcal{F}})$. *Then* $\widetilde{\mathcal{F}}_n\{\tau < \infty\} \overset{P}{=} \widetilde{\mathcal{F}}_{(\tau)} \cap \{\tau < \infty\}$

Proof. 1) Let $\widetilde{\mathcal{F}}' = \{B \in \widetilde{\mathcal{F}} \cap \{\tau < \infty\} : (\exists B_1 \in \widetilde{\mathcal{F}}_{(\tau)} \cap \{\tau < \infty\})\, P(B \Delta B_1) = = 0\}$. Evidently $\widetilde{\mathcal{F}}'$ is a σ-algebra on $\{\tau < \infty\}$ and $\widetilde{\mathcal{F}}' \subset \widetilde{\mathcal{F}} \cap \{\tau < \infty\}$. But $\widetilde{\mathcal{F}} \cap \{\tau < \infty\}$ is generated by events of the form $\{\tilde\tau_m = \infty,\ \tau < \infty\}$ and $\{\tilde\tau_m < \infty,\ \pi_{\tilde\tau_m} \in S,\ \tau < \infty\}$ $(S \in \mathcal{B}(X))$, where $\tilde\tau_m = \tilde\tau_\delta^m$, $\delta \in (\delta_n)_1^\infty$, $\tilde\tau_\delta^m$ is defined by closures of sets which define the τ_δ^m. Let $\tau_k = \tau_{(\Delta_1, \ldots, \Delta_k)}$ and $\tilde\tau_k = = \tau_{([\Delta_1], \ldots, [\Delta_k])}$ $(k = 1, \ldots, m)$. We have

$$\{\tilde\tau_m = \infty,\ \tau < \infty\} = \bigcup_{k=0}^{m-1} \{\tilde\tau_k \leqslant \tau < \tilde\tau_{k+1},\ \tilde\tau_m = \infty,\ \tau < \infty\} =$$

$$(1) = \bigcup_{k=0}^{m-1} \{\tilde\tau_k \leqslant \tau < \tilde\tau_{k+1},\ \tau_{([\Delta_{k+1}], \ldots, [\Delta_m])} \circ \theta_\tau = \infty,\ \tau < \infty\} \overset{P}{=}$$

$$(2) \overset{P}{=} \bigcup_{k=0}^{m-1} \{\tau_k \leqslant \tau < \tau_{k+1},\ \tau < \infty\} \cap \theta_\tau^{-1}\{\tau_{([\Delta_{k+1}], \ldots, [\Delta_m])} = \infty\} =$$

$$(3) = \bigcup_{k=0}^{m-1} \alpha_\tau^{-1}\{\tau_k < \infty,\ \tau_{k+1} = \infty\} \cap \theta_\tau^{-1}\{\tau_{([\Delta_{k+1}], \ldots, [\Delta_m])} = \infty\} \cap \{\tau < \infty\} \subset$$

$$\subset \widetilde{\mathcal{F}}_{(\tau)} \cap \{\tau < \infty\},$$

where (1) through (3) are justified as follows: (1) $(\forall t \in [\tilde\tau_k, \tilde\tau_{k+1}))\ \tilde\tau_m = t \dotplus \dotplus \tau_{([\Delta_{k+1}], \ldots, [\Delta_m])}$; (2) P-a.s. correct exit from corresponding open sets implies that $P\left(\bigcap_{k=1}^{m} \{\tau_k = \tilde\tau_k\}\right) = 1$; (3): $(\forall t \in R_+)\ \{\tau_k \leqslant t\} = \{\tau_k \circ \alpha_t < \infty\} \equiv \alpha_t^{-1}\{\tau_k < \infty\}$ and $\{\tau_k > t\} = \{\tau_k \circ \alpha_t = \infty\} \equiv \alpha_t^{-1}\{\tau_k = \infty\}$.

Further, $\{\tilde\tau_m < \infty,\ \pi_{\tilde\tau_m} \in S,\ \tau < \infty\} = \bigcup_{k=0}^{m-1} \{\tilde\tau_k \leqslant \tau < \tilde\tau_{k+1},\ \tau_{([\Delta_{k+1}], \ldots, [\Delta_m])} \circ \theta_\tau < < \infty,\ \pi_{\tau_{([\Delta_{k+1}], \ldots, [\Delta_m])}} \circ \theta_\tau \in S\} \cup \{\tilde\tau_m \leqslant \tau < \infty,\ \pi_{\tilde\tau_m} \in S\} \overset{P}{=} \bigcup_{k=0}^{m-1} (\{\tau_k \leqslant \tau < \tau_{k+1}\} \cap \cap\, \theta_\tau^{-1}\{\tau_{([\Delta_{k+1}], \ldots, [\Delta_m])} < \infty,\ \pi_{\tau([\Delta_{k+1}], \ldots, [\Delta_m])} \in S\}) \cup \{\tau_m \leqslant \tau < \infty,\ \pi_{\tau_m} \in S\} =$

$$= \left(\bigcup_{k=0}^{m-1} \alpha_\tau^{-1}\{\tau_k < \infty,\ \tau_{k+1} = \infty\} \cap \theta_\tau^{-1}\{\tau_{([\Delta_{k+1}], \ldots, [\Delta_m])} < \infty,\ \pi_{\tau_{([\Delta_{k+1}], \ldots, [\Delta_m])}} \in S\} \cup \right.$$

$$\left. \cup\, \alpha_\tau^{-1}\{\tau_m < \infty,\ \pi_{\tau_m} \in S\}\right) \cap \{\tau < \infty\} \subset \widetilde{\mathcal{F}}_{(\tau)} \cap \{\tau < \infty\}.$$

Therefore $\widetilde{\mathcal{F}} \cap \{\tau < \infty\} = \widetilde{\mathcal{F}}'$.

2) Let $\tilde\tau$ be defined by the closures of the spheres which define $\tau \equiv \tau_r^k$ We have $\theta_{\tilde\tau}^{-1}\{\tilde\tau_m = \infty\} = \{\tilde\tau < \infty,\ \tilde\tau_m \circ \theta_{\tilde\tau} = \infty\} = \{\tilde\tau > \infty,\ \tilde\tau \dotplus \tilde\tau_m = \infty\} \in \in \widetilde{\mathcal{F}}$ and $\theta_\tau^{-1}\{\tilde\tau_m < \infty,\ \pi_{\tilde\tau_m} \in S\} = \{\tilde\tau < \infty,\ \tilde\tau_m \circ \theta_{\tilde\tau} > \infty,\ \pi_{\tilde\tau_m} \circ \theta_{\tilde\tau} \in S\} = \{\tilde\tau \dotplus$

$\dot{+}\tilde{\tau}_m < \infty,\ \pi_{\tilde{\tau}\dot{+}\tilde{\tau}_m} \in S\} \in \tilde{\mathcal{F}}$. Consequently $\theta_{\tilde{\tau}}$ is a $\tilde{\mathcal{F}}$-measurable mapping $\{\tilde{\tau} < \infty\} \to \mathfrak{D}$. Further, $\tilde{\mathcal{F}}_{(\tau)} \cap \{\tau < \infty)$ is generated by events of the form $\alpha_\tau^{-1} A \cap \theta_\tau^{-1} B \cap \{\tau < \infty\}$, where $A = \{\tau_m = \infty\}$ or $\{\tau_m < \infty,\ \pi_{\tau_m} \in S\} (S \in \in \mathcal{B}(X),\ m \geqslant 0,\ \tau_0 = 0)$ and $B \in \tilde{\mathcal{F}}$. We have $\alpha_\tau^{-1}\{\tau_m = \infty\} \cap \{\tau < \infty\} = = \{\tau_m \circ \alpha_\tau = \infty,\ \tau < \infty\} = \{\tau < \tau_m\} \overset{P}{=} (4)\ \{\tilde{\tau} < \tilde{\tau}_m,\ \tau < \infty\} \in (5)\ \tilde{\mathcal{F}} \cap \{\tau < \infty\}$ where (4) follows form the equality $P(\tau = \tilde{\tau}) = 1$, and (5) from $\{\tilde{\tau} < \tilde{\tau}_m\} = \bigcup_{N=1}^{\infty} \bigcap_{n \geqslant N} \{\tilde{\tau} \circ L_{\tilde{\delta}_n} < \tilde{\tau}_m \circ L_{\tilde{\delta}_n}\} \in \tilde{\mathcal{F}}$, where $\tilde{\delta}_n$ is composed from the closures of the sets which compose δ_n; evidently, $\tilde{\delta}_n \in \mathrm{OS}(\mathcal{A})$ and $\{\tilde{\tau} \circ L_{\tilde{\delta}_n} < \tilde{\tau}_m \circ L_{\tilde{\delta}_n}\} \in \in \tilde{\mathcal{F}}$. Moreover, $\theta_\tau^{-1} B \overset{P}{=} \theta_{\tilde{\tau}}^{-1} B \in \tilde{\mathcal{F}}$. The theorem is proved.

We will investigate the conditional distribution $P_x(\cdot \mid \tilde{\mathcal{F}})$ which depends on the parameter $x \in X$. From countable generating of $\tilde{\mathcal{F}}$ and $\mathcal{B}(X)$-measurability with respect to the Borel σ-algebra $\mathcal{B}(X)$ of the family of measures $(P_x)_{x \in X}$ (i.e. $(\forall B \in \tilde{\mathcal{F}})\ P.(B)$ is a $\mathcal{B}(X)$-measurable function) the existence of a $\mathcal{B}(X) \otimes \tilde{\mathcal{F}}$-measurable version of the conditional distribution which depends on the parameter (i.e. $(\forall B \in \tilde{\mathcal{F}})\ P.(B \mid \tilde{\mathcal{F}})$ is a $\mathcal{B}(X) \otimes \tilde{\mathcal{F}}$-measurable function) follows. Since π_0 is an $\tilde{\mathcal{F}}$-measurable function, the functions $P_{\pi_0}(B \mid \tilde{\mathcal{F}}) (B \in \tilde{\mathcal{F}})$ and $E_{\pi_0}(f \mid F)$ (f is an $\tilde{\mathcal{F}}$-measurable function) are $\tilde{\mathcal{F}}$-measurable.

Theorem 3 (Regenerative property of conditional probabilities). *Let* $(P_x) \in SM$, f_1 *a* $\mathcal{B}(\bar{R}_+) \otimes \tilde{\mathcal{F}}$*-mesurable function and* f_2 *an* $\tilde{\mathcal{F}}$*-measurable function. Then* $(\forall x \in X)$ *there exists a sequence* $(r_n)_1^\infty (r_n \downarrow 0)$ *such that* $(\forall k,n \in N)\ E_{\pi_0}(f_1(\tau, \alpha_\tau) \cdot f_2(\theta_\tau) \mid \tilde{\mathcal{F}}) = E_{\pi_0}(f_1(\tau, \alpha_\tau) \mid \tilde{\mathcal{F}}) \cdot E_{\pi_0}(f_2 \mid \tilde{\mathcal{F}}) \circ \theta_\tau$, P_x*-a.s. on the set* $\{\tau_{r_n}^k < < \infty\}$, *where* $\tau = \tau_{r_n}^k$.

Proof. Let $x \in X$. Choose $(r_n)_1^\infty$ such that $P_x\left(\bigcap_{n=1}^{\infty} \Pi(r_n)\right) = 1$. It is required to prove that $(\forall A_1 \in \tilde{\mathcal{F}})(\forall \tau \in \{\tau_{r_n}^k,\ k,\ n \in N\})\ E_x(E_{\pi_0}(f_1(\tau, \alpha_\tau) \times \times f_2 \theta(\tau) \mid \tilde{\mathcal{F}}),\ A_1 \cap \{\tau < \infty\}) = E_x(E_{\pi_0}(f_1(\tau, \alpha_\tau) \mid \tilde{\mathcal{F}}) \cdot E_{\pi_0}(f_2 \mid \tilde{\mathcal{F}}) \circ \theta_\tau,\ A_1 \cap \{\tau < \infty\})$. According to Theorem 2 we can replace A_1 by $\alpha_\tau^{-1} \cap \theta_\tau^{-1} B$ where $A \in \tilde{\mathcal{F}}$, $B \in \tilde{\mathcal{F}}$. Further

$E_x(E_{\pi_0}(f_1(\tau, \alpha_\tau) \cdot f_2(\theta_\tau) \mid \tilde{\mathcal{F}}),\ \alpha_\tau^{-1} A \cap \theta_\tau^{-1} B \cap \{\tau < \infty\}) =$

$(6) = E_x\,(f_1(\tau, \alpha_\tau) \cdot f_2(\theta_\tau),\ \alpha_\tau^{-1} A \cap \theta_\tau^{-1} B \cap \{\tau < \infty\}) =$

$(7) = E_x(f_1(\tau, \alpha_\tau) \cdot E_{\pi_\tau}(f_2, B),\ \alpha_\tau^{-1} A \cap \{\tau < \infty\}) =$

$(8) = E_x(f_1(\tau, \alpha_\tau)\, E_{\pi_\tau}(E_{\pi_0}(f_2 \mid \tilde{\mathcal{F}}),\ B),\ \alpha_\tau^{-1} A \cap \{\tau < \infty\}) =$

$(9) = E_x(f_1(\tau, \alpha_\tau) \cdot E_{\pi_0}(f_2 \mid \tilde{\mathcal{F}}) \circ \theta_\tau,\ \alpha_\tau^{-1} A \cap \theta_\tau^{-1} B \cap \{\tau < \infty\}) =$

$(10) = E_x(E_{\pi_0}(f_1(\tau, \alpha_\tau) \mid \tilde{\mathcal{F}}) \cdot E_{\pi_0}(f_2 \mid \tilde{\mathcal{F}}) \circ \theta_\tau,\ \alpha_\tau^{-1} A \cap \theta_\tau^{-1} B \cap \{\tau < \infty\}).$

where (6) through (10) are justified as follows: (6) For some set $A_1 \in \tilde{\mathcal{F}} \cap \cap \{\tau < \infty\}$, $\alpha_\tau^{-1} A \cap \theta_\tau^{-1} B \cap \{\tau < \infty\} \overset{P_x}{=} A_1$; (7) The semi-Markov property; (8) $(\forall x \in X)\ E_x(f_2,\ B) = E_x(E_{\pi_0}(f_2 \mid \tilde{\mathcal{F}}),\ B)$ on account of the $\mathcal{B}(X) \otimes \otimes \tilde{\mathcal{F}}$-measurability of the function $E.\,(f_2 \mid \tilde{\mathcal{F}})$; (9) The semi-Markov property; (10) Replace $\alpha_\tau^{-1} A \cap \theta_\tau^{-1} B \cap \{\tau < \infty\}$ by the equivalent $A_1 \in \tilde{\mathcal{F}} \cap \{\tilde{\tau} < \infty\}$

and θ_τ by $\theta_{\tilde{\tau}}$, where $\tilde{\tau}$ is an iterated first exit time from the closures of the spheres which define τ. The choice of (r_n) implies $P_x(\tau=\tilde{\tau})=1$. Then, according to Theorem 2, $E_{\pi_0}(f_2|\widetilde{\mathcal{F}})\circ\theta_{\tilde{\tau}}$ is an $\widetilde{\mathcal{F}}$-measurable function. The theorem is proved.

3. Degeneracy of the conditional distribution of passage times for a Markov process

Denote $\mathfrak{M}_\tau = E_{\pi_0}(\tau|\widetilde{\mathcal{F}})$ and $\exp(-b_\tau(\lambda)) = E_{\pi_0}(e^{-\lambda\tau}|\widetilde{\mathcal{F}})$ $(\lambda>0)$ where $\tau\in\mathcal{T}_+$. On account of Theorem 3 it follows that $(\forall x\in X)(\exists(r_n),\ r_n\downarrow 0)$ $(\forall k,n\in N)(\forall\tau\in\mathcal{T}_+)$ $\mathfrak{M}_{\tau_1\dot{+}\tau}=\mathfrak{M}_{\tau_1}+\mathfrak{M}_\tau\circ\theta_{\tau_1}$, $b_{\tau_1\dot{+}\tau}(\lambda)=b_{\tau_1}(\lambda)+b_\tau(\lambda)\circ\theta_{\tau_1}$ $(\tau_1\equiv$ $\equiv\tau^k_{r_n}<\infty)$, and $\mathfrak{M}_{\tau_1}=b_{\tau_1}(\lambda)=\infty$ P_x-a.s. on the set $\{\tau_1=\infty\}$.

Theorem 4 (Conditional degeneracy). *Let* $(P_x)\in SM$, $\Delta\in\mathcal{H}$, $\lambda>0$, *and* $A^r_\lambda 1(x)\to-\lambda$ *uniformly in* $x\in\Delta$. *Let* $x\in\Delta$ *and* $(\exists r>0)$ $P_x(\tau_\Delta{}^{+r}<\infty)=1$, $m_{\tau_\Delta{}^{+r}}(x)<\infty$. *Then* $b_{\tau_{[\Delta]}}(\lambda)=\lambda\mathfrak{M}_{\tau_{[\Delta]}}$, P_x-a.s.

Proof. Let $r_n\downarrow 0$ and $P_x\left(\bigcap_{n=1}^{\infty}\Pi(r_n)\right)=1$. We have $(\forall\xi\in\mathfrak{D})$ $\tau_{[\Delta]}(L_r\xi)\downarrow$ $\downarrow\tau_{[\Delta]}\xi$ $(r\downarrow 0)$. Therefore $E_x(\exp(-b_{\tau_{[\Delta]}\circ L_r}(\lambda)))=E_x(\exp(-\lambda\tau_{[\Delta]}\circ L_r))\uparrow$ $\uparrow E_x(\exp(-\lambda\tau_{[\Delta]}))=E_x(\exp(-b_{\tau_{[\Delta]}}(\lambda)))$. Further $(\forall r\in(r_n)_1^\infty)$

$$(11)\quad 0\leqslant E_x(\exp(-b_{\tau_{[\Delta]}\circ L_r}(\lambda))-\exp(-\lambda\mathfrak{M}_{\tau_{[\Delta]}\circ L_r}))=$$

$$=\sum_{k=1}^{\infty}E_x(\exp(-b_{\tau_r^k}(\lambda))-\exp(-\lambda\mathfrak{M}_{\tau_r^k}),\ \tau_{[\Delta]}\circ L_r=\tau_r^k)=$$

$$(12)\quad =\sum_{k=1}^{\infty}E_x\left(\prod_{i=1}^{k}\exp(-b_{\tau_r}(\lambda)\circ\theta_{\tau_r^{i-1}})-\prod_{i=1}^{k}\exp(-\lambda\mathfrak{M}_{\tau_r}\circ\theta_{\tau_r^{i-1}}),\ \tau_{[\Delta]}\circ L_r=\tau_r^k\right)\leqslant$$

$$\leqslant\sum_{k=1}^{\infty}\sum_{i=1}^{k}E_x((\exp(-b_{\tau_r}(\lambda))-\exp(-\lambda\mathfrak{M}_{\tau_r}))\circ\theta_{\tau_r^{i-1}},\ \tau_{[\Delta]}\circ L_r=\tau_r^k)=$$

$$(13)\quad =\sum_{i=1}^{\infty}E_x((\exp(-b_{\tau_r}(\lambda))-\exp(-\lambda\mathfrak{M}_{\tau_r}))\circ\theta_{\tau_r^{i-1}},\ \tau_{[\Delta]}\circ L_r>\tau_r^{i-1})=$$

$$(14)\quad =\sum_{i=1}^{\infty}E_x(E_{\pi(\tau_r^{i-1})}(\exp(-b_{\tau_r}(\lambda))-\exp(-\lambda\mathfrak{M}_{\tau_r})),\ \tau_{[\Delta]}\circ L_r>\tau_r^{i-1})\leqslant$$

$$(15)\quad \leqslant\varepsilon\sum_{i=1}^{\infty}E_x(m_{\tau_r}\circ\pi_{\tau_r^{i-1}},\ \tau_{[\Delta]}\circ L_r>\tau_r^{i-1})=$$

$$(16)\quad =\varepsilon\sum_{i=1}^{\infty}E_x(\tau_r\circ\theta_{\tau_r^{i-1}},\ \tau_{[\Delta]}\circ L_r>\tau_r^{i-1})=$$

$$(17)\quad =\varepsilon\sum_{k=1}^{\infty}\sum_{i=1}^{k}E_x(\tau_r\circ\theta_{\tau_r^{i-1}},\ \tau_{[\Delta]}\circ L_r=\tau_r^k)=$$

$$=\varepsilon\sum_{k=1}^{\infty}E_x(\tau_r^k,\ \tau_{[\Delta]}\circ L_r=\tau_r^k)=\varepsilon E_x(\tau_{[\Delta]}\circ L_r)\leqslant\varepsilon m_{\tau_\Delta{}^{+r}}(x),$$

where (11) through (17) are justified as follows: (11) Jensen's inequality; (12) Theorem 3; (13) Change of the summation order; (14) The semi-Markov property; (15) Jensen's inequality and the uniform convergence

$$E_x(1 - \exp(-\lambda\tau_r))/m_{\tau_r}(x) = E_x(1 - \exp(-b_{\tau_r}(\lambda)))/m_{\tau_r}(x) \to \lambda,$$

imply the uniform convergence

$$E_x(1 - \exp(-\lambda \mathfrak{M}_{\tau_r}))/m_{\tau_r}(x) \to \lambda,$$

(where $E_x\ (1 - \exp(-\lambda \mathfrak{M}_{\tau_r}))/m_{\tau_r}(x) \leqslant \lambda E_x(\mathfrak{M}_{\tau_r})/m_{\tau_r}(x) = \lambda$); therefore $(\forall \varepsilon > 0)(\exists r' \in (r_n)_1^\infty)(\forall r < r')$ the stated inequality holds; (16) The semi-Markov property, (17) Change of the summation order.

Consequently, $\lim_{r\to 0} E_x(\exp(-b_{\tau_{[\Delta]}\circ L_r}(\lambda)) - \exp(-\lambda \mathfrak{M}_{\tau_{[\Delta]}\circ L_r})) = E_x(\exp(-b_{\tau_{[\Delta]}}(\lambda)) - \exp(-\lambda \lim_{r\to 0} \mathfrak{M}_{\tau_{[\Delta]}\circ L_r})) = 0$. Since $\mathfrak{M}_{\tau_{[\Delta]}} \leqslant \mathfrak{M}_{\tau_{[\Delta]}\circ L_r}$ and $\exp(-b_{\tau_{[\Delta]}}(\lambda)) \geqslant \exp(-\lambda \mathfrak{M}_{\tau_{[\Delta]}}) \geqslant \exp(-\lambda \lim_{r\to 0} \mathfrak{M}_{\tau_{[\Delta]}\circ L_r})$ P_x-a.s. we have $b_{\tau_{[\Delta]}}(\lambda) = \lambda \mathfrak{M}_{\tau_{[\Delta]}} = \lambda \lim_{r\to 0} \mathfrak{M}_{\tau_{[\Delta]}\circ L_r}$, P_x-a.s. The theorem is proved.

Remarks. 1. The Wiener process on $X = R^n (n \geqslant 1)$ is an example of a strong Markov process for which the conditions of Theorem 4 are satisfied. **2.** The property $E_x(e^{-\lambda\tau}|\widetilde{\mathcal{F}}) = \exp(-\lambda E_x(\tau|\widetilde{\mathcal{F}}))$ proved in Theorem 4 can naturally be interpreted as degeneracy of the conditional distribution of passage times given the σ-algebra $\widetilde{\mathcal{F}}$. **3.** Let $\mathcal{F}_{(L)} = \{B \in \mathcal{F}: (\xi \in B,\ L\xi_1 = L\xi) \Rightarrow \xi_1 \in B\}$ and $\mathcal{F}_{(l)} = \{B \in \mathcal{F}: (\xi \in B,\ l\xi_1 = l\xi) \Rightarrow \xi_1 \in B\}$. Then $\widetilde{\mathcal{F}} \subset \mathcal{F}_{(l)} \subset \mathcal{F}_{(L)}$. It is natural to denote by $P_x(\cdot \mid L)$ and $P_x(\cdot \mid l)$ the conditional distributions given $\mathcal{F}_{(L)}$ and $\mathcal{F}_{(l)}$ (i.e., given a sequence of states and a sequence of essential states, respectively). Let $\widetilde{\mathcal{F}} \subset \mathcal{F}' \subset \mathcal{F}$. From Jensen's inequality we have $E_x(e^{-\lambda\tau}|\mathcal{F}') \geqslant \exp(-\lambda E_x(\tau|\mathcal{F}'))(P_x\text{-a.s.})$. Let $E_x(e^{-\lambda\tau}|\mathcal{F}') > \exp(-\lambda E_x(\tau|\mathcal{F}'))$ on some set $A \in \mathcal{F}'$ such that $P_x(A) > 0$. Then

$$E_x(E_x(e^{-\lambda\tau}|\mathcal{F}')) > E_x(\exp(-\lambda E_x(\lambda|\mathcal{F}'))) = E_x(E_x(\exp(-\lambda E_x(\tau|\mathcal{F}))|\widetilde{\mathcal{F}})) \geqslant$$

$$\geqslant E_x(\exp(-\lambda E_x(E_x(\tau|\mathcal{F}')|\widetilde{\mathcal{F}}))) = E_x(\exp(-\lambda E_x(\tau|\widetilde{\mathcal{F}}))) = \tag{18}$$

$$= E_x(e^{-\lambda\tau}) = E_x(E_x(e^{-\lambda\tau}|\mathcal{F}')),$$

where (18) follows from Jensen's inequality. Thus we reached a contradiction. Consequently, $E_x(e^{-\lambda\tau}|\mathcal{F}') = \exp(-\lambda E_x(\tau|\mathcal{F}'))$, P_x-a.s. Therefore the distribution $P_x(\cdot \mid L)$ is degenerated, too. This can be interpreted as expressing the degeneracy of the conditional distribution of passage times given a sequence of states. **4.** The condition of Theorem 4 concerning the finiteness of $m_{\tau_{[\Delta]}}(x)$ is a technical one and might be removed by using a sharper estimate of the difference $\exp(-b_\tau(\lambda)) - \exp(-\lambda \mathfrak{M}_\tau)$.

REFERENCES

1. Gihman, I. I., Skorohod, A. V., *Theory of Stochastic Processes, Vol. 2*. Moscow, Nauka, 1976. (Russian)
2. Kolmogorov, A. N., Fomin, S. V., *Elements of Function Theory and Functional Analysis*. Moscow, Nauka, 1972. (Russian)

3. Koroljuk, V. S., Turbin, A. F., *Semi-Markcv Processes.* Kiev, Naukova Dumka, 1976. (Russian)
4. Širjaev, A. N., *Statistical Sequential Analysis.* Moscow, Nauka, 1976. (Russian)
5. Harlamov, B. P., *Stochastic processes with semi-Markov streams of first entrances.* Zap. Naučn. Sem. Leningrad. Otdel. Mat. Inst. Steklov (LOMI), **41**, 139 – 164 (1974). (Russian)
6. Harlamov, B. P., *On connections between random curves, time changes and regeneration times.* Zap. Naučn. Sem. Leningrad. Otdel. Mat. Inst. Steklov (LOMI), **55**, 128 – 164 (1976). (Russian)
7. Harlamov, B. P., *A "correct exit" property and a limit theorem for semi-Markov processes.* Zap. Naučn. Sem. Leningrad. Otdel. Mat. Inst. Steklov (LOMI), **72**, 186 – 202. (1977). (Russian)
8. Harlamov, B. P., *A criterion for the Markov property for continuous semi-Markov processes.* Teor. Verojatnost. i Primenen., **25**, 535 – 548 (1980). (Russian)

OCTAVIAN IORDACHE, CORNEL BERBENTE

A new stochastic model of turbulence

This work presents a model intended to describe the fully developed turbulence. We start by considering that the turbulent system's evolution is described probabilistically by the forward type equation $u_{n+1} = f(u_n)$ or by the backward type equation $u_n = f(u_{n+1})$. Using the framework of the random system with complete connections theory, some results concerning ergodicity are emphasized.

1. Classical theories of turbulence postulate a direct cascade that descends toward smaller eddies. Some experiments in the limit of infinite Reynolds number evidence ascending or inverse cascades [8]. Such a cascade of noisy periods is expected after the cascade of period doublings in the Feigenbaum theory of turbulence (see Eckmann [2]).

2. Our model starts by considering that in the fully developed turbulence a fluid element switches at random between direct and inverse cascades. The turbulence field results by the stochastic superposition of the two types of evolution that characterize the two types of cascades.

Let us consider that the difference equation

$$V_{n+1} = f(V_n) \tag{1}$$

describes the direct cascade i.e. the "forward" mode of evolution. Here V_n denotes the value of V (vorticity, the square of the spectral component of the field, velocity, etc.) at moment n. The inverse cascade or the "backward" mode of evolution does not introduce new possible states, changing only the temporal order of their appearance. Consequently, the backward evolution is characterized by the equation

$$V_n = f(V_{n+1}) \tag{2}$$

obtained by interchanging V_n and V_{n+1} in (1). Evidently we can write

$$V_{n+1} = g(V_n) \tag{3}$$

Successive applications at random of f and g generate the value V_{n+1} of V at moment $n + 1$.

In many applications the functions f and g depend on controllable quantities. Examples of such parameters are the Reynolds or Rayleigh number, the amount of friction, heat or mass added per time unit, etc.

3. A difficulty in turbulence theory is the inability of the Navier-Stokes equations to describe the vortex formation. Such equations describe the decay of vortices rather than their formation. The vortex appearance by inverse cascades, e.g. in the presence of solid walls, is easily understood in our model.

A fundamental problem in turbulence theory is the necessity of dealing with a system that is not in thermal equilibrium. The impossibility to reach equilibrium for infinite wave numbers is one of the characteristics of turbulence. In our model such a problem is avoided in the case where the asymptotic states are of null probability.

In fact the direct and inverse cascade represents the elements of a sequential adaptation of the fluid energy to the flow and to the boundary conditions acting in chaotic or random environments.

4. Let us consider the one dimensional form of the Navier-Stokes equation

$$\frac{\partial V}{\partial t} + V\frac{\partial V}{\partial y} = \nu\frac{\partial^2 V}{\partial y^2} - \frac{1}{\rho}\frac{\partial p}{\partial y}. \tag{4}$$

Here V denotes the velocity, ν the kinematic viscosity, y the spatial coordinate, ρ the density, p the pressure. A discrete counterpart of (4) is

$$V_{n+1} = aV_n^2 + bV_n + c = f(V_n). \tag{5}$$

Here a, b and c are parameters. Of course, turbulence phenomena could be modeled using more complicated systems of equations. Equation (5) may contain many parameters but in fact even a one-parameter family of maps will be relevant. Probably no new phenomena can be found by using more parameters or equations. An equation of type (5) is obtained if instead of Navier-Stokes equations the corresponding Fourier transforms are considered [8].

The backward mode of evolution is characterized by an equation having $-\frac{\partial V}{\partial t}$ instead of $\frac{\partial V}{\partial t}$ in (4). By interchanging V_n and V_{n+1} in (5) we obtain an equation of type (3).

We observe that if a, b and c are constant and nonrandom $g = f^{-1}$ if f has an inverse.

5. This model can be discussed in terms of the random system with complete connections (R.S.C.C.) theory [4]. In this special case the space W of the values of V_n is R, $X = \{F, B\}$ where F abbreviates the forward evolution (direct cascades) while B abbreviates the backward evolution (inverse cascade). We denote also the probabilities of transition from W to X by $P(V_n, F) = \Pi_F$ and $P(V_n, B) = \Pi_B$ where $\Pi_F + \Pi_B = 1$. The operator $u: W \times X \to W$ is given by $u(V_n, F) = f(V_n)$ and $u(V_n, B) = g(V_n)$.

6. A case of special interest appears when only the forward type of evolution is followed by the system. In such a case $X = \{F\}$ and $\Pi_F = 1$. A model with $f(V_n) = aV_n^2 + 1$ was used to describe the onset of turbulence (see for a review Collet and Eckmann [1]).

Similar results are obtained using either $f(V_n) = aV_n(1 - V_n)$ or the Lorenz model for which

$$f(V_n) = \begin{cases} \frac{V_n}{2} \text{ if } 0 \leqslant V_n \leqslant \frac{1}{2} \\ \frac{1}{2} - \frac{V_n}{2} \text{ if } \frac{1}{2} < V_n \leqslant 1 \end{cases} \tag{6}$$

(see also [3], p. 191).

7. Models with two equations of types (1) and (3) have been studied by Kaijser [7, 8] (see [3] p. 194). The X-valued variable is a chain of infinite order if Π_F and Π_B depend on V. In fact, the memory effects on the dynamics of turbulence are well-known and the non-Markovian nature of such phenomena is pointed out [8].

A fundamental problem refers to the existence of a limiting distribution of V_n. According to Kaijser [6] such limiting distributions exist if

$$\Pi_F \log \frac{\partial f}{\partial V_n} + \Pi_B \log \frac{\partial g}{\partial V_n} < 0 \tag{7}$$

For special results of the same type see [4], p. 271.

If $f = g^{-1}$ we have

$$\lim_{n \to \infty} V_n = \begin{cases} V_F \text{ if } \Pi_F > \Pi_B \\ V_B \text{ if } \Pi_F < \Pi_B \end{cases} \tag{8}$$

Here V_F and V_B are defined by $f(V_F) = V_F$ and $g(V_B) = V_B$. Moreover, if $\Pi_F = \Pi_B = \frac{1}{2}$, then V_n oscillates between V_F and V_B. Despite this oscillation there is a limiting distribution for V_n. It is concentrated at V_F and V_B.

8. Let us consider that $a = c = 0$ in (5). In such a case the two equations (1) and (3) reduce to

$$V_{n+1} = b_n V_n. \tag{9}$$

Assuming the b_n to be independent identically distributed random variables taking on the values b and $1/b$ with probabilities Π_F and Π_B, respectively, equation (9) offers a support of the log-normal hypothesis of Kolmogorov (see also [7]). Actually, from $\frac{V_{n+1} - V_n}{V_n} = b_n - 1$ we obtain $\sum_n \frac{V_{n+1} - V_n}{V_n} = \sum_n (b_n - 1)$, and finally $\ln V_n \sim \sum_n (b_n - 1)$. Since the $(b_n - 1)$ are independent identically distributed random variables we obtain using the central limit theorem that $\ln V_n$ is asymptotically normal and consequently V_n is lognormal.

9. Asymptotic distributions as (8) explains the presence of large and small scale eddies and the extremely intermittent character which is typical of fully developed turbulence in free boundaries. Oscillations between V_F and V_B can be seen in many physical experiments.

Some Monte-Carlo simulations performed with models of the type described in § 6 evidence limit distributions having a great number of singularities in the histogram. Singular distributions of the Cantor type, occurring even in linear models of types (1) and (3) seem to be frequent in turbulence systems. It should be pointed out that the study of such limit distributions suggests a wide range of possible applications in the domain of energy conversion.

REFERENCES

1. Collet, P., Eckmann, J. P., *Iterated Maps on the Interval as Dynamical Systems.* Progress in Physics, Vol. 1. Basel — Boston — Stuttgart, Birkhäuser, 1980.
2. Eckmann, J. P., *Roads to turbulence in dissipative dynamical systems.* Rev. Mod. Phys., **53**, 643—654 (1981).
3. Grigorescu, S., Iosifescu, M., *Dependence with Complete Connections and Its Applications.* București, Ed. științifică și enciclopedică, 1982. (Romanian)
4. Iosifescu, M., Theodorescu, R., *Random Processes and Learning.* Berlin, Springer, 1969.
5. Kaijser, T., *Some limit theorems for Markov chains with applications to learning models and product of random matrices* Ph. D. Thesis, Institute Mittag-Leffler, Djursholm (Sweden), 1972.
6. Kaijser, T., *On a new contraction condition for random systems with complete connections.* Rev. Roumaine Math. Pures Appl., **26**, 1075—1117 (1981).
7. Mandelbrot, B., *Intermittent turbulence in self-similar cascades: Divergence of high moments and dimension of the carrier.* J. Fluid Mech., **62**, 331—358 (1974).
8. Rose, H. A., Sulem, P. L., *Fully developed turbulence and statistical mechanics.* J. Physique, **39**, 441—484 (1978).

MARIUS IOSIFESCU

On the random Riemann hypothesis*

The aim of this paper is to explain why conclusions like 'the Riemann hypothesis (about the complex zeros of the zeta function) is true with probability one' are not at all absurd but, moreover, they give 'good' reasons for believing the Riemann hypothesis. In fact, as usual with statistical reasoning, the real Riemann hypothesis is imbedded in a probabilistic framework. This is done in Section 5, after giving in the first four sections prerequisites, historical remarks and some motivation. Section 6 collects a few final remarks.

1. Introduction

In his famous paper *Ueber die Anzahl der Primzahlen unter einer gegebenen Grösse* [Monatsberichte d. Preuss. Akad. d. Wissens. (Berlin: 1859; 1860), 671—680; *Gesammelte mathematische Werke*, 2. Aufl., 145—153, Teubner, Leipzig, 1892 (reprinted by Dover Books, New York, 1953)] Riemann considered the function ζ of the complex variable $s = \sigma + it$ defined by the equation

(1) $$\zeta(s) = \frac{\Gamma(1-s)}{2\pi i} \int_{+\infty}^{+\infty} \frac{(-x)^s}{e^x - 1} \frac{dx}{x},$$

where Γ is Euler's gamma function, and the limits of integration are intended to indicate a path of integration which begins at $+\infty$, moves to the left down the positive real axis, circles counterclockwisely the origin once and returns up the positive real axis to $+\infty$. The definition of $(-x)^s$ is $(-x)^s = \exp[s \log(-x)]$, where the definition of $\log(-x)$ conforms to the usual definition of $\log z$ for z not on the negative real axis as the branch which is real for positive real z; thus $(-x)^s$ is not defined on the positive real axis and, strictly speaking, the path of integration must be taken to be slightly above the real axis as it descends from $+\infty$ to 0 and slightly below the real axis as it goes from 0 back to $+\infty$.

* This paper is a revised version of Preprint No. 500, Institut für Angewandte Mathematik der Universität Bonn, Sonderforschungsbereich 72. Financial support from the Deutsche Forschungsgemeinschaft in the period November 1981 — February 1982 is gratefully acknowledged.

Formula (1) defines a function $\zeta(s)$ which is analytic at all points of the complex s-plane except for a simple pole with residue 1 at $s = 1$. This function coincides with the sum of the Dirichlet series

$$\sum_{n \geqslant 1} \frac{1}{n^s}$$

for Re $s > 1$.

The function $\zeta(s)$ defined by (1) is known as the *Riemann zeta function.* The real zeros of $\zeta(s)$ are the integers —2, —4, ... and the complex zeros are located in the strip $0 <$ Re $s < 1$, lying symmetrically about the real line and the line Re $s = 1/2$. In his paper, Riemann says that he considers it "sehr wahrscheinlich" that the complex zeros of $\zeta(s)$ all have real part 1/2, but that he has been unable to prove the truth of this statement, which is now known as the *Riemann hypothesis.* Hilbert included the problem of proving the Riemann hypothesis (R.H.) in his well known list of the most important unsolved problems confronting mathematics in 1900. He is reputed to have said (cf. Bellman (1980), p. 76) "If I woke after a thousand years, the first question I would ask is 'Has the R.H. been settled yet'?" In his 1900 talk he said that so much had been done on the theory of entire functions that it was clear that the R.H. would be settled in about twenty years at most. In spite of the efforts of many of the best mathematicians of our century just partial results have been obtained. In 1914, G.H. Hardy proved that infinitely many zeros of the zeta function have real part equal to 1/2. By laborious computations it has been shown (January 1982) that $\zeta(s)$ has exactly 200,000,001 zeros of the form $\sigma + it$ in the region $0 < t < 81{,}702{,}130.19$, all these zeros are simple and lie on the line $\sigma = 1/2$. In 1974, N. Levinson proved that more than one third of the zeros lie on the line Re $s = 1/2$. Later 'one third' was successively improved to 0.3474 (by Levinson), 0.35 (by Lou), and 0.3658 (by Conrey). For details and proofs the reader is referred to Brent & al (1982), Edwards (1974), Levinson (1974, 1975), Lou (1981), and the review MR **82** a:10043.

The fact that the first 200,000,001 complex zeros of the zeta function are on the line Re $s = 1/2$ is not at all a decisive reason for believing that the R.H. should be true. For, as put by Good and Churchhouse (1968), p. 857, "... in the theory of the zeta function, and in the closely allied theory of the distribution of prime numbers, the iterated logarithm log log is often involved in asymptotic formulae*, and this function increases extremely slowly. The first zero off the line Re $s = 1/2$, if there is one, might have an imaginary part whose iterated logarithm is, say, as large as 10, and if so, it might never be practicable to find this zero by calculation". (If log log $t = 10$, then t is approximately $10^{10{,}000}$.)

The aim of this paper, mainly expository, is to explain as clearly as possible that conclusions like 'the R.H. is true *with probability one*' are not at all preposterous, and, moreover, they give 'good reasons' for believing the R.H. (Compare with Edwards (1974), pp. 164—166 and 268—269.)

---

* A well known example is the estimate

$$\sum_{p \leqslant x} \frac{1}{p} = \log\log x + B + O\left(\frac{1}{\log x}\right), \quad x \geqslant 2,$$

where the sum is over all the primes not exceeding x, and B is a constant.

It seems that Denjoy (1931) was the first author to state explicitly that the R.H. is true with probability one. But, at least in an implicit manner, such a statement should also be found in papers by N. Wiener, R.E.A.C. Paley and N. Wiener, and H. Cramér in the thirties.

Remark. As these proceedings are dedicated to Professor Octav Onicescu it is adequate to show that in his paper *Sur les zéros des fonctions* $\zeta(s)$ written in 1921 and published in 1922 in Bull. Sect. Sci. Acad. Roumaine 7,117—124 (1920/21) he proposed a method for proving the R.H. It was later noticed by himself that the method failed due to the fact that a certain series is not uniformly convergent.

2. Some values of the zeta function

It is known that

$$\zeta(-n) = (-1)^n \frac{B_{n+1}}{n+1}, \tag{2}$$

$$\zeta(2n) = \frac{(2\pi)^{2n}(-1)^{n+1}B_{2n}}{2 \cdot (2n)!} \tag{3}$$

for $n = 0, 1, 2, \ldots$. Here the B_n are the Bernoulli numbers defined by the equation

$$\frac{z}{e^z - 1} = \sum_{n \geqslant 0} \frac{B_n z^n}{n!}, \quad |z| < 2\pi.$$

In particular, $B_1 = -1/2$, $B_{2n+1} = 0$, $n \geqslant 1$, $B_0 = 1$, $B_2 = 1/6$, $B_4 = -1/30$, $B_6 = 1/42$, $B_8 = -1/30, \ldots$, and the even Bernoulli numbers can be obtained from the recursion

$$\binom{n+1}{0} B_0 + \binom{n+1}{1} B_1 + \ldots + \binom{n+1}{n} B_n = 0.$$

Hence

$$\zeta(2) = \pi^2/6,\ \zeta(4) = \pi^4/90,\ \zeta(6) = \pi^6/945, \ldots$$

Thus by (2) (which is due to Lipschitz (1884)) all the values $\zeta(n)$, $n = 0, -1, -2, \ldots$ are rational and by (3) (which is due to Euler) all the values $\zeta(2n)$, $n = 1, 2, \ldots$ are irrational. There are no useful analogous closed evaluations of ζ at odd positive arguments. Finding the arithmetic nature of $\zeta(2n+1)$, $n = 1, 2, \ldots$ constitutes one of the outstanding problems of the theory of the zeta function. It was therefore something of a sensation when in June 1978 at the Journées Arithmétiques de Marseille-Luminy, R. Apéry (see Apéry (1979)) presented an elementary proof of the fact that $\zeta(3)$ is irrational. For a detailed discussion of Apéry's paper see van der Poorten (1978/79, 1980); see also Beukers (1979, 1981).

It is interesting to note that the values $\zeta(n)$, $n = 2\ 3, \ldots$ occur in formulae expressing probabilities of various arithmetic properties.

Thus if $P_k(n)$ denotes the probability that k positive integers, chosen at random from $\{1, 2, \ldots, n\}$ (i.e. assuming the uniform distribution on the set of the first n positive integers), are relatively prime then (Lehmer (1900), Nymann (1972))

$$P_2(n) = 1/\zeta(2) + O((\log n)/n), \; P_k(n) = 1/\zeta(k) + O(1/n), \; k \geqslant 3.$$

Hence for $k \geqslant 2$

$$\lim_{n\to\infty} P_k(n) = \frac{1}{\zeta(k)}. \tag{4}$$

The last result for the special case $k = 2$ goes back to Césaro (1881, 1883) and Sylvester (1883). Césaro (1884) obtains (4) for arbitrary $k \geqslant 2$.

More generally, if $T_k^r(n)$ denotes the number of k-tuples of positive integers $(x_1, \ldots, x_k)$ with $1 \leqslant x_i \leqslant n$, $1 \leqslant i \leqslant k$, and $(x_1, \ldots, x_k)_r = 1$ (this means that the largest r-th power dividing all the x_i equals 1) then (Benkoski (1976)) for $kr \geqslant 2$

$$\lim_{n\to\infty} \frac{T_k^r(n)}{n^k} = \frac{1}{\zeta(rk)}. \tag{5}$$

(Clearly, $T_k^1(n) = n^k P_k(n)$, $k \geqslant 2$.)

Note that $T_1^r(n)$ is the number of r-free integers not exceeding n (an integer is said to be r-free if the largest r-th power dividing it equals 1). For this special case equation (5) can be improved (Lehmer (1900)) to

$$T_1^r(n) = \frac{n}{\zeta(r)} + o(n^{1/r}), \; r \geqslant 2. \tag{6}$$

The last result except for the error estimate has been also obtained by Césaro (1884).

We conclude this section by noticing that the zeta function can be used (Golomb (1970)) to define a probability distribution on the natural integers. For any real $s = \sigma > 1$ consider the probability distribution $\mathbf{P}_\sigma = \{\mathrm{pr}_\sigma(n)\}$ defined by

$$\mathrm{pr}_\sigma(n) = n^{-\sigma}/\zeta(\sigma), \; n = 1, 2, \ldots$$

The probability that an integer chosen at random from $\mathbf{P}_\sigma$ is divisible by m is

$$\sum_{n\geqslant 1} \mathrm{pr}_\sigma(mn) = \frac{1}{\zeta(\sigma)} \sum_{n\geqslant 1} (mn)^{-\sigma} = \frac{m^{-\sigma}\zeta(\sigma)}{\zeta(\sigma)} = m^{-\sigma}.$$

(Consequently, the probability that a random integer is *not* divisible by m is $1 - m^{-\sigma}$.) It follows that if $(m_1, m_2) = 1$, then divisibility by m_1 is statistically independent of divisibility by m_2.

Many heuristic probability arguments based on the fictitious *uniform distribution on the natural integers* become rigorous statements under $\mathbf{P}_\sigma$.

3. A few Dirichlet series connected with the zeta function, and the Euler factorization formula

Let us list a certain number of equations involving the zeta function we shall need in the following. Their proofs are to be found in the books by Hardy and Wright (1962) and Titchmarsh (1951).

The (arithmetical) Möbius function $\mu(n)$ is defined as follows

(i) $\mu(1) = 1$;

(ii) $\mu(n) = 0$ if n has a repeated factor;

(iii) $\mu(p_1 \dots p_k) = (-1)^k$ if all the primes $p_1, \dots, p_k$ are different.

The (arithmetical) Liouville function $\lambda(n)$ is defined as follows

(i) $\lambda(1) = 1$;

(ii) $\lambda(n) = (-1)^k$, where k is the total number of prime factors of $n > 1$, multiple factors being counted multiply.

The (arithmetical) function $q_k(n)$, $k \geqslant 2$, is defined as follows

(i) $q_k(1) = 1$;

(ii) $q_k(n) = 0$ if n has a k-th power as a factor;

(iii) $q_k(n) = 1$ if n has not a k-th power as a factor (i.e. if n is k-free).

Obviously, $q_2(n) = |\mu(n)|$, $n \geqslant 1$.

The equations below are all valid for Re $s > 1$

(7) $$\frac{1}{\zeta(s)} = \sum_{n \geqslant 1} \frac{\mu(n)}{n^s},$$

(8) $$\frac{\zeta(2s)}{\zeta(s)} = \sum_{n \geqslant 1} \frac{\lambda(n)}{n^s},$$

(9) $$\frac{\zeta(s)}{\zeta(ks)} = \sum_{n \geqslant 1} \frac{q_k(n)}{n^s}, \quad k \geqslant 2.$$

The essence of the relationship between $\zeta(s)$ and prime numbers is the Euler factorization formula

(10) $$\zeta(s) = \prod_p (1 - p^{-s})^{-1}, \quad \text{Re } s > 1,$$

where the product on the right is over the monotone sequence of primes. This is, in fact, a special case of the following result (Ingham (1932), p. 16):

Let $f(n)$ be a multiplicative function on the set of natural integers, i.e. one which has the property that $f(mn) = f(m)f(n)$ whenever m and n are relatively prime, and $f(1) = 1$*. Then

$$\sum_{n \geqslant 1} f(n) = \prod_p \{1 + f(p) + f(p^2) + \dots\},$$

provided that the series on the left is absolutely convergent, in which case the product is also absolutely convergent.

If $f(n)$ is completely multiplicative, i.e. $f(mn) = f(m)\, f(n)$ for all m and n, relatively prime or not, and $f(1) = 1$**, then

(11) $$\sum_{n \geqslant 1} f(n) = \prod_p \frac{1}{1 - f(p)}.$$

Equation (10) is obtained from (11) for $f(n) = n^{-s}$.

---

* μ and q_k are examples of multiplicative functions.

** λ is an example of a completely multiplicative function.

An immediate probabilistic proof of (10) for a real $s > 1$ (therefore for any complex s with Re $s > 1$, by analytic continuation) is as follows. It is clear that, under P_σ (see the last lines of the previous section), we have

$$\mathrm{pr}_\sigma(1) = \text{probability that a random integer has no prime factors.}$$

But $\mathrm{pr}_\sigma(1) = 1/\zeta(\sigma)$ and

$$\text{probability that a random integer has no prime factors} =$$

$$= \lim_{x\to\infty} \prod_{p<x} (1 - p^{-\sigma}) = \prod_p (1 - p^{-\sigma}), \text{ q.e.d.}$$

Clearly, (10) implies

(12)
$$\frac{\zeta(s)}{\zeta(2s)} = \prod_p (1 + p^{-s}), \text{ Re } s > 1.$$

(This is also obtained from (11) for $f(n) = \lambda(n)n^{-s}$ on account of (8)).

We conclude this section with the following remark.
Equation (6) can be written as

(13)
$$q_k(1) + \dots + q_k(n) - \frac{n}{\zeta(k)} = o(n^{1/k})$$

It follows from (9) that

(14)
$$\zeta(s) = \frac{\zeta(k)\,\zeta(ks)}{\zeta(k) - \zeta(ks)} \sum_{n\geqslant 1} \frac{q_k(n) - (1/\zeta(k))}{n^s}, \quad \text{Re } s > 1.$$

But on account of (13) it is easily seen that the series on the right in (14) converges for Re $s > 1/k$. Therefore, by analytic continuation, equation (14) furnishes a representation of the zeta function valid in the halfplane Re $s > 1/k$, except for $s = 1$. Since $k = 2, 3, \dots$ we have, in fact, a representation of the zeta function valid in the halfplane Re $s > 0$, except for $s = 1$. Equation (14) is to be associated with the well-known equation

$$\sum_{n\geqslant 1} \frac{(-1)^{n-1}}{n^s} = (1 - 2^{1-s})\,\zeta(s)$$

valid for Re $s > 0$.

4. A few statements equivalent to the Riemann hypothesis

There are several equivalent formulations of the R.H. We proceed to describe some of them.

Let

$$M(x) = \sum_{n\leqslant x} \mu(n), \quad L(x) = \sum_{n\leqslant x} \lambda(n)$$

for any $x \geqslant 1$, and put $M(x) = L(x) = 0$ for $0 \leqslant x < 1$. On account of the obvious inequalities $|M(x)| \leqslant x$ and $|L(x)| \leqslant x$ it is easily seen that equations (7) and (8) yield

$$(15) \qquad \frac{1}{\zeta(s)} = s\int_0^\infty M(x)\,x^{-s-1}\,dx, \text{ Re } s > 1,$$

and

$$\frac{\zeta(2s)}{\zeta(s)} = s\int_0^\infty L(x)\,x^{-s-1}\,dx, \text{ Re } s > 1.$$

It was shown by J.E. Littlewood in 1912 (see Edwards (1974), p. 261) that the R.H. is equivalent to the statement that $M(x) = O(x^{1/2+\varepsilon})$ for every $\varepsilon > 0$. (The implication $M(x) = O(x^{1/2+\varepsilon}) \Rightarrow$ R.H. is obvious from (15).)

A careful examination of Edwards' proof shows that a similar result holds true with L in place of M, i.e. the R.H. is equivalent to the statement that $L(x) = O(x^{1/2+\varepsilon})$ for every $\varepsilon > 0$.

It follows easily from the above statements that the R.H. is equivalent to convergence for Re $s > 1/2$ of either series (7) or (8) to $1/\zeta(s)$ or $\zeta(2s)/\zeta(s)$, respectively.

It is perhaps interesting to note that the truth of any one of the four inequalities

$$M(x) < Kx^{1/2},\; M(x) > -Kx^{1/2},\; L(x) < Kx^{1/2},\; L(x) > -Kx^{1/2},$$

for all sufficiently large x, where K is a constant, would imply both the R.H. and the fact that all the zeros (on the line Re $s = 1/2$) are simple (see Ingham (1942)).

Pólya (1919) conjectured that $L(x) \leqslant 0$ for $x \geqslant 2$, which on account of the above remark should imply the R.H. Although verified up to about $x = 600{,}000$, Pólya's conjecture was disproved by Haselgrove (1958), who made essential use of results of Ingham (1942).

Stieltjes claimed in 1885 and in 1887 (see Stieltjes (1905)) to have succeeded in proving that $M(x) = O(x^{1/2})$ (clearly, a stronger statement than $M(x) = O(x^{1/2+\varepsilon})$ for every $\varepsilon > 0$.) It is unclear whether he had a complete proof. In his own words: «Ma démonstration est bien pénible; je tâcherai, lorsque je reprendrai ces recherches, de la simplifier encore» (1885), and «Mais la démonstration de ce lemme est purement arithmétique et très difficile et je ne l'obtiens que comme résultat de toute une série de propositions préliminaires. J'espère que cette démonstration pourra encore être simplifiée, mais en 1885 j'ai déjà fait de mon mieux en envisageant encore la question d'une autre manière et en remplaçant ce lemme par un autre, d'une nature pareille toutefois» (1887). (We could not help to compare this with the well known Fermat's assertion about the proof of his great theorem.) Moreover, Stieltjes suggested that $|M(x)| < x^{1/2}$ for $x > 1$ (this assertion is now known as Mertens' conjecture; see Mertens (1897)). Stieltjes-Mertens' conjecture has been neither proved nor disproved. See, e.g., te Riele (1979).

In fact, all that has been proved (see Landau (1909), pp. 613 and 620, Titchmarsh (1951), p. 316, Diamond and McCurley (1982)) does not go much beyond the estimates

$$(16) \qquad M(x) = o(x),\; L(x) = o(x)$$

as $x \to \infty$. (Both these estimates are equivalent to the prime number theorem.)

We close by giving two equivalent formulations of the R.H., which are different in nature from those discussed above.

1) The R.H. is equivalent to the statement that $\zeta'(s)$ has no zeros in $0 < \operatorname{Re} s < 1/2$.

This result essentially due to Speiser (1934) was rediscovered by Spira (1973) (see Levinson and Montgomery (1974)).

2) The R.H. is equivalent to the statement that for every $\varepsilon > 0$ the relative error in the prime number theorem is less that $x^{-(1/2)+\varepsilon}$ for all sufficiently large x:

$$\left| \frac{\pi(x)}{\operatorname{Li}(x)} - 1 \right| < x^{(-1/2)+\varepsilon}$$

(If the R.H. is true, then the relative error in the prime number theorem is in fact $O(x^{-1/2} (\log x)^2$.)

Here $\pi(x)$ denotes the number of primes not exceeding x and

$$\operatorname{Li}(x) = \lim_{a \to 0+} \left(\int_0^{1-a} + \int_{1+a}^x \right) \frac{du}{\log u} = \int_2^x \frac{du}{\log u} + 1.04 \ldots$$

is the logarithmic integral. For a detailed discussion see Edwards (1974), pp. 88—91.

For other statements equivalent to or implying the R.H. and certain consequences of the R.H. the reader should consult the books by Edwards (1974) and Titchmarsh (1951). For a recent result see Fogels (1980).

5. Two probabilistic frameworks for the Riemann hypothesis

We are now proceeding to describe two probabilistic approaches to the R.H. As usual with probabilistic reasoning we shall imbed (in quite natural a manner) the (real) R.H. in an ensemble of like situations and replace statements about it by statements about the ensemble. For a similar treatment of other problems see Erdös and Ulam (1971), and Kendall (1948).

5.1. *The series approach.* Consider the equation

$$\frac{1}{\zeta(s)} = \sum_{n \geqslant 1} \frac{\mu(n)}{n^s}, \quad \operatorname{Re} s > 1.$$

Clearly, the Möbius function μ is deterministic: once a number n is given, then $\mu(n)$ is 0 or $(-1)^k$ according as n has any repeated factor or, if it has no repeated factors, the number of factors equals k. But a look at a table of values of μ conveys us a feeling of 'randomness' in the sense that it appears to be chaotic, with no discernible pattern or regularity, except for the fact that the values 1 and —1 occur with almost equal frequencies. In fact, equation (6) with $r = 2$ shows that

$$(6') \qquad \frac{\#\{k\colon 1 \leqslant k \leqslant n,\ \mu(k) \neq 0\}}{n} = \frac{6}{\pi^2} + o(n^{-1/2})$$

and on account of the first equation (16), which can be written as

$$\frac{\#\{k: 1 \leqslant k \leqslant n, \mu(k) = 1\} - \#\{k: 1 \leqslant k \leqslant n, \mu(k) = -1\}}{n} = o(1),$$

we deduce that

$$\frac{\#\{k: 1 \leqslant k \leqslant n,\ \mu(k) = 1\}}{n} = \frac{3}{\pi^2} + o(1),$$

$$\frac{\#\{k: 1 \leqslant k \leqslant n,\ \mu(k) = -1\}}{n} = \frac{3}{\pi^2} + o(1)$$

as $n \to \infty$. In other words, the asymptotic densities of the values $-1, 0, 1$ of μ are equal to

$$\frac{3}{\pi^2},\ 1 - \frac{6}{\pi^2},\ \frac{3}{\pi^2}, \tag{17}$$

respectively.

The above remarks suggest the following probabilistic framework. Let $(\mu_n)_{n \geqslant 1}$ be a sequence of independent identically distributed random variables taking on the values -1, 0, 1 with probabilities (17). A natural realization of the probability space on which the μ_n are defined and of its probability measure is as follows. Consider the set I_3 of the points of the unit interval which are not of the form $m/3^n$, $1 \leqslant m < 3^n$, $n = 1, 2, \ldots$. Any $\omega \in I_3$ has a unique infinite 3-adic expansion

$$\omega = 0.\, \omega_1 \omega_2 \ldots \omega_n \ldots,$$

where each $\omega_n = \omega_n(\omega)$ is either 0, 1, or 2 (the ternary digits). Define on I_3 a distribution function F by

$$F(\omega) = b_{\omega_1(\omega)} + \sum_{m>1} \left(\prod_{n=1}^{m-1} a_{\omega_n(\omega)} \right) b_{\omega_m(\omega)}.$$

where $a_0 = a_2 = 3/\pi^2$, $a_1 = 1 - 6/\pi^2$, $b_0 = 0$, $b_1 = 3/\pi^2$, $b_2 = 1 - 3/\pi^2$, and consider on $\mathcal{B}(I_3)$, the class of Borel sets in I_3, the probability measure P_F defined by $\mathsf{P}_F([0, \omega) \cap I_3) = F(\omega)$, $\omega \in I_3$. (It can be shown (Chatterji (1964)) that P_F is singular with respect to Lebesgue measure.) Then (see Spătaru (1978), p. 88) if we define $\mu_n(\omega) = \omega_n(\omega) - 1$, $n \geqslant 1$, $\omega \in I_3$, the sequence $(\mu_n)_{n \geqslant 1}$ on $(I_3, \mathcal{B}(I_3), \mathsf{P}_F)$ has the desired properties.

Now consider the random series

$$\sum_{n \geqslant 1} \frac{\mu_n(\omega)}{n^s}, \quad \omega \in I_3,\ \mathrm{Re}\ s > 1. \tag{18}$$

Notice that series (7) is obtained as a special case of (18) for

$$\omega = \sum_{n \geqslant 1} \frac{\mu(n) + 1}{3^n} = 0.681994\ldots. \tag{19}$$

Since

$$\sum_{n\geq 1} \operatorname{var}\left(\operatorname{Re}\frac{\mu_n}{n^s}\right) = \frac{6}{\pi^2}\sum_{n\geq 1}\frac{\cos^2(t\log n)}{n^{2\sigma}},$$

$$\sum_{n\geq 1} \operatorname{var}\left(\operatorname{Im}\frac{\mu_n}{n^s}\right) = \frac{6}{\pi^2}\sum_{n\geq 1}\frac{\sin^2(t\log n)}{n^{2\sigma}},$$

with $s = \sigma + it$, on account of the well-known Kolmogorov criterion for convergence of series of independent random variables we deduce that for any fixed $\sigma > 1/2$ almost all (with respect to P_F) series (18) are convergent. If this is applied to $\sigma = k^{-1} + 1/2$, where $k = 1, 2, \dots$, it follows, by adding a sequence of ω-sets of P_F-measure 0, that almost all series (18) converge in the halfplane $\sigma = \operatorname{Re} s > 1/2$. Thus (see Section 4), if ω defined by (19) does not belong to the excluded set of those ω-values for which (18) need not converge for $\operatorname{Re} s > 1/2$, the R.H. should be true!

An alternative method for establishing convergence of almost all series (18) for $\operatorname{Re} s > 1/2$ is available. This is the one used by Denjoy (1931). It follows from the classical law of iterated logarithm (in fact this is a result of Hausdorff going back to the first decade of this century) that for any $\varepsilon > 0$ there exists a positive constant $C = C(\varepsilon)$ such that setting $M_\omega(n) = \mu_1(\omega) + \dots$ $\dots + \mu_n(\omega)$ one has

$$|M_\omega(n)| \leqslant Cn^{1/2+\varepsilon} \tag{20}$$

for P_F-almost all $\omega \in I_3$. Now, similarly to (15), series (18) can be written as

$$s\int_0^\infty M_\omega(x)x^{-s-1}\,dx \ (\operatorname{Re} s > 1),$$

with an obvious definition of $M_\omega(x)$ for a non integer $x \geqslant 0$. On account of Section 4, if ω defined by (19) does not belong to the excluded set of those ω-values for which (20) need not hold, the R.H. should be true. Moreover, on account of Ingham's result given in the same Section 4, all the zeros (on the line $\operatorname{Re} s = 1/2$) should be simple.

All that has been said above applies mutatis mutandis to the equation

$$\frac{\zeta(2s)}{\zeta(s)} = \sum_{n\geq 1}\frac{\lambda(n)}{n^s}, \quad \operatorname{Re} s > 1.$$

On account of the second equation (16), which can be written as

$$\frac{\#\{k: 1 \leqslant k \leqslant n,\ \lambda(k) = 1\} - \#\{k: 1 \leqslant k \leqslant n,\ \lambda(k) = -1\}}{n} = o(1)$$

we deduce that

$$\frac{\#\{k: 1 \leqslant k \leqslant n,\ \lambda(k) = 1\}}{n} = \frac{1}{2} + o(1),$$

$$\frac{\#\{k: 1 \leqslant k \leqslant n,\ \lambda(k) = -1\}}{n} = \frac{1}{2} + o(1)$$

as $n \to \infty$. This suggests to consider a sequence $(\lambda_n)_{n\geq 1}$ of independent identically distributed random variables taking on the values -1 and 1 with equal probabilities and define the random Dirichlet series

$$\sum_{n\geq 1} \frac{\lambda_n(\omega)}{n^s}. \tag{21}$$

The probability space on which the λ_n are defined and its probability measure can be constructed as follows. Let I_2 denote the set of the points of the unit interval which are not of the form $m/2^n$, $1 \leqslant m < 2^n$, $n = 1, 2, \ldots$. Any $\omega \in I_2$ has a unique infinite dyadic expansion

$$\omega = 0 \cdot \omega_1 \omega_2 \ldots \omega_n \ldots,$$

where each $\omega_n = \omega_n(\omega)$ is either 0 or 1 (the binary digits). Consider the Lebesgue measure Λ on $\mathcal{B}(I_2)$, the class of Borel sets in I_2, and define $\lambda_n(\omega) = 2\omega_n(\omega) - 1$, $n \geqslant 1$, $\omega \in I_2$. On account of a classical result of Borel the sequence $(\lambda_n)_{n\geq 1}$ on $(I_2, \mathcal{B}(I_2), \Lambda)$ has the desired properties. Note that now series (8) is obtained as a special case of (21) for

$$\omega = \sum_{n\geq 1} \frac{\lambda(n) + 1}{2^{n+1}} = 0.5811 \ldots$$

Remark. The Bernoullian character of the λ_n allows an elementary proof of the fact that almost all series (21) converge in the halfplane $\operatorname{Re} s > 1/2$ via a direct study of the sum $L_\omega(n) = \lambda_1(\omega) + \ldots + \lambda_n(\omega)$. See Denjoy (1964).

5.2. *The infinite product approach.* The series approach we have just discussed fully disregards the fact that both Dirichlet series (7) and (8) have coefficients which are multiplicative, i.e. $\mu(mn) = \mu(m)\,\mu(n)$, $\lambda(mn) = \lambda(m)\lambda(n)$ whenever n and m are relatively prime, and $\mu(1) = \lambda(1) = 1$. This arithmetical restriction (stressed by Lévy (1931), p. 155) is equivalent (see Wintner (1944b)) to the assumption that, corresponding to (10), the Dirichlet series (18) be representable (at least for $\operatorname{Re} s > 1$) in the Eulerian product form $\prod_p (1 \pm p^{-s})$, where p runs through the monotone sequence of all primes. It is easy to show that this happens only for a set of ω-points of P_F-measure 0.

The above difficulty was elegantly overcome by Wintner (1944a) who considered on I_2 the random Eulerian product

$$\frac{1}{\zeta_\omega(s)} = \prod_{n\geq 1} \left(1 - (2\omega_n(\omega) - 1)p_n^{-s}\right), \quad \operatorname{Re} s > 1, \tag{22}$$

where p_n denotes the n-th prime, $n \geqslant 1$. (The set I_2 and the $\omega_n(\omega)$, $\omega \in I_2$, have been defined at the end of the preceding subsection.) The product (22) can be rearranged into a Dirichlet series which, corresponding to (7), can be denoted by

$$\frac{1}{\zeta_\omega(s)} = \sum_{n\geq 1} \frac{\mu_\omega(n)}{n^s}, \quad \operatorname{Re} s > 1, \tag{23}$$

where $\mu_\omega(\cdot)$ *is* a multiplicative function on the set of natural integers for *all* $\omega \in I_2$.

Using function-theoretical facts concerning Dirichlet series, Wintner proved the following results:

I. For almost all $\omega \in I_2$, not only the Eulerian product (22) but also its Dirichlet series (23) is convergent in the half plane Re $s > 1/2$.

II. For almost all $\omega \in I_2$, the line Re $s = 1/2$ is a natural boundary of the function defined by (22) or (23).

II'. If $M_\omega(n)$ denotes the sum $\sum_{k=1}^{n} \mu_\omega(k)$, then, for any fixed $\varepsilon > 0$, both $M_\omega(n) = O(n^{(1/2)+\varepsilon})$ and $M_\omega(n) \neq O(n^{(1/2)-\varepsilon})$ hold for almost all $\omega \in I_2$.

III. For suitable $\omega \in I_2$, the line Re $s = 1$ is a natural boundary of the function represented for Re $s > 1$ by (22) and (23).

IV. The limit $f(\omega) = \lim_{n\to\infty} M_\omega(n)/n$ exists for *every* $\omega \in I_2$. More precisely $f(\omega) > 0$ or $= 0$ according as

$$\sum_{\{n:\omega_n(\omega)=1\}} \frac{1}{p_n}$$

diverges or converges.

It is clear that **I** above shows that, for almost all $\omega \in I_2$, the R.H. is true for $\zeta_\omega(s)$. The value of ω corresponding to the (real) zeta function is now

$$\omega = \sum_{n\geqslant 1} \frac{1}{2^n} = 1.$$

Again, as before, the falsity or truth of the (real) R.H. amounts to whether the point $\omega = 1$ does or does not belong to the set of ω-points excluded in **I**.

Remark. It is not difficult to prove that the sets of ω-points excluded in **I**, **II** and **II'** are of the second category.

We conclude this section by mentioning that Jessen (1932, 1935) considered a random zeta function of the form

$$\prod_{n\geqslant 1} (1 - \varepsilon_n p_n^{-s})^{-1}, \quad \text{Re } s > 1, \tag{24}$$

where the ε_n are independent identically distributed complex valued random variables of modulus 1. He proved that almost all functions (24) are regular and $\neq 0$ in the half-plane Re $s > 1/2$, their natural boundary being the line Re $s = 1/2$. Clearly, Jessen's result contains Wintner's results **I** and **II**.

6. Final remarks

6.1. The truth with probability one of the random R.H. does not settle in any way the real R.H., which is just an element of a 'homogeneous' set in which a probability measure introduces a regularity pattern. In spite of this, on account of a basic probability principle, one could not help thinking the real R.H. as entirely plausible. In other words, the truth with probability one of the random R.H. is a 'good reason' for believing the truth of the real R.H.

It has been often said that if the R.H. is true, then there should be some deep relation, at present hidden, allowing to prove it. It is the falsity of the R.H. that would also imply the existence of some bad relation assigning a special importance to e.g. the point (19).

6.2. In connection with the equivalence of the R.H. to a more precise statement of the prime number theorem (see Section 4) it should be noted that such a statement holds true for the so-called 'random primes' constructed according to the Hawkins sieving procedure. For details see Neudecker and Williams (1974).

6.3. It is interesting to speculate why Denjoy returned in 1964 to the problem he had taken up 33 years earlier. Did he intuit that the probability measure P_F involved in his 1931 paper is singular? (A set of P_F-measure 0 might comprise a whole interval in [0, 1], thus contradicting our intuition of a negligible set.)

6.4. On account of the remarks made at the beginning of Subsection 5.1, Good and Churchhouse (1968) conjectured that

A. The sums of $\mu(n)$ in blocks of length N, where N is large, have asymptotically a normal distribution with mean zero and variance $6N/\pi^2$.

Note that Conjecture A is much stronger than the condition $M(x) = O(x^{(1/2)+\varepsilon})$. Even so, numerical evidence strongly supports Conjecture A.

A second conjecture by the same authors is constructed by formally applying the law of the iterated logarithm to the (nonrandom) sequence $(\mu(n))_{n\geq 1}$:

$$\text{B.} \qquad \limsup_{x\to\infty} \{M(x)\,(x \log\log x)^{-1/2}\} = \sqrt{12}/\pi.$$

Clearly, Conjecture B (for which it is difficult to obtain numerical support) contradicts Stieltjes-Mertens' conjecture that $|M(x)| < x^{1/2}$ even in the extended form $|M(x)| < Cx^{1/2}$ for any constant C. (This modification of Stieltjes-Mertens' conjecture was refuted numerically for $C = 1/2$ (at e.g. $x = 7.76 \times 10^9$) by Neubauer (1963).) Haselgrove (1958) asserted that his method of refuting Pólya's conjecture $L(x) \leqslant 0$ for $x \geqslant 2$ could be applied, with 1000 times as much calculation, to disprove Stieltjes-Mertens' conjecture. Good and Churchhouse conclude that the latter "is improbable".

We are inclined to believe that, nevertheless, Stieltjes was right. In fact, his claimed estimate $M(x) = O(x^{1/2})$ is equivalent on account of (6') to

$$\frac{\#\{k:\, 1\leqslant k\leqslant n,\ \mu(k) = u\}}{n} = \frac{3}{\pi^2} + O(n^{-1/2}) \tag{25}$$

for $u = -1$ or 1, i.e. to the fact that the order of the error terms for the asymptotic densities of square-free integers with an odd or an even number of prime factors is $O(n^{-1/2})$, which is even worse than the (genuine) order of the error term in (6').

We cannot see any reason against (25). As Einstein put it, "God is subtle, but not malicious".

[Let us note that, assuming the R.H., the order of the error term in (6') can be lowered to $O(n^{-0.68+\varepsilon})$ for any $\varepsilon > 0$ (see Graham (1981)). It is also known (without assuming the R.H.) that the error term is *not* $o(n^{-3/4})$ (see Evelyn and Linfoot (1931))].

6.5. In Subsection 5.1 the random variables μ_n, as well as the random variables λ_n, have been supposed to be independent. At least Denjoy's treatment carries over to certain classes of dependent random variables, e.g. Markov chains (see Iosifescu (1980)). For checking Markovness (as well as independence), we need know the following asymptotic densities (assuming that they do exist):

$$\lim_{n\to\infty} \frac{\#\{k\colon 1\leqslant k\leqslant n,\ \lambda(k)=u,\ \lambda(k+1)=-u\}}{n}$$

for $u = 1$ or $u = -1$, and

$$\lim_{n\to\infty} \frac{\#\{k\colon 1\leqslant k\leqslant n,\ \mu(k)=u,\ \mu(k+1)=v\}}{n}$$

for $u = -1$, 0 or 1, and $v = u^2 - 1$ or u^2.

We are unaware whether the above quantities have been already calculated.

6.6. We have included in the bibliography a number of historical references. Results known for a long time are again and again rediscovered by authors unaware of their availableness. This illustrates once more the Fejér-Erdös dictum 'Everybody writes, nobody reads'. (Or, as put by Montherlant: «Au fond, les gens ne lisent pas; ou s'ils lisent, ils ne comprennent pas; quant à ceux qui comprennent, ils oublient.»)

REFERENCES

1. Apéry, R., *Irrationalité de ζ(2) et ζ(3)*. In: *Journées Arithmétiques de Luminy.* (Colloq. Internat. CNRS, Centre Univ. Luminy, Luminy, 1978). Astérisque, 61, pp. 11–13. Paris, Soc. Math. France, 1979.
2. Bellman, R., *Analytic Number Theory: An Introduction.* Reading, Mass., The Benjamin/Cummings Publ. Co., 1980.
3. Benkoski, S. J., *The probability that k positive integers are relatively r-prime.* J. Number Theory, **8**, 218–223 (1976).
4. Beukers, F., *A note on the irrationality of ζ(2) and ζ(3).* Bull. London Math. Soc., **11**, 268–272 (1979).
5. Beukers, F., *Padé-approximations in number theory.* In: *Padé Approximation and Its Applications, Amsterdam* 1980. Lecture Notes in Math., Vol. 888, pp. 90–99. Berlin–Heidelberg–New York, Springer, 1981.
6. Brent, R. P., van de Lune, J., te Riele, H. J. J., Winter, D. T., *On the zeros of the Riemann zeta function in the critical strip, II.* Preprint NW 121/82, Mathematical Centre, Amsterdam, 1982.
7. Césaro, E., *Question proposée* 75. Mathesis, **1**, 184 (1881).
8. Césaro, E., *Question 75 (solution).* Mathesis, **3**, 224–225 (1883).
9. Césaro, E., *Probabilités de certains faits arithmétiques.* Mathesis, **4**, 150–151 (1884).
10. Chatterji, S. D., *Certain induced measures and the fractional dimensions of their "supports".* Z. Wahrsch. Verw. Gebiete, **3**, 184–192 (1964).
11. Denjoy, A., *L'hypothèse de Riemann sur la distribution des zéros de ζ(s), reliée à la théorie des probabilités.* C. R. Acad. Sci. Paris, **192**, 656–658 (1931).
12. Denjoy, A., *Probabilités confirmant l'hypothèse de Riemann sur les zéros de ζ(s).* C. R. Acad. Sci. Paris, **259**, 3143–3145 (1964).
13. Diamond, H. G., McCurley, K. S., *Constructive elementary estimates for M(x).* In: *Analytic Number Theory.* Lecture Notes in Math., Vol. 899, pp. 239–253. Berlin – Heidelberg–New York, Springer, 1981.
14. Edwards, H. M., *Riemann's Zeta Function.* New York, Academic Press, 1974.
15. Erdös, P., Ulam, S., *Some probabilistic remarks on Fermat's last theorem.* Rocky Mountain J. Math., **1**, 613–616 (1971).

16. Evelyn, C. J. A., Linfoot, E. H., *On a problem in the additive theory of numbers. IV.* Ann, of. Math. (2), **32**, 261–270 (1931).
17. Fogels, È. K., *On the zeros of Riemann's zeta function.* Latvijas PSR Zinātņ. Akad. Vēstis, no. 4 (393), 136–144, 160(1980). (Russian)
18. Golomb, S., *A class of probability distributions on the integers.* J. Number Theory, **2**, 189–192 (1970).
19. Good, I. J., Churchhouse, R. F., *The Riemann hypothesis and pseudorandom features of the Möbius sequence.* Math. Comp., **22**, 857–861 (1968).
20. Graham, S. W., *The distribution of squarefree numbers.* J. London Math. Soc. (2), **24**, 54–64 (1981).
21. Hardy, G. H., Wright, E. M., *An Introduction to the Theory of Numbers, Corrected Fourth Ed.* Oxford, Clarendon Press, 1962.
22. Haselgrove, C. B., *A disproof of a conjecture of Pólya.* Mathematika, **5**, 141–145 (1958).
23. Ingham, A. E., *The Distribution of Prime Numbers.* Cambridge Tracts in Math. and Math. Phys., No. 30. London, Cambridge Univ. Press, 1932.
24. Ingham, A. E., *On two conjectures in the theory of numbers.* Amer. J. Math., **64**, 313–319 (1942).
25. Iosifescu, M., *Finite Markov Processes and Their Applications.* Chichester & București, Wiley & Ed. Tehnica, 1980.
26. Jessen, B., *Eine Integrationstheorie fur Funktionen unendlich vieler Veränderlichen, mit Anwendung auf das Wertverteilungsproblem für fastperiodische Funktionen, insbesondere fur die Riemannsche Zetafunktion.* Mat. Tidsskrift B, 59–65 (1932).
27. Jessen, B., *Some analytical problems relating to probability.* J. Math. Phys., **14**, 24–27 (1935).
28. Kendall, D. G., *On the number of lattice points inside a random oval.* Quart. J. Math., Oxford Ser., **19**, 1–26 (1948).
29. Landau, E., *Handbuch der Lehre von der Verteilung der Primzahlen. 2. Band.* Leipzig, Teubner, 1909.
30. Lehmer, D. N., *An asymptotic evaluation of certain totient sums.* Amer. J. Math., **22**, 293–355 (1900).
31. Levinson, N., *More than one third of zeros of Riemann's zeta function are on $\sigma = 1/2$.* Adv. in Math., **13**, 383–436 (1974).
32. Levinson, N., *Deduction of semi-optimal mollifier for obtaining lower bound for $N_0(T)$ for Riemann's zeta function.* Proc. Nat. Acad. Sci. USA, **72**, 294–297 (1975).
33. Levinson, N., Montgomery, H. L., *Zeros of the derivatives of the Riemann zeta function.* Acta Math., **133**, 49–65 (1974).
34. Lévy, P., *Sur les séries dont les termes sont des variables éventuelles indépendantes.* Studia Math., **3**, 119–155 (1931).
35. Lipschitz, R., *Beiträge zu der Kenntniss der Bernoullischen Zahlen.* J. Reine Angew. Math., **96**, 1–16 (1884).
36. Lou, Shi-Tuo, *A lower bound for the number of zeros of Riemann's zeta-function on $\sigma = 1/2$* In: Halberstam, H., Hooley, C. (Eds.), *Recent Progress in Analytic Number Theory Vol.* 1, pp. 319–325. London, Academic Press, 1981.
37. Mertens, F., *Über eine zahlentheoretische Funktion.* Sitzungsber. Kaiser. Akad. Wiss. Wien Math.-Natur. Cl., (2a) **106**, 761–830 (1897).
38. Neubauer, G., *Eine empirische Untersuchung zur Mertensschen Funktion.* Numerische Math., **5**, 1–13 (1963).
39. Neudecker, W., Williams, D., *The 'Riemann Hypothesis' for the Hawkins random sieve.* Compositio Math., **29**, 197–200 (1974).
40. Nyman, J. E., *On the probability that k positive integers are relatively prime.* J. Number Theory, **4**, 469–473 (1972).
41. Pólya, G., *Verschiedene Bemerkungen zur Zahlentheorie.* Jahresbericht d. deutschen Math-Ver., **28**, 31–40 (1919).
42. van der Poorten, A., *A proof that Euler missed... Apéry's proof of the irrationality of ζ (3). An informal report.* Math. Intelligencer, **1**, 195–203 (1978/79).
43. van der Poorten, A. J., *Some wonderful formulae ... footnotes to Apéry's proof of the irrationality of ζ (3).* In: *Séminaire Delange-Pisot-Poitou, 20e année 1978/1979. Théorie des nombres,* Fasc. 2, Exp. No. 29, 7 pp. Paris, Secrétariat Math., 1980.
44. te Riele, H. J. J., *Computations concerning the conjecture of Mertens.* J. Reine Angew. Math **311/312**, 356–360 (1979).
45. Spătaru, A., *Stochastic independence of the digits of an f-expansion.* Stud. Cerc. Mat., **30** 75–117 (1978). (Romanian)

46. Speiser, A., *Geometrisches zur Riemannschen Zetafunktion*. Math. Ann. **110**, 514–521 (1934).
47. Spira, R., *Zeros of $\zeta'(s)$ and the Riemann hypothesis*. Illinois J. Math., **17**, 147–152 (1973).
48. Stieltjes, T. J., *Lettre du 11 juillet 1885 à Hermite; Lettre du 15 avril 1887 à Mittag-Leffler*. In: Baillaud, B., Bourget, H., (Éditeurs) *Correspondance d'Hermite et de Stieltjes, Tome I*, pp. 160–164; *Tome II*, pp. 449–457. Paris, Gauthier-Villars, 1905.
49. Sylvester, J. J., *On the number of fractions in their lowest terms whose numerators and denominators are limited not to exceed a certain number*. John Hopkins University Circulars II., 44–45 (1883). (Also in Sylvester, J. J., *The Collected Mathematical Papers of James Joseph Sylvester, Vol. III*, pp. 672–676. London, Cambridge Univ. Press, 1900.)
50. Titchmarsh, E. C., *The Theory of the Riemann Zeta-Function*. Oxford, Clarendon Press, 1951.
51. Wintner, A., *Random factorizations and Riemann's hypothesis*. Duke Math. J., **11**, 267–275 (1944 a).
52. Wintner, A., *Eulerian products and analytic continuation*. Duke Math. J., **11**, 277–285 (1944 b).
53. Wintner, A., *The singularities in a family of zeta-functions*. Duke Math. J., **11**, 287–291 (1944 c).

FRANZ KONECNY

On nonlinear filtering of Poisson driven dynamical systems

1. Introduction

In this paper we are concerned with a one-dimensional stochastic dynamical system which is governed by a stochastic differential equation of the type

$$\mathrm{d}X_t = f(t, X_t)\mathrm{d}t + \int c(t, X_t, u)\, \tilde{\nu}(\mathrm{d}t, \mathrm{d}u), \tag{1.1}$$

where $\tilde{\nu}(.,.)$ is a centered Poisson measure on $[0, T] \times \mathbb{R}$. The problem is to estimate the state X_t given an observable process Y_t of the form

$$\mathrm{d}Y_t = h(X_t)\, \mathrm{d}t + \mathrm{d}W_t, \quad Y_0 = 0. \tag{1.2}$$

The noise W_t is a standard Wiener process independent of X_t.

H. Kwakernaak (1975) studied linear systems driven by Poisson noise with Gaussian observation noise having in mind applications to a water pollution problem. Assuming the existence of the conditional density he formally derived a generalized version of the Kushner-Stratonovich equation.

We write $\pi_t(\varphi) := E(\varphi(X_t) \mid Y_s;\ s \leqslant t)$ for a bounded smooth function φ. The general filter equation (c.f. Fujisaki et al. (1972), Liptser-Shiryayev (1977)) gives the following representation of the measure valued process

$$\pi_t(\varphi) = \pi_0(\varphi) \int_0^t \pi_s(A_s\varphi)\, \mathrm{d}s + \int_0^t \pi_s(\varphi(h - \pi_s(h)))\, \mathrm{d}I_s \tag{1.3}$$

where A_t is the extended generator of X_t (see Section 2) and

$$\mathrm{d}I_t = \mathrm{d}Y_t - \pi_t(h)\, \mathrm{d}t \tag{1.4}$$

is the so-called innovation process.

Instead of using the filter equation (1.3) directly, it is more convenient to perform a measure transformation, so that under the new measure X_t and Y_t are independent. This approach was first suggested by Zakai (1969). By this tool he derived an equation for the nonnormalized conditional density of diffusion process (see also Pardoux (1977, 1979)).

Stating the necessary definitions and assumptions in Section 2 we state in Section 3 the equation for the nonnormalized density of the law of X_t.

In Section 4 we establish the existence of a smooth conditional density in the case where the noise term in (1.1) does not depend on the state variable and $\Pi(\mathbb{R}) < \infty$. The latter assumption means that the jump intensity during an infinitesimal time period is finite.

2. Definitions and assumptions

Let $(\Omega, \mathbf{F}, P)$ be a probability space on which the processes X_t and Y_t are defined.

Let $\nu(dt, du)$ denote a homogeneous Poisson measure on $[0, T]\times\mathbb{R}$, i.e. $E\nu(dt, du) = dt\,\Pi(du)$ where $\Pi(.)$ is a σ-finite measure with support $\mathbb{R}\setminus\{0\}$ and $\Pi(|u| > \varepsilon) < \infty$ for any $\varepsilon > 0$. Set $\tilde{\nu}(dt, du) = \nu(dt, du) - dt\,\Pi(du)$.

We assume that

a) The functions f and c fulfil a Lipschitz and a growth condition so that (1.1) has a strong unique solution which is Markov (see Gihman-Skorohod (1972)).

b) h is a bounded measurable function.

Let $\varphi(x)$ be a continuously differentiable function satisfying

$$\int_0^T\int[\varphi(x+c(t, x, u)) - \varphi(x) - \varphi'(x)\,c(t, x, u)]\,\Pi(du)\,dt < \infty$$

and

$$\int_0^T\int|\varphi(x+c(t, x, u)) - \varphi(x)|^2\,\Pi(du)\,dt < \infty.$$

Let

$$A_t\varphi(x) := f(t, x)\,\varphi'(x) + \int[\varphi(x+c(t, x, u)) - \varphi(x) - \varphi'(x)\,c(t, x, u)]\,\Pi(du). \tag{2.1}$$

On account of Itô's formula

$$\varphi(X_t) - \varphi(X_0) - \int_0^t A_s\varphi(X_s)\,ds$$

is a local martingale with respect to $\sigma(X_s; s\leqslant t)$. We call A_t the *extended generator* of (X_t).

Let $\mathbf{F}_t := \sigma(X_s, Y_s; s\leqslant t)$ and

$$Z_t = \exp\left\{\int_0^t h(X_s)\,dY_s - \frac{1}{2}\int_0^t h(X_s)^2\,ds\right\}. \tag{2.2}$$

Define the probability P_0 on $(\Omega, \mathbf{F}_\infty)$ by $\left.\frac{dP_0}{dP}\right|_{\mathbf{F}_t} = Z_t^{-1}$, $0\leqslant t\leqslant T$. Under P_0 the law of (X_t) is unchanged but (Y_t) becomes a standard Wiener process.

3. The Zakai equation

We introduce now the following transformation

$$\sigma_t(\varphi) = \pi_t(\varphi)\exp\left\{\int_0^t \pi_s(h)\,dI_s + \frac{1}{2}\int_0^t \pi_s(h)^2\,ds\right\}. \tag{3.1}$$

By the Itô formula, we obtain Zakai's equation:

$$\sigma_t(\varphi) = \sigma_0(\varphi) + \int_0^t \sigma_s(A_s\varphi)\,ds + \int_0^t \sigma_s(h\varphi)\,dY_s. \tag{3.2}$$

σ_t is a measure-valued process with $\pi_t(\varphi) = \sigma_t(\varphi)/\sigma_t(1)$.

If the nonnormalized conditional law σ_t has a sufficiently smooth density, Zakai's equation can be used to derive a characterization for it. The following result can be found in Konecny (1982)

Theorem 3.1. *Let the assumptions of Section 2 be satisfied. Moreover we assume that*

i) *σ_t has a (nonnormalized) density $q(t, x)$ with respect to Lebesgue measure.*

ii) *The derivatives*

$$\frac{\partial}{\partial x}[f(t, x)q(t, x)] \tag{3.3}$$

and

$$\frac{\partial}{\partial x}\int_{|u|>\varepsilon} c(t, x, u)\,\Pi(du) = \int_{|u|>\varepsilon} \frac{\partial c}{\partial x}(t, x, u)\,\Pi(du) \tag{3.4}$$

exist for each $t \in [0, T]$.

iii) *$\frac{\partial c}{\partial x}(t, x, u) > -1$, the function $y = x + c(t, x, u)$ is invertible for any t and u and the inverse function $x = y - \gamma(t, y, u)$ is continuously differentiable.*

iv) *If we set*

$$J(t, x, u) := \left(1 + \frac{\partial}{\partial x} c(t, x - \gamma(t, x, u); u)\right)^{-1} \tag{3.5}$$

and

$$A_t^* q(t, x) := -\frac{\partial}{\partial x}[f(t, x)q(t, x)] + \tag{3.6}$$

$$+ \int\left\{q(t, x) - \gamma(t, x, u))\, J(t, x, u) - q(t, x) + \frac{\partial}{\partial x}[q(t, x)\, c(t, x; u)]\right\} \Pi(du),$$

then for any twice continuously differentiable function $\varphi(x)$ with compact support

$$\int_0^T\int \varphi(x)\, A_t^* q(t, x)\,dx\,dt < \infty \text{ a.s.} \tag{3.7}$$

and

$$E\int_0^T \left\{\int |\varphi(x)\, q(t, x)\, h(x)|\,dx\right\}^2 dt < \infty. \tag{3.8}$$

Under these assumptions the nonnormalized density $q(t, x)$, $0 \leqslant t \leqslant T$, satisfies the stochastic integro-differential equation

$$dq(t, x) = A_t^* q(t, x)\,dt + h(x)\, q(t, x)\,dY_t \tag{3.9}$$

a.e. in x with probability one.

Let $\hat{q}(t, x)$ be a measurable function which satisfies (3.7), (3.8) *and equation* (3.9). *Then $\hat{q}(t, x)$ is a.s. a version of the conditional density of σ_t.*

Remark. In the case where the function $c(t, x, u)$ does not depend on x we have

$$A_t^* q(t, x) = -\frac{\partial}{\partial x}[f(t, x)\, q(t, x)] + \int\left[q(t, x - c(t, u)) - q(t, x) + \frac{\partial q}{\partial x} c(t, u)\right]\Pi(\mathrm{d}u).$$

4. Existence of the conditional density

In this section we consider a system governed by the stochastic differential equation

$$\mathrm{d}X_t = f(X_t)\,\mathrm{d}t + \int u\tilde{\nu}(\mathrm{d}t, \mathrm{d}u), \tag{4.1}$$

$0 \leqslant t \leqslant T$, where the random measure $\nu(\mathrm{d}t, \mathrm{d}u)$ represents a compound Poisson process on $\mathbb{R}$ with jumps U_n having the common density function $g(u)$. Hence we have

$$E\nu([0, t], A) = \lambda t \int_A g(u)\,\mathrm{d}u \tag{4.2}$$

with a parameter $\lambda > 0$.

Under the assumptions of Section 2 the ordinary differential equation $\dot{x} = f(x)$ has a unique global solution. Let $\Phi(t, x)$ denote the flow of this differential equation.

Theorem 4.1. *In addition to the general conditions of Section 2 we make the following assumptions*

i) *The partial derivatives $\frac{\partial \Phi}{\partial x}$ and $\frac{\partial^2 \Phi}{\partial x^2}$ exist and are bounded (by a constant K_1). Moreover $\left|\frac{\partial \Phi}{\partial x}\right| \geqslant c > 0$.*

ii) *The functions f and h are bounded together with their derivatives f' and h'.*

iii) *The density $g \in C^1(\mathbb{R})$ possesses a derivative $g' \in L^1(\mathbb{R})$.*

iv) *The random variable X_0 is independent of the system noise and the Wiener process. Moreover X_0 has a density $\rho_0 \in C^1(\mathbb{R})$ which is bounded together with ρ_0'.*

Then the conditions of Theorem 3.1 *are fulfilled and there exists a smooth nonnormalized conditional density which is the solution of equation* (3.9) *with $q(0, x) = \rho_0(x)$ and*

$$A_t^* q(t, x) = -\frac{\partial}{\partial x}[\tilde{f}(x)\, q(t, x)] + \lambda \int [q(t, x-u) - q(t, x)]\, g(u)\,\mathrm{d}u, \tag{4.3}$$

where

$$\tilde{f}(x) = f(x) - \int u g(u) \, du.$$

Before proving Theorem 4.1 we prove a proposition concerning the absolute continuity of the law of X_t.

Lemma 4.1. *Under the assumptions of Theorem* 4.1 *the law of* X_t, $t \in [0, T]$ *possesses a smooth density which is bounded on* $[0, T] \times \mathbb{R}$.

Proof. We suppose $\frac{\partial \Phi}{\partial x} \geqslant c > 0$. For each $t \in [0, T]$ there exists a function $\Psi(t, x)$ inverse to $\Phi(t, x)$. Let N_t denote the number of jumps on $[0, t]$, τ_n the random time when the nth jump occurs and U_n the amplitude of the nth jump. Then $(N_t;\ t \in [0, T])$ is a homogeneous Poisson process with intensity λ and we have

$$P(X_t \leqslant x) = \sum_{n=0}^{\infty} e^{-\lambda t} \frac{(\lambda t)^n}{n!} P(X_t \leqslant x \mid N_t = n). \tag{4.4}$$

Let $N_t = 0$. We see that $P(X_t \leqslant x \mid N_t = 0) = P(\Phi(t, X_0) \leqslant x)$ and $\Phi(t, X_0$ has the density

$$p_0(t, x) = \rho_0(\Psi(t, x)) \left[\frac{\partial \Phi}{\partial x} (t, \Psi(t, x)) \right]^{-1}. \tag{4.5}$$

As a consequence $X_{\tau_1} = \Phi(\tau_1, X_0) + U_1$ has the density

$$\rho_1(x) = \lambda \int_0^{\infty} \int e^{-\lambda t} p_0(t, v) \, g(x - v) \, dv \, dt. \tag{4.6}$$

Let $\rho_n(x \mid t)$ denote the conditional density of X_{τ_n} given $\tau_n = t$. We find $P(X_t \leqslant x \mid N_t = n) = P(\Phi(t - \tau_n, X_{\tau_n}) \leqslant x \mid \tau_n \leqslant t)$ and the conditional density of X_t is

$$p_n(t, x) = \frac{\lambda}{(n-1)!} \int_0^t (\lambda s)^{n-1} e^{-\lambda s} \frac{\rho_n(\Psi(t - s, x) \mid s)}{\frac{\partial \Phi}{\partial x} (t - s, \Psi(t - s, x))} \, ds. \tag{4.7}$$

If we use (4.4) and apply the monotone convergence theorem we obtain

$$p(t, x) = \sum_{n=0}^{\infty} \frac{(\lambda t)^n}{n!} e^{-\lambda t} p_n(t, x). \tag{4.8}$$

We introduce now the norm $\| \psi \| = \sup \ \{ | \psi(t, x) | : t \in [0, T], \quad x \in \mathbb{R} \}$. Since $\rho_n(. \mid t) = p_{n-1}(t, .) * g$, we have

$$\| p_n \| \leqslant \frac{1}{c} \| p_{n-1} \| \leqslant \frac{1}{c^n} \| \rho_0 \| \tag{4.9}$$

and as a consequence

$$\| p \| \leqslant \frac{\| \rho_0 \|}{c} e^{\lambda T \left(1 - \frac{1}{c}\right)}. \tag{4.10}$$

Moreover for each n the partial derivative $\partial p_n/\partial x$ exists and a similar estimation as above yields that the series

(4.11) $$\sum_{n=0}^{\infty} \frac{(\lambda t)^n}{n!} e^{-\lambda t} \frac{\partial p_n}{\partial x}$$

converges uniformly in x for each $t \in [0, T]$.

Now we proceed to the proof of the existence of the conditional density. Zakai (1969) established the following result: *If X_t has the density $p(t, x)$, then the a posteriori law $P(X_t \leqslant x | Y_s;\ s \leqslant t)$ has the nonnormalized density*

(4.12) $$q(t, x) = p(t, x)\, E_0\left(\exp\left\{\int_0^t h(X_s)\, \mathrm{d}Y_s - \frac{1}{2}\int_0^t h(X_s)^2\, \mathrm{d}s\right\} \middle| X_t = x\right).$$

Let $(X_s^{tx};\ 0 \leqslant s \leqslant t)$ denote the process whose law coincides with the reverse conditional distribution of X which is given by $P(s, \cdot, t, x) = P(X_s \in \cdot \mid X_t = x)$, $s \leqslant t$. Set

(4.13) $$v(t, x) = v(t, x, Y) = \tilde{E}\left(\exp\left\{\int_0^t h(X_s^{tx})\, \mathrm{d}Y_s - \frac{1}{2}\int_0^t h(X_s^{tx})\, \mathrm{d}s\right\}\right),$$

where $\tilde{E}$ denotes the expectation with respect to the law of X_t.

Lemma 4.2. *There exists a subset $\Omega_0 \subseteq \Omega$ with $P(\Omega_0) = 1$, so that for each $\omega \in \Omega_0$ the function $x \to X_s^{tx}(\omega)$ belongs to $C^1(\mathbb{R})$.*

Proof (see Konecny (1982)). Using Kolmogoroff's forward equation for $p(t, x)$ a "backward stochastic differential equation" can be found so that (X_s^{tx}) is a solution of this equation. Thus our assertion can be shown in the same way as the corresponding results for "forward equations" (see Métivier (1982)).

Lemma 4.3. *Under the conditions of Theorem* 4.1 *for almost all sample paths of Y the function $x \to v(t, x)$, $t \in [0, T]$ belongs to $C^1(\mathbb{R})$.*

Proof. We denote the expression under the expectation in formula (4.13) by $Z_t(x, Y)$. By Itô's formula

(4.14) $$Z_t(x, Y) = 1 + \int_0^t Z_s(x, Y)\, h(X_s^{tx})\, \mathrm{d}Y_s.$$

Using Lemma 4.2 we obtain $x \to Z_t(x, Y) \in C^1(\mathbb{R})$ for almost all sample paths of Y(see, e.g., Meyer (1981)). The continuity of $\partial Z_t/\partial x$ and Fubini's theorem yield the assertion.

Proof of Theorem 4.1. Since under P_0 the processes X and Y are independent we have $q(t, x) = p(t, x)\, v(t, x)$. As $p(t, x)$ is bounded on $[0, T] \times \mathbb{R}$ and $h(x)$ is bounded on $\mathbb{R}$, by a result of Zakai (1969)

(4.15) $$E\int_0^T \int q^2(t, x)\, \mathrm{d}x\, \mathrm{d}t < \infty.$$

This fact, Lemma 4.1 and Lemma 4.3 imply the conclusion of the theorem.

REFERENCES

1. Fujisaki, M., Kallianpur, G., Kunita, H., *Stochastic differential equations for the nonlinear filtering problem.* Osaka J. Math., **9**, 19–40 (1972).
2. Gihman, I. I., Skorohod, A. V., *Stochastic Differential Equations.* Berlin, Springer, 1972.
3. Konecny, F., *On nonlinear filtering of Poisson driven dynamical systems.* Preprint, Universität für Bodenkultur, Wien, 1982.
4. Kwakernaak, H., *Filtering for systems excited by Poisson white noise.* In: Bensoussan, A., Lions, J. L. (Eds.), *Control Theory, Numerical Methods and Computer Systems Modelling.* Lecture Notes in Economics and Math. Systems, Vol. 107, pp. 468–492. Berlin, Springer, 1975.
5. Liptser, R. S., Shiryayev, A. N., *Statistics of Random Processes I: General Theory.* New York, Springer, 1977.
6. Métivier, M., *Pathwise differentiability with respect to a parameter of solutions of stochastic differential equations.* In: *Séminaire de Probabilités XVI.* Lecture Notes in Math., Vol. 920, pp. 490–502. Berlin, Springer, 1982.
7. Meyer, P. A., *Flot d'une équation différentielle stochastique.* In: *Séminaire de Probabilités XV.* Lecture Notes in Math., Vol. 850, pp. 103–117. Berlin, Springer, 1981.
8. Pardoux, E., *Filtrage de diffusions avec conditions frontières: Caractérisation de la densité conditionnelle.* In: Dacunha-Castelle, D., van Cutsem, B. (Eds.), *Journées de Statistique des Processus Stochastiques.* Lecture Notes in Math., Vol. 636, pp. 163–188. Berlin, Springer, 1978.
9. Pardoux, E., *Stochastic partial differential equations and filtering of diffusion processes.* Stochastics, **3**, 127–167 (1979).
10. Zakai, M., *On the optimal filtering of diffusions processes.* Z. Wahrsch. Verw. Gebiete, **11**, 230–243 (1969).

NORBERT LENZ

Convolution inverses and branching random walk

1. Introduction

In the branching random walk an initial ancestor forms the 0th generation and has position 0 on the real line. He produces children which form the 1st generation. Their positions on the real line are given by a point process Z_1. Every individual i of the nth generation with position $z_{n,i}$ produces children independently of the others. The positions of these children are given by a point process ${}_iZ_{n+1}$, which is an independent version of Z_1 shifted by $z_{n,i}$.

Let us define $Z_n(a) = \sum_i \mathbf{1}_{(-\infty,a]}(z_{n,i})$ and $G(a) = EZ_1(a)$. Then we have $EZ_n(a) = G^{*n}(a) =: G_n(a)$. We will assume that G is concentrated on $[0, \infty)$ with $G(0) > 1$ and $m = G(\infty) < \infty$. G is continuous from the right; it is the distribution function of a finite positive measure concentrated on $[0, \infty)$.

The question of the almost sure convergence and the convergence in rth mean, $r \in (1, 2]$, of $Z_n(a)/G_n(a)$, $a \in [0, \infty]$ is treated by Lenz [6] and Asmussen and Kaplan [1]. Only in the cases $a = 0$ and $a = \infty$, $\{Z_n(a)\}$ is a supercritical Galton-Watson process. Then $Z_n(a)/G_n(a)$ is a nonnegative and thus almost surely convergent martingale.

Here we will construct a martingale by norming $Z_n(.)$ with a convolution inverse of $G_n(.)$. We prove that the function

$$F_n(x) = G(0)^{-n} \sum_{k=0}^{\infty} (-1)^k \binom{n+k-1}{k} J^{*k}(x) \tag{1}$$

satisfies $F_n * G_n = \mathbf{1}_{[0,\infty)}$, where

$$J(x) = G(x)/G(0) - \mathbf{1}_{[0,\infty)}(x) \tag{2}$$

and $J^{*0} = \mathbf{1}_{[0,\infty)}$. Note that

$$U_n(a) = F_n * Z_n(a) \tag{3}$$

is a martingale for every $a \in [0, \infty)$. If $EZ_1(\infty)^r$ exists for some $r \in (1,2]$, then $U_n(a)$ converges almost surely and in rth mean to a random variable $U(a)$. $U(\cdot)$ is continuous from the right in rth mean. Let

$$R(s) = \int e^{-sx} \, dG(x) \tag{4}$$

be the Laplace transform of G. Biggins [3] has shown that

$$W_n(s) = R(s)^{-n} \int e^{-sx} \, dZ_n(x) \tag{5}$$

is a nonnegative martingale and thus $W_n(s)$ converges almost surely to a random variable $W(s)$. We prove that for all t sufficiently large $W(t) = t \int U(x) e^{-tx} \, dx$ holds almost surely.

2. Convolution inverses

We define $\mathcal{M} = \{f\colon \mathbb{R} \to \mathbb{R};\ f \cdot \mathbf{1}_{(-\infty,0)} \equiv 0,\ f$ is of bounded variation on every finite interval and continuous from the right$\}$ and denote the total variation of $f \in \mathcal{M}$ on $(-\infty, x]$ by $V_f(x)$. Clearly we have $G \in \mathcal{M}$, $J \in \mathcal{M}$. For the following theorem it suffices to assume $G(0) > 0$.

Theorem 1. *Let $G \in \mathcal{M}$ be the distribution function of a finite positive measure with $G(0) > 0$. Then for every $n \in \mathbb{N}$ holds:*

(i) *For all $x \in \mathbb{R}$ the series $F_n(x)$ as defined in* (1) *is absolutely convergent. If $s \in [0, \infty)$ is so large that $R(s) < 2G(0)$, then*

$$V_{F_n}(x) \leq e^{sx}(2G(0) - R(s))^{-n}. \tag{6}$$

Therefore we have $F_n \in \mathcal{M}$.

(ii) *$F_n = F^{*n}$ and F_n is the unique function in $\mathcal{M}$ satisfying $F_n * G_n = \mathbf{1}_{[0,\infty)}$.*

For the proof we need the following simple

Lemma 2. *Let $H \in \mathcal{M}$ be the distribution function of a finite positive measure and $L(\lambda) = \int e^{-\lambda x} \, dH(x)$, $\lambda \geq 0$. Then*

$$H^{*n}(a) \leq (L(\lambda))^n e^{\lambda a}. \tag{7}$$

Proof.

$$(L(\lambda))^n = \int e^{-\lambda x} \, dH^{*n}(x) \geq \int_{-\infty}^{a} e^{-\lambda x} \, dH^{*n}(x) \geq e^{-\lambda a} \cdot H^{*n}(a).$$

Proof of Theorem 1. (i) We define $L(s) = \int e^{-sx} \, dJ(x)$. Because of (2) and (4) $L(s) = R(s)/G(0) - 1 < 1$ holds if $R(s) < 2G(0)$. Since $\lim_{s\to\infty} R(s) = G(0)$, there exists some $s_0 \geq 0$ such that $R(s) < 2G(0)$ for all $s > s_0$. With (7) we obtain

$$|F_n(x)| \leq V_{F_n}(x) \leq G(0)^{-n} \sum_{k=0}^{\infty} \binom{n+k-1}{k} J^{*k}(x) \leq G(0)^{-n} e^{sx} \sum_{k=0}^{\infty} \binom{n+k-1}{k} \left(\frac{R(s)}{G(0)} - 1\right)^k = e^{sx}(2G(0) - R(s))^{-n}.$$

(ii) Define

$$f_{n,1}(x) = G(0)^{-n} \sum_{k \in \{0,2,4,\ldots\}} \binom{n+k-1}{k} J^{*k}(x)$$

and

$$f_{n,2}(x) = f_{n,1}(x) - F_n(x).$$

By the generalized convergence theorem for Laplace transforms (see Feller ([4], p. 433, Theorem 2a))

$$\int_0^\infty e^{-st}\, df_{n,1}(t) = G(0)^{-n} \sum_{k \in \{0,2,4,\ldots\}} \binom{n+k-1}{k} L(s)^k$$

is convergent for all $s > s_0$. An analogous result holds for $f_{n,2}$. Thus we have for $s > s_0$:

$$\text{(8)} \qquad \int_0^\infty e^{-st}\, dF_n(t) = G(0)^{-n} \sum_{k=0}^\infty (-1)^k \binom{n+k-1}{k} L(s)^k =$$

$$= \left(\int_0^\infty e^{-st}\, dF(t)\right)^n = R(s)^{-n} = 1 \Bigg/ \left(\int_0^\infty e^{-st}\, dG_n(t)\right).$$

By Widder ([8], Ch. II, Theorems 11.5 and 6.3) the proposition follows.

Example 3. For some discrete distributions the convolution inverses have a particularly simple form (in the cases (i), (ii) and (iii) the frequency functions are also indicated).

(i) Bernoulli distribution:

Let G be given by $G\{0\} = p$, $G\{1\} = 1 - p = q$.

Then $\qquad G_n\{k\} = \binom{n}{k} p^{n-k} q^k \qquad$ and

$$F_n\{k\} = \frac{1}{p^n}\left(-\frac{q}{p}\right)^k \binom{n+k-1}{k}, \quad n \in \mathbb{N},\ k \in \mathbb{N}_0.$$

(ii) Geometric distribution:

Let G be given by: $G\{k\} = pq^k$, $k \in \mathbb{N}_0$.

Then $\qquad G_n\{k\} = p^n q^k \binom{n+k-1}{k} \qquad$ and

$$F_n\{k\} = \binom{n}{k} \frac{1}{p^{n-k}}\left(-\frac{q}{p}\right)^k, \quad n \in \mathbb{N},\ k \in \mathbb{N}_0.$$

(iii) Poisson distribution:

Let G be given by $G\{k\} = e^{-\lambda} \lambda^k / k!$, $k \in \mathbb{N}_0$.

Then $\qquad G_n\{k\} = \frac{(n\lambda)^k}{k!} e^{-n\lambda} \qquad$ and

$$F_n\{k\} = \frac{(-n\lambda)^k}{k!} e^{n\lambda}, \quad n \in \mathbb{N},\ k \in \mathbb{N}_0.$$

(iv) Exponential distribution with an atom at $\{0\}$:

Let
$$G(x) = \begin{cases} 0 & \text{for } x < 0 \\ p + q(1 - e^{-\alpha x}) & \text{for } x \geq 0. \end{cases}$$

Then
$$F_1(x) = \begin{cases} 0 & \text{for } x < 0 \\ 1 + \frac{q}{p} e^{-\frac{\alpha}{p} x} & \text{for } x \geq 0. \end{cases}$$

Remark. An essential assumption for the uniqueness of F_n is $F_n \cdot \mathbf{1}_{(-\infty,0)} \equiv \equiv 0$. The following example shows that there may exist several functions F which satisfy $F * G = \mathbf{1}_{[0,\infty)}$ for some distribution function G and which are of finite local variation and continuous from the right.

Example 4. Let G be defined as in Example 3 (i) with $p = q = 1/2$

Let
$$F(x) = \begin{cases} 2 & \text{for } x \in [2n - 1, 2n),\ n \in \mathbb{Z} \\ 0 & \text{for } x \in [2n, 2n + 1),\ n \in \mathbb{N}_0 \\ -2 & \text{for } x \in [2n, 2n + 1),\ -n \in \mathbb{N}. \end{cases}$$

Then
$$(G * F)(x) = \frac{1}{2}(F(x) + F(x - 1)) = \mathbf{1}_{[0,\infty)}(x).$$

Another theorem concerning the existence of convolution inverses can be found in Gelfand, Raikow, and Schilow ([5], p. 252).

3. The processes $\{U_n(a)\}$

Let $U_n(a)$ be defined as in (3). We put $U_n(0) = Z_n(0)/G(0)^n$. Let the σ-algebra generated by the births of the first n generations be denoted by $\mathfrak{F}_n$.

Theorem 5. *For all $a \in \mathbb{R}$, $\{U_n(a)\}$ is a martingale.*

Proof.
$$E(U_n(a) \mid \mathfrak{F}_{n-1}) = \sum_i E((F_n * {}_iZ_n)(a) \mid \mathfrak{F}_{n-1}) =$$
$$= \sum_i \int F_n(a - y)\, dG(y - z_{n-1,i}) = \sum_i (F_n * G)(a - z_{n-1,i}) =$$
$$= (F_{n-1} \,\bar{*}\, Z_{n-1})(a) = U_{n-1}(a).$$

As mentioned in the introduction we assume $G(0) > 1$, $m = G(\infty) < \infty$ and $EZ_1(\infty)^r < \infty$ for some $r \in (1, 2]$.

Theorem 6. *$U_n(a)$ converges almost surely and in rth mean to a random variable $U(a)$. For all $s \in [0, \infty)$ sufficiently large we have*

$$\rho = \frac{R(s)}{(2G(0) - R(s))^r} < 1$$

With $c = 4\,(E(Z_1(\infty))^r + m^r)$ we have

$$E\,|U(a) - U_k(a)|^r \leq \frac{c\, e^{sar}}{1 - \rho}\, \rho^{k+1}. \tag{9}$$

Proof. First we will show

(10) $$E\,|Z_n(a)-(G*Z_{n-1})(a)|^r\leqslant 2(E(Z_1(\infty))^r+m^r)\,G_{n-1}(a).$$

Let the atoms of ${}_iZ_n$ be denoted by $\{{}_iz_{n,j}\}$. Since G is concentrated on $[0,\infty)$, we have

(11) $${}_iZ_n(a)=\sum_j \mathbf{1}_{(-\infty,a]}({}_iz_{n,j})\leqslant\sum_j \mathbf{1}_{(-\infty,a]}(z_{n-1,i})=\mathbf{1}_{(-\infty,a]}(z_{n-1,i})\cdot{}_iZ_n(\infty).$$

By inequality (4) of von Bahr and Esseen [2] and (11)

$$E\,|Z_n(a)-(G*Z_{n-1})(a)|^r=EE\left(\left|\sum_i({}_iZ_n(a)-G(a-z_{n-1,i}))\right|^r\,\middle|\,\mathfrak{F}_{n-1}\right)\leqslant$$

$$\leqslant 2\,E\left(\sum_i E(({}_iZ_n(a))^r+(G(a-z_{n-1,i}))^r\,\middle|\,\mathfrak{F}_{n-1}\right)\leqslant$$

$$\leqslant 2\,E\sum_i \mathbf{1}_{(-\infty,a]}(z_{n-1,i})\,(E(Z_1(\infty))^r+m^r)=2(E(Z_1(\infty))^r+m^r)\,G_{n-1}(a),$$

thus (10) is proved.

If s is large enough that $R(s)<2G(0)$, then we get with Jensen's inequality, (10), (1), (6) and (7)

$$E|U_n(a)-U_{n-1}(a)|^r\leqslant E\left(\int_0^a|Z_n(a-y)-(G*Z_{n-1})(a-y)|\,\mathrm{d}V_{F_n}(y)\right)^r\leqslant$$

$$\leqslant(V_{F_n}(a))^{r-1}\int_0^a\frac{c}{2}G_{n-1}(a-y)\,\mathrm{d}V_{F_n}(y)\leqslant(V_{F_n}(a))^{r-1}\frac{c}{2}\sum_{k=0}^{\infty}\binom{n+k-1}{k}(G_{n-1}*J^{*k})(a)\leqslant$$

$$\leqslant\left(\frac{\mathrm{e}^{sa}}{(2G(0)-R(s))^n}\right)^{r-1}\frac{c}{2}\,\frac{R(s)^{n-1}\cdot\mathrm{e}^{sa}}{(2G(0)-R(s))^n}\leqslant\frac{c}{2}\,\mathrm{e}^{sar}\,\rho^n.$$

Since $R(s)$ decreases monotonically to $G(0)>1$, $\rho<1$ for all sufficiently large s. Using again (4) we get

(12) $$E\,|U_n(a)-U_k(a)|^r\leqslant 2\sum_{j=k+1}^{n}E\,|U_j(a)-U_{j-1}(a)|^r\leqslant$$

$$\leqslant c\,\mathrm{e}^{sar}\sum_{j=k+1}^{n}\rho^j\leqslant\frac{c\,\mathrm{e}^{sar}}{1-\rho}\,\rho^{k+1}$$

Therefore $\sup_{n\in\mathbf{N}}E\,|U_n(a)|^r<\infty$ and because of the martingale property (see Loève ([7], p. 60)) $U_n(a)$ converges almost surely and in rth mean. By applying Fatou's lemma to (12), (9) is proved.

Example 7. Let $G(x)=G(0)\,\mathbf{1}_{[0,\infty)}(x)+(m-G(0))\,\mathbf{1}_{[1,\infty)}(x)$ with $1<G(0)<m$. Applying Theorems 2.5 and 3.5 of Lenz [6] we get

$$\lim_{n\to\infty}\left(\frac{Z_n(1)}{nG(0)^n}-\frac{Z_n(0)}{G(0)^n}\,\frac{m-G(0)}{G(0)}\right)=0$$

almost surely and in rth mean. Using Theorem 6 and Example 3 (i) we obtain

$$\lim_{n\to\infty}\left(\frac{Z_n(1)}{G(0)^n}-n\frac{Z_n(0)}{G(0)^n}\,\frac{m-G(0)}{G(0))}\right)=U(1)$$

almost surely and in rth mean.

Remark. If $m^{1/r}+m<2G(0)$, then we can put $s=0$ in inequality (9) and thus the inequality is independent of a.

Theorem 8. *$U(\cdot)$ is continuous from the right in rth mean on $[0, \infty)$.*

Proof. Let $a\in[0, \infty)$, $b>a$ and $y\in(a,b]$. Because of (9) we obtain

$$\lim_{y\downarrow a} E\,|U(y)-U(a)|^r\leqslant\lim_{y\downarrow a} 3^{r-1}\,(E\,|U(y)-U_k(y)|^r+E\,|U_k(y)-U_k(a)|^r+$$
$$+E\,|U_k(a)-U(a)|^r)\leqslant\text{const.}\ (\rho^k+\lim_{y\downarrow a} E\,|U_k(y)-U_k(a)|^r)$$

with $\rho<1$. Therefore it suffices to prove

$$\lim_{n\to\infty} E\,|U_k(y_n)-U_k(a)|^r=0$$

for all sequences $\{y_n\}\subset(a, b]$ with $\lim_{n\to\infty} y_n=a$. Because Z_k and F_k are both continuous from the right, U_k also is right continuous. Therefore $\lim_{n\to\infty}|U_k(y_n)-U_k(a)|=0$ holds almost surely. Since

$$|U_k(y_n)-U_k(a)|^r\leqslant 2^{r-1}\left(\left(\int_0^{y_n} V_{F_k}\,(y_n-y)\,\mathrm{d}Z_k(y)\right)^r+\left(\int_0^{a} V_{F_k}(a-y)\,\mathrm{d}Z_k(y)\right)^r\right)\leqslant$$
$$\leqslant 2^r(V_{F_k}(b))^r\,(Z_k(\infty))^r.$$

is P-integrable, we can complete the proof by applying the theorem of dominated convergence.

Let us now take a look at the random variables $W_n(s)$ defined in (5). Biggins [3] has proved $\lim_{n\to\infty} W_n(s)=W(s)$ almost surely. If s is large enough so that $R(s)<2G(0)$, then by (8), $\int_0^\infty \mathrm{e}^{-sx}\,\mathrm{d}F_n(x)=R(s)^{-n}$ is convergent. Therefore the empirical Laplace transform of the signed measure $U_n(\cdot)$ exists:

$$\int_0^\infty \mathrm{e}^{-sx}\,\mathrm{d}U_n(x)=R(s)^{-n}\int_0^\infty \mathrm{e}^{-sx}\,\mathrm{d}Z_n(x)=W_n(s).$$

Corollary 9. *Let $s\in[0, \infty)$ satisfy $R(s)<(2G(0)-R(s))^r$. Then for all $t>s$*

$$W(t)=t\int_0^\infty U(x)\,\mathrm{e}^{-tx}\,\mathrm{d}x$$

almost surely.

Proof. Assume $t>s$. By (6) we obtain

$$|\mathrm{e}^{-tx}U_n(x)|\leqslant\mathrm{e}^{-tx}\,Z_n(\infty)\,\mathrm{e}^{sx}\,(2G(0)-R(s))^{-n}\xrightarrow[x\to\infty]{}0$$

almost surely. Using partial integration we have

$$\int_0^\infty \mathrm{e}^{-tx}\,\mathrm{d}U_n(x)=t\int_0^\infty \mathrm{e}^{-tx}U_n(x)\,\mathrm{d}x.$$

Because of (9)

$$E\left|\int_0^\infty \mathrm{e}^{-tx}\,(U_n(x)-U(x))\,\mathrm{d}x\right|\leqslant\int_0^\infty \mathrm{e}^{-tx}\,\mathrm{e}^{-sx}\,\mathrm{d}x\cdot\text{const}\cdot\tilde{\rho}^{n+1}$$

is summable since $\tilde{\rho} < 1$. We obtain

$$W(t) = \lim_{n\to\infty} W_n(t) = \lim_{n\to\infty} t\int_0^\infty e^{-tx} U_n(x)\,dx = t\int_0^\infty e^{-tx} U(x)\,dx$$

almost surely.

REFERENCES

1. Asmussen, S. Kaplan, N., *Branching random walks I.* Stochastic Process. Appl., **4**, 1–13 (1976).
2. von Bahr, B., Esseen, C.-G., *Inequalities fcr the rth absolute mcment of a sum of random variables,* $1 \leqslant r \leqslant 2$. Ann. Math. Statist., **36**, 299–303 (1965).
3. Biggins, J. D., *Martingale convergence in the branching randcm walk.* J. Appl. Probab., **14**, 25–37 (1977).
4. Feller, W., *An Introduction to Probability Theory and Its Applications, Vol. 2, 2nd Ed.* New York, Wiley, 1971.
5. Gelfand, I. M., Raikow, D. A., Schilow, G. E., *Kcmmutative ncrmierte Algebren.* Berlin, Deutscher Verlag der Wissenschaften, 1964.
6. Lenz, N., *Konvergenzsätze für Branching Random Walks.* Dissertation, Universität Mainz, 1981.
7. Loève, M., *Probability Theory II, 4th Ed.* Berlin, Springer, 1978.
8. Widder, D. V., *The Laplace Transform.* Princeton, Princeton University Press, 1946.

ARISTIDE LEONTE

On the ergodicity of the tensor product of two stochastic systems

Let $(S, \mathcal{A}, \mu)$ be a measure space and let P be a transition probability function on $(S, \mathcal{A})$, invariant with respect to μ. For x from $L^1(\mu)$ we put $(Tx)(s) = \int P(s, ds')\, x(s')$. Obviously Tx belongs to $L^1(\mu)$ and the L^1-norm of T is one. It follows that T is an endomorphism of $L^p(\mu)$ for each $p > 1$.

Definition ([4], p. 94). P is said to be μ-*ergodic* if

$$\lim_n \frac{1}{n} \sum_{k=1}^{n} T^k x = E_\mu(x) \qquad \mu\text{-a.e.}$$

for each x from $L^1(\mu)$.

It is proved that P is (μ-)ergodic iff

$$\lim_n \frac{1}{n} \sum_{k=1}^{n} \int_A \mu(ds)\, P_k(s, B) = \mu(A)\, \mu(B)$$

for each pair of sets A, B from $\mathcal{A}$ ([4], p. 97). (In this statement P_k is the usual convolution product of k copies of P.)

Equivalently, P is (μ-)ergodic iff

$$\lim_n \frac{1}{n} \sum_{k=1}^{n} \langle T^k x, y\rangle = \langle x, 1\rangle \langle 1, y\rangle$$

for each pair of complex functions x, y from $L^2(\mu)$.

In this context it seems rather natural to call the transition probability P (μ-)*weakly mixing* if

$$\lim_n \frac{1}{n} \sum_{k=1}^{n} \left| \int_A \mu(ds)\, P_k(s, B) - \mu(A)\, \mu(B) \right| = 0$$

for each pair of sets A, B from $\mathcal{A}$.

Equivalently, P is (μ-)weakly mixing iff

$$\lim_n \frac{1}{n} \sum_{k=1}^{n} |\langle T^k x, y\rangle - \langle x, 1\rangle \langle 1, y\rangle| = 0$$

for each pair of complex functions x, y from $L^2(\mu)$.

Obviously, if P is (μ-)weakly mixing, then P is (μ-)ergodic.

Let P_j ($j = 1, 2$) be two transition probability functions on $(S_j, \mathcal{A}_j)$ By the formula

$$(P_1 \otimes P_2)((s_1, s_2), A) = \int P_1(s_1, du_1)\, P_2(s_2, A_{u_1}) =$$

$$= \int P_2(s_2, du_2)\, P_1(s_1, A_{u_2})$$

one defines unambiguously the tensor product of P_1, P_2 in $(S_1, \mathcal{A}_1) \otimes (S_2, \mathcal{A}_2)$ (see [2], p. 6, for the case of kernels).

If P_1 and P_2 are invariant with respect to μ_1 and μ_2 respectively, then $P_1 \otimes P_2$ is invariant with respect to $\mu_1 \otimes \mu_2$.

Proposition. *If P_j is (μ_j-)weakly mixing, $j = 1, 2$, then $P_1 \otimes P_2$ is ($\mu_1 \otimes \mu_2$-)ergodic.*

Proof. Let x_j belong to $L^2(\mu_j)$ ($j = 1, 2$) and let $(x_1 \otimes x_2)(s_1, s_2) = x_1(s_1)\, \overline{x_2(s_2)}$. Clearly, $x_1 \otimes x_2$ belongs to $L^2(\mu_1 \otimes \mu_2)$. We put

$$T(x_1 \otimes x_2)(s_1, s_2) = (T_1 x_1)(s_1) \cdot \overline{(T_2 x_2)(s_2)}.$$

There exists a unique bounded linear operator on $L^2(\mu_1 \otimes \mu_2)$ so that if $x_j \in L^2(\mu_j)$ ($j = 1, 2$), then

$$T(x_1 \otimes x_2) = T_1 x_1 \otimes T_2 x_2$$

([1], p. 45). It follows that $T^k = T_1^k \times T_2^k$ and consequently

$$\frac{1}{n}\left|\sum_{k=1}^{n} \langle T^k(x_1 \otimes x_2), y_1 \otimes y_2\rangle - \langle x_1 \otimes x_2, 1\rangle \langle 1, y_1 \otimes y_2\rangle\right| =$$

$$= \frac{1}{n}\left|\sum_{k=1}^{n} \langle T_1^k x_1, y_1\rangle \langle T_2^k x_2, y_2\rangle - \langle x_1, 1\rangle \langle x_2, 1\rangle \langle 1, y_1\rangle \langle 1, y_2\rangle\right|.$$

But

$$\lim_n \frac{1}{n} \sum_{k=1}^{n} |\langle T_j^k x_j, y_j\rangle - \langle x_j, 1\rangle \langle 1, y_j\rangle| = 0$$

for $j = 1, 2$. Then

$$\lim_n \frac{1}{n}\left|\sum_{k=1}^{n} \langle T^k(x_1 \otimes x_2, y_1 \otimes y_2)\rangle - \langle x_1 \otimes x_2, 1\rangle \langle 1, y_1 \otimes y_2\rangle\right| = 0$$

Finally $\overline{\mathrm{sp}}\{x_1 \otimes x_2;\, x_j \in L^2(\mu_j)\} = L^2(\mu_1 \otimes \mu_2)$.

Theorem. *P is (μ-)weakly mixing if and only if $P \otimes P$ is ($\mu \otimes \mu$)-ergodic.*

Proof. On account of the above proposition, if P is (μ-)weakly mixing then $P \otimes P$ is ($\mu \otimes \mu$-)ergodic.

Conversely, $P \otimes P$ is ($\mu \otimes \mu$-)ergodic iff

$$\lim_n \frac{1}{n} \sum_{k=1}^{n} \langle (T \otimes T)^k x, y\rangle = 0 \tag{1}$$

for every x and y from $L^2(\mu \otimes \mu)$ with $\langle x, 1\rangle = \langle y, 1\rangle = 0$. In particular, for $x = x_1 \otimes x_2$, $y = y_1 \otimes y_2$, from (1) one obtains that

$$\lim_n \frac{1}{n} \sum_{k=1}^{n} \langle T^k x_1, y_1\rangle \langle y_2, T^k x_2\rangle = 0$$

for every x_j, y_j $(j = 1, 2)$ from $L^2(\mu)$ with $\langle x_j, 1\rangle = \langle y_j, 1\rangle = 0$. If one takes $x_1 = x_2$ and $y_1 = y_2$ it follows that

$$\lim_n \frac{1}{n} \sum_{k=1}^{n} |\langle T^k x, y\rangle|^2 = 0,$$

hence

$$\lim_n \frac{1}{n} \sum_{k=1}^{n} |\langle T^k x, y\rangle| = 0$$

for each pair of complex functions x, y from $L^2(\mu)$ with $\langle x, 1\rangle = \langle y, 1\rangle = 0$. So P is (μ-)weakly mixing.

Corollary. ([3], p. 45). *Let (S, A, P, T) be a measure dynamical system (i.e., T is a μ-invariant endomorphism of $(S, \mathcal{A})$. T is weakly mixing iff $T \otimes T$ is ergodic.*

Proof. The function $P(., A) = (1_A \circ T)(.)$ is a transition probability function, which is invariant with respect to μ. T is weakly mixing (ergodic) iff P is (μ-)weakly mixing (ergodic).

REFERENCES

1. Halmos, P. R., *Lectures on Ergodic Theory*. Tokyo, Math. Soc. of Japan, 1956.
2. Halmos, P. R., Sunder, V. S., *Bounded Integral Operators on L^2 Spaces*. Berlin, Springer, 1978.
3. Kornfeld, I. P., Sinai, Ia. G., Fomin, S. V., *Ergodic Theory*. Moskva, Nauka, 1980. (Russian)
4. Rosenblatt, M., *Markov Processes. Structure and Asymptotic Behaviour*. Berlin, Springer, 1971.

E. MARINESCU

Two remarks on the Skorohod representation

Using a technique from the theory of time reversal, two simple results connected with the Skorohod representation are proved.

The setting of this note is the general frame of a standard Markov process X, having finite lifetime ζ, for which the filtration is given by a family of killing operators (the filtration is completed in the usual way). For notation and definitions we refer to [1] and [2].

Let only recall (see [1] that $\mathcal{H}_g^\circ$ is the σ-algebra of the random sets generated by the left- continuous processes Y, whose supports are in the stochastic interval $]0, \zeta]$, and for which $Y_{s+t} = Y_s \circ \theta_t$ on $\{0 < s + t \leqslant \zeta\}$, $\forall\, s, t \in R_+$. Let μ be a positive σ-finite measure on the state space $(E, \mathcal{E})$. $\mathcal{H}_g^\mu$ is the σ-algebra of the random sets that are P^μ-indistinguishable from a $\mathcal{H}_g^\circ$-measurable set.

Following [1], a random variable τ taking values in $R_+ \cup \{-\infty\}$ is said to be a "temps de retour algébrique" if its graph is included in $]0, \zeta]$ and $\forall$ $t \in R_+$, $\tau \circ \theta_t$ equals $\tau - t$ on $\{\tau > t\}$ and $-\infty$ on $\{\tau \leqslant t\}$. We need also the following.

Lemma ([1]). *Let X be a standard Markov process and τ a "temps de retour algébrique". The function $f_\tau : E \to R_+$ defined by $f_\tau(x) = P^x(\tau > 0)$ is an excessive function for X.*

Proof. It consists of the following simple computations

$$P_t f_\tau(x) = E^x(f_\tau \circ X_t) = E^x(P^{X_t}(\tau > 0)) = E^x(\tau \circ \theta_t) =$$
$$= P^x(\tau - t > 0) \leqslant P^x(\tau > 0) = f_\tau(x),$$
$$\lim_{t \downarrow 0} P_t f_\tau(x) = \lim_{t \downarrow 0} P^x(\tau > t) = P^x(\tau > 0) = f_\tau(x).$$

Theorem 1. *Let μ and ν be two positive σ-finite measures on the state space $(E, \mathcal{E})$ of a standard Markov process such that $\mu \prec \nu$. If A and A' are two random sets in $\mathcal{H}_g^\circ$ that are P^μ-indistinguishible, then they are also P^ν-indistinguishable.*

Proof. Obviously, it is enough to prove that if A is P^μ-vanishing, it is also P^ν-vanishing. Suppose $\alpha = P^\nu(\pi(A)) > 0$ (π is the projection on Ω, the space of trajectories). Using a section theorem we obtain that $\forall \varepsilon > 0$, $\varepsilon < \alpha$, $\exists \tau_\varepsilon$ "temps de retour algébrique" such that $[\tau_\varepsilon] \subset A$ and

$$P^\nu(\tau_\varepsilon > 0) + \varepsilon \geqslant P^\nu(\pi(A)) \Rightarrow \int \nu(\mathrm{d}x)\, P^x(\tau_\varepsilon > 0) > 0.$$

Note that $\mu \prec \nu$ implies $\int \nu(dx) f(x) \leqslant \int \mu(dx) f(x)$ for all excessive functions f (see [3]). Since $x \to f_{\tau_\varepsilon}(x) = P^x(\tau_\varepsilon > 0)$ is an excessive function, we get

$$0 < P^\nu(\tau_\varepsilon > 0) = \int \nu(dx) f_{\tau_\varepsilon}(x) \leqslant \int \mu(dx) f_{\tau_\varepsilon}(x) = P^\mu(\tau_\varepsilon > 0) \leqslant P^\mu(\pi(A))$$

so $P^\mu(\pi(A)) > 0$. This contradicts the property of A to be P^μ-vanishing.

Remark. A similar result is to be found in [1], where the argument relies on Meyer's (L) hypothesis.

Corollary. $\mu \prec \nu \Rightarrow \mathcal{H}_g^\mu \subset \mathcal{H}_g^\nu$

Proof. It is an immediate consequence of Theorem 1.

For the next result we need the following.

Definition ([1]) Let μ be a positive σ-finite measure on the state space $(E, \mathcal{E})$ of a standard Markov process X. A set $\Gamma \in \mathcal{E}$ is said to be μ-left-polar (μ-polaire-à gauche) if the random set $\{(\omega, t) \mid t > 0, X_{t^-}(\omega) \in \Gamma\}$ is P^μ-vanishing.

Theorem 2. *Let μ, ν be positive measures on $(E, \mathcal{E})$ such that $\mu \prec \nu$. If $\Gamma \in \mathcal{E}$ is μ-left-polar then Γ is ν-left-polar.*

Proof. Since the random set $\{(\omega, t) \mid t > 0, X_{t^-}(\omega) \in \Gamma\} \in \mathcal{H}_g^\circ$ (see [1]), the theorem is a consequence of Theorem 1.

REFERENCES

1. Azéma, J., *Théorie générale des processus et retournement du temps.* Ann. Sci. École Norm. Sup. (4), 6, 459–519 (1973).
2. Blumenthal, R.M., Getoor, R.K., *Markov Processes and Potential Theory.* New York–London, Academic Press, 1968.
3. Rost, H., *The stopping distribution of a Markov process.* Invent. Math., 4, 1–16 (1971).

MAGDA PELIGRAD

On convergence rates of the strong law for stationary mixing sequences

This note summarizes recent results obtained in evaluation of the convergence rate in the Marcinkiewicz–Zygmund strong law for stationary mixing sequences.

Let $\{X_n\}_n$ be a sequence of random variables and set $\mathbf{F}_n^m = \sigma(X_i; n \leqslant \leqslant i \leqslant m)$. The sequence X_n is said to be φ-*mixing* (cf. Ibragimov) if $\varphi(n) \to 0$, where

$$\varphi(n) = \sup_m \sup_{\{A \in \mathbf{F}_1^m,\ B \in \mathbf{F}_{n+m}^\infty,\ P(A) \neq 0\}} |P(A \mid B) - P(B)|.$$

The sequence $\{X_n\}_n$ is said to be *strong mixing* (cf. Rosenblatt) if $\alpha(n) \to 0$, where

$$\alpha(n) = \sup_m \sup_{\{A \in \mathbf{F}_1^m,\ B \in \mathbf{F}_{n+m}^\infty\}} |P(A \cap B) - P(A)P(B)|.$$

The sequence $\{X_n\}_n$ is said to be ρ-*mixing* if $\rho(n) \to 0$, where the definition of ρ involves the maximal coefficient of correlation:

$$\rho(n) = \sup_m \sup_{\{f \in L_2(\mathbf{F}_1^m),\ g \in L_2(\mathbf{F}_{n+m}^\infty)\}} \frac{|E(f - Ef)(g - Eg)|}{\|f - Ef\|_2 \|g - Eg\|_2}.$$

This mixing coefficient is equivalent (cf. Bradley [2]) to the coefficient

$$\lambda(n) = \sup_m \sup_{\{A \in \mathbf{F}_1^m,\ B \in \mathbf{F}_{n+m}^\infty,\ P(A) \neq 0,\ P(B) \neq 0\}} \frac{|P(A \cap B) - P(A)\,P(B)|}{[P(A)\,P(B)]^{1/2}}.$$

Let $S_n = \sum_{k=1}^{n} X_k$. The problem in many limit theorems for stationary sequences is to find necessary and sufficient conditions for sequences $\{P(|S_n - ES_n| > \varepsilon n^\alpha)\}_n$ and $\{P(\max_{k \leqslant n} |S_k - ES_k| > \varepsilon n^\alpha)\}_n$ to converge to zero at specified rates (here $\alpha > 1/2$). Baum and Katz (1965) estimated the rate of convergence of the strong law, for partial sums S_n of i.i.d. random variables. Their well-known theorem is as follows.

Theorem 1. *Let $\{X_n\}_n$ be a sequence of i.i.d. random variables, $\alpha > 1/2$, $p\alpha > 1$ and $EX_1 = 0$ if $\alpha \leqslant 1$. Then the following statements are equivalent*

a) $E|X_1|^p < \infty$;

b) $\sum_n n^{p\alpha-2} P(\max_{k\leqslant n} |S_k| > \varepsilon n^\alpha) < \infty$;

c) $\sum_n n^{p\alpha-2} P(|S_n| > \varepsilon n^\alpha) < \infty$.

This result generalizes earlier theorems of Hsu and Robbins [5] and Erdős [3], the proof being based on the truncation method of P. Erdos, and exploits higher moments of partial sums of truncated random variables. Motivated by applications to sequential analysis of time series, Lai [7] extended this theorem from the i.i.d. case to some classes of stationary strong mixing sequences of random variables. In order to handle events which are not too far apart, he needed to reinforce the mixing condition by requiring that the bivariate probabilities of random variables which are sufficiently far apart decrease faster then the univariate tail probabilities. More precisely, he assumed that *there exist $\beta > 1$ and a positive integer m such that*

(C) $$\sup_{i\geqslant m} P(|X_1| > x, |X_i| > x) = O(P^\beta(|X_1| > x)).$$

as $x \to \infty$.

Under this condition and mixing rates of the type $\varphi(n) = O(n^{-g})$ and $\alpha(n) = O(n^{-f})$, where g and f are some functions of p, α and β, T.L. Lai proved that a) is equivalent to b). In order to establish this theorem, he used a combination of both above mentioned methods and a different approach based on a stopping time technique due to I.S. Chow and himself. In another theorem he used instead of Condition (C) a stronger moment condition, namely $E|X_1|^{p+\delta} < \infty$ for some $\delta > 0$, and got with the same mixing coefficients as in the former theorem that

$$E|X_1|^{p+\delta} < \infty \Rightarrow \sum_n n^{p\alpha-2} P(\sup_{n\leqslant k} |S_k| > \varepsilon n^\alpha) < \infty.$$

Later on Hipp [4] using the classical Erdös-Katz method obtained the same conclusion as Lai under Condition (C) and a φ-mixing rate satisfying $\Sigma[\varphi(n)]^{1/\theta} > \infty$ for some $\theta > 1$, where θ does not depend on β.

In Lai [7] there appears the following problem: *Is it possible to obtain an analogue of the Baum-Katz theorem for mixing sequences without using Condition (C)?* In [8] and [9] this question is answered positively, the equivalence a) $\Leftrightarrow$ b) being obtained for some classes of φ-and ρ-mixing sequences. The proofs of the theorems are based on a refinement of the method involving the estimation of the moments of S_n. In fact, for ρ-mixing sequences we have the following result [9].

Proposition 2. *Let $\{X_n\}_n$ be a stationary ρ-mixing sequence of random variables such that $E|X_1|^k < \infty$, where $k \geqslant 2$ is an integer and $\sum_i \rho^{2/k}(2^i) < \infty$.*

Then

$$\|S_n\|_k = O(n^{1/k} \|X_1\|_k + n^{1/(k-1)} \|X_1\|_{k-1} + \dots + n^{1/2} \|X_1\|_2 + n\,|EX_1|).$$

In [8] we prove

Theorem 3. *Let $\{X_n\}_n$ be a strictly stationary φ-mixing sequence of random variables, $\alpha p > 1$, $\alpha > 1/2$, $EX_1 = 0$ for $\alpha \leqslant 1$.*

If

$$\sum_i \rho^{1/k}(2^i) < \infty, \tag{1}$$

where

$$k = \begin{cases} 0 & \text{for } p < 1, \\ 2 & \text{for } 1 \leqslant p < 2, \\ \left[\dfrac{\alpha p - 1}{\alpha - 1/2}\right] + 1 & \text{for } p \geqslant 2, \end{cases}$$

then a) $\Leftrightarrow$ b) $\Leftrightarrow$ c') *with* c') *defined as*

c') $\sum_n n^{p\alpha - 2} \max_{k \leqslant n} P(|S_k| > \varepsilon n^\alpha) < \infty$.

The result in [9] is as follows.

Theorem 4. *Let $\{X_n\}_n$ be a strictly stationary ρ-mixing sequence of random variables, $p\alpha > 1$, $\alpha > 1/2$ and $EX_1 = 0$ for $\alpha \leqslant 1$.*

If

$$\sum_i \rho^{2/k}(2^i) < \infty, \tag{2}$$

where k is defined above then a) $\Rightarrow$ b).

If

$$\sum_i \rho^m(i) < \infty, \tag{3}$$

where

$$m = \begin{cases} \left[\dfrac{2(p\alpha - 1)}{2 - p\alpha}\right] & \text{for } 1 < p < 4/3, \\ 1 & \text{for } p > 4/3, \end{cases}$$

then b) $\Rightarrow$ a).

We note that while Theorem 3 and condition (2) of Theorem 4 assume a logarithmic rate, (3) uses a stronger rate. The problem is whether it is also possible in this case to weaken the rate to a logarithmic one.

REFERENCES

1. Baum, L.E., Katz, M., *Convergence rates in the law of large numbers.* Trans. Amer. Math. Soc. **120**, 108–123 (1965).
2. Bradley, C., *Equivalent measures of dependence.* J. Multivariate Anal., **13**, 167–176 (1983).
3. Erdös, P., *On a theorem of Hsu and Robbins.* Ann. Math. Statist., **20**, 286–291 (1949).
4. Hipp, C., *Convergence rates of the strong law for stationary mixing sequences.* Z. Wahrsch. Verw. Gebiete, **49**, 49–62 (1979).
5. Hsu, P.L., Robbins, H., *Complete convergence and the law of large numbers.* Proc. Nat. Acad. Sci. USA, **33**, 25–31 (1947).
6. Katz, M., *The probability in the tail of a distribution.* Ann. Math. Statist., **34**, 312–318 (1963).
7. Lai, T.L., *Convergence rates and r-quick versions of the strong law for stationary mixing sequences.* Ann. Probab., **5**, 693–706 (1977).
8. Peligrad, M., *Convergence rates of the strong law for stationary φ-mixing sequences.* INCREST Preprint Series in Math., No. 112. Bucureşti, 1981.
9. Peligrad, M., *Convergence rates of the strong law for stationary ρ-mixing sequences.* Preprint, Centre of Mathematical Statistics. Bucureşti, 1982.

D. PLACHKY

A characterization of product-measurability of Radon-Nikodym derivatives by separability

Let $(\Omega, \mathfrak{A})$ be a measurable space and denote the set of all probability measures P on $\mathfrak{A}$ by $\mathfrak{K}$. Furthermore, on $\mathfrak{K}$ the smallest σ-algebra $\mathfrak{A}^*$ is introduced such that all mappings $\mathfrak{K} \to \mathbf{R}$ defined by $P \to P(A)$, $A \in \mathfrak{A}$, are measurable.

If $\mathfrak{A}$ is separable, i.e. $\mathfrak{A}$ is generated by a countable set $\{G_1, G_2, \ldots\}$, the Radon-Nikodym derivative $\mathrm{d}P_Q/\mathrm{d}Q$, where P_Q denotes the Q-continuous part of P of the Lebesgue-decomposition of $P \in \mathfrak{K}$ with respect to $Q \in \mathfrak{K}$, can be constructed in the following way. Denoting by P_n resp. Q_n the restrictions of P resp. Q to the σ-algebra $\mathfrak{A}_n$ generated by $\{G_1, \ldots, G_n\}$, $n = 1, 2, \ldots$, a version of $\mathrm{d}P_{nQ_n}/\mathrm{d}Q_n$ is given by

$$\sum_{i=1}^{m_n} I_{A_n^{(i)}} I_{\{\mu \in \mathfrak{K} : \mu(A_n^{(i)}) > 0\}} (P) \frac{Q(A_n^{(i)})}{P(A_n^{(i)})}$$

where $A_n^{(i)}$, $i = 1, \ldots, m_n$ are the atoms of $\mathfrak{A}_n$. According to [8], Proposition III-1-5, the sequence $(\mathrm{d}P_{nQ_n}/\mathrm{d}Q_n)$ $n = 1, 2, \ldots$ is a positive supermartingale converging Q - a.e. to a version of $\mathrm{d}P_Q/\mathrm{d}Q$. Since the version of $\mathrm{d}P_{nQ_n}/\mathrm{d}Q_n$ given above is $\mathfrak{A} \otimes \mathfrak{A}^* \otimes \mathfrak{A}^*$-measurable one obtains the following

Lemma 1. *If $\mathfrak{A}$ is separable then there exists an $\mathfrak{A} \otimes \mathfrak{A}^* \otimes \mathfrak{A}^*$-measurable version of $\mathrm{d}P_Q/\mathrm{d}Q$, $P, Q \in \mathfrak{K}$.*

Remark. From Lemma 1 one obtains easily the $(\mathfrak{A}^* \otimes \mathfrak{A}^*, \mathfrak{A}_*)$-measurability of the mapping $(P, Q) \to P_Q$ and hence the $(\mathfrak{A}^*, \mathfrak{A}_*)$-measurability of the Lebesgue-decomposition $P \to P_Q$, where $\mathfrak{A}_*$ denotes the smallest σ-algebra of subsets of the set of all finite measures $\mathfrak{M}$ on $\mathfrak{A}$ such that all mappings $\mu \to \mu(A)$, $A \in \mathfrak{A}$ $(\mu \in \mathfrak{M})$ are measurable (see also [2]).

Denoting the property of a sub-σ-algebra $\mathfrak{A}_0$ of $\mathfrak{A}$ such that for every $A \in \mathfrak{A}$ there exists an $A_0 \in \mathfrak{A}_0$ with $P(A \Delta A_0) = 0$ by $\mathfrak{A}_0 = \mathfrak{A}[P]$ it will be proved that the existence of a separable sub-σ-algebra $\mathfrak{A}_0$ of $\mathfrak{A}$ with $\mathfrak{A}_0 = \mathfrak{A}[P]$ for every $P \in \mathfrak{K}$ implies the $\mathfrak{A} \otimes \mathfrak{A}^* \otimes \mathfrak{A}^*$-measurability of a version of $\mathrm{d}P_Q/\mathrm{d}Q$.

In order to show that the condition for the $\mathfrak{A} \otimes \mathfrak{A}^* \otimes \mathfrak{A}^*$-measurability of a version of $\mathrm{d}P_Q/\mathrm{d}Q$ described in the second part of the remark above is necessary and sufficient, the following characterization of the property $\mathfrak{A}_0 = \mathfrak{A}[P]$ of a sub-σ-algebra $\mathfrak{A}_0$ of $\mathfrak{A}$ is given.

Lemma 2. *$\mathfrak{A}_0 = \mathfrak{A}[P]$ holds if and only if P is an extremal point in the set of all $Q \in \mathfrak{K}$ with $Q \mid \mathfrak{A}_0 = P \mid \mathfrak{A}_0$.*

Proof. If $\mathfrak{A}_0 = \mathfrak{A}[P]$ holds, $P = \frac{1}{2} P_1 + \frac{1}{2} P_2$ with $P_i \in \mathfrak{K}$, $P_i \mid \mathfrak{A}_0 = P|\mathfrak{A}_0$, $i = 1, 2$, implies $P = P_1 = P_2$ and, therefore, P is an extremal point. In order to show that the property of P to be an extremal point implies $\mathfrak{A}_0 = \mathfrak{A}[P]$ one observes that for any fixed $\mathfrak{A}$-measurable real valued function f with $0 < m \leqslant f \leqslant M$ ($m, M \in \mathbf{R}$) and for any $Q \in \mathfrak{K}$

$$Q_f(A) = \int_A \frac{f}{E_Q(f \mid \mathfrak{A}_0)} \, dQ, \; A \in \mathfrak{A},$$

defines an element Q_f of $\mathfrak{K}$ with $Q_f| \mathfrak{A}_0 = Q \mid \mathfrak{A}_0$. Choosing $f = 1 + I_{A_1}$ for $A_1 \in \mathfrak{A}$ fixed yields $Q_f \leqslant 2Q$. For $Q = P$ one obtains $P = \frac{1}{2} P_f + \frac{1}{2} (2P - P_f)$ which implies $P = P_f$ since P is an extremal point. Therefore, $I_{A_1} = E_P(I_{A_1} \mid \mathfrak{A}_0)$ holds P-a.e. which yields $P(A_1 \Delta A_0) = 0$ with $A_0 = \{E_P(I_{A_1} \mid \mathfrak{A}_0) = 1\} \in \mathfrak{A}_0$.

Remark. In Lemma 2 the condition that P is an extremal point in the set of all $Q \in \mathfrak{K}$ with $Q \mid \mathfrak{A}_0 = P \mid \mathfrak{A}_0$ can be replaced by the condition that P is an extremal point in the set of all $Q \in \mathfrak{K}$ with $Q | \mathfrak{A}_0 = P | \mathfrak{A}_0$ and $Q \ll P_0$, where P_0 is a fixed element of $\mathfrak{K}$ with $P \ll P_0$. Furthermore, the condition $\mathfrak{A}_0 = \mathfrak{A}[P]$ for every $P \in \mathfrak{K}$ does not imply in general that $\mathfrak{A}$ is contained in $\bigcap_{P \in \mathfrak{K}} \mathfrak{A}_0(P)$ where $\mathfrak{A}_0$ (P) denotes the completion of $\mathfrak{A}_0$ with respect to $P \mid \mathfrak{A}_0$. Consider for example the σ-algebra $\mathfrak{A}$ of all subsets of a set Ω with cardinality $\aleph_1$ and introduce $\mathfrak{A}_0$ as the σ-algebra of subsets of Ω generated by $\{\{\omega\} \mid \omega \in \Omega \setminus \{\omega_0\}\}$ where $\omega_0 \in \Omega$ is fixed. Then $P_0 = \delta_{\omega_0}$ is uniquely determined by its restriction to $\mathfrak{A}_0$ since according to a result of Ulam only discrete probability measures exist on $\mathfrak{A}$ and $\mathfrak{A}_0 = \{A \subset \Omega\,;\, A$ or A^c is a countable subset of $\Omega \setminus \{\omega_0\}\}$. The last equation implies $\mathfrak{A}_0(P_0) = \mathfrak{A}_0[P_0]$.

Now all preparations are made in order to prove the main result.

Theorem. *There exists an $\mathfrak{A} \otimes \mathfrak{A}^* \otimes \mathfrak{A}^*$-measurable version of* dP_Q/dQ $P, Q \in \mathfrak{K}$, *if and only if there exists a separable sub-σ-algebra $\mathfrak{A}_0$ of $\mathfrak{A}$ with $\mathfrak{A}_0 = \mathfrak{A}[P]$ for all $P \in \mathfrak{K}$.*

Proof. In order to prove sufficiency one observes that for any $\mu \in \mathfrak{M}$ and $Q \in \mathfrak{K}$ with $\mu \ll Q$

$$\mu^{(Q)}(A) = \int_A \frac{d(\mu \mid \mathfrak{A}_0)}{d(Q \mid \mathfrak{A}_0)} \, dQ, \; A \in \mathfrak{A},$$

defines an element $\mu^{(Q)}$ of $\mathfrak{M}$ with $\mu^{(Q)} \mid \mathfrak{A}_0 = \mu \mid \mathfrak{A}_0$. Choosing $\mu = P$, where P satisfies $\mathfrak{A}_0 = \mathfrak{A}[P]$, yields

$$\frac{dP}{dQ} = \frac{d(P \mid \mathfrak{A}_0)}{d(Q \mid \mathfrak{A}_0)}$$

Q-a.e. for any $Q \in \mathfrak{K}$ with $P \ll Q$. Furthermore the condition that $\mathfrak{A}_0 = \mathfrak{A}[P]$ holds for any $P \in \mathfrak{K}$ implies that all $P \in \mathfrak{K}$ are uniquely determined by $P \mid \mathfrak{A}_0$ from which one obtains $P_Q \mid \mathfrak{A}_0 = (P \mid \mathfrak{A}_0)_{Q|\mathfrak{A}}$ for any $P \in \mathfrak{K}$ and $Q \in \mathfrak{K}$ since P_Q is the greatest lower bound (with respect to the natural order of $\mathfrak{M}$) of all $\nu \in \mathfrak{M}$ with $\nu \ll P$ and $\nu \ll Q$. Therefore there exists an $\mathfrak{A} \otimes \mathfrak{A}^* \otimes \mathfrak{A}^*$-measurable version of dP_Q/dQ namely

$$\frac{d(P \mid \mathfrak{A}_0)_{Q \mid \mathfrak{A}_0}}{d(Q \mid \mathfrak{A}_0)}$$

according to Lemma 1. In order to show that the existence of an $\mathfrak{A} \otimes \mathfrak{A}^* \otimes \mathfrak{A}^*$-measurable version of dP_Q/dQ implies the existence of a separable sub-σ-algebra $\mathfrak{A}_0$ of $\mathfrak{A}$ with $\mathfrak{A}_0 = \mathfrak{A}[P]$ for all $P \in \mathfrak{K}$ one observes that a set belonging to a σ-algebra generated by a system $\mathfrak{G}$ of subsets is an element of a sub-σ-algebra generated by a countable subset of $\mathfrak{G}$. This yields the existence of a separable sub-σ-algebra $\mathfrak{A}_0$ of $\mathfrak{A}$ such that a version of dP_Q/dQ is $\mathfrak{A}_0 \otimes \mathfrak{A}^* \otimes \mathfrak{A}^*$-measurable (see also [10]). This implies that $\mathfrak{A}_0$ is pairwise sufficient for $\mathfrak{A}$ according to the factorization criterion for sufficiency (see [5]). Therefore, from $P_1 \mid \mathfrak{A}_0 = P_2 \mid \mathfrak{A}_0$ for $P_i \in \mathfrak{K}$, $i = 1, 2$, the equality $P_1 = P_2$ follows. This means that every $P \in \mathfrak{K}$ is uniquely determined by its restriction $P \mid \mathfrak{A}_0$, i.e. every $P \in \mathfrak{K}$ is an extremal point in the set of all $Q \in \mathfrak{K}$ with $Q \mid \mathfrak{A}_0 = P \mid \mathfrak{A}_0$. Lemma 2 yields the assertion $\mathfrak{A}_0 = \mathfrak{A}[P]$ for every $P \in \mathfrak{K}$.

Remark. The condition $\mathfrak{A}_0 = \mathfrak{A}[P]$ for every $P \in \mathfrak{K}$ with $\mathfrak{A}_0$ as a separable sub-σ-algebra does not in general imply the separability of $\mathfrak{A}$. Choose for example for $\mathfrak{A}$ the set of all universally measurable subsets of $\mathbf{R}$ and for $\mathfrak{A}_0$ the σ-algebra of Borel subsets of $\mathbf{R}$. In [11] it has been proved that in this case $\mathfrak{A}$ is not separable. Here is a proof for this fact according to J.P.R. Christensen: Assuming that $\mathfrak{A}$ is generated by $\{G_1, G_2, \ldots\}$ the mapping of E. Marczewski $f\colon \mathbf{R} \to \mathbf{R}$ defined by $f = \sum_{n=1}^{\infty} 3^{-n} I_{G_n}$ is a Borel isomorphism between $(\mathbf{R}, \mathfrak{A})$ and $(f(\mathbf{R}), \mathfrak{A}_0 \cap f(\mathbf{R}))$. The theorem of Luzin implies the existence of a compact subset C of $f(\mathbf{R})$ which is not countable and $f \mid C\colon C \to \mathbf{R}$ is continuous. This yields $C \cap \mathfrak{A} = C \cap \mathfrak{A}_0$. According to [11], p. 43, the set C contains a Cantor-like subset which implies card $(C \cap \mathfrak{A}) = 2^{\mathfrak{c}}$ whereas card $(C \cap \mathfrak{A}_0) = \mathfrak{C}$ holds. Finally it should be remarked that it seems to be an open problem (in ZFC) whether card $(C \cap \mathfrak{A}) = 2^{\mathfrak{c}}$ holds (see [3]). However, the argument used above works (in ZFC) since $C \cap \mathfrak{A}$ is equal to the universally measurable subsets of C and, as well known, different from the Borel subsets of C.

The proof of the theorem above together with the remark following Lemma 2 yield the following generalization: Let $\mathfrak{K}_1$ be a non-void subset of $\mathfrak{K}$ and $\mathfrak{K}_0 = \bigcup_{Q \in \mathfrak{K}_1} \mathfrak{K}_Q$ where $\mathfrak{K}_Q$ denotes the set of all $P \in \mathfrak{K}$ with $P \ll Q$ ($Q \in \mathfrak{K}$ fixed), then the existence of an $\mathfrak{A} \otimes (\mathfrak{A}^* \cap \mathfrak{K}_0) \otimes (\mathfrak{A}^* \cap \mathfrak{K}_1)$-measurable version of dP_Q/dQ, $P \in \mathfrak{K}_0$, $Q \in \mathfrak{K}_1$, is equivalent to the existence of a separable sub-σ-algebra $\mathfrak{A}_0$ of $\mathfrak{A}$ with $\mathfrak{A}_0 = \mathfrak{A}[P]$ for every $P \in \mathfrak{K}_0$.

Finally it should be pointed out that hints concerning the importance of separable σ-algebras for the product-measurability of Radon-Nikodym derivatives resp. for integral representations can be found in [9], [10], [12], [13] resp. in [6], [7]. Investigations concerning probability measures on $\mathfrak{A}^*$ can be found in [4].

Acknowledgements. I would like to thank J.P.R. Christensen, Z. Lipecki, D. Mussmann and C. Portenier for valuable remarks.

REFERENCES

1. Christensen, J. P. R., *Topology and Borel Structure*. Amsterdam, North-Holland, 1974.
2. Dubins, L., Freedman, D., *Measurable sets of measures*. Pacific J. Math., **14**, 1211 – 1222 (1964).
3. Grzegorek, E., Ryll-Nardzewski, C., *A remark on absolute measurable sets*. Bull. Acad. Polon. Sci. Sér. Sci. Math., **28**, 229–239 (1980).
4. Jiřina, M., *Convergence in distribution of random measures*. Ann. Math. Statist., **43**, 1727–1731 (1972).
5. Lehmann, E.L., *Testing Statistical Hypotheses*. New York, Wiley, 1959.
6. Meyer, P.A., *Représentation intégrale des fonctions excessives, résultats de Mokobodzki*. In: *Séminaire de Probabilités V*. Lecture Notes in Math., Vol. 191, pp. 196–208. Berlin–Heidelberg–New York, Springer, 1971.
7. Mokobodzki, G., *Noyaux absolument mesurables ou basiques et opérateurs nucléaires*. In: *Séminaire Goulaouic–Schwartz*, 1971–1972, Exp. No. 6, 21 pp. Palaiseau, École Polytech., 1972.
8. Neveu, J., *Discrete Parameter Martingales*. Amsterdam, North-Holland, 1975.
9. Nölle, G., *Produktmessbarkeit von Funktionen*. Z. Wahrsch. Verw. Gebiete, **17**, 163–167 (1971).
10. Pfanzagl, J., *On the existence of product measurable densities*. Sankhyā Ser. A, **31**, 13–18 (1969).
11. Rao, B.V., *Remarks on analytic sets*. Fund. Math., **66**, 237–239 (1970).
12. Roy, K.K., Ramamoorthi, R.V., *Relationship between Bayes, classical and decision theoretic sufficiency*. Sankhyā Ser. A, **41**, 48–58 (1979).
13. Seheult, A.H.S., Quesenberry, C.P., *On unbiased estimation of density functions*. Ann. Math. Statist., **42**, 1434–1438 (1971).

ILEANA POPESCU, GH. POPESCU

Biased estimates for the solution of an integral equation *

Rezolvarea prin metoda Monte Carlo a ecuației (2) este bine cunoscută ([1], [8]). Ecuația (1) este o generalizare a ecuației (2). Scopul acestui articol este de a găsi estimatori deplasați pentru produsul scalar $<\varphi, z>$ unde φ este o funcție dată iar z este soluția ecuației (1). Este examinată de asemenea metoda selecției după importanță în vederea aplicării eficiente a metodei Monte Carlo. Cazul estimatorilor nedeplasați va fi prezentat într-un alt articol.

1. Introduction

We consider the integral equation

$$z(x) = \int_{G_2} K(x, y)\, z(u(x, y))\mathrm{d}y + f(x), \tag{1}$$

where G_i is a Borel subset of $\mathbb{R}^{s_i}$, $i = 1, 2$; $K(\cdot, \cdot)$ is a Borel function from $G_1 \times G_2$ to $\mathbb{R}$; $u(\cdot,\cdot)$ is a Borel function from $G_1 \times G_2$ to G_1, $f(\cdot)$ is a Borel function from G_1 to $\mathbb{R}$. These are given elements and z: $G_1 \to \mathbb{R}$ is an unkown function.

In the particular case when $G_1 = G_2$ and $u(x, y) = y$ equation (1) becomes

$$z(x) = \int_{G_1} K(x, y)\, z(y)\, \mathrm{d}y + f(x) \tag{2}$$

The Monte-Carlo method of solving equation (2) is based on the construction of certain functionals of the trajectories of a Markov process which are biased or unbiased estimates for the scalar product $<\varphi, z>$, where φ is a given function. The main results concerning equation (2) are given in [1], [8].

The aim of this paper is to prove similar results for equation (1) which should allow the Monte Carlo approach to this equation.

Examples leading to equation (1) are given in [7].

2. Preliminaries

Let $B(G_1)$ be the normed space of all bounded Borel functions $f : G_1 \to \mathbb{R}$ with the norm given by $\|f\| = \sup_{x \in G_1} |f(x)|$

Suppose that

$$\text{A)} \qquad \sup_{x \in G_1} \int_{G_2} |K(x, y)| \, dy = M < \infty.$$

Then the equation

$$(3) \qquad (Kf)(x) = \int_{G_2} K(x, y) f(u(x, y)) \, dy$$

defines a bounded linear operator K from $B(G_1)$ to $B(G_1)$ for which $\| K \| \leqslant M$.

Now equation (1) may be written in the form

$$(4) \qquad z = Kz + f$$

It is well known that if $\| K \| \leqslant 1$ and $\| K^l \| < 1$ for some $l \in N^*$, then for each $f \in B(G_1)$ equation (4) has a unique solution in $B(G_1)$. This solution can be found by the method of successive approximations.

Let us now consider the normed space $L_2(G_1)$ of all Borel, square integrable functions $f : G_1 \to \mathbb{R}$ with the norm

$$\|f\|_2 = \left[\int_{G_1} f^2(x) dx\right]^{1/2}$$

Suppose that

$$\text{B}_1) \qquad \int_{G_1 \times G_2} K^2(x, y) \, dx \, dy = Q^2 < \infty$$

$$\text{B}_2) \qquad \lambda_2 \circ u^{-1}(x, \cdot) \ll \lambda_1, \quad \frac{d\lambda_2 \circ u^{-1}(x, \cdot)}{d\lambda_1} \leqslant c^2$$

for each $x \in G_1$ (here λ_i is the Lebesgue measure on G_i $i = 1,2$).

One can prove (see [6]) that equation (3) defines a bounded linear operator K from $L_2(G_1)$ to $L_2(G_1)$ for which $\| K \|_2 \leqslant cQ$.

Similarly, if $\| K \|_2 \leqslant 1$ and $\| K^l \|_2 < 1$ for some $l \in N^*$, then for each $f \in L_2(G_1)$ equation (4) has a unique solution in $L_2(G_1)$, which can be found by the method of successive approximations.

Obviously, conditions **A**) and $\| K \| \leqslant 1$, $\| K^l \| < 1$ for some $l \in N^*$ (or $\mathbf{B}_1$, $\mathbf{B}_2$ and $\| K \|_2 \leqslant 1$, $\| K^l \|_2 < 1$ for some $l \in N^*$) are sufficient for the existence and uniqueness of the solution of equation (4) in $B(G_1)$ (or $L_2(G_1)$).

In what follows we assume that equation (1) has a solution which may be expressed as a series $z = \sum_{n=0}^{\infty} K^n f$ which is supposed to be convergent with respect to $\|\cdot\|$ and $\|\cdot\|_2$ respectively.

It is known that this condition is necessary and sufficient for convergence of the sequence given by the method of successive approximations to the solution of equation (1). This sequence is defined by $z^{(0)} = \psi$ and $z^{(n)} = Kz^{(n-1)} = f + Kf + \ldots + K^{n-1}f + K\psi$ where $\psi \in B(G_1)$ or $\psi \in L_2(G_1)$ respectively.

The stochastic processes used to construct estimates for $\langle \varphi, z \rangle$ in the case of equation (1) are random systems with complete connections (R.S.C.C.) (see [3]) and absorbing R.S.C.C. (see [5]). The computer-simulation of these processes is considered in [5]. For a better understanding of the underlying facts we recall here the definition of these processes as required in the sequel.

We consider the following elements

i) $u_0(\cdot, \cdot)$ — a Borel function from $G_1 \times G_2$ to G_1;

ii) $q(\cdot)$ — a probability density function on G_1;

iii) $p(\cdot, \cdot)$ — a transition density function from G_1 to G_2 i.e. for each $x \in G_1$, $p(x, \cdot)$ is a probability density function on G_2 and for each $y \in G_2$, $p(\cdot, y)$ is a real Borel function.

$\{G_1, G_2, q, p, u_0\}$ are the elements defining a R.S.C.C. With these elements one can associate in a standard way (see [3]) a probability space (Ω, K, P) and two sequences of random vectors $(X_n)_{n \in N}$ and $(Y_n)_{n \in N^*}$ taking values in G_1 and G_2 respectively, such that:

a) for any Borel set $A \subset G_1$, $\mathsf{P}(X_0 \in A) = \int_A q(x)\mathrm{d}x$

b) for any Borel set $B \subset G_2$,

$$\mathsf{P}(Y_{n+1} \in B \mid X_0 = x_0, Y_1 = y_1, \ldots Y_n = y_n) = \int_B p(x_n, y)\mathrm{d}y,$$

where $x_n = u_0(x_{n-1}, y_n) = u_0(x_0, y^{(n)}),\ y^{(n)} = (y_1, \ldots, y_n)$;

c) $X_n = u_0(X_{n-1}, Y_n) = u_0(X_0, Y^{(n)}),\ Y^{(n)} = (Y_1, \ldots, Y_n)$

In the following we find biased estimates for $\langle \varphi, z \rangle = \int_{G_1} \varphi(x)\, z(x)\, \mathrm{d}x$, where $\varphi \in L_1(G_1)$ if $z \in B(G_1)$ and $\varphi \in L_2(G_1)$ if $z \in L_2(G_1)$. The method of importance sampling (see [4]) is also considered. The construction for equation (1) of a class of unbiased estimates generalizing the results obtained by Hisamutdinov for equation (2) (see [2]) will be given elsewhere.

3. Biased estimates for $\langle \varphi, z \rangle$

The construction of biased estimates for $\langle \varphi, z \rangle$ is based on the following lemma, whose proof is very simple and therefore will be omitted.

Lemma 1. a) *If the series* $\sum_{n=0}^{\infty} K^n f$ *converges in* $B(G_1)$ *then for any* $\varphi \in L_1(G_1)$

$$\left\langle \varphi, \sum_{n=0}^{m} K^n f \right\rangle \to 0 \quad \textit{as } m \to \infty.$$

b) *If the series* $\sum_{n=0}^{\infty} K^n f$ *converges in* $L_2(G_1)$ *then for any* $\varphi \in L_2(G_1)$

$$\left\langle \varphi, \sum_{n=0}^{m} K^n f \right\rangle \to 0 \quad \textit{as } m \to \infty.$$

Hence for sufficiently large m, $\left\langle \varphi, \sum_{n \in 0}^{m} K^n f \right\rangle \cong \langle \varphi, z\rangle$ and any unbiased estimate for $\left\langle \varphi, \sum_{n \in 0}^{m} K^n f \right\rangle$ is a biased estimate for $\langle \varphi, z\rangle$.

Suppose that the R.S.C.C. $\{G_1, G_2, q, p, u_0\}$ satisfies the following admissibility conditions:

j) $q(x) > 0$ for any $x \in G_1$, such that $\varphi(x) \neq 0$;

jj) $p(x, y) > 0$ for any $(x, y) \in G_1 \times G_2$ such that $K(x, y) \neq 0$.

It was shown in [6] that under these conditions the function

$$I_{n,\varphi}(X_0, Y_1, \dots, X_{n-1}, Y_n) = \begin{cases} \dfrac{\varphi(X_0)}{q(X_0)} W_n(X_0, Y_1, \dots, X_{n-1}, Y_n) f(u(X_0, Y^{(n)})) & \text{if } n \geqslant 1 \\ \dfrac{\varphi(X_0)}{q(x_0)} f(X_0) & \text{if } n=0 \end{cases}$$

is an unbiased estimate for $\langle \varphi, K^n f\rangle$, where the weight W_n is defined by

$$W_n(X_0, Y_1, \dots, X_{n-1}, Y_n) = \frac{K(X_0, Y_1) \cdot K(u(X_0, Y_1), Y_2) \cdot \dots \cdot K(u(X_0, Y^{(n-1)}), Y_n)}{p(X_0, Y_1)\, p(X_1, Y_2) \dots p(X_{n-1}, Y_n)}$$

Analogously,

$$J_{m,\varphi}(X_0, Y_1, \dots X_{m-1}, Y_m) = \sum_{n=0}^{m-1} I_{n,\varphi}(X_0, Y_1, \dots X_{n-1}, Y_n) +$$

$$+ \frac{\varphi(X_0)}{q(X_0)} W_m(X_0, Y_1, \dots, X_{m-1}, Y_m)\, \psi(u(X_0, Y^{(m)}))$$

is an unbiased estimate for $\langle \varphi, z^{(m)}\rangle$, where $z^{(m)} = f + Kf + \dots K^{m-1}f + K^m\psi$ is the mth successive approximation of z, the initial approximation being $z^{(0)} = \psi$.

Therefore $J_{m,\varphi}(X_0, Y_1, \dots, X_{m-1}, Y_m)$ is a biased estimate for $\langle \varphi, z\rangle$. Of course the larger m, the better the approximation of $< \varphi, z >$.

The only conditions to be verified by the R.S.C.C. $\{G_1, G_2, q, p, u_0\}$ in order to get an unbiased estimate for $\langle \varphi, z\rangle$ are the admissibility conditions; otherwise q, p, u_0 are at our choice. Nevertheless the variance of this estimate and consequently the accuracy of the method depends upon the choice of q, p, u_0.

In order to obtain a smaller variance it is necessary to choose, whenever possible, the density function q "proportional" to the absolute value of the integrand. This method introduced by Kahn [4] is called the importance sampling. Accordingly, on those parts of the domain where the absolute value of the integrand is large, whose contribution to the value of the integral is determinant, q must be also large in order to provide more sampling values belonging to those parts.

In the theorem below an R.S.C.C. is defined, for which the variance of $J_{m,\varphi}(X_0, Y_1, \dots, X_{m-1}, Y_m)$ is minimum.

Theorem 1. *Suppose that $K(x, y) \geqslant 0$ and $z(x) \geqslant 0$. If the initial approximation $z^{(0)}$ is the solution z of equation (1), then for the R.S.C.C. $\{G_1, G_2, q, p, u_0\}$ defined by*

$$u_0(x, y) = u(x, y) \quad (\forall)(x, y) \in G_1 \times G_2$$

$$q(x) = \frac{|\varphi(x)|\, z(x)}{\langle |\varphi|, z\rangle} \quad (\forall) x \in G_1$$

$$p(x, y) = K(x, y) \frac{z(u(x, y))}{(Kz)(x)} \quad (\forall)(x, y) \in G_1 \times G_2$$

we have

$$D^2 J_{m,\varphi}(X_0, Y_1, \ldots, X_{m-1}, Y_m) = \langle |\varphi|, z^{(m)}\rangle^2 - \langle \varphi, z^{(m)}\rangle^2 = D^2_{\min}.$$

Proof. Taking into account the assumptions of the theorem we have

$$J_{0,\varphi}(X_0) = \frac{\varphi(X_0)}{q(X_0)} z^{(0)}(X_0) = \frac{\varphi(X_0) z(X_0) \langle |\varphi|, z\rangle}{|\varphi(X_0)|\, z(X_0)} =$$

$$= \operatorname{sgn} \varphi(X_0) \langle |\varphi|, z\rangle$$

Hence

$$\mathsf{E}(J^2_{0,\varphi}(X_0)) = \langle |\varphi|, z\rangle$$

Suppose that $J_{m,\varphi}(X_0, Y_1, \ldots X_{m-1}, Y_m) = J_{0,\varphi}(X_0)$. Then the proof is concluded as follows

$$D^2(J_{m,\varphi}(X_0, Y_1, \ldots, Y_{m-1}, X_m)) = \mathsf{E}(J^2_{m,\varphi}(X_0, Y_1, \ldots X_{m-1}, Y_m)) -$$

$$- [\mathsf{E}(J_{m,\varphi}(X_0, Y_1, \ldots, X_{m-1}, Y_m))]^2 = \mathsf{E}(J^2_{0,\varphi}(X_0)) - [\mathsf{E}(J_{0,\varphi}(X_0))]^2 =$$

$$= \langle |\varphi|, z\rangle^2 - \langle \varphi, z\rangle^2 = \langle |\varphi|, z^{(m)}\rangle^2 - \langle \varphi, z^{(m)}\rangle^2,$$

since $z^{(m)} = z^{(0)} = z$ and $J_{0,\varphi}(X_0)$ is an unbiased estimate of $\langle \varphi, z^{(0)}\rangle$.

Let us now prove by induction that $J_{m,\varphi}(X_0, Y_1, \ldots, X_{m-1}, Y_m) = J_{0,\varphi}(X_0)$. We have

$$J_{1,\varphi}(X_0, Y_1) = \frac{\varphi(X_0)}{q(X_0)} \left[f(X_0) + \frac{K(X_0, Y_1)\, z^{(0)}(u(X_0, Y_1))}{p(X_0, Y_1)} \right] =$$

$$= \frac{\varphi(X_0) \langle |\varphi|, z\rangle}{|\varphi(X_0)|\, z(X_0)} \left[f(X_0) + \frac{K(X_0, Y_1)\, z(u(X_0, Y_1))\, (Kz)(X_0)}{K(X_0, Y_1)\, z(u(X_0, Y_1))} \right] =$$

$$= \frac{\operatorname{sgn} \varphi(X_0) \langle |\varphi|, z\rangle}{z(X_0)} [f(X_0) + (Kz)(X_0)] =$$

$$= \frac{\operatorname{sgn} \varphi(X_0) \langle |\varphi|, z\rangle}{z(X_0)} z(X_0) = \operatorname{sgn} \varphi(X_0) \langle |\varphi|, z\rangle.$$

For $m \geqslant 2$ we have

$$J_{m-1}, \varphi(X_0, Y_1, \ldots, X_{m-2}, Y_{m-1}) - J_{m,\varphi}(X_0, Y_1, \ldots, X_{m-1}, Y_m) =$$

$$= \frac{\varphi(X_0)}{q(X_0)} [W_{m-1}(X_0, Y_1, \ldots, X_{m-2}, Y_{m-1})\, z(u(X_0, Y^{(m-1)}) -$$

$$- W_{m-1}(X_0, Y_1, \ldots, X_{m-2}, Y_{m-1}) f(u(X_0, Y^{(m-1)})) -$$

$$- W_m(X_0, Y_1, \ldots, X_{m-1}, Y_m)\, z\,(u(X_0, Y^{(m)}))] =$$

$$= \frac{\varphi(X_0)}{q(X_0)} W_{m-1}(X_0, Y_1, \ldots, X_{m-2}, Y_{m-1}) \Big[z(u(X_0, Y^{(m-1)})) -$$

$$- f(u(X_0, Y^{(m-1)})) - \frac{K(u(X_0, Y^{(m-1)}), Y_m)}{p(X_{m-1}, Y_m)} z\,(u\,(X_0, Y^{(m)})) \Big] =$$

$$= \frac{\operatorname{sgn} \varphi(X_0) \langle |\varphi|, z \rangle}{z(X_0)} W_{m-1}(X_0, Y_1, \ldots, X_{m-2}, Y_{m-1}) \Big[z(X_{m-1}) -$$

$$- f(X_{m-1}) - \frac{K(X_{m-1}, Y_m)\,(Kz)\,(X_{m-1})}{K(X_{m-1}, Y_m)\, z(u(X_{m-1}, Y_m))} z(X_m) \Big] =$$

$$= \frac{\operatorname{sgn} \varphi(X_0) \langle |\varphi|, z \rangle}{z(X_0)} W_{m-1}\,(X_0, Y_1, \ldots, X_{m-2}, Y_{m-1})\, [z(X_{m-1}) -$$

$$- f(X_{m-1}) - (Kz)(X_{m-1})] = 0,$$

taking into account the fact that $u(\cdot, \cdot) = u_0(\cdot, \cdot)$, hence $u(X_0, Y^{(m-1)}) = X_{m-1}$ and $u(X_{m-1}, Y_m) = X_m$.

Remark. A direct application of this theorem is practically impossible, the above R.S.C.C. being constructed by means of the unknown solution z and the integral $\langle |\varphi|, z \rangle$. As a matter of fact, the computation of $\langle |\varphi|, z \rangle$ is in general as difficult as that of $\langle \varphi, z \rangle$.

However, under certain stability conditions, if we know a sufficiently good approximation $z^{(0)}$ of z and consider the R.S.C.C. for which

$$u_0(\cdot,\cdot) = u(\cdot,\cdot),\ q(x) = \frac{|\varphi(x)|\, z^{(0)}(x)}{\langle |\varphi|, z^{(0)} \rangle},\ p(x, y) = \frac{K(x, y)\, z^{(0)}(u(x, y))}{(Kz^{(0)})(x)}$$

then the variance of $J_{m,\varphi}(X_0, Y_1, \ldots, X_{m-1}, Y_m)$ will be close to $D^2_{\min}$.

We mention that the estimates $J_{m,\varphi}$ have the advantage that in their computation only finite trajectories are involved but they have the disadvantage of being biased.

REFERENCES

1. Ermakov, S.M., *Metod Monte Carlo i smežnye voprosy*. Moskva, Nauka, 1971.
2. Hisamutdinov, A.I., *Ediničnyi klass očenok dlja vyčislenya po metodu Monte Carlo funkcionalov ot rešenya integralnogo uravnenjy vtorogo roda.* Ž. Vyčisl. Mat. i Mat. Fiz., **10**, 1269–1280 (1970).
3. Iosifescu, M., Theodorescu, R., *Random Processes and Learning*, Berlin, Springer, 1969.
4. Kahn, H., *Use of different Monte Carlo sampling techniques.* In: Meyer, H.A. (Ed.), *Symposium on Monte Carlo Methods*, pp. 145–191. New York, Wiley, 1956.
5. Popescu, Ileana, *Generarea cu calculatorul a sistemelor aleatoare cu legături complete.* Stud. Cerc. Mat., **30**, 69–76 (1978).
6. Popescu, Ileana, *Calcul des intégrales multiples par la méthode de Monte Carlo.* Rev. Roumaine Math. Pures Appl., **26**, 619–629 (1981).
7. Popescu, Ileana, Popescu, Gh., *Modele matematice care conduc la o ecuaţie integrală.* Stud. Cerc. Mat., **34**, 468–485 (1982).
8. Sobol', I.M., *Čislennye metody Monte Carlo.* Moskva, Nauka, 1973.

V. RADU

On a family of metrics for the distribution functions

0. Using the method of the so called T-conjugate transforms, in [6] is introduced a (deterministic) metric which generates the (ε, λ)-topology in Menger spaces under an Archimedean t-norm. This metric is given in terms of a multiplicative generator of the t-norm T.

Considering a modified form of this metric, in terms of an additive generator of the t-norm, we shall give a direct simpler proof of the above result.

Thereafter we consider the particular case of the t-norm T_m and we show how this case can always be used in order to obtain a metric for the (ε, λ)-topology (also in the Archimedean case). Our formula can be considered as a direct generalization of the Fréchet - Ky Fan metric for convergence in probability.

Using the results of [1] we then obtain a family of metrics on the set of distribution functions of nonnegative random variables. In the case of the t-norm T_m these metrics are seen to be uniformly equivalent to the metric of P. Lévy.

The terminology and the notation not given here are as in [9,10].

1. Let $(S, \mathcal{F}, T)$ be a Menger space under an Archimedean t-norm. It is known that the (ε, λ)-topology is induced by a metrizable uniformity on S generated by the random metric $\mathcal{F}$ [11]. In [6] the following is proved.

Theorem A. *For any multiplicative generator h of T and for any positive real number z, the mapping d_z defined on $S \times S$ by*

$$d_z(p, q) = -\ln \sup_{x>0} e^{-zx}\, hF_{pq}(x) \tag{1}$$

is a metric on S, which generates the (ε, λ)-topology. Moreover the metrics d_z are uniformly equivalent.

This result is proved by using the properties of the so called T-conjugate transform.

Let us observe that if h is a multiplicative generator of T then the function $f : [0, 1] \to [0, \infty)$, $f = -\ln h$ is an additive generator for T. Moreover f is continuous, decreasing, $f(1) = 0$ and any other additive generator of T is of the form $g = af$, where a is a positive number.

Using this remark, we restate the above Theorem A in terms of f and so we can give a direct simpler proof. Namely we have

Theorem B. *Let f be an additive generator of T and define the mapping*

$$\rho_f(p, q) = \inf_{t>0} \{t + f \circ F_{pq}(t)\}, \; p, q \in S. \tag{2}$$

Then

(i) ρ_f *is a metric on* S;

(ii) *The uniformity generated by* ρ_f *is the* (ε, λ)*-uniformity;*

(iii) *If* a *is a positive real number then* ρ_f^a *defined by*

$$\rho_f^a(p, q) = \inf_{t>0} \{at + f \circ F_{pq}(t)\} \tag{2a}$$

has the properties (i)—(ii);

(iv) *For each* $a \in (0, 1]$ *one has*

$$a\, \rho_f \leqslant \rho_{af} \leqslant \rho_f$$

and so all ρ_f *are uniformly equivalent.*

Proof. Since $\lim_{t\to\infty} F_{pq}(t) = 1$, then we see that ρ_f is finite and, clearly, nonnegative.

If we suppose that $\rho_f(p, q) = 0$, then for each n there exists $t_n < \frac{1}{n}$ such that $t_n + f(F_{pq}(t_n)) < \frac{1}{n}$. Therefore $t_n \to 0$ and $f(F_{pq}(t_n)) \to 0$. Therefore $F_{pq}(0+) = 1$, which is equivalent to $p = q$. It is obvious that $\rho_f(p, p) = 0$ for any $p \in S$, and that ρ_f is symmetric.

In order to prove the triangle inequality, we observe [9, Chapter 2, Theorem 11] that

$$f \circ F_{pq}(x + y) \leqslant f \circ F_{pr}(x) + f \circ F_{rq}(y), \; \forall\, p, q, r \in S, \forall\, x, y > 0. \tag{3}$$

By adding $x + y$ in both sides and taking the infimum we obtain that (i) holds.

Now let $\varepsilon < \delta = f(0)$. If $\rho_f(p, q) < \varepsilon$ then there exists $t < \varepsilon$ such that $f \circ F_{pq}(t) < \varepsilon - t$, which implies that $F_{pq}(t) \geqslant f^{(-1)}(\varepsilon - t) > f^{(-1)}(\varepsilon)$. Therefore $F_{pq}(\varepsilon) > 1 - (1 - f^{(-1)}(\varepsilon))$, and so $V(\varepsilon) \subset U(\varepsilon, 1 - f^{(-1)}(\varepsilon))$.

Conversely, if $(p, q) \in U(\varepsilon, \lambda)$, that is $F_{pq}(\varepsilon) > 1 - \lambda$, then $\varepsilon + f \circ F_{pq}(\varepsilon) < \varepsilon + f(1 - \lambda)$ and so $\rho_f(p, q) < \varepsilon + f(1 - \lambda)$ which implies that $U(\varepsilon, \lambda) \subset V(\varepsilon + f(1 - \lambda))$ and (ii) is proved.

In order to prove (iii) we observe that $g = \frac{1}{a} f$ is also an additive generator of T and we can write

$$\rho_f^a(p, q) = \inf_{t>0} \{at + f \circ F_{pq}(t)\} = \inf_{t>0} \{at + ag \circ F_{pq}(t)\} = a\, \rho_g(p, q),$$

and so (iii) follows from (i) and (ii).

Now for $a \leqslant 1$, we have

$$a\rho_f(p, q) = \inf_{t>0} \{at + af \circ F_{pq}(t)\} \leqslant \inf_{t>0} \{t + af \circ F_{pq}(t)\} =$$

$$= \rho_{af}(p, q) \leqslant \inf_{t>0} \{t + f \circ F_{pq}(t)\} = \rho_f(p, q).$$

Since for every additive generator g of T there exists $a \in (0, 1]$ such that $g = af$ or $f = ag$ then (iv) holds and the theorem is proved.

Remark 1. It is easy to see that ρ_f^g in (2a) and d_z in (1) are identical for $f = -\ln h$.

2. Although the following result is a very particular case of the above results we will see that it can be used in order to obtain an equivalent of Theorem B.

Proposition 1. (The Fréchet-Ky Fan metric). *If $(S, \mathcal{F}, T_m)$ is a Menger space then the two-place function k defined by*

$$k(p, q) = \inf_{t>0} \{t + 1 - F_{pq}(t)\} \tag{4}$$

is a metric on S which induces the (ε, λ)-topology.

In particular in every E-space, the function k metrizes the topology of the convergence in probability.

Proof. It suffices to observe that the Archimedean t-norm T_m has the additive generator f_m given by $f_m(t) = 1 - t$.

Now Proposition 1 can be used in the general case of the Archimedean t-norms. Namely we have

Theorem C. *If $(S, \mathcal{F}, T)$ is a Menger space under an Archimedean t-norm T then there exists an increasing bijection $\varphi : [0, 1] \to [0, 1]$ such that the two-place function k defined by*

$$k_\varphi(p, q) = \inf_{t>0} \{t + 1 - \varphi \circ F_{pq}(t)\} \tag{5}$$

is a metric on S which metrizes the (ε, λ)-uniformity. The same is true for any t-norm T not weaker than T_m.

Proof. Since for all a, b in $[0, 1]$ we have that $ab \geqslant a + b - 1$ then from [5] we obtain that for every Archimedean t-norm T, there exists φ as above such that

$$T(a, b) \geqslant \varphi^{-1}(T_m(\varphi(a), \varphi(b)) = T_\varphi(a, b). \tag{6}$$

But this shows that $(S, \varphi \circ \mathcal{F}, T_m)$ is a Menger space. Now if we apply formula (4) for $\varphi \circ \mathcal{F}$, then we obtain (5). Finally we observe that the (ε, λ)-uniformities generated by $\mathcal{F}$ and $\varphi \circ \mathcal{F}$ are identical, and the theorem is proved. We note also that the above t-norm T has the additive generator $f = f_m \circ \varphi = 1 - \varphi$, and so formula (5) is equivalent to (2)

Remark 2. If $(S, \mathcal{F}, \mathrm{Min})$ is a Menger space then formula (5) gives us a metric for *any* increasing bijection φ; the same is true for *any* f in (2), since Min is the strongest t-norm.

Remark 3. Obviously formula (4) can be used in any PM-space in order to obtain a symmetric positively defined mapping on $S \times S$, which is *not* necessarily a metric. But if it *is* a metric, then it gives the (ε, λ)-uniformity on S. The same is true for formula (2). The above theorems A, B, C give us sufficient conditions for the corresponding two-place functions to be metrics. These are best possible in the sense of the following

Theorem D. *Suppose that the t-norm T_1 is weaker than the Archimedean t-norm T. Then there exists a Menger space $(S, \mathcal{F}, T_1)$ such that ρ_f given by (2) is not a metric for any additive generator f of T.*

Proof. Let $S = \{p, q, r\}$ be a set of three distinct points and choose $a, b \in (0, 1)$ such that $T_1(a, b) < T(a, b)$. If we set

$$F_{pr}(t) = \begin{cases} 0, & t \leqslant a \\ 1, & t > a \end{cases}, \quad F_{rq}(t) = \begin{cases} 0, & t \leqslant b \\ 1, & t > b \end{cases}, \quad F_{pq}(t) = \begin{cases} 0, & t \leqslant T_1(a, b) \\ 1, & t > T_1(a, b) \end{cases}$$

then $(S, \mathcal{F}, T_1)$ is a Menger space under T_1 and by simple calculation we obtain that

$$\rho_f(p, r) = f(a), \quad \rho_f(r, q) = f(b), \quad \rho_f(p, q) = f(T_1(a, b)).$$

If we suppose that ρ_f is a metric then $f(T_1(a, b)) \leqslant f(a) + f(b)$, which implies that $T_1(a, b) \leqslant f^{(-1)}(f(a) + f(b)) = T(a, b)$, evidently a contradiction.

3. In this section we consider the problem of the construction of a quasi-norm on a random normed space.

Let $(S, \mathcal{F}, T)$ be a random normed space, that is [12] S is a linear space, $\mathcal{F}$ is a mapping of S into the family D of distribution functions F with $F(0) = 0$, T is a t-norm not weaker than T_m and the following axioms are verified ($\mathcal{F}(p)$ is denoted by F_p)

I. $F_p = E_0 \Leftrightarrow p = 0 \qquad (E_0 \in D,\ E_0(x) = 1 \text{ for } x > 0)$;

II. $F_{\lambda p}(x) = F_p\left(\frac{x}{|\lambda|}\right), \ x > 0, \ p \in S, \ \lambda$ scalar;

III. $F_{p+q}(x + y) \geqslant T(F_p(x), F_q(y)), \ p, q \in S, \ x, y > 0$.

It is known [12] that endowed with the (ε, λ)-topology, S is a linear metric space. Also the function given by

$$\|p\| = \sum_{n=0}^{\infty} \frac{1}{2^n} \frac{1 - F_p\left(\frac{1}{n}\right)}{2 - F_p\left(\frac{1}{n}\right)} \tag{7}$$

is a quasi-norm on S, which gives the (ε, λ)-topology.

Since by definition the t-norm T is not weaker than T_m then as in Theorem C we can obtain also a quasi-norm on S.

Theorem E. *If $(S, \mathcal{F}, T)$ is a random normed space, then the mapping ν defined on S by*

$$\nu(p) = \inf_{t>0} \{t + 1 - F_p(t)\} \tag{8}$$

is a quasi-norm on S which induces the (ε, λ)-topology.

Proof. Clearly ν is positively defined and "monotone". The triangle inequality results from the fact that $(S(p, q) \to \mathcal{F}(p - q), T_m)$ is a Menger space and $\nu(p) = k(p, 0)$. The last part of the theorem can be obtained in a similar manner.

Remark 4. Clearly, if T is Archimedean, then we can use the formula (2) to obtain a quasi-norm on S

$$\nu_f(p) = \inf_{t>0} \{t + f \circ F_p(t)\} \tag{9}$$

For example let $(S, \|\cdot\|)$ be a normed space. If we set $F_p(x) = \begin{cases} 0, & x \leqslant \|p\| \\ 1, & x > \|p\| \end{cases}$ then it is known that $(S, p \to F_p, \text{Min})$ is a random normed space. Since $T_m \leqslant \text{Prod} \leqslant \text{Min}$ then we can take $f = -\ln$ in (9) and we obtain that $\nu_f(p) = \|p\|$. From (8) we obtain $\nu(p) = \min\{1, \|p\|\}$. In the case of the t-norm Min, the formula (9) can be used with *any* f.

Remark 5. In the axiom (III) of a random normed space it is supposed that T is not weaker than T_m. Remark 4 shows that we can use any other (not weaker than an) Archimedean t-norm in the role of T_m.

4. Let T be a left-continuous t-norm and consider the corresponding t-function τ_T([10, 12]). In [1] the following is proved

Lemma. *There exists a mapping* $\mathcal{F}_T : D \times D \to D$ *such that* $(D, \mathcal{F}_T, T)$ *is a Menger space. More precisely,*

$$\mathcal{F}_T(F, G) = \sup\{H \in D,\ \tau_T(F, H) \leqslant G,\ \tau_T(G, H) \leqslant F\}. \tag{10}$$

From this lemma and Theorem B we obtain

Theorem F. *Let* $f : [0,1] \to [0,\infty]$ *be a continuous and nonincreasing function such that* $f(1) = 0$. *If* T *is the Archimedean t-norm generated by* f *then*

$$L_f(F, G) = \inf_{t>0}\{t + f \circ \mathcal{F}_T(F, G)(t)\} \tag{11}$$

defines a metric on D.

The case where $f(t) = 1 - t$ can be considered as an extension of the Fréchet - Ky Fan metric. We shall see that it is sufficient in order to obtain all the topologies generated by the L_f.

Theorem G. *Let* $\mathcal{K} : D \times D \to D$ *be defined by*

$$\mathcal{K}(F, G) = \sup_{K \in D}\{K\}, \tag{12}$$

where K *is such that*

$$K(t) \leqslant \text{Min}\{G(x) - F(x - t) + 1,\ F(x) - G(x - t) + 1\} \quad \forall t \leqslant x,\ \forall x > 0 \tag{12'}$$

Then L_k *defined on* $D \times D$ *by*

$$L_k(F, G) = \inf_{t>0}\{t + 1 - \mathcal{K}(F, G)(t)\} \tag{13}$$

is a metric on D. *Moreover*

$$L(F, G) \leqslant L_k(F, G) \leqslant 2L(F, G), \tag{14}$$

where L *is the P. Lévy metric.*

Proof. We will prove only the inequalities (14). If $L_k(F, G) < h$, then let $t < h$ such that $t + 1 - \mathcal{K}(F, G)(t) < h$. Therefore, for $x \geqslant t$ we have that $G(x) - F(x - t) > t - h$ and $F(x) - G(x - t) > t - h$, which imply

$$F(x - h) - h \leqslant G(x) \leqslant F(x + h) + h,$$

that is $L(F, G) \leqslant h$.

Conversely, suppose that $L(F, G) < h$, and let $K_1(t) = \begin{cases} 0, & t \leqslant h \\ 1-h, & t > h \end{cases}$.
Then for $x \geqslant t > h$, we have that $G(x) - F(x-t) + 1 \geqslant -h + 1$, $F(x) - G(x-t) + 1 \geqslant -h + 1$. Therefore, $\operatorname{Min}(G(x) - F(x-t) + 1, F(x) - G(x-t) + 1) \geqslant K_1(t)$. This shows that $\mathcal{K}(F, G) \geqslant K_1$, which implies that

$$L_k(F, G) \leqslant \inf_{t>0} \{t + 1 - K_1(t)\} \leqslant 2h$$

and the required inequalities follow.

Corollary. *Let f be as in Theorem* **F**. *Then there exists an increasing bijection $\varphi : [0, 1] \to [0, 1]$ such that L_φ defined on $D \times D$ by*

(15) $$L_\varphi(F, G) = L_k(\varphi \circ F, \varphi \circ G)$$

is a metric on D which generates the same uniformity as L_f and

(16) $$L_\varphi \leqslant L_f.$$

REFERENCES

1. Höhle, U., *Probabilistic metrics on the set of nonnegative distribution functicns.* Aequationes Math., **18**, 345–356 (1978).
2. Kingman, J.F.C., *Metrics for Wald spaces*, J. London Math. Soc. (2), **39**, 129–130 (1964).
3. Ling, C.H., *Representation of associative functions.* Publ. Math. Debrecen, **12**, 189–212 (1965).
4. Menger, K., *Statistical metrics.* Prcc. Nat. Acad. Sci. U.S.A., **28**, 535–537 (1942).
5. Mostert, P.S., Shields, A.L., *On the structure of semigroups on a compact manifold with boundary.* Ann. of Math. (2), **65**, 117–143 (1957).
6. Moynihan, R., Schweizer, B., *Betweenness relations in PM-spaces.* Pacific J. Math., **81**, 175–196 (1979).
7. Radu, V., *On the metrizability of Menger spaces.* Math. Balkanica, **4**, 497–498 (1974).
8. Radu, V., *On some metrics for the (ε, λ)-topology in PM-spaces.* Seminarul de Teoria Probabilităţilor şi Aplicaţii, nr. 61, Univ. Timişoara, 1982.
9. Schweizer, B., *Lectures on PM-spaces.* Univ. of Arizona, 1975. (Unpublished)
10. Schweizer, B., Sklar, A., *Statistical metric spaces.* Pacific J. Math., 10, 313–334 (1960).
11. Schweizer, B., Sklar, A., Thorp, E., *The metrizaticn of PM-spaces.* Pacific J. Math., **10**, 73–75 (1960).
12. Serstnev, A.N., *Sluchainyie normirovannyie prostranstva, voprosy polnoty.* Učen. Zap. Kazan Gos. Univ., **122**, *no.* 4, 3–20 (1962).
13. Sibley, D.A., *A metric for weak convergence of distribution functions.* Rocky Mountain J. Math., 1, 427–430 (1971).
14. Wald, A., *On a statistical generalization of metric spaces.* Proc. Nat. Acad. Sci. U.S.A., **29**, 196–197 (1943).

NICOLAE ROMANOF

Application of the orthonormal expansion of random functions to turbulent diffusion

1. Introduction

The turbulent diffusion is the fundamental mechanism which ensures the transfer of mass, heat and momentum in turbulent flow. In many cases, the molecular transfer is much smaller than the turbulent transfer and therefore the classical diffusion theory is not useful in turbulent diffusion.

In the description of turbulence and turbulent diffusion there are used random functions coupled by nonlinear dynamical equations and these lead to "closing problems" for the moment equations [Monin and Jaglom (1965), (1967)].

As it is well known the first closure is the K-theory in which the turbulent flows are represented in analogy with molecular flows. Recently, the K-theory was generalized to the so called K-spectral theory [Bercovicz and Prahm (1979)].

A nonlinear integro-differential equation for turbulent diffusion was found by Roberts (1961), who based his considerations on direct interaction approximation [Kraichnan (1961)].

Saffman (1969) used the Wiener-Hermite expansion for random functions to study the diffusion of a passive scalar in homogeneous and stationary turbulence and established a linear integro-differential equation. A generalization of Saffman's equation to the boundary layer was made by Romanof [6] in Gaussian approximation. Saffman's equation was also obtained by Romanof [5] using orthonormal expansions for Gaussian random functions. In this paper, we use the orthonormal expansion of random functions (with finite second order moments) to study the turbulent diffusion of a passive scalar and establish an "exact" linear integro-differential equation for turbulent diffusion which is not local in space and time. Any linear equation for turbulent diffusion is a particular case of this equation.

2. The mathematical formalism

Let $\{\Omega, \mathcal{F}, P\}$ be a probability space, $u(x, t)$ a vectorial random field, $x \in R^s$, $t \in R$, $\mathcal{F}(u)$ the σ-algebra generated by u, $L_2(u) = L_2(\Omega, \mathcal{F}(u), P)$ the Hilbert space of random variables which are $\mathfrak{F}(u)$-measurable and have finite second order moments, $H(u)$ the closed vectorial subspace of $L_2(u)$ generated by $u(x, t)$. The sample paths of the random field u are assumed sufficiently smooth to ensure the validity of the equations to be written.

Let the system of differential equations

$$\frac{dx(t, \omega)}{dt} = u(x, t, \omega) \tag{2.1}$$

with the initial condition

$$x(t_0, \omega) = x_0 \tag{2.2}$$

One fundamental problem in the turbulent diffusion is: given the statistical properties of the random field $u(x, t)$, which are the statistical properties of the family of random processes $\{X(t, \omega, t_0, x_0),\ x_0 \in R^s,\ t_0 \in R\}$?

In the special case $s = 3$, this is an unsolved Eulerian-Lagrangian relationship problem.

To apply the orthonormal expansions method it is useful to replace (2.1) and (2.2) by

$$\frac{\partial C}{\partial t} + \sum_{i=1}^{s} u_i \frac{\partial C}{\partial x_i} = 0 \tag{2.3}$$

$$C(x, t_0) = \delta(x - x_0), \tag{2.4}$$

where δ is the Dirac function and $C(x, t; x_0, t_0, \omega)$ is the concentration random field [Monin and Jaglom (1965), Ch. 5].

In the following it is assumed that the u-field is divergenceless:

$$\sum_{i=1}^{s} \frac{\partial u_i}{\partial x_i} = 0 \tag{2.5}$$

In such a case, equation (2.3) becomes

$$\frac{\partial C}{\partial t} + \sum_{i=1}^{s} \frac{\partial}{\partial x_i} (u_i\, C) = 0 \tag{2.6}$$

If $s = 3$, equation (2.6) expresses the mass conservation law of a passive scalar C which is released in u-flow at $t = t_0$ in $x = x_0$ with strength unity.

In what follows our aim is to establish an equation for the transition probability

$$P(x, t; x_0, t_0) = EC(x, t; x_0, t_0) = \bar{C}.$$

Let $\{1; \xi_\alpha, \alpha \in I\}$ be an orthonormal basis in $L_2(u)$; generally, I is not a denumerable set. If u is Gaussian, then $H(u)$ is a separable Hilbert space and it is possible to build up a denumerable orthonormal basis in $L_2(u)$ [Neveu (1968), Ch. 3,7].

The random fields u and C are represented in the orthonormal basis as

$$u_i = \bar{u}_i + \sum_{\alpha \in I} u_{\alpha i} \xi_\alpha, \quad 1 \leqslant i \leqslant s, \tag{2.7}$$

$$C = \bar{C} + \sum_{\alpha \in I} C_\alpha\, \xi_\alpha \tag{2.8}$$

Because $u \in H(u)$, in (2.7) $u_\alpha \neq 0$ only for those α for which $\xi_\alpha \in H(u)$.

If we use these expansions in equation (2.6), multiply by 1 and ξ_α and then take the average, we finally obtain an infinite system of differential equations with $\bar{C}$ and C_α as unknowns. The system is

$$\frac{\partial \bar{C}}{\partial t} + \sum_{i=1}^{s} \frac{\partial}{\partial x_i} (\bar{u}_i \bar{C}) + \sum_{i=1}^{s} \frac{\partial}{\partial x_i} \sum_{\alpha \in I} u_{\alpha i} C_\alpha = 0 \tag{2.9}$$

$$\frac{\partial C_\alpha}{\partial t} + \sum_{i=1}^{s} \frac{\partial}{\partial x_i} \left[\bar{u}_i C_\alpha + \sum_{\beta, \gamma \in I} \omega_{\alpha\beta\gamma} u_{\beta i} C_\gamma \right] = -\sum_{k=1}^{s} u_{\alpha k} \frac{\partial \bar{C}}{\partial x_k}, \tag{2.10}$$

where

$$\omega_{\alpha\beta\gamma} = \overline{\xi_\alpha \xi_\beta \xi_\gamma}$$

The solution of equation (2.10) can be symbolically represented using the Green matrix as

$$C_\alpha(x, t) = -\int_{t_0}^{t} \mathrm{d}t' \int_{R^s} \mathrm{d}x' \sum_{\beta \in I} G_{\alpha\beta}(x, t; x', t') \sum_{k=1}^{s} u_{\beta k}(x', t') \frac{\partial \bar{C}(x', t')}{\partial x'_k} \tag{2.11}$$

Using (2.11) and replacing $\bar{C}$ by P, (2.9) becomes

$$\frac{\partial P}{\partial t} + \sum_{i=1}^{s} \frac{\partial}{\partial x_i} (\bar{u}_i P) = \sum_{i=1}^{s} \frac{\partial}{\partial x_i} \int_{t_0}^{t} \mathrm{d}t' \int_{R^s} \mathrm{d}x' \sum_{k=1}^{s} D_{ik}(x, t; x', t') \frac{\partial P(x', t')}{\partial x'_k}, \tag{2.12}$$

where D_{ik} is defined as

$$D_{ik}(x, t; x', t') = \sum_{\alpha, \beta \in I} u_{\alpha i} (x, t) G_{\alpha\beta}(x, t; x', t') u_{\beta k}(x', t') \tag{2.13}$$

We call D_{ik} the *generalized diffusion tensor*. It depends only on the statistical properties of the u-field, which are included in $\bar{u}$ and u_α.

If D_{ik} has the form

$$D_{ik}(x, t; x', t') = K_{ik}(x, t) \, \delta(x - x') \, \delta(t - t') \tag{2.14}$$

then, equation (2.12) becomes the second Kolgomorov equation with transfer coefficients $\bar{u}_i + \sum_{k=1}^{s} \frac{\partial K_{ik}}{\partial x_k}$ and diffusion coefficients K_{ik}.

3. Application to the turbulent diffusion

If $s = 3$ and the random field $u = (u_i, i = 1, 2, 3)$ is the velocity field of a turbulent flow, then we have the real turbulent diffusion problem. In this case, equation (2.12) is an "exact" equation of turbulent diffusion. The particle motion in the turbulent flow is a process with memory due to the existence of a space-time correlation of the velocity field. The memory of this process is reflected by the non-local term in the right side of equation (2.12).

If D_{ik} has the form (2.14), equation (2.12) becomes the equation of turbulent diffusion established in the K-theory

$$\frac{\partial P}{\partial t}+\sum_{i=1}^{3}\frac{\partial}{\partial x_i}(\bar{u}_i P)=\sum_{i=1}^{3}\frac{\partial}{\partial x_i}\sum_{k=1}^{3}K_{ik}\frac{\partial P}{\partial x_k} \tag{3.1}$$

In what follows we restrict ourselves to the homogeneous and stationary turbulence, $\bar{u}_i = 0$. The transition probability $P(x, t;\ x_0,\ t_0)$ depends only on $x - x_0, t - t_0$ and therefore it may be expected that $D_{ik}(x, t; x', t')$ depends only on $x - x'$, $t - t'$.

Equation (2.12) becomes

$$\frac{\partial P}{\partial t}=\sum_{i=1}^{3}\frac{\partial}{\partial x_i}\int_{t_0}^{t}\mathrm{d}t'\int_{R^3}\mathrm{d}x'\sum_{k=1}^{3}D_{ik}(x-x',\ t-t')\frac{\partial P(x',\ t')}{\partial x'_k} \tag{3.2}$$

If $D_{ik}(x-x', t-t') = B_{ik}(x-x')\ \delta(t-t')$, then equation (3.2) is identical to the equation established by Bercovicz and Prahm (1979) on the basis of the spectral diffusivity concept.

If $D_{ik}(x-x', t-t') = R_{ik}(t-t')\delta(x-x')$ and R_{ik} is the time correlation of the velocity field, then equation (3.2) becomes Saffman's equation [Saffman (1969)]. Saffman's equation can be also found if we assume that u and C are Gaussian fields, therefore $\omega_{\alpha\beta\gamma} = 0$ [equation (2.10)].

The Laplace-Fourier ($\mathcal{L}\mathcal{F}$) transform of equation (3.2) is

$$s\tilde{P}(s,\ \varkappa) - 1 = -\sum_{i=1}^{3}\sum_{k=1}^{3}\varkappa_i\varkappa_k\tilde{D}_{ik}(s,\ \varkappa)\ \tilde{P}(s,\ \varkappa) \tag{3.3}$$

where $\tilde{f} = \mathcal{L}\mathcal{F} f$.

Let $\tilde{P}_B$ be the $\mathcal{L}\mathcal{F}$-transform of the transition probability P_B of the Brownian motion process. It is easy to see that $\tilde{P}_B(s, \varkappa)$ satisfies an equation like (3.3) in which $\tilde{D}_{ik}(s, \varkappa) = \delta_{ik}D$. This remark suggests that turbulent diffusion in homogeneous and stationary turbulence may be considered as a generalization of the Brownian motion.

REFERENCES

1. Berkovicz, R., Prahm, L.P., *Generalizaticn of K-thecry for turbulent diffusicn. Part I: Spectral turbulent diffusivity concept.* J. Appl. Meteorology, 18, 266—272 (1979).
2. Kraichnan, R.H., *Dynamics of non-linear stcchastic systems.* J. Math. Phys., 2, 124—148 (1961).
3. Monin, A.S., Jaglom, A.M., *Statistical Hydrcmechanics. I, II.* Moskva, Nauka, 1965, 1967. (Russian)
4. Neveu, J., *Processus Aléatoires Gaussiens.* Montréal, Les Presses de l'Université de Montréal, 1968.
5. Romanof, N., *Applications of the orthonormal developments for Gaussian random functions to the pollutant diffusion emitted by instantaneous sources.* Meteorology and Hydrology, 2, 25-31 (1981).
6. Romanof, N., *Applications of the Wiener-Hermite expansions to the turbulent diffusion.* Meteorology and Hydrology, 2, 17-21 (1982),
7. Roberts, P.H., *Analytical theory of turbulent diffusicn.* J. Fluid Mech., 11, 257—283 (1961).
8. Saffman, P.G., *Application of the Wiener-Hermite expansion to the diffusicn of a passive scalar in homogeneous turbulent flow.* Phys. Fluids, 12, 1786—1798 (1969).

CONSTANTIN TUDOR

On infinite dimensional continuous square integrable martingales with differentiable characteristics

We treat continuous martingales with one parameter and strong martingales with two parameters, which take values in infinite dimensional spaces, in the case where they have differentiable characteristics. These martingales are described as solutions of stochastic integral equations with the abstract Wiener process as martingale differential, and as exponential martingales.
Existence theorems for the above mentioned martingales are given.

1. Preliminaries

Throughout this paper H is a real separable Hilbert space with norm $|\cdot|$ and inner product $<.,.>$. The time parameter domain is a set $D \subset R^k$ with $k = 1, 2$. We write ∂D for the boundary of D.

If $X = (X_z)_{z \in D}$ is an H-valued stochastic process then we put $\partial X =$ the restriction of X to ∂D, $\square_z X_{z'} = X_{z'} - X_z$ if $k = 1$ and $\square_z X_{z'} = X_{s',t'} - X_{s',t} - X_{s,t'} + X_{s,t}$ if $k = 2$, $z = (s, t)$, $z' = (s', t')$.

By $\mathcal{L}_1(H)$ (resp. $\mathcal{L}_2(H)$) we denote the collection of all trace (resp. Hilbert-Schmidt) operators on H with norms $\|\cdot\|_1$, $\|\cdot\|_2$.

Next, let $(\Omega, \mathcal{F}, P)$ be a probability space and $(\mathcal{F}_z)_{z \in D}$ be a stochastic basis on Ω. We define $\overline{\mathcal{F}}_z = \mathcal{F}_z$ if $k = 1$ and $\overline{\mathcal{F}}_z = \mathcal{B}(\mathcal{F}_{u,v};\ u \leqslant s \text{ or } v \leqslant t)$ if $k = 2$ and $z = (s, t)$.

Definition. An H-valued process X defined on $(\Omega, \mathcal{F}, P)$ is a *k-parameter $\square$-martingale* if it is an $\mathcal{F}_z$-martingale if $k=1$ and an $\mathcal{F}_z$-strong martingale if $k = 2$.

We write $\mathcal{M}_2^c(\mathcal{F}_z, H)$ for the collection of all continuous square integrable k-parameter $\square$-martingales X with $\partial X = 0$.

If (e_n) is an orthonormal basis in H and $X \in \mathcal{M}_2^c(\mathcal{F}_z, H)$ then for every n the process $<e_n, X_z>$ is a real k-parameter $\square$-martingale, so there is the quadratic variation $<X^n>_z$ of it (see [3]).

Definition. The characteristic operator of X is the $\mathcal{L}_1(H)$-valued process $A = (A_z)$ which satisfies the conditions

a) A is $\mathcal{F}_z$-adapted and has continuous paths;

b) For every (z, ω) the operator $A_z(\omega)$ is symmetric, nonnegative definite and satisfies the equality $\| A_z(\omega) \|_1 = \sum_{n=1}^{\infty} (<X^n>_z)(\omega)$.

Denote by B the completion of H with respect to a measurable norm $\| \cdot \|$. The triple (i, H, B), where $i : H \to B$ is the natural injection, is called an *abstract Wiener space.*

Let p_t be the Wiener measure on B with variance parameter t (see Gross [6], Kallianpur [8] or Kuo [9]).

There exists a Markov process X with state space B, with $\partial X = 0$ and which has continuous paths, the transition function $p_t(x, A) = p_t(A - x)$ if $k = 1$ and $p_{s,t}(x, y, z, A) = p_{st}(A + x - y - z)$ if $k = 2$.

Such a process X is called a *k-parameter abstract Wiener process with state space B.*

Definition. Let ξ be an $\mathcal{L}_2(H)$-valued process which is $\mathcal{F}_z$-nonanticipating and satisfies $E\left(\int_0^z \| \xi_u \|_2^2 \, du\right) < \infty$ for each z. By the *stochastic integral of ξ with respect to an abstract Wiener process X* we mean the k-parameter process $I(\xi)$ of the class $\mathcal{M}_2^c(\mathcal{F}_z, H)$ with $\int_0^z \xi_u \xi_u^* \, du$ a characteristic operator (the construction of this integral is similar to that of the finite dimensional case; see [2], [9] for the stochastic integral with respect to a 1-parameter abstract Wiener process and [7], [12], [15] for the stochastic integral with respect to a 1-parameter infinite dimensional martingale).

We also use the notation $\int_0^z \xi_u dX_u$ for $I(\xi)\,(z)$.

2. Connection with exponential martingales and stochastic integral equations

In the following we restrict ourselves to $\square$-martingales with characteristic operators absolutely continuous with respect to the Lebesgue measure.

Theorem 2.1. *Let X be a continuous and adapted H-valued process with $\partial X = 0$. Let (a_z) be an $\mathcal{L}_1(H)$-valued nonanticipating process such that for every z, ω, $a_z(\omega)$ is symmetric, nonnegative definite and moreover assume that $\sup_{z,\omega} \| a_z(\omega) \|_1 < \infty$. Then the following are equivalent*

(1) $X \in \mathcal{M}_c^2(\mathcal{F}_z, H)$ *and* $\int_0^z a_u \, du$ *is its characteristic operator.*

(2) *For each* $z < z'$ *and* $h \in H$

$$E\left[\exp\left(<h, \square_z X_{z'}> - 1/2 \int_z^{z'} <a_u h, h> du\right) \mid \mathcal{F}_z\right] = 1.$$

Proof. First observe that (1) is equivalent to the fact that for every $h \in H$ the process $<h, X_z>$ is a real k-parameter $\square$-martingale having $\int_0^z < a_u h, h > du$ as quadratic variation.

Now the theorem follows from the corresponding result for the finite dimensional case (see[1; Proposition 87, p. 72] if $k = 1$ and [14; Theorem 1.8, p. 512] if $k = 2$).

Theorem 2.2. *Suppose that* $\sigma: D\times H \to \mathcal{L}_2(H)$ *and* $b: D\times H \to H$ *are continuous and bounded. Then the following are equivalent*

(1) *On a probability space* $(\Omega, \mathcal{F}, P)$ *with a filtration* $(\mathcal{F}_z)$ *there is a process* X *such that* $Y_z = X_z - \int_0^z b(u, X_u)\, du \in \mathcal{M}_c^2(\mathcal{F}_z, H)$ *and* Y *has* $\int_0^z (\sigma\sigma^*)(u, X_u)\, du$ *as characteristic operator.*

(2) *The stochastic integral equation*

$$\square_0 X_z = \int_0^z \sigma(u, X_u)\, dw_u + \int_0^z b(u, X_u)\, du$$

has a weak solution (X, w), *with* w *an abstract Wiener process.*

Proof. Assume (1). Let T be an injective operator from $\mathcal{L}_2(H)$ and define $\|x\| = |Tx|$ for $x\in H$. The norm $\|\cdot\|$ is measurable, so that there exist the projections P_n that increase to the identity operator such that we have $\gamma_Q(\|Qx\| > 1/2^n) < 1/2^n$ for every n and for every finite dimensional projection Q with $Q \perp P_n$, where γ_Q is the standard Gaussian measure in $Q(H)$.

Let (e_n) be an orthonormal basis in H such that $(e_1, \dots, e_{k_n})$ is a basis in $P_n(H)$. From hypotheses and Theorem 2.1 it follows that

$$E\left[\exp(<P_n h, \square_z Y^n_{z'}> - 1/2\int_z^{z'} <(P_n\sigma\sigma^* P_n)(u, X_u) P_n h, P_n h> du \,|\, \mathcal{F}_z\right] = 1$$

for every n, $z < z'$, $h \in H$, where $Y^n_z = P_n X_z - \int_0^z P_n b(u, X_u)\, du$.

By using a result from the finite dimensional case (see [13; Theorem 4.5.1, p. 108]) we find a k_n-dimensional Brownian motion w^n such that

$$\square_0 Y^n_z = \int_0^z (P_n\sigma P_n)(u, X_u)\, dw^n_u \tag{2.1}$$

(the proof of this result for $k = 2$ is similar to that for $k = 1$).

From $P_{n+1} - P_n \perp P_n$ and the construction of w^n it follows that

$$w^n_z = \sum_{i=1}^{k_n} <w^n_z, e_i> e_i = \sum_{i=1}^{k_n} <w^{n+1}_z, e_i> e_i.$$

Then

$$P(\|w^{n+1}_z - w^n_z\| > 1/2^n) = \gamma_{(P_{n+1}-P_n)(H)}(\|P_{n+1}x - P_n x\| > 1/2^n) < 1/2^n$$

so that w^n_z converges in probability for each z.

Since

$$P(\|\square_z w^n_{z'}\| > \varepsilon) \leqslant 1/\varepsilon^2 E(\|\square_z w^n_{z'}\|^2) = 1/\varepsilon^2 E(|\square_z T w^n_{z'}|^2) =$$

$$= 1/\varepsilon^2 \sum_{i=1}^{k_n} |T^* e_i|^2 \lambda((z, z']) \leqslant 1/\varepsilon^2 \|T\|_2^2 \lambda((z, z'])$$

(λ is the Lebesgue measure) it follows that

$$\lim_{\delta\to 0} \sup_{n} \sup_{|z-z'|<\delta} P(\| w^n_z - w^n_{z'} \| > \varepsilon) = 0.$$

By a result of Skorohod [10; Theorem 6, p. 9] there exists a sequence of processes $(\overline{Y}^n_z, \overline{w}^n_z)_{n\geq 0}$ with state space $H\times B$ (B is the completion of H with respect to the norm $\|\cdot\|$) such that

(i$_1$) For every $n\geq 0$ the process $(\overline{Y}^n, \overline{w}^n)$ has continuous paths;

(i$_2$) For every $n\geq 1$ the process $(\overline{Y}^n, \overline{w}^n)$ is identical in law with (X, w^n);

(i$_3$) $\overline{w}^n_z \xrightarrow{P} \overline{w}^0_z$, $\overline{Y}^n_z \xrightarrow{P} \overline{Y}_z$ for every z.

Because for $h\in B^*$ and $z<z'$

$$E\left[\exp\{<h, \Box_z \overline{w}^n_{z'}> - 1/2\, |P_n h|^2\, \lambda((z, z'])\}\,|\, \overline{\mathcal{F}}\right] = 1,$$

the same equality for $n=0$ follows. Hence $\overline{w}^0$ is a k-parameter abstract Wiener process with state space B.

Also from (2.1) and (i$_2$) it follows that

$$\Box_0 P_n \overline{Y}^n_z = \int_0^z P_n\sigma(u, \overline{Y}^n_u)\, P_n\, d\overline{w}^n_u + \int_0^z P_n b(u, \overline{Y}^n_u)\, du. \tag{2.2}$$

It is easy to see that $P_n \overline{Y}^n_z \xrightarrow{P} \overline{Y}^0_z$, $\int_0^z P_n b(u, \overline{Y}^n_u)\, du \xrightarrow{P} \int_0^z b(u, \overline{Y}^0_u)\, du$.

If we prove that

$$\int_0^z P_n\sigma(u, \overline{Y}^n_u)\, P_n\, d\overline{w}^n_u \xrightarrow{P} \int_0^z \sigma(u, \overline{Y}^0_u)\, d\overline{w}^0_u, \tag{2.3}$$

then from (2.2), by passing to the limit, will follow that $(\overline{Y}^0, \overline{w}^0)$ is a weak solution of the stochastic equation from (2). To this end let $0 = z_0 < \ldots < z_r = z$ be a partition of $[0, z]$ such that $\max_i (z_{i+1} - z_i) \to 0$ (we consider only the case $k=1$; the proof is identical for $k=2$). We have

$$P\left(\left|\int_0^z P_n\sigma(u, \overline{Y}^n_u)\, P_n\, d\overline{w}^n_u - \int_0^z \sigma(u, \overline{Y}^u_0) d\overline{w}^0_u\right| > \varepsilon\right) \leq$$

$$\leq P\left(\left|\int_0^z P_n\sigma(u, \overline{Y}^n_u)\, P_n\, d\overline{w}^n_u - \sum_{i=0}^{r-1} (P_n\sigma(z_i, \overline{Y}^n_{z_i})\, P_n)\, (\Box_{z_i} \overline{w}^n_{z_{i+1}})\right| > \varepsilon/3\right) +$$

$$+ P\left(\left|\int_0^z \sigma(u, \overline{Y}^0_u)\, d\overline{w}^0_u - \sum_{i=0}^{r-1} \sigma(z_i, \overline{Y}^n_{z_i})\, (\Box_{z_i} \overline{w}^0_{z_{i+1}})\right| > \varepsilon/3\right) +$$

$$+ P\left(\left|\sum_{i=0}^{r-1} P_n\sigma(z_i, \overline{Y}^n_{z_i})\, P_n\, (\Box_{z_i} \overline{w}^n_{z_{i+1}}) - \sum_{i=0}^{r-1} \sigma(z_i, \overline{Y}^0_{z_i})\, (\Box_{z_i} \overline{w}^0_{z_{i+1}})\right| > \varepsilon/3\right) =$$

$$= I_1 + I_2 + I_3.$$

Clearly $I_3 \to 0$ as $n\to\infty$.

If we put $\overline{\sigma}(u) = \sigma(z_i, \overline{Y}^0_{z_i})$ for $u\in[z_i, z_{i+1})$ then we have

$$I_2 \leq 1/\varepsilon^2 E\left(\left|\int_0^z [\sigma(u, \overline{Y}^0_u) - \overline{\sigma}(u)]\, d\overline{w}^0_u\right|^2\right) = 1/\varepsilon^2 E\left(\int_0^z \|\sigma(u, \overline{Y}^0_u) - \overline{\sigma}(u)\|_2^2\, du\right)$$

$\to 0$ as $\max (z_{i+1} - z_i) = \delta \to 0$.

It remains to show that $\lim_{\delta\to 0}\lim_{n\to\infty} I_1 = 0$. We have

$$I_1 \leqslant 9/\varepsilon^2 E\left(\int_0^z \operatorname{Trace}\{[P_n(\sigma-\bar\sigma^n)]^* P_n\sigma\sigma^* P_n(\sigma-\bar\sigma^n)\}\,du\right) \leqslant$$

$$\leqslant \text{const. } E\left(\int_0^z \|\sigma-\bar\sigma^n\|_2^2\,du\right)$$

$\to 0$ as $n\to\infty$ and $\delta\to 0$ (in this order) where $\bar\sigma^n(u) = \sigma(z_i, \bar Y^n_{z_{i+1}})$ if $u\in[z_i, z_{i+1})$.

The implication (2) $\Rightarrow$ (1) is clear.

The proof of the theorem is complete.

Remark. We do not know whether in assertion (2) of the theorem we may take an arbitrary measurable norm.

By $C(D, H)$ we denote the collection of all H-valued continuous functions on D. $C(D, H)$ is a separable complete metric space relative to the metric

$$\rho(f, g) = \sum_{k=1}^{\infty} 2^{-k} \min\left(\sup_{z\leqslant k} |f(z)-g(z)|, 1\right).$$

We denote $\mathcal{F}_z^H =$ the Borel field generated by the sets of the type $\{f; f(u)\in C\}$, where $C\in\mathcal{B}_H$, $u\leqslant z$; $x_z^H\colon C(D, H)\to H$ the function defined by $x_z^H(f) = f(z)$. Note that $\mathcal{B}_{C(D,H)} = \mathcal{B}(x_z^H; z\in D)$.

Theorem 2.3. *Suppose we are given*

(1) $\sigma\colon D\times C(D, H)\to \mathcal{L}_2(H)$ and $b\colon D\times C(D,H)\to H$, which are bounded and $\mathcal{F}_z^H$-nonanticipating;

(2) A weak solution (X, w) of the stochastic integral equation

$$\square_0 X_z = \int_0^z \sigma(u, X)\,dw_u + \int_0^z b(u, X)\,du \tag{2.4}$$

such that for every T we have

$$E(\sup_{\substack{z\leqslant T\\ z\in\partial D}} |X_z|^{2(k+1)}) = C_1(T) < \infty \tag{2.5}$$

$$E(\sup_{\substack{z,z'\in\partial D\\ 0\leqslant z'-z\leqslant h\\ z,z'\leqslant T}} |X_{z'}-X_z|^{2(k+1)}) \leqslant C_2(T)\,h^{k+1} \tag{2.6}$$

(if $k=2$ by T, h we mean $(T, T), (h, h)$).

Then there exists a constant $C_3(T)$ depending only on the bounds of σ, b and $C_1(T), C_2(T)$ such that

(i) $E(\sup_{z\leqslant T} |X_z|^{2(k+1)}) \leqslant C_3(T) < \infty$,

(ii) $E(\sup_{\substack{|z'-z|\leqslant h\\ z,z'\leqslant T}} |X_{z'}-X_z|^{2(k+1)}) \leqslant C_3(T)h$.

Moreover, assume that for a completely continuous operator S which is symmetric and positive definite and a constant $c(T)$ the following conditions are satisfied

$$E(\sup_{\substack{z\leqslant T\\ z\in\partial D}} |S^{-1}X_z|^{2(k+1)}) \leqslant c(T) < \infty, \tag{2.7}$$

$$\sup_{z,\omega} \|S^{-1}\sigma(z,\omega)\|_2^2 + \sup_{z,\omega} |S^{-1}b(z,\omega)|^2 \leqslant c(T). \tag{2.8}$$

Then

$$E(\sup_{z\leqslant T} |S^{-1}X_z|^2)\leqslant c_1(T) < \infty \tag{2.9}$$

for a constant $c_1(T)$ depending only on $c(T)$ and S.

Proof. The case $k = 1$ is treated in [11]. Now we give the proof for $k=2$. Let $\sup_{z,\omega} \|\sigma(z, \omega)\|_2^2 = K$. The assertion will follow if we prove that

$$E\left(\left|\int_0^T \sigma(z)\, dw_z\right|^6\right)\leqslant \bar{C}_3(T, K). \tag{2.10}$$

Consider the process $y_s = \int_0^s\int_0^T \sigma(z)\, dw_z$. This process is a 1-parameter martingale with the characteristic operator $\int_0^s A_u\, du$ where $A_s = \int_0^T (\sigma\sigma^*)(s, u)\, du$.

By applying the Ito formula to y_s and to the function $f(x) = |x|^6$ (see [2] or [15]) and then taking expectation we obtain

$$E(|y_T|^6) = \frac{1}{2} E\left(\int_0^T \mathrm{Trace}\,(A(u)\, D^2f(y_u))\, du\right).$$

But $\langle D^2f(x)\, h, h\rangle = 3\,|x|^4\,|h|^2 + 12\,|x|^2 \langle x, h\rangle \leqslant 15\,|x|^4\,|h|^2$, whence

$$\mathrm{Trace}\ (A(u)\ D^2f(y_u)) \leqslant 15\,|y_u|^4\ \mathrm{Trace}\ A(u) \leqslant 15\,KT\,|y_u|^4.$$

Then $E(|y_T|^6) \leqslant \frac{15}{2} KT\,E\left(\int_0^T |y_u|^4\, du\right) \leqslant \frac{15}{2} KT^2[E(|y_T|^6)]^{2/3}$, hence $E(|y_T|^6) \leqslant \leqslant \left(\frac{15}{2}\right)^3 (KT^2)^3$ and this proves (2.10).

The above calculus shows that in general for $z < z'$ the following inequality hol ds

$$E\left(\left|\int_{[0,z']\setminus[0,z]} \sigma(u)\, dw_u\right|^6\right)\leqslant C_3(K)\,[\lambda((0, z]) - \lambda((0, z])]^3. \tag{2.11}$$

Assertion (ii) follows from (2.11) (see [4]; p. 80, 81).
Finally, (2.9) follows from the equation

$$\square_0 S^{-1}X_z = \int_0^z S^{-1}\sigma(u, X)\, dw_u + \int_0^z S^{-1}b(u, X)\, du$$

by an application of the Schwartz inequality and of the properties of stochastic integral.

3. Existence theorems

In this section we give three results concerning the construction of a k-parameter $\square$-martingale with a given absolutely continuous characteristic operator. In this direction we mention [13] for the finite dimensional case and $k = 1$ and [4], [5], [14], [16] for $k=2$. Also we mention the paper [11] which is devoted to the 1-parameter and the infinite dimensional case.

Theorem 3.1 *Suppose the coefficients* σ *and* b *in the stochastic equation* (2.4) *satisfy the following conditions:*

There exists a Radon measure μ *on* D *such that for each* $T>0$ *there is* $L_T>0$ *such that*

$$(3.1)\qquad \|\sigma(z,f)-\sigma(z,g)\|_2^2+|b(z,f)-b(z,g)|^2\leqslant L_T\left\{|f(z)-g(z)|^2+\right.$$
$$\left.+\int_0^z|f(u)-g(u)|^2\,d\mu(u)\right\}.$$

$$(3.2)\qquad \|\sigma(z,f)\|_2^2+|b(z,f)|^2\leqslant L_T\left\{|f(z)|^2+1+\int_0^z|f(u)|^2\,d\mu(u)\right\}$$

for all $z\leqslant T$ *and* f, g.

Let (w_z) *be a k-parameter* $\mathcal{F}_z$*-abstract Wiener process defined on the probability space* $(\Omega,\mathcal{F},P)$ *and let* $(\varphi_z)_{z\in\partial D}$ *be an* $\mathcal{F}_z$*-adapted and continuous process with values in* H *such that for every* $T>0$ *we have* $\sup_{z\leqslant T}E(|\varphi_z|^2)<\infty$.

Then on $(\Omega,\mathcal{F},P)$ *there exists an* H*-valued process* (X_z) *such that* (X,w) *is a solution of* (2.4) *with* $\partial X=\varphi$ *and* $\sup_{z\leqslant T}E(|X_z|^2)<\infty$. *Moreover, the solution of* (2.4) *is pathwise unique, i.e. whenever there are two solutions* (X,w) *and* (Y,w) *such that* $\partial X=\partial Y$ *then* X *and* Y *are indistinguishable.*

Proof. Since we use the well known method of succesive approximations we omit the proof (the reader may find details in [16]).

Theorem 3.2. *Let* $a\colon R_+\times H\to\mathcal{L}_1(H)$ *and* $b\colon R_+\times H\to H$ *be two functions such that*

(1) *For every* $h\in H$ *the function* $a(t,x)h$ *is continuous,* a *is bounded, symmetric, positive definite and* $\lim_{n\to\infty}\sup_{t,x}\sum_{i=n}^{\infty}<a(t,x)\,e_i,e_i>=0$ *for an orthonormal basis* (e_i) *in* H.

(2) b *is continuous, bounded and* $\lim_{n\to\infty}\sup_{t,x}\sum_{i=n}^{\infty}<b(t,x),e_i>^2=0$.

Then for each $x\in H$ *there exists on* $\mathcal{B}_{C(R+,H)}$ *a probability measure* Q *such that* $x_t^H-x-\int_0^t b(u,x_u^H)\,du\in\mathcal{M}_2^c(\mathcal{F}_z^H,H)$ *(with respect to* Q*) and with* $\int_0^t a(u,x_u^H)\,du$ *as its characteristic operator.*

Proof. Let P_n be the projection on the span of $(e_1,\ldots,e_n)$ and let us denote $a_n=P_naP_n$, $b_n=P_nb$, $x_n=P_nx$, $H_n=P_nH$.

According to a result of Stroock and Varadhan (see [13; Theorem 6.1.7, p. 144]) we can find a probability measure Q_n on $C(R_+,H)$ such that if we consider $x_t^n\colon C(R_+,H_n)\to H_n$ defined by $x_t^n(f)=f(t)$ then

$$(3.3)\qquad Q_n(x_0^n=x_n)=1,$$

$$(3.4)\qquad \exp\left(<h,x_t^n-x_n-\int_0^t b_n(u,x_u^n)\,du>-\frac{1}{2}\int_0^t<a_n(u,x_u^n)h,h>du\right)$$

is a Q_n-martingale for every $h\in H_n$.

Let $\tilde{Q}_n=Q_n\circ i^{-1}$, where $i\colon C(R_+,H_n)\to C(R_+,H)$ is the inclusion.

Then

$$(3.5) \qquad \tilde{Q}_n(x_0^H = x_n) = 1,$$

$$(3.6) \qquad \exp\left(<h, x_t^H - x_n - \int_0^t b_n(u, x_u^H)\, du> - 1/2 \int_0^t <a_n(u, x_u^H)\, h, h> du\right)$$

is a $\tilde{Q}_n$-martingale for every $h \in H$.

Now from Theorem 2.3 it follows that

$$(3.7) \qquad \int |x_t^H - x_s^H|^4 \, d\tilde{Q}_n \leqslant C_T(t-s)^2 \quad \text{if} \quad 0 \leqslant s \leqslant t \leqslant T.$$

Next, if we put $r_n^2(x) = \sum_{i=n}^{\infty} < x, e_i >^2$ then we have

$$\sup_n \int r_N^2\left(x_t^H - x_n - \int_0^t b_n(u, x_u^H)\, du\right) d\tilde{Q}_n = \sup_n \int \left(\sum_{j=N}^{\infty} < a_n(u, x_u^H)\, e_j, e_j > du\right) d\tilde{Q}_n \leqslant$$

$$\leqslant t \cdot \sup_{u,x} \sum_{j=N}^{\infty} < a(u, x)\, e_j, e_j > \to 0 \quad \text{as} \quad N \to \infty,$$

$$\sup_n r_N^2(x_n) = \sum_{j=N}^{\infty} < x, e_j >^2 \to 0 \quad \text{as} \quad N \to \infty,$$

$$\sup_n \int r_N^2 \left(\left(\int_0^t b_n(u, x_u^H)\, du\right) d\tilde{Q}_n \leqslant t \cdot \sup_{u,x} \sum_{j=N}^{\infty} < b(u, x), e_j >^2 \to 0 \quad \text{as} \quad N \to \infty,\right.$$

whence

$$(3.8) \qquad \lim_{N \to \infty} \sup_n \int r_N^2(x_t^H)\, d\tilde{Q}_n = 0.$$

The relations (3.7) and (3.8) imply that the sequence $(\tilde{Q}_n)$ is precompact in the space of probability measures on $C(R_+, H)$. Therefore we can suppose that $\tilde{Q}_n$ converges weakly to a probability measure Q. Then it is easy to see that Q satisfies the requirements.

Remark. The theorem holds for $k = 2$, too, if we suppose in addition that $< a(t, x)\, h, h > \geqslant C\, |h|^2$ for each h and for some positive constant C(one uses Theorem 4.5 from [14]).

Theorem 3.3. *Let* $a: D \times C(D, H) \to \mathcal{L}_1(H)$ *and* $b: D \times C(D, H) \to H$ *be such that*

(1) *a and b are $(\mathcal{F}_z^H$-nonanticipating;*

(2) *There exists a completely continuous operator S which is symmetric, positive definite and there exists a constant $C(T)$ such that*

$$\sup_{z,f} \operatorname{Trace} S^{-2} a(z, f) + \sup_{z,f} |S^{-1} b(z, f)|^2 \leqslant C(T);$$

(3) $\int_0^z a(u, f)\, du$ *and* $\int_0^{z]} b(u, f)\, du$ *are continuous in (z, f) relative to the norms* $\|\cdot\|_1$, $|\cdot|$.

Then we have the same conclusion as in Theorem 3.2.

Proof. The case $k=1$ is treated in [11]. The case $k=2$ is based on an application of Theorem 2.3 and Theorem 3.1 in a similar manner as in case $k=1$ (see[4] in connection with the finite dimensional case).

REFERENCES

1. Ciucu, G., Tudor, C., *Probabilities and Stochastic Processes, Volume II*. Bucureşti, Ed. Acad. R.S. Romania, 1979. (Romanian)
2. Dalec'kiĭ, Ju.L., *Infinite-dimensional elliptic operators and parabolic equations connected with them*. Uspehi Mat. Nauk, 22, 4, 3–54 (1967). (Russian)
3. Gihman, I.I., *Square integrable difference martingales*. Teor. Verojatnost. i Mat. Statist., **15**, 21–30 (1976). (Russian)
4. Gihman, I.I., Pjasec'ka, T.E., *A class of stochastic differential equations that contain two parameter white noise*. In: *Limit Theorems for Random Processes*, pp. 71–92. Tashkent, Fan, 1977. (Russian)
5. Gihman, I.I., *Existence of weak solutions for a class of hyperbolic systems that contain two parameter white noise*. In: *Theory of Random Processes, No.* 6, pp. 39–48. Kiev, Naukova, Dumka 1978. (Russian)
6. Gross, L., *Abstract Wiener space*. In: *Proc. 5th Berkeley Sympos. Math. Statist. Probab., Vol.* 2, pp. 31–42. Berkeley, Univ. California Press, 1965.
7. Kunita, H., *Stochastic integrals based on martingales taking values in Hilbert space*. Nagoya Math. J., **38**, 41–52 (1970).
8. Kallianpur, G., *Abstract Wiener processes and their reproducing kernel Hilbert space*. Z. Wahrsch. Verw. Gebiete, **17**, 113–123 (1971).
9. Kuo, H.H., *Gaussian Measures in Banach Spaces*. Moscow, Mir, 1979. (Russian)
10. Skorohod, A.V., *Studies in the Theory of Random Processes*. Reading, Mass., Addison-Wesley, 1965.
11. Skorohod, A.V., *K-martingales and stochastic equations*. In: *Proc. School-Seminar Theory Random Processes (Druskininkai*, 1974), *Part II*, pp. 195–234. Vilnius, 1975. (Russian).
12. Skorohod, A.V., *Random Linear Operators*. Kiev, Naukova Dumka, 1978. (Russian)
13. Stroock, D.V., Varadhan, S.R., *Multidimensional Diffusion Processes*. Berlin–Heidelberg–New York, Springer, 1979.
14. Tudor, C., *Stochastic integral equations in the plane*. Rev. Roumaine Math. Pures Appl., **26**, 507–538 (1981).
15. Tudor, C., *On the martingale problem in abstract Wiener space*. Rev. Roumaine Math. Pures Appl. (to appear)
16. Yeh, J., *Existence of strong solutions for stochastic differential equations in the plane*. Pacific J. Math., **97**, 217–247 (1981).

SIEGFRIED WEBER

An integral for ⊥-decomposable measures and applications

I consider normed ⊥-decomposable measures m in the sense that $m(A \cup B) = m(A) \perp m(B)$ with respect to some t-conorm ⊥. Using Ling's representation theorem for Archimedean t-conorms, an integral with respect to those measures is defined as extension of Lebesgue's one. This integral can be taken as measure of fuzzy sets extending so Zadeh's concept. As another application, this integral can serve as power function in robust test problems between two approximately known simple hypotheses, which is formalized by composite hypotheses containing ⊥-decomposable measures, alternatively to the approach by Huber and Strassen.

1. Introduction

Mathematical models using only additive measures can be too rough. The theory of capacities by Choquet [1] presents one way of weakening the additivity, and Choquet's integral is an extension of Lebesgue's integral. This concept is used in robust statistics, see [2]. Sugeno [7] has considered monotonous measures instead of additive ones and defined an integral for this more general situation. Sugeno's integral is used in the theory of fuzzy sets, founded by Zadeh [10]. But it does not reduce to Lebesgue's integral if the underlying measure reduces to be additive. This is obviously a drawback in those cases where the underlying measure has a special form as that proposed in [7]

(1) $$m(A \cup B) = m(A) + m(B) + \lambda \cdot m(A) \cdot m(B), \quad \lambda > -1$$

where $A \cup B$ means $A \cap B = \emptyset$.

Therefore the author [8] has suggested to look at normed measures m on a measurable space $(\Omega, \mathcal{B})$, which are ⊥-decomposable,

(2) $$m(A \cup B) = m(A) \perp m(B),$$

by means of t-conorms ⊥, and to define an integral in the Archimedean case, i.e. when ⊥ can be written as

(3) $$a \perp b = g^{(-1)} \, (g(a) + g(b))$$

with additive generator g and pseudoinverse $g^{(-1)}$. Originally t-conorms were introduced to formulate triangular (briefly: t-) inequalities for probabilistic metric spaces [6]. This new integral is an extension of Lebesgue's one, advantageously compared with Sugeno's concept. On the other hand, the linear structure induced by (2) is inherited to the integral, which is an advantage compared with Choquet's concept. It can be seen, that Sugeno's λ-measures (1) are of form (2) with (3).

Section 2 introduces t-conorms, for more details see [4], [6], [8].

Section 3 deals with the basic results of the theory of decomposable measures from [8], especially for the three Archimedean cases: (S) $\perp$ strict and therefore $g \circ m$ an infinite additive measure; (NSA) $\perp$ non strict and $g \circ m$ a finite additive measure; (NSP) $\perp$ non strict but $g \circ m$ a finite pseudoadditive measure.

Section 4 notes the fundamentals for the integral from [8], restricting to the cases (S) and (NSA), which can be considered as generalizations of Lebesgue's theory.

In Section 5 I take this new integral $\mathbf{I}_{(m)}$ with underlying decomposable measure m as measure $\tilde{m}$ for fuzzy subsets φ of Ω.

$$\tilde{m}(\varphi) = \mathbf{I}_{(m)}\varphi \tag{4}$$

In the case of a probability measure $m=P$, construction (4) reduces to that by Zadeh [11]. For more details see [9].

In Section 6 I suggest an approach to robust test problems between two approximately known simple hypotheses, where the known hypotheses are probabilities P_0 and P_1, and the approximations are described by the composite hypotheses, $\mathcal{M}_0$ and $\mathcal{M}_1$, each $\mathcal{M}_i$ contains $\perp$-decomposable measures of the form

$$m_{i,\lambda} = g_{i,\lambda}^{-1} \circ P_i, \tag{5}$$

which correspond to case (NSA) with normed generators. The integral will be used to define the power function.

2. t-conorms

2.1. **Definition** ([6]). A function $\perp : [0,1]\times[0,1] \to [0,1]$ will be called a *t-conorm*: $\Leftrightarrow$ $\perp$ is non decreasing in each argument, commutative, associative, and has 0 as unit.

A t-conorm $\perp$ will be called *Archimedean*: $\Leftrightarrow$ $\perp$ is continuous, and $\perp(a, a) > a$ for all $a \in (0,1)$.

An Archimedean t-conorm $\perp$ will be called *strict*: $\Leftrightarrow$ $\perp$ is (strictly) increasing.

2.2. **Theorem** ([4]). *A function $\perp$: $[0,1]\times[0,1] \to [0,1]$ is an Archimedean t-conorm $\Leftrightarrow$ There exists an increasing and continuous function g : $[0,1] \to [0, \infty]$ with $g(0) = 0$ so that*

$$\perp(a,b) = g^{(-1)}(g(a)+g(b))$$

where $g^{(-1)}$ is the pseudoinverse of g, defined by

$$g^{(-1)}(y) := g^{-1}(\min(y, g(1))).$$

Moreover: $\perp$ strict $\Leftrightarrow$ $g(1) = \infty$.

The functions g from 2.2 are called *(additive) generators* of $\perp$. They are unique except for multiplication with positive numbers. In the non strict case we will call the generator with $g(1) = 1$ the *normed generator*.

2.3. *Examples.* a) $\perp(a,b) = a \vee b = \max(a, b)$ is the weakest t-conorm, continuous, but non Archimedean.

b) $\perp_s(a,b) := \begin{cases} a & \text{if } b = 0 \\ b & \text{if } a = 0 \\ 1 & \text{otherwise} \end{cases}$ is the strongest t-conorm, non continuous.

c) $\perp_0(a,b) := a + b - ab$ is strict with generator $g_0(x) = -\ln(1 - x)$.
d) $\perp_m(a,b) := \min(a+b, 1)$ is non strict with normed generator $g_m(x) = x$.
e) $U_\lambda(a, b) := \min(a + b + \lambda ab, 1)$ for $\lambda > -1$ is non strict with normed generator

$$g_\lambda(x) = \frac{\ln(1+\lambda x)}{\ln(1+\lambda)} \text{ with } g_\lambda^{-1}(y) = ((1+\lambda)^y - 1)/\lambda \text{ for } y \leq 1.$$

The U_λ increase between the weakest $\lim\limits_{\lambda \to -1} U_\lambda = \perp_0$ and the strongest $\lim\limits_{\lambda \to \infty} U_\lambda = \perp_s$, including also $U_0 = \perp_m$.

For more details and references I refer to [4], [6], [8].

3. $\perp$-decomposable measures

Let $(\Omega, \mathcal{B})$ be some measurable space, and $\perp$ denote any t-conorm.

3.1. **Definition** ([8]). A function $m : \mathcal{B} \to [0,1]$ with $m(\emptyset) = 0$ and $m(\Omega) = 1$ will be called *$\perp$-or σ-$\perp$-decomposable measure*: $\Leftrightarrow m(A \cup B) = m(A) \perp m(B)$ or $m\left(\bigcup\limits_{k=1}^{\infty} A_k\right) = \mathop{\perp}\limits_{k=1}^{\infty} m(A_k)$ resp, where $\mathop{\perp}\limits_{k=1}^{\infty}$ means $\lim\limits_{N\to\infty} \mathop{\perp}\limits_{k=1}^{N}$.

3.2. **Theorem** ([8]). *a) m $\perp$-decomposable $\Rightarrow$ m monotonuos,*
b) m σ- $\perp$-decomposable $\Leftrightarrow$ m $\perp$-decomposable and continuous from below.

3.3. **Remark.** There are σ-$\perp$-decomposable measures, which are not continuous from above. For the case $\perp = \vee$ recently Puri and Ralescu [5] have proved this for all non trivial possibility measures, introduced by Zadeh [12]. On the other hand, the author [8] has proved that for Archimedean $\perp$ always follows continuity from above, with an obvious restriction.

3.4. **Theorem** ([8]). *Let m be a (σ-)$\perp$-decomposable measure with respect to an Archimedean t-conorm $\perp$ with generator g.*

$$\Rightarrow (g \circ m)\left(\bigcup_k A_k\right) = \min\left(\sum_k (g \circ m)(A_k),\ g(1)\right),$$

where we have to take a finite or enumerable number resp. of A_k.

In view of this result we can distinguish between three cases:
(S) $\perp$ strict and therefore $g \circ m$ an infinite (σ-)additive measure,
(NSA) $\perp$ non strict and $g \circ m$ a finite (σ-) additive measure,
(NSP) $\perp$ non strict and $g \circ m$ finite *pseudo* (σ-) *additive*, i.e.

$$(g \circ m)\left(\bigcup_k A_k\right) = g(1) < \sum_k (g \circ m)(A_k)$$

is possible.

Clearly, always $(g \circ m)(\Omega) = g(1)$.

In Section 6 we need the following reverse result.

3.5. **Theorem.** *Let μ be a (σ-) additive measure and g some generator or an Archimedean t-conorm $\perp$ with $g(1) = \mu(\Omega)$.*
$\Rightarrow m := g^{-1} \circ \mu$ is a (σ-) $\perp$-decomposable measure of type (S) or (NSA) if $\perp$ is strict or non strict respectively.

3.6. *Example.* In 3.5 take as $\mu = P$ some probability measure and g_λ from 2.3e), then we obtain Sugeno's λ-measure

$$m_\lambda(A) = ((1+\lambda)^{P(A)} - 1)/\lambda,$$

including $m_0 = P$.
The result by Kruse [3] follows immediately.

More about decomposable measures can be found in [8].

4. Integrals for the Archimedean cases

Let m be a σ-$\perp$-decomposable measure on $(\Omega, \mathcal{B})$ with respect to an Archimedean t-conorm $\perp$ with generator g. The integrands $\varphi, \psi, \ldots$ are measurable functions from Ω to $[0,1]$.

4.1. **Definition** ([8]). The *integral* of φ over $A \in \mathcal{B}$ is defined for cases (S) and (NSA) as

$$\int_A \varphi \perp m := g^{-1}\left(\int_A \varphi \mathrm{d}\,(g \circ m)\right).$$

Write $\int_\Omega \varphi \perp m$ as $\mathbf{I}\varphi$ for short or as $\mathbf{I}_{(m)}\varphi$ if necessary. We can introduce a multiplication between "scalar" $c \in [0,1]$ equipped with $+$ and "vector" x equipped with $\perp$ by

$$c * x := g^{-1}(c \cdot g(x)).$$

4.2. **Theorem** ([8]). *The functional $\mathbf{I}$ is*

an extension of m	: $\mathbf{I}(1_A) = m(A)$ *for* $A \in \mathcal{B}$,
normed	: $\mathbf{I}(0) = 0,\ \mathbf{I}(1) = 1$,
monotonuous	: $\varphi \leqslant \psi \Rightarrow \mathbf{I}\varphi \leqslant \mathbf{I}\psi$,
semi-linear	: $\mathbf{I}(\varphi + \psi) = \mathbf{I}\varphi \perp \mathbf{I}\psi$ *for* $\varphi + \psi \leqslant 1$,
	$\mathbf{I}(c \cdot \varphi) = c * \mathbf{I}\varphi$ *for* $c \in [0,1]$,

continuous from below: $\varphi_n \uparrow \varphi$ *a.e.* $\Rightarrow \mathbf{I}\varphi_n \uparrow \mathbf{I}\varphi$.

4.3. *Example* (continuation from 3.6). Write $\mathbf{I}_{(m_\lambda)}$ as $\mathbf{I}_\lambda$, then

$$\mathbf{I}_\lambda \varphi = ((1+\lambda)^{\mathbf{I}_0\varphi} - 1)/\lambda \text{ with } \mathbf{I}_0\varphi = \int_\Omega \varphi \,\mathrm{d}P.$$

Clearly, $\mathbf{I}_\lambda\varphi \to \mathbf{I}_0\varphi$ if $\lambda \to 0$.

More details can be found in [8], including also the case (NSP) and a comparison with the integrals by Choquet and Sugeno.

5. Measures of fuzzy sets

5.1. **Definition** ([10]). *A fuzzy (sub-) set* of (the ordinary set) Ω is a function $\varphi: \Omega \to [0,1]$.

Here $\varphi(\omega)$ will mean the *grade of membership* of ω to φ. From this interpretatorical point of view, φ will be called often the *membership function* of the fuzzy set, e.g. the (fuzzy-) set of all feal numbers ω which are "much greater than zero" can be defined by, say,

$$\varphi(\omega) = \begin{Bmatrix} 0 & \text{if} & \omega < 0 \\ 10^{-4} \cdot \omega^2 & \text{if} & 0 \leqslant \omega \leqslant 100 \\ 1 & \text{if} & \omega > 100 \end{Bmatrix}.$$

In connection with 5.1 an ordinary subset A of Ω is identified with its characteristic function $\varphi = 1_A$.

Analogously to the ordinary case, measurable fuzzy sets are called *fuzzy events*. For those we can define a measure, following the idea by Zadeh [11] for probability measures.

5.2. **Definition.** Let m be a (σ-) $\perp$-decomposable measure for events $A \in \mathcal{B}$, then a measure $\tilde{m}$ for fuzzy events φ is given by

$$\tilde{m}(\varphi) := \mathbf{I}_{(m)}\varphi.$$

By 4.2, $\tilde{m}$ extends m, is normed by $\tilde{m}(0) = 0$ and $\tilde{m}(1) = 1$, and is (σ-)-$\perp$ decomposable in the sense that

$$\tilde{m}\left(\sum_k \varphi_k\right) = \mathop{\perp}_k \tilde{m}(\varphi_k) \quad \text{for} \quad \sum_k \varphi_k \leqslant 1.$$

For more details see the author's forthcoming paper [9].

6. A robust testing model

Alternatively to the approach by Huber and Strassen [2] I suggest to formalize the approximation of a known probability P as a neighborhood of P containing $\perp$-decomposable measures m with induced measure $g \circ m = P$. In the following I will give the first steps in this direction.

6.1. *Assumption.* For $i = 1,2$ let be P_i two distinct probabilities on $(\Omega, \mathcal{B})$, $\Lambda_i \subseteq \mathbb{R}$, $g_{i,\lambda}$ normed generators of non strict Archimedean t-conorms $\perp_{i,\lambda}$, and

$$\mathcal{M}_i = \{m_{i,\lambda} \mid m_{i,\lambda} = g_{i,\lambda}^{-1} \circ P_i,\ \lambda \in \Lambda_i\}.$$

6.2. **Definition.** Let $\varphi: \mathbb{R}^n \to [0,1]$ be the critical function of a *test* and $X: \Omega \to \mathbb{R}^n$ a (measurable) *observation*. Then define the:

(**I**-) *power function* of φ as $\beta_\varphi(i, \lambda) := \mathbf{I}_{(m_{i,\lambda})}(\varphi X)$, especially:

(**I**-) *size* of φ as $\sup_{\lambda \in \Lambda_0} \beta_\varphi(0, \lambda)$

(**I**-) *power* of φ *against* $\lambda \in \Lambda_1$ as $\beta_\varphi(1, \lambda)$.

A test φ is said to be (**I**-) UMP of (**I**)-size α for testing $\mathcal{M}_0$ against $\mathcal{M}_1$: $\Leftrightarrow$ φ is of (**I**-) size α, and any other test ψ of (**I**-) size at most α there is

$$\beta_\psi(1, \lambda) \leqslant \beta_\varphi(1, \lambda) \text{ for all } \lambda \in \Lambda_1.$$

Write for the classical case as usual $\mathbf{E}_i(\varphi X) := \int_\Omega \varphi X \, \mathrm{d}P_i$, then we can write the

(**I**-) power function of φ as $\beta_\varphi(i, \lambda) = g_{i,\lambda}^{-1} \circ \mathbf{E}_i(\varphi X)$, i.e. in terms of the classical power function.

6.3. **Theorem** (Neyman-Pearson-Lemma for $\perp$-decomposable measures). *Let P_i have density f_i and suppose that there exists $\lambda^* \in \Lambda_0$ with* $\sup_{\lambda \in \Lambda_0} g_{0,\lambda}^{-1} = g_{0,\lambda^*}^{-1}$.

a) *Any test*

$$\varphi(x) = \begin{cases} 1 & \text{if } f_1(x) > k \cdot f_0(x) \\ \gamma(x) & \text{if } \quad = \\ 0 & \text{if } \quad < \end{cases},$$

for some $k \geqslant 0$ and $0 \leqslant \gamma(x) \leqslant 1$, is (**I**-) *UMP of its* (**I**-) *size for testing $\mathcal{M}_0$ against $\mathcal{M}_1$. Corresponding to $k = \infty$, the test*

$$\varphi(x) = \begin{cases} 1 & \text{if } f_0(x) = 0 \\ 0 & \text{if } \quad > \end{cases}$$

is (**I**-)*UMP of* (**I**-)*size zero.*

b) *The test φ of* (**I**-) *size α is equivalent to the (classical) Neyman-Pearson-test of (classical) size $g_{0,\lambda^*}(\alpha)$.*

c) *For every $\alpha \in (0,1)$ there exists a test of the form above with constant γ and* (**I**-)*size α, and is unique.*

Proof. a) Take any test ψ with $\sup_{\lambda \in \Lambda_0} \beta_\psi(0, \lambda) \leqslant \sup_{\lambda \in \Lambda_0} \beta_\varphi(0, \lambda)$. The additional assumption leads to $\mathbf{E}_0(\psi X) \leqslant \mathbf{E}_0(\varphi X)$. Now, the (classical) N-P-lemma tells us that any test of the given form is best of its (classical) size $\mathbf{E}_0(\varphi X)$. Therefore $\mathbf{E}_1(\psi X) \leqslant \mathbf{E}_1(\varphi X)$, which implies that $\beta_\psi(1, \lambda) \leqslant \beta_\varphi(1, \lambda)$ for all $\lambda \in \Lambda_1$.

6.4. *Example.* If $\{g_{0,\lambda}^{-1}\}_{\lambda \in \Lambda_0}$ is non increasing, then $\lambda^* = \inf_{\lambda \in \Lambda_0} \lambda$. This is the case, e.g., for $g_{0,\lambda} = g_\lambda$ from 2.3e). Furthermore, we see that for the choice of $\Lambda_0 = [0, \lambda_0]$ with arbitrary $\lambda_0 \geqslant 0$, (**I**-) size and (classical) size coincide.

REFERENCES

1. Choquet, G., *Theory of capacities.* Ann. Inst. Fourier (Grenoble), **5**, 131–292 (1953/54).
2. Huber, P.J., Strassen, V., *Minimax tests and the Neyman-Pearson lemma for capacities.* Ann. Statist., **1**, 251–263 (1973).
3. Kruse, R., *A note on λ-additive fuzzy measures.* Fuzzy Sets and Systems, **8**, 219–222 (1982).
4. Ling, C.H., *Representation of associative functions.* Publ. Math. Debrecen, **12**, 189–212 (1965).

5. Puri, M.L., Ralescu, D., *A possibility measure is not a fuzzy measure.* Fuzzy Sets and Systems, 7, 311–313 (1982).
6. Schweizer, B., Sklar, A., *Associative functions and statistical triangle inequalities.* Publ. Math. Debrecen, 8, 169–186 (1961).
7. Sugeno, M., *Theory of fuzzy integrals and its applications.* Ph. D. Thesis, Tokyo Institute of Technology, 1974.
8. Weber, S., *⊥-decomposable measures and integrals fcr Archimedean t-conorms* ⊥. J. Math. Anal. Appl. (to appear).
9. Weber, S., *Measures of fuzzy sets and fuzziness.* (to appear).
10. Zadeh, L., *Fuzzy sets.* Inform. and Control, 8, 338–353 (1965).
11. Zadeh, L., *Probability measures of fuzzy events.* J. Math. Anal. Appl., 23, 421–427 (1968).
12. Zadeh, L., *Fuzzy sets as a basis for a theory of possibility.* Fuzzy Sets and Systems, 1, 3–28 (1978).

G. ZBĂGANU

Two inequalities concerning centered moments

Let $(\Omega, \mathbf{K}, P)$ be a probabilized space. Let also $\mathbf{F} \subset \mathbf{K}$ be a σ-algebra, $f \in L^p(\Omega, \mathbf{K}, P)$ with some $1 < p < \infty$. In this short note two inequalities are established between $||f - E^{\mathbf{F}} f||_p$ and $||f||_p^p - ||E^{\mathbf{F}} f||_p^p$. The aim is to prove that if f_n is a sequence of L^p-functions, $||f_n||_p = 1$ and $\lim_n ||E^{\mathbf{F}}(f_n)||_p = 1$, then $\lim_n ||f_n - E^{\mathbf{F}}(f_n)||_p = 0$.

Proposition 1. (i) *If* $p \geqslant 2$, *then*

$$E(|f - E^{\mathbf{F}} f|^p) \leqslant 2^{p-1}(E(|f|^p) - E(|E^{\mathbf{F}} f|^p)) \tag{1}$$

(ii) *If* $f_n \in L^p$ *and* $\lim(||f_n||_p^p - ||E^{\mathbf{F}}(f_n)||_p^p) = 0$, *then* $\lim ||f_n - E^{\mathbf{F}} f_n|| = 0$.

Proof. (i) We shall use the elementary inequality $|x-a|^p \leqslant |x^p - a^p|$ if a, $x \geqslant 0$ and $p > 1$. Consider first the case $f \geqslant 0$. Then $E^{\mathbf{F}} f \geqslant 0$ and $E(|f - E^{\mathbf{F}} f|^p) = E((|f - E^{\mathbf{F}} f|^{p/2})^2) \leqslant E(|f^{p/2} - (E^{\mathbf{F}} f)^{p/2}|)^2 = E(f^p + (E^{\mathbf{F}} f)^p - 2f^{p/2}(E^{\mathbf{F}} f)^{p/2}) \leqslant E(f^p) - E((E^{\mathbf{F}} f)^p)$ (since $x \mapsto x^{p/2}$ is a convex function, Jensen's inequality implies that $E^{\mathbf{F}}(f^{p/2}) \geqslant (E^{\mathbf{F}} f)^{p/2}$, hence $E(f^{p/2}(E^{\mathbf{F}} f)^{p/2}) = E(E^{\mathbf{F}}(f^{p/2})(E^{\mathbf{F}} f)^{p/2}) \geqslant E((E^{\mathbf{F}} f)^p))$).

In the general case $f = f_+ - f_-$, $|f|^p = f_+^p + f_-^p$. Therefore

$$E(|f - E^{\mathbf{F}} f|^p) = E(|f_+ - f_- - E^{\mathbf{F}}(f_+) + E^{\mathbf{F}}(f_-)|^p) \leqslant$$

$$\leqslant E((|f_+ - E^{\mathbf{F}}(f_+)| + |f_- - E^{\mathbf{F}}(f_-)|)^p) \leqslant 2^{p-1}(||f_+ - E^{\mathbf{F}}(f_+)||_p^p + ||f_- - E^{\mathbf{F}}(f_-)||_p^p)$$

(due to the elementary inequality $(a+b)^p \leqslant 2^{p-1}(a^p + b^p)$ if $a, b \geqslant 0$)

$$\leqslant 2^{p-1}(E(f_+^p) - E((E^{\mathbf{F}}(f_+))^p) + E(f_-^p) - E((E^{\mathbf{F}}(f_-))^p)) =$$

$$= 2^{p-1}(E(|f|^p) - E(E^{\mathbf{F}}(f_+)^p + E^{\mathbf{F}}(f_-)^p)).$$

But the functions $x \mapsto x_+$ and $x \mapsto x_-$ are convex; a new application of Jensen's inequality yields $E^{\mathbf{F}}(f_+) \geqslant (E^{\mathbf{F}} f)_+$ and $E^{\mathbf{F}}(f_-) \geqslant (E^{\mathbf{F}} f)_-$ hence $E^{\mathbf{F}}(f_+)^p \geqslant (E^{\mathbf{F}} f)_+^p$, $E^{\mathbf{F}}(f_-)^p \geqslant (E^{\mathbf{F}} f)_-^p$ and $E^{\mathbf{F}}(f_+)^p + E^{\mathbf{F}}(f_-)^p \geqslant (E^{\mathbf{F}} f)_+^p + (E^{\mathbf{F}} f)_-^p = |E^{\mathbf{F}} f|^p$.

Therefore the last term in the above evaluations is less or equal to $2^{p-1}(E(|f|^p) - E(|E^{\mathbf{F}} f|^p))$ and (1) is proved.

As for the assertion (ii), it is a straightforward application of the inequality (1). The proof is complete.

We are going to prove now that no such evaluation is possible when $1 < p < 2$.

Proposition 2. (i) *Let* $1 < p < 2$ *and* $f, g \in L^2 (\subset L^p)$. *Suppose that* $Ef = Eg$. *Then* $\lim_{t\to\infty} E((f+t)^p - (g+t)^p) = 0$

(ii) *If* $f \geq 0, f \in L^2$, *then* $\lim_{t\to\infty} E((f+t)^p - (E^{\mathbf{F}}f + t)^p) = 0$

(iii) *It is not possible to obtain an inequality of the form*

$$\|f - E^{\mathbf{F}}f\|_p \leq h(\|f\|_p^p - \|E^{\mathbf{F}}f\|_p^p)$$

where $h: R_+ \to R_+$ *is some continuous function with* $h(0) = 0$.

Proof. Clearly, (ii) and (iii) are trivial consequences of (i). To check it, set $x = 1/t$. It is no lack of generality in agreeing that $x < 1$. Now, we have $\lim_{t\to\infty} (E((f+t)^p) - E((g+t)^p)) = \lim_{x\to 0} \frac{E((1+xf)^p - (1+xg)^p)}{x^p} =$ $= \lim_{x\to 0} \frac{E(f(1+xf)^{p-1} - g(1+xg)^{p-1})}{x^{p-1}} = \lim_{x\to 0} E\left(x^{2-p}\left(\frac{f^2}{(1+xf)^{2-p}} - \right.\right.$ $\left.\left. - \frac{g^2}{(1+xg)^{2-p}}\right)\right) = 0$. In these computations L'Hospital's theorem was used. This is allowed because $((1+xf)^p)' = pf(1+xf)^{p-1} \leq p(1+f)^p \in L^1$ and Lebesgue's dominated convergence theorem permits the differentiation under the expectation sign.

Remark 1. The above result points out in a dramatical manner that not always what is true in L^2 can be "approximated" to be true in L^p, $1 < p < \infty$, as it seemed, due to uniform convexity properties of these spaces.

Remark 2. We can ask ourselves if it were possible that a majorization of the form $E(|f - E^{\mathbf{F}}f|^p) \leq C_p \|E^{\mathbf{F}}f\|_p^{pa}(\|f\|_p^p - \|E^{\mathbf{F}}f\|_p^p)^b$ should hold with some $a, b > 0$. The answer is negative, as the following example points out: $f = c1_A + (c+x)1_{A^c}$ with $c, x \geq 0$, $\mathbf{F}$ = the trivial σ-algebra, $P(A) = q$ and $P(A^c) = 1 - q = q'$. Let also $h_{ab}(c, x, p) = \frac{E(|f - Ef|^p)^{1/b}}{(Ef)^{pa/b}(E(|f|^p) - (Ef)^p)}$.
The problem is if we can choose a and b such that $\sup\{h_{ab}(c, x, p): c, x \geq 0, 0 < q < 1\}$ be finite. After some easy computations we get

$$h_{ab}(c, x, p) = \frac{x^{p/b}(qq'^p + q'q^p)^{1/b}}{(qc^p + q'(c+x)^p - (c+q'x)^p)(c+q'x)^{pa/b}}.$$

Setting $c = 0$, it turns out that $a + b = 1$ (otherwise either $h_{ab}(0, 0, p) = \infty$ or $h_{ab}(0, \infty, p) = \infty$). We obtain quickly a contradiction observing that the request that $\sup\{h_{ab}(0, 1, p): 0 < q < 1\}$ be finite implies that $pa + b < 1$.

The study of this example made us feel that *an inequality of the form* $\|f - E^{\mathbf{F}}f\|_p^{2p} \leq C_p \|f\|_p^p(\|f\|_p^p - \|E^{\mathbf{F}}f\|_p^p)$ *could hold*, at least when $\mathbf{F}$ is trivial. A simulation on the computer makes us believe that $C_p \leq 2/(p-1)$. But, in the absence of any proof of that conjecture, we shall confine ourselves to prove a much weaker inequality, still good enough for our purpose. From now on, $\|f\|$ will denote $\|f\|_p$ as p will be fixed.

Proposition 3. *Let $1 < p < 2$, $q = p/(p-1)$ be the conjugate of p, $f \geqslant 0$ from $L^p(\Omega, \mathbf{K}, P)$ and $\mathbf{F} \subset \mathbf{K}$ be a sub σ-algebra of $\mathbf{K}$. Let also $Nf = (E^{\mathbf{F}}(f^p))^{1/p}$ and $\varepsilon > 0$. Then the following inequalities hold:*

(i) $||f - Nf||^p \leqslant \varepsilon^p ||f||^p + 2q(1 + \varepsilon^{-1})^{2-p} ||f||^{p-1}||Nf - E^{\mathbf{F}}f||$,

(ii) $||f - E^{\mathbf{F}}f|| \leqslant 2 ||f - Nf||$,

(iii) $||f - Ef|| \leqslant 2 ||f|| \left(\varepsilon^p + 2q(1 + \varepsilon^{-1}) \left(1 - \dfrac{Ef}{||f||}\right)\right)^{1/p}$

Proof. (i). The function $g(x) = x^p - px + p - 1$ is convex and has a single minimum point $x = 1$, $g(1) = 0$. Expanding g in Taylor's series in a neighbourhood of $x = 1$, we get $g(x) = p(p-1)(x-1)^2(1 + t(x-1))^{p-2}/2$ with some $0 < t < 1$. Therefore

$$(2)\qquad p(p-1)\max(1, x^{p-2})(x-1)^2 \geqslant 2(x^p - px + p - 1) \geqslant \geqslant p(p-1)\min(1, x^{p-2})(x-1)^2$$

In these inequalities, set $x = f/Nf$ and apply the conditional expectation. Then

$$(3)\qquad (p-1)E^{\mathbf{F}}((1-(f/Nf))^2 \max(1,(f/Nf)^{p-2}) \geqslant 2(1 - (E^{\mathbf{F}}f/Nf)) \geqslant \geqslant (p-1)E^{\mathbf{F}}((1-(f/Nf))^2 \min(1, (f/Nf)^{p-2}))$$

Let now $\varepsilon > 0$ be arbitrary but fixed, and set $A_\varepsilon = \left\{\left|\dfrac{f}{Nf} - 1\right| \geqslant \varepsilon\right\}$. Since for $|x-1| \geqslant \varepsilon$, the elementary inequality $\min(1, x^{p-2})(x-1)^2 \geqslant \geqslant (1 + \varepsilon^{-1})^{p-2} |x-1|^p$ holds, it turns out that

$$E^{\mathbf{F}}(|1 - f/Nf|^p) \leqslant E^{\mathbf{F}}(|1 - f/Nf|^p 1_{A_\varepsilon^c}) + E^{\mathbf{F}}(|1 - f/Nf|^p 1_{A_\varepsilon}) \leqslant$$
$$\leqslant \varepsilon^p E^{\mathbf{F}}(1_{A_\varepsilon^c}) + (1 + \varepsilon^{-1})^{2-p} E^{\mathbf{F}}(\min(1,(f/Nf)^{p-2})(1 - f/Nf)^2 1_{A_\varepsilon}) \leqslant$$
$$\leqslant \varepsilon^p + (1 + \varepsilon^{-1})^{2-p} E^{\mathbf{F}}(\min(1, (f/Nf)^{p-2})(1 - f/Nf)^2)$$

and, using (3) the last quantity is further less or equal to

$$\varepsilon^p + \frac{2}{p-1}(1 + \varepsilon^{-1})^{2-p}(1 - E^{\mathbf{F}}f/Nf).$$

Or, otherwise written

$$(4)\qquad E^{\mathbf{F}}(|f - Nf|^p) \leqslant \varepsilon^p (Nf)^p + \frac{2}{p-1}(1 + \varepsilon^{-1})^{2-p}(Nf - E^{\mathbf{F}}f)(Nf)^{p-1} \leqslant \leqslant \varepsilon^p(Nf)^p + 2q(1 + \varepsilon^{-1})^{2-p}(Nf - E^{\mathbf{F}}f)(Nf)^{p-1}$$

Taking the expectation in (4) and remarking that $E((Nf)^p) = E(f^p)$ and also that $E((Nf - E^{\mathbf{F}}f)(Nf)^{p-1}) \leqslant ||(Nf)^{p-1}||_q ||Nf - E^{\mathbf{F}}f||$ (Hölder!) $= ||f||^{p-1}||Nf - E^{\mathbf{F}}f||$, the inequality (i) immediately follows.

(ii). $||f - E^{\mathbf{F}}f|| \leqslant ||f - Nf|| + ||Nf - E^{\mathbf{F}}f||$ and $||Nf - E^{\mathbf{F}}f|| = = ||E^{\mathbf{F}}(f - Nf)|| \leqslant ||f - Nf||$.

(iii). This is a particular case of the inequality (i) when $\mathbf{F}$ is trivial. Then $Nf = ||f||$ and $E^{\mathbf{F}}f = Ef$. The proof is complete.

Proposition 4. *Let* $1 < p < \infty$ *and* $(f_n)_{n\geqslant 1} \in L^p$ *such that* $||f_n|| = 1$. *Suppose that* $||E^F f_n|| \to 1$. *Then* $||f_n - E^F f_n|| \to 0$.

Proof. If $p \geqslant 2$, the proof is already made. We suppose that $1 < p < 2$. Consider first the case when $f_n \geqslant 0$ for every n. Then the inequality (i) from the above Proposition becomes

$$||f_n - Nf_n||^p \leqslant \varepsilon^p + 2q(1 + \varepsilon^{-1})^{2-p}||Nf_n - E^F f_n||. \tag{5}$$

We shall check that $||Nf_n - E^F f_n|| \to 0$. Indeed, $Nf_n \geqslant E^F f_n$ (Jensen) and $1 = Nf_n = \lim E^F(f_n)$. Therefore $1 = E((Nf_n)^p) = E(E^F(f_n) + (Nf_n - E^F(f_n)))^p \geqslant$ $\geqslant E((E^F f_n)^p) + E((Nf_n - E^F f_n)^p)$ hence $E((Nf_n - E^F f_n)^p) \leqslant 1 - E((E^F f_n)) \to$ $\to 0$. This fact further implies that limsup $||f_n - Nf_n|| \leqslant \varepsilon$ i.e. $\lim ||f_n - Nf_n|| =$ $= 0$ since ε is arbitrarily small. Using (ii) from Proposition 3, it follows that $||f_n - E^F f_n|| \to 0$.

In the general case, write $f_n = f_{n+} - f_{n-}$, $|f_n| = f_{n+} + f_{n-}$. As $E^F(|f_n|)$ tends to 1, it is clear that $|||f_n| - E^F(|f_n|)|| \to 0$. Take now an ε arbitrarily small and divide the sequence $(f_n)_n$ into two subsequences $(f_i, i \in I)$, $(f_j, j \in J)$ where $I = \{n \in N / ||f_{n+}|| \leqslant \varepsilon$ or $||f_{n-}|| \leqslant \varepsilon\}$ and $J = I^c$. Without lack in generality, we may suppose that both I and J are infinite.

Let $i \in I$. Suppose, say, that $||f_{i-}|| \leqslant \varepsilon$. Then $||E^F(f_{i-})|| \leqslant \varepsilon$ and $||f_i - E^F(f_i)|| \leqslant ||f_{i+} + f_{i-} - E^F(f_{i+}) - E^F(f_{i-})|| + 2(||f_{i-}|| + ||E^F(f_{i-})||) \leqslant$ $\leqslant ||f_i - E^F(|f_i|)|| + 4\varepsilon$. The same relation holds also if $||f_{i+}|| < \varepsilon$.

Let now $j \in J$. We shall verify that

$$\frac{||E^F(f_{j+})||}{||f_{j+}||} \to 1; \quad \frac{||E^F(f_{j-})||}{||f_{j-}||} \to 1 \text{ as } j \to \infty. \tag{6}$$

Suppose ad absurdum that, for instance $||E^F(f_{j+})|| \leqslant m||f_{j+}||$ i.o. with some $0 < m < 1$. Take into account that $E((E^F(f_{j-}))^p) \leqslant E((f_{j-})^p)$, $E^F(f_{j+}) \geqslant$ $\geqslant (E^F f_j)_+$ and $E^F(f_{j-}) \geqslant (E^F f_j)_-$. Then we get $||E^F f_j||^p = E(|E^F f_j|^p) \leqslant$ $\leqslant m^p E(f^p_{j+}) + E(f^p_{j-}) \leqslant 1 - (1 - m^p)$ i.o. and that is absurd because $||E^F f_j|| \to$ $\to 1$ as $j \to \infty$.

It is clear now that (6) implies $||f_{j+} - E^F(f_{j+})|| \to 0$, $||f_{j-} - E^F(f_{j-})|| \to$ $\to 0$ hence $||f_j - E^F f_j|| \to 0$ as $j \to \infty$. Joining together the two cases it follows that limsup $||f_n - E^F f_n|| \leqslant 4\varepsilon$ and we are done, since ε is arbitrarily small.

Remark 3. If $p = 1$ or $p = \infty$ and $||f||_p = ||E^F f||_p$, one cannot infer that $f = E^F f$ (however, it is not difficult to prove that $||f||_1 = ||E^F f||_1 \Leftarrow \{f >$ $> 0\} \in \mathbf{F}$!); that's why, in these cases Proposition 5 cannot hold.

Remark 4. We are sure that in the right side of (1), 2^{p-1} is *not* the best constant. We conjecture that the best one is $2^{p-1}/p$, but, so far, our methods seemed to be fruitless.

Acknowledgements. The author is deeply indebted to his colleagues M. Dragomirescu and F. Gonzacenco for their assistance and fruitful ideas.

REFERENCES

1. Lindenstrauss, J., Tzafriri, L., *Classical Banach Spaces*. Lecture Notes in Math, Vol., 338. Berlin–Heidelberg–New York, Springer, 1973.
2. Neveu, J., *Martingales à Temps Discret*. Paris, Masson, 1972.

H. AHRENS, R. PINCUS, J. SANCHEZ

Unbalancedness and efficiency in estimating components of variance

Several papers have been published regarding comparisons of estimation procedures for components of variance, including MINQUE and ANOVA estimators (Ahrens, Kleffe and Tenzler (1979), Hess (1979), Ahrens (1978), La Motte (1976), Corbeil and Searle (1976), Townsend and Searle (1971)).
The basis of these comparisons are MSE functions of the estimators. Beside of the parameters these functions essentially depend on the n_j — the numbers of observations in each group. Assuming a pattern of the n_j it is possible to compare several estimation procedures. But this makes sense only if the results remain valid for other similar designs, where "similar" is understood as "equally unbalanced".
This paper discusses measures of unbalancedness and shows that for a given statistical procedure they may serve as measures of efficiency of a design. They also allow to compare for example estimation methods for variance components in designs with a fixed level of unbalancedness.

1. Introduction

We use the one-way classification random model as a basic model to study the relations between unbalancedness and efficiency for different MINQ-and ANOVA-estimation procedures.

Ahrens and Pincus (1981) defined measures of unbalancedness in the one-way classification model. Using one of these we are able to define classes of unbalanced designs for a given level of unbalancedness and within each class the most unbalanced design. See also Ahrens and Sanchez (1982). Based on these measures we compare by means of maximum and minimum efficiencies several MINQ-estimators and also the ANOVA-estimator. Moreover we define an efficiency concept of a given design as the relation of the MSE-functions of balanced and the given unbalanced design. For the minimum efficiency of the MINQ-estimators

(i) MINQE (I) — invariant and biased
(ii) MINQE (I, +) — invariant, biased and positive
(iii) MINQE (U, I) — unbiased and invariant

in the case $\rho_0 = \infty$ we derive upper and lower bounds which are simultaneously valid for the three procedures. We include in our considerations only the estimator $\hat{\sigma}_a^2$, statements for $\hat{\sigma}_e^2$ are similar and more simple.

2. Measures of unbalancedness

We consider the one-way model

$$y_{jk} = \mu + a_j + e_{jk} \quad (j = 1, 2, ..., m; \; k = 1, 2, ..., n_j),$$

$N = \Sigma n_j$, with μ fixed, the e_{jk} independently $N(0, \sigma_e^2)$ distributed and the a_j' either

— independently $N(0, \sigma_a^2)$ distributed (random effects), or

— fixed but unknown, with reparametrization condition $\Sigma a_j = 0$ (fixed effects).

By N we denote a design

$$N := \{n_1, ..., n_m\}, \; n_j \geqslant 1, \; n_1 \leqslant ... \leqslant n_m$$

and will introduce $\bar{n} = \frac{1}{m} \Sigma n_j$. A measure of unbalancedness should

(i) be a simple function of the n_j, symmetric in its arguments,

(ii) be invariant under k-fold replicated designs

$$1_k \otimes N := \{n_1, ..., n_m, ..., n_1, ..., n_m\}$$

as well as under proportionate increase $\{kn_1, ..., kn_m\}$ of observations in each group,

(iii) reflect in a specified way properties of statistical analyses.

A measure which fulfils at least the first two requirements is given by

$$\gamma(N) = m \Big/ \left(\bar{n} \Sigma \frac{1}{n_j}\right).$$

A second one

$$\nu(N) = 1 \Big/ \left(m \Sigma \left(\frac{n_j}{N}\right)^2\right)$$

is simply a function of the well known coefficient of variability $c = \sqrt{\frac{1}{m} \Sigma(n_j - \bar{n})^2 / \bar{n}}$, namely $\nu = 1/(1 + c^2)$. Since these two measures are the squared cosinus between the vectors $(n_1^{1/2}, ..., n_m^{1/2})'$ and $(n_1^{-1/2}, ..., n_m^{-1/2})'$ or between $(\bar{n}, ..., \bar{n})'$ and $(n_1, ..., n_m)'$, respectively, we have

$$0 < \gamma(N) \leqslant 1, \; 0 < \nu(N) \leqslant 1,$$

and γ and ν equal 1 iff the design N is balanced. More precisely we have even $\frac{1}{m} < \nu \leqslant 1$.

3. The measure of unbalancedness as a measure of inefficiency

Beside of other advantages, like simplicity and uniqueness of statistical analyses in balanced models, the balanced case has—compared with unbalanced models, with the same number of observations N—also optimum properties w.r.t. variance and power of certain estimates and tests. Often the measures $\gamma(N)$ and $\nu(N)$ may express the efficiency of an unbalanced design in a simple manner. We shall formally denote the design $\{\bar{n}, \ldots, \bar{n}\}$ 'balanced', although $\bar{n}$ is not necessarily an integer.

Estimation of the mean. If in a random one-way model we estimate μ by

$$\bar{y}_{..} = \frac{1}{N} \Sigma\, n_j \bar{y}_{j.},$$

we get an unbiased estimate with variance

$$\mathrm{var}(\bar{y}_{..}) = \sigma_a^2 \Sigma \left(\frac{n_j}{N}\right)^2 + \sigma_e^2 \frac{1}{N}$$

which is minimal in the balanced case and reduces to

$$\mathrm{var}(\bar{y}_{..}) = \sigma_a^2 \frac{1}{m} + \sigma_e^2 \frac{1}{N}\cdot$$

If we as usual define the efficiency of N as the ratio $\mathrm{var}_{\mathrm{bal}}/\mathrm{var}_N$ we get

$$\nu(N) \leqslant \mathrm{Eff}\,(N) \leqslant 1,$$

so that ν has an interpretation as 'guaranteed efficiency'. An alternative estimator would be the unweighted mean

$$\bar{y}_0 = \frac{1}{m} \Sigma \bar{y}_{j.},$$

which leads to an efficiency

$$\mathrm{Eff}(N) = \frac{\sigma_a^2 \dfrac{1}{m} + \sigma_e^2 \dfrac{1}{N}}{\sigma_a^2 \dfrac{1}{m} + \sigma_e^2 \dfrac{1}{m^2}\, \Sigma \dfrac{1}{n_j}}$$

Here we have

$$\gamma(N) \leqslant \mathrm{Eff}(N) \leqslant 1.$$

We note that in the fixed effect model with reparametrization $\Sigma a_j = 0$ only the second estimator is senseful.

Testing the mean in a fixed effect model. Under the reparametrization $\Sigma a_j = 0$ the F-test for the hypothesis

$$H : \mu = 0$$

is based on the statistic

$$\frac{m^2}{\Sigma \dfrac{1}{n_j}} \left(\frac{1}{m} \Sigma \bar{y}_{j.}\right)^2 \Big/ S^2$$

which is $F_{1,N-m}$-distributed with non-centrality $N\gamma\mu^2/\sigma^2$. The non-centrality is maximal in the balanced case, and the efficiency, defined as the ratio of the noncentrality parameters, is

$$\mathrm{Eff}(N) = \gamma(N).$$

After the general illustration of the relations between unbalancedness and efficiency we consider now the same for components of variance.

4. Some useful results about MSE(.)'s for MINQ-estimators

Using the random effect model

$$y_{jk} = \mu + a_j + e_{jk}, \quad j = 1, \ldots, m; \; k = 1, \ldots, n_j$$

with the common assumptions about μ, a_j and e_{jk} and the components of variance σ_a^2 and σ_e^2 we obtain the MSE (.)' s of different MINQ-estimator procedures (Ahrens and Sanchez (1982)).

All MINQE's are based on the two quadratic forms

$$Q_1 = \sum_j \frac{n_j^2}{(n_j \rho_0 + 1)^2} (y_{j\cdot} - \tilde{y}_{\cdot\cdot})^2$$

$$Q_2 = \sum_{j,k} (y_{jk} - y_{j\cdot})^2 + \sum_j \frac{n}{(n_j \rho_0 + 1)^2} (y_{j\cdot} - \tilde{y}_{\cdot\cdot})^2,$$

where

$$y_{j\cdot} = \frac{1}{n_j} \sum_{k=1}^{n_j} y_{jk}$$

$$\tilde{y}_{\cdot\cdot} = \left(\sum_j \frac{n_j}{(n_j \rho_0 + 1)} \right)^{-1} \cdot \sum_j \frac{n_j}{(n_j \rho_0 + 1)} y_{j\cdot}$$

and

$$\rho_0 = \sigma_a^2 / \sigma_e^2.$$

The estimators for the different MINQE's are given in general by

$$\hat{\sigma}_a^2 = a_1 Q_1 + a_2 Q_2$$

$$\hat{\sigma}_e^2 = e_1 Q_1 + e_2 Q_2$$

and therefore the MSE (.)'s of the estimator $\hat{\sigma}_a^2$ include the second moments of Q_1 and Q_2 [Ahrens (1978)]. MSE ($\hat{\sigma}_a^2$(I)) for $\rho_0 \to \infty$ in the case MINQE (I). Each MSE(.) is a quadratic function in σ_a^2 and σ_e^2 depending on a parameter

$$\rho_0 : 0 \leq \rho_0 \leq \infty.$$

A comparison on the basis of MSE(.)'s depends on the choice of ρ_0. Therefore we eliminate the dependency choosing the MSE(.)'s for $\rho_0 \to \infty$.

$$\lim_{\rho_0 \to \infty} \text{MSE}\,(\hat{\sigma}_a^2(\text{I})) = \frac{2m-1}{m^2} \sigma_a^4 + \frac{2(2m-1)}{m^3} \Sigma \frac{1}{n_j} \sigma_a^2 \sigma_e^2 +$$

$$+ \frac{1}{m} \left\{ \frac{2(m-2)}{m^2} \Sigma \frac{1}{n_j^2} + \frac{1}{m} \left(\Sigma \frac{1}{n_j} \right)^2 \left[\frac{3}{m^2} + \frac{2}{\Sigma n_j - m} \right] \right\} \sigma_e^4.$$

For the balanced case $n_j = n \; \forall j$ we obtain therefore

$$\text{MSE}\,(\hat{\sigma}_a^2(\text{I})) = \frac{2m-1}{m^2} \sigma_a^4 + \frac{2(2m-1)}{m^2 n} \sigma_a^2 \sigma_e^2 + \frac{2mn - n + 1}{m^2 n^2 (n-1)} \sigma_e^4.$$

MSE $(\hat{\sigma}_a^2(\mathrm{I}, +))$ *for* $\rho_0 \to \infty$ *in the case* MINQ(I, +).

In the general unbalanced case we obtain

$$\lim_{\rho_0\to\infty} \mathrm{MSE}\,(\hat{\sigma}_a^2(\mathrm{I}, +)) = \frac{2m-1}{m^2}\sigma_a^4 + \frac{2(m-1)}{m^3}\Sigma\frac{1}{n_j}\sigma_a^2\sigma_e^2 +$$

$$+\frac{1}{m}\left\{\frac{2(m-2)}{m^2}\Sigma\frac{1}{n_j^2}+\frac{1}{m}\left(\Sigma\frac{1}{n_j}\right)^2\left[\frac{2}{m^2}+\frac{(m-1)^2}{m^2}\right]\right\}\sigma_e^4.$$

For the balanced case $n_j = n\ \forall j$ we obtain:

$$\mathrm{MSE}\,(\hat{\sigma}_a^2(\mathrm{I}, +)) = \frac{2m-1}{m^2}\sigma_a^4 + \frac{2(m-1)}{m^2 n}\sigma_a^2\sigma_e^2 + \frac{m^2-1}{n^2m^2}\sigma_e^4.$$

MSE $(\hat{\sigma}_a^2(\mathrm{U}, \mathrm{I}))$ *for* $\rho_0 \to \infty$ *in the case* MINQ(U, I).

This estimator is unbiased (U) and invariant (I) [Rao (1971)]. Its explicit form is given by Ahrens (1978). Because of the unbiasedness MSE(.) = = Var (.). We obtain

$$\lim_{\rho_0\to\infty} \mathrm{Var}\,(\hat{\sigma}_a^2(\mathrm{U}, \mathrm{I})) = \frac{2}{m-1}\sigma_a^4 + \frac{4}{m(m-1)}\Sigma\frac{1}{n_j}\sigma_a^2\sigma_e^2 +$$

$$+\frac{2}{m}\left\{\frac{m-2}{(m-1)^2}\Sigma\frac{1}{n_j^2}+\frac{1}{m}\left(\Sigma\frac{1}{n_j}\right)^2\left[\frac{1}{(m-1)^2}+\frac{1}{\Sigma n_j - m}\right]\right\}\sigma_e^4.$$

In the balanced case we have MINQE(U, I) = ANOVA:

$$\mathrm{Var}\,(\hat{\sigma}_a^2(\mathrm{U}, \mathrm{I})) = \frac{2}{m-1}\sigma_a^4 + \frac{4}{n(m-1)}\sigma_a^2\sigma_e^2 +$$

$$+\frac{2(mn-1)}{n^2m(n-1)(m-1)}\sigma_e^4 = \mathrm{Var}\,(\hat{\sigma}_a^2(\mathrm{AN}))$$

5. Classes of unbalanced designs

For a given $N = \Sigma n_j$ and m (m—number of groups (samples) in a one-way classification) we denote with

$$N := \{n_1, n_2, \ldots, n_m\}$$

a design where without any limitation of generality $n_1 \leqslant n_2 \leqslant \ldots \leqslant n_m$.

Let us use a measure of unbalancedness as it is given in Ahrens and Pincus (1981)

$$\gamma = m\bigg/\left(\bar{n}\Sigma\frac{1}{n_j}\right), \quad \bar{n} = \frac{1}{m}\Sigma n_j$$

with $$0 < \gamma(N) \leqslant 1.$$

To form classes of unbalanced designs under fixed N and m we take into account that

$$N_b := \{\bar{n}, \bar{n}, \ldots, \bar{n}\}$$

is the most balanced one, e.g. $\gamma(N_b) = 1$.

If $m|N$ then $\bar{n}$ is integer, if $m\nmid N$ then $N = mp + q$; $p, q \in \mathfrak{N}$ and the most balanced design has the following form

$$N_i = \{\underbrace{p, \ldots, p,}_{(m-q)\text{-times}} \underbrace{(p+1), \ldots, (p+1)}_{q\text{-times}}\}$$

For a given m and N all designs $N := \{n_1, n_2, \ldots, n_m\}$ are obtained by starting with $n_1 = 1$ and choosing the lexicographic order such that $n_1 \leqslant n_2 \leqslant \ldots \leqslant n_m$ holds. The process continues with $n_1 = 2$ and so on, where n_1 runs from 1 to $[N/m]$.

Example. $N = 10$, $m = 3$; then $10 = 3 \times 3 + 1$, $p = 3$, $q = 1$.

Designs:			
	(1,1,8)	(2,2,6)	(3,3,4)
	(1,2,7)	(2,3,5)	
	(1,3,6)	(2,4,4)	
	(1,4,5)		

Corollary. *In the class of all designs starting with n_1 the following relation holds:*

$$\gamma(\{n_1, \ldots, n_1, N - (m-1)n_1\}) \leqslant \gamma(\{n_1, n_2, \ldots, n_m\}).$$

By means of this corollary we define classes of designs starting with n_1. Within each class we choose the most unbalanced design as the design for the comparison, that means we choose the "worst case", let us denote it by $N_{n_1} := \{n_1, \ldots, n_1, N - (m-1)\, n_1\}$.

According to the considerations made for the most balanced design out of a given combination ("best case") it is easy to see that for a class beginning with any n_j, the most balanced design within it will be:

$$N^*_{n_j} := \{n_j, \underbrace{p, \ldots, p,}_{(m-q-1)\text{-times}} \underbrace{(p+1), \ldots, (p+1)}_{q\text{-times}}\},$$

where

$$N - n_j = (m-1)\, p + q.$$

In Ahrens, Kleffe and Tenzler (1981) a real design with $N = 379$ observations in $m = 36$ groups (data for the daily growth in weight of descendants (bulls) of $m = 36$ males [G. Herrendörfer (1978)]) is used. By means of this design some numerical results are given in Ahrens and Sanchez (1982).

6. The efficiency of an unbalanced design

Let

$$N_u := \{n_1, n_2, \ldots, n_m\}$$

be any unbalanced design, m and $N = \Sigma\, n_j$ are given. Let

$$N_b := \{\bar{n}, \bar{n}, \ldots, \bar{n}\}$$

be the corresponding balanced design with $\bar{n} = \frac{1}{m} \Sigma\, n_j$. The influence of unbalancedness will be measured as the deviation of the given design N_u

from the design N_b. This will be derived for the MSE of $\hat{\sigma}_a^2$ for MINQE procedures.

Let $\hat{\sigma}^2_{a(u)}$ be the estimator by using N_u and $\hat{\sigma}^2_{a(b)}$, the corresponding by using N_b. The efficiency of a given (unbalanced) design will be defined as

$$\text{Eff}\,(N) = \text{MSE}\ (\hat{\sigma}^2_{a(b)})/\text{MSE}\ (\hat{\sigma}^2_{a(u)})$$

for any of the three MINQE's (at $\rho_0 = \infty$). This function Eff (N) depends on the "true" σ_a^2's and σ_e^2's. Then the minimum for $(\sigma_a^2, \sigma_e^2) = (0, 0)$ is given by

$$\min_{\sigma_a^2,\, \sigma_e^2 \geqslant 0} \text{Eff}\ (N)$$

which fulfils simultaneously for MINQE (U, I), MINQE (I, +) and MINQE(I) the following relation

$$\left(\frac{\min n_j}{\max n_j}\right)^2 \leqslant \min_{\sigma_a^2, \sigma_e^2 \geqslant 0} \text{Eff}(N) \leqslant \gamma^2,$$

where γ is the measure of unbalancedness.

REFERENCES

1. Ahrens, H., *MINQUE and ANOVA estimator for one-way classification — a risk comparison.* Biometrical J., 20, 535–556 (1978).
2. Ahrens, H., Kleffe, J., Tenzler, R., *Mean square error comparison for MINQUE, ANOVA and two alternative estimators under the unbalanced one-way random model.* Biometrical J., 23, 323–342 (1981).
3. Ahrens, H., Pincus, R., *On two measures of unbalancedness in a one-way model and their relation to efficiency.* Biometrical J., 23, 227–235 (1981).
4. Ahrens, H., Sanchez, J., *Unbalancedness and efficiency in estimating components of variance: MINQUE and ANOVA procedure.* Biometrical J., 24, 649–661 (1982).
5. Corbeil, R.R., Searle, S.R., *A comparison of variance component estimators.* Biometrics, 32, 779–791 (1976).
6. Herrendörfer, G., Private communication, 1978.
7. Hess, J.L., *Sensitivity of MINQUE with respect to a priori weights.* Biometrics, 35, 645–649 (1979).
8. LaMotte, L.R., *Invariant quadratic estimators in the random one-way ANOVA model.* Biometrics, 32, 793–804 (1976).
9. Rao, C.R., Kleffe, J., *Estimation of variance components.* Tech. Report No. 79–1, Inst. Statist. and Applications, Dept. Math. and Statist., Univ. of Pittsburgh, 1979.
10. Rao, C.R., *Estimation of variance and covariance components—MINQUE theory.* J. Multivariate Anal., 1, 257–275 (1971).

ADNAN M. AWAD*, MOHAMMAD FAYOUMI

Estimate of P(Y < X) in case of the double exponential distribution

Recently a large number of papers considered the problem of estimating $R = P(Y < X)$ in case the random variables X and Y are independent and normally distributed or exponentially distributed and in some other cases. In this paper we estimate $P(Y < X)$ in case the random variables X and Y are independent and have double exponential distributions with parameters θ and μ respectively. We have proved that if $\mu - \theta = \rho$ then

$$R = P(Y < X) = \begin{cases} \frac{1}{4}(2+\rho)\exp(-\rho) & \text{if } \rho \geqslant 0 \\ 1 - \frac{1}{4}(2-\rho)\exp(\rho) & \text{if } \rho < 0 \end{cases}$$

and therefore it is enough to consider only the case $\rho \geqslant 0$. In applications, having obtained the sample $x_1, \ldots, x_n, y_1, \ldots, y_n$, we obtain the maximum likelihood estimates $\hat{\theta}$ and $\hat{\mu}$ of θ and μ which are in fact the sample medians respectively and then use

$$\hat{R} = \begin{cases} \frac{1}{4}(2+\hat{\rho})\exp(-\hat{\rho}) & \text{if } \hat{\rho} \geqslant 0 \\ 1 - \frac{1}{4}(2-\hat{\rho})\exp(\hat{\rho}) & \text{if } \hat{\rho} < 0 \end{cases}$$

to estimate R.
Simulation study shows that if $\theta < \mu$ then in about 95% of the cases it is also true that $\hat{\theta} < \hat{\mu}$. Therefore, it appears that the procedure suggested to estimate $P(Y < X)$ is reasonable.

1. Introduction

Recently a number of papers dealt with the problem of estimating $R = P(Y < X)$ when X and Y are independent normal variates. Church and Harris (1970) derived the maximum likelihood estimator (MLE) of R. The problem of estimating R when X and Y are independent exponential variates has been considered by Kelley, Kelley and Schucany (1976). Awad, Azzam and Hamdan (1981) derived the MLE of R when X and Y have a bivariate

* Department of Statistics, Yarmouk University, Irbid, Jordan.

exponential distribution. In this paper, we derive the MLE of R when X and Y are two independent double exponential variates.

A distribution of theoretical and some practical importance is that of a random variable X with probability density function

$$f(x|\theta) = \frac{1}{2} \exp(-|x-\theta|), \quad -\infty < x < \infty, \quad -\infty < \theta < \infty.$$

This distribution is called double (or bilateral) exponential distribution. It is also known as Laplace's first law of error. A survey of the properties and the practical importance of this distribution is given in Chapter 23 of Johnson and Kotz (1970).

Consider a system with stress Y and strength X. Assume that X and Y are independent random variables with double exponential densities

$$f(x|\theta) = \frac{1}{2} \exp(-|x-\theta|)$$

and

$$f(y|\mu) = \frac{1}{2} \exp(-|y-\mu|).$$

The probability that the stress is less than the strength, $P(Y < X)$, is called the *mechanical reliability* of the system. Note that

$$P(Y < X|X) = \begin{cases} \int_{-\infty}^{\mu} f(y|\mu)\, dy + \int_{\mu}^{X} f(y|\mu)\, dy & \text{if } \mu \leqslant X \\ \int_{-\infty}^{X} f(y|\mu)\, dy & \text{if } \mu > X \end{cases}$$

$$= \begin{cases} 1 - \frac{1}{2} \exp(\mu - X) & \text{if } \mu \leqslant X \\ \frac{1}{2} \exp(X - \mu) & \text{if } \mu > X \end{cases}$$

Since

$$P(Y < X) = E(P(Y < X \mid X)),$$

we get

$$P(Y < X) = \int_{-\infty}^{\mu} P(Y < X|X=x) f(x|\theta)\, dx + \int_{\mu}^{\infty} P(Y < X|X=x) f(x|\theta)\, dx.$$

Now, if $\theta \leqslant \mu$ then

$$P(Y < X) = \frac{1}{4} \int_{-\infty}^{\theta} \exp(2x - \mu - \theta)\, dx + \frac{1}{4} \int_{\theta}^{\mu} \exp(\theta - \mu)\, dx +$$

$$+ \int_{\mu}^{\infty} \left\{ \frac{1}{2} \exp(\theta - x) - \frac{1}{4} \exp(\mu + \theta - 2x) \right\} dx = \frac{1}{4} (2 + \mu - \theta) \exp(\theta - \mu).$$

Similarly, if $\theta > \mu$ then

$$P(Y<X) = \frac{1}{4}\int_{-\infty}^{\mu} \exp(2x-\theta-\mu)\,dx + \int_{\mu}^{\theta}\left\{\frac{1}{2}\exp(x-\theta) - \frac{1}{4}\exp(\mu-\theta)\right\}dx +$$

$$+\int_{\theta}^{\infty}\left\{\frac{1}{2}\exp(\theta-x) - \frac{1}{4}\exp(\mu+\theta-2x)\right\}dx = 1 - \frac{1}{4}(2+\theta-\mu)\exp(\mu-\theta).$$

Setting $\mu - \theta = \rho$ we have

$$P(Y<X) = \begin{Bmatrix} \frac{1}{4}(2+\rho)\exp(-\rho) & \text{if } \rho \geqslant 0 \\ 1 - \frac{1}{4}(2-\rho)\exp(\rho) & \text{if } \rho < 0 \end{Bmatrix} = g(\rho), \text{ say.}$$

It is interesting to notice that $P(Y < X)$ depends only on ρ where

$$\rho = \mu - \theta = E(Y) - E(X).$$

It is possible to find the difference between the strength mean and the stress mean to achieve a given reliability of the system. For example, the reliability is about 90% when $\rho = -2.4$, and about 10% when $\rho = 2.4$, i.e. if the strength mean is greater than the stress mean by 2.4 units then the probability that the strength is greater than the stress is about 0.90, and if the strength mean is smaller than the stress mean by 2.4 units then the probability that the strength is greater than the stress is about 0.10.

It can be shown that the first two derivatives of $g(\rho)$ are

$$g'(\rho) = -\frac{1}{4}(1+|\rho|)\exp(-|\rho|) < 0,$$

and

$$g''(\rho) = \frac{1}{4}\rho\exp(-|\rho|).$$

Hence $g(\rho)$ is a decreasing function in ρ with an inflection point at $\rho = 0$. The following figure gives the graph of this function.

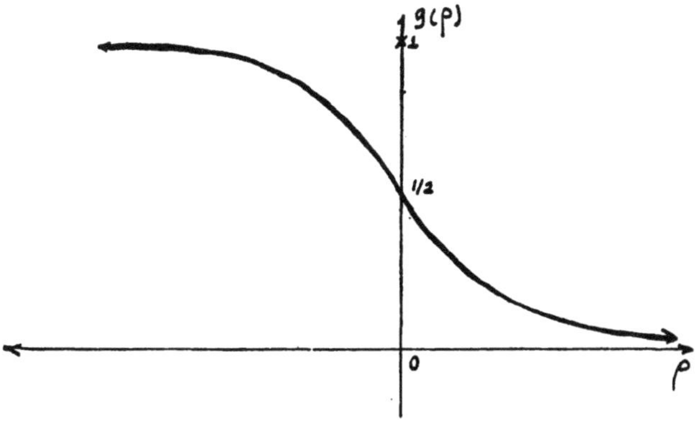

Fig. 1.

Since $\theta = E(X)$ and $\mu = E(Y)$, this graph shows that the reliability of the system decreases when ρ = mean of stress — mean of strength increases.

It is also clear that $g(-\rho) = 1 - g(\rho)$. Hence it is enough to consider only the case $\rho \geqslant 0$.

2. Estimate of R

Let $x_1, ..., x_n$ be a random sample from X and let $y_1, ..., y_n$ be a random sample from Y. Assume that the two samples are independent. It is well known that the maximum likelihood estimates of θ, μ and ρ are

$$\hat{\theta} = M(X) = \text{sample median of } X,$$

$$\hat{\mu} = M(Y) = \text{sample median of } Y,$$

and $\hat{\rho} = \hat{\mu} - \hat{\theta}$ respectively. Since $g(\rho) = P(Y < X)$ is continuous in ρ, the maximum likelihood estimate of $g(\rho)$ is $g(\hat{\rho})$, i.e.

$$\hat{R}_1 = \begin{cases} \frac{1}{4}(2 + M(Y) - M(X)) \exp(M(X) - M(Y)) & \text{if } \theta \leqslant \mu \\ 1 - \frac{1}{4}(2 + M(X) - M(Y)) \exp(M(Y) - M(X)) & \text{if } \theta > \mu. \end{cases}$$

Since θ and μ are unknown, $\hat{R}_1$ cannot be used to estimate R. However, one may suggest the following procedure to estimate R:

1. Given the sample $x_1, ..., x_n$ from X, find $\hat{\theta} = M(X)$.
2. Given the sample $y_1, ..., y_n$ from Y, find $\hat{\mu} = M(Y)$.
3. Estimate $\hat{\rho}$ by $\hat{\mu} - \hat{\theta}$.
4. Estimate R by

$$\hat{R} = \begin{cases} \frac{1}{4}(2 + \hat{\rho}) \exp(-\hat{\rho}) & \text{if } \hat{\rho} \geqslant 0 \\ 1 - \frac{1}{4}(2 - \hat{\rho}) \exp(\hat{\rho}) & \text{if } \hat{\rho} < 0 \end{cases}$$

In order to justifiy this procedure we have done a simulation study. A BASIC program (obtainable from the authors) was used as follows:

(i) Random samples of size $n = 20(10)60$ from double exponential distributions with means $\theta = 5$ and $\mu = 5.1(0.1)6$ were generated.

(ii) These samples were used to estimate θ, μ and $g(\rho)$ depending on a Monte Carlo study of 5000 runs. The mean square errors of these estimates were also calculated in order to evaluate the goodness of these estimates.

(iii) Since $\rho = \mu - \theta > 0$ in all the cases which have been used in the calculations, we have counted the percentage of the cases in which $\hat{\rho}$ is positive in order to justify the validity of the estimation procedure.

This program was carried on Wang VS 2200 at Yarmouk University, Jordan. The output of this program shows that

(i) The MLE of θ and μ underestimate these parameters, i.e. $\hat{\theta} < \theta$ and $\hat{\mu} < \mu$.

(ii) For fixed sample size, the percentage of the positive values of $\hat{\rho} = \hat{\mu} - \hat{\theta}$ increases when ρ increases in almost all cases. This percentage approaches 100% when ρ is about 0.5 or more.

(iii) $g(\hat{\rho})$ seems to be a good estimate for $g(\rho)$ because the MSE $(g(\hat{\rho}))$ is very small.

(iv) For fixed values of θ and μ, the percentage of positive values of $\hat{\rho}$ increases when the sample size increases. Moreover the MSE$(g(\hat{\rho}))$ is a decreasing function in the sample size.

(v) For a given n, the MSE$(g(\hat{\rho}))$ is a decreasing function in ρ. This means that when the two distributions become well separated then we can estimate $P(Y < X)$ more precisely.

These remarks validate our procedure.

REFERENCES

1. Awad, A.M., Azzam, M.M., Hamdan, M.A., *Some inference results on $Pr(X < Y)$ in the bivariate exponential model.* Comm. Statist. A-Theory Methods, **10**, 2515–2525 (1981).
2. Church, J.D., Harris, B., *The estimation of reliability from stress – strength relationships.* Technometrics, **12**, 49–54 (1970).
3. Kelley, G.D., Kelley, J.A., Schucany, W.R., *Efficient estimation of $P(Y < X)$ in the exponential case.* Technometrics, **18**, 359 – 360 (1976).

DRAGOMIRA BAZ, VALERIU BĂDIN

Quelques considérations statistiques concernant la loi normale généralisée

En partant de certains résultats connus [3] sur la loi normale généralisée, on donne une propriété de stabilité de cette loi et on construit certaines statistiques χ^2 généralisées qu'on utilise pour vérifier des hypothèses concernant la matrice moyenne lorsqu'on connaît la variance généralisée.

1. Analyse préliminaire

Soient $(E, \mathcal{X}, P)$ un espace de probabilité, $R^{q \times N}$ l'espace matriciel de dimension $q \times N$ et $\mathcal{B}^{q \times N}$ la tribu borélienne de $R^{q \times N}$. L'application $X: E \to R^{q \times N}$ s'appelle *matrice aléatoire* si elle est une application $\mathcal{X}$-mesurable. En posant $X = (X_{ij})$, $1 \leqslant i \leqslant q$, $1 \leqslant j \leqslant N$, les éléments X_{ij} de la matrice aléatoire X sont des variables aléatoires.

La matrice aléatoire X a une répartition normale si sa densité de probabilité s'exprime par la formule

$$\frac{1}{(2\pi)^{q \times N/2} |\Sigma|^{N/2} |B|^{q/2}} e^{-1/2 \operatorname{tr} B^{-1}(x-U)' \Sigma^{-1}(x-U)} \tag{1}$$

où $\Sigma = (\sigma_{ij})$ représente la matrice des covariances entre les lignes de X et $B = (b_{km})$ représente la matrice des covariances entre les colonnes de X, $1 \leqslant i, j \leqslant q$, $1 \leqslant k, m \leqslant N$.

On sait [3], [2] que $E(X) = U$, $D^2(X) = \Sigma \otimes B$ et

$$\varphi_X(T) = E(e^{\operatorname{tr} i T'X}) = e^{\operatorname{tr}(iT'U - 1/2\, BT'\Sigma T)} = e^{\operatorname{tr}(iT'U - 1/2\, W(T) \Sigma \otimes B)}$$

où $\Sigma \otimes B$ représente le produit Kronecker [1] des matrices Σ et B, T est une matrice $q \times N$ et $W(T)$ une matrice dont les éléments sont des matrices et qui est définie comme il suit: $W(T) = (t_i t'_j)$ avec $T' = (t_1, \ldots, t_i, \ldots, t_q)$ et $t_i t'_j = (t_{ik} t_{jm})$, $1 \leqslant i, j \leqslant q$, $1 \leqslant k, m \leqslant N$.

Proposition 1. *Si* $X_\alpha \sim N(U_\alpha, \Sigma_\alpha \otimes B_\alpha)$, $1 \leqslant \alpha \leqslant s$, *sont des matrices aléatoires indépendantes* $q \times N$ *et* c_α *des scalaires quelconques, on a*

$$\sum_{\alpha=1}^{s} c_\alpha X_\alpha \sim N\left(\sum_{\alpha=1}^{s} c_\alpha U_\alpha, \sum_{\alpha=1}^{s} c_\alpha^2 (\Sigma_\alpha \otimes B_\alpha)\right).$$

Démonstration. Nous établirons d'abord la propriété suivante:

(2) $$\sum_{\alpha=1}^{s} X_\alpha \sim N\left(\sum_{\alpha=1}^{s} U_\alpha, \sum_{\alpha=1}^{s} (\Sigma_\alpha \otimes B_\alpha)\right).$$

En effet, en posant $Y = \sum_{\alpha=1}^{s} X_\alpha$ on a

$$\varphi_Y(T) = E(e^{\operatorname{tr} i\, T'Y}) = E\left(e^{\operatorname{tr} i\, T' \sum_{\alpha=1}^{s} X_\alpha}\right) = \prod_{\alpha=1}^{s} E(e^{\operatorname{tr} i\, T'X_\alpha}) =$$

$$= \prod_{\alpha=1}^{s} e^{\operatorname{tr}(i\, T'U_\alpha - 1/2\, W(T)\, (\Sigma_\alpha \otimes B_\alpha))} = e^{\operatorname{tr}\left(i\, T' \sum_{\alpha=1}^{s} U_\alpha - 1/2\, W(T) \sum_{\alpha=1}^{s} (\Sigma \otimes B_\alpha)\right)}.$$

On a aussi

(3) $$cX \sim N(cU, c^2\, \Sigma \otimes B), \qquad c \in R.$$

En effet

$$\varphi_{cX}(T) = E(e^{\operatorname{tr} i\, T'(cX)}) = E(e^{\operatorname{tr} i\, (T'c)X}) = e^{\operatorname{tr}(i\, T'(cU) - 1/2\, W(T) c^2 \Sigma \otimes B)}.$$

Les relations (2) et (3) démontrent la proposition 1.

Corollaire 1. *La loi normale généralisée est une loi stable.*

Corollaire 2. *Si* $X_i \sim N(U, \Sigma \otimes B)$ *sont des matrices aléatoires indépendantes, la matrice* $\bar{X} = \frac{1}{n} \sum_{i=1}^{n} X_i$ *suit une loi normale, c'est-à-dire* $\bar{X} \sim N\left(U, \frac{1}{n} \Sigma \otimes B\right)$.

Proposition 2. [3] *Si* $Y \sim N(0, \Sigma \otimes B)$, *alors la trace* $\operatorname{tr} B^{-1}Y'\Sigma^{-1}Y = \operatorname{tr} \Sigma^{-1}YB^{-1}Y'$ *suit la loi de* χ^2_{qN}.

2. Hypothèses concernant la moyenne de la loi normale généralisée ($\Sigma \otimes$ B connue)

Proposition 3. *Soit* $X_1, X_2, \ldots, X_n$ *un échantillon sur la matrice aléatoire* $X \sim N(U, \Sigma \otimes B)$ *avec la variance généralisée* $\Sigma \otimes B$ *connue. Alors la statistique*

$$n \operatorname{tr} B^{-1}(\bar{X} - U_0)'\, \Sigma^{-1}(\bar{X} - U_0)$$

suit la loi de χ^2_{qN} *lorsque* $U = U_0$ *et de* $\chi^2_{qN,\delta}$ *non centrée lorsque* $U \neq U_0$, *où le paramètre de non-centricité* δ *a l'expression*

$$\delta^2 = n \operatorname{tr} B^{-1}(U - U_0)'\, \Sigma^{-1}(U - U_0).$$

Démonstration. Posons $Y = \sqrt{n}(\bar{X} - U_0)$. En tenant compte du Corollaire 2 on a $\bar{X} \sim N\left(U, \frac{1}{n} \Sigma \otimes B\right)$ et donc $E(Y) = 0$ et $D^2(Y) =$

$= \Sigma \otimes B$. Par conséquent, les hypothèses de la proposition 2 sont vérifiées et ainsi la proposition 3 est démontrée.

Le résultat précédent nous permet de déterminer une région de confiance pour U lorsque $\Sigma \otimes B$ est connue. Cette région représente un ellipsoïde dans l'espace $R^{q \times N}$ ayant le centre $\bar{X}$ et qui est défini par la relation:

$$n \operatorname{tr} B^{-1}(\bar{X} - U)' \Sigma^{-1}(\bar{X} - U) \leqslant \chi^2_{qN, p},$$

où p représente la probabilité de confiance donnée.

On peut aussi construire un test pour l'hypothèse $H_0 : U = U_0$ contre l'alternative $H_1 : U \neq U_0$.

La région critique de l'hypothèse est définie par l'inégalité

$$n \operatorname{tr} B^{-1}(\bar{X} - U_0)' \Sigma^{-1}(\bar{X} - U_0) > \chi^2_{qN, \alpha},$$

α étant le risque de première espèce.

Pour déterminer le risque β de deuxième espèce, on supposera vraie l'hypothèse H_1 lorsque $\delta > \delta_0$ où δ_0 représente une constante positive dont le choix est imposé par des considérations pratiques. On a $\beta = H_{qN; \delta_0}(\chi^2_{qN, \alpha})$ où $H_{qN; \delta_0}$ est la fonction de répartition de la variable aléatoire $\chi^2_{qN, \alpha}$.

Proposition 4. *Soient* $X_1^{(1)}, \ldots, X_{n_1}^{(1)}$ *et* $X_1^{(2)}, \ldots, X_{n_2}^{(2)}$ *deux échantillons indépendants effectués respectivement sur les matrices aléatoires* $X^{(1)} \sim N(U^{(1)}, \Sigma \otimes B)$ *et* $X^{(2)} \sim N(U^{(2)}, \Sigma \otimes B)$ *et* $\bar{X}^{(1)} = \frac{1}{n_1} \sum_{i=1}^{n_1} X_i^{(1)}$ *et* $\bar{X}^{(2)} = \frac{1}{n_2} \sum_{i=1}^{n_2} X_i^{(2)}$.

Alors la statistique

$$\frac{n_1 n_2}{n_1 + n_2} \operatorname{tr} B^{-1}(\bar{X}^{(1)} - \bar{X}^{(2)})' \Sigma^{-1}(\bar{X}^{(1)} - \bar{X}^{(2)})$$

suit la loi de χ^2_{qN} *lorsque* $U^{(1)} = U^{(2)}$. *Lorsque* $U^{(1)} \neq U^{(2)}$, *la même statistique suit la loi de* $\chi^2_{qN, \delta}$ *où* δ *a l'expression*

$$\delta^2 = \frac{n_1 n_2}{n_1 + n_2} \operatorname{tr} B^{-1}(U^{(1)} - U^{(2)})' \Sigma^{-1}(U^{(1)} - U^{(2)}).$$

Démonstration. En posant $Y = \sqrt{\frac{n_1 n_2}{n_1 + n_2}} (\bar{X}^{(1)} - \bar{X}^{(2)})$ on voit que les hypohèses de la proposition 2 sont satisfaites et par conséquent la proposition 4 est démontrée.

En observant que la statistique

$$\frac{n_1 n_2}{n_1 + n_2} \operatorname{tr} B^{-1}[(\bar{X}^{(1)} - \bar{X}^{(2)}) - (U^{(1)} - U^{(2)})]' \Sigma^{-1}[(\bar{X}^{(1)} - \bar{X}^{(2)}) - (U^{(1)} - U^{(2)})]$$

est une variable aléatoire χ^2_{qN}, on peut déterminer une région de confiance pour la différence $U^{(1)} - U^{(2)}$ lorsque $\Sigma \otimes B$ est connue. Cette région de l'espace $R^{q \times N}$ est représentée par l'intérieur de l'ellipsoïde défini par la relation

$$\frac{n_1 n_2}{n_1 + n_2} \operatorname{tr} B^{-1}[(\bar{X}^{(1)} - \bar{X}^{(2)}) - (U^{(1)} - U^{(2)})]' \Sigma^{-1}[(\bar{X}^{(1)} - \bar{X}^{(2)}) - (U^{(1)} - U^{(2)})] \leqslant \chi^2_{qN, p}$$

et dont le centre est $\bar{X}^{(1)} - \bar{X}^{(2)}$ (p = la probabilité de confiance).

La région critique du test de l'hypothèse $H_0 : U^{(1)} = U^{(2)}$ contre l'alternative $H_1 : U^{(1)} \neq U^{(2)}$ est définie par l'inégalité

$$\frac{n_1 n_2}{n_1 + n_2} \operatorname{tr} B^{-1}(\overline{X}^{(1)} - \overline{X}^{(2)})' \, \Sigma^{-1}(\overline{X}^{(1)} - \overline{X}^{(2)}) > \chi^2_{qN,\alpha}$$

et, comme dans le cas de la proposition 3, la probabilité β a l'expression $\beta = H_{qN;\,\delta_0}(\chi^2_{qN,\alpha})$.

BIBLIOGRAPHIE

1. Anderson, T.W., *An Introduction to Multivariate Statistical Analysis*. New York, Wiley, 1958.
2. Baz, Dragomira, *Contribuţii la analiza statistică multidimensională cu aplicaţii la controlul statistic al calităţii*. Teză de doctorat, Centrul de statistică matematică, Bucureşti, 1980.
3. Perez Gonzales, A.L., *Studiul statistico-matematic al calităţii şi fiabilităţii în cazul multidimensional*. Teză de doctorat, Univ. Bucureşti, 1979.
4. Theodorescu, R., Văduva, I., *Multivariate statistical quality control*. Rev. Roumaine Math. Pures Appl., 12, 237–265 (1967).

ANCA CREANGĂ-SÂMBOAN, DOINA STOICA

Tests of hypotheses based on maximal contrasts

The paper considers a p-variate normal population with mean μ and covariance matrix Σ. Under certain special assumptions about Σ some procedures based on maximal contrasts for testing the hypothesis H_0: $\mu_1 = \mu_2 = \ldots = \mu_p$ are proposed. The corresponding confidence intervals are constructed, too. These present the advantage of being exact and needing no knowledge of any latent root.

1. Introduction

Let $X^\alpha = (X_{\alpha 1}, X_{\alpha 2}, \ldots, X_{\alpha p})'$, $\alpha = 1,2, \ldots, N$, be a random sample of size N from a p-variate normal population with mean $\mu = (\mu_1, \mu_2, \ldots, \mu_p)'$ and positive definite unknown covariance matrix Σ.

It is known that the hypothesis H_0: $\mu_1 = \mu_2 = \ldots = \mu_p$ is equivalent to $H_0' : C\mu = 0$, where C is a $(p-1) \times p$ contrast matrix.

Roy's step-down procedure [2] tests the hypothesis H_0 avoiding all latent roots of a covariance matrix calculation. One uses there only the dominant latent root. But the confidence intervals for the parametrical functions obtained by this procedure are not exact, but approximate.

In this paper we propose some procedures for testing the hypothesis H_0, which lead to exact confidence intervals for maximal contrasts, with no knowledge of any latent root of the covariance matrix.

2. The univariate case

Let X_α, $\alpha = 1, 2, \ldots, n$ be one-dimensional random variables distributed independently and normally with the same variance σ^2 and expectations given by

$$EX = A\theta + M\beta = \mu \tag{1}$$

where $X = (X_1, \ldots, X_n)'$, $\theta_{m\times 1}$ and $\beta_{q\times 1}$ are unknown parameters and $A_{n\times m}$, $M_{n\times q}$ are matrices of known constants with rank $A = r$, rank$(A:M) = r + q$, $n > r + q$.

Let

$$\Phi_{t\times 1} = B_{t\times m}\, \theta_{m\times 1} \tag{2}$$

be a set of t linearly independent estimable linear functions and $\hat{\Phi}_{t\times 1}$ the minimum variance unbiased estimator of Φ. Let $D\sigma^2$ be the covariance matrix of $\hat{\Phi}$, where $D_{t\times t}$ is positive definite.

If s^2 is the usual variance unbiased estimator with $(n-q-r)$ degrees of freedom, then

$$F = \frac{(\hat{\Phi}-\Phi)' D^{-1}(\hat{\Phi}-\Phi)/t}{s^2/(n-q-r)} \tag{3}$$

is F-distributed with t and $(n-q-r)$ degrees of freedom.

Therefore with probability not less than $(1-\alpha)$ the inequality

$$(\hat{\Phi}-\Phi)'D^{-1}(\hat{\Phi}-\Phi) \leqslant tfs^2/(n-q-r) \tag{4}$$

holds, where f is the upper 100α percent point of the $F_{t;\, n-q-r}$ distribution.

This equation defines a confidence interval for Φ.

The step-down procedure replaces this interval by an approximate one, given by

$$(\hat{\Phi}-\Phi)'(\hat{\Phi}-\Phi) \leqslant ts^2 f\, \lambda_{\max}(D)/(n-q-r), \tag{5}$$

where $\lambda_{\max}(D)$ is the dominant latent root of D.

For a vector $a = (a_1, a_2, \ldots, a_t)'$ and a subset $T = (j_1, j_2, \ldots, j_u)$ of the integers $\{1, 2, \ldots, t\}$, where $j_1 < j_2 < \ldots < j_u$, $u < t$, let us consider the T-norm defined in [2] as

$$T[a] = \{a_{j_1}^2 + a_{j_2}^2 + \ldots + a_{j_u}^2\}^{1/2},$$

and, for a matrix $B_{t\times t}$, the T-submatrix defined in [2] as the $u\times u$ submatrix $B_{(T)}$ obtained from B by eliminating the rows and columns of indices $i \neq j_k$ $(i = 1, 2, \ldots, t;\ k = 1, 2, \ldots, u)$.

Considering the T-norms of Φ and $\hat{\Phi}$ we obtain from (5) the following confidence interval for $T[\Phi]$.

$$T[\hat{\Phi}] - t^{1/2}sf^{1/2}\lambda_{\max}^{1/2}(D_{(T)})/(n-q-r)^{1/2} \leqslant T[\Phi] \leq T[\hat{\Phi}] + t^{1/2}sf^{1/2}\lambda_{\max}^{1/2}(D_{(T)})/(n-q-r)^{1/2}. \tag{6}$$

We reject the hypothesis H_0 if (4) or (5) do not contain the origin.

Let us consider now the orthogonal contrast Helmert matrix $C_{(t-1)\times t}$

$$C = \begin{pmatrix} \frac{1}{\sqrt{1\cdot 2}} & -\frac{1}{\sqrt{1\cdot 2}} & 0 & & 0 \\ \frac{1}{\sqrt{2\cdot 3}} & \frac{1}{\sqrt{2\cdot 3}} & -\frac{2}{\sqrt{2\cdot 3}} & & 0 \\ \vdots & \vdots & \vdots & & \vdots \\ \frac{1}{\sqrt{t(t-1)}} & \frac{1}{\sqrt{t(t-1)}} & \frac{1}{\sqrt{t(t-1)}} & \cdots & -\frac{t-1}{\sqrt{t(t-1)}} \end{pmatrix}$$

and the functions Φ (see equation (2)) of the form $\Phi = C\theta$.

For the model (2), if $\theta = (\theta_1, \ldots, \theta_t)'$, then $\hat{\theta}_i = X_{i\cdot}$, $i = 1, 2, \ldots, k$; $j = 1, 2, \ldots, k$, where

$$X_{i\cdot} = \frac{1}{k} \sum_{j=1}^{k} X_{ij}.$$

Therefore $\hat{\Phi}_i = CX_{i\cdot}$.

Since $\mathrm{Var}(X_{i\cdot}) = \sigma^2/k$ and $\mathrm{cov}(X_{i\cdot}, X_{j\cdot}) = 0, i \neq j$, it follows that

$$\Sigma_{\hat{\Phi}} = C\Sigma_{X_{\cdot}}C' = C\frac{\sigma^2}{k} IC' = \frac{\sigma^2}{k} I$$

and

$$(\Sigma_{\hat{\Phi}})^{-1} = \frac{k}{\sigma^2} I$$

Now denoting, as in (4), $D = k^{-1}I$, $D^{-1} = kI$, the test statistic is given by

$$F = \frac{k \sum_{i=1}^{t} (\hat{\Phi}_i - \Phi_i)^2/(t-1)}{s^2/(n-q-r)}$$

Because

$$C'C = \begin{pmatrix} \frac{t-1}{t} & -\frac{1}{t} & & -\frac{1}{t} \\ -\frac{1}{t} & \frac{t-1}{t} & \cdots & \frac{1}{t} \\ \vdots & \vdots & & \vdots \\ -\frac{1}{t} & -\frac{1}{t} & & \frac{t-1}{t} \end{pmatrix}$$

the hypothesis H_0 is rejected at level of significance α if

$$\mathcal{F} = \frac{k \sum_{i=1}^{t} X_{i\cdot}(X_{i\cdot} - X_{\cdot\cdot})/(t-1)}{s^2/(n-q-r)} > f_\alpha,$$

where

$$X_{\cdot\cdot} = \frac{1}{t} \sum_{i=1}^{t} X_{i\cdot}$$

Thus, the confidence interval (4) for Φ becomes

(7) $$(\hat{\Phi} - \Phi)'(\hat{\Phi} - \Phi) \leqslant (t-1) f_\alpha s^2/k(n-q-r),$$

where f_α is the upper 100α point of the $F_{(t-1),(n-q-r)}$ distribution.

3. The p-variate case

Let $X_{n\times p}$ be a matrix of random variables such that the rows are independently distributed each having a p-variate normal distribution with the same unknown covariance matrix $\Sigma_{p\times p}$. We suppose that the expected values are given by $EX = A\,\theta$, where $A_{n\times m}$ is a matrix of known constants with rank $A = r \leqslant (n-p)$ and $\theta_{m\times p}$ is the matrix of unknown parameters.

Applying the step-down procedure developed in [2] for testing the hypothesis H_0 or H_0', let us consider the contrasts $\Phi_i = C\,\theta_i$, $i = 1, 2, \ldots, p$.

The statistics corresponding to (3) are

$$\mathcal{F}_i = \frac{k(\hat{\Phi}_i - \Phi_i)'\,(\hat{\Phi}_i - \Phi_i)/(t-1)}{s_i^2/(n-r-i+1)}, \quad i = 1, 2, \ldots, p.$$

If the first $(i-1)$ columns of the matrix $X_{n\times p}$ are fixed $(i = 2,3, \ldots, p)$, $\mathcal{F}_i$ is F-distributed with $t-1$ and $(n-r-i+1)$ degrees of freedom. The conditional distribution of $\mathcal{F}_i$, given the first $(i-1)$ columns of $X_{n\times p}$, does not involve these columns, and hence $\mathcal{F}_{i-1}, \mathcal{F}_{i-2}, \ldots, \mathcal{F}_1$. Therefore the statistics $\mathcal{F}_1, \mathcal{F}_2, \ldots, \mathcal{F}_p$ are independent.

Then, for the given constants α_i, $0 < \alpha_i < 1$, $i = 1, 2, \ldots, p$, the inequalites

$$(\hat{\Phi}_i - \Phi_i)'\,(\hat{\Phi}_i - \Phi_i) \leqslant (t-1)\, f_i s_i^2/k(n-r-i+1),$$

where f_i are the upper 100 α_i percent points of $F_{i\,;\,(t-1),\,(n-r-i+1)}$, hold simultaneously with probability not less than $\prod_{i=1}^{p} (1-\alpha_i)$.

Introducing the T-norms of Φ_i and $\hat{\Phi}_i$ we also obtain approximate simultaneous confidence intervals of type (6)

$$T[\hat{\Phi}_i] - (t-1)^{1/2} s_i f_i^{1/2}\, \lambda_{\max}^{1/2}(D_{i(T)})/(n-r-i+1)^{1/2} \leqslant$$

$$\leqslant T[\Phi_i] \leqslant T[\hat{\Phi}_i] + (t-1)^{1/2} s_i f_i^{1/2} \lambda_{\max}^{1/2}(D_{i(T)})(n-r-i+1)^{1/2}$$

with confidence level $\prod_{i=1}^{p} (1-\alpha_i)$.

But the hypothesis H_0 is true if and only if the hypotheses $\Phi_i = 0$, $i = 1, 2, \ldots, p$, are true. Thus we can decompose H_0 in p component hypotheses and test them successively.

The introduction of the T-norm permits to test the hypothesis H_0 in the special case where the random variables are ordered according to their importance.

REFERENCES

1. Giri, N.C., *Multivariate Statistical Inference.* New York, Academic Press, 1977.
2. Roy, J., *Step-down procedure in multivariate analysis.* Ann. Math. Statist. 29, 1177—1187 (1958).
3. Scheffé, H., *The Analysis of Variance.* New York, Wiley, 1959.

I. DEÁK

On random number generation procedures

Several general procedures for generating pseudo-random numbers are known and we treat here the connection between the methods. It is shown that both the acceptance-complement method and the economical method are special cases of the composition technique.

1. Introduction

Several general methods and algorithms for generating pseudo-random numbers from a given distribution are known and still others are emerging [2], [3], [7]. There are some methods known since long time, as the inversion technique, the acceptance-rejection method and the composition method. Other recently proposed techniques are the acceptance-complement method [5], the alias method [8] and the economical method [1].

The *composition technique* is as follows (we describe only the case that is most frequently used in practice). Let $f(x)$ be the density function of the target distribution and let us assume that we have a decomposition of $f(x)$ of the form

$$f(x) = p_1 g_1(x) + p_2 g_2(x) + \dots + p_k g_k(x),$$

where $p_1, \dots, p_k \geqslant 0$, $\sum_{i=1}^{k} p_i = 1$, and $g_i(x)$, $i = 1, \dots, k$, are density functions. Then the generation of numbers with density $f(x)$ is accomplished by the following procedure

1. [Generate a random index.] Generate a sample i from that discrete distribution $P\{\xi = j\} = p_j$, $j = 1, \dots, k$.

2. Generate x with density $g_i(x)$ and deliver it.

As far as the *alias method* is considered, it has been shown that it can be viewed as a special kind of the composition technique [4]. In more detail the alias method considers an n-point discrete distribution of a random variable ξ for which

$$P\{\xi = x_i\} = p_i > 0, \quad i = 1, \dots, n, \quad p_1 + p_2 + \dots + p_n = 1,$$

and gives an algorithm for decomposing this distribution into a weighted sum of n two-point discrete distributions of random variables η_i determined by the probabilities p_{i1}, p_{i2}, i.e. $P\{\eta_i = x_{ij}\} = p_{ij}$, $j = 1, 2$, $i = 1, \dots, n$.

The weight of each two-point distribution is $1/n$, i.e. all the component distributions are equally probable. Thus an algorithm realizing the alias method runs as follows.

1. [Generate a uniformly distributed index i.] Generate u uniformly in $[0,1)$ compute $s \leftarrow nu$, $i \leftarrow [s] + 1$ and preserve $u \leftarrow s - i + 1$ (u is now uniformly distributed between 0 and 1).

2. [Generate a sample from the ith distribution.] If $u \leqslant p_{i1}$ then deliver x_{i1}, otherwise deliver x_{i2}.

In the next sections of the present paper we show that both the economical method and the acceptance-complement method are in fact instances of the composition method, where special choices are made for the weights and the sampling technique.

2. The economical method

Here we describe the economical method only in a simple form. A more general description of the method can be found in [1].

Our aim is to generate random samples with density function $f(x)$, $x \in [a, b]$. Let us assume that we are given another density function $h(x)$, $x \in [a, b]$, which is "simple", that is we can "easily" produce random numbers with density h, and there is a number $c \in [a, b]$ for which the relations $f(x) < h(x)$, $x \in [a, c)$ and $f(x) \geqslant h(x)$, $x \in [c, b]$ hold. Define the following surplus and shortage intervals, respectively,

$$I^+ = \{x \mid h(x) > f(x)\} = [a, c),$$

$$I^- = \{x \mid h(x) \leqslant f(x)\} = [c, b].$$

Let us furthermore assume that we know a transformation $T(x)$ from I^+ to I^- with the property that the equilibrum equation

$$\int\limits_{\substack{-\infty \\ z \in I^+}}^{x} [h(z) - f(z)]\, \mathrm{d}z = \int\limits_{\substack{-\infty \\ z \in I^-}}^{T(x)} [f(z) - h(z)]\, \mathrm{d}z \tag{2.1}$$

holds for all $x \in I^+$, i.e. for a given amount of surplus we can determine the same amount of shortage by the aid of T. With this notation the economical method for generating numbers with density f is as follows.

1. Generate x with density h.
2. If $x \in I^-$ then deliver x (finish).
3. [Transfer x to I^- if necessary.] Generate u. If $f(x)/h(x) \leqslant u$, then deliver $T(x)$, otherwise deliver x.

In order to describe this method as a special kind of the composition technique some more notation is required. Set

$$p_1 = \int_{x \in I^+} h(x)\, \mathrm{d}x, \quad p_2 = 1 - p_1 = \int_{x \in I^-} h(x)\, \mathrm{d}x,$$

$$q_1 = \int_{x \in I^+} f(x)\, \mathrm{d}x, \quad q_2 = 1 - q_1 = \int_{x \in I^-} f(x)\, \mathrm{d}x.$$

The probability of the delivery in Step 2. is p_2, thus making use of the density

$$h_2(x) = \begin{cases} \frac{1}{p_2} h(x), & x \in I^- \\ 0, & x \notin I^-, \end{cases}$$

we can state that in Steps 1. and 2. we generate a random number according to the density h_2 with probability p_2.

If x is in the surplus area I^+, then acceptance of x in Step 3. occurs with probability $P\{f(x)/h(x) \geqslant u\} = q_1/p_1$. This means that this delivery happens with probability $p_1(q_1/p_1) = q_1$ (the probability of the condition $x \in I^+$ is p_1). This part of the algorithm is simply an acceptance-rejection technique for generating numbers with density $f_1(x) = q_1^{-1}f(x)$, $x \in I^+$, by using a majorizing function $q_1^{-1}h(x)$ and a density function h_1 proportional to the majorizing function, given by

$$h_1(x) = \begin{cases} \frac{1}{p_1} h(x), & x \in I^+, \\ 0, & x \notin I^+. \end{cases}$$

According to our assumptions

$$f_1(x) = \frac{1}{q_1} f(x) \leqslant \frac{1}{q_1} h(x) \text{ and } \frac{1}{p_1} h(x) \leqslant \frac{1}{q_1} h(x).$$

The probability of delivering $T(x)$ in Step 3. of the algorithm is $1 - p_2 - q_1 = p_1 - q_1$. Let us denote the density function of the delivered values $T(x)$ by $h_3(x)$, $x \in I^-$. Since Step 3. of the algorithm produces random numbers with density function $f(x)$ — for a proof see [1] — we can write

$$f(x) = p_2h_2(x) + q_1f_1(x) + (p_1 - q_1)\, h_3(x). \tag{2.2}$$

Hence we can derive the density $h_3(x)$ as follows

$$h_3(x) = \begin{cases} \frac{1}{p_1 - q_1}[f(x) - h(x)], & x \in I^+ \\ 0, & x \notin I^+. \end{cases}$$

Thus according to (2.2) we can say that Step 3. of the economical method is a composition method with the following special features: (i) the target density is decomposed into three densities; (ii) the choice among the densities is carried out by observing the placement of x in I^+ or I^- and then u; (iii) for sampling with density $h_3(x)$ a special transformation $T(x)$, given by (2.1), is used.

3. The acceptance-complement method

This method was presented as a variant of the acceptance-rejection technique, where no rejection occurs. The original formulation of the method is due to Kronmal and Peterson [5] and is as follows.

Decompose the target density $f(x)$ into two subdensities $g_1(x)$, $g_2(x)$ ($g_1(x) \geqslant 0$, $g_2(x) \geqslant 0$) as

(3.1) $$f(x) = g_1(x) + g_2(x)$$

and denote

$$p_1 = \int g_1(x)\, dx, \qquad p_2 = \int g_2(x)\, dx.$$

Choose a density function h_1 that dominates the subdensity g_1, i.e. $h_1(x) \geqslant g_1(x)$. Then the algorithm for generating random numbers with density f is as follows.

1. Generate u uniformly in [0,1) and generate x with density h_1.
2. If $u \leqslant g_1(x)/h_1(x)$, then deliver x (finish).
3. Generate y with density $h_2(x) = p_2^{-1} g_2(x)$ and deliver y.

In the first two steps of the algorithm with probability

$$P\{u \leqslant g_1(x)/h_1(x)\} = \int_{-\infty}^{+\infty} \int_0^{g_1(x)/h_1(x)} h_1(x)\, du\, dx =$$

$$= \int_{-\infty}^{+\infty} g_1(x)\, dx = p_1$$

we accept a sample with density h_1. Steps 1. and 2. are clearly recognized as generating a random variable with density $f_1(x) = p_1^{-1} g_1(x)$ by the usual acceptance—rejection technique by making use of the majorizing function $p_1^{-1} h_1(x)$ and the density function $h_1(x)$, proportional to this majorizing function, going then to Step 3. of the algorithm when rejection occurs. The main point here is that the choice of the first term of the sum in (3.1) has the same probability as the probability of acceptance in Step 2. Similarly, rejection occurs with the same probability $1—p_1 = p_2$ as that for the choice of the second density function $p_2^{-2} g_2(x)$ in the sum (3.1) Thus the desired form of the decomposition is

(3.2) $$f(x) = p_1 \left[\frac{1}{p_1} g_1(x) \right] + p_2 \left[\frac{1}{p_2} g_2(x) \right],$$

where the functions in square brackets are density functions.

Consequently, from (3.2) we can deduce that the acceptance-complement method is a composition method where: (i) the target density is decomposed into two densities; (ii) the choice among these densities is carried out by the same procedure as generating a sample with the first density; (iii) for sampling with the first density we use an acceptance-rejection technique with a slight modification (see above).

REFERENCES

1. Deák, I., *An economical method for random number generation and a normal generator.* Computing, **27**, 113–121 (1981).
2. Deák, I., Benne, B., *Random number generation: A bibliography.* MTA SzTAKI Working Paper MO/5, 53 pp., 1979.
3. Dieter, U., Ahrens, J.H., *Pseudorandom Numbers.* Graz, 1974.
4. Kronmal, R.A., Peterson, A.V., Jr., *On the alias method for generating random variables from a discrete distribution.* Amer. Statist., **33**, 214–218 (1979).
5. Kronmal, R.A., Peterson, A.V., Jr., *A variant of the acceptance-rejection method for computer generation of random variables.* J. Amer. Statist. Assoc., **76**, 446–451 (1981).
6. Peterson, A.V., Jr., Kronmal, R.A., *A representation for discrete distribution by equiprobable mixtures.* J. Appl. Probab., **17**, 102–111 (1098).
7. Văduva, I., *Computer generation of random vectors based on transformaticn of uniformly distributed vectors.* These Proceedings, pp. 589–598.
8. Walker, A.J., *An efficient method for generating discrete random variables with general distributions.* ACM Trans. Math. Software, **3**, 253–256 (1977).

J. KLEFFE, B. SEIFERT

Matrix free computation of C. R. Rao's MINQUE and its sample variance under unbalanced nested classification models

This paper presents an efficient algorithm for computation of MINQUE under arbitrarily large nested classification models which is based on explicitely inverting the covariance matrix of the data vector. It does not require storage of matrices and by obvious modifications may be used for computation of various other types of optimal estimates (surveyed in Rao and Kleffe (1980)).

0. Introduction

Rao's Minimum Norm Quadratic Unbiased Estimators (MINQUE) for variance components own a number of desirable properties. They are to this field what best linear unbiased estimates are to estimation of fixed effects. But their calculation requires inversion of matrices as high in dimension as the total number of observations.

In special cases as considered in analysis of variance, inversion of these high dimensional matrices is easily avoided by expressing them in terms of much less dimensional matrices (see e.g. Kleffe (1980a)). However, also storage of the inverse puts heavy limitations on the size of the model. Therefore, explicit or partly explicit expressions for MINQUE have been developed by Ahrens (1978), Giesbrecht and Burrows (1978), Kleffe (1980b) and Swallow and Searle (1978) in special cases. Hemmerle (1979) reports on a comprehensive matrix free algorithm for computation of MINQUE when prior values of all variance components, other than the overall experimental error, are zero.

This paper presents an efficient algorithm for computation of MINQUE under arbitrary nested classification models, which does not require storage of matrices. The average computational effort is about the same as in Hemmerle's algorithm. A detailed outline of the algorithm including proofs and hints for organization of computations is going to be published.

1. General results

MINQUE-theory for estimating variance components is based on the linear model

$$Y = X\beta + U_1\xi_1 + \dots + U_p\xi_p + \xi_{p+1} \tag{1.1}$$

where Y is an N-vector of observations, X, $U_1, \ldots, U_p$ are known matrices and $\xi_1, \ldots, \xi_p, \xi_{p+1}$ are independently distributed random vectors with expectations zero and covariance matrices $E\,\xi_i\xi_i' = \sigma_i^2 I_{c_i}$. The k-vector β and the variance components $\sigma_1^2, \ldots, \sigma_p^2, \sigma_{p+1}^2$ are unknown parameters. C. R. Rao's method for estimation of the variance components (later on called MINQUE(U,I) in Rao and Kleffe (1980)) is now as follows. Fix prior rations $\alpha_i = \sigma_i^2/\sigma_{p+1}^2$ and define $U_{p+1} = I$, $V = \sum_{i=1}^{p+1} \alpha_i U_i U_i'$ and $R = V^{-1} - V^{-1}X(X'V^{-1}X)^- X'V^{-1}$. Based on R let us compute sums of squares

$$q_i = Y'RU_iU_i'RY, \quad i = 1, \ldots, p+1 \tag{1.2}$$

and traces

$$h_{ij} = \operatorname{tr} U_i'RU_jU_j'RU_i \quad i, j = 1, \ldots, p+1 \tag{1.3}$$

Then, MINQUE (U, I) for $\sigma = (\sigma_1^2, \sigma_2^2, \ldots, \sigma_{p+1}^2)'$ is given by any solution $\hat\sigma = (\hat\sigma_1^2, \hat\sigma_2^2, \ldots, \hat\sigma_{p+1}^2)'$ to the system of linear equations

$$\begin{pmatrix} h_{11} & & h_{1,p+1} \\ \cdot & & \cdot \\ \cdot & & \cdot \\ \cdot & & \cdot \\ h_{p+1,1} & \cdots & h_{p+1,p+1} \end{pmatrix} \begin{pmatrix} \hat\sigma_1^2 \\ \cdot \\ \cdot \\ \cdot \\ \hat\sigma_{p+1}^2 \end{pmatrix} = \begin{pmatrix} q_1 \\ \cdot \\ \cdot \\ \cdot \\ q_{p+1} \end{pmatrix} \tag{1.4}$$

It provides a unique solution iff all variance components are unbiasedly estimable by quadratic forms $Y'AY$ with $AX=0$. More generally, if $f'\sigma = \sum_{i=1}^{p+1} f_i\sigma_i^2$ admits such an estimator, $f'\hat\sigma$ is unique for all solutions $\hat\sigma$ to the system (1.4). Computation of sample variances for the estimates σ_i^2, $i = 1, \ldots, p+1$, requires knowledge of the variances and covariances of the quadratic forms q_i, $i = 1, \ldots, p+1$. They are given by

Theorem 1. *Assume Y to be multivariate normal and define for $i, j, k, \lambda = 1, \ldots, p$*

$$c_{k\lambda}^{ij} = \operatorname{tr} U_j'RU_kU_k'RU_iU_i'RU_\lambda U_\lambda'RU_j \tag{1.5}$$

$$c_k^{ij} = \operatorname{tr} U_j'RU_kU_k'RU_iU_i'RU_j. \tag{1.6}$$

Then, for $i, j = 1, \ldots, p+1$,

$$\begin{aligned} \operatorname{cov}(q_i, q_j) &= 2\sigma_{p+1}^4 h_{ij} + 4\sigma_{p+1}^2 \sum_{k=1}^{p} (\sigma_k^2 - \alpha_k\sigma_{p+1}^2)\, c_k^{ij} \\ &+ 2 \sum_{k,\lambda=1}^{p} (\sigma_k^2 - \alpha_k\sigma_{p+1}^2)(\sigma_\lambda^2 - \alpha_\lambda\sigma_{p+1}^2)\, c_{k\lambda}^{ij}, \end{aligned} \tag{1.7}$$

where

$$\begin{aligned} c_{k\lambda}^{i,p+1} &= c_k^{i\lambda} - \sum_{j=1}^{p} \alpha_j c_{k\lambda}^{ij}, \quad c_k^{i,p+1} = h_{ik} - \sum_{j=1}^{p} \alpha_j c_k^{ij}, \\ c_{k\lambda}^{p+1,p+1} &= c_k^{\lambda,p+1} - \sum_{j=1}^{p} \alpha_j c_{k\lambda}^{j,p+1}, \quad c_k^{p+1,p+1} = h_{p+1,k} - \sum_{j=1}^{p} \alpha_j c_k^{j,p+1}. \end{aligned} \tag{1.8}$$

It is important to notice that c_k^{ij} is entirely invariant under permutation of indices and

$$c_{k\lambda}^{ij} = c_{k\lambda}^{ji} = c_{\lambda k}^{ij} = c_{\lambda k}^{ji} = c_{ij}^{k\lambda} = c_{ij}^{\lambda k} = c_{ji}^{k\lambda} = c_{ji}^{\lambda k}$$

2. A general nested classification model

The linear model for a general nested classification is given by

$$Y_{i_1 \dots i_p i_{p+1}} = a + a_{i_1} + a_{i_1 i_2} + \dots + a_{i_1 \dots i_p} + \varepsilon_{i_1 \dots i_p i_{p+1}}, \tag{2.1}$$

where indices $i_1, \dots, i_p$ describe levels of p factors characterizing experimental conditions under which the observations $Y_{i_1 \dots i_p i_{p+1}}$ for $i_{p+1} = 1, \dots, n_{i_1 \dots i_p}$ have been made. The index i_{p+1} describes replication.

Model (2.1) is called a *random model* if a is an unknown constant and $a_{i_1}, a_{i_1 i_2}, a_{i_1 i_2 i_3}, a_{i_1 \dots i_p}, \varepsilon_{i_1 \dots i_p i_{p+1}}$ are all independently distributed random variables with zero expectations and different variances $\sigma_1^2, \dots, \sigma_{p+1}^2$.

More generally, in the mixed model case, the first r effects are considered fixed so that $a_{i_1}, a_{i_1 i_2}, \dots, a_{i_1 \dots i_r}$ are all unknown constants. Then, it is well-known that not all of these constants are identifiable unless additional restrictions are placed on them. For the sake of estimating variance components $\sigma_{r+1}^2, \dots, \sigma_p^2, \sigma_{p+1}^2$ model (2.1) becomes equivalent to

$$Y_{i_1 \dots i_p i_{p+1}} = a_{i_1 \dots i_r} + a_{i_1 \dots i_{r+1}} + \dots + a_{i_1 \dots i_p} + \varepsilon_{i_1 \dots i_p i_{p+1}} \tag{2.2}$$

and MINQE (U, I) is easily obtained by pooling MINQE (U, I) estimators for model (2.2) given $i_1, \dots, i_r$. This method of pooling is extensively discussed in Kleffe (1980) so that we confine on random models.

In order to write model (2.1) in a form like (1.1) let us arrange observations $Y_{i_1 \dots i_p i_{p+1}}$ in lexical order to a column vector

$$Y = \mathrm{col}_{i_1} \mathrm{col}_{i_2} \dots \mathrm{col}_{i_p} \mathrm{col}_{i_{p+1}} (Y_{i_1 \dots i_p i_{p+1}}).$$

The operator $\mathrm{col}_{i_{p+1}}$ stands for arranging observations $Y_{i_1 \dots i_p i_{p+1}}$ one below each other for given $i_1 \dots i_p$ and as $i_{p+1} = 1, \dots, n_{i_1 \dots i_p}$. In this way vectors are obtained which depend on $i_1, \dots, i_p$ and are of different dimensions. The operator col_{i_p} stands for arranging these vectors one below the other to a larger column according to the subscript i_p. This operation is followed by $\mathrm{col}_{i_{p-1}}$ and so on. Using this notation the matrix U_k associated with the kth variance component σ_k^2 is given by

$$U_k = \mathrm{Diag}_{i_1} \dots \mathrm{Diag}_{i_k} \mathrm{col}_{i_{k+1}} \dots \mathrm{col}_{i_p} \mathrm{col}_{i_{p+1}} (1) \tag{2.3}$$

and $X = \mathrm{col}_{i_1} \dots \mathrm{col}_{i_{p+1}} (1)$ is the vector with all components 1. Diag_{i_k} means arranging the vectors or matrices obtained so far in block diagonal form according to index i_k. The arranging operations "col" and "Diag" have to be performed in the given order from the right to the left.

The transpose of (2.3) is

$$U_k' = \mathrm{Diag}_{i_1} \dots \mathrm{Diag}_{i_k} \mathrm{col}_{i_{k+1}}' \dots \mathrm{col}_{i_p}' \mathrm{col}_{i_{p+1}}' (1),$$

where col'_{i_k} denotes arrangement in a row. Further, the covariance matrix of the random vector Y given $\sigma^2_{p+1} = 1$, $\sigma^2_i = \alpha_i$, $i = 1, \ldots, p$, may be written as

$$V = \text{Diag}_{i_1} (\alpha_1 1_{n_1} 1'_{n_1} + \text{Diag}_{i_2} (\alpha_2 1_{n_2} 1'_{n_2} + \text{Diag}_{i_3} (\ldots \tag{2.4}$$

$$\ldots \text{Diag}_{i_p} (\alpha_p 1_{n_p} 1'_{n_p} + \text{Diag}_{i_{p+1}} (1))\ldots))).$$

This special form of V admits derivation of a closed form expression of its inverse, by repeatedly using Rao (1973), p. 33.

Lemma 1. *Set* $e^p_{i_1 \ldots i_p i_{p+1}} = 1$ *and compute recursively for* $\mu = p-1$, $p-2, \ldots, 1$

$$e^{\mu}_{i_1 \ldots i_p i_{p+1}} = \frac{e^{\mu+1}_{i_1 \ldots i_p i_{p+1}}}{1 + \alpha_{\mu+1} e^{\mu+1}_{i_1 \ldots i_1 i_{\mu+1}\cdot}}, \tag{2.5}$$

where $e^{\mu}_{i_1, \ldots, i_\mu \cdot} = \sum_{i_{\mu+1}, \ldots, i_p, i_{p+1}} e^{\mu}_{i_1 \ldots i_p i_{p+1}}$. *Then,*

$$V^{-1} = I - \sum_{\mu=1}^{p} \text{Diag}_{i_1} \ldots \text{Diag}_{i_\mu} \frac{\alpha_\mu}{1 + \alpha_\mu e^{\mu}_{i_1 \ldots i_\mu}} b_\mu b'_\mu, \tag{2.6}$$

$$b_\mu = \text{col}_{i_{\mu+1}} \ldots \text{col}_{i_p} \text{col}_{i_{p+1}} (e^{\mu}_{i_1 \ldots i_p i_{p+1}}).$$

3. Co putation of quadratic forms

For computation of the quadratic forms $q_1, \ldots, q_p, q_{p+1}$ defined by (1.2) let us introduce

$$e^0_{i_1 \ldots i_{p+1}} = \frac{e^1_{i_1 \ldots i_{p+1}}}{1 + \alpha_1 e^1_{i_1 \cdot}}, \quad e^0_{\cdot} = \sum_{i_1, \ldots, i_{p+1}} e^0_{i_1 \ldots i_{p+1}} \tag{3.1}$$

by continuing (2.5). Further let us define the totals

$$\tilde{Y}_{i_1 \ldots i_p \cdot} = \sum_{i_{p+1}} Y_{i_1 \ldots i_{p+1}}$$

and recursively

$$\tilde{Y}_{i_1 i \quad \mu \cdot} = \sum_{i_{\mu+1}} \frac{\tilde{Y}_{i_1 \ldots i_{\mu+1} \cdot}}{1 + \alpha_{\mu+1} e^{\mu+1}_{i_1 \ldots i_{\mu+1} \cdot}} \tag{3.2}$$

for $\mu = p-1$, $p-2, \ldots, 1, 0$ leading to

$$\tilde{Y}_{\cdot} = \sum_{i_1} \frac{\tilde{Y}_{i_1 \cdot}}{1 + \alpha_1 e^1_{i_1 \cdot}}.$$

Theorem 2. *The quadratic forms* q_k, $k = 1, \dots, p$, *are given by* $q_k = \|t_k\|^2$, *where* $t_k = \mathrm{col}_{i_1} \dots \mathrm{col}_{i_k}(t^k_{i_1 \dots i_k})$ *has components*

$$t^k_{i_1 \dots i_k} = \frac{e^k_{i_1 \dots i_k \cdot}}{1 + \alpha_k e^k_{i_1 \dots i_k \cdot}} \Bigg(\tilde{\tilde{Y}}_k - \tilde{\tilde{Y}}_{k-1} + \frac{1}{1 + \alpha_{k-1} e^{k-1}_{i_1 \dots i_{k-1} \cdot}} \Bigg(\tilde{\tilde{Y}}_{k-1} - \tilde{\tilde{Y}}_{k-2} +$$

(3.3)

$$+ \dots + \frac{1}{1 + \alpha_2 e^2_{i_1 i_2 \cdot}} \Bigg(\tilde{\tilde{Y}}_2 - \tilde{\tilde{Y}}_1 + \frac{1}{1 + \alpha_1 e^1_{i_1 \cdot}} (\tilde{\tilde{Y}}_1 - \tilde{\tilde{Y}}_{\cdot})\Bigg) \cdots \Bigg)\Bigg)$$

with $\tilde{\tilde{Y}}_\mu = \tilde{\tilde{Y}}_{i_1 \dots i_\mu} = \tilde{Y}_{i_1 \dots i_\mu} / e^\mu_{i_1 \dots i_\mu \cdot}$.
The quadratic form q_{p+1} *is given by*

$$q_{p+1} = \mathrm{SSR} + \sum_{i_1 \dots i_p} (t^p_{i_1 \dots i_p})^2 / n_{i \dots i_p} \tag{3.4}$$

where SSR *is the ordinary residual sum of squares*

$$\mathrm{SSR} = \sum_{i_1, \dots, i_p, i_{p+1}} (Y_{i_1, \dots, i_p, i_{p+1}} - \tilde{\tilde{Y}}_{i_1 \dots i_p})^2.$$

Formula (3.3) offers an interesting computational scheme for simultaneous computation of $q_1 \dots q_{p+1}$. We first compute the totals $\tilde{Y}_{i_1 \dots i_p \cdot} \to \tilde{Y}_{i_1 \dots i_{p-1} \cdot} \to \dots \tilde{Y}_{i_1 \cdot} \to \tilde{Y}_{\cdot}$, recursively, based on (3.5), and simultaneously the weight $e^p_{i_1 \dots i_p \cdot} \to \dots \to e^0_{\cdot}$. Afterwards we are able to compute, backwards, all the vectors $t_1 \to t_2 \to \dots \to t_{p-1} \to t_p$ according to (3.3). Sums of squares in corresponding blocks yield $q_k = \|t_k\|^2$, $k = 1, \dots, p$, and the vector t_p is simultaneously used to obtain q_{p+1} by (3.4).

4. Computation of expectations and sample variances

Theorem 2 is also employed to compute the coefficients (1.3). For this let $U_j^{i_1^* \dots i_j^*}$ denote the $i_1^* \dots i_j^*$th column of U_j and introduce $t_k(U_j^{i_1^* \dots i_j^*})$ to be the vector (3.3) obtained when Y is replaced by $U_j^{i_1^* \dots i_j^*}$. Then from (1.3)

$$h_{kj} = \sum_{i_1^*, \dots, i_j^*} \|t_k (U_j^{i_1^* \dots i_j^*})\|^2 = \sum_{i_1^*, \dots, i_j^*} q_k(U_j^{i_1^* \dots i_j^*}) \tag{4.1}$$

for $k, j = 1, \dots, p$.
It is worth noting that the totals $Y_{i_1 \dots i_k}$ computed for the $i_1^* \dots i_k^*$th column of U_k, i.e. $U_k^{i_1^* \dots i_k^*}$, are $e^k_{i_1^* \dots i_k^*}$ in the $i_1^* \dots i_k^*$th place and zero elsewhere. This fact admits considerable shortenings in computation along (3.3).
Slightly different is the computation of $h_{p+1,k}$ for $k = 1, \dots, p+1$. We have

$$h_{p+1,k} = \sum_{i_1^*, \dots, i_p^*} q_k (U^{p i_1^* \dots i_p^*}) / n_{i_1^* \dots i_p^*} \qquad \text{for } k = 1, \dots, p,$$

(4.2) $$h_{p+1,p+1} = N - n + \sum_{\overset{*}{i}_1, \ldots, \overset{*}{i}_p} q_{p+1}(U_p^{\overset{*}{i}_1 \ldots \overset{*}{i}_p})/n_{\overset{*}{i}_1 \ldots \overset{*}{i}_p},$$

where $N = \sum_{i_1, \ldots, i_p} n_{i_1 \ldots i_p}$ and $n = \sum_{i_1, \ldots, i_p} 1$ are the numbers of observations and cells, respectively. The coefficients $h_{p+1\,k}$ and $h_{p+1,\,p+1}$ are simultaneously obtained when computing h_{pk} for $k = 1, \ldots, p$.

The computation of coefficients $c_{k\lambda}^{ij}$ and c_k^{ij} defined by (1.5) and (1.6) is also based on Theorem 2. For this let us observe that

(4.3) $$c_{k\lambda}^{ij} = \sum_{\overset{*}{i}_1, \ldots, \overset{*}{i}_j} (t_i(U_\lambda t_\lambda(U_j^{\overset{*}{i}_1 \ldots \overset{*}{i}_j})))' (t_i(U_k t_k(U_j^{\overset{*}{i}_1 \ldots \overset{*}{i}_j})))$$

and

(4.4) $$c_k^{ij} = \sum_{\overset{*}{i}_1, \ldots, \overset{*}{i}_j} (t_i(U_j^{\overset{*}{i}_1 \ldots \overset{*}{i}_j}))' (t_i(U_k t_k(U_j^{\overset{*}{i}_1 \ldots \overset{*}{i}_j}))).$$

The vectors $t_i(U_k t_k(U_j^{\overset{*}{i}_1 \ldots \overset{*}{i}_j}))$ are obtained by repeated use of the basic computational procedure of Theorem 2.

Notice that by using (3.6) the relation $q_{p+1} + \sum_{i=1}^{p} \alpha_i q_i = Y'RY$ turns to

(4.5) $$q_{p+1} + \sum_{i=1}^{p} \alpha_i q_i = \mathrm{SSR} + \sum_{i_1, \ldots, i_p} \tilde{\tilde{Y}}_{i_1 \ldots i_p} t_{i_1 \ldots i_p}^p.$$

This provides an alternative way of computing the quadratic form q_{p+1} and hence a check against miscalculation.

Similarly, each row of the coefficient matrix H in (1.4) admits such checks based on the equations $(k = 1, \ldots, p)$

(4.6) $$h_{k,\,p+1} + \sum_{i=1}^{p} \alpha_i h_{ki} = \operatorname{tr} U_k' R U_k = \sum_{i_1, \ldots, i_k} t_{i_1 \ldots i_k}^k (U_k^{i_1 \ldots i_k})$$

The terms of the last row in (1.4) have to satisfy

(4.7) $$h_{p+1,\,p+1} + \sum_{i=1}^{p} \alpha_i h_{p+1,i} = N - n + \sum_{i_1, \ldots, i_p} t_{i_1 \ldots i_p}^p (U_p^{i_1 \ldots i_p})/n_{i_1 \ldots i_p},$$

where N and n are as in (4.2).

Some checks for the coefficients $c_{k\lambda}^{ij}$ and c_k^{ij} are based on the direct computation of $c_{k\lambda}^{i,p+1}$ and $c_k^{i,p+1}$. We have

(4.8) $$c_{k\lambda}^{i,p+1} = \sum_{i_1, \ldots, i_p} (t_k(U_\lambda t_\lambda(U_p^{i_1 \ldots i_p})))' (t_k(U_i t_i(U_p^{i_1 \ldots i_p})))/n_{i_1 \ldots i_p}$$

and

(4.9) $$c_k^{i,\,p+1} = \sum_{i_1, \ldots, i_p} (t_k(U_p^{i_1 \ldots i_p}))' (t_k(U_i t_i(U_p^{i_1 \ldots i_p})))/n_{i_1 \ldots i_p},$$

where the vectors in (4.8) and (4.9) all appear during computation of $c_{k\lambda}^{ij}$ and c_k^{ij}. Further, it is easy to establish that the matrices $((h_{ij}))_{i,j=1,\ldots,p+1}$, $((c_{kk}^{ij}))_{i,j=1,\ldots,p}$, $((c_{k\lambda}^{ii}))_{k,\lambda=1,\ldots,p}$, and $((c_k^{ij}))_{i,j=1,\ldots,p}$ are all nonnegative definite and that (1.4) is a consistent system of linear equations, so that any failure in these points is due to miscalculation.

REFERENCES

1. Ahrens, H., *MINQUE and ANOVA estimator for one-way- classification — a risk comparison.* Biometrical J., 20, 535—556 (1978).
2. Giesbrecht, F.G., Burrows, P.M., *Estimating variance components in hierarchical structures using MINQUE and restricted maximum likelihood.* Comm. Statist. A, 7, 891—904 (1978).
3. Hemmerle, W.J., *A comprehensive, matrix free algorithm for analysis of variance.* Institute of Statistics Mimeo Ser., Nos. 1210, 1211, 1212. Raleigh, N.C., North Carolina State Univ., 1979.
4. Kleffe, J., *C.R. Rao's MINQUE for replicated and multivariate observations.* In: *Math. Statist. and Probab. Theory* (Proc. Sixth Internat. Conf., Wisła, 1978). Lecture Notes in Statist., Vol. 2, pp. 188—200. New York, Springer, 1980.
5. Kleffe, J., *On recent progress of MINQUE theory—nonnegative estimation, consistency, asymptotic normality and explicit formulae.* Math. Operationsforsch. Statist. (Statist.), 11, 563—588 (1980 a).
6. Kleffe, J., *C.R. Rao's MINQUE under four two-way ANOVA models.* Biometrical J., 22, 93—104 (1980 b).
7. Rao, C.R., *Linear Statistical Inference and Its Applications, 2nd Ed.*. New York, Wiley, 1973.
8. Rao, C.R., Kleffe, J., *Estimation of variance components.* In: Krishnaiah, P.R. (Ed.), *Analysis of Variance.* Handbook of Statistics, Vol. 1, pp. 1—40. Amsterdam—New York, North-Holland, 1980.
9. Swallow, W.H., Searle, S.R., *Minimum variance quadratic unbiased estimation (MINQUE) of variance components.* Technometrics, 20, 265—272 (1978).

EUGEN MĂRGĂRITESCU

Sur une extension de la méthode T de comparaison multiple

T-confidence intervals are given for the class of all contrasts among k means $\{\theta_i\}$ under the underlying assumptions (1.3), with a and b satisfying the inequalities $-\frac{a^2}{k-1} \leqslant b < a^2$.

1. Introduction

Sous les hypothèses fondamentales

$$(1.1)\quad \begin{cases} \{\hat{\theta}_i\}_{1\leqslant i\leqslant k} \text{ sont des variables aléatoires mutuellement indépendantes,} \\ \hat{\theta}_i \sim N(\theta_i, a^2\sigma^2),\ i = 1, \ldots, k, \text{ où } a > 0 \text{ est une constante connue.} \\ s^2 \text{ est un estimateur quadratique indépendant de } \sigma^2 \text{ à } \nu \text{ degrés de liberté,} \\ \text{à savoir } \frac{\nu s^2}{\sigma^2} \sim \chi^2_\nu \text{ et la variable } s^2 \text{ est indépendante de } \{\hat{\theta}_i\} \end{cases}$$

est valable la relation probabiliste suivante ([1], [4]):

$$P\left\{\sum_i c_i\hat{\theta}_i - aq_{\alpha:k,\nu}\, s\left(\frac{1}{2}\sum_i |c_i|\right) \leqslant \sum_i c_i\theta_i \leqslant \sum_i c_i\hat{\theta}_i + \right.$$

$$(1.2)\qquad \left. + aq_{\alpha:k,\nu}\, s\left(\frac{1}{2}\sum_i |c_i|\right), \forall\{c_i\} : \sum_i c_i = 0\right\} = 1 - \alpha,$$

qui constitue le support de la méthode T (Tukey [3]) de comparaison multiple.

Une extension du modèle (1.1), considérée dans [2], traite le même problème pour le cas des variables $\{\hat{\theta}_i\}$ corrélées de la manière suivante:

$$(1.3)\quad \begin{cases} \hat{\theta} = (\hat{\theta}_1, \ldots, \hat{\theta}_k)' \text{ est distribué suivant une loi normale.} \\ E(\hat{\theta}_i) = \theta_i,\ D(\hat{\theta}_i) = a^2\sigma^2,\ i = 1, \ldots, k; \operatorname{Cov}(\hat{\theta}_i, \hat{\theta}_{i'}) = b\sigma^2, \forall i \neq i', \\ \text{où } a > 0 \text{ et } b \text{ sont des constantes connues.} \\ \frac{\nu s^2}{\sigma^2} \sim \chi^2_\nu \text{ et la variable } s^2 \text{ est independante de } \{\hat{\theta}_i\}. \end{cases}$$

Si $\Sigma_{\hat{\theta}}$ désigne la matrice des covariances du vecteur $\hat{\theta}$, on a

$$\Sigma_{\hat{\theta}}/\sigma^2 = \begin{pmatrix} a^2 & b & \cdots & b & b \\ \vdots & \vdots & & \vdots & \vdots \\ b & b & \cdots & b & a^2 \end{pmatrix}$$

Parce que la matrice $\Sigma_{\hat{\theta}}/\sigma^2$ est non négativement définie, on déduit la non-négativité de ses racines caractéristiques a^2-b (d'ordre de multiplicité $k-1$) et $a^2+(k-1)b$, ce qui entraîne $-\dfrac{a^2}{k-1} \leqslant b \leqslant a^2$.

Sous les hypothèses (1.3) et pour $-\dfrac{a^2}{k-1} \leqslant b \leqslant 0$ dans [2] on détermine des intervalles communs de confiance pour l'ensemble des contrastes dans les moyennes inconnues $\{\theta_i\}$ à l'aide des nouvelles variables aléatoires $\tilde{\theta}_i = \hat{\theta}_i + x (i = 1, ..., k)$, où $x \sim N(0, -b\sigma^2)$ est une variable auxiliaire indépendante de $\{\hat{\theta}_i\}$ et s^2. Il est facile à voir que la restriction $b \leqslant 0$ n'est pas nécessaire. Elle en est imposée par le procédé de démonstration.

Dans le présent article on étudie le cas général $-\dfrac{a^2}{k-1} < b < a^2$ (la matrice $\Sigma_{\hat{\theta}}/\sigma^2$ est positivement définie), en montrant que les inférences concernant les intervalles communs de confiance pour des contrastes, obtenues dans [2], sont également valables si $0 < b < a^2$.

2. Intervalles communs de confiance pour l'ensemble des contrastes

Il est bien connu que pour la matrice symétrique $\Sigma_{\hat{\theta}}/\sigma^2$ il existe une matrice orthogonale B de sorte que $B'(\Sigma_{\hat{\theta}}/\sigma^2)B$ soit diagonale, les éléments non nuls représentant les racines caractéristiques de la matrice $\Sigma_{\hat{\theta}}/\sigma^2$. À partir de cette remarque, par une transformation lineaire convenable appliquée aux variables $\{\hat{\theta}_i\}$, nous allons introduire k variables non corrélées. Soit par exemple la matrice

$$B = \begin{pmatrix} b_1' \\ \vdots \\ b_k' \end{pmatrix}$$

définie de la façon suivante:

$$(2 \cdot 1) \quad b_i' = \begin{cases} \Big(\underbrace{\frac{1}{\sqrt{i(i+1)}}, \dots, \frac{1}{\sqrt{i(i+1)}}}_{i \text{ composantes}}, -\frac{i}{\sqrt{i(i+1)}}, 0, \dots, 0\Big) & \text{si } 1 \leqslant i \leqslant k-1, \\ \left(\frac{\lambda}{\sqrt{k}}, \dots, \frac{\lambda}{\sqrt{k}}\right) & \text{si } i = k, \end{cases}$$

où $\lambda = \left(\dfrac{a^2-b}{a^2+(k-1)b}\right)^{1/2}$*.

Si l'on pose $\hat{\beta} = (\hat{\beta}_1, \dots, \hat{\beta}_k)' = B\hat{\theta}$, alors $\Sigma_{\hat{\beta}} = \sigma^2(a^2-b)I$, I désignant la matrice identité d'ordre k.

---

* $\lambda \in R^+$. Pour $\lambda = 1$ le modèle (1.3) se réduit à (1.1).

Il est facile à montrer que *toute matrice $B^{*k\times k}$ qui conduit à la matrice des covariances $\sigma^2(a^2-b)I$ a la forme $B^* = PB$, où P désigne une matrice orthogonale quelconque.*

Considérons maintenant les variables $\{\hat{\beta}_i^*\}_{1\leqslant i\leqslant k}$ définies par $\hat{\beta}^* = B^*\hat{\theta}$, où $B^* = PB$. Alors on a $\beta^* = E(\hat{\beta}^*) = B^*\theta$ et évidemment la variable s^2 est indépendante de $\{\hat{\beta}_i^*\}$. Puisque, comme l'on a déjà montré, $\Sigma_{\hat{\beta}^*}$ est une matrice diagonale et le vecteur aléatoire $\hat{\beta}^*$ suit une loi normale, il en resulte l'indépendance des variables $\{\hat{\beta}_i^*\}$. Par conséquent, pour $-\dfrac{a^2}{k-1} < b < a^2$ les hypothèses fondamentales (1.3) peuvent être écrites de la manière équivalente:

$$(2.2)\quad \begin{cases} \{\hat{\beta}_i^*\}_{1\leqslant i\leqslant k} \text{ sont des variables mutuellement indépendantes,} \\ \hat{\beta}_i^* \sim N(\beta_i^*, (a^2-b)\sigma^2),\ i=1, \ldots, k, \text{ où } a^2-b>0 \text{ est une} \\ \text{constante connue.} \\ \dfrac{\nu s^2}{\sigma^2} \sim \chi^2_\nu \text{ et la variable } s^2 \text{ est indépendante de } \{\hat{\beta}_i^*\}. \end{cases}$$

Sous les hypothèses (2.2) peuvent être déterminés des intervalles communs de confiance pour tous les contrastes dans les nouveaux paramètres $\{\beta_i^*\}$. Mais, en général, le retour aux paramètres initiaux $\{\theta_i\}$ ne conduit pas aux contrastes dans ces paramètres. Le lemme qui suit permet de déterminer les matrices $B^* = PB$ qui satisfont à une telle propriété d'invariance.

Lemme. *L'égalité $c'PB = c'$ où B est la matrice définie dans* (2.1) *et $P = (p_1, \ldots, p_k)$ une matrice orthogonale, est satisfaite pour tout vecteur $c = (c_1, \ldots, c_k)'$ pour lequel $\sum_i c_i = 0$ si et seulement si $p_i = b_i (i = 1, \ldots, k-1)$,*

$$p_k = \pm \frac{1}{\lambda} b_k.$$

En considérant maintenant le modèle (2.2), il en résulte que pour l'ensemble des contrastes $\{\sum_i c_i\beta_i^*\}$ est valable la relation probabiliste (1.2) dont les quantités $\{\theta_i\}$, $\{\hat{\theta}_i\}$ et a sont remplacées respectivement par $\{\beta_i^*\}$, $\{\hat{\beta}_i^*\}$ et $(a^2-b)^{1/2}$. Si l'on choisit dans la transformation $\hat{\beta}^* = PB\hat{\theta}$ comme matrice P l'une de celles déterminées à l'aide du lemme, l'on déduit que si $\sum_i c_i = 0$ alors $\sum_i c_i\beta_i^* = \sum_i c_i\theta_i$ et $\sum_i c_i\hat{\beta}_i^* = \sum_i c_i\hat{\theta}_i$. On peut donc énoncer le résultat suivant:

Sous les hypothèses fondamentales (1.3), *la probabilité que tous les contrastes $\sum_i c_i\theta_i (\sum_i c_i = 0)$ satisfassent simultanément aux inégalités:*

$$\sum_i c_i\hat{\theta}_i - (a^2 < b)^{1/2} q_{\alpha;k,\nu} s\left(\frac{1}{2}\sum_i |c_i|\right) \leqslant \sum_i c_i\theta_i \leqslant \sum_i c_i\hat{\theta}_i + $$
$$+ (a^2-b)^{1/2} q_{\alpha;k,\nu} s\left(\frac{1}{2}\sum_i |c_i|\right)$$

où $-\dfrac{a^2}{k-1} \leqslant b < a^2$, *est égale à* $1-\alpha$.

BIBLIOGRAPHIE

1. Miller, Jr., R.G., *Simultaneous Statistical Inference*. New York, McGraw-Hill, 1966.
2. Scheffé, H., *The Analysis of Variance*. New York, Wiley, 1959.
3. Tukey, J.W., *The Problem of Multiple Comparisons*. Princeton, Princeton Univ. Press, 1953.
4. Văduva, I., *Analiză dispersională*. Bucureşti, Ed. tehnică, 1970.

A. N PHILIPPOU, C. GEORGHIOU [1], G. N. PHILIPPOU [2]

A generalized geometric distribution and some of its properties

A generalized geometric distribution is introduced and briefly studied. First it is shown that it is a proper probability distribution. Then its probability generating function, mean and variance are derived. Finally, the probabillity distribution of the sum of r independent random variables distributed as generalized geometric is obtained. This distribution is a generalized negative binomial distribution.

1. Introduction and summary

In the sequel, k and r are fixed integers greater than or equal to 1, unless otherwise stated, x_i $(1 \leqslant i \leqslant k)$ and x are nonnegative integers as specified and p is a real number in the interval (0,1). Motivated by the recent work of Philippou and Muwafi [2], we presently introduce in Section 2 the geometric distribution of order k with parameter p. We denote this distribution by $G_k(.\,;p)$. First we observe that $G_1(.\,;p)$ is the usual geometric distribution, which shows that $G_k(.\,;p)$ is a generalized geometric distribution. Next we establish that $G_k(.\,;p)$ is a proper probability distribution, and we derive its probability generating function, mean and variance. Finally, in Section 3, we obtain the probability distribution of r independent random variables distributed as geometric of order k. We call this probability distribution the negative binomial distribution of order k with parameter vector (r, p).

We note in ending this section that the present work generalizes well known results on the usual geometric distribution (see, e.g. Zehna [4], pp. 129—130 and Chung [1], pp. 88 and 181).

2. The geometric distribution of order k

In this section, we introduce the geometric distribution of order k and briefly study some of its properties.

---

[1] University of Patras, Patras, Greece;
[2] Higher Technical Institute Patras, Patras, Greece.

Definition 2.1. A random variable X is said to have the *geometric distribution of order k with parameter p*, to be denoted by $G_k(x;p)$, if

$$P[X=x]=\sum_{x_1,\dots,x_k}\binom{x_1+\dots+x_k}{x_1,\dots,x_k}p^x\left(\frac{q}{p}\right)^{x_1+\dots+x_k}, \qquad x\geqslant k,$$

where the summation is over all nonnegative integers $x_1, \ldots, x_k$ such that $x_1+2x_2+\ldots+kx_k=x-k$, and $q=1-p$.

If we denote by $G(.;p)$ the geometric distribution, it follows from Definition 2.1 that $G_1(x;p)=G(x;p)$. Therefore, $G_k(.;p)$ is a generalized geometric distribution.

We now proceeed to show that $G_k(.;p)$ is a proper probability distribution.

Lemma 2.1. *Let X be a random variable distributed as* $G_k(x;p)$. *Then*

$$\sum_{x=k}^{\infty}P[X=x]=1.$$

Proof. We observe that by Definition 2.1

$$\sum_{x=k}^{\infty}P[X=x]=\sum_{x=k}^{\infty}\sum_{\substack{x_1,\dots,x_k\\ x_1+2x_2+\dots+kx_k=x-k}}\binom{x_1+\dots+x_k}{x_1,\dots,x_k}p^x\left(\frac{q}{p}\right)^{x_1+\dots+x_k}=$$

$$=p^k\sum_{x=0}^{\infty}\sum_{\substack{x_1,\dots,x_k\\ x_1+2x_2+\dots+kx_k=x}}\binom{x_1+\dots+x_k}{x_1,\dots,x_k}p^x\left(\frac{q}{p}\right)^{x_1+\dots+x_k}\text{(by replacing } x \text{ by } x+k)=$$

$$=p^k\sum_{n=0}^{\infty}\sum_{\substack{n_1,\dots,n_k\\ n_1+\dots+n_k=n}}\binom{n_1+\dots+n_k}{n_1,\dots,n_k}p^{n_1+2n_2+\dots+kn_k}\left(\frac{q}{p}\right)^{n_1+\dots+n_k},$$

by setting $x_i=n_i(1\leqslant i\leqslant k)$ and $x=n+\sum_{i=1}^{k}(i-1)n_i$,)

$$=p^k\sum_{n=0}^{\infty}\left[\left(\frac{q}{p}\right)(p+p^2+\dots+p^k)\right]^n \text{ (by the multinomial theorem)}$$

$$=p^k\sum_{n=0}^{\infty}(1-p^k)^n=1, \text{ QED.}$$

Remark 2.1. In particular, Lemma 2.1 states that

$$\sum_{x=1}^{\infty}(1-p)^{x-1}p=1,$$

which is well known.

Lemma 2.1. will be utilized in establishing

Lemma 2.2. *Let X be a random variable distributed as* $G_k(x;p)$. *Then its probability generating function, to be denoted by* $\gamma_k(s)$, *is given by*

$$\gamma_k(s)=p^ks^k(1-ps)/[1-s+qp^ks^{k+1}], \quad |s|\leqslant 1.$$

Proof. By definition,

$$\gamma_k(s) = E(s^X) = \sum_{x=k}^{\infty} s^x P[X = x],$$

so that $\gamma_k(s)$ converges absolutely for $|s| \leqslant 1$, by Lemma 2.1. Next we proceed as follows. We have

$$\gamma_k(s) = \sum_{x=k}^{\infty} s^x P[X = x] = \sum_{x=0}^{\infty} s^{x+k} P[X = x + k] \text{ (by replacing } x \text{ with } x + k)$$

$$= \sum_{x=0}^{\infty} \sum_{\substack{x_1,\ldots,x_k \\ x_1+2x_2+\ldots+kx_k=x}} \binom{x_1 + \ldots + x_k}{x_1, \ldots, x_k} (ps)^{x+k} \left(\frac{q}{p}\right)^{x_1+\ldots+x_k} \text{ (by Definition 2.1.)}$$

$$= p^k s^k \sum_{n=0}^{\infty} \sum_{\substack{n_1,\ldots,n_k \\ n_1+\ldots+n_k=n}} \binom{n_1 + \ldots + n_k}{n_1, \ldots, n_k} (p\,s)^{n_1+2n_2+\ldots+kn_k} \left(\frac{q}{p}\right)^{n_1+\ldots+n_k},$$

$$\left(\text{by setting } x_i = n_i \ (1 \leqslant i \leqslant k) \text{ and } x = n + \sum_{i=1}^{k} (i-1)n_i, \right)$$

$$= p^k s^k \sum_{n=0}^{\infty} \left[\left(\frac{q}{p}\right)(ps + p^2s^2 + \ldots + p^ks^k)\right]^n \text{ (by the multinomial theorem)}$$

$$= p^k s^k \frac{1}{1 - \dfrac{q}{p}(ps + p^2s^2 + \ldots + p^ks^k)} \text{ (since } |s| \leqslant 1)$$

$$= \frac{p^k s^k (1 - ps)}{1 - s + qp^k s^{k+1}}, \text{ QED.}$$

We can now employ Lemma 2.2 to get the mean and variance of the geometric distribution of order k.

Proposition 2.1. *Let X be a random variable distributed as $G_k(x; p)$. Then $E(X) = (1 - p^k)/qp^k$ and $\sigma^2(X) = [1 - (2k + 1)qp^k - p^{2k+1}]/q^2p^{2k}$.*

Proof. We observe that

$$E(X) = \frac{\partial \gamma_k(s)}{\partial s}\bigg|_{s=1} = \frac{(kq - p)\,p^k - (k + 1)\,qp^k + 1}{qp^k} = \frac{1 - p^k}{qp^k} \tag{2.1}$$

and

$$E[X(X-1)] = \frac{\partial^2 \gamma_k(s)}{\partial s^2}\bigg|_{s=1} = \frac{-2kp^k}{qp^k} - \frac{2[-1 + (k + 1)qp^k](1 - p^k)}{q^2p^{2k}} =$$

$$= \frac{2\,[1 - p^k + qp^{2k} - (k + 1)\,qp^k)}{q^2p^{2k}}. \tag{2.2}$$

It follows that

$$\sigma^2(X) = E[X(X-1)] + E(X) - [E(X)]^2 = \frac{1 - (2k + 1)qp^k - p^{2k+1}}{q^2p^{2k}}, \tag{2.3}$$

by means of (2.1), (2.2) and some algebra. Relations (2.1) and (2.3) establish the proposition.

For $k=1$, Proposition 2.1 reduces to the following well known result (see, e.g Zehna [4], pp. 129—130).

Corollary 2.1. *Let X be a random variable distributed as $G(x;p)$. Then*

$$E(X) = \frac{1}{p} \text{ and } \sigma^2(X) = \frac{q}{p^2}.$$

We end this section by noting the following proposition which describes how the geometric distribution of order k may arise.

Proposition2.2. *Let N_k be a random variable denoting the number of trials until the occurrence of the kth consecutive success in independent trials with success probability p. Then*

(a) N_k *is* $G_k(n;p)$;

(b) $E(s^{N_k}) = p^k s^k(1-ps)/[1-s+qp^k s^{k+1}]$, $|s| \leqslant 1$;

(c) $E(N_k) = (1-p^k)/qp^k$ *and* $\sigma^2(N_k) = [1-(2k+1)qp^k - p^{2k+1}]/q^2p^{2k}$.

Proof. (a) For $k=1$, it is well known that N_1 is $G(n;p)$. But $G(n;p) = G_1(n;p)$, by Definition 2.1, so that N_1 is $G_1(n;p)$. For $k \geqslant 2$, (a) holds true, because of Theorem 3.1 of Philippou and Muwafi [2] and Definition 2.1. Part (b) follows from (a) and Lemma 2.2. Finally, (c) follows from (a) and Proposition 2.1.

Remark 2.2. Shane [3] was able to derive (b) and (c) by developing a recursion formula for $P[N_k = n]$, $n \geqslant k$, in terms of his polynacci polynomials of order k in p.

3. The negative binomial distribution of order k

In this section we derive the probability distribution of the sum of r independent random variables distributed as geometric of order k. This probability distribution is a generalized negative binomial distribution.

Theorem 3.1. *Let $X_1, \ldots, X_r$ be independent random variables distributed as $G_k(x;p)$ and set $Y_r = \sum_{i=1}^{r} X_i$. Then*

$$P[Y_r = y] = \sum_{y_1,\ldots,y_k} \binom{y_1+\ldots+y_k+r-1}{y_1, \ldots, y_k, r-1} p^y \left(\frac{q}{p}\right)^{y_1+\ldots+y_k}, \quad y \geqslant kr,$$

where the summation is over all nonnegative integers $y_1, \ldots, y_k$ such that $y_1 + 2y_2 + \ldots + ky_r = y - kr$.

Proof. We observe that by the definition of Y_r and the independence of $X_1, \ldots, X_r$,

$$\sum_{y=kr}^{\infty} s^y P[Y_r = y] = E(s^{Y_r}) = \prod_{i=1}^{r} E(s^{X_i}) =$$

$$= \left[\frac{p^k s^k(1-ps)}{1-s+qp^k s^{k+1}}\right]^r = \frac{p^{kr}s^{kr}}{\left[1-\frac{q}{p}(ps+p^2s^2+\ldots+p^ks^k)\right]^r} \text{ (by Lemma 2.2)}$$

$$= p^{kr}s^{kr}\sum_{m=0}^{\infty}\binom{-r}{m}\left[-\frac{q}{p}(ps+p^2s^2+\ldots+p^ks^k)\right]^m$$

$$= p^{kr}s^{kr}\sum_{m=0}^{\infty}\binom{m+r-1}{m}\left(\frac{q}{p}\right)^m[ps+p^2s^2+\ldots+p^ks^k]^m$$

$$= p^{kr}s^{kr}\sum_{m=0}^{\infty}\binom{m+r-1}{m}\left(\frac{q}{p}\right)^m \sum_{\substack{m_1,\ldots,m_k\\ m_1+\ldots+m_k=m}}\binom{m_1+\ldots+m_k}{m_1,\ldots,m_k}(ps)^{m_1+2m_2+\ldots+km_k},$$

(by the multinomial theorem)

$$= p^{kr}s^{kr}\sum_{m=0}^{\infty}\sum_{\substack{m_1,\ldots,m_k\\ m_1+\ldots+m_k=m}}\binom{m_1+\ldots+m_k+r-1}{m_1,\ldots,m_k,\ r-1}\left(\frac{q}{p}\right)^{m_1+\ldots+m_k}(ps)^{m_1+2m_2+\ldots+km_k},$$

$$= p^{kr}s^{kr}\sum_{y=kr}^{\infty}\sum_{\substack{y_1,\ldots,y_k\\ y_1+2y_2+\ldots+ky_k=y-kr}}\binom{y_1+\ldots+y_k+r-1}{y_1,\ldots,y_k,\ r-1}\left(\frac{q}{p}\right)^{y_1+\ldots+y_k}(ps)^{y-kr},$$

$\left(\text{by setting } m_i = y_i\ (1\leqslant i\leqslant k) \text{ and } m = y-kr-\sum_{i=1}^{k}(i-1)y_i\right)$

$$= \sum_{y=kr}^{\infty} s^y \sum_{\substack{y_1,\ldots,y_k\\ y_1+2y_2+\ldots+ky_k=y-kr}}\binom{y_1+\ldots+y_k+r-1}{y_1,\ldots,y_k,\ r-1}p^y\left(\frac{q}{p}\right)^{y_1+\ldots+y_k}.$$

It follows that

$$P[Y_r = y] = \sum_{\substack{y_1,\ldots,y_k\\ y_1+2y_2+\ldots+ky_k=y-kr}}\binom{y_1+\ldots+y_k+r-1}{y_1,\ldots,y_k,\ r-1}p^y\left(\frac{q}{p}\right)^{y_1+\ldots+y_k} \quad y\geqslant kr,$$

which was to be shown.

We say that the random variable Y_r has the negative binomial distribution of order k with parameter vector (r, p) and denote it by $NB_k(y; r, p)$. Let $NB(.;r,p)$ denote the negative binomial distribution with parameter vector (r,p). Then $NB_1(y; r, p) = NB(y; r, p)$, which shows that the negative binomial distribution of order k is a generalized negative binomial distribution.

REFERENCES

1. Chung, K.L., *Elementary Probability Theory with Stochastic Processes.* New York, Springer, 1975.
2. Philippou, A.N., Muwafi, A.A., *Waiting for the kth consecutive success and the Fibonacci sequence of order k.* Fibonacci Quart., **20**, 28–32 (1982).
3. Shane, H.D., *A Fibonacci probability function.* Fibonacci Quart., **11**,517–522 (1973).
4. Zehna, P.W., *Probability Distributions and Statistics.* Boston, Allyn and Bacon, 1970.

VIORICA POSTELNICU

General criteria for positive dependence

1. Introduction

The concept of stochastic independence is defined by equality relations and the absence of it implies dependence. This negative way of describing dependence hardly provides a clue about the nature of dependence. Of particular importance are conditions of "positive dependence".

Intuitively one could say that random variables are positively dependent if they "hang together" in the sense that large values of any one of the variables tend to accompany large values of the other variables and similarly for small values.

Motivated by the necessities in reliability theory and simultaneous inference, many authors defined types of positive dependence. The main types of such dependence were introduced in [6], [4].

Definition 1. [Lehman (1966)] The random variables $U_1, \ldots, U_n$ are said to be *positively quadrant dependent* (PQD) if

$$P[U_1 \leqslant u_1, \ldots, U_n \leqslant u_n] \geqslant \prod_1^n P[U_i \leqslant u_i]$$

for all real numbers $u_1, \ldots, u_n$.

Definition 2. [Esary, Proschan and Walkup (1967)] The random variables $T_1, \ldots, T_n$ are said to be *associated* if

$$\text{Cov}[f(T_1, \ldots, T_n), g(T_1, \ldots, T_n)] \geqslant 0$$

for all nondecreasing functions f and g for which $Ef(\overline{T})$, $Eg(\overline{T})$, $Ef(\overline{T})g(\overline{T})$ exist.

Other types of positive dependence are defined in [1], [2], [3].

Various concepts of positive dependence can be unified to some extent. The aim of the present paper is to do this by means of the concept of dependence functional..

2. Dependence functionals

Let us denote by $\mathbf{F}^{(n)}$ the set of all n-dimensional distribution functions

$$\mathbf{F}^{(n)} = \{F_{X_1,\dots,X_n}(x_1, \dots, x_n) \mid F_{X_1,\dots,X_n}(x_1, \dots, x_n) = P[X_1 \leqslant x_1, \dots, X_n \leqslant x_n]\}$$

and by $\overline{\varphi}$ an n-dimensional vector of functions $\overline{\varphi} = (\varphi_1, \dots, \varphi_n)$ where $\varphi_i(u_1, \dots, u_n)$ are nondecreasing functions in each argument.

For each $\overline{\varphi}$ define the application

$$\alpha_{\overline{\varphi}} : \mathbf{F}^{(n)} \to \mathbf{F}^{(n)}$$
$$\alpha_{\overline{\varphi}}(F_{X_1,\dots,X_n}) = F(\varphi_1(X_1, \dots, X_n), \dots, \varphi_n(X_1, \dots, X_n))$$

Definition 3. An application $\delta : \mathbf{F}^{(n)} \to R$ is said to be a *dependence functional* if for each $\overline{\varphi}$

1) $\delta(\alpha_{\overline{\varphi}}(F)) \geqslant \delta(F)$

2) $\delta(F_{X_1,\dots,X_n}) \geqslant 0$ if $X_1, \dots, X_n$ are independent.

Definition 4. Let Δ be a set of dependence functionals. $F \in \mathbf{F}^{(n)}$ is said to be *positive dependent* with respect to Δ (*Δ-positive dependent*) if $\delta(F) \geqslant 0$ for each $\delta \in \Delta$.

The random variables $X_1, \dots, X_n$ are said to be *Δ-positive dependent* if $F_{X_1,\dots,X_n}(x_1, \dots, x_n) = P[X_1 \leqslant x_1, \dots, X_n \leqslant x_n]$ is a Δ-positive dependent distribution function.

It follows immediately from Definition 4 that:

— any subset of Δ-positive dependent random variables is also Δ-positive dependent;

— if F is Δ-positive dependent then $\alpha_{\overline{\varphi}}(F)$ is Δ-positive dependent.

When two distribution functions satisfy a condition of positive dependence it is natural to ask whether either exhibits this dependence in a stronger form than the other.

According to Definition 4 positive dependence with respect to Δ is stronger than positive dependence with respect to Δ' if $\Delta' \subset \Delta$ or if the positivity of all functionals in Δ implies the positivity of all functionals in Δ'.

The first natural step to do now is to check if the known types of positive dependence are Δ-positive dependences and to establish the corresponding sets Δ. This is done by means of the following two propositions for the types defined by Defintions 1 and 2.

Proposition 1. *For each distribution function with marginals $F_1(x), \dots, F_n(x)$, $\delta_I(F) = \inf\limits_{(x_1,\dots,x_n) \in R^n} [F(x_1, \dots, x_n) - F_1(x_1) \dots F_n(x_n)]$ is a dependence functional.*

Positive dependence with respect to $\Delta = \{\delta_I\}$ is the positive quadrant dependence.

The proposition can be proved via Theorems 3 and 6 of Sklar [8].

Proposition 2. a) *If X, Y are random variables with the bivariate distribution function $F(x, y)$ then $C(F) = \operatorname{cov}[X, Y]$ is a dependence functional.*

b) *If $X_1, \dots, X_n$ are random variables with the n-dimensional distribution $F(x_1, \dots, x_n)$ then for each pair of nondecreasing functions f and g for which $Ef(X_1, \dots, X_n)$, $Eg(X_1, \dots, X_n)$, $Ef(X_1, \dots, X_n)\, g(X_1, \dots, X_n)$ exist $C_{f,g}(F) = \operatorname{cov}[f(X_1, \dots, X_n), g(X_1, \dots, X_n)]$ is a dependence functional.*

Positive dependence with respect to $\Delta = \{C_{f,g}\}$, *f and g binary nondecreasing, is equivalent to association.*

Proof. To prove a) is it sufficient to observe that all possible nondecreasing functions $\varphi : \{0, 1\}^2 \to \{0, 1\}$ can be listed, so that it is easy to see by direct computation that $\mathrm{cov}\,[\varphi(X, Y), \psi(X, Y)] \geqslant \mathrm{cov}\,[X, Y]$ if φ and ψ are such functions and X, Y are binary random variables. This inequality may be extended to general φ, ψ, X and Y using an analog of an identity attributed to Hoeffding (presented in [6]), which completes the proof of the first part of the proposition.

b) is a consequence of a) and of Theorem 3.1. in (4).

3. The dependence model

We use the following notation

1) Let $\mathbf{A} \subset \mathbf{P}(\{1, 2, \ldots, n\})$, $\mathbf{A} = \{A_1, A_2, \ldots, A_{N(\mathbf{A})}\}$, $\bigcup_{i=1}^{N(\mathbf{A})} A_i = \{1, 2, \ldots, n\}$ and let $k(i)$ be the number of the A_j, $1 \leqslant j \leqslant N(\mathbf{A})$ such that $i \in A_j$.

2) Let $\{S_A\}_{A \in \mathbf{A}}$ be a collection of independent random variables defined on a probability space $(\Omega, \mathbf{K}, P)$.

3) Let $\{\varphi_i\}_{i=1,\ldots,n}$ be a collection of functions $\varphi_i : \Omega^{k(i)} \to \Omega$ nondecreasing in each of their arguments.

Then the random variables $U_i = \varphi_i(S_{A^{(i)}_{j_1}} \ldots, S_{A^{(i)}_{j_{k(i)}}})$, where $i \in A^{(i)}_{j_u}$, $A^{(i)}_{j_u} \neq A^{(i)}_{j_v}$, and $\{A^{(i)}_{j_1}, \ldots, A^{(i)}_{j_{k(i)}}\} \subset \mathbf{A}$ for $1 \leqslant u \neq v \leqslant k(i)$, $1 \leqslant i \leqslant N(\mathbf{A})$, are generated by the dependence model

$$(\mathbf{A}, \ \{S_A\}_{A \in \mathbf{A}}, \ \{\varphi_i\}). \tag{1}$$

Theorem. *The random variables* $\{U_i\}$ *generated by the dependence model* (1) *are positive dependent with respect to any class of dependence functionals.*

The statement follows from the fact that for independent random variables $\delta(F) \geqslant 0$ for each dependence functional.

It follows from this theorem that the random variables generated by a dependence model provide the strongest form of Δ-positive dependence.

In many cases it is useful to solve a converse problem: given a Δ-positive dependent distribution function F, find a dependence model generating the random variables $X_1, \ldots, X_n$ with joint distribution F.

An example of such a dependence model is given by Theorem 3.2 of Marshall and Olkin [7]: A random vector $U = (U_1, \ldots, U_n)$ is multivariate exponential if and only if it is generated by a dependence model of the form $(\mathbf{A}, \ \{S_A\}_{A \in \mathbf{A}}, \ \{\min\,[S_{A^{(i)}_{j_{ki}}}, \ldots, S_{A^{(i)}_{j_{k(i)}}}]\})$.

Other examples of dependence models are given in [5].

REFERENCES

1. Alan, K., Wallenius, K.T., *Positive dependence and monotonicity in conditional distributions*. Comm. Statist. A, **5**, 525–534 (1976).
2. Dykstra, R.L., Hewitt, J.E., *Positive dependence of the roots of a Wishart matrix*. Ann. Statist. **6**, 235–238 (1978).
3. Esary, J.D., Proschan, F., *Relations among some concepts of bivariate dependence*. Ann. Math. Statist., **43**, 651–655 (1972).

4. Esary, J.D., Proschan, F., Walkup, D.W., *Association of random variables with applications.* Ann. Math. Statist., **38**, 1466–1474 (1967).

5. Lanberg, N., Proschan, F., Quinzi, A.J., *Converting dependent models into independent ones, with applications in reliability.* In: Tsokos, C.P., Shimi, I.N. (Eds.), *The Theory and Applications of Probability,* pp. 259–276. San Francisco–London, Academic Press, 1977.

6. Lehmann, E.L., *Some concepts of dependence.* Ann. Math. Statist., **37**, 1137–1173 (1966).

7. Marshall, A.W., Olkin, I., *A multivariate exponential distribution.* J. Amer. Statist. Assoc., **62**, 30–44 (1967).

8. Sklar, A., *Random variables, joint distribution functions and copulas.* Kybernetika, **9**, *6*, 449–460 (1973).

VASILE PREDA

The generalized gamma distribution and the principle of maximum entropy

It is shown that the generalized gamma distribution can be obtained by means of the principle of maximum entropy.

1. Introduction

The principle of maximum entropy was established, independently, by Ingarden (1963), Jaynes (1957), Kullback and Leibler (1951). According to this principle we choose the probability distribution which maximizes the entropy compatible to some set of restraints. Using this principle, Jaynes (1957), Kampé de Fériet (1963), Ingarden and Kossakowski (1971), Preda (1982 a, b, 1984) obtained some usual probability distributions. In this paper another probability distribution is obtained.

2. Statement of the problem

Let $(\Omega, \mathbf{K}, P)$ be a probability space, $T=(0, +\infty)$, $\mathbf{T}$ the class of Lebesgue measurable sets in T, μ Lebesgue measure on $\mathbf{T}$. If X is a random variable on $(\Omega, \mathbf{K}, P)$ and taking values in $(T, \mathbf{T})$, then its probability distribution is $\nu_X = P \circ X^{-1}$. If $\nu_X \ll \mu$ let $\rho(x) = (d\nu_X/d\mu)(x)$ be the probability density function of X and $H(X) = -E_{\nu_X} \ln \rho(X)$ the entropy of the random variable X, where E_{ν_X} denotes expectation under ν_X.

Let $\alpha>0$, $\beta>0$, $\gamma \neq 0$ be real numbers and $\mathcal{X}(\alpha, \beta, \gamma)$ the set of all random variables X so that the following conditions are verified:

(1) $\nu_X \ll \mu$,

(2) $E_{\nu_X} X = \alpha \cdot \beta$,

(3) $E_{\nu_X} \ln X = A(\alpha, \beta, \gamma)$,

where

$$A(\alpha, \beta, \gamma) = |\gamma|^{-1} \cdot \left(\int_0^1 (1 - y^{\alpha-1}) \cdot (1-y)^{-1} \, dy + \ln \beta - C \right)$$

and C is Euler's constant.

Thus, we have the following optimization problem:

(4) $$\sup \{H(X) \mid X \in \mathfrak{X}(\alpha, \beta, \gamma)\}.$$

In the following section we shall prove the following

Theorem. *The solution of the problem* (4) *is given by a random variable X with generalized gamma probability density function with parameters* α, β, γ.

3. Proof of the theorem

Using Lagrange's multipliers method, we get

$$H(X) - u - v \cdot \alpha \cdot \beta - w \cdot A(\alpha, \beta, \gamma) = E_{\nu_x} \ln X^{-w} \exp(-u - v \cdot X^{\gamma})(\rho(X))^{-1} \leqslant$$
$$\leqslant E_{\nu_x} X^{-w} \cdot \exp(-u - v \cdot X^{\gamma})(\rho(X))^{-1} - 1,$$

with equality if and only if

(5) $$\rho(X) = X^{-w} \cdot \exp(-u - v \cdot X^{\gamma}), \; X \in \mathfrak{X}(\alpha, \beta, \gamma).$$

Because $\rho(x)$ is a probability density function, according to Lemma 3 of [8] from (5) we obtain

(6) $$|\gamma| \exp(u) = v^{(w-1)/\gamma} \cdot \Gamma((1-w)/\gamma)$$

and $(1-w)/\gamma > 0$, $v > 0$, where $\Gamma(.)$ is the gamma function. From (2) and (5) it results

(7) $$\alpha \cdot \beta |\gamma| \exp(u) = v^{(w-\gamma-1)/\gamma} \cdot \Gamma((\gamma + 1 - w)/\gamma).$$

Now, from (6) and (7) we have

(8) $$v = (1-w) \cdot (\alpha \cdot \beta \cdot \gamma)^{-1}.$$

From (3) and (5), using a change of variable, we get

(9) $$\gamma^2 \cdot A(\alpha, \beta, \gamma) \cdot \exp(u) = v^{(w-1)/\gamma} \cdot [\Gamma'((1-w)/\gamma) - \Gamma((1-w)/\gamma) \ln v],$$

where $\Gamma'(a) = \mathrm{d}\Gamma(a)/\mathrm{d}a$.

Now, from (6), (8) and (9) we obtain

(10) $$\Gamma'(t)/\Gamma(t) - \ln(t(\alpha \cdot \beta)^{-1}) = \int_0^1 (1 - y^{\alpha-1}) \cdot (1-y)^{-1} \, \mathrm{d}y + \ln \beta - C,$$

where $t = (1-w) \cdot \gamma^{-1}$. By using Lemma 1 of [8] and equation (10) we have

(11) $$\int_0^1 (y^{\alpha-1} - y^{t-1})(1-y)^{-1} \, \mathrm{d}y - \ln(t \cdot \alpha^{-1}) = 0.$$

But, according to Lemma 3 of [8], equation (11) has a unique solution $t = \alpha$. Hence

(12) $$w = 1 - \alpha \cdot \gamma, \; v = \beta^{-1} \text{ and } \exp(u) = |\gamma|^{-1} \cdot \beta^{\alpha} \cdot \Gamma(\alpha).$$

From (5) and (12) it results

$$\rho(x) = |\gamma|/(\Gamma(\alpha) \cdot \beta^{\alpha}) \cdot x^{\alpha \cdot \gamma - 1} \cdot \exp(-x^{\gamma}/\beta), \quad x > 0$$

i.e. the generalized gamma distribution.

REFERENCES

1. Ingarden, R.S., *Information theory and variational principles in statistical theories.* Bull. Acad. Polon. Sci. Sér. Sci. Math. Astronom. Phys., **11**, 541–547 (1963).
2. Ingarden, R.S., Kossakowski, A., *Poisson probability distribution and information thermodynamics.* Bull. Acad. Polon. Sci. Sér. Sci. Math. Astronom. Phys., **19**, 83–86 (1971).
3. Jaynes, E.T., *Information theory and statistical mechanics. I, II.* Phys. Rev., **106**, 620–630; **108**, 171–190 (1957).
4. Kampé de Fériet, J., *Théorie de l'Information. Principe du Maximum de l'Entropie et ses Applications à la Statistique et à la Mécanique.* Lille, Lab. Calcul Fac. Sci. Univ. Lille, 1963.
5. Kullback, S., Leibler, R.A., *On information and sufficiency.* Ann. Math. Statist., 22, 79–86 (1951).
6. Preda, V., *Binomial distributions and information thermodynamics.* Bull. Acad. Polon. Sci. Sér. Sci. Math., **30**, 569–573 (1982a).
7. Preda, V., *The Student distribution and the principle of maximum entropy.* Ann. Inst. Statist. Math., **34**, 335–338 (1982b).
8. Preda, V., *Sur la distribution gamma et le principe du maximum de l'information.* (A paraître 1984).

NICOLAE ROMANOF, VICTOR ION PESCARU

Numerical simulation of turbulent diffusion for a one-dimensional model

1. Introduction

The numerical simulation of turbulent diffusion is a useful method to test some theoretical models [Kraichnan (1970)] and it can be also used in practical applications [Hanna (1982)].

Let $x(t)$ be a random process defined by the differential equation $\frac{dx}{dt} = u(x, t)$, where $u(x, t)$ is the random velocity field of turbulent flow.

One of the problems of turbulent diffusion is to relate the statistical properties of the random field u to the statistical properties of the random process $x(t)$ [Monin and Jaglom (1965), Ch. 5]. This problem, known as "Eulerian-Lagrangian relationship" leads to high mathematical difficulties and until now the results obtained are very poor.

Patterson and Corrsin (1966) simulated the turbulent diffusion using a binary random velocity field in a one-dimensional model. It was found that at small times the distribution of the particles is similar to the solution of the telegraph equation.

Kraichnan (1970) made a computer experiment in which the velocity field was generated along the particle's trajectory, using the Fourier representation.

In this paper we report on numerical simulation of the motion of a particle in a one-dimensional Gaussian random velocity field with mean 0 and covariance $\exp(-|\Delta x| - |\Delta t|)$.

2. Some results from statistical theory of turbulent diffusion

We consider a homogeneous and stationary turbulent flow.

Let $x(t)$ and $V(t) = \frac{dx}{dt}$ be the position and velocity of a particle which starts at $t = 0$ from $x = 0$. In this case, $V(t)$ is a stationary random process [Monin and Jaglom (1965), Ch. 5]. We assume that the length, time and velocity are normalized with some characteristic scales such that $EV^2 = 1$.

The mean square of the position is given by the Taylor formula

$$Ex^2 = \sigma^2 = 2\int_0^t d\tau \int_0^\tau R_L(\zeta)\, d\zeta, \tag{2.1}$$

where $R_L(\zeta) = EV(0)\, V(\zeta)$ is the Lagrangian autocorrelation function.

At small time ($t \ll 1$), $\sigma^2 = t^2$ and at large time ($t \gg 1$), $\sigma^2 = 2tT_L$, where $T_L = \int_0^\infty R_L(\zeta)\, d\zeta$.

Some experimental as well as theoretical results suggest that $x(t)$ has a Gaussian distribution at any time [Monin and Jaglom (1965), Ch. 5] and the transition probability is

$$P(x, t) = \frac{1}{\sqrt{2\pi}\,\sigma} \exp\left(-\frac{x^2}{2\sigma^2}\right). \tag{2.2}$$

The transition probability satisfies the turbulent diffusion equation

$$\frac{\partial P}{\partial t} = K(t) \frac{\partial^2 P}{\partial x^2}, \tag{2.3}$$

where $K(t) = \frac{1}{2} \frac{d\sigma^2}{dt}$ is known as the *apparent eddy diffusivity* (the turbulent diffusion coefficient). At small times $K(t) = t$, and at large times $K(t) = T_L$. Therefore, the classical diffusion equation is true at large times; at small times the diffusion is an accelerated process due to the increasing of K with time.

3. Results and conclusions

In our experiment one thousand trajectories were generated and the following statistical quantities estimated: dispersion $\sigma^2(t) = Ex^2(t)$, Lagrangian autocorrelation function $R_L(t) = EV(0)\, V(t)$, turbulent diffusion coefficient $K(t) = Ex(t)\, V(t)$ and transition probability $P(x, t)$.

Based on results of Hanna (1979) in simulation in time of Eulerian and Lagrangian velocity, we generated the velocity field along the trajectory by the equation

$$u(x + \Delta x,\ t + \Delta t) = Ru(x, t) + \sqrt{1 - R^2}\zeta, \tag{3.1}$$

where ζ is a Guassian random variable with $E\zeta = 0$, $E\zeta^2 = 1$ and $R(\Delta x, \Delta t) = \exp\,(-|\Delta x| - |\Delta t|)$ is the covariance of the velocity field.

The random variable ζ was generated by the rejection method [Văduva (1975)].

In each sample the differential equation $dx/dt = u(x, t)$, which describes the particle path, was integrated by means of finite differences ($\Delta t = 0.1$) in a time interval [0,10]. The transition probability histogram was built in the interval [—10,10] using classes with length 0.2. The results are presented in Figures 1—4. The numerical experiment outlined the principal characteristics of the turbulent diffusion under homogeneous and stationary turbulence:

1) At small times ($t \ll 1$) $\sigma^2 \sim t^2$, and at large times ($t \gg 1$) $\sigma^2 \sim t$ (Figure 1).

2) The turbulent diffusion coefficient increases with time until $t \cong 2$, then it becomes quite constant (Figure 4).

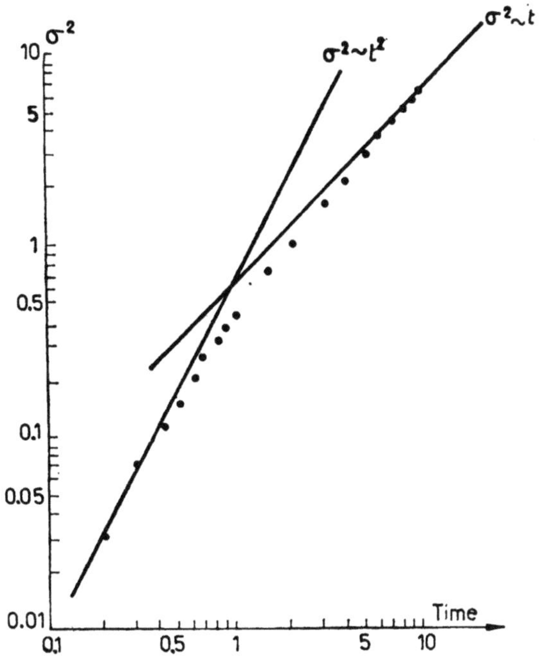

Fig. 1. Dotted line is dispersion σ^2 of the spread of 1000 particles; full lines correspond to the theoretical dispersion at small time ($\sigma^2 \sim t^2$) and large time ($\sigma^2 \sim t$).

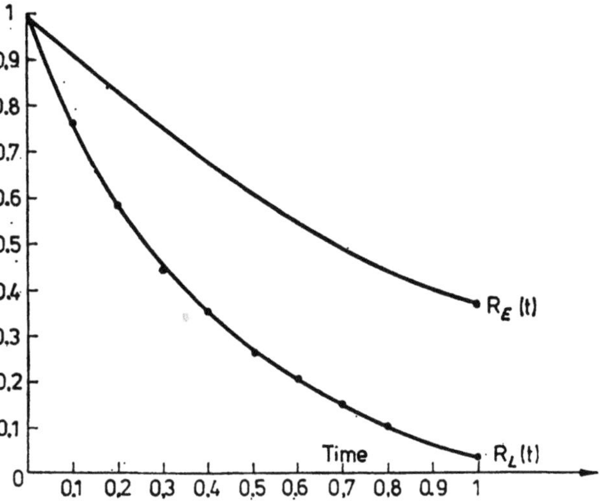

Fig. 2. Variation of Eulerian (R_E) and Lagrangian (R_L) correlation functions with time.

3) The transition probability histogram (Figure 3) shows a Gaussian shape at different times. No statistical test was made but the flatness factor $F = \frac{\overline{x^4}}{3(\overline{x^2})^2}$ has values close to 1.

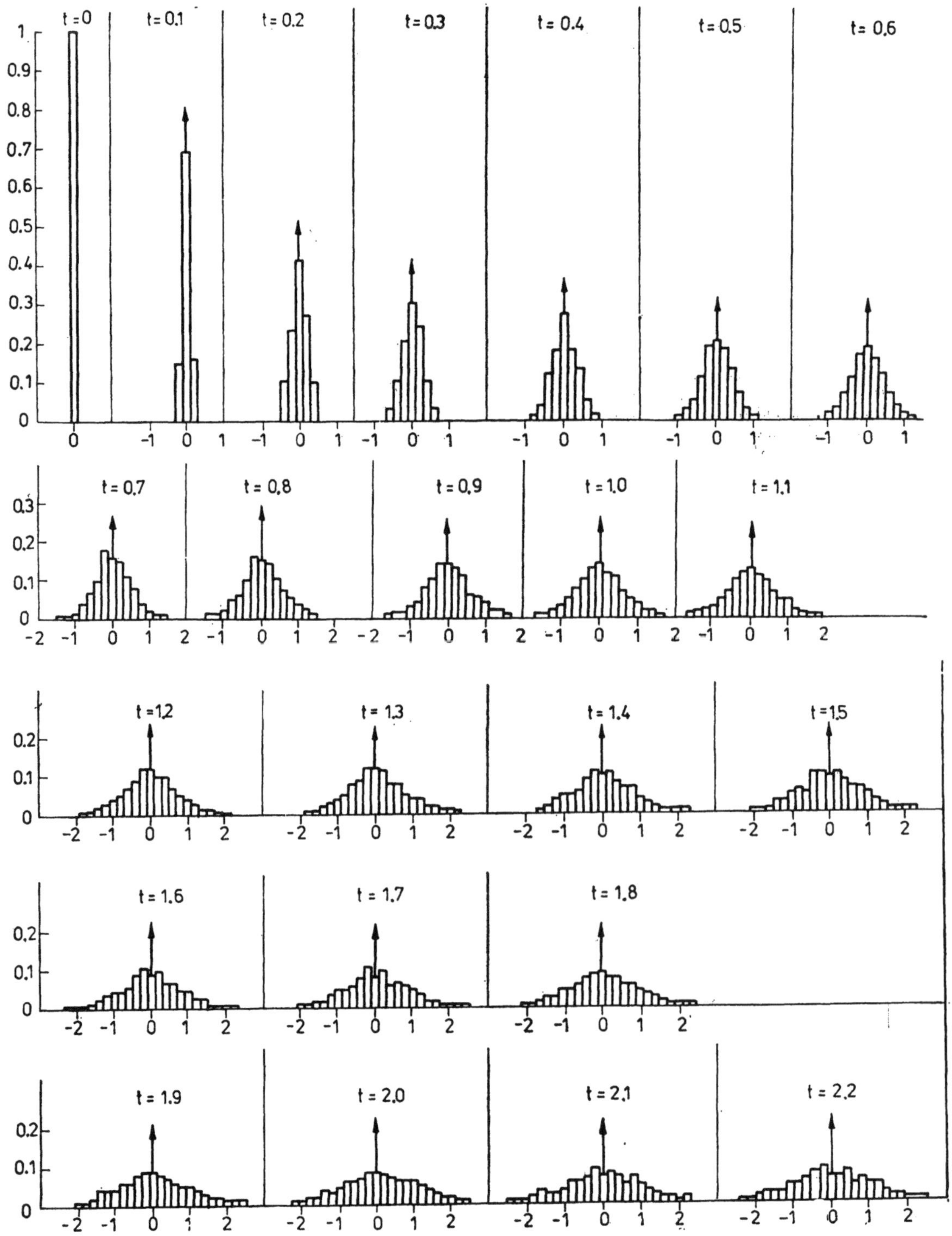

Fig. 3. The transition probability histogram at different times.

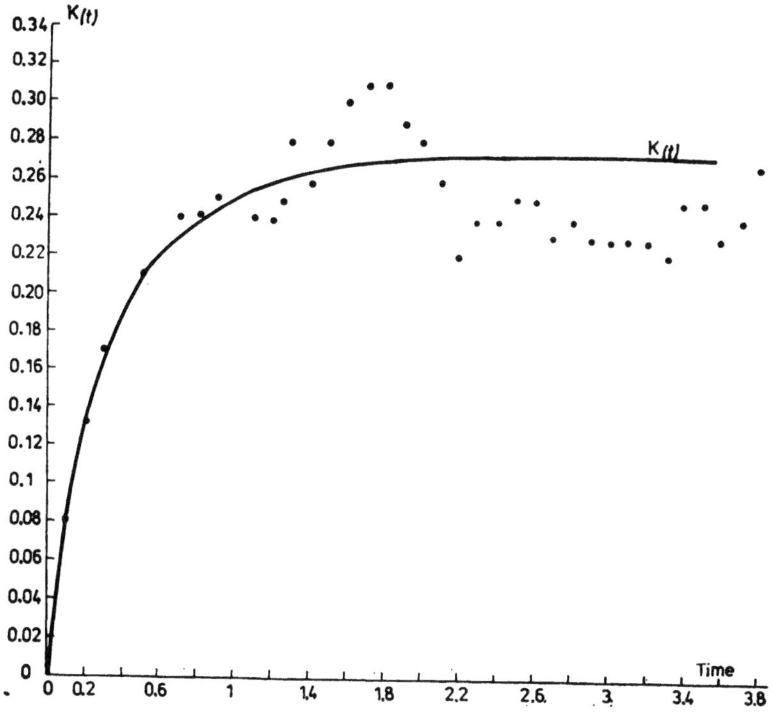

Fig. 4. Variation of the diffusivity coefficient with time.

It was found that the Lagrangian autocorrelation function $R_L(t)$ decreases with time faster than the Eulerian $R_E(t) = \exp(-t)$ (Figure 4).

These preliminary results show that it is possible to find by numerical experiments some of the characteristics of turbulent diffusion. Nevertheless it is necessary to improve the simultation technique and to extend the results to several dimensions.

REFERENCES

1. Kraichnan, R.H., *Diffusion by a random velocity field.* Phys. Fluids, **13**, 22–31 (1970).
2. Hanna, S.R., *Some statistics of Lagrangian and Eulerian wind fluctuations.* J. Appl. Meteorology, **18**, 518–525 (1979).
3. Hanna, S.R., *Review of atmospheric diffusion models for regulatory applications.* W.M.O. Technical Note No. 177. Geneva, 1982.
4. Monin, A.S., Jaglom, A.M., *Statistical Hydromechanics, I.* Moskow, Nauka, 1965 (Russian).
5. Patterson, G.S., Corrsin, S., *Computer experiments on random walks and both Eulerian and Lagrangian statistics.* In: Pai, S.I. (Ed.), *Dynamics of Fluids and Plasmas*, pp. 275–307. New York–London, Academic Press, 1966.
6. Văduva, I., *Simulation Models for Computer Applications.* București, Ed. tehnică, 1977. (Romanian).

GÁBOR J. SZÉKELY

Multiplicative infinite divisibility of standard normal and gamma distributions

The infinite divisibility of the standard normal and the gamma distributions is proved when the operation corresponds not to the addition but to the multiplication of independent random variables. On the contrary, the Poisson distributions are irreducible under multiplicative convolution.

Besides the additive convolution $*$ of probability distributions corresponding to the addition of independent random variables we may also define the multiplicative convolution $*$ corresponding to the multiplication of independent random variables. This multiplicative convolution has been investigated in the last decades by several authors (see e.g. [3] and [10]). Here we deal with the $*$-infinite divisibility of some special distributions.

Denote by N a random variable having standard normal distribution. Since the distribution of sign N is obviously $*$-infinitely divisible, for the $*$-infinite divisibility of N it is enough to prove that the distribution of $|N|$, i.e. the χ_1 distribution, is also $*$-infinitely divisible. (If the expectation of the normally distributed N were negative then the distribution of sign N would not be $*$-infinitely divisible, moreover it would not even have a $*$-square root.) The χ_1 distribution is $*$-infinitely divisible if and only if its square has this property. But χ^2 distributions are special cases of gamma distributions (whose probability densities are proportional to $x^{a-1}e^{-\lambda x}$ if $x>0$ and 0 otherwise; here the parameters a and λ can be arbitrary positive numbers), thus the $*$-infinite divisibility of the standard normal distribution is implied by the following

Theorem. *The gamma distributions are $*$-infinitely divisible.*

Denote by $\beta_{a,b}$ the probability density function of the beta distribution with parameters a and $b-a$, i.e. let $\beta_{a,b}(x)$ be proportional to $x^{a-1}(1-x)^{b-a-1}$ if $0<x<1$ and 0 otherwise (a and $b-a$ are supposed to be positive). Denote further by $\gamma_{a,\lambda}$ the probability density function of the gamma distribution with parameters a and λ.

Lemma. $\gamma_{a,\lambda}=\gamma_{b,\lambda} * \beta_{a,b}$ *if* $0<a<b$.

This lemma can be proved by a simple integration but we can also refer to the "easy direction" of the following characterization theorem (see [4] and [6]). Let X and Y be independent non-degenerated random variables. Then $X+Y$ is independent of $X/(X+Y)$ if and only if X and Y or $-X$ and $-Y$

have gamma distributions with the same scale parameter λ. Thus if X has the density $\gamma_{a,\lambda}$ and Y has the density $\gamma_{b-a,\lambda}$, then $X+Y$ is independent of $X/(X+Y)$. But $X+Y$ resp. $X/(X+Y)$ are known to have the densities $\gamma_{b,\lambda}$ resp. $\beta_{a,b}$.

Proof of the theorem. By the lemma

$$\gamma_{a,\lambda} = \gamma_{b,\lambda} * \beta_{a,a_1} * \beta_{a_1,a_2} * \dots * \beta_{a_n,b}$$

for arbitrary sequences $0<a=a_0< a_1<a_2<\dots<a_{n+1} = b$. Since the expectation of $\gamma_{b,\lambda}$ is equal to b/λ and its variance is b/λ^2, a $\gamma_{b,\lambda}$-distributed random variable divided by b/λ tends stochastically to 1 as $b\to\infty$ (by the Čebyshev inequality). On the other hand the expectation of $\beta_{a_i,a_{i+1}}$ is a_i/a_{i+1} and its variance is $a_i(a_{i+1}-a_i)\,a_{i+1}^{-2}(a_{i+1}+1)^{-1}$, thus a $\beta_{a_i,a_{i+1}}$-distributed random variable divided by a_i/a_{i+1} $(i=0,1,\dots,n)$ tends stochastically to 1 as $a_i\to\infty$. Therefore any $\gamma_{a,\lambda}$-distributed random variable divided by a/λ can be decompossed into the $*$-convolution of arbitrary many random variables which are uniformly close to 1. This implies (see e.g. [5] Corollary to Theorem 5.6.1) the $*$-infinite divisibility of $\gamma_{a,\lambda}$.

Remarks. (i) Our theorem is obviously equivalent to the $*$-infinite divisibility of the logarithm of gamma variables. Using somewhat less elementary results one can also prove the self-decomposability of these variables [7].

(ii) The $*$-infinite divisibility of the standard normal distribution is a dual of Thorin's theorem [9] (see also [1] and [2]) on the $*$-infinite divisibility of the lognormal distribution.

(iii) Our lemma is enlightened from another point of view in [8].

(iv) Besides the normal distributions the Poisson distributions are the most typical $*$-infinitely divisible distributions, thus it is interesting to note that the Poisson distributions are not $*$-infinitely divisible. Moreover, with one of my students, András Zempléni, we have submitted for publication to The American Mathematical Monthly the following theorem. *If X and Y are independent nonnegative integer valued random variables and XY has a Poisson distribution, then either X or Y takes at most two values, zero and one, with probability one. In this sense the Poisson distribution is irreducible.*

Problems. (i) Is it true that a normal distribution is $*$-infinitely divisible if and only if its expectation is 0?

(ii) Find the $*$-irreducible components of the standard normal distribution.

REFERENCES

1. Bondesson, L., *A general result on infinite divisibility.* Ann. Probab., **7**, 965–979 (1979).
2. Bondesson, L., *Classes of infinitely divisible distributions and densities.* Z. Wahrsch. Verw. Gebiete, **57**, 39–72 (1981).
3. Lévy, P., *Esquisse d'une théorie de la multiplication des variables aléatoires.* Ann. Sci. École Norm. Sup., **76**, 59–82 (1959).
4. Lukács, E., *A characterization of the gamma distribution.* Ann. Math. Statist., **26**, 319–324 (1955).

5. Lukács, E., *Characteristic Functions*. London, Griffin, 1960.

6. Marsaglia, G., *Extension and applications of Lukács' characterization of gamma distribution*. In: *Proc. Sympos. Statist. Related Topics (Carleton Univ., Ottawa, Ont., 1974)*, Paper No. 9, 13 pp. Ottawa, Ont., Carleton Univ., 1975.

7. Shanbhag, D.N., Sreehari, M., *On certain self-decomposable distributions*. Z. Wahrsch. Verw. Gebiete, **38**, 217–222 (1977).

8. Stuart, A., *Gamma-distributed products of independent random variables*. Biometrika, **49**, 564–565 (1962).

9. Thorin, O., *On the infinite divisibility of the lognormal distribution*. Scand. Actuar. J., 121–148 (1977).

10. Zolotarev, V.M., *On a general theory of multiplication of independent random variables*. Dokl. Akad. Nauk SSSR, **142**, 788–791 (1962). (Russian)

EMILIANA URSIANU

Multiple comparison techniques for dependent variables

Simultaneous confidence intervals for all linear parametric functions are derived under the general case of dependence using the studentized maximum absolute deviate from the population mean of normally distributed variables. A comparison with other competitors is made.

1. Introduction

The data obtained from various types of research may be highly correlated.

Various methods of simultaneous inference have been suggested for obtaining simultaneous confidence intervals for means or for linear contrasts among means.

Confidence intervals for the true means of dependent normally distributed variables are derived in Dunn [1] when these are of interest rather than confidence intervals for contrasts or linear functions among means.

The confidence intervals can be derived for the general case of dependence using the studentized maximum absolute deviate variable from the population mean in normal samples (the h-method) [17].

This general case includes in particular the result of Dunn regardless of the correlations among the variables, as well as the result of Schilling and Ott [12][14] to compare the individual means of k samples of equal size with the overall grand mean.

An approximate simultaneous confidence intervals for the differences on the effect of the ith factor level α_i and the general mean μ in the one-way fixed effect model ANOVA are given by Enderlein [5].

The exact confidence intervals of overall level $1-\alpha$ are obtained by Dunn [3] for some special form of the correlation matrix and also $k=2,3$.

In this paper the confidence intervals for linear functions among means are derived using the above mentioned distribution and a comparison with the serious competitors is made.

2. Notation

Let the general model Ω be:

(2.1).
$\{\hat{\theta}_i\}$ are random variables $N(\theta_i, a^2\sigma^2)$, $i = 1, \dots, k$; $\{a\}$ is a known positive constant; θ_i and σ^2 are unknown parameters;
$\text{Cov}(\hat{\theta}_i, \hat{\theta}_j) = b\sigma^2$, $i \neq j$, b is a known constant;
s^2 is an independent quadratic estimate of σ^2 with ν d.f., i.e. $\nu s^2/\sigma^2$ has a Chi-square distribution with ν d.f. and is statistically independent of the $\{\hat{\theta}_i\}_{1 \leqslant i \leqslant k}$

Let $\{x_i\}_{1 \leqslant i \leqslant k}$ be i.i.d. $N(\mu, \sigma^2)$ and let s^2 be an estimate of σ^2 which is independent of $\{x_i\}$ and such that $\nu s^2/\sigma^2$ is χ^2_ν; then the random variable

$$H_{k,\nu} = \left\{\max_i |x_i - \bar{x}|\right\} / s_x \tag{2.2}$$

has the studentized maximum absolute deviate from the sample mean distribution, i.e.

$$P\left(H_{k,\nu} = \left\{\max_i |x_i - \bar{x}|\right\} / s_x > H_{\alpha;k,\nu}\right) = \alpha \tag{2.3}$$

The values $H_{\alpha;k,\nu}$ are given by Halperin et al. [6]. The values $h_{\alpha;k,\nu} = H_{\alpha;k,\nu}\sqrt{k/(k-1)}$ given by Schilling [14] are upper bounds for studentized maximum absolute deviate from the population mean μ, namely

$$P\left(h_{k,\nu} = \left\{\max_i |x_i - \mu|\right\} / s_x > h_{\alpha;k,\nu}\right) = \alpha, \tag{2.4}$$

and are, in fact, a slight modification of the $(1-\alpha/2k)$-points of the Student t distribution given by Dunn [1], [3]. Extended tables of factors $h_{\alpha;k,\nu}$ are given by Nelson [11].

3. Confidence intervals for linear functions among means

The studentized maximum absolute deviate from the population mean technique (the h-method) can be extended to include confidence intervals for linear functions among means using the following

Theorem 1. (Ext h) *Under the general assumptions* Ω (2.1) *the probability is at least* $1-\alpha$ *that all linear functions* $\psi = \sum_{i=1}^{k} c_i\theta_i$ *among a set of means* $\{\theta_i\}_{1 \leqslant i \leqslant k}$ *simultaneously satisfy*

$$\sum_{i=1}^{k} c_i\hat{\theta}_i - s \cdot a \cdot h_{\alpha;k,\nu} \sum_{i=1}^{k} |c_i| \leqslant \psi \leqslant \sum_{i=1}^{k} c_i\hat{\theta}_i + s \cdot a \cdot h_{\alpha;k,\nu} \sum_{i=1}^{k} |c_i|, \quad \forall\psi \tag{3.1}$$

The *proof* lies entirely on the following statements:

a). The result attributed to Tukey [8], [13] as follows: Let us suppose that $\theta_1, \dots, \theta_k$ are unknown parameters and $(\underline{\theta}_1, \overline{\theta}_1), \dots, (\underline{\theta}_k, \overline{\theta}_k)$ are $100(1-$

$-\alpha/k)\%$ (one-tail test) or $100(1-\alpha/2k)\%$ (two-tail test) confidence intervals for $\theta_1, \ldots, \theta_k$. Then the probability is at least $1-\alpha$ that these confidence intervals simultaneously contain $\theta_1, \ldots, \theta_k$, respectively.

b) The following lemma [7], [13]: Let $u_1, \ldots, u_k$ be random variables. Then $\max_i \{|u_i|\} \leqslant c$ if and only if: $\left|\sum_{i=1}^{k} c_i u_i\right| \leqslant c \sum_{i=1}^{k} |c_i|$ for every set $\{c_i\}$ of real numbers, $i = 1, \ldots, k$.

c) The following double implication: If the random variables $\hat{\gamma}_i = \hat{\theta}_i - \theta_i$, $i = 1, \ldots, k$ are $N(0, a^2\sigma^2)$ and are statistically independent of $\nu a^2 s^2/a^2\sigma^2$ which is χ^2_ν, then

$$\left|\frac{\hat{\gamma}_i}{a \cdot s}\right| = \left|\frac{\hat{\theta}_i - \theta_i}{a \cdot s}\right| \leqslant c, \; i = 1, \ldots, k$$

if and only if

$$h_{k,\nu} = \left\{\max_i |\hat{\gamma}_i|\right\} / as \leqslant c = h_{\alpha;k,\nu}.$$

Then using Boole inequality the following relations are valid with probability at least $1-\alpha$:

$$(3.1) \qquad P\left(\max_i |\hat{\gamma}_i| \leqslant a\, s\, h_{\alpha;k,\nu}\right) = P(|\hat{\gamma}_i| \leqslant a \cdot s \cdot h_{\alpha;k,\nu}, \; \forall i) \geqslant$$

$$\geqslant (1-\alpha/k)^k > 1-\alpha,$$

where

$$(3.2) \qquad P(|\hat{\gamma}_i| \leqslant a\, s\, h_{\alpha;k,\nu}) = 1 - \alpha/k.$$

Using $\{\hat{\gamma}_i\}$ instead of $\{u_i\}$ and $a\, s\, h_{\alpha;k,\nu}$ instead of $\{c\}$ in the above lemma, the following inequalities should be true:

$$(3.3) \qquad P\left(\max_i |\hat{\gamma}_i| \leqslant a\, s\, h_{\alpha;k,\nu}\right) \geqslant 1-\alpha$$

$$(3.4) \quad P\left(\left|\sum_{i=1}^{k} c_i \hat{\gamma}_i\right| \leqslant a\, s\, h_{\alpha;k,\nu} \cdot \sum_{i=1}^{k} c_i, \; \forall \{c_i\} \text{ real numbers}, \; i = 1, \ldots, k\right) \geqslant 1-\alpha.$$

Then

$$(3.5) \quad P\left(\left|\sum_{i=1}^{k} c_i(\hat{\theta}_i - \theta_i)\right| \leqslant a \cdot s \cdot h_{\alpha;k,\nu} \cdot \sum_{i=1}^{k} |c_i|, \; \forall \{c_i\} \text{ real numbers}\right) \geqslant 1-\alpha,$$

or

$$(3.6) \quad P\left(\hat{\psi} - a \cdot s \cdot h_{\alpha;k,\nu} \sum_{i=1}^{k} |c_i| \leqslant \psi \leqslant \hat{\psi} + a \cdot s \cdot h_{\alpha;k,\nu} \sum_{i=1}^{k} |c_i|, \; \forall \psi\right) \geqslant 1-\alpha,$$

where $\hat{\psi}$ is an l.s.-estimator of linear parametric function ψ.

Remark. When the $\{\hat{\theta}_i\}$ are independently distributed, then the exact confidence intervals for linear parametric functions among means can be obtained using the Ext m-method [7], [10].

4. Comparison of some techniques

Under the model Ω (2.1) the S-method due to Scheffé [13] can be applied using the transformed underlying assumptions Ω':

$$(4.1)\quad \begin{cases} \{\hat{\beta}_i\} \text{ are random variables with } \hat{\beta}_i = \hat{\theta}_i + \hat{\theta}_0; \\ \hat{\theta}_0 \text{ is } N(0, \sigma_0^2) \text{ and is independent of } \{\hat{\theta}_i\} \text{ and } s^2; \\ \sigma_0^2 \text{ is a known positive constant;} \\ E\hat{\beta}_i = \theta_i; \\ \text{Var } \hat{\beta}_i = (a^2 - b)\sigma^2; \\ \text{Cov }(\hat{\beta}_i, \hat{\beta}_j) = \text{Cov }(\hat{\theta}_i, \hat{\theta}_j) + \text{Var } \hat{\theta}_0 = b\sigma^2 + \sigma_0^2, \\ \text{where } \sigma_0^2 = -b\sigma^2, \ -a^2/(k-1) \leqslant b \leqslant 0, \ i \neq j. \end{cases}$$

The S-method is based on the following theorem: *The probability is* $1 - \alpha$ *that the relations*

$$\hat{\psi} - (a^2 - b)^{1/2}(k \cdot F_{\alpha;k,\nu})^{1/2} \cdot s \cdot \left(\sum_{i=1}^{k} c_i^2\right)^{1/2} \leqslant \psi \leqslant$$

$$\leqslant \hat{\psi} + (a^2 - b)^{1/2} \ (k \cdot F_{\alpha;k,\nu})^{1/2} \cdot s \cdot \left(\sum_{i=1}^{k} c_i^2\right)^{1/2}$$

be simultaneously satisfied for every estimable function $\psi = \sum_{i=1}^{k} c_i\theta_i$.

In order to compare the relative efficiency of the Ext h-method and the S-method the ratio R of the squared lenghts of the corresponding confidence intervals should be studied.

The ratio R can be expressed as

$$(4.2)\qquad R_{\text{Ext } h, S} = h^2_{\alpha;k,\nu}/kF_{\alpha;k,\nu} \cdot a^2/(a^2 - b) \frac{\left(\sum_{i=1}^{k} |c_i|\right)^2}{\sum_{i=1}^{k} c_i^2}$$

Numerical tabulations show that

$$h^2_{\alpha;k,\nu}/kF_{\alpha;k,\nu} < 1$$

for every k and $\alpha = 0.01$ or 0.05.

The following criterion for relative efficiency of the Ext h-method and the S-method for the linear functions among means is thus obtained:

Criterion. The Ext h-method is better than the S-method for linear functions among means if $R < 1$.

Taking into account that $b \leqslant 0$ it follows that $a^2/(a^2 - b) \leqslant 1$ or $(k-1)/k \leqslant$ $\leqslant a^2/(a^2 - b) \leqslant 1$.

Therefore the inequality $R < 1$ holds for every k. Thus the Ext h-method gives confidence intervals shorter than the S-confidence intervals for linear functions among means.

It is only when k is considerably bigger i.e., for a large number of populations that the reverse inequality $R > 1$ holds and the S-method gives shorter intervals than the Ext h-intervals.

Under model Ω' the Ext T-method can be applied for linear functions among means and is based on the following theorem: *The probability is* $1-\alpha$ *that the relations*

$$\hat{\psi} - s(a^2 - b)^{1/2}\, q'_{\alpha;k,\nu} \cdot \max\left(\sum_{i\in I^+} c_i, -\sum_{i\in I^-} c_i\right) \leqslant \psi \leqslant$$

$$\leqslant \hat{\psi} + s(a^2 - b)^{1/2}\, q'_{\alpha;k,\nu} \cdot \max\left(\sum_{i\in I^+} c_i, -\sum_{i\in I^-} c_i\right)$$

be simultaneously satisfied, where $q'_{\alpha;k,\nu}$ *are percentage points of the augmented studentized range distribution,* $I^+ = \{i;\ c_i > 0\}$, $I^- = \{i;\ c_i < 0\}$, $n^+ =$ $=$ card I^+, $n^- =$ card I^- [15].

A similar criterion of relative efficiency can be established. Taking into account that $a^2/(a^2-b) \leqslant 1$ and $h^2_{\alpha;k,\nu}/q'^2_{\alpha;k,\nu} < 1$ it follows that the Ext h-method is better than the Ext T-method for the linear functions among means if:

$$\frac{\left(\sum_{i=1}^{k} |c_i|\right)^2}{\max\left(\left(\sum_{i\in I^+} c_i\right)^2, \left(-\sum_{i\in I^-} c_i\right)^2\right)} < 1. \tag{4.3}$$

The following cases of (4.3) will be considered:

1) $n^+ \geqslant 1$, $n^- = 0$

The R ratio becomes:

$$R_{\text{Ext } h,\ \text{Ext } T} = \frac{h^2_{\alpha;k,\nu}}{q'^2_{\alpha;k,\nu}}\ \frac{a^2}{a^2-b} < 1$$

Therefore the Ext h-method gives shorter intervals than the Ext T-method for all parametric functions with positive coefficients.

2) $n^+, n^- \geqslant 1$

The ratio R becomes:

$$R_{\text{Ext. } h\ \text{Ext } T} = h^2_{\alpha;k,\nu}/q'^2_{\alpha;k,\nu} \cdot a^2/(a^2-b)\ \frac{\left(\sum_{i=1}^{k} |c_i|\right)^2}{\left(\sum_{i\in I^+} c_i\right)^2}$$

It is only when $\left(\sum_{i=1}^{k} |c_i|\right)$ is considerably bigger than $\left(\sum_{i\in I^+} c_i\right)^2$ that the Ext T-method is better than the Ext h-method.

This conclusion holds even if $\sum_{i\in I^-} c_i \geqslant \sum_{i\in I^+} c_i$.

Remark. Confidence intervals for a preassigned number of linear combinations among means (i.e the set of linear functions among means which are to be estimated must be planned in advance) are derived by Dunn [4].

The main difference between the h-intervals given here and Dunn intervals is that the Ext h-method gives intervals for all possible linear combinations of k means and they must be selected after studying the data.

REFERENCES

1. Dunn, O.J., *Estimation of the means of dependent variables.* Ann. Math. Statist., **29**, 1095–1111 (1958).
2. Dunn, O.J., *Confidence intervals for the means of dependent normally distributed variables.* J. Amer. Statist. Assoc., **54**, 613–621 (1959).
3. Dunn, O.J., *Multiple comparison among means.* J. Amer. Statist. Assoc., **56**, 52–64 (1961).
4. Dunn, O.J., Massey, F.J., *Estimation of multiple contrasts using t-distribution.* J. Amer. Statist. Assoc., **60**, 573–583 (1965).
5. Enderlein, G., *Multiple comparisons of means with the total mean.* In: *Progress in Statistics* (Proc. European Meeting of Statisticians, Budapest, 1972), Colloq. Math. Soc. J. Bolyai, Vol. 9, pp. 191–197. Amsterdam, North-Holland, 1974.
6. Halperin, M., Greenhouse, S.W., Cornfield, J., Zalokar, J., *Tables of percentage points for the studentized maximum absolute deviate in normal samples.* J. Amer. Statist. Assoc., **50**, 185–195 (1955).
7. Lehmann, E.L., Shaffer, Juliet, *On a fundamental theorem in multiple comparisons.* J. Amer. Statist. Assoc., **72**, 576–578 (1977).
8. Mărgăritescu, E., *Metode de comparaţie multiplă.* Bucureşti, Ed. Acad. R.S. România, 1981.
9. Mărgăritescu, E., *The studentized maximum modulus method of multiple comparison.* Rev. Roumaine Math. Pures Appl., **26**, 89–94 (1981).
10. Miller, R.G., *Simultaneous Statistical Inference.* New York, McGraw-Hill, 1966.
11. Nelson, L.S., *Factor for the analysis of means.* J. Quality Tech., **6**, 175–181 (1974).
12. Ott, E.R., *Analysis of means – a graphical procedure.* Indust. Quality Control, **24**, 101–109 (1967).
13. Scheffé, H., *The Analysis of Variance.* New York, Wiley, 1959.
14. Schilling, E.G., *A systematic approach to the analysis of means. Part I.* J. Quality Tech., **5**. 93–108 (1973).
15. Stoline, M.R., *Tables of the studentized augmented range and applications to problems of multiple comparison.* J. Amer. Statist. Assoc. **73**, 656–660 (1978).
16. Ursianu, Emiliana, *On a method of multiple comparison among means.* Paper presented at the 2nd Pannonian Sympos. Math. Statist. (Bad Tatzmannsdorf, 1981).
17. Ursianu, Emiliana, Sâmboan, Anca, *On the efficiency of some methods of multiple comparison.* Bull. Math. Soc. Sci. Math. R.S. Roumanie (N.S.), **28** (**76**), 81–86 (1984).

ION VĂDUVA

Computer generation of random vectors based on transformation of uniformly distributed vectors

0. Introduction and summary

The paper [3] presents a method for computer generation of a gamma random variable as a ratio of two random variables which have a joint uniform distribution over some domain $D \subset R^2$. The papers [1, 2] present other similar methods for computer generation of different types of random variables by transforming uniform random vectors over some domain $D \subset R^2$. In this paper we generalize results from [1, 2, 3] and give methods for generating random vectors X, $X \in R^k$, by transforming random vectors over some domain $D \subset R^{k+1}$. The general method is briefly the following:

Given a random vector $X = (X_1, \ldots, X_k)'$ (a column vector) which has the probability density function (p.d.f) $f(x)$, $x \in R^k$ in the form

$$(1) \qquad f(x) = \frac{1}{H} h(x), \; x \in R^k, \; H = \int_{R^k} h(x)\, \mathrm{d}x > 0$$

one chooses (in a suitable manner) a transform φ: $R^{k+1} \to R^k$ and a constant $d = d(k)$ such as the domain $D \subset R^{k+1}$ is defined as

$$(2) \qquad D = \{(v_0, v_1, \ldots, v_k)' ;\; \gamma(v_0, v_1, \ldots, v_k) \leq 0\},$$

where

$$(2') \qquad \gamma(v_0, v_1, \ldots, v_k) = \ln\, v_0 - d \ln\, h(\varphi(v_0, v_1, \ldots, v_k)).$$

The domain D is assumed to be bounded in R^{k+1}. If V is a random vector uniformly distributed over D and φ and d are properly chosen, then $X = \varphi(V)$ has the p.d.f. $f(x)$. In this paper we give some propositions which specify particular forms of φ and d, and build up computer algorithms for generating X. Applications to generate multivariate normal vectors, Dirichlet vectors and multivariate t-distributed vectors are finally presented.

These distributions are very important in multivariate statistical analysis and the generation algorithms can be used in various simulation models or Monte Carlo calculations.

1. Fundamental propositions

In this section we present four propositions which specify particular forms of φ and d.

Theorem 1. *If $f(x)$ is in the form* (1) *and D is bounded and in the form* (2) *with $d = d(k) = \dfrac{1}{k+1}$ and $\varphi : R^{k+1} \to R^k$,*

$$\varphi(v_0, v_1, \dots, v_k) = \left(\frac{v_1}{v_0}, \frac{v_2}{v_0}, \dots, \frac{v_k}{v_0}\right)' \tag{3}$$

and if V is a random vector uniformly distributed over D, then $X = \varphi(V)$ has the p.d.f. $f(x)$.

The *proof* of this theorem is to be found in [6] and is similar to the proof of the next theorem. From the proof it results that

$$\operatorname{mes} D = \frac{H}{k+1} \tag{4}$$

Note also that

$$\gamma(v_0, v_1, \dots, v_k) = \ln v_0 - \frac{1}{k+1} \ln h\left(\frac{v_1}{v_0}, \dots, \frac{v_k}{v_0}\right). \tag{4'}$$

Theorem 2. *If $f(x)$ is in the form* (1) *and D is bounded and in the form* (2) *with $d(k) = \dfrac{1}{k+1}$ and $\varphi(v_0, v_1, \dots, v_k) = \left(\dfrac{v_0+v_1}{v_0}, \dfrac{v_0+v_2}{v_0}, \dots, \dfrac{v_0+v_k}{v_0}\right)'$ and if V is uniformly distributed over D then $X = \varphi(V)$ has the p.d.f. $f(x)$.*

Proof. Note that in this case

$$D = \left\{(v_0, v_1, \dots, v_k)' \in R^{k+1}; 0 \leqslant v_0 \leqslant h^{\frac{1}{k+1}}\left(\frac{v_0+v_1}{v_0}, \dots, \frac{v_0+v_k}{v_0}\right)\right\}. \tag{5}$$

Since the random vector $V = (V_0, V_1, \dots, V_k)'$ is uniformly distributed over D its p.d.f. is

$$u(v_0, v_1, \dots, v_k) = \begin{cases} \dfrac{1}{\operatorname{mes} D} & \text{if } (v_0, v_1, \dots, v_k) \in D \\ 0 & \text{otherwise,} \end{cases} \tag{4''}$$

where

$$\operatorname{mes} D = \int \dots \int_D \mathrm{d}v_0 \, \mathrm{d}v_1 \dots \mathrm{d}v_k < \infty, \tag{6}$$

because D is bounded.

If in the integral (6) we make the transform

$$x_0 = v_0, \; x_i = \frac{v_0 + v_i}{v_0}, \quad 1 \leqslant i \leqslant k, \tag{7}$$

which has the Jaccobian

$$(7')\qquad \frac{D(v_0, v_1, \ldots, v_k)}{D(x_0, x_1, \ldots, x_k)} = \begin{vmatrix} 1 & 0 & \ldots & 0 \\ x_1 - 1 & x_0 & \ldots & 0 \\ \cdot & & \ldots & \\ \cdot & & \ldots & \\ \cdot & & \ldots & \\ x_k - 1 & 0 & \ldots & x_0 \end{vmatrix} = x_0^k,$$

and denote by D^* the new domain (transformed) i.e.

$$(5')\qquad D^* = \{(x_0, x_1, \ldots, x_k)';\ 0 \leqslant x_0 \leqslant h^{\frac{1}{k+1}}(x_1, \ldots, x_k)\},$$

then we obtain

$$(6')\qquad \text{mes } D =$$

$$= \int \underset{D^*}{\ldots} \int x_0^k \, dx_0 \ldots dx_k = \int \underset{R^k}{\ldots} \int\int_0^{h^{\frac{1}{k+1}}(x_1, \ldots, x_k)} x_0^k \, dx_0 \, dx_1 \ldots dx_k =$$

$$= \int \underset{R^k}{\ldots} \int \frac{h(x)\, dx}{k+1} = \frac{H}{k+1}$$

Put

$$(8)\qquad Y = (Y_1, \ldots, Y_k)',\ Y_i = \frac{V_0 + V_i}{V_0},\quad 1 \leqslant i \leqslant k.$$

In order to prove the theorem we must prove that

$$(9)\qquad P(Y_1 < x_1, \ldots, Y_k < x_k) = \frac{1}{H} \int_{-\infty}^{x} \ldots \int h(z)\, dz = F(x)$$

where $F(x)$ is the cummulative distribution function (c.d.f.) of X and $x = (x_1 \ldots, x_k)'$.

We have

$$P\,(Y_1 < x_1, \ldots, Y_k < x_k) = \frac{1}{\text{mes } D} \int \underset{D \cap \Delta}{\ldots} \int dv_0 \ldots dv_k,$$

where

$$\Delta = \left\{ (v_0, v_1, \ldots, v_k)';\ \frac{v_0 + v_1}{v_0} < x_1, \ldots, \frac{v_0 + v_k}{v_k} < x_k \right\}.$$

If in the last integral we introduce the transform (7)and use (7'), we obtain

$$P(Y_1 < x_1, \ldots, Y_k < x_k) = \frac{k+1}{H} \int \underset{\Delta^*}{\ldots} \int z_0^k \, dz_0 \, dz_1 \ldots dz_k =$$

$$= \frac{1}{H} \int_{-\infty}^{x} \ldots \int h(z)\, dz = F(x),$$

where

$$\Delta^* = \{z_1 < x_1, \ldots, z_k < x_k,\ 0 \leqslant z_0 \leqslant h^{\frac{1}{k+1}}(z_1, \ldots, z_k)\}.$$

The theorem is proved.

Theorem 3. *If $f(x)$ is in the form* (1), *D is bounded, $d(k) = \frac{2}{k+2}$, $\varphi(v_0, v_1, \ldots, v_k) = \left(\frac{v_1}{\sqrt{v_0}}, \ldots, \frac{v_k}{\sqrt{v_0}}\right)'$ and V is uniformly distributed over D, then $X = \varphi(V)$ has the p.d.f. $f(x)$.*

The *proof* of this theorem is similar to that of Theorem 2. From the proof it results that

$$\text{mes } D = \frac{2H}{k+2} \tag{10}$$

Furthermore the function γ used to define D is

$$\gamma(v_0, v_1, \ldots, v_k) = \ln v_0 - \frac{2}{k+2} \ln h\left(\frac{v_1}{\sqrt{v_0}}, \ldots, \frac{v_k}{\sqrt{v_0}}\right). \tag{10'}$$

Theorem 4. *Let X be a random vector which has the p.d.f. $f(x)$ in the form* (1) *with $x \in I \subset R^k$, $I = [a_1, b_1] \times \ldots \times [a_k, b_k]$, $-\infty < a_i < b_i < \infty$, $1 \leq i \leq k$ and $0 \leqslant h(x) < M < \infty$, $x \in I$; $h(x) = 0$ for $x \notin I$. Assume that the domain D is in the form* (2), *$d(k) = 1$ and $\varphi(v_0, v_1, \ldots, v_k) = (v_0 + v_1, \ldots, v_0 + v_k)'$. If V is a random vector uniformly distributed over D, then $X = \varphi(V)$ has the p.d.f. $f(x)$.*

The *proof* of this theorem is also similar to that of Theorem 2. From the proof it results that

$$\text{mes } D = H \tag{11}$$

and the function γ which defines D in the form (2) is

$$\gamma(v_0, v_1, \ldots, v_k) = \ln v_0 - \ln h(v_0 + v_1, \ldots, v_0 + v_k). \tag{12}$$

Remarks. 1. In the above proofs the condition "D is bounded" was not yet directly involved. As the random vector V must be uniformly distributed over D, this condition becomes compulsory.

2. We do not know a priori that D is bounded. Therefore, we need a technique to establish this, whenever we use one of the previous theorems for generating X. This technique is presented in the next section and is also necessary for designing an algorithm for generating V.

2. A general algorithm

If X is a random vector whose p.d.f. $f(x)$ satisfies one of the above theorems, then the following algorithm can be applied for computer generation of X.

Algorithm GENER

Step 0. Initializations.

Step 1. Generate a random vector $V = (V_0, V_1, \ldots, V_k)'$ uniformly distributed over $D \subset R^{k+1}$.

Step 2. Calculate $Y_i = \varphi_i(V_0, \ldots, V_k)$ where $\varphi(V) = (\varphi_1(V), \ldots, \varphi_k(V))'$.

Step 3. Take $X = (Y_1, \ldots, Y_k)'$. Stop.

In this algorithm the function φ is one of those involved in the above theorems.

Note that the generation of X reduces to the generation of V, and, when D is not an interval of R^{k+1}, this is not a simple thing. The known procedure for generating V is the following (see [4, 6]).

Algorithm GENUNV

Step 0. Determine an interval $I^* \subset R^{k+1}$ such as $I^* \supset D$. Denote

$$I^* = [v_0^0, v_0^1] \times \ldots \times [v_k^0, v_k^1].$$

Step 1. Generate the random vector V^* uniformly distributed over I^*. If RND is a uniform random number generator (i.e. which produces numbers U uniformly distributed over [0,1]), then V^* may be generated as follows:

1.1. Generate $U_0, U_1, \ldots, U_k$ uniformly distributed over [0,1] and stochastically independent. (U's are generated by RND).

1.2. Calculate $V_i^* = v_i^0 + (v_i^1 - v_i^0)U_i,\ 0 \leqslant i \leqslant k$.

1.3. Take $V^* = (V_0^*, V_1^* \ldots, V_k^*)'$. Stop Step 1.

Step 2. If $V^* \notin D$ then go to Step 1. (Reject V^*.)

Step 3. (When V^* is accepted.) Take $V = V^*$. Stop.

Note that GENUNV is a *rejection* algorithm [5]. The performance of this algorithm is characterized by the probability to accept V^*, in Step 2 of the previous algorithm. This probability is

$$p_a = \frac{\text{mes } D}{\prod\limits_{i=1} (v_i - v_i^0)} \tag{13}$$

and it must be as large as possible. The largest p is obtained when I^* is the *minimum* interval which contains $D(I^* \supset D)$.

An important practical problem is now to determine I^*, when D is in the form (2). This problem reduces [6] to solve the following problems:

$$v_i^0 = \min_{v \in D} v_i,\ v_i^1 = \max_{v \in D} v_i,\ v = (v_0, v_1, \ldots, v_k)',\ 0 \leqslant i \leqslant k. \tag{14}$$

By applying the Lagrange multipliers method to problems (14) the following optimum problems are derived:

$$\min_{v \in D} F_i(v)\,;\ \max_{v \in D} F_i(v),\ 0 \leqslant i \leqslant k, \tag{14'}$$

$$F_i(v) = v_i + \lambda_i \cdot \gamma(v).$$

If for a fixed i, $0 \leqslant i \leqslant k$ denote by $(v^{(i)*}, \lambda_i^*)$, $(v^{(i)**}, \lambda_i^{**})$ the optimum points of the two problems (14'), respectively, then we have

$$v_i^0 = v_i^{(i)*},\ v_i^1 = v_i^{(i)**} \tag{15}$$

where $v^{(i)*} = (v_0^{(i)*}, v_1^{(i)*}, \ldots, v^{(i)*})'$

$$v^{(i)**} = (v_0^{(i)**}, v_1^{(i)**}, \ldots, v_k^{(i)**})'.$$

Therefore, Step 0 in the algorithm GENUNV consists now in solving the optimum problems (14') which lead to the interval I^* having limits (15).

Step 1 in GENUNV requires that D (and therefore I^*) be bounded. This happens when

$$-\infty < v_i^0 < v_i^1 < +\infty, \quad 0 \leq i \leq k. \tag{16}$$

Therefore the solution of optimum problems (14), (14') show also whether D is bounded (when (16) takes place) or not. In the case of Theorem 4 the domain D is obviously bounded. Furthermore, if in each of the other three theorems there is a finite interval I such as $f(x) = 0$, $x \in I$ and $0 \leq f(x) \leq M < \infty$, then the domain D is also bounded. The discussion on computer generation of a random vector X which has the p.d.f. in the form (1) is therefore terminated.

3. Applications

Many of the classical statistical distributions such as: normal, beta, gamma, lognormal, arcsin, logistic, Snedecor, t, etc., satisfy one of the above theorems for $k = 1$ (see [1, 2, 3, 6]) and therefore the algorithm GENER (combined with GENUNV) can be adapted to generate sampling values on them.

Here we shall give applications of these theorems for generating multivariate normal random vectors, Dirichlet random vectors and multivariate t random vectors.

Multivariate normal distribution. Let X be a normal $N(0, \Sigma)$ random vector, i.e. its p.d.f. is

$$\begin{aligned} f(x) &= \frac{1}{H} h(x), \quad H = (2\pi)^{\frac{k}{2}} (\det \Sigma)^{\frac{1}{2}}, \\ h(x) &= \exp\left\{-\frac{1}{2} x' \Sigma^{-1} x\right\}, \quad x \in R^k \end{aligned} \tag{17}$$

where $\Sigma = (\sigma_{ij})$ is the positive definite covariance matrix. If we apply Theorem 3, the domain D is the in form (2) with

$$\gamma(v_0, \ldots, v_k) = \gamma(v_0, v) = \ln v_0 + \frac{1}{k+2} \frac{1}{v_0} v' \Sigma^{-1} v, \quad v = (v_1, \ldots, v_k)'. \tag{18}$$

The Lagrangians F_i in (14') are

$$F_i(v_0, v; \lambda_i) = v_i + \lambda_i \left[\ln v_0 + \frac{1}{k+2} \cdot \frac{1}{v_0} \cdot v' \Sigma^{-1} v\right], \quad 0 \leq i \leq k. \tag{18'}$$

The optimum conditions for F_0 lead to the following equations:

$$\begin{cases} \dfrac{\partial F_0}{\partial v_0} = 1 + \lambda_0 \cdot \left[\dfrac{1}{v_0} - \dfrac{1}{v_0^2} \cdot \dfrac{1}{k+2} \cdot v' \Sigma^{-1} v\right] = 0, \\ \dfrac{\partial F_0}{\partial v_m} = 0 \Leftrightarrow \lambda_0 \cdot \left[\dfrac{1}{v_0} \cdot 2 \cdot \Sigma^{-1} v\right] = 0, \quad 1 \leq m \leq k, \\ \dfrac{\partial F_0}{\partial \lambda_0} = \gamma(v_0, v) = 0. \end{cases} \tag{19}$$

The second vector equation (19) gives $\Sigma^{-1}v = 0$, i.e. $v = 0$ (since Σ^{-1} is positive definite) and from the first equation (19) we get $v_0 = -\lambda_0$. Using now the last equation (19) combined with (18) we obtain $\lambda_0 = -1$. Therefore

$$v_0^0 = 0,\ v_0^1 = 1. \tag{20}$$

The optimum conditions for F_n, $1 \leqslant n \leqslant k$, give the following systems of equations:

$$\begin{cases} \dfrac{\partial F_n}{\partial v_0} = \lambda_n \cdot \left[\dfrac{1}{v_0} - \dfrac{1}{v_0^2} \cdot \dfrac{1}{k+2} v' \Sigma^{-1} v\right] = 0 \\ \dfrac{\partial F_n}{\partial v_m} = 0 \Leftrightarrow \xi_n + \lambda_n \left[\dfrac{1}{v_0} \dfrac{2}{k+2} \Sigma^{-1} v\right] = 0 \\ \dfrac{\partial F_n}{\partial \lambda_n} = \gamma(v_0, v) = 0,\ 1 \leqslant n \leqslant k, \end{cases} \tag{21}$$

where

$$\xi_n = (0, \ldots, 1, \ldots, 0)';\ \xi_{ni} = \begin{cases} 1 \text{ for } i = n \\ 0 \text{ for } i \neq n. \end{cases}$$

The solution of system (21) is

$$v_n^0 = -\sqrt{\frac{(k+2)\sigma_{nn}}{e}},\ v_n^1 = \sqrt{\frac{(k+2)\sigma_{nn}}{e}},\ 1 \leqslant n \leqslant k. \tag{22}$$

Therefore, the limits of I^* are defined by (20) and (22) and the algorithm GENER for generating $x \sim N(0, \Sigma)$ is completely specified. Finally, we note that the acceptance probability p_a is

$$\begin{aligned} p_a &= \frac{2H}{k+2} \cdot \frac{1}{\prod_{i=1}^{k} \sqrt{\frac{(k+2)\sigma_{ii}}{e}}} = \\ &= 2\left(\frac{2\pi e}{2}\right)^{\frac{k}{2}} \cdot \frac{(\det \Sigma)^{\frac{1}{2}}}{(k+2)^{\frac{k+2}{2}} \cdot \left(\prod_{i=1}^{k} \sigma_{ii}\right)^{\frac{1}{2}}} < \frac{2}{k+2}\left(\frac{\pi e}{2(k+2)}\right)^{\frac{1}{2}}. \end{aligned} \tag{23}$$

From (23) we have

$$\lim_{k \to \infty} p_a = 0 \tag{23'}$$

which shows that the algorithm GENER is not performant for a large dimension k. (Unfortunately, this is expected from any algorithm for generating random vectors.)

Dirichlet distribution. Assume that $X \in R^k$ has a Dirichlet distribution $D(\nu_1, \ldots, \nu_{k+1})$, [7], which has the p.d.f. in the form (1) with

$$H = \frac{\Gamma(\nu_1) \ldots \Gamma(\nu_{k+1})}{\Gamma(\nu_1 + \ldots + \nu_{k+1})}, \tag{24}$$

$$h(x_1, \ldots, x_k) = x_1^{\nu_1 - 1} \ldots x_k^{\nu_k - 1}(1 - x_1 \ldots - x_k)^{\nu_{k+1} - 1}.$$

Denoting

$$(24')\qquad \delta = k+1,\ \alpha_i = \nu_i - 1,\ 1 \leqslant i \leqslant \delta,\ \alpha = \sum_{i=1}^{\delta} \alpha_i,$$

and intending to apply Theorem 1, we have the function γ which defines the domain D in the form

$$(25)\qquad \begin{aligned} \gamma(v_0, v_1, \ldots, v_k) &= \frac{\alpha+\delta}{\delta} \ln v_0 - \sum_{i=1}^{k} \frac{\alpha_i}{\delta} \ln v_i - \frac{\alpha_{k+1}}{\delta} \ln (v_0 - v), \\ v &= \sum_{i=1}^{k} v_i. \end{aligned}$$

As we shall see below, we must assume ν_i-integers, $\nu_i > 1$, $1 \leqslant i \leqslant k+1$.

In order to define the interval I^*, $I^* \supset D$, $D \subset R^{k+1}$, we use again the mentioned technique based on the optimum problems (14') related to the Lagrangians F_i, $0 \leqslant i \leqslant k$.

The optimum conditions for F_0 or F_n, $1 \leqslant n \leqslant k$, lead to systems of equations similar to (19) or (21). By solving these systems one obtains finally

$$(26)\qquad v_0^0 = 0,\ v_0^1 = \prod_{m=1}^{\delta} \left(\frac{\alpha_m}{\alpha}\right)^{\frac{\alpha_m}{\delta}}$$

$$(27)\qquad v_i^0 = 0,\ v_i^1 = \left(\frac{\delta+\alpha_i}{\alpha_i}\right)^{\frac{\alpha_i}{\delta}} \cdot \prod_{m=1}^{\delta} \left(\frac{\alpha_m}{\delta+\alpha_i}\right)^{\frac{\alpha_m}{\delta}} \cdot \left(\frac{\delta+\alpha_i}{\alpha+\delta}\right)^{\frac{\alpha+\delta}{\delta}},\quad 1 \leqslant i \leqslant k.$$

Therefore the interval $I^* = [v_0^0, v_0^1] \times \ldots \times [v_k^0, v_k^1]$ is defined by (26) and (27) and the algorithm GENER for generating the Dirichlet random vector $X \sim D(\nu_1, \ldots, \nu_{k+1})$ is completely specified.

The acceptance probability p_a for this rejection algorithm is

$$p_a = \frac{H}{k+1} \frac{1}{\prod_{i=0}^{k} v_i^1},\quad P = \frac{H}{k+1} = \frac{\Gamma(\nu_1) \ldots \Gamma(\nu_{k+1})}{(k+1) \cdot \Gamma(\nu_1 + \ldots + \nu_{k+1})}$$

which after some calculations becomes

$$(28)\qquad p_a = P \cdot \frac{(\alpha+\delta)^{\frac{k(\alpha+\delta)}{\delta}} \cdot \alpha^{\frac{\alpha}{\delta}}}{\alpha_{k+1}^{\frac{\alpha_{k+1}}{\delta}} \prod_{i=1}^{k} (\delta+\alpha_i)^{\frac{\alpha_i+\delta}{\delta}} \left[\prod_{m=1}^{\delta} (\alpha_m)^{\frac{\alpha_m}{\delta}}\right]^k}.$$

Now, applying Stirling's formula for large ν_i, $1 \leqslant i \leqslant \delta$, in (28) we have

$$(28')\qquad p_a = \frac{(2\pi)^{k/2} \cdot e^k \cdot (\alpha+\delta)^{\frac{k(\alpha+\delta)}{\delta}} \cdot \alpha^{\frac{\alpha}{\delta}} \cdot \alpha_{k+1}^{1/2}}{\delta(\alpha+k)^{\alpha+k+1/2} \cdot \prod_{m=1}^{k} (\delta+\alpha_m)^{1/2}} \prod_{m=1}^{k} \left(\frac{\alpha_m}{\delta+\alpha_m}\right)^{\frac{\alpha_m}{\delta}+\frac{1}{2}}$$

Since

$$\lim_{\alpha_i \to \infty} p_a = 0$$

the algorithm is not performant for large values of α_i(e.g. ν_i, $1 \leqslant i \leqslant \delta$); neither is the algorithm performant when $k = \delta - 1$ is large, which is actually to be expected.

Multivariate t-distribution. Let X be a random vector which has a multivariate t-distribution [4], i.e. its p.d.f. is in the form (1) with

$$\text{(29)} \qquad \frac{(\det A)^{\frac{1}{2}} \cdot (n\pi)^{\frac{k}{2}} \Gamma\left(\frac{n}{2}\right)}{\Gamma\left(\frac{n+k}{2}\right)},$$

$$h(x_1, \ldots, x_k) = h(x) = \left(1 + \frac{1}{n} x' A x\right)^{-\frac{n+k}{2}},$$

where $n > 1$ is the number of degrees of freedom and $\Sigma = A^{-1}$ is the covariance (correlation) matrix.

We apply again Theorem 1 with D given by (2) and

$$\text{(30)} \qquad \gamma(v_0, v_1, \ldots, v_k) = -\frac{n-1}{k+1} \ln v_0 + \frac{n+k}{2(k+1)} \ln\left(v_0^2 + \frac{1}{n} v' A v\right),$$

$$v = (v_1, \ldots, v_k)'.$$

By applying the usual technique based on the optimum problems (14') we obtain [6] for the interval I^* the following limits

$$v_0^0 = 0, \ v_0^1 = 1,$$

$$\text{(31)} \qquad v_i^0 = -\sqrt{\frac{n(k+1)}{n-1}}, \ v_i^1 = \sqrt{\frac{n(k+1)}{n-1}}, \ 1 \leqslant i \leqslant k.$$

Thus, the algorithm GENER for generating the random vector X which has a multivariate t-distribution is completely specified.

The acceptance probability p_a, after some calculations, can be put in the form

$$\text{(32)} \qquad p_a = \frac{\pi^{k/2} \cdot \Gamma\left(\frac{n}{2}\right)}{(\det \Sigma)^{1/2} \cdot \Gamma\left(\frac{n+k}{2}\right) \cdot 2^k (k+1)} \cdot \left(\frac{n-1}{k+2}\right)^{\frac{k}{2}}$$

which again gives

$$\text{(32')} \qquad \lim_{n \to \infty} p_a = 0, \ \lim_{k \to \infty} p_a = 0$$

showing a bad behaviour of the algorithm for large values of k and/or n

4. Final remarks

Any of the above Theorems 1—3 can be applied to build up algorithms for generating one of the mentioned random vectors, and these algorithms have almost the same behaviour as illustrated by the acceptance probability p_a. In order to test the method, computer tests were performed on algorithms based on Theorem 1 [6]. The tests followed the following line: samples of large size N were produced and the mean value vector and covariance matrix were estimated in each case and they were compared with the theoretical figures. On the other hand, the computer time was compared with the time obtained for other known methods of generating normal, Dirichlet and t-random vectors. These tests conclude that for small values of k ($k \leqslant 4$) *better results* are obtained when the method shown in this paper is used. For the normal and t-distributed random vectors good results were obtained even for $k \leqslant 15$, while for the Dirichlet distribution satisfactory results were obtained for $k \leqslant 5$, and $\nu_i \leqslant 6, 1 \leqslant i \leqslant 6$. As the Dirichlet distribution satisfies Theorem 4, it is expected that the corresponding algorithm will give good results in this case, but the computer tests are not yet done.

Finally, we note that Theorems 1 and 4 lead to very good generation algorithms for normal, lognormal, t, F, and Cauchy distributions (see [1, 2]).

REFERENCES

1. Barbu, Gh., *On computer generation of random variables by transformations of uniform variables.* Bull. Math. Soc. Sci. Math. R.S. Roumanie (N.S.), **26** (74), 129—139(1982).
2. Barbu, Gh., *On computer generation of random variables as ratios of uniform random variables.* Econom. Comput. Econom. Cybernet. Stud. Res., **7**, 2, 49—65 (1982).
3. Cheng, R.C.H., Feast, G.M., *Gamma variate generators with increased shape parameters range.* Comm. ACM, **23**, 389—394 (1980).
4. Johnson, N.L., Kotz, S., *Distributions in Statistics: Multivariate Distributions, Vol.* 4. New York, Wiley, 1972.
5. Văduva, I., *Simulation Models for Computers.* București, Ed. tehnică, 1977. (Romanian)
6. Văduva, I., *On computer generation of random variables and random vectors as ratios of uniform variables.* Unpublished paper presented at the Meeting INFO-Iași, 1981. (Romanian)
7. Wilks, S., *Mathematical Statistics.* New York, Wiley, 1962.

RĂZVAN ANDONIE

A generative picture processing device used in two-dimensional cutting problems

It is for the first time that syntactic methods are used in two-dimensional cutting problems. For this purpose, we define a new type of context-free picture processing grammars. Then we give a heuristic algorithm for the generation of admissible solutions of a two-dimensional cutting problems class. The admissible solutions of a cutting problem are words accepted by an ordered picture processing grammar. The cutting problems discussed are considered with the restrictions specific to guillotine-type cutting of iron tables in the heavy industry. Our aim was to rest in the area of syntactic picture processing, so that we only touch the optimization aspects.

1. Introduction

Solving an optimal two-dimensional cutting problem means — generally speaking — the achievement of a given plan of two-dimensional components by using a minimal cutting surface, or — which means the same — by obtaining a minimal rest. An optimal cutting problem can be divided into two parts:

— the generation of admissible solutions, which is a problem of combinatorics;

— the search of an optimal solution, which is an optimization problem.

There are several syntactic possibilities to generate complexes of adjacent rectangles:

1. Shaw's PDL [10]. The primitives can be, for example, two orthogonal unit vectors.

2. The same language can be considered as in [4]. In this case, the primitives are oriented rectangles.

3. Web grammars, tree grammars and graph grammars [5].

4. Coordinate grammars [3].

5. Siromoney's picture languages with array rewriting rules [11].

We shall define a new generative picture processing device which is a class of string-grammars, for which every accepted word (a string) can be represented as a complex of adjacent rectangles.

2. Definition of the cutting problem

In the following, we aim to define the terms which are proper to our cutting problems.

Resource — a sort of rectangular plate;

Component — a rectangular required table-piece;

η = *Minimal shearing* — the distance between the cutting knife and the guard;

Cutting — we consider only guillotine-type rectangular cuttings. The width of the cut is not considered. A cut can be effected only if:

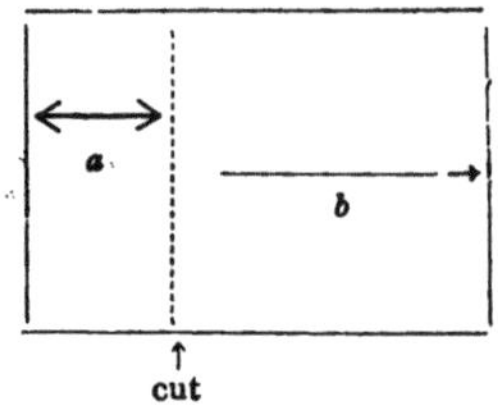

$a > \eta$ or $b > \eta$

Coupon — utilizable shearing (for which the length or the width is greater than η);

Residue — nonutilizable shearing (for which the length and the width are less than or equal to η);

Positioning — some componentes have to be positioned to the length (width) of the fibre because of the subsequent processings;

Cutting formula — cutting type of a resource-piece;

Admissible solution of a cutting problem — a linear combination of cutting formulas which satisfy a set of restrictions and generate the planned components;

Optimal solution of a cutting problem — the admissible solution which uses the minimal number of resource-tables;

c = *Complexity* — a measure of the physical effort done by the worker at the cutting operation.

A *cutting problem* is defined as follows: A set of components (which can be different) of the same material and the same thickness have to be obtained under the following restrictions:

1°. Only one resource is to be used, in unlimited quantity.

2°. The number of different components for each cutting formula must be less than or equal to m^* (usually, $m^* = 5$).

3°. The minimal shearing is η^* (usually, $\eta^* = 7$ cm).

4°. Some components may have positioning restrictions.

5°. The number of cutting formulas must be less than or equal to n^* (usually, $n^* = 3$).

6°. The complexity for each cutting formula must be less than or equal to c^*.

The solution obtained must be optimal (or as optimal as possible in case of a heuristic procedure).

3. The mathematical model

In the following we shall define a syntactic device for the generation of admissible solutions by observing the restrictions 1°, 4° and 6°.

Consider the resource (n, m), $n \leqslant m$, and the components $R_j(l_j, q_j)$, $l_j \leqslant q_j$, $i = \overline{1, s}$, where $n, l_1, \ldots, l_s$ are widths and $m, q_1, \ldots, q_s$ are lengths, all of them natural numbers (in our cases, millimeters). It is obvious that we are reducing the cutting problems to discrete ones. We have to obtain the components in the quantities $n(R_1), \ldots, n(R_s)$.

Definition 1. We shall call *cutting grammar with one resource and minimal shearing* η^*, $(\eta^* \in \mathbb{N})$, a system $G = (\Sigma, N, P, S(n, m))$, where $\Sigma = \{\{\overset{k}{\underset{j}{[}} \mid j = \overline{1, s},\ k = \overline{1, 10}\},,,],\ \{\$ (l, q) \mid l \leqslant n,\ q \leqslant m,\ l > \eta^* \text{ or } q > \eta^*\},$ $\{*(l, q) \mid l \leqslant n,\ q \leqslant m, l \leqslant \eta^*,\ q \leqslant \eta^*\}\}$ is the set of primitives, $N = \{S(l, q) \mid l \leqslant n, q \leqslant m\}$ is the set of nonprimitives, $S(n, m)$ is the initial symbol, $P = P_1 \cup P_2 \cup P_3 \cup P_4$ is the set of productions. Here

$$P_1\colon S(l, q) \to \overset{k}{\underset{j}{[}} S\ \delta_j^{k(1)}(l, q),\ S\ \delta_j^{k(2)}(l, q)],$$

where $\delta_j^k\colon \mathbb{N}^2 \to \mathbb{N}^2 \times \mathbb{N}^2$ is an application $\delta_j^k(l, q) = (\delta_j^{k(1)}(l, q), \delta_j^{k(2)}(l, q))$, $j = \overline{1, s}$, $k = \overline{1, 4}$, $l > l_j$, $q > q_j$, defined as follows:

(1) 

$$\delta_j^1(l, q) = ((l - l_j, q), (l_j, q - q_j)),$$

(2) 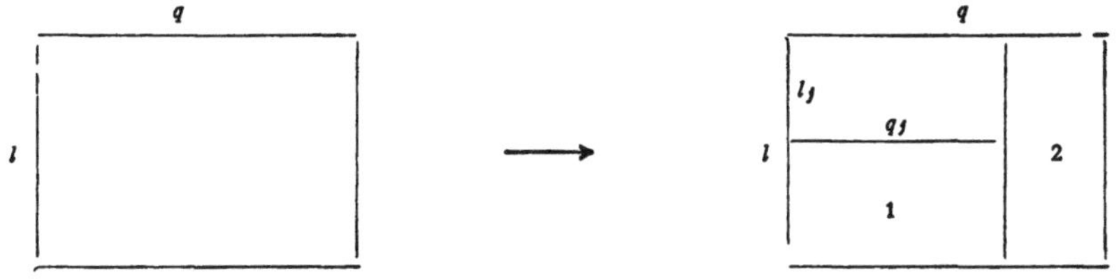

$$\delta_j^2(l, q)) = ((l - l_j, q_j), (l, q - q_j)),$$

(3) 

$$\delta_j^3(l, q) = ((l - q_j, l_j),\ (l, q - l_j)),$$

(4)

$$\delta_j^4(l, q) = ((l - q_j, q), \ (q_j, q - l_j));$$

P_2:

(5)

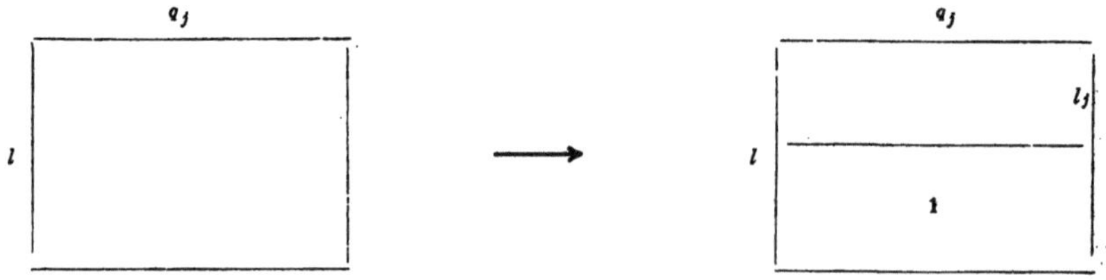

$$S(l, q_j) \to \overset{5}{\underset{j}{[}}\, S(l - l_j, q_j)], \ j = \overline{1, s}, \ l > l_j,$$

(6)

$$S(l_j, q) \to \overset{6}{\underset{j}{[}}\, S(l_j, q - q_j)], \ j = \overline{1, s}, \ q > q_j,$$

(7)

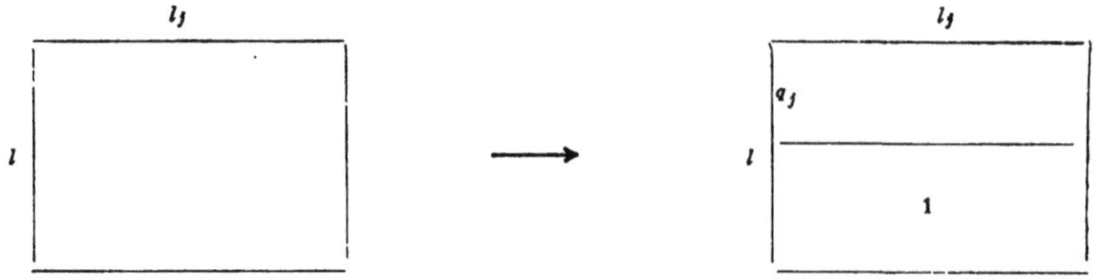

$$S(l, l_j) \to \overset{7}{\underset{j}{[}}\, S(l - q_j, l_j)], \ j = \overline{1, s}, \ l > q_j,$$

(8)

$$S(q_j, q) \to \overset{8}{\underset{j}{[}}\, S(q_j, q - l_j)], \ j = \overline{1, s}, \ q > l_j;$$

P_3:

$$(9) \qquad S(l_j, q_j) \to \underset{j}{\overset{9}{[}}\ \ ], \; j = \overline{1, s},$$

$$(10) \qquad S(q_j, l_j) \to \underset{j}{\overset{10}{[}}\ \ ], \; j = \overline{1, s};$$

P_4:

$$(11) \qquad S(l, q) \to \$(l, q), \text{ for: } l > \eta^* \text{ or } q > \eta^*,$$

$$(12) \qquad S(l, q) \to *(l, q), \text{ for: } l \leqslant \eta^* \text{ and } q \leqslant \eta^*.$$

For the positioned components we define only the productions

$$\begin{cases} \{(1), (2)\} \text{ or } \{(3), (4)\} \text{ in } P_1 \\ \{(5), (6)\} \text{ or } \{(7), (8)\} \text{ in } P_2 \\ (9) \text{ or } (10) \text{ in } P_3 \\ P_4 \end{cases}$$

For the components for which $l_j = q_j$ we do not define the productions (3), (4) and (10).

G is a picture processing grammar which generates strings that can be represented as bidimensional pictures. The pictures are cutting formulas. We attach to them, in a unique way the following tree:

1. A piece may be considered as a strip of subpieces which can be detached by applying exactly one set of parallel guillotine-type cuttings. In case that it is possible to apply a set of vertical cuttings, but also a set of horizontal cuttings, we choose the set with fewer cuttings.

2. We attach the root to the initial piece (a resource-piece).

3. For every node we attach an offspring for every subpiece. We respect the normal order in labelling: from the left to the right and top-down.

4. The terminal nodes are residues, coupons and components.

Definition 2. The *complexity* of a cutting formula is equal to the number of nonterminal nodes of the attached tree.

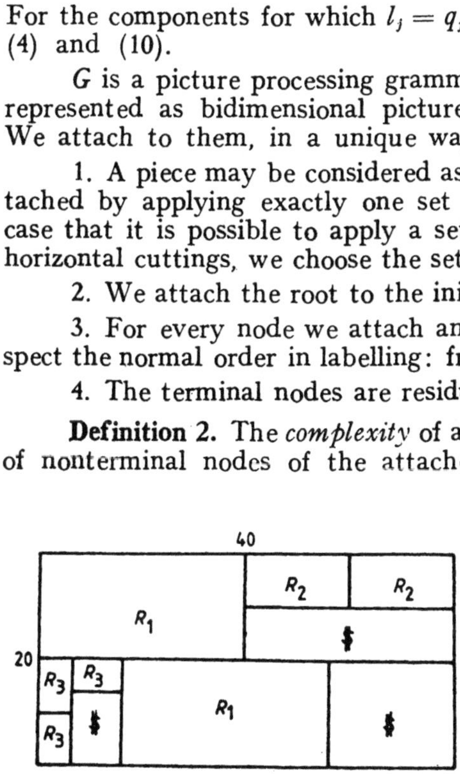

Fig. 1 a

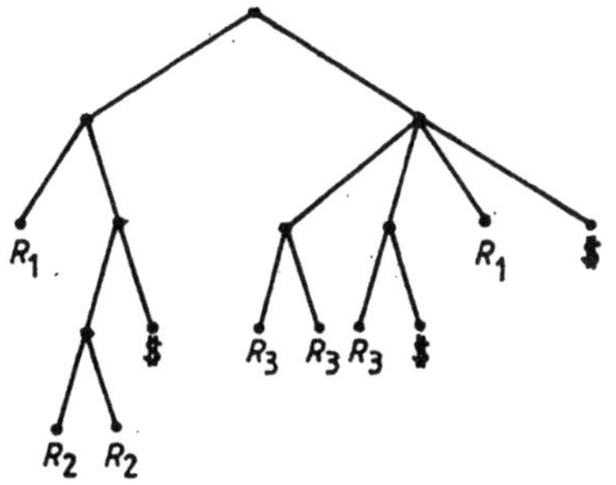

Fig. 1 b

Example. We have the resource (20, 40) and the components $R_1(10, 20)$, $R_2(5, 10)$, $R_3(3,5)$. We assume that we have no positioning. The string

$$\underset{1}{\overset{1}{[}}\underset{3}{\overset{3}{[}}\underset{3}{\overset{10}{[}}\,], \; \underset{3}{\overset{2}{[}}\$(7,5), \underset{1}{\overset{6}{[}}\$(10, 12)]]], \; \underset{2}{\overset{1}{[}}\$(5,20), \underset{2}{\overset{9}{[}}]]]$$

can be decoded as the cutting formula in Fig. 1 a, and the corresponding tree is in Fig. 1 b. We have $c = 7$.

We shall extend now the generative capacity of the grammar G:

1. We add the nonprimitives: $S(n, i \cdot m)$, $i = \overline{2, T}$, $T = \sum_{j=1}^{s} n(R_j)$.
2. We consider $S(n, T \cdot m)$ as initial symbol.
3. We add the following rule:

P_0:

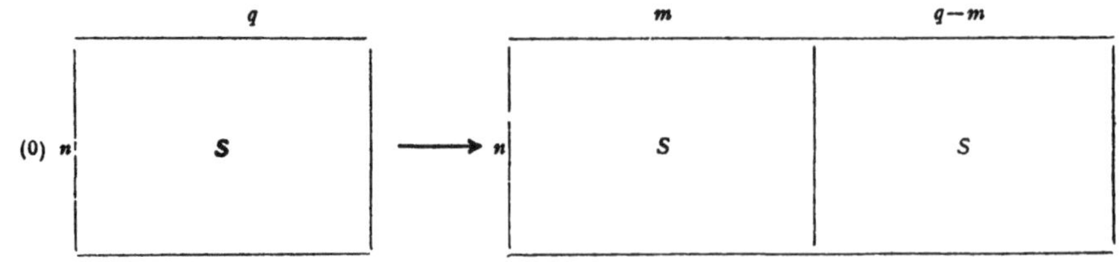

$$S(n, q) \rightarrow \underset{0}{[}S(n, m), S(n, q—m)], \ q>m$$

and to P_4 we add the rule:

(13) $$S(n, m) \rightarrow \#(n, m)$$

4. We add $\#(n, m)$ to the primitive set. We shall write more concisely $\#$ and we mean by this symbol an unused resource-piece.

Definition 3. We shall call *generalized cutting grammar with one resource and minimal shearing* η^*, ($\eta^* \in \mathbb{N}$), the system

$$G' = (\Sigma', N', P', S(n, T \cdot m)).$$

G' is a picture processing grammar which generates strings that can be represented as admissible solutions for the cutting problem (linear combinations of cutting formulas). The complexity may be defined in the same manner as for G. The grammar G' generates also the solution "one component over one resource-piece".

Definition 4. A *cutting problem* is defined as follows. Find $w^* \in L(G')$ for which

(A) $\begin{cases} \text{— the appearance number of "[" symbols = planned number} \\ \text{of } R_j \text{ components } (j = \overline{1, s}), \\ \text{— the appearance number of "\#" symbols} \rightarrow \max. \end{cases}$

We observe that $L(G)$ and $L(G')$ are finite.

Property 1. *G and G' are parenthesis grammars.*

It follows that the equivalence and the inclusion problems are solvable for G and G' (see Salomaa [9], p. 287).

Property 2. *G and G' are invertible.*

It follows that G and G' are unambiguous (see Mc Naughton[7]).

A pictorial representation of a string is the result of the parsing of the string. We have

Property 3. 1°. *For every string from $L(G)$ or $L(G')$ we have a unique pictorial representation.*

2°. *For an arbitrary pictorial representation of a cutting formula (or an admissible solution of the cutting problem) we may have several string representations.*

The bottom-up parsing can be done by using

Theorem 1. *G and G' are operatorial precedence grammars.*

We aim to give an efficient algorithm for the generation of solutions of problem (A).

We may generalize the definition of complexity also for sentential forms. The terminal nodes of the attached trees are in this case nonprimitives of the grammars.

Lemma 1. *Consider a sentential form of the grammar G, with complexity c. Then:*

1. *By applying a production from P_3 or P_4, the complexity remains constant.*
2. *By applying a production from P_2, the new complexity c' is $c \leqslant c' \leqslant c+1$.*
3. *By applying a production from P_1, we have $c \leqslant c' \leqslant c+2$.*

We give the following heuristic algorithm for the generation of a cutting formula in the grammar G:

1°. If there exists an unexhausted component for which we can apply a production from P_3, then we do so. Go to 1°.

2°. If there exists an unexhausted component for which we can apply a production from P_2, then we do so. Go to 1°.

3°. If there exists an unexhausted component for which we can apply a production from P_1, then we do so. Go to 1°.

4°. We apply rules from P_4 as many times as possible.

The algorithm generates cutting formulas with approximately little complexity.

And now we give a heuristic algorithm for the generation of admissible solutions for the cutting problem:

a) We apply the production (0) as many times as possible.

b) We apply 1°, 2°, 3° from the above algorithm.

c) We apply the production (13) as many times as possible.

d) We apply 4° from the above algorithm.

This algorithm leads us to admissible solutions with (approximately) little complexity. The grouping into strips is favoured.

We define a partial ordering relation in the set P (and P'):

$$(0) > P_3 > P_2 > P_1 > (13) > P_4.$$

We obtain the *ordered grammars* $(G, >)$ and $(G', >)$.

Property 4. *$L(G, >)$ = the set of cutting formulas with (approximately) little complexity; $L(G' >)$ = the set of admissible solutions for the cutting problem, with (approximately) little complexity.*

Our new generative device for admissible solutions of the cutting problem is an *ordered picture processing grammar.*

Subsequently we may eliminate from the obtained set of admissible solutions those solutions with a complexity greater than c^*.

Another possibility is to complicate the generation algorithm so as to obtain only admissible solutions with a complexity less than or equal to c^*.

We can restrict the obtained set of solutions by applying (in case of the sets P_2 and P_3) with priority those productions which ensure the cutting of the largest components. We are close then to the heuristic algorithm proposed by Albano and Orsini [1], [2], and to the algorithm of Weiss, Duţă and Fabian [12].

4. Conclusions

The cutting problem (A) is complete if we consider also the restrictions 2°, 3° and 5°. These restrictions may be introduced in the derivations of the grammars G and G'. Another possibility is to build for each derivation the corresponding grammar and to intersect the obtained languages with $L(G)$ or $L(G')$. Such a procedure was used by Păun [8] in the study of economic processes. We have to ensure first that the intersection is a closed operation.

Restriction 1° is a particular case of the more general situation when we use more resources. In this latter case, we have to modify our grammars by considering for each of them a set of initial symbols (and not only one symbol).

Our syntactic approach to the cutting problem leads us to the known fact that the most habitual way in which we understand and express things is a linguistic one. The development of Mathematical Linguistics as a pilot science [6] is a necessity. Its applications to economic and industrial processes is still at the beginning, because of their complexity. But, stepwise, we are finding syntactic order in these domains, too.

REFERENCES

1. Albano, A., Orsini, R., *A tree search approach to the M-partition and knapsack problem.* Comput. J., **23**, 256–261 (1980).
2. Albano, A., Orsini, R., *A heuristic solution of the rectangular cutting stock problem.* Comput. J., **23**, 338–343 (1980).
3. Anderson, R.H., *Two-dimensional mathematical notations.* In: Fu, K.S. (Ed.), *Syntactic Pattern Recognition, Applications*, pp. 147–178. Berlin, Springer, 1977.
4. Ghioca, G., *Model matematic al structurilor arhitecturale.* Stud. Cerc. Mat., **28**, 417–438(1976).
5. Gonzales, R.C., Thomason, M.G., *Syntactic Pattern Recognition.* Reading, Mass., Addison-Wesley, 1978.
6. Marcus, S., *Linguistics as a pilot science.* In: Sebeok, T. (Ed.), *Linguistics and Adjacent Arts and Sciences.* Current Trends in Linguistics, Vol. 12, pp. 2871–2887. The Hague–Paris, Mouton, 1974.
7. Mc Naughton, R., *Parenthesis grammars.* J. Assoc. Comput. Mach., **14**, 490–500 (1967).
8. Păun, Gh., *Mecanisme generative ale proceselor economice.* Bucureşti, Ed. tehnică, 1981.
9. Salomaa, A., *Formal Languages.* New York–London, Academic Press, 1973.
10. Shaw, A.C., *A formal picture description scheme on a basis of picture processing systems.* Inform. and Control, **14**, 9–52 (1969).
11. Siromoney, G., Siromoney, R., Krithivasan, K., *Picture languages with array rewriting rules.* Inform. and Control, **22**, 447–470 (1973).
12. Weiss, Gh., Duţă, D.L., Fabian, Cs., *Consideraţii asupra problemelor de croire. I–III.* Stud. Cerc. Calcul Econom. Cibernet. Econom. **6**, *3*, 33–38 (1971); **7**, *1*, 45–59 (1972); **7**, *4*, 5–21 (1972).

BERNARD BEREANU

The world nuclear weapons system is self-activating

Firstly, results of Bereanu ([2], [3]) are briefly reviewed. In [2] a stochastic model of a nuclear weapons race, essentially different form the mathematical models of arms races existing in the literature, was introduced. This model allowed, among others, to reveal a basic peculiarity of this type of arms competition which appears for the first time in the history of arms races. Namely two antagonistic nuclear weapons systems engaged in an arms race merge into one world, randomly self-activiting system, and the race leads with probability 1 to self-activation (accidental nuclear war).

In the second part of the paper the effects of both recent launching doctrines and demands of antinuclear movements on the probability of self-activation are investigated.

0. Introduction

There is a rich literature on the causes of wars based on historical experience and the theories generated by it (see Singer [12] for a review). There has been also considerable research on arms races stimulated especially by Richardson's mathematical model of such a race [11]. Richardson's classic model written as a matrix differential equation [14] is given by

$$\frac{d\mathbf{x}}{dt} = \mathbf{A}\mathbf{x} + \mathbf{g}, \quad \text{where} \quad \mathbf{x} = \begin{pmatrix} x_1 \\ x_2 \end{pmatrix}, \mathbf{A} = \begin{pmatrix} -a & l \\ k & -b \end{pmatrix}, \mathbf{g} = \begin{pmatrix} g_1 \\ g_2 \end{pmatrix}.$$

Here $x_1(t)$, $x_2(t)$ are the arms levels (usually measured as military expenditures) of the two powers engaged in an arms race, l and k are "threat" parameters, a and b are "economic fatigue" parameters and g_1, g_2 represent historic "grievances" or hostility of the two nations for one another [17]. There are various generalizations of this linear model, according to which the driving force behind the fluctuations in military expenditures is a process of action and reaction between two rival states. See [14] for such generalizations prompted especially by the failure of the original model to explain the arms expenditures of great powers during the 19th and 20th centuries. There is another school of thought in this field which explains, based on empirical data, the dynamics of military expenditure through purely internal factors. An outstanding example of such research is contained in [10]. An interesting self-activation mechanism of the technological arms race based on a "mirror image" phenomenon in which one protagonist attributes to his antagonist technological developments, which appear as possibilities from his own achievements and blueprints, is disclosed in [7]. Naturally enough the relation between the escalation of conflicts to wars and arms races received due attention in the specialized literature mainly in relation to the doubtful relevance to modern times of the political wisdom of the fourth century Roman general Vegetius ("if you seek peace, prepare for war"), who apparently is the forerunner of today theorists of "peace through strength". This problem was considered by Wallace (see for instance [15] and [16]). Thus in [16] the author investigates serious great powers disputes since 1816. Using mathematical statistics he concludes among others that "the existence of an arms race between the two powers prior to the dispute was very strongly associated with its subsequent escalation to full-scale hostilities". Testing of empirical data using statistical methods to investigate various factors accounting for wars or the relevance of various models of arms races and the relations of such races to war escalation has been stimulated in the last decade by the Correlates of War project at Michigan which has collected information about wars (including colonial wars) any place in the world since the Congress of Vienna (1814—1815), and has provided coding rules and resulting data including summary statistics (see Singer and Small [13]).

Of course the research based on the study of conventional wars provide interesting insights into the causes of arms conflicts of the prenuclear era and international military conflicts in general but cannot reveal the qualitative changes brought about by the advent of the present day nuclear weapons systems — combining weapons of awesome destructive power with extraordinary rapid delivery vehicles and computer programs of great complexity impossible to "completely" test—and which, for all practical purposes, control all operational aspects of these systems. But these circumstances, neglected in the literature on nuclear weapons, are of paramount importance in unleashing a world nuclear war. Indeed as stressed in [16, p. 50] the ever-growing power of destruction of nuclear weapons leads to the inhibition of the factors which used to escalate conflicts into wars in the prenuclear era. *Thus the self-activation property of two antagonistic nuclear weapons systems engaged in an arms race due to inherent technical factors and unavoidable randomness and erratic behaviour of such systems becomes the overriding potential cause of nuclear war.* Hence as underlined in [4] the adage that "mankind will have to *live* with *nuclear weapons*" while continuing the nuclear arms race is contradictory.

1. The mathematical model of a nuclear arms race

We shall use the notations and definitions introduced in [2].

1.1. *Notations and assumptions.* Let $\mathbf{S}_{1t}$ and $\mathbf{S}_{2t}$ be two antagonistic nuclear weapons systems (**NWS**) at time t each composed of two subsystems i.e., $\mathbf{S}_{it} = (\mathbf{O}_{it}, \mathbf{W}_{it})$, $i = 1, 2$. Here $\mathbf{O}_{it}$ is the operational component (the set of nuclear weapons together with their delivery vehicles), while $\mathbf{W}_{it}$ is the corresponding Early Warning System (**EWS**). In [3] details are given about these components based on published information on existing **NWS**s.

Each component $\mathbf{O}_{it}$ and $\mathbf{W}_{it}$ and the system $\mathbf{S}_{it}$ can move from the normal state (preparedness) to an "activated" state. We write $\bar{\mathbf{L}}_{it}$ for activated $\mathbf{L}_{it}$ and $(\mathbf{L}_{it} \rightarrow \bar{\mathbf{L}}_{it})$ for the process of activation. ($\mathbf{O}_{it}$ is activated iff one of its subsets is launched towards targets, while activation of $\mathbf{W}_{it}$ takes place when it records, perhaps erroneously, an activation of $\mathbf{O}_{jt}$, $j \neq i$). The activations of $\mathbf{W}_{it}$ results in nuclear alarms of various degrees. In what follows we shall be interested in false nuclear alarms only. Let $\{T_{in}\}$, $i = 1, 2$ and $n = 1, 2, 3, \ldots$ be two random sequences (random times) at which false nuclear alarms take place in the corresponding **NWS**s. At the time $T_{in} = \tau$, $\mathbf{W}_{i\tau}$ moves into one of the $h(i)$ active states $\bar{\mathbf{W}}^q_{i\tau}$, $q = 1, \ldots, h(i)$, i.e., $\mathbf{W}_{i\tau}$ is q-activated, with probabilities $p_{i1}, \ldots, p_{ih(i)}$, $p_{iq} > 0, \sum_{q=1}^{h(i)} p_{iq} = 1$. Furthermore there are given two sequences ξ_{in} of independent, identically distributed positive random variables with probability distribution functions $F_i(.)$, $i = 1, 2$. The random variable ξ_{in} represents the required length of time to discover the fault which generated the activation of $\mathbf{W}_{it}$ at the time $t = T_{in}$. We shall denote by x_{in} a realization of ξ_{in}.

Let the positive number δ_{it} be the minimum time to activate $\mathbf{S}_{it}$, or perhaps an average of it. In [2] δ_{it} could include also the time of actually producing the elements of $\mathbf{O}_{it}$. We denote by r the number of types of nuclear weapons delivery vehicles (the launchers of the same type of the two systems differ in general by technical characteristics and their proportions in the respective stockpile). If $\Delta_{ik}(t)$, $k = 1, \ldots, r$ is the average time of reaching its targets of the fastest delivery vehicle of type k in O_{it}, then the Efficient Checking Time (**ECT**) corresponding to the q-active state of $\mathbf{W}_{it}$, $\bar{\mathbf{W}}^q_{it}$ and denoted by $d_{iq}(t)$ is given by

$$(1) \qquad d_{iq}(t) \equiv \min_{k \in \mathbf{K}_q} \{\Delta_{jk}(t)\} - \delta_{it}, \quad j \neq i, \quad q = 1, 2, \ldots, h(i).$$

In (1) $\mathbf{K}_q$ is the subset of r types of launchers of $\mathbf{O}_{jt}$, the activation of which is implied by $\bar{\mathbf{W}}^q_{it}$.

Because the introduction of new systems of delivery vehicles in $\mathbf{O}_{it}$ with shorter delivery times (due either to the technological improvement or/and to their deployment nearer to targets), takes place at discrete times, $\Delta_{ik}(t)$, $i = 1, 2$; $k = 1, \ldots, r$ are piecewise constant nonincreasing functions of t and from (1) it follows that $d_{ik}(t)$ are functions of the same type.

It is reasonable to assume that alarms of higher degree will involve faster launchers, i.e., $(q_2 > q_1) \Rightarrow (d_{iq_2}(t) < d_{iq_1}(t))$, $t \geqslant 0$, $q = 1, \ldots, k(i)$, $i = 1, 2$.

In [2] and [3] there were taken into consideration two main sources of randomness inherent in the functioning of $\mathbf{W}_{it}$ (based on published official information on existing **NWS**s cited there). Namely

1) Errors due to the software of the computer systems of $\mathbf{W}_{it}$.

2) Failures of components of $\mathbf{W}_{it}$.

Of course the same types of sources of randomness affect $\mathbf{O}_{it}$ *after* activation, but we shall not consider them in the model since the main purpose of the paper is to *investigate the process of activation.*

It is shown in [2] using results in [1], that under "optimistic" assumptions on **EWS**s (i.e., assumptions under which the false nuclear alarms in the model are less frequent than in reality) the times $\{T_{in}\}$, $i = 1, 2$ are the arrival times of two Poisson processes, each of them the superposition of two Poisson processes. Let λ_i, $i = 1, 2$ be the arrival rates of these Poisson processes.

1.2. *Activation of systems and subsystems.* The activation of the systems $\mathbf{S}_{it}$, $i = 1, 2$ and their subsystems take place in accordance with the following rules.

$$(A_1) \quad (\mathbf{S}_{it} \to \overline{\mathbf{S}}_{it}) \Leftrightarrow (\mathbf{O}_{it} \to \overline{\mathbf{O}}_{it}),$$

$$(A_2) \quad (\mathbf{S}_{it} \to \overline{\mathbf{S}}_{it}) \Rightarrow (\mathbf{S}_{jt} \to \overline{\mathbf{S}}_{jt},\ j \neq i),$$

$$(A_3) \quad (\tau(i) = T_{in}) \to (\mathbf{W}_{i\tau(i)} \to \overline{\mathbf{W}}_{i\tau(i)})_{p_{iq}},\ i = 1, 2;\quad q = 1, \dots, h(i).$$

(A_1) simply stipulates that the operational part of $\mathbf{S}_{it}$ is its subsystem $\mathbf{O}_{it}$.

(A_2) corresponds to existing most cautious doctrines on the use of nuclear weapons and in fact to their raison d'être and underlines the tandem peculiarity of the two systems $\mathbf{S}_{it}$.

In (A_3), $(L \to L^q)_{p_{iq}}$ means that the activation takes place in one of the states q with probability p_{iq}. In [2], [3] more details are given about the technical interpretation of all various types of activation.

1.3. *Self-activation of the global system.* The activation of $\mathbf{W}_{it}$ in our model is a necessary, but not a sufficient condition of the activation of $\mathbf{O}_{it}$. However this may cease to be the case if $d_{iq}(t)$ decreases to zero as will be clear from what follows. We further discuss this case in [4].

As mentioned in 1.1 we are interested in false nuclear alarms only. We give now the rule of activation of $\mathbf{O}_{it}$ as a result of activation of $\mathbf{W}_{it}$.

$$(A_4) \quad (\tau(i) = T_{in},\ (\mathbf{W}_{i\tau(i)} \to \overline{\mathbf{W}}^q_{i\tau(i)}) \text{ and } (x_n^i - d_{iq}(\tau(i)) > 0) \Rightarrow$$

$$\Rightarrow (\mathbf{O}_{i\tau(i)} \to \overline{\mathbf{O}}_{i\tau(i)}).$$

From (A_4) via (A_2) and (A_1) it follows that $\mathbf{S}_{it}$, $i = 1, 2$ merge into a single random system from the point of view of its activation. We shall denote it by $\mathbf{S}_t = (\mathbf{O}_t, \mathbf{W}_t)$, where $\mathbf{O}_t = \mathbf{O}_{1t} \cup \mathbf{O}_{2t}$ and $\mathbf{W}_t = \mathbf{W}_{1t} \cup \mathbf{W}_{2t}$ and $\cup$ stands as usual for the union of sets.

Since the activation of $\mathbf{S}_t$ does not imply the initiation of the process by a decision maker but simply results from "retaliation" to a false nuclear alarm, we call it self-activation.

From (A_4) it is seen that the term Efficient Checking Time (**ECT**) used for $d_{iq}(t)$ is justified since $d_{iq}(t)$ represent the maximum checking time to verify if the activation of $\mathbf{W}_{it}$ is due to an error, while the system is to remain "efficient" in accordance with (A_2), i.e. fit to retaliate.

For the launching doctrines advocated at present ("launch on warning", or as some writers on the subject prefer, "launch under attack"), the rule (A_4) is replaced by

$$(A_4)' \quad (\tau(i) = T_{in},\ (\mathbf{W}_{i\tau(i)} \to \overline{\mathbf{W}}^q_{i\tau(i)})) \Rightarrow (\mathbf{O}_{i\tau(i)} \to \overline{\mathbf{O}}_{i\tau(i)}).$$

2. The probability of self-activation of the world nuclear weapons system

Let $\Pi(t)$ be the probability that the next false nuclear alarm in $\mathbf{S}_t$ will result in the self-activation of $\mathbf{S}_t$. In [2] it was shown that a lower bound for $\Pi(t)$ is given by

$$(2) \qquad \overline{\Pi}(t) = (\lambda_1 + \lambda_2)^{-1} \sum_{i,q} p_{iq} \lambda_i \overline{F}_i(d_{iq}(t)), \; q = 1, \ldots, k(i), \; i = 1, 2,$$

where

$$\overline{F}_i(d_{iq}(t)) = 1 - F_i(d_{iq}(t))$$

As remarked in 1.1, the functions $d_{iq}(t)$, $i = 1, 2$, $q = 1, \ldots, k(i)$, are piecewise constant. According to existing information on the deployment of new types of nuclear weapons vehicles (when the **ECTs** change) and on false nuclear alarms in $\mathbf{S}_t$ (see information cited in [3]) the interarrival times of false nuclear alarms are much shorter than the periods of time the **ECTs** given by $d_{iq}(t)$ remain unchanged. It follows from the derivation of formula (2) in [3] that in general $\overline{\Pi}(t)$ provides the probability of self-activation (**PSA**) itself.

From formula (2) a number of results can be derived. Here are some.

2.1. **Proposition.** *The probability that* $\mathbf{S}_{it}$ *is activated due to an erroneous alarm, i.e. that the antagonistic system* $\mathbf{S}_{jt}$, $j \neq i$ *suffers a first strike by error increases when the delivery times of the latter decrease.*

For the proof see [2, p. 515].

2.2. **Proposition.** *Let* $\Delta_i(t) = \min_l \{\Delta_{il}(t), \; l = 1, \ldots, r\}$ *and* t_i^* *defined by* $\Delta_j(t_i^*) \leqslant \delta_i$, $i = 1, 2$, *where* $\delta_i = \delta_{it_i^*}$. *Let further put* $\bar{t} = \min\{t_i^*\}$. *Then if at any moment* $t > \bar{t}$, *there is in* S_t *a false nuclear alarm involving the fastest delivery vehicle of the system, then* S_t *will be activated with probability* 1.

For the proof see [3, p. 656].

2.3. **Remark.** In [3] there are given some results which may be used to minimize the **PSA** of $\mathbf{S}_t$ for $t \geqq t_1$, subject to various restictions on $\Delta_{ik}(t)$ of the type $\Delta_{ik}(t) = \Delta_{ik}(t_1)$, $k \in \mathbf{K} \subset \{1, 2, \ldots, r\}$, which would appear during negotiations on eliminating certain types of nuclear weapons vehicles from $\mathbf{O}_{it}$, $i = 1, 2$.

2.4. **Remark.** Most of the results of [2] and [3] including Propositions 2.1 and 2.2 remain valid if the interarrival times $(T_{i,n+1} - T_{in})$, $i = 1, 2$ are two general renewal processes, as shown in [9].

3. The PSA, present launching doctrines, and some proposals of antinuclear movements

As seen in 1.3 in the present paper the launching doctrine embodied in (A_4) advocates the maximum time for checking the early warning of a nuclear attack, compatible with the principle of retaliating in kind to such an attack. However since on one hand the new **NWSs** planned to be deployed reduce the flight time of nuclear weapons vehicles and on the other hand the doctrine of first disarming strike seems to be at the basis of these projects of deployment, various launching policies have been proposed, taking into consideration these circumstances.

It is safe to say that never were the dangers of the continuation of the nuclear arms race for the very existence of mankind and in general of life on our planet so widely recognized as at present. A number of proposals for acting against these dangers have found wide support.

In this paragraph such proposals as well as lauching doctrines are discussed from the point of view of their effect on the **PSA.**

3.1. *"Launch-on-warning" or "Launch under attack"*. Although there are some differences in the formulation of the two launching doctrines (see [8]), for all practical purposes there are no essential differences except semantic ones in the present developments.

Let us denote by E_{iq}, $i = 1, 2$; $q = 1, 2, ..., k(i)$ the event {the next nuclear alarm will take place in $\mathbf{S}_{it}$ and will be of degree q} and by B the event {$\mathbf{S}_t$ is activated}. Then the probability $\Pi(t)$, that the next erroneous nuclear alarm in $\mathbf{S}_t$ results in the self-activation of $\mathbf{S}_t$ is given by

$$\Pi(t) = \sum_{i,q} P\{E_{iq}\} \cdot P\{B | E_{iq}\}, \quad i = 1, 2; \quad q = 1, 2, ..., k(i).$$

It is seen that if the "launch-on-warning" policy is adopted, then in $\Pi(t)$ $P\{B \mid E_{iq}\}$ is replaced by 1 or (A_4) is replaced by (A_4)'. One may imagine intermediate measure in which the activation takes place only after alarms of certain degree.

3.2. *"Launch unarmed / arm in flight" and "Launch armed / disarm in flight"*. Under these launching policies (see [8]) the nuclear weapons vehicles are launched on warning (under attack) and armed (disarmed) in flight by a radio signal if the attack proves to be (not to be) real. Since the antagonistic **EWS** makes no difference between armed or unarmed nuclear weapons delivery vehicles it will generate correctly a nuclear alarm and following (A_4) the system will be activated.

It follows that both these launching policies lead with probability 1 to self-activation. It should be however underlined that simply by continuation of the nuclear arms race and consequent reduction of **ECT**, the "launch-on-warning" policy becomes a technical result of the dynamics of such race.

3.3. *No-first-use option.* According to this proposal both antagonistic global nuclear powers should pledge not to use first nuclear weapons. Of course such a measure would reduce the subjective component of the probability of self-activation which exists always in the decision process related to the activation of $\mathbf{S}_t$, especially when one protagonist is committed through policy statements to the first-use option.

3.4. *The freeze demand.* According to this demand, widely supported, no new **NWS** should be deployed or developed, and this measure would be hopefully followed by the mutual reduction of **NWS**s until their total elimination.

Such a measure would "freeze" the functions $d_{iq}(t)$, i.e. the **ECT**s. Thus, if the freeze will take place at time t_1, then $d_{iq}(t) \geqslant d_{iq}(t_1)$, $i = 1, 2$; $q = 1, 2, ..., k(i)$, $t \geqslant t_1$. The inequality will be strict and the **ECT**s will increase if the freeze will be followed by reductions of the **EWS**s of the type discussed in [3], which minimize the **PSA**, subject to certain restrictions. The freeze would especially avoid to reach the critical time of Proposition 2.2 which may be brought about by the planned deployment of Cruise and Pershing 2 missiles in Europe. Indeed according to the existing information, the launching of Cruise missiles cannot be detected by **EWS**s, while the flight time of Pershing 2 would be less than four times shorter than that of the existing land based nuclear weapons vehicles which can reach strategic objectives of the two global nuclear powers.

REFERENCES

1. Barlow, R.E., Proschan, F., *Statistical Theory of Reliability and Life Testing*. New York, Holt, Rinehart and Winston, 1975.
2. Bereanu, B., *On the probability of self-activation of the world nuclear weapons stockpile*. Methods Oper. Res., **44**, 507–520 (1981).
3. Bereanu, B., *The self-activation property of the world nuclear weapons stockpile. A minimum risk approach*. Rev. Roumaine Math. Pures Appl., **27**, 647–662 (1982).
4. Bereanu, B., *The self-activation of the world nuclear system*. J. Peace Research, **20**, 49–57 (1983).
5. Çinlar, E., *Introduction to Stochastic Processes*. Englewood Cliffs, New Jersey, Prentice-Hall, 1975.
6. Feller, W., *An Introduction to Probability Theory and Its Applications, Vol. II*. New York, Wiley, 1966.
7. Forsberg, R., *What is the driving force behind the technological arms race?* Proc. American Philos. Soc., **124**, 266–270 (1980).
8. Garwin, R.L., *Launch under attack to redress Minuteman vulnerability?* International Security, **4**, 117–139 (1980).
9. Ghiţă, A., *On the self-activation property of the world nuclear weapons stockpile*. INCREST, Preprint Series in Mathematics, No. 30. Bucureşti, 1982.
0. Nincic, M., Cusack, T.R., *The political economy of U.S. military spending*. J. Peace Research, **16**, 101–115 (1979).
11. Richardson, L.F., *Arms and Insecurity*. Chicago, Quadrangle Press, 1960.
12. Singer, J.D., *Accounting for international war: The state of the discipline*. J. Peace Research, **18**, 1–18 (1981).
13. Singer, J.D., Small, M., *The Wages of War: A Statistical Handbook, 1816–1965*. New Yorh, Wiley, 1972.
14. Wallace, M.D., Wilson, M.J., *Non-linear arms race models*. J. Peace Research, **15**, 175–192 (178).
15. Wallace, M.D., *Arms races and escalation: Some new evidence*. J. Conflict Resolution, **23**, 3–16 (1979).
16. Wallace, M.D., *Armaments and escalation: Two competing hypotheses*. Internat. Stud. Quart., **26**, 37–56 (1982).
17. Ward, M.D., Guetzkov, H., *Toward integrated global models: From economic engineering to social science modelling*. J. Policy Modelling, **3**, 445–464 (1979).

ALEXANDER MEHLMANN, RUDOLF WILLING

Ein Differentialspiel zu Goethes Faust

Are there any questions that cannot be clarified by a mathematical formulation? Here we propose a differential game model for Goethe's tragedy Faust in the confident expectation that the answer is no. By using the method of phase plane analysis we are able to describe and understand the dynamic conflicting situation of Faust's *compact* with Mephistopheles.

Werd ich zum Augenblicke sagen:
Verweile doch! du bist so schön!
Dann magst du mich in Fesseln schlagen,
Dann will ich gern zugrunde gehn!
Faust 1, ii, 1699—1702

1. Einleitung

Eine bereits traditionelle Vorgangsweise auf dem Gebiet der dynamischen Optimierungstheorie, besteht darin, Konfliktsituationen, deren Realitätsbezug nicht so sehr auf der Hand zu liegen scheint, mathematisch zu formulieren und zu analysieren. Für die Präsentation von eher trockenen mathematischen Inhalten hat sich dieses Mittel gut bewährt. Ein Beispiel für diese Art von Modellen wäre das Dieb-Polizei Spiel (Feichtinger [3], Sethi [12]) oder Hartl und Mehlmann's [5] Transsylvanisches Problem erneuerbarer Ressourcen. Die bekannten pursuer-evader Spiele nach Isaacs haben andererseits oft deshalb märchenhafte Züge, um den militaristischen Hintergrund abzuschwächen. In diese Kategorie gehören Spiele wie "lady in the lake" und "the princess and the monster" (Basar und Olsder [1]).

In der vorliegenden Arbeit wird ein spieltheoretisches Modell der Wette zwischen Faust und Mephisto aufgestellt. Das literarisch anspruchsvolle Thema dient jedoch nicht allein als Transportmittel für technische Fragestellungen, wie die der Phasendiagrammanalyse optimaler Lösungen. Die mathematische Formulierung wurde weitgehend an Goethes Tragödienfassung des Fauststoffes angepaßt, um deren Handlung durch die Dynamik des Spiels erklären zu können — eine Bereicherung der Palette interpretativer literaturwissenschaftlicher Arbeiten, die die Überprüfung der Rationalität der dramatischen Spielführung ebenfalls mit quantitativen Argumenten vornehmen. Man vergleiche zum Beispiel die Begründung der Walpurgisnachtszene in [4]. Darüber hinaus kann unser, ketzerisch anmutende, Beitrag durchaus im Sinne der von Marcus [9] definierten „mathematischen Poetik" verstanden werden, eine Disziplin, die es sich zum Ziel gesetzt hat die Kritik literarischer Texte und die Beschreibung dramatischer Situationen mit den Mitteln der mathematischen Linguistik, Topologie und Statistik zu unterstützen.

2. Das Modell

Grau, teurer Freund ist alle Theorie,
Und grün des Lebens goldner Baum.
Faust 1, ii, 2038–2039

Das ursprüngliche Faustmotiv enthält eine zeitliche Begrenzung der Wette im Ausmaß von 24 Jahren, ein Intervall das wir unseren Überlegungen ebenfalls zugrunde legen wollen. Das wesentliche Kriterium des Spiels ist das vorzeitige Ende der Wette, das im Augenblick der Bereitschaft Faustens erfolgt die in der Einleitung zitierten Worte auszusprechen. Es handelt sich dabei um einen eher zufälligen Zeitpunkt, dessen Eintreten von beiden Gegnern verschieden eingeschätzt wird.

Die wesentlichen Größen, die die momentane Spielsituation beschreiben, sind die akkumulierten Maßnahmen M des Teufels und die Schuld S, die Faust auf sich geladen hat. Letztere entstand aus nicht bereutem Unterliegen der Versuchung. Die Systemdynamik kann nun wie folgt beschrieben werden, wobei jetzt und in der Folge bei Größen, die nur implizit von der Zeit abhängen, das Argument t unterdrückt wird

$$\dot{M} = m, \quad M(0) = 0. \tag{1}$$

$$\dot{S} = m - r, \quad S(0) = 0. \tag{2}$$

Mephisto glaubt, daß er das Streben Faustens mit seinen Machinationen m ersticken kann *(Ein solcher Auftrag schreckt mich nicht, Mit solchen Schätzen kann ich dienen;* Faust 1, ii, 1688—1689). Faust hingegen bezweifelt dies *(Was willst du armer Teufel geben? Ward eines Menschen Geist, in seinem hohen Streben, Von deinesgleichen je gefaßt;* Faust 1, ii, 1675—1677) und wirkt der Vergrößerung der Schuld durch tätige Reue r entgegen.

Die Zielfunktionale der beiden Spieler sind gegeben durch

$$\operatorname*{Min}_{m} \left\{ \int_0^{24} \exp(-\alpha M)\, \delta(m, r)\, dt - (1 - \exp(-\alpha M(24))\, L \right\} \tag{3}$$

$$\operatorname*{Max}_{r} \left\{ \int_0^{24} \exp(-\beta\,(M - S))\, \Phi(m, r)\, dt \right\} \tag{4}$$

wobei $\delta(m, r)$ den Disnutzen Mephistos und $\Phi(m, r)$ den Nutzen Faustens zum Zeitpunkt t bezeichnet. L ist Faustens Seelenwert.

Die Diskontierungen $\exp(-\alpha M)$ und $\exp(-\beta(M-S))$ ergeben sich als Wahrscheinlichkeiten, daß die Wette, aus der Sicht des jeweiligen Kontrahenten, noch im Gange ist, und lassen sich wie folgt ableiten:

Mephisto vermutet, das es ein $\hat{M}$ gibt, so das Faust befriedigt ist und die ominösen Worte ausspricht. Er kennt diesen Wert jedoch nicht genau und versucht deshalb seine subjektiven Erwartungen in Form einer Wahrscheinlichkeitsverteilung für eine Zufallsgröße $\hat{M}^*$ zusammenzufassen. Faust, seinerseits, ist sich bewußt, daß der entscheidende Augenblick von hoher Schuld verzögert wird und daß es eine Differenz $(\widehat{M-S})$ gibt, deren Realisation das Ende der Wette bedeutet. Er stützt sich ebenfalls auf eine Wahrscheinlichkeitsverteilung für eine Zufallsgröße $(\widehat{M-S})^*$. Aus Gründen der vereinfachten Darstellung wählen wir für beide Spieler Exponentialverteilungen mit den Ausfallsraten α und β. Das Erreichen von $\hat{M}$ beziehungsweise von $\widehat{M-S}$ erfolgt zu zufälligen Zeitpunkten T_M^* und T_F^*, deren Dichte durch

$$\frac{\dot{M}}{\alpha} \exp(-\alpha M) \quad \text{bzw.} \quad \frac{(\dot{M}-\dot{S})}{\beta} \exp(-\beta(M-S))$$

gegeben ist. Das sieht man durch Ableiten von

$$W(T_M^* \leqslant t) = W(M(t) \geqslant M^*) = \int_0^{M(t)} \frac{1}{\alpha} \exp(-\alpha M)\, dM$$

nach t ein, analog für T_F^*.

Mephisto minimiert somit seinen erwarteten Aufwand

$$E^{T_M^*} \int_0^t \delta(m, r)\, ds - W(T_M^* \leqslant 24)\, L$$

denn mit Wahrscheinlichkeit $W(T_M^* \leqslant 24)$ gewinnt er die Wette, was seinen Aufwand um L verkleinert *(Mir ist ein großer, einziger Schatz entwendet: Die hohe Seele, die sich mir verpfändet;* Faust 2, v, 11828—11829). Faust maximiert seinen erwarteten Nutzen

$$E^{T_F^*} \int_0^t \Phi(m, r)\, ds.$$

Im Gegensatz zu Mephisto verknüpft er keinerlei Erwartungen an das Jenseits *(Das Drüben kann mich wenig kümmern;* Faust 1, ii, 1660), was die Nichtberücksichtigung eines Endwertes bei der Maximierung verständlich macht. (3) und (4) ergeben sich durch Auswertung der Erwartungswerte mittels partieller Integration.

Für die spezielle Wahl der Funktionen $\delta(m, r)$ und $\Phi(m, r)$ lassen sich einschlägige Kommentare (siehe [4]) des literarischen Originals heranziehen

$$\delta(m, r) = d_1 m^2 + d_2 r \tag{5}$$

$$\Phi(m, r) = f_1 m(\bar{m} - m) + f_2 r(2m - r). \tag{6}$$

Mephisto versucht seinen quadratischen Aufwand m^2 *(Ein großer Aufwand, schmählich! ist vertan;* Faust 2, v, 11837) und seinen aus Faustens Reue entstandenen Disnutzen zu minimieren. Aus Freudscher Sicht lassen sich die Komponenten des Zielfunktionals Faustens den verschiedenen Schichten der faustischen Seele zuordnen. Das rationale Ich ist sich bewußt, daß der entscheidende Augenblick von hoher Schuld verzögert wird. Das hedonistische Es bezieht konkaven Nutzen aus m, wobei $\bar{m}$ eine natürliche obere Schranke des ihm möglichen individuellen Glücksempfindens (Edgeworth [2]) bedeutet. Das moralische Überich kann maximal din momentane Verführung bereuen. d_1, d_2, f_1 und f_2 sind zeitunabhängige Gewichte.

3. Das Nash-Gleichgewicht und das Pontryaginsche Maximumprinzip

Hier wittert's nach der Hexenküche.
Faust 2, i, 6229

Die mathematische Theorie der Differentialspiele soll nun zur Lösung des im vorangegangenen Abschnitt beschriebenen Problems herangezogen werden. Der interessierte Leser sei auf die Standardarbeiten von Basar und Olsder [1], Isaacs [6] und Starr und Ho [13], [14] verwiesen. Eines der grundlegenden Probleme besteht in der Auswahl eines Lösungskonzeptes. Im vorliegenden Fall werden Nash- (Gleichgewichts)-Lösungen gesucht. Das sind Politiken m^* und r^*, so daß für alle anderen zulässigen Politiken m und r gilt:

$$J_1(m, r^*) \geqslant J_1(m^*, r^*), \quad J_2(m^*, r) \leqslant J_2(m^*, r^*)$$

wobei $J_i(.,.)$ das Mephisto ($i = 1$) und Faust ($i = 2$) zugeordnete Zielfunktional ist. Als zulässig mögen alle (closed-loop) Strategien $m(t, M, S) \geqslant 0$, $r(t, M, S) > 0$ gelten, die stetig in den Argumenten und gleichmäßig Lipschitzstetig in den Zuständen (M, S) sind. Zur Bestimmung von m^* und r^* läßt sich eine Variante des Pontryaginschen Maximumprinzips heranziehen. Dazu definieren wir die Hamiltonfunktionen beider Spieler:

$$7) \qquad H^{\mathrm{Me}} = \exp(-\alpha M)\,(d_1 m^2 + d_2 r) + \mu_1 m + \mu_2 (m - r)$$

$$8) \qquad H^{\mathrm{Fa}} = \exp(-\beta(M - S))\,(f_1 m(\bar{m} - m) + f_2 r(2m - r)) + \lambda_1 m + \lambda_2 (r - m).$$

Faßt man nunmehr die Wahrscheinlichkeiten, daß das Spiel, aus der Sicht des jeweiligen Spielers, noch im Gange ist, als neue Zustände $x = \exp(-\alpha M)$, $y = \exp(-\beta(M-S))$ mit den Zustandsgleichungen $\dot{x} = -\alpha m x$ und $\dot{y} = -\beta r y$ auf, so ergeben sich folgende notwendige Gestalten für ein optimales Lösungspaar:

$$(9) \qquad m^* = \frac{\alpha \mu_1}{2 d_1}, \quad r^* = m^* - \frac{\beta \lambda_2}{2 f_2}$$

wobei μ_1 und λ_2 durch folgende Differentialgleichungen gegeben sind:

$$(10) \qquad \begin{aligned} \dot{\mu}_1 &= -(d_1 m^2 + d_2 r) + \alpha m \mu_1; \quad \mu_1(24) = L \\ \dot{\lambda}_2 &= -(f_1 m(\bar{m} - m) + f_2 r(2m - r)) + \beta r \lambda_2; \quad \lambda_2(24) = 0. \end{aligned}$$

Offensichtlich haben wir bei dieser Zustandstransformation eine entartete closed-loop Lösung erhalten. Für diese sind auch die hinreichenden Optimalitätsbedingungen von Leitmann und Stalford [8] *trivialerweise erfüllt* da die maximierte (minimierte) Hamiltonfunktion konkav (konvex) in den Zuständen ist. Reinganum [11] leitet für eine mit (1)—(4) verwandte Klasse von Differentialspielen entartete closed-loop Lösungen mit Hilfe der Hamilton-Jacobi-Bellmanschen Gleichung. Ein beträchtlicher Nachteil ihres Ansatzes liegt jedoch im übersehen zahlreicher nichtdegenerierter closed-loop Kandidaten die wir für unser Modell durch Anwendung beliebiger von $x = \exp(-\alpha M)$ und $y = \exp(-\beta(M-S))$ abweichender Zustandstransformationen erhalten. Für eine detailierte Untersuchung sei auf Mehlmann und Willing [10] verwiesen.

Aus (9) und (10) ergeben sich nunmehr Differentialgleichungen für das optimale Lösungspaar;

(11) $$\dot{m} = \frac{\alpha}{2}(m^2 - \mathrm{d}r),\ \dot{r} = \frac{\beta}{2}r^2 - \frac{\alpha \mathrm{d}}{2}r + \left(\frac{\alpha}{2} - \frac{\beta}{2}f\right)m^2 + \frac{\beta}{2}f\overline{m}m$$

wobei $f = f_1/f_2$ und $d = d_2/d_1$ ist.

Eine bereits klassische Methode zur Analyse von Systemen zweier Differentialgleichungen besteht in der Erstellung eines Phasendiagramms. Die Isokline $\dot{m} = 0$ ist für das obige System stets eine Parabel, während $\dot{r} = 0$ ein beliebiger Kegelschnitt sein kann. Bis zu 4 stationäre Punkte (Schnittpunkte der Isoklinen) sind möglich, wobei der Koordinatenursprung (m_1, r_1) und ein uns nicht interessierender Punkt (m_2, r_2) im 2. Quadranten stets gesichert sind. Befindet sich kein explodierendes Gleichgewicht im 1. Quadranten, so würden sich die dazugehörigen Lösungstrajektorien zum Teil im unzulässigen Bereich bewegen und man müßte einen Lagrangeschen Multiplikatorenansatz zur Lösung heranziehen. Der nun folgende Satz gibt Bedingungen für die Existenz eines explodierenden Gleichgewichtspunktes im 1. Quadranten an.

Theorem 1. *Ist die Ungleichung*

(12) $$d(f/3)^{1/2} \geqslant 3/2\,\overline{m}$$

strikt erfüllt, dann existieren zwei zusätzliche Gleichgewichtspunkte

$$0 < m_3 < d(f/3)^{1/2} < m_4,\quad 0 < r_3 < df/3 < r_4$$

die im Fall, daß (12) *zur Gleichung entartet, zusammenfallen.*

Der Beweis erfolgt mittels Kurvendiskussion.

(12) läßt sich wie folgt interpretieren: *Je höher das individuelle Glücksempfinden Faustens $\overline{m}$ angesetzt wird, desto weniger darf das Verführen Mephisto kosten, respektive die Reue Faust Genuß verschaffen.*

Theorem 2. *Die den stationären Punkten (m_i, r_i) $i \neq 1$ zugeordnete Jacobideterminante hat folgende Gestalt:*

(13) $$\Delta = \frac{\alpha\beta f d}{2}\left(m - \frac{3}{2}\overline{m}\right).$$

Beweis.

(14) $$\Delta = \partial\dot{m}/\partial m\ \partial\dot{r}/\partial r - \partial\dot{m}/\partial r\ \partial\dot{r}/\partial m = \alpha\beta(rm - fdm/2 + fd\overline{m}/4).$$

Für Schnittpunkte der Isoklinen erhält man aus der Kurvendiskussion aber folgende Gleichung

(15) $$rm - dfm + df\overline{m} = 0$$

so daß Δ dann die angegebene Gestalt hat.

Korrolar 2.1. *Der Gleichgewichtspunkt (m_4, r_4) kann nie ein Sattelpunkt sein.*

Beweis. $\Delta > 0$ da $m_4 > d(f/3)^{1/2} > 3/2\,\overline{m}$.

Proposition 1. *Gilt*

(16) $$\alpha m_4 + \beta r_4 - \alpha d/2 > 0$$

was zum Beispiel für

(17) $$(f/3)^{1/2} > 1/2$$

stets erfüllt ist, dann ist der Gleichgewichtspunkt (m_4, r_4) *instabil, ist zusätzlich*

(18) $$(\alpha m_4 + \beta r_4 - \alpha d/2)^2/4 > \alpha\beta f d/2(m_4 - 3/2\bar{m})$$

gegeben, so ist (m_4, r_4) *ein Knotenpunkt.*

Beweis. Man berechnet die Eigenwerte der Jacobimatrix in (m_4, r_4) und berücksichtigt die bekannten Gesetzmäßigkeiten.

(17) *bedeutet, daß die Gewichtung des Nutzens aus der Verführung zumindest* 75% *der Gewichtung des durch die Reue entstehenden Gewinns beträgt.*

Proposition 2. *Ist* $f_1 > 3f_2$ *so gilt* $m_4 < r_4$, *andernfalls gilt*

(19) $$m_4 > r_4 \Leftrightarrow d(1 - f) + f\bar{m} > 0.$$

Beweis. $f_1 > 3f_2 \Rightarrow m_4 > d(f/3)^{1/2} > d \Rightarrow r_4 = m_4^2/d > m_4$. Aus $f_1 < 3f_2$ und der rechten Ungleichung in (21) folgt mittels Kurvendiskussion $m_4 < d$.

Da $m_4 > \bar{m}$ gilt, bringt das starke Verführen Mephistos Faust Disnutzen. Ist also die Gewichtung dessen zu groß, so wird er in der Nähe dieses Gleichgewichtes mehr bereuen als er sündigt.

4. Ein spezielles Szenario

In bunten Bildern wenig Klarheit
Faust, Vorspiel auf dem Theater, 170

Zur Illustration des Verhaltens der Kontrollpfade soll folgender spezieller Parametersatz bei variierendem Seelenwert L betrachtet werden: $d = 16$; $f = 2$; $\alpha = \beta = 1$; $\bar{m} = 4$.

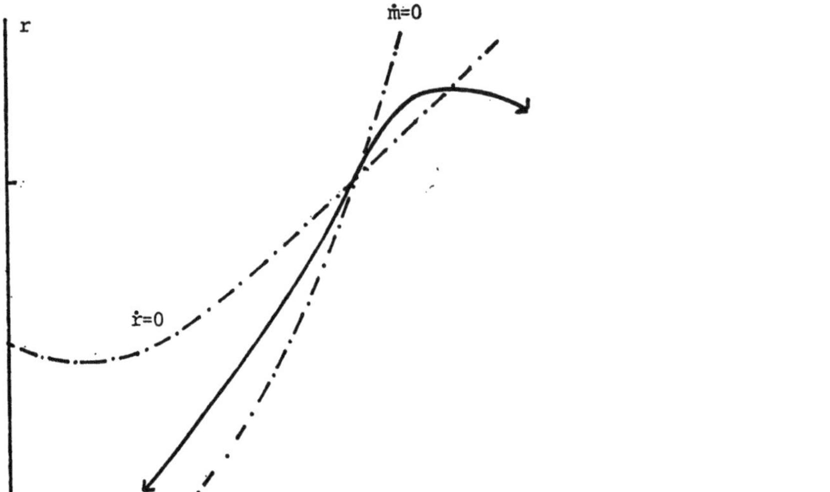

Figur 1: $\alpha = \beta = 1, m = 4, d = 16, f = 2.$

Überprüft man die Aussagen des vorangegangenen Abschnittes für den obigen Fall, so ergeben sich für die stationären Punkte folgende Charakteristika:

(m_1, r_1) stabiler Strudel
(m_3, r_3) Sattelpunkt
(m_4, r_4) instabiler Knoten.

Die interessanten Politiken gehen alle vom Knoten (m_4, r_4) aus. Die längste Zeit über befinden sie sich in seiner Nähe, um dann im letzten halben Jahr die Endbedingungen zu erfüllen.

Da den Teufel die Reue seines Opfers so sehr stört $(d_2 : d_1 = 16)$, muß er auf ein baldiges Ende der Wette dringen. Er setzt deshalb seine Machinationen für Faust viel zu hoch $(m_4 \gg \bar{m})$ an und senkt sie nur dann ab, wenn der Seelenwert L für ihn nicht verlockend genug ist.

Proposition 3. *Je mehr Mephisto die Reue irritiert und je höher er den Erwartungswert von $\hat{M}^* = 1/\alpha$ ansetzt, desto höher muß die Seelenprämie L sein, damit er am Schluß seine Aktivitäten nochmals steigert.*

Beweis. $\dot{m}(24) = \alpha/2(m^2(24) - dr\,(24)) = \alpha/2(\alpha^2 L^2/(4d_1^2) - d_2 \alpha L/(2d_1^2)) =$

$$= \alpha^3 L(L - 2d_2/\alpha)/(8d_1^2),$$

(20) $$\dot{m}(24) > 0 \Leftrightarrow L > 2d_2/\alpha.$$

Faust bezieht also in Wirklichkeit gar keinen Nutzen aus dem, was der Teufel ihm bietet. So muß Mephisto unserem Doktor zweimal das schönste Weib der griechischen Antike, Helena, anbieten, obwohl diesem bereits ein einfaches deutsches Mädchen genügen würde. In unserem Beispiel ist $\Phi(m_4, r_4) \sim -315$, da der Disnutzen aus der übertriebenen Verführung doppelt so stark ins Gewicht fällt. Faust hat daher selbst ein großes Interesse, die Wette durch Überbereuen $(r_4 > m_4)$ abzukürzen. Dies mag auch der Grund sein, weshalb es in Faust 2 zu einem für Mephisto überraschenden Ausgang kommt. Er erkennt, daß seine Sicht des Spiels falsch war *(Ihn sättigt keine Lust, ihm genügt kein Glück;* Faust 2, v. 11587) und er der Betrogene ist. Dauert das Spiel die vollen 24 Jahre, so würde Faust im Fall, daß $L > 2d_2/\alpha$, kurz vor Verstreichen der Frist vorerst exzesiv bereuen, eine Art Torschußpanik, um dann am Ende doch zu resignieren. Gilt die letzte Ungleichung mit anderen Vorzeichen, so senken beide Spieler ihre Kontrollen ab, da es um einen zu geringen Einsatz geht.

Mathematisch interessant ist der Fall eines beinahe vernachlässigbaren Seelenwertes $(L \sim 0.1)$, bei dem es dann zu einem Strudeln um den Ursprung und somit zu einem Ablenken in unzulässige Bereiche kommt. Ist überhaupt kein Seelenwert vorhanden $(L = 0)$, so tun beide die ganze Zeit über nichts. Die beiden letzten Fälle sind jedoch sehr unrealistisch, da es unter diesen Bedingungen nie zu einem Teufelspakt gekommen wäre. Der historische Dr. Faustus (Georg und nicht Heinrich) weist durch seine Behauptung (siehe [4]), der Teufel habe in der Hölle den Vertrag bestätigen lassen müssen, offensichtlich darauf hin, daß die Unterwelt eine nach kapitalistischen Gesichtspunkten ausgerichtete Institution ist.

LITERATURVERZEICHNIS

1. Basar, T., Olsder, G. J., *Dynamic Noncooperative Games.* London—New York—Paris, Academic Press, 1982.
2. Edgeworth, F.Y., *Mathematical Psychics: An Essay on the Application of Mathematics to the Moral Sciences.* London, Kegan Paul, 1881.

3. Feichtinger, G., *A differential game solution to a model of competitcn between a thief and police.* Management Sci., **29**, 686 – 699 (1983).
4. Friedrich, T., Scheithauer, L. J., *Kommentare zu Goethes Faust.* Stuttgart, Phillipp Reclam Jun., 1980.
5. Hartl, R., Mehlmann, A., *The Transylvanian problem of renewable resources.* RAIRO Rech. Opér., **16**, 379 – 390 (1982).
6. Isaacs, R., *Differential Games, 2nd Edition.* Huntington-New York, Kruger Publ. Co., 1975.
7. Goethe, J.W., *Faust: der Tragödie beider Teile.* Suttgart, Phillipp Reclam Jun., 1980.
8. Leitmann, G., Stalford, H., *Sufficiency for optimal strategies in Nash equilibrium games.* In: Balakrishnan, A.V. (Ed.), *Techniques of Optimization.* New York-London, Academic Press, 1972.
9. Marcus, S., *Poetica matematică.* București, Ed. Acad. R.S. România, 1970.
10. Mehlmann, A., Willing, R., *Some results about Reinganum's class of differential games.* Preprint, Institut für Ökonometrie und Operations Research, Technische Universität Wien, 1982.
11. Reinganum, J.F., *A class of differential games for which the closed-loop and open-loop Nash equilibria coincide.* J. Optim. Theory Appl., **36**, 253–262 (1982).
12. Sethi, S.P., *Optimal pilfering policies for dynamic continuous thieves.* Management Sci., **25**, 535–542 (1979).
13. Starr, A.W., Ho, Y.C., *Nonzero-sum differential games.* J. Optim. Theory Appl., **3**, 184–206 (1969).
14. Starr, A.W., Ho, Y.C., *Properties of nonzero-sum differential games.* J. Optim. Theory Appl., **3**, 207–219 (1969).

MIHAI FLORIN POPESCU

Efficiency of some approximate implicit enumeration algorithms

Algoritmii de tip enumerare implicită sînt metode de rezolvare a problemelor de optimizare combinatorială şi de programare în numere întregi. În lucrare este analizat modul cum depinde eficienţa algoritmilor aproximativi (ε-optimali) de acest tip de funcţia de toleranţă utilizată (ε). Se arată întîi că nu orice algoritm aproximativ este mai eficient decît algoritmul exact corespunzător. Sînt studiate apoi două clase de algoritmi pentru care efortul de calcul este monoton descrescător relativ la „mărimea" funcţiei de toleranţă utilizate.

1. Introduction

Implicit enumeration algorithms are among the simplest methods used to solve integer programming and combinatorial problems. Sometimes the practical problems are so difficult that the quantity of computer resources required exceeds the existing possibilities. Then, in order to speed up the convergence of the procedures, instead of obtaining an exact optimal solution of the input problem, one searches for a suboptimal solution with a guaranteed accuracy. For primal methods, such as the implicit enumeration procedures, a usual way to this goal is to modify the improvement-validation of the optimality phase by the aid of an allowance function $a(P, z) \geqslant 0$, specifying the allowable deviation of a suboptimal value from the exact optimal value. Then, whenever the currently inspected element can not improve the best available value by a quantity greater than the value of the allowance function, it is discarded.

The approximate implicit enumeration procedures have essentially four components: an enumeration, a skip function, a rejection test and an allowance function. The principle they are based on is the following. Assume that the objective function $f_P(x)$ of the input problem P is to be minimized. The elements considered to be potential solutions for P are enumerated one by one by the aid of the enumeration. Using the rejection test and the allowance function one determines whether the currently inspected element x is a feasible solution of P and whether $f_P(x) < z_a(P, x) - a(P, z_a(P, x))$, where $z_a(P, x)$ is the best available value when x is reached by the algorithm (i.e. $z_a(P, x) = f_P(y)$, y being the best feasible solution already examined). If a solution occurs, which improves the best available value with a quantity greater than the value of the allowance function, it is memorized and one continues the verification with one of the remaining elements. If the test rejects the current element x, this means that there is no feasible solution x', starting with x up to the value of the skip function, which improves the best available value with a quantity greater than the value of the allowance function. If such is the case, one skips over this set and the verification is pursued with the element given by the skip function. The above procedure is repeated until the domain of the potential solutions is exhausted.

The aim of this paper is to clarify the way the efficiency of the above approximate methods depends on the allowance function used. In order to estimate the efficiency we use here two natural measures of the computing effort: the total number of elements tested until the algorithm stops and the number of elements tested until the final suboptimal solution is attained. The approach described in this paper is close to the similar study of Ibaraki [1], for the Branch and Bound methods.

In Section 2 a formal description of the components of an approximate implicit enumeration algorithm is given. Two types of such procedures are introduced and their finite convergence and correctness are proved.

The effort measures are discussed in Section 3. An equivalence relation among the allowance functions is introduced with the property that the approximate algorithms using equivalent allowance functions have identical computing efforts. Some properties of the allowance functions belonging to the same class of equivalence are given and the set of equivalence classes is characterized.

Lower and upper bounds of the computing effort of an approximate procedure are given in Section 4. Although it seems intuitively obvious that solving approximately a problem must be more advantageous than searching for an exact optimal solution, this is not always true, as it is shown by an example. Along the same line, the approximate algorithm using the "less restrictive" allowance function is not always the most efficient. However, if one imposes supplementary conditions for the components of the procedure, these desirable properties can be obtained. Two such classes of algorithms are analyzed here: the procedures with skip functions having nested jumps and those with nondecreasing skip functions and whose tests satisfy a certain regularity condition.

In Section 5 two relations of order among the allowance functions are considered and the monotonicity of the effort measures with respect to these orders is studied. The first relation is the natural order among the sizes of the allowances: $a \preccurlyeq a'$ iff $a(P, z) \leqslant a'(P, z)$ for any problem P and real z. It is shown that, contrary to the intuitive belief, the efficiency does not always increase with respect to the size of the allowance functions even if only the algorithms in Section 4 are considered. The second order relation depends on the other components of the approximate method: $a \preccurlyeq^* a'$ iff $z_{a'}(P, x) - a'(P, z_{a'}(P, x)) \leqslant z_a(P, x) - a(P, z_a(P, x))$, for any problem P and element x. Although the effort measures are generally not monotonous functions with respect to this order they become nonincreasing functions for the algorithms in Section 4.

Some consequences of the results obtained are discussed in the conclusions section; therein the relation with the similar results of Ibaraki [1] are also clarified.

2. Approximate implicit enumeration algorithms

Let $\mathbf{C}$ be a class of integer programming problems $P = (f_P, r_P, D_P)$, of the form

$$\begin{aligned} \min \quad & f_P(x) \\ & r_P(x) \qquad\qquad \text{(GIP)} \\ & x \in D_P \end{aligned}$$

where $f_P : N^{n(P)} \to R$ is a real valued function, $r_P \subseteq N^{n(P)}$ a relation equivalent to the set of the proper constraints of the problem and $D_P \subseteq N^{n(P)}$ a finite set of vectors with integer components from the elements of which one has to find a (sub)optimal solution of P. (We denoted here by $n(P)$ the number of variables of the problem P.)

An *enumeration* for a class $\mathbf{C}$, $\alpha(P) = (x_{\alpha(P,1)}, succ_\alpha(P, x))$ is a function $\alpha(P): \mathbf{C} \to \bigcup_{P \in \mathbf{C}} (N^{n(P)} \times (N^{n(P)})^{(N^{n(P)})})$ which satisfies the conditions:

(e1) $x_{\alpha(P,1)} \in D_P$ for every problem $P \in \mathbf{C} \cdot x_{\alpha(P,1)}$ is the *first element* of the set D_P with respect to $\alpha(P)$;

(e2) $succ_\alpha(P, x)$: $N^{n(P)} \to N^{n(P)}$, which is said to be the *successor function* of the enumeration, has the property:

$$\min \ \{n \in N \mid succ_\alpha^n(P, x_{\alpha(P,1)}) \notin D_P\} = d(P)$$

for every problem $P \in \mathbf{C}$ (Here $d(P)$ is the cardinal of the set D_P).

In other words α is an enumeration for the class $\mathbf{C}$ if one can generate the elements of the sets D_P by making use of the successor function $succ_\alpha(P, x)$ and beginning from the first element $x_{\alpha(P,1)}$. The conditions (e1), (e2) assure that the elements of the sets D_P are nonredundantly enumerated so that they can be represented uniquely as values of the $succ_\alpha(P, x)$ function: $x_{\alpha(P,1)}, x_{\alpha(P,2)}, \ldots, x_{\alpha(P,d(P))}$, where $x_{\alpha(P,j)} = succ_\alpha(P, x_{\alpha(P,j-1)}) = succ_\alpha^{j-1}(P, x_{\alpha(P,1)})$.

An enumeration introduces in every set D_P a total order relation "$\leqslant_{\alpha(P)}$", defined by $x_{\alpha(P,i)} \leqslant_{\alpha(P)} x_{\alpha(P,j)}$ iff $i \leqslant j$. We denote as usually by $[x, y]_{\alpha(P)}$ and $(x, y)_{\alpha(P)}$ the closed intervals, respectively the open ones with respect to this order and by $x_{\alpha(P,d(P)+1)} = succ_\alpha(P, x_{\alpha(P,d(P))}) \in N^{n(P)} \setminus D_P$ the least upper bound of D_P. If P is a feasible problem let also denote by x^P the first feasible solution of P with respect to the order $\leqslant_{\alpha(P)}$.

A function $sk_\alpha(P, x) : \mathbf{C} \times \bigcup_{P \in \mathbf{C}} N^{n(P)} \to \bigcup_{P \in \mathbf{C}} N^{n(P)}$ is said to be a *skip function* for the class $\mathbf{C}$ compatible with the enumeration α if it satisfies the condition

(s) $$(\forall x \in D_P)\ (sk_\alpha(P, x) \in [succ_\alpha(P, x), x_{\alpha(P,d(P)+1)}]_{\alpha(P)})$$

for any problem $P \in \mathbf{C}$.

A *rejection test* (or an *exclusion test*) for the class $\mathbf{C}$ compatible with α and sk_α is a three-valued function $T(P, x, z)$: $\mathbf{C} \times (\bigcup_{P \in \mathbf{C}} N^{n(P)}) \times R^\cdot \to \{0, 1, 2\}$ which has the properties:

(t1) $T(P, x, z) = 2$ iff $(r_P(x)$ and $f_P(x) < z)$

(t2) $T(P, x, z) = 1 \to (\forall y \in [x, sk_\alpha(P, x))_{\alpha(P)})$ (non $r_P(y)$ or $f_P(y) \geqslant z)$)

(t3) $T(P, x, z) = 1 \to (\forall z' < z)\ (T(P, x, z') = 1)$

for any problem $P \in \mathbf{C}$, $x \in D_P$ and $z \in R^\cdot$, where $R^\cdot = R \cup \{+\infty\}$.

If the test takes the value 1 one says that it *rejects* for this argument and if its value is 2 that it *quashes* the argument. From the condition (t1) it follows that T quashes iff x is a solution of the problem P for which the objective function has a better value than the bound z. The condition (t2) implies that if T rejects, $T(P, x, z) = 1$, then in the interval $[x, sk_\alpha(P, x))_{\alpha(P)}$ there is no solution of P for which the objective function takes a better value than z. (t3) is a natural regularity condition.

Finally a function $a(P, z) : \mathbf{C} \times R \to R_+$ is an *allowance function* for the class $\mathbf{C}$ if (according to Ibaraki [1]):

(a) $z \leqslant z' \to z - a(P, z) \leqslant z' - a(P, z')$

for any $P \in \mathbf{C}$ and $z \in R$ (R_+ = the nonnegative real numbers). Let us denote by $\mathbf{A}(\mathbf{C})$ the set of the allowance functions for the class $\mathbf{C}$.

The previously introduced components — the enumeration, the skip function, the rejection test and the allowance function — can be assembled so that they produce approximate algorithms for the classes of problems $\mathbf{C} \subseteq \subseteq$ GIP. As for the exact methods which solve the problems of the form GIP, two types of approximate implicit enumeration algorithms are discussed below. The program schema of the procedures of the first type is given in Figure 1.

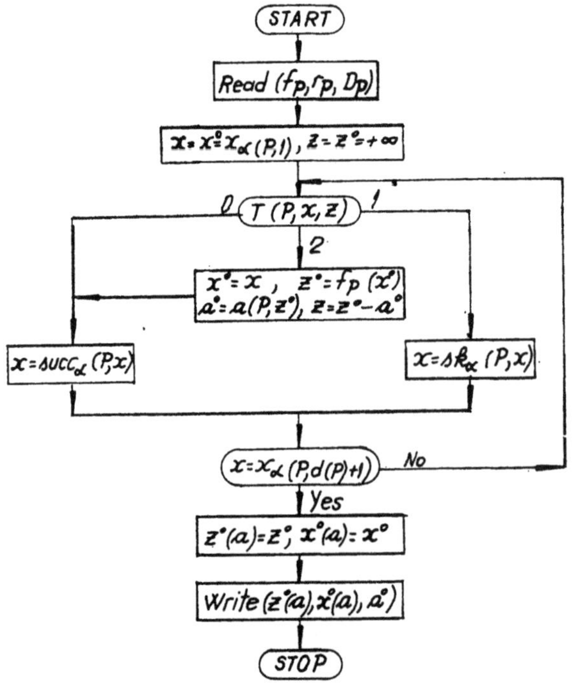

Figure 1. The program schema of the first-type approximate algorithms.

Consider a programming language and a class $\mathbf{C} \subseteq$ GIP. Let α, sk_α and T be an enumeration of $\mathbf{C}$, a skip function and a rejection test which are compatible and a an allowance function. Assume that the data of the problems (f_P, r_P, D_P), $P \in \mathbf{C}$ and the functions $succ_\alpha$, sk_α, T and a can be expressed in some manner in the given programming language. On the basis of the above program schema one obtains a procedure in the programming language having α, sk_α, T and a as components. The input of the procedure consists of the data describing a problem P within the class $\mathbf{C}$. The output, denoted by $(z^0(a), x^0(a), a^0)$ has the usual meaning: $z^0(a) = +\infty$, if the current problem has no solutions, otherwise $x^0(a) =$ a suboptimal feasible solution of P, $z^0(a) = f_P(x^0(a))$, the corresponding suboptimal value and $a^0 = a(P, f_P(x^0(a)))$, the final value of the allowance function.

With these specifications we can state now the algorithm in an ALGOL-like programming language.

S1(Input): Read $P = (f_P, r_P, D_P)$;

S2(Initialization): $x = x^\circ = x_{\alpha(P,1)}$, $z = z^0 = +\infty$;

S3(Test): Compute the value $y = T(P, x, z)$ and go to:

S4 if $y = 2$,

S5 if $y = 1$,

S6 if $y = 0$;

S4 (Improvement): $x^0 = x$, $z^0 = f_P(x^0)$, $a^0 = a(P, z^0)$, $z = z^0 - a^0$;

go to S6;

S5(Skip): $x = sk_\alpha(P, x)$, go to S7;

S6 ; $x = succ_\alpha(P, x)$;

S7 : If $x = x_{\alpha(P,d(P)+1)}$ then go to S8, otherwise go to S3;

S8 (Output): $z^0(a) = z^0$, $x^0(a) = x^0$; Write $(z^0(a), x^0(a), a^0)$; STOP.

We shall consider that an algorithm of this kind is completely determined by its components and let us denote it by $(\alpha, sk_\alpha, T, a)_1$.

In particular, if one uses the identical null allowance function $0(P, z) \equiv 0$, one obtains the corresponding exact implicit enumeration algorithm $(\alpha, sk_\alpha, T, 0)_1$, which solves the problems $P \in \mathbf{C}$ optimally. (These exact procedures are studied in [5], [6], [7]).

For an algorithm $(\alpha, sk_\alpha, T, a)_1$ and a problem $P \in \mathbf{C}$ denote by $x_a(P, x)$ and $z_a(P, x)$ the functions:

$$x_a(P, x) : \mathbf{C} \times \bigcup_{P \in \mathbf{C}} (D_P \cup \{x_{\alpha(P,d(P)+1)}\}) \to D_P \cup \{x_{\alpha(P,d(P)+1)}\},$$

$$z_a(P, x) : \mathbf{C} \times \bigcup_{P \in \mathbf{C}} (D_P \cup \{x_{\alpha(P,d(P)+1)}\}) \to D_P \cup \{x_{\alpha(P,d(P)+1)}\},$$

which take as values for an argument (P, x) the solution respectively its value computed in the step S4 by $(\alpha, sk_\alpha, T, a)_1$ and which are available when x is attained:

$$x_a(P, x) = \begin{cases} x_{\alpha(P,1)}, \text{if } P \text{ has no solution in the interval } [x_{\alpha(P,1)}, x)_{\alpha(P)} \\ x^\circ, \text{otherwise} \end{cases}$$

$$z_a(P, x) = \begin{cases} +\infty, \text{ if } P \text{ has no solution in the interval } [x_{\alpha(P,1)}, x)_{\alpha(P)} \\ f_P(x^0), \text{ otherwise} \end{cases}$$

where $x^0 <_{\alpha(P)} x$ is the last solution of P for which the step S4 was performed.

$z_a(P, x)$ is a nonincreasing function and $x_a(P, x)$ a nondecreasing function in the x variable. The output values of the algorithm are $z^0(a) = z_a(P, x_{\alpha(P,d(P)+1)})$, $x^0(a) = x_a(P, x_{\alpha(P,d(P)+1)})$. In particular $z^0(0) = z_0(P, x_{\alpha(P,d(P)+1)})$ and $x^0(0) = x_0(P, x_{\alpha(P,d(P)+1)})$ are the exact optimal value of P, respectively its optimal solution given by $(\alpha, sk_\alpha, T, 0)_1$.

If in addition to the conditions (e1), (e2), (s), (t1) — (t3) the components α, sk_α, T satisfy also the condition:

(sc) $r_P(x) \to (\forall y \in (x, sk_\alpha(P, x))_{\alpha(P)}$ (non $r_P(y)$ or $f_P(y) \geqslant f_P(x))$ and $(\forall z \in R)$ $(T(P, x, z) \neq 0)$,

for any problem $P \in \mathbf{C}$ and $x \in D_P$, then α, sk_α and T are said to be *strongly compatible (s-compatible)*.

The condition (sc) implies that whenever a test quashes a solution x, $T(P, x, z) = 2$, then in the interval $(x, sk_\alpha(P, x))_{\alpha(P)}$ there is no feasible solution of P better than x. Moreover, this condition assures that if x is a feasible solution of P then the test T either rejects or quashes for all the arguments of the form (P, x, z), $z \in R$. (The above condition (sc) differs slightly from the corresponding condition (sc) given in [6] in order to assure the correctness of the approximate algorithms).

If the components α, sk_α, T are s-compatible then besides the algorithms $(\alpha, sk_\alpha, T, a)_1$, $a \in \mathbf{A}(\mathbf{C})$, of the first type, one can construct approximate methods of a second type. Their program schema differs from that given in Figure 1 only on the branch where the test has the value 2. In this case the computing process will continue also with a skip, after the current solution x^0 and the variables z^0, a^0 are updated. We denote a procedure of the second type by $(\alpha, sk_\alpha, T, a)_2$. In the above programming language, it differs from the corresponding first-type algorithm $(\alpha, sk_\alpha, T, a)_1$ only in the last instruction of the step S4 which is here replaced by: "go to S5". Note that both methods supply the same output and the functions $x_a(P, x)$, $z_a(P, x)$ associated with these algorithms are respectively identical. However these procedures differ from the point of view of the computing effort.

For a problem $P \in \mathbf{C}$ and a procedure $(\alpha, sk_\alpha, T, a)_1$ denote by:

$M_1(P, a) =$ the set of elements $x \in D_P$ tested in step S3 prior to the termination of the algorithm.

The same notations with the index 2 instead of 1 will be used to denote the corresponding sets for the second-type procedures.

Note that if the components α, sk_α, T are s-compatible then the functions $x_a(P, x)$, $z_a(P, x)$ associated with the algorithm $(\alpha, sk_\alpha, T, a)_2$ are respectively identical with those corresponding to $(\alpha, sk_\alpha, T, a)_1$ and therefore we do not use any indexing for these functions. In the subsequent discussion we will use the same notations with index 1 or 2 depending on the type of the method. Moreover, we will omit the index whenever similar results hold true for both types of algorithms (e.g. $(\alpha, sk_\alpha, T, a)$ stands for $(\alpha, sk_\alpha, T, a)_1$ and $(\alpha, sk_\alpha, T, a)_2$, $M(P, a)$ for $M_1(P, a)$ and $M_2(P, a)$, etc.).

Proposition 2.1. *Let α be an enumeration for a class $\mathbf{C}$, sk_α a skip function, T a rejection test which are (s) compatible and $a \in \mathbf{A}(\mathbf{C})$. Then*

(i) $x_a(P, x) \leqslant_{\alpha(P)} x_0(P, x)$ *and*

(ii) $z_a(P, x) - a(P, z_a(P, x)) \leqslant z_0(P, x) \leqslant z_a(P, x)$

for any problem $P \in \mathbf{C}$ and $x \in D_P \cup \{x_{\alpha(P,d(P)+1)}\}$.

(iii) $y \in (x_a(P, x), x)_{\alpha(P)} \rightarrow y$ *unfeasible or* $f_P(y) \geqslant z_a(P, x) - a(P, z_a(P,x))$;

(iv) *If P has solutions in the interval* $[x_{\alpha(P,1)}, x)_{\alpha(P)}$ *then* $x_a(P, x)$ *is the feasible solution with the property* $x_a(P, x) = \min \{x_a(P, y) \mid z_a(P, y) - a(P, z_a(P, y)) \leqslant z_0(P, x)\}$.

Proof. We prove simultaneously the inequalities (i) and (ii) by induction on x for the algorithms of the first type. For $x_{\alpha(P,1)}$ we have $x_a(P, x_{\alpha(P,1)}) = x_0(P, x_{\alpha(P,1)}) = x_{\alpha(P,1)}$ hence (i) is satisfied. Then $z_a(P, x_{\alpha(P,1)}) = z_0(P, x_{\alpha(P,1)}) = +\infty$ and there is nothing to prove for (ii) The induction step: Assume that (i) and (ii) hold true for all $y \leqslant_{\alpha(P)} x$ and consider the case of $succ_\alpha(P, x)$. Let $\bar{y} \in M_1(P, a)$, $\bar{y} \leqslant_{\alpha(P)} x$, be the last element less than or equal to x for which T was evaluated when the algorithm $(\alpha\ sk_\alpha, T, a)_1$ is performed. The following cases are possible.

Case 1. Assume $T(P, \bar{y}, z_a(P, \bar{y}) - a(P, z_a(P, \bar{y}))) = 2$. Then $\bar{y} = x$, according to the program schema step S4 is performed and $x_a(P, succ_\alpha(P, x)) = x$, $z_a(P, succ_\alpha(P, x)) = f_P(x)$. The condition (t1) and the induction hypothesis imply that x is a feasible solution of P for which $f_P(x) < z_a(P, x) - a(P, z_a(P, x)) \leqslant z_0(P, x) \leqslant z_a(P, x)$. Therefore according to (t1) we have also $T(P, x, z_0(P, x)) = 2$, $x \in M_1(P, 0)$ and $x_0(P, succ_\alpha(P, x)) = x$, $z_0(P, succ_\alpha(P, x)) = f_P(x)$. The inequality (i) is satisfied and (ii) follows immediately since $a(P, z) \geqslant 0$.

Case 2. Assume $T(P, \bar{y}, z_a(P, \bar{y}) - a(P, z_a(P, \bar{y}))) = 0$. Then $\bar{y} = x$, according to the program schema step S4 is not performed by $(\alpha, sk_\alpha, T, a)_1$ and $x_a(P, succ_\alpha(P, x)) = x_a(P, x)$, $z_a(P, succ_\alpha(P, x)) = z_a(P, x)$. On the other hand consider the exact procedure $(\alpha, sk_\alpha, T, 0)_1$. If $T(P, x, z_0(P, x)) \neq 2$ then $x_0(P, succ_\alpha(P, x)) = x_0(P, x)$, $z_0(P, succ_\alpha(P, x)) = z_0(P, x)$ and the inequalities (i), (ii) follow from the induction hypothesis. Next assume $T(P, x, z_0(P, x)) = 2$. Then $x_0(P, succ_\alpha(P, x)) = x$, $z_0(P, succ_\alpha(P, x)) = f_P(x)$. The inequality (i) follows from the induction hypothesis and the fact that $x_0(P, x) < _{\alpha(P)} x$. On the other hand the above assumptions and the condition (t1) imply that x is a feasible solution of P and $z_a(P, succ_\alpha(P, x)) - a(P, z_a(P, succ_\alpha(P, x)) = z_a(P, x) - a(P, z_a(P, x)) < f_P(x) = z_0(P, succ_\alpha(P, x)) < z_0(P, x) \leqslant z_a(P, x) = z_a(P, succ_\alpha(P, x))$.

Case 3. Assume $T(P, \bar{y}, z_a(P, \bar{y}) - a(P, z_a(P, \bar{y}))) = 1$. Then $\bar{y} \leqslant_{\alpha(P)} x < _{\alpha(P)} sk_\alpha(P, \bar{y})$, hence $succ_\alpha(P, x) \leqslant_{\alpha(P)} sk_\alpha(P, \bar{y})$ and $x_a(P, succ_\alpha(P, x)) = x_a(P, \bar{y})$, $z_a(P, succ_\alpha(P, x)) = z_a(P, \bar{y})$. The inequalities (i), (ii) follow from the induction hypothesis if $T(P, x, z_0(P, x)) \neq 2$. In the case $T(P, x, z_0(P, x)) = 2$ holds then $x_0(P, succ_\alpha(P, x)) = x$, $z_0(P, succ_\alpha(P, x)) = f_P(x)$. The inequality (i) is obviously satisfied. Then the conditions (t1),(t2) and the above assumptions imply $z_a(P, succ_\alpha(P, x)) - a(P, z_a(P, succ_\alpha(P, x)) = z_a(P, \bar{y}) - a(P, z_a(P, \bar{y})) \leqslant f_P(x) = z_0(P, succ_\alpha(P, x)) < z_0(P, x) \leqslant z_a(P, \bar{y}) = z_a(P, succ_\alpha(P, x))$.

The proof of the inequalities (i) and (ii) for the methods of the second type is similar. (The algorithms compute the same functions $x_a(P, x)$, $z_a(P, x)$). Parts (iii) and (iv) of the proposition follow immediately. Q.E.D.

Proposition 2.2. *Let α be an enumeration for a class* **C**, sk_α *a skip function and T a rejection test which are (s) compatible and an allowance function* $a \in \mathbf{A}(\mathbf{C})$. *The algorithm* $(\alpha, sk_\alpha, T, a)$ *halts in a finite number of steps for every problem* $P \in \mathbf{C}$ *as input and solves it suboptimally with the allowance* a^0, *i.e.*: $z^0(a) = +\infty$ *if P has no solutions, otherwise* $x^0(a) =$ *a feasible solution of P*, $z^0(a) = f_P(x^0(a))$, $a^0 = a(P, z^0(a))$ *and* $z^0(a) - a^0 \leqslant z^0(0) \leqslant z^0(a)$.

Moreover $x^0(a) \leqslant_{\alpha(P)} x^0(0)$ *holds. (Here* $x^0(0)$ *and* $z^0(0)$ *are the output optimal values supplied by the exact algorithm* $(\alpha, sk_\alpha, T, 0)$.

Proof. Finiteness. Let $P \in \mathbf{C}$ be an input problem for the algorithm $(\alpha, sk_\alpha, T, a)$ and $x_1 = x_{\alpha(P,1)}, x_2, \ldots,$ the resulting sequence of elements for which the test T is evaluated in step S3 during the computation. From the property (s) of the skip function and from the statement of the algorithm it follows that all the elements in the sequence are from the finite set D_P. Then note that these elements are strictly increasing with respect to the $\leqslant_{\alpha(P)}$- order. Therefore the sequence is finite and the algorithm stops.

Validity. Let $(z^0(a)), x^0(a), a^0)$ be the output supplied by the algorithm. First assume P has no solutions. Then according to (t1) $T(P, x, z) \neq 2$ for any $x \in D_P$ and $z \in R^0$. Step S4 is never performed and the algorithm will supply the output $z^0(a) = +\infty$. Next assume P has solutions. Then from Proposition 2.1 it follows $x^0(a) = x_a(P, x_{\alpha(P,d(P)+1)}) =$ a feasible solution of P, $z^0(a) = f_P(x^0,(a))$, $a^0 = a(P, z^0(a))$ and $z^0(a) - a^0 \leqslant z_0(0) \leqslant z^0(a)$. Q.E.D.

Note that the above propositions do not imply that $y = x^0(a)$ is the first feasible solution of P with respect to $\leqslant_{\alpha(P)}$ which satisfies the suboptimality condition $z^0(0) \leqslant f_P(y) \leqslant z^0(0) + a^0$.

In the integer programming literature the reader can find many examples of implicit enumerations algorithms for general and particular problems. We cite here only the method of Lawler and Bell [3], that of Mao and Wallingford [4], LEXS of Rödder [8], a part of the algorithms discussed by Korte, Krelle and Oberhoffer [2], and the procedures constructed by Popescu [5]. The most widespread enumerations are the lexicographic ones. The skip functions and the rejection tests are usually constructed on the basis of the monotonicity of the functions which are contained in the data of the problems with respect to the natural order on each of the components. Concerning the allowance functions, two types are often used. The first one, of the form $a(P,z) = \varepsilon(P) > 0$, assures a suboptimal solution $x^0(a)$ with its absolute error from the optimal value at most $\varepsilon(P)$: $z^0(a) - z^0(0) \leqslant \varepsilon(P)$. The second type, of the form $a(P, z) = r(P)|z|$, with $0 < r(P) < 1$, guarantees that the exact optimal value $z^0(0)$ is within $r(P) \times 100\ \%$ relative error from the suboptimal value: $|(z^0(a) - z^0(0))/z^0(a)| \leqslant r(P)$.

3. Computing effort of the approximate algorithms

In practice, the most time-consuming part of an implicit enumeration procedure is the evaluation of the rejection test. Therefore, when one solves a problem P by an algorithm $(\alpha, sk_\alpha, T, a)$ — exact or approximate —, the efficiency is usually estimated by (one of) the following effort measures:

$m(P, a)$ = the number of executions of step S3 (the evaluation of the test) prior to the algorithms halts (i.e. the cardinal number of the set $M(P, a)$);

$n(P, a)$ = the number of executions of step S3 until the last execution of step S4 (the last modification of the variables x^0, z^0, a^0) if P has solutions and $n(P, a) = m(P, a)$ otherwise.

Both measures count the number of the elements for which the test T is evaluated. The parameter $m(P, a)$ is related to the total computation time. The measure $n(P, a)$ is useful when the computing process may be cut off prior to the natural termination of the algorithm. This situation can occur due to the insufficiency of the available computer time or to a more complex usage of the algorithm. $n(P, a)$ is related to the computation time required until the final suboptimal solution $x^0(0)$ is attained, that is until the output is obtained.

Remark 3.1. Note that for a given problem P the measures m and n do not simultaneously increase or decrease from one algorithm to another, as it was shown in [6]. Therefore both measures m, n must be considered explicitly.

Remark also that if one solves P by two procedures $(\alpha, sk_\alpha, T, a)$, $(\alpha, sk_\alpha, T, a')$ then $x^0(a) \leqslant_{\alpha(P)} x^0(a')$ does not always imply $m(P, a) \leqslant m(P, a')$ or $n(P, a) \leqslant n(P, a')$.

Let α be an enumeration for a class $\mathbf{C} \subseteq$ GIP, sk_α a skip function and T a rejection test which are compatible. Denote by $a \sim a'$ the following relation in the set $\mathbf{A}(\mathbf{C})$ of the allowance functions:

$$a \sim a' \text{ iff } T(P, x, z_a(P, x) - a(P, z_a(P, x))) =$$

$$= T(P, x, z_{a'}(P, x) - a'(P, z_{a'}(P, x)))$$

for any feasible problem $P \in \mathbf{C}$ and $x \in M_1(P, a)$, $x^P <_{\alpha(P)} x$.

Proposition 3.1. *The relation* $\sim$ *is an equivalence relation in the set* $\mathbf{A}(\mathbf{C})$ *and* $a \sim a'$ *implies*

$$x_a(P, x) = x_{a'}(P, x), \quad z_a(P, x) = z_{a'}(P, x),$$

$$m(P, a) = m(P, a'), \quad n(P, a) = n(P, a'),$$

for any $P \in \mathbf{C}$ *and* $x \in D_P \cup \{x_{\alpha(P, d(P)+1)}\}$.

Proof. If the input problem P is unfeasible or if $x^P = x_{\alpha(P, d(P)+1)}$, then the result is obvious. Otherwise the proof goes by induction on the elements of the set $M_1(P, a)$. Q.E.D.

Denote by $\tilde{a}$ the equivalence class of a and by $\inf \tilde{a}$, $\sup \tilde{a}$ the bounds of the set $\tilde{a}$:

$$\inf \tilde{a}(P, z) = \inf \{a'(P, z) \mid a' \in \mathbf{A}(\mathbf{C}) \text{ and } a' \sim a\},$$

$$\sup \tilde{a}(P, z) = \sup \{a'(P, z) \mid a' \in \mathbf{A}(\mathbf{C}) \text{ and } a' \sim a\}.$$

Then $\inf \tilde{a}(P, z) \leqslant a'(P, z) \leqslant \sup \tilde{a}(P, z)$ for any $a' \in \tilde{a}$, $P \in \mathbf{C}$ and $z \in R$. Obviously the bounds $\inf \tilde{a}(P, z)$, $\sup \tilde{a}(P, z)$ are allowance functions for $\mathbf{C}$, if their values are finite. A result which clarifies how the content of an equivalence class is, is the following.

Proposition 3.2. *Let* $a, a' \in \mathbf{A}(\mathbf{C})$. *If* $\inf \tilde{a}(P, z_a(P, x)) < a'(P, z_a(P, x)) < \sup \tilde{a}(P, z_a(P, x))$ *for any feasible problem* $P \in \mathbf{C}$ *and* $x \in M_1(P, a)$, $x^P <_{\alpha(P)} x$, *then* $a' \sim a$.

Proof. Let $P \in \mathbf{C}$ be a feasible problem with $x^P <_{\alpha(P)} x_{\alpha(P, d(P))}$ and $x \in M_1(P, a)$, $x^P <_{\alpha(P)} x$. There exists $a_1, a_2 \in \tilde{a}$ with the property $a_1(P, z_a(P, x)) \leqslant a'(P, z_a(P, x)) \leqslant a_2(P, z_a(P, x))$.

If $T(P, x, z_a(P, x) - a(P, z_a(P, x))) = 2$ then from the inequality $z_a(P, x) - a_2(P, z_a(P, x)) \leqslant z_a(P, x) - a'(P, z_a(P, x))$ and from the condition (t1) it follows $T(P, x, z_a(P, x) - a'(P, z_a(P, x))) = 2$.

If $T(P, x, z_a(P, x) - a(P, z_a(P, x))) = 1$ then the inequality $z_a(P, x) - a'(P, z_a(P, x)) \leqslant z_a(P, x) - a_1(P, z_a(P, x))$ and the condition (t2) imply $T(P, x, z_a(P, x) - a'(P, z_a(P, x))) = 1$.

Finally if $T(P, x, z_a(P, x) - a(P, z_a(P, x))) = 0$ then an analysis similar to the above ones shows that $T(P, x, z_a(P, x) - a'(P, z_a(P, x))) = 0$.

Therefore $M_1(P, a) = M_1(P, a')$ and $z_a(P, x) = z_{a'}(P, x)$, hence $a \sim a'$. Q.E.D.

From the conditions (t1) — (t3) one deduces the following relations for the values of the bounds $\inf \tilde{a}$, $\sup \tilde{a}$ in the tested points $x \in M_1(P, a)$, $x^P <_{\alpha(P)} x$:

1) If $T(P, x, z_a(P, x) - a(P, z_a(P, x))) = 2$ then $\sup \tilde{a}(P, z_a(P, x)) \leqslant z_a(P, x) - f_P(x)$;

2) If $T(P, x, z_a(P, x) - a(P, z_a(P, x))) = 1$ then $\inf \tilde{a}(P, z_a(P, x)) \geqslant z_a(P, x) - \min \{f_P(y) | r_P(y)$ and $y \in [x, sk_\alpha(P, x))_{\alpha(P)}\} \geqslant z_0(P, sk_\alpha(P, x))$;

3) If $T(P, x, z_a(P, x) - a(P, z_a(P, x))) = 0$ and x is a feasible solution of P then $\inf \tilde{a}(P, z_a(P, x)) \geqslant z_a(P, x) - f_P(x)$.

For an arbitrary problem $P \in \mathbf{C}$, denote by $\mathbf{A}(\{P\}) / \sim$ the set of equivalence classes of the allowance functions $a \in \mathbf{A}(\{P\})$. The set $\mathbf{A}(\{P\})/\sim$ characterizes all the possible ways in which the problem P can be solved by approximate algorithms $(\alpha, sk_\alpha, T, a)$, $a \in \mathbf{A}(\{P\})$.

First, remark that $\mathbf{A}(\{P\})/\sim$ is finite. Indeed, for P fixed any equivalence class $\tilde{a}$ is determined by a system of values $\{T(P, x, z_a(P, x) - a(P, z_a(P, x)))\}_{x \in M_1(P,a) \text{ and } x^P <_{\alpha(P)} x}$. The rejection test T can take only the values 0, 1, 2 and the set of indices $\{x \mid x \in M_1(P, a)$ and $x^P <_{\alpha(P)} x\}$ is finite, since D_P is always finite. Therefore there is only a finite number of different systems of values which define an equivalence class.

Note also that if P is unfeasible or if $x^P = x_{\alpha(P,d(P))}$ then $\mathbf{A}(\{P\})/\sim$ has a single element.

The allowance function $u(P)$. In the set $\mathbf{A}(\mathbf{C})$, outside the identical null allowance function $0(P, z) = 0$, there is another one $u(P)$, with a special rôle. Let

$$l(P, x) = \begin{cases} \sup \{z \mid T(P, x, z) = 1\}, & \text{if there is such a } z\text{-value} \\ f_P(x), & \text{otherwise} \end{cases}$$

and

$$l(P) = \begin{cases} \min\{l(P, x) \mid x \in D_P \text{ and } x^P <_{\alpha(P)} x\}, & \text{if } P \text{ is feasible and } x^P <_{\alpha(P)} x_{\alpha(P,d(P))} \\ +\infty, & \text{otherwise.} \end{cases}$$

Define

$$u(P) = \begin{cases} f_P(x^P) - l(P) + \varepsilon, & \text{if } l(P) \neq +\infty \\ 0, & \text{otherwise} \end{cases}$$

where $\varepsilon > 0$ is arbitrary.

As it was defined, $\tilde{u}(P)$ is the less restrictive class of allowance functions. That is, any time a problem P has solutions then the output suboptimal solution supplied by $(\alpha, sk_\alpha, T, u)$ is x^P. Moreover, from the point of view of the efficiency, the algorithm $(\alpha, sk_\alpha, T, u)$ has the property that any time the test can reject in a point in principle, then the value of $u(P)$ is sufficiently large such that the approximate procedure $(\alpha, sk_\alpha, T, u)$ will reject the corresponding argument.

On the other hand $\sup \tilde{u}(P, z) = +\infty$ holds for any $P \in \mathbf{C}$. Indeed, let $r > 0$ be arbitrary, and consider the allowance function $u(P) + r$. If P is unfeasible then obviously $u(P) \sim u(P) + r$. Next, assume P has solutions. Then from the definition of $l(P)$ it follows by induction $T(P, x, z_u(P, x) - u(P)) = T(P, x, z_{u+r}(P, x) - u(P) - r)$, for any $x \in M_1(P, u)$, $x^P <_{\alpha(P)} x$, hence $u(P) \sim u(P) + r$. Therefore $\sup \tilde{u}(P, z) = +\infty$.

4. Lower and upper bounds of the computing effort

A first question related to the efficiency of the approximate algorithms is whether solving suboptimally a problem is really more advantageous than solving it by the corresponding exact procedure.

Remark 4.1. The use of an arbitrary approximate algorithm instead of the exact one does not imply the decrease of the effort when solving the problems in the corresponding class (even $x^0(a) \leqslant_{\alpha(P)} x^0(0)$ holds for any $a \in \mathbf{A(C)}$ and $P \in \mathbf{C}$). That is, the inequalities $m(P, a) \leqslant m(P, 0)$, $n(P, a) \leqslant n(P, 0)$ do not always hold for any α, sk_α, T which are (s) compatible, $a \in \mathbf{A(C)}$ and $P \in \mathbf{C}$, where $\mathbf{C}$ is the class of the problems solved by the methods $(\alpha, sk_\alpha, T, a)$, $a \in \mathbf{A(C)}$. Indeed, consider the following example.

Let $P \in \mathrm{GIP}$ be a problem with $d(P) = 6$, $\mathbf{C} = \{P\}$, an enumeration α of the elements of the set D_P in the order $x_1, x_2, \ldots, x_6$ and sk_α defined by its graph: $\{(x_1, x_2), (x_2, x_4), (x_3, x_6), (x_4, x_5), (x_5, x_6)\}$. sk_α is a skip function compatible with α. Assume also that x_1, x_6 are the feasible solutions of P and the objective function $f_P(x)$ is such that $f_P(x_6) < f_P(x_1)$. Define T by:

$$T(P, x_i, z) = \begin{cases} 2, \text{ for } i = 1 \text{ and } z > f_P(x_1), \\ \qquad i = 6 \text{ and } z > f_P(x_6); \\ 0, \text{ for } i = 2 \text{ and } z \geqslant f_P(x_1); \\ 1, \text{ otherwise.} \end{cases}$$

T is a rejection test for the class $\mathbf{C}$ and the components α, sk_α, T are s-compatible. Let $a(P, z) \equiv \varepsilon > 0$. If we solve the problem P by the exact method $(\alpha, sk_\alpha, T, 0)$ and by the approximate one $(\alpha, sk_\alpha, T, a)$ we obtain: $m(P, a) = n(P, a) = 5$, $m(P, 0) = n(P, 0) = 4$.

The solving process by the above procedures is illustrated in Figure 2, where the upper part stands for the exact algorithm and the lower part for the approximate one.

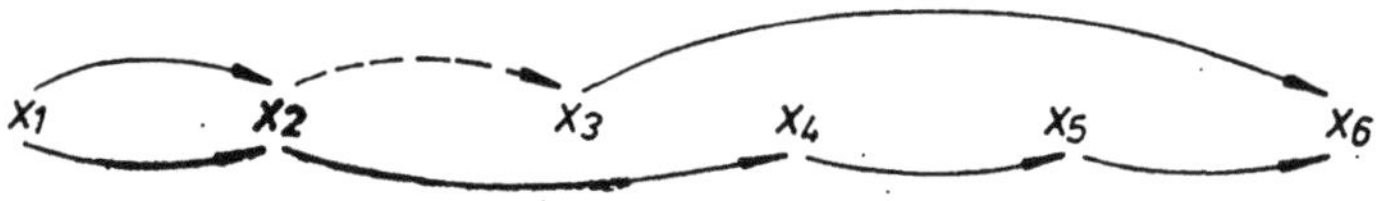

Figure 2. The computation process by the algorithms used in Remark 4.1.

Remark 4.2. A similar question arises with respect to the algorithm $(\alpha, sk_\alpha, T, u)$. In the previous section it was shown that any time the test T can reject in a point in principle, this procedure rejects effectively. Moreover $x^0(u) = x^P \leqslant_{\alpha(P)} x^0(a)$ for any $a \in \mathbf{A(C)}$ and $P \in \mathbf{C}$. Therefore it seems very natural that $(\alpha, sk_\alpha, T, u)$ will be the most efficient among the approximate algorithms $(\alpha, sk_\alpha, T, a)$, $a \in \mathbf{A}(\mathbf{C})$. However this is not always true. Using the same example as in the previous remark it results that neither do the inequalities $m(P, a) \geq m(P, u)$, $n(P, a) \geq n(P, u)$ always hold for any α, sk_α, T, which are (s) compatible, $a \in \mathbf{A}(\mathbf{C})$ and $P \in \mathbf{C}$.

In spite of the above results there is a wide class of approximate methods for which the upper and the lower bounds of their computing effort are given by the exact algorithm, respectively by the procedure using the allowance function $u(P)$. Two such classes are given below.

Algorithms whose skip functions are with nested jumps. Let sk_α be a skip function for a class $\mathbf{C}$ compatible with an enumeration $\alpha \cdot sk_\alpha$ is *with nested jumps* if

$$(\forall x \in D_P)\ (\forall y \in (x, sk_\alpha(P, x))_{\alpha(P)})\ (sk_\alpha(P, y) \leq_{\alpha(P)} sk_\alpha(P, x))$$

for all problems $P \in \mathbf{C}$.

For instance, the skip function in Lawler and Bell's algorithm [3], and that in the method of Mao and Wallingford [4], are with nested jumps.

Proposition 4.1. *Let α be an enumeration for a class $\mathbf{C}$, sk_α a skip function with nested jumps and T a rejection test which are (s) compatible. Then*

$$n(P, u) \leq n(P, a) \leq n(P, 0),$$

$$m(P, u) \leq m(P, a) \leq m(P, 0),$$

for any allowance function $a \in \mathbf{A}(\mathbf{C})$ and problem $P \in \mathbf{C}$.

Proof. If the input problem $P \in \mathbf{C}$ is unfeasible or if $x^P = x_{\alpha(P,d(P))}$ then all the approximate algorithms proceed identically, hence the inequalities are satisfied. Next, let $P \in \mathbf{C}$ be a feasible problem with $x^P <_{\alpha(P)} x_{\alpha(P,d(P))}$.

Proof of the inequality $m_1(P, a) \leq m_1(P, 0)$, for the procedures of the first type. Denote by $V_1(P, a) = D_P \setminus M_1(P, a)$ the set of skipped elements when one solves P by $(\alpha, sk_\alpha, T, a)_1$. The inequality follows immediately provided we show the inclusion $V_1(P, 0) \subseteq V_1(P, a)$. If $V_1(P, 0) = \emptyset$ the result is obvious. Let $x \in V_1(P, 0) \neq \emptyset$. If $x <_{\alpha(P)} x^P$ then $x \in V_1(P, a)$, since both algorithms $(\alpha, sk_\alpha, T, 0)_1$ and $(\alpha, sk_\alpha, T, a)_1$ proceed identically until x^P. Next assume $x^P <_{\alpha(P)} x$ $(x^P \notin V_1(P, 0))$. There exists an element $y \in M_1(P, 0)$, $x^P \leq_{\alpha(P)} y <_{\alpha(P)} x <_{\alpha(P)} sk_\alpha(P, y)$, such that $T(P, y, z_0(P, y)) = 1$. The inequality $z_a(P, y) - a(P, z_a(P, y)) \leq z_0(P, y)$, which follows from Proposition 2.1, and the property (t3) of the tests imply $T(P, y, z_a(P, y) - a(P, z_a(P, y))) = 1$. If $y \in M_1(P, a)$ then $x \in V_1(P, a)$ and this ends the proof. Next assume $y \notin M_1(P, a)$ and let $y' \in M_1(P, a)$, $x^P \leq_{\alpha(P)} y' <_{\alpha(P)} y$, be the greatest element less than y tested by $(\alpha, sk_\alpha, T, a)_1$. (Such an element does exist since $x^P <_{\alpha(P)} y$.) Then, as shown below, one must have $T(P, y', z_a(P, y') - a(P, z_a(P, y'))) = 1$ and $y <_{\alpha(P)} sk_\alpha(P, y')$, the other cases leading to a contradiction. Indeed:

a) Assume $T(P, y', z_a(P, y') - a(P, z_a(P, y'))) = 0$. Then $succ_\alpha(P, y') = y$ is the next element in the set $M_1(P, a)$, which contradicts the assumption on y.

b) Assume $T(P, y', z_a(P, y') - a(P, z_a(P, y'))) = 1$ and $sk_\alpha(P, y') <_{\alpha(P)} y$. The case $sk_\alpha(P, y') = y$ can be rejected by an argument similar to that in a). On the other hand the situation $sk_\alpha(P, y') <_{\alpha(P)} y$ contradicts the definition of y'.

c) Assume the value of the test when $(\alpha, sk_\alpha, T, a)_1$ reach y' is 2. This situation is identical to that in a).

Therefore $y <_{\alpha(P)} sk_\alpha(P, y')$ and $T(P, y', z_a(P, y') - a(P, z_a(P, y'))) = 1$ hold. Then $x \in [y', sk_\alpha(P, y'))_{\alpha(P)}$ since sk_α is with nested jumps. Hence $x \in V_1(P, a)$.

All the other inequalities in the statement of the proposition can be proved similarly for the first-type procedures as well as for those of the second type. Q.E.D.

Another class of approximate methods which are always more efficient than the corresponding exact procedures is the following.

Algorithms with nondecreasing skip functions and regular tests. Let α be an enumeration for a class **C**, sk_α a skip function and T a rejection test which are compatible. The test T is said to be *regular* (with respect to α and sk_α) if whenever it rejects for an element x and a value z it also rejects for all the elements y in the interval $(x, sk_\alpha(P, x))_{\alpha(P)}$ and the bound z. More precisely, T is regular if it satisfies the condition.

$$T(P, x, z) = 1 \rightarrow (\forall y \in (x, sk_\alpha(P, x))_{\alpha(P)} \; (T(P, y, z) = 1)$$

for any problem $P \in \mathbf{C}$, $x \in D_P$ and $z \in R^{\cdot}$.

Generally the tests constructed on the basis of the piecewise monotonicity of the functions are all regular.

As the next result shows, the computing effort of the approximate algorithms using regular tests and skip functions which are nondecreasing with respect to the order introduced by the enumerations is upper and lower bounded by the effort of the exact procedure, respectively by the effort of the approximate methods using the allowance function $u(P)$.

Proposition 4.2. *Let α be an enumeration for a class* **C**, *sk_α a nondecreasing skip function with respect to the $\leqslant_{\alpha(P)}$-order and T a regular rejection test which are (s) compatible. Then*

$$n(P, u) \leqslant n(P, a) \leqslant n(P, 0),$$

$$m(P, u) \leqslant m(P, a) \leqslant m(P, 0)$$

for any allowance function $a \in \mathbf{A}(\mathbf{C})$ and problem $P \in \mathbf{C}$.

Proof. If P is unfeasible or if $x^P = x_{\alpha(P, d(P))}$ the inequalities are satisfied obviously. Next assume P is feasible and $x^P <_{\alpha(P)} x_{\alpha(P, d(P))}$.
Proof of the inequality $m_1(P, a) \leqslant m_1(P, 0)$, for the first type algorithms. If $M_1(P, a) \setminus M_1(P, 0) = \emptyset$ the result is obvious. Assume $M_1(P, a) \setminus M_1(P, 0) = \{x_1 <_{\alpha(P)} x_2, \ldots, <_{\alpha(P)} x_k\} \neq \emptyset$. In this case the inequality follows immediately provided we show that there exists an injective function $g(P, x)$, $g(P, .)\colon M_1(P, a) \setminus M_1(P, 0) \rightarrow M_1(P, 0) \setminus M_1(P, a)$.

For every x_i, $i = 1, \ldots, k$, there exists an element $y_i \in M_1(P, 0)$, $x^P < <_{\alpha(P)} y_i <_{\alpha(P)} sk_\alpha(P, y_i)$, such that $T(P, y_i, z_0(P, y_i)) = 1$. From the property (t3) and Proposition 2.2 it follows $T(P, y_i, z_a(P, y_i) - a(P, z_a(P, y_i))) = 1$. Then from this equality and from $y_i <_{\alpha(P)} x_i <_{\alpha(P)} sk_\alpha(P, y_i)$ it results $y_i \notin M_1(P, a)$, since otherwise $x_i \notin M_1(P, a)$, which is absurd.

Next let us show that $x_{i-1} <_{\alpha(P)} y_i$, for any $i \neq 1$. Indeed, assume $y_i \leqslant_{\alpha(P)} x_{i-1}$ holds for some i. Then $x_{i-1} \in [y_i, sk_\alpha(P, y_i))_{\alpha(P)}$, since $x_{i-1} < <_{\alpha(P)} x_i <_{\alpha(P)} sk_\alpha(P, y_i)$. Therefore from the regularity of the test it follows $T(P, x_{i-1}, z_a(P, x_{i-1}) - a(P, z_a(P, x_{i-1}))) = 1$. Hence $sk_\alpha(P, x_{i-1}) \leqslant_{\alpha(P)} x_i$, since $x_{i-1}, x_i \in M_1(P, a)$. On the other hand the nondecreasing monotonicity of the function sk_α and $y_i \leqslant_{\alpha(P)} x_{i-1}$ imply $x_i <_{\alpha(P)} sk_\alpha(P, y_i) \leqslant_{\alpha(P)} sk_\alpha(P, x_{i-1})$, which contradicts the previous conclusion.

Therefore for every x_i there exists an $y_i \in M_1(P, 0) \setminus M_1(P, a)$ such that $x_{i-1} <_{\alpha(P)} y_i$. If one defines $g(P, x_i) = y_i$ then the function $g(P, x)$ is injective and the inequality is proved.

The proofs of the other inequalities in the statement of the proposition are analogous for the first type procedures as well as for those of the second type. Q.E.D.

5. Monotonicity of the effort measures with respect to the allowance function

In this section two relations of order among the allowance functions are introduced and the monotonicity of the effort measures $m(P, a)$, $n(P, a)$ with respect to these orders is studied.

Let $a, a' \in \mathbf{A}(\mathbf{C})$. Denote by "$a \preccurlyeq a'$" the natural order relation between the sizes of the allowance functions:

$$a \preccurlyeq a' \text{ iff } (\forall P \in \mathbf{C})(\forall z \in R)(a(P, z) \leqslant a'(P, z))$$

and by "$a \prec a'$" the strict inequality "$a \preccurlyeq a'$ and $a' \not\preccurlyeq a$".

Lemma 5.1. *$\{\mathbf{A}(\mathbf{C}), \preccurlyeq\}$ is a distributive lattice, 0 is its minimum element and u has the maximality property $a \succcurlyeq u \rightarrow a \sim u$.*

Proof. The reflexivity, antisymmetry and transitivity properties of the relation $\preccurlyeq$ are obvious, hence it is a partial order. If $a, a' \in \mathbf{A}(\mathbf{C})$, define:

$$\inf (a, a')(P, z) = \min \{a(P, z), a'(P, z)\},$$

$$\sup (a, a')(P, z) = \max \{a(P, z), a'(P, z)\}.$$

inf (a, a'), sup $(a, a') \in \mathbf{A}(\mathbf{C})$, since they satisfy the condition (a) from the definition of the allowance functions. Then inf $(a, a') \preccurlyeq a, a' \preccurlyeq$ sup (a, a') and obviously they are the greatest lower bound, respectively the least upper bound of a, a'. The distributivity of the operations inf (a, a'), sup (a, a'), one to each other, follows from the similar property of the min, max operations among the real numbers. The property of maximality of the allowance function $u(P)$ follows from the equality sup $\tilde{u}(P, z) = +\infty$, shown in Section 3. Q.E.D.

Remark 5.1. Although it seems natural that the greater the size of the allowance function, the more efficient the algorithm, this is not always true, even if only the procedures in Section 4 are considered. Indeed, let us consider the following example.

Let $P \in$ GIP be a problem with $d(P) = 7$, $\mathbf{C} = \{P\}$, α an enumeration of the elements of the set D_P in the order $x_1, x_2, \ldots, x_7$ and sk_α defined by its graph: $\{(x_1, x_2), (x_2, x_4), (x_3, x_4), (x_4, x_7), (x_5, x_7), (x_6, x_7)\}$. sk_α is a skip function with nested jumps compatible with α (sk_α is also a nondecreasing function with respect to $\leqslant_{\alpha(P)}$).

Let x_1, x_3 and x_7 be the feasible solutions of P and suppose $f_P(x_1) > f_P(x_3) > f_P(x_7)$. Let $a, a' \in \mathbf{A}(\mathbf{C})$, $a' \prec a$, be two allowance functions satisfying the following inequalities:

$$f_P(x_7) < f_P(x_3) - a'(P, f_P(x_3)) < f_P(x_1) - a(P, f_P(x_1)) < f_P(x_3) < f_P(x_1) - a'(P, f_P(x_1)).$$

Now let T have the following values:

$$T(P, x_i, z) = \begin{cases} 2, \text{ for: } i = 1 \text{ and } z > f_P(x_1), \\ \qquad\quad i = 3 \text{ and } z > f_P(x_3), \\ \qquad\quad i = 7 \text{ and } z > f_P(x_7); \\ 1, \text{ for: } i = 1 \text{ and } z \leqslant f_P(x_1), \\ \qquad\quad i = 2, 3 \text{ and } z \leqslant f_P(x_3), \\ \qquad\quad i = 4, 5, 6 \text{ and } z \leqslant f_P(x_3) - a'(P, f_P(x_3)); \\ 0, \text{ orherwise.} \end{cases}$$

As it was defined, T satisfies the conditions (t1)—(t3), (sc) and the regulatrity condition. Now if one solves the problem P by the algorithms $(\alpha, sk_\alpha, T, a)$, $(\alpha, sk_\alpha, T, a')$ one obtains: $m(P, a) = n(P, a) = 6 > m(P, a') = n(P, a') = 5$.

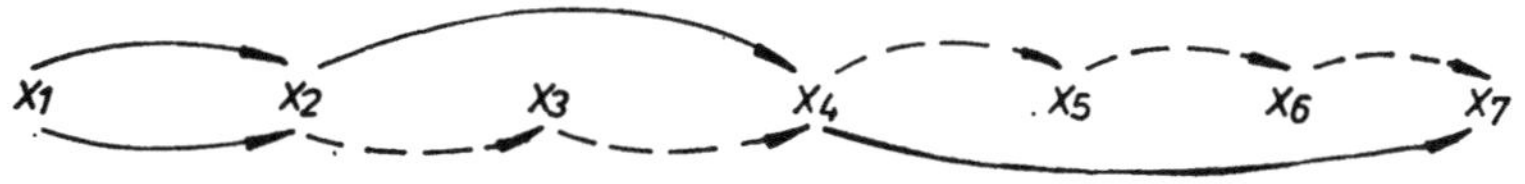

Figure 3. The computation process by the algorithms used in Remark 5.1.

The solving process is illustrated in Figure 3, where the upper part stands for $(\alpha, sk_\alpha\ T, a)$ and the lower one for the algorithm $(\alpha, sk_\alpha, T, a')$.

Remark 5.2. Note that $a \prec a'$ does not imply $x^0(a') \leqslant_{\alpha(P)} x^0(a)$ even for the algorithms in Section 4 (where $x^0(a')$ and $x^0(a)$ are the output suboptimai solutions supplied by $(\alpha, sk_\alpha, T, a')$, $(\alpha, sk_\alpha, T, a)$, respectively). Indeed, consider the following example. Let $P \in$ GIP with $d(P) = 3$, α an enumeration the elements of the set D_P in the order x_1, x_2, x_3 and $sk_\alpha = succ_\alpha$. Assume also that the elements in D_P are all feasible. Let $r, r' \in R$, $0 < r < r' < 1$, and f_P satisfy the following inequalities:

$$0 < f_P(x_3) < f_P(x_2) < f_P(x_1),$$

$$f_P(x_2) - rf_P(x_2) \leqslant f_P(x_3) < f_P(x_1) - r'f_P(x_1) \leqslant f_P(x_2) < f_P(x_1) - rf_P(x_1).$$

Define T by:

$$T(P, x, z_i) = \begin{cases} 2, \text{ for } i = 1 \text{ and } z > f_P(x_1), \\ \qquad\quad i = 2 \text{ and } z > f_P(x_2), \\ \qquad\quad i = 3 \text{ and } z > f_P(x_3); \\ 1, \text{ otherwise.} \end{cases}$$

Then T is a rejection test compatible with α and sk_α. Note also that $succ_\alpha$ is with nested jumps and non-decreasing and T is regular. If one solves P by the approximate algorithms $(\alpha, sk_\alpha, T, a')$, $(\alpha, sk_\alpha, T, a)$, where $a(P, z) = rz$, $a'(P, z) = r'z$ and $a \prec a'$, one obtains: $x^0(r') = x_3$, $x^0(r) = x_2$.

Another order relation among the allowance functions which this time depends on the other components of the approximate procedure, is the following. Let α be an enumeration for a class **C**, sk_α a skip function and T a rejection test which are compatible. Denote by "$a \preccurlyeq^* a'$" the relation in the set **A**(**C**) defined by: "$a \preccurlyeq^* a'$ iff $(z_a, (P, x) - a'(P, z_a, (P, x)) \leqslant z_a(P, x) - a(P, z_a(P, x))$ for any feasible problem $P \in \mathbf{C}$ and $x \in D_P$, $x^P <_{\alpha(P)} x)$.

Denote also by "$a \approx a'$" the relation "$a \preccurlyeq *a'$ and $a' \preccurlyeq *a$". Then we have

Lemma 5.2. *The relation $\preccurlyeq^*$ is a pseudo-order in the set* $\mathbf{A}(\mathbf{C})$; 0 *is the minimum element of* $\mathbf{A}(\mathbf{C})$ *with respect to $\preccurlyeq^*$ and u has the maximality property:* $a^* \succcurlyeq u \rightarrow a \sim u$. *Moreover:*

$$a \preccurlyeq^* a' \rightarrow x^0(a') \leqslant_{(\alpha P)} x^0(a)$$

and

$$a \approx a' \rightarrow a \sim a'.$$

Proof. The reflexivity and the transitivity of the relation $\preccurlyeq^*$ are obvious, hence it is a pseudo-order. The fact that 0 is the minimum element of $\mathbf{A}(\mathbf{C})$ follows from Proposition 2.1. Next, let $a \in \mathbf{A}(\mathbf{C})$, $a \,{}^*\!\succcurlyeq u$. It follows by induction on the elements of $M(P, u)$ that $M(P, u) = M(P, a)$, $z_u(P, x) = z_a(P, x)$ and $a(P, z_a(P, x)) \geqslant u(P)$. Hence $a \sim u$.

Now assume $P \in \mathbf{C}$ is a feasible problem and consider the implication $a \preccurlyeq^* a' \rightarrow x^0(a') \leqslant_{\alpha(P)} x^0(a)$. Assume by way of contradiction $x^0(a) <_{\alpha(P)} x^0(a')$. Then $a \preccurlyeq^* a'$ and $T(P, x^0(a'), z_{a'}(P, x^0(a')) - a'(P, z_{a'}(P, x^\circ(a')))) = 2$ imply $T(P, x^0(a'), z_a(P, x^0(a')) - a(P, z\,(P, x^0(a')))) = 2$. Therefore $x^0(a') \in$ $\in M(P, a)$ and $f_P(x^0(a')) < f_P(x^0(a)) - a(P, f_P(x^0(a)))$ which contradicts the assumption that $x^0(a)$ is the output suboptimal solution supplied by $(\alpha, sk_\alpha, T, a)$. Hence $x^0(a') \leqslant_{\alpha(P)} x^0(a))$.

The implication $a \approx a' \rightarrow a \sim a'$ can be proved by induction on the elements of the set $M(P, a)$. Q.E.D.

It follows from the previous lemma that $\approx$ is an equivalence relation and $\preccurlyeq^*$ is a partial order in the set $\mathbf{A}(\mathbf{C}) / \approx$. Note that the set $\mathbf{A}(\mathbf{C})/\sim$ is strictly included in $\mathbf{A}(\mathbf{C})/\approx$, even if one considers allowance functions which are constant with respect to z. As in the case of the order $\preccurlyeq$ one can show by an example that the effort measures m and n are not monotonous with respect to the $\preccurlyeq^*$-order over the set of all the approximate algorithms. However, these measures become nonincreasing functions if one considers the procedures in Section 4.

Proposition 5.1. *Let α be an enumeration for a class* $\mathbf{C}$, *sk_α a skip function with nested jumps and T a rejection test which are (s) compatible. If a, $a' \in$* $\mathbf{A}(\mathbf{C})$ *are two allowance functions satisfying $a \preccurlyeq^* a'$ then*

$$n(P, a') \leqslant n(P, a), m(P, a') \leqslant m(P, a)$$

for any problem $P \in$ $\mathbf{C}$.

Proof. If the input problem $P \in \mathbf{C}$ is unfeasible or if $x^P = x_{\alpha(P, d(P))}$ then all the approximate algorithms proceed identically when solving P, hence the inequalities are satisfied.

Next assume P is feasible and $x^P <_{\alpha(P)} x_{\alpha(P,\, d(P))}$. We prove the second inequality for the second-type methods. The inequality $m_2(P, a') \leqslant m_2(P, a)$ follows immediately provided we show the inclusion $V_2(P, a) \subseteq V_2(P, a')$. If $V_2(P, a) = \varnothing$ the result is obvious. Let $x \in V_2(P, a) \neq \varnothing$. If $x <_{\alpha(P)} x^P$ then $x \in V_2(P, a')$. Next assume $x^P <_{\alpha(P)} x (x^P \in M_2(P, a) \cap M_2(P, a')$ and therefore $x \neq x^P$). There exists an element $y \in M_2(P, a)$, $x^P \leqslant_{\alpha(P)} y <_{\alpha(P)} x <_{\alpha(P)} sk_\alpha\,(P, y)$, such that $T(P, y, z_a(P, y) -$ $- a(P, z_a(P, y))) \neq 0$.

Case 1. Assume $T(P, y, z_a(P, y) - a(P, z_a(P, y))) = 1$. From the inequality $a \preccurlyeq^* a'$, and the property (t3) it follows $T(P, y, z_{a'}(P, y) - a'(P, z_{a'}(P, y))) = 1$. If $y \in M_2(P, a')$ then $x \in V_2(P, a')$. Next assume $y \notin M_2(P, a')$ and let $y' \in M_2(P, a')$, $x^P \leqslant_{\alpha(P)} y' <_{\alpha(P)} y$, be the greatest element less than y which is tested by the procedure $(\alpha, sk_\alpha, T, a')_2$. (Such an element does exist since $x^P \in M_2(P, a')$ and $x^P <_{\alpha(P)} y$.) Then, as shown below, $x^P \neq y'$, $y <_{\alpha(P)} sk_\alpha(P, y')$ and $T(P, y', z_{a'}(P, y') - a'(P, z_{a'}(P, y'))) = 1$. Indeed, all the other cases lead to a contradiction:

a) Assume $T(P, y', z_{a'}(P, y') - a'(P, z_{a'}(P, y'))) = 0$. Then $succ_\alpha(P, y') = y \in M_2(P, a')$, which contradicts the assumption on y.

b) Assume $T(P, y', z_{a'}(P, y') - a'(P, z_{a'}(P, y'))) = 1$ and $sk_\alpha(P, y') \leqslant_{\alpha(P)} y$. The case $sk_\alpha(P, y) = y'$ can be rejected by an argument similar to that in a). On the other hand the situation $sk_\alpha(P, y') <_{\alpha(P)} y$ contradicts the definition of y'.

c) The value of the test T the moment $(\alpha, sk_\alpha, T, a')_2$ reaches y' is 2. First note that one must have $y' \neq x^P$ since $x^P \in M_2(P, a)$. Then remark that this situation can be rejected as in case a).

Therefore $y <_{\alpha(P)} sk_\alpha(P, y')$ and then $y' <_{\alpha(P)} x <_{\alpha(P)} sk_\alpha(P, y')$, since sk_α is with nested jumps. Hence $x \in V_2(P, a')$.

Case 2. Assume $T(P, y, z_a(P, y) - a(P, z_a(P, y))) = 2$. The condition (sc) implies $T(P, y, z_{a'}(P, y) - a'(P, z_{a'}(P, y))) \neq 0$. If $y \in M_2(P, a')$ then obviously $x \in M_2(P, a')$. Next, assume $y \notin M_2(P, a')$ and let $y' \in M_2(P, a')$, $x^P <_{\alpha(P)} y' <_{\alpha(P)} y$, be the greatest element less than y which is tested by $(\alpha, sk_\alpha, T, a')_2$. One shows as in Case 1 that $T(P, y', z_{a'}(P, y') - a'(P, z_{a'}(P, y'))) \neq 0$ and $y <_{\alpha(P)} sk_\alpha(P, y')$. Therefore it results $x \in V_2(P, a')$, since the function sk_α is with nested jumps.

The proof of the analoguos inequality for the corresponding first-type procedures as well as that of the first inequality in the statement of the proposition are similar. Q.E.D.

For the approximate procedures using nondecreasing skip functions and regular rejection tests a similar result of monotonicity can be proved.

Proposition 5.2. *Let α be an enumeration for a class* **C**, *sk_α a skip function nondecreasing with respect to the $\leqslant_{\alpha(P)}$-order and T a regular rejection test which are (s) compatible. If $a, a' \in$* **A(C)** *are two allowance functions satisfying $a \preccurlyeq^* a'$ then*

$$n(P, a') \leqslant n(P, a), \; m(P, a') \leqslant m(P, a)$$

for any problem $P \in$ **C**.

Proof of the inequality $m_2(P, a') \leqslant m_2(P, a)$, for the second-type algorithms. The results is obvious if $M_2(P, a') \setminus M_2(P, a) = \emptyset$. Next assume $M_2(P, a') \setminus M_2(P, a) = \{x_1 <_{\alpha(P)} x_2, \ldots, <_{\alpha(P)} x_k\} \neq \emptyset$. In this case the inequality follows immediately provided we show that there exists an injective function $g(P, x)$, $g(P, .): M_2(P, a') \setminus M_2(P, a) \to M_2(P, a) \setminus M_2(P, a')$.

Let $x_i \in M_2(P, a') \setminus M_2(P, a)$. There exists an element $y_i \in M_2(P, a)$, $x^P <_{\alpha(P)} y_i <_{\alpha(P)} x_i <_{\alpha(P)} sk_\alpha(P, y_i)$, such that $T(P, y_i, z_a(P, y_i) - a(P, z_a(P, y_i))) \neq 0$.

Case 1. Assume $T(P, y_i, z_a(P, y_i) - a(P, z_a(P, y_i))) = 1$. The relation $a \preccurlyeq^* a'$ and the property (t3) of the tests imply $T(P, y_i, z_{a'}(P, y_i) - a'(P, z_{a'}(P, y_i))) = 1$. Then, from this equality and from $y_i <_{\alpha(P)} x_i <_{\alpha(P)} sk_\alpha(P, y_i)$, it follows $y_i \notin M_2(P, a')$, since otherwise $x_i \notin M_2(P, a')$.

Case 2. Assume $T(P, y_i, z_a(P, y_i) - a(P, z_a(P, y_i))) = 2$. The condition (sc) implies $T(P, y_i, z_{a'}(P, y_i) - a'(P, z_{a'}(P, y_i))) \neq 0$. We must have also $T(P, y_i, z_{a'}(P, y_i) - a'(P, z_{a'}(P, y_i))) = 1$ and $y_i \notin M_2(P, a')$, since otherwise $x_i \notin M_2(P, a')$.

Therefore for every $x_i \in M_2(P, a') \setminus M_2(P, a)$ there exists an element $y_i \in M_2(P, a) \setminus M_2(P, a')$ such that $T(P, y_i, z_{a'}(P, y_i) - a'(P, z_{a'}(P, y))) = 1$. Next let us show that $x_{i-1} <_{\alpha(P)} y_i$ for any $i \neq 1$. Indeed, assume $y_i \leqslant_{\alpha(P)} x_{i-1}$ for some $i \neq 1$. The regularity of the test and $x_{i-1} \in [y_i, sk_\alpha(P, y_i))_{\alpha(P)}$ would imply $T(P, x_{i-1}, z_{a'}(P, x_{i-1}) - a'(P, z_{a'}(P, x_{i-1}))) = 1$. Then $sk_\alpha(P, x_{i-1}) < <_{\alpha(P)} x_i$, since $x_{i-1}, x_i \in M_2(P, a')$ and $x_{i-1} <_{\alpha(P)} x_i$. On the other hand $y_i \leqslant_{\alpha(P)} x_{i-1}$ and the nondecreasing monotonicity of the function sk_σ imply $x_i <_{\alpha(P)} sk_\alpha(P, y_i) \leqslant_{\alpha(P)} sk_\alpha(P, x_{i-1})$, which contradicts the above conclusion. Therefore if one defines $g(P, x_i) = y_i$ then the function $g(P, x)$ is injective, hence the inequality is proved.

The proof of the analogous inequality for the procedures of the first type as well as that of the first inequality are similar. Q.E.D.

The fact that the effort measures are not monotonous with respect to the relation $\preccurlyeq$ but they become nonincreasing functions with respect to the $\preccurlyeq^*$ order for the algorithms in Section 4, can be explained as follows. The conditions defining the two classes of procedures in Section 4 do not imply any relation between the difference of the sizes of the allowance functions, on the one hand, and the difference of the values of the best available solutions of the corresponding algorithms on the other hand.

In particular one can have two algorithms $(\alpha, sk_\alpha, T, a)$, $(\alpha, sk_\alpha, T, a')$ for a class $\mathbf{C}$ with $a \prec a'$, a problem $P \in \mathbf{C}$ and an element $x \in D_P$ such that $z_a(P, x) \ll z_{a'}(P, x)$. Therefore $z_a(P, x) - a(P, z_a(P, x)) < z_{a'}(P, x) - -a'(P, z_a(P, x))$. Then the algorithm $(\alpha, sk_\alpha, T, a)$ skips in the moment it reaches the element x or in the next step but $(\alpha, sk_\alpha, T, a')$ does not. The relation $a \preccurlyeq^* a'$ prevents precisely such a situation: If $a \preccurlyeq^* a'$ then either $z_{a'}(P, x) \leqslant \leqslant z_a(P, x)$ and $a(P, z_a(P, x)) \leqslant a'(P, z_{a'}(P, x) + (z_a(P, x) - z_{a'}(P, x))$ or $z_{a'}(P, x) > z_a(P, x)$ and in this case $a'(P, z_{a'}(P, x)) > a(P, z_a(P, x)) + + (z_{a'}(P, x) - z_a(P, x))$. Hence in both situations if $(\alpha, sk_\alpha, T, a)$ skips, then $(\alpha, sk_\alpha, T, a')$ does it as well.

However from Propositions 5.1 and 5.2 one can deduce a monotonicity result w.r.t. the relation $\preccurlyeq$ for two important types of allowance functions. First consider the case of the ε-type allowance functions which guarantee that the value of the suboptimal solution has an absolute error less than ε(P).

Corollary 5.1. *Let α be an enumeration for a class* $\mathbf{C}$, *sk_α a skip function and T a rejection test which are (s) compatible. Let be also $\varepsilon(P), \varepsilon'(P) \geqslant 0$ and $\Delta(P, \varepsilon, \varepsilon') \geqslant 0$ such that $|z(P, x) - z'(P, x)| \leqslant \Delta(P, \varepsilon, \varepsilon')$ for any $P \in \mathbf{C}$ and $x \in D_P \cap \{x_{\alpha(P, d(P)+1)}\}$. If $\varepsilon'(P) \geqslant \varepsilon(P) + \Delta(P, \varepsilon, \varepsilon')$ and*

(i) *sk_α is with nested jumps*

or

(ii) *sk_α is nondecreasing w.r.t. $\leqslant_{\alpha(P)}$ and T regular*

then $n(P, \varepsilon') \leqslant n(P, \varepsilon)$ and $m(P, \varepsilon') \leqslant m(P, \varepsilon)$ for any problem $P \in \mathbf{C}$ (Here ε, ε' denote the corresponding allowance functions $\varepsilon(P, z) = \varepsilon(P)$, $\varepsilon'(P, z) = = \varepsilon'(P)$ and $z(P, x)$, $z'(P, x)$ the functions z_a associated with ε, ε').

Proof. We show that we have $\varepsilon \preccurlyeq^* \varepsilon'$. Indeed, $z'(P, x) - \varepsilon'(P) = = z(P, x) - \varepsilon(P) - ((\varepsilon(P) - \varepsilon'(P)) - (z'(P, x) - z(P, x)) \leqslant z(P, x) - \varepsilon(P)$. Therefore depending on the condition (i) or (ii) one of the Propositions 5.1 or 5.2 applies. Q.E.D.

Put

$$m(P) = \min \{|z_0(P, x)| \mid x \in D_P \cup \{x_{\alpha(P, d(P)+1)}\}\}$$

$$M(P) = \max \{|z_0(P, x)| \mid x \in D_P \cup \{x_{\alpha(P, d(P)+1)}\}\}.$$

For the allowance functions of the r-type which assure that the output suboptimal value has a relative error less than $r(P)$ we have the following result.

Corollary 5.2. *Let α be an enumeration for a class* **C**, *sk_α a skip function and T a rejection test which are (s) compatible. Let be also $r(P)$, $r'(P)$, $0 < r(P), r'(P) < 1$ and $\Delta(P, r, r')$ such that $|z(P, x) - z'(P, x)| \leqslant \Delta(P, r, r')$ for any $P \in$* **C** *and $x \in D_P \cup \{x_{\alpha(P, d(P)+1)}\}$. If*

$$r'(P) \geqslant r(P) \frac{M(P)}{m(P)} + \frac{\Delta(P, r, r')}{m(P)}$$

and

(i) *sk_α is with nested jumps*

or

(ii) *sk_α is nondecreasing w.r.t. $\leqslant_{\alpha(P)}$ and T regular then $n(P, r') \leqslant n(P, r$ and $m(P, r') \leqslant m(P, r)$ for any problem $P \in$* **C**. *(Here r, r' denote the corresponding allowance functions $r(P, z) = r(P)\,|z|$, $r'(P, z) = r'(P)\,|z|$ and z, z' the functions z_α associated with r, r'.)*

Proof. The inequality in the statement implies $r \preccurlyeq^* r'$. Indeed $r'(P)\,|z'(P, x)| - r(P)\,|z(P, x)| \geqslant r'(P)\, m(P) - r(P)\, M(P) \geqslant \Delta(P, r, r') \geqslant$ $\geqslant z'(P, x) - z(P, x)$. Then this inequality together with one of the hypotheses (i) or (ii) imply the desired monotonicity. Q.E.D.

6. Conclusions

In practice the use of the approximate algorithms instead of the exact ones is justified only if they produce a decrease of the computing efforts. As it was shown in the present paper this improvement of the efficiency holds at least for two important classes of implicit enumeration methods: the procedures using skip functions with nested jumps and those having nondecreasing skip functions and regular rejection tests. Fortunately these classes of algorithms contain the largest part of the implicit enumeration methods used in practice: [3], [4], [7], etc.

In the paper general bounds for the computing effort of the considered algorithms were given. These bounds depend on all the components of the procedure and therefore, in order to improve the efficiency the choice of each of the components α, sk_α, and T is crucial. Along this line the results in the papers [6], [7] are useful.

Finally, let us explain the relation among our results and the similar ones, obtained by Ibaraki [1], for B & B methods. Implicit enumeration algorithms cannot be considered as a particular case of B & B methods from the *constructive point of view* as well as from the *formal point of view* (if we look at the formalization of B & B methods which stands behind the definitions in [1]). If we try to put an i.e. procedure into the frame of B&B schema given by Ibaraki the natural definition of the dominance relation D from [1], is $P_i D P_j = (x_i = x_j$ or $(x_i <_\alpha x_j <_\alpha sk_\alpha(P, x_i)$ and $T(P, x_i, Z(P, x_i)) = 1)$, where any input problem P has a unique row of subproblems P_i, defined by "$x = x_i$". If so defined, relation D depends on the enumeration and on the skip function. Moreover D is not a partial order among subproblems even for i.e. procedures with sk_α nondecreasing and T regular. For this type of procedures generally D is not consistent with g, where, in this case, g is constructed such that $g(x_i) \leqslant \min \{f_P(y) \mid x_i \leqslant_\alpha y <_\alpha sk_\alpha(P, x_i)\}$. However, for i.e. algorithms with skip function with nested jumps D becomes a partial order which is consistent with g.

Acknowledgements. The author wishes to thank dr. C. Bergthaller, dr. Mioara Buiculescu and dr. S. Grigorescu for helpful suggestions given during the preparation of the paper.

REFERENCES

1. Ibaraki, T., *Computational efficiency of approximate Branch and Bound algorithms.* Math. Oper. Res., **1**, 287–299 (1976).
2. Korte, B., Krelle, W., Oberhoffer, W., *Ein lexikographischer Suchalgorithmus zur Lösung allgemeiner ganzzähliger Programmierungsaufgaben. I, II.* Z. Oper. Res. Ser. A–B, **13**, 73–98, 171–192 (1969).
3. Lawler, E.L., Bell, M.D., *A method for solving discrete optimization problems.* Oper. Res., **14**, 1098–1112 (1966).
4. Mao, J.C.T., Wallingford, B.A., *An extension of Lawler and Bell's method of discrete optimizaion with examples from capital budgeting.* Management Sci., **15**, B52–B60 (1968).
5. Popescu, M.F., *Méthodes d'énumération implicite basées sur la monotonie par morceaux.* In: *Proc. Sixth Conference Probab. Theory (Brașov, 1979)*, pp. 443–454. București Ed. Acad. R. S. România, 1981.
6. Popescu, M.F., *Efficiency of implicit enumeration algorithms and rejection power of the tests.* Rev. Roumaine Math. Pures Appl., **28**, 875–892 (1983).
7. Popescu, M.F., *Efficiency of implicit enumeration algorithms and the power of the skip functions,* (In preparation)
8. Rödder, W., *Ein lexikographischer Suchalgorithmus zur ganzzähligen Programmierung : LEXS.* Z. Oper. Rss. Ser. A–B, **20**, 209–217 (1976).

A. ŞTEFĂNESCU, MARIA V. ŞTEFĂNESCU

General stochastic games

A unified theory of stochastic games is proposed starting from an adequate definition of the global pay-off function.
A stochastic game may be described as a sequential decision process with n players in which the global pay-off of each player is a resultant of the partial utilities received in all the stages of the process. In our definition of the general stochastic game the global pay-off is assumed to be dependent on the partial utilities, but not in an explicit particular form.

1. Stochastic game-structure and global strategies

Definition 1. A *discrete stochastic game-structure* is a collection

$$\gamma = \{T;\, I_n;\, I;\, \{X_i^k\}_{i\in I_n,\, k\in I'};\, \{u_i^k\}_{i\in I_n,\, k\in I'};\, \pi^0,\, \{\pi(x)\}_{x\in X}\},$$

where $T = \{1, 2, ...\}$ is the set of the *stages*, $I_n = \{1, 2, ..., n\}$ is the set of the *players*, $I = \{0, 1, ...\}$ is the set of the *states*, $I' = \{1, ...\}$ is the set of *"transient"* states, while 0 is the *"stopping state"*, the X_i^k are finite sets (X_i^k is the set of the *pure strategies* of player i in state k), u_i^k is a real valued function defined on $X^k = \underset{i\in I_n}{\times} X_i^k$ named the *partial utility* of player i in state k, $\pi^0 = (\pi_k)_{k\in I}$, $\pi_k \geqslant 0$, $\sum \pi_k = 1$ is the *initial distribution* of the states, $\pi(x) = (\pi_{kl}(x))_{kl\in I}$ is a stochastic matrix for each $x \in X = \underset{k\in I'}{\times} X^k$, with the properties: $\pi_{kl}(x) = \pi_{kl}(x^k) \geqslant 0$, $\sum_{l\in I} \pi_{kl}(x) = 1$,

$$\pi_{0l}(x) = \begin{cases} 0 & \text{if } l \neq 0 \\ 1 & \text{if } l = 0. \end{cases}$$

Definition 2. If π^0 is the degenerate probability $\pi_k = 1$, $\pi_l = 0$, $l \neq k$, we say that the stochastic game-structure has *initial state* k (denoted γ^k).

Let us denote by S_i^k the set of all probabilities on X_i^k and put $S^k = \{s^k = \underset{i\in I_n}{\otimes} s_i^k \mid s_i^k \in S_i^k,\ i \in I_n\}$, $S_i = \{s_i = \underset{k\in I'}{\otimes} s_i^k \mid s_i^k \in S_i^k\}$ and $S = \{s = \underset{k\in I'}{\otimes} s^k \mid s^k \in S^k\}$ $(S = \underset{i\in I_n}{\otimes} s_i \mid s_i \in S_i\})$.

We agree to consider that $X_i^k \subset S_i^k$, thus to identify each $x_i^k \in X_i^k$ with the degenerate probability $\varepsilon_{x_i^k}$ whose support is x_i^k.

In the following we shall use the symbols $s^k\setminus r_i^k$, $s_i^k\setminus x_i^k$, $s\setminus r_i$ for the product probabilities $\bigotimes_{j\in I_n\setminus\{i\}} s_j^k\otimes r_i^k$, $\bigotimes_{j\in I_n\setminus\{i\}} s_j^k\otimes \varepsilon_{x_i^k}$, $\bigotimes_{j\in I_n\setminus\{i\}} s_j\otimes r_i$, respectively.

By $u_i^k(s^k)$ we shall mean the expected value of u_i^k with respect to the probability $s^k\in S^k$.

Now, let us write $\Delta = \bigcup_{k\in I'}(k, S^k)$ and $\Delta^t = \underbrace{\Delta\times\Delta\times\ldots\times\Delta}_{t \text{ times}}$, $t\in T$.

Definition 3. A *decision function* of player i at stage t is a mapping $\delta_i^t : \Delta^{t-1}\times I' \to \bigcup_{k\in I'} S_i^k$ such that $\delta_k^t(d^{t-1}, k)\in S_i^k$ if $d^{t-1}\in\Delta^{t-1}$ $k\in I'$, $t\geqslant 2$ (for $t = 1$, the term Δ^0 will be omitted).

Definition 4. A *global strategy* of player i is a sequence $\delta_i = (\delta_i^t)_{t\in T}$, where δ_i^t is a decision function at stage t.

Denoting by $\mathbf{D}_i$ the set of the global strategies of player i, an element $\delta = (\delta_1,\ldots,\delta_n)\in\mathbf{D} = \times_{i\in I_n}\mathbf{D}_i$ is said to be a *global strategy* in the stochastic game-structure γ.

Definition 5. The global strategy δ_i is named *stationary* if for every $t, \tau\in T$, $k\in I'$, $d^{t-1}\in\Delta^{t-1}$, $d^{\tau-1}\in\Delta^{\tau-1}$ we have

$$\delta_i^t(d^{t-1}, k) = \delta_i^\tau(d^{\tau-1}, k) \text{ (denoted } \delta_i(k)).$$

Remark. A stationary strategy δ_i may be identified with an element s_i of S_i (taking $s_i^k = \delta_i(k)$, $k\in I'$).

2. General stochastic games

A *stochastic game* Γ is defined by a stochastic game-structure and a set of global pay-off functions $\{f_i^0\}_{i\in I_n}$ of the n players.

If the stochastic game-structure has initial state k, then we denote by f_i^k the global pay-off function of player i.

Obviously, if the stochastic game Γ has initial distribution of the states π^0, then $f_i^0 = \sum_{k\in I'}\pi_k f_i^k$, $i\in I_n$. Consequently, we agree to represent a general stochastic game by the ensamble $\Gamma = \{\gamma, f_i^k, i\in I_n, k\in I'\}$, where γ is a stochastic game-structure and f_i^k are real valued functions defined on $\mathbf{D}$.

Definition 6. A general stochastic game Γ is of the *first category* if

$$(1)\qquad f_i^k(s) = u_i^k(s^k) + g_i^k(s^k, f_i(s)),\ i\in I_n,\ k\in I',\ s\in S$$

(here $f_i = (f_i^k)_{k\in I'}$) for some given applications g_i^k defined on $S^k\times l_\infty$.

Now, for each $i\in I_n$ and $s\in S$ let us consider the system

$$(H_i(s))\quad w + v^k = \sup_{r_i^k\in S_i^k}[u_i^k(s^k\setminus r_i^k) + g_i^k(s^k\setminus r_i^k, v)],\quad k\in I',$$

where $\{g_i^k\}_{k\in I'}$, is a set of applications defined on $S^k\times l_\infty$, w is a real variable and $v = (v^k)_{k\in I'}\in l_\infty$.

We say that $\{g_i^k\}_{k\in I', i\in I_n}$ is a *family of generators* of the game Γ, if for each $i\in I_n$, $s\in S$, the system $(H_i(s))$ has at least a solution $(w_i(s), v_i(s))$, which is a continuous function of s.

Definition 7. A general stochastic game is of the *second category* if $f_i^k = f_i^l$ every $k, l \in I'$, $i \in I_n$, and,

$$f_i^0(s) = h(s, u_i(s)), i \in I_n, \ s \in S \tag{2}$$

(here $u_i = (u_i^k)_{k \in I'}$), where h is an application defined on $S \times l_\infty$ satisfying the equality,

$$h(s, g_i(s, v)) = h(s, v), \tag{3}$$

for every $s \in S$, $v \in l_\infty$ and for each family of generators of the game $(g_i = (g_i^k)_{k \in I'})$.

3. The existence of the stationary equilibrium points in a general stochastic game

The optimal behaviour of the players in a stochastic game is defined by the equilibrium point of the normalized form of the game.

Definition 8. The *normalized form* of the stochastic game $\Gamma(\Gamma^k)$ is the game in the normal form $\{\mathbf{D}_i; f_i^0, i \in I_n\}$ $(\{\mathbf{D}_i; f_i^k, i \in I_n\})$.

In several particular cases it was shown that stationary strategies are sufficient, and, therefore, the central problem of the stochastic game theory is to establish sufficient conditions for the existence of equilibrium points in the game.

Definition 9. The *reduced normalized form* of the stochastic game $\Gamma(\Gamma^k)$ is $\{S_i, f_i^0; i \in I_n\}$ $(\{S_i, f_i^k; i \in I_n\})$.

The following result concerning stochastic games of the first category has been proved in [5].

Theorem 1. *Let a general stochastic game of the first category satisfy the following properties:*

(4) g_i^k, f_i^k *are continuous functions for each* $i \in I_n$, $k \in I'$;

(5) $g_i^k(s^k \setminus r_i^k, v)$ *are linear functions in the variable* r_i^k *for each* $s^k \in S^k$ *and* $v \in l_\infty$;

(6) $g_i^k(s^k, .)$ *are linear functions on* l_∞ *for each* $s^k \in S^k$;

(7) $g_i^k(s^k, v) \leqslant \beta \sup_l v^l$ *for every* $s^k \in S^k$ *and* $v \in l_\infty$, *where* β *is given in the unit interval;*

(8) $|g_i^k(s^k, v)| \leqslant g_i^k(s^k, |v|)$ *for every* $s^k \in S^k$, $v \in l_\infty$ $(|v| = (|v^k|)_{k \in I'})$

Then there exists $\bar{s} \in S$ *such that* $f_i^k(\bar{s}) \geqslant f_i^k(\bar{s} \setminus s_i)$ *(and, consequently,* $f_i^0(\bar{s}) \geqslant f_i^0(\bar{s} \setminus s_i)$*) for every* $s_i \in S_i$, $i \in I_n$, $k \in I'$.

We can establish a similar result for the games of the second category.

Theorem 2. *Consider a general stochastic game of the second category with a family of generators* $\{g_i^k\}_{k \in I', i \in I_n}$ *such that properties* (4)—(7) *are fulfilled. If,*

(9) $h(s, \vec{c}) = c$ *for each* $s \in S$, $\vec{c} \in R$ *(*$c \in l_\infty$ *is the constant sequence with general term* c*);*

(10) $h(s, .)$ *is a strictly increasing linear function on* l_∞ *for each* $s \in S$,

then there exists $\bar{s} \in S$ *such that*

$$f_i^0(\bar{s}) \geqslant f_i^0(\bar{s} \setminus s_i), \ s_i \in S_i, \ i \in I_n.$$

Lemma 1. *For each* $i \in I_n$, $s \in S$, *the system*

(11) $$w + v^k = \max_{r_i^k \in S_i^k} [u_i^k(s^k \backslash r_i^k) + g_i^k(s^k \backslash r_i^k, v)], \; k \in I'$$

has solutions, and if (w, v), $(\bar{w}, \bar{v})$ *are two distinct solutions then* $w = \bar{w}$.

Proof. The first assertion of the lemma follows from the continuity of the functions u_i^k, g_i^k and from the definition of the family of generators.

Now, let be $\bar{s}_i \in S_i$ such that

$$\bar{w} + \bar{v}^k = u_i^k(s^k \backslash \bar{s}_i^k) + g_i^k(s^k \backslash \bar{s}_i^k, \bar{v}), \; k \in I'.$$

Clearly,

$$w + v^k \geqslant u_i^k(s^k \backslash \bar{s}_i^k) + g_i^k(s^k \backslash \bar{s}_i^k, v), \; k \in I'$$

and, then, for each $k \in I'$,

$$\bar{w} - w + \bar{v}^k - v^k \leqslant g_i^k(s^k \backslash \bar{s}_i^k, \bar{v} - v) \leqslant \beta \sup_l (\bar{v}^l - v^l)$$

Hence, $\bar{w} - w \leqslant 0$. By the same argument we conclude that $w - \bar{w} \leqslant 0$ and, consequently, $\bar{w} = w$.

Lemma 2. *Let* $(w_i(s), v_i(s))$ *be a solution of the system* (11) *(with* $w_i(s)$ *uniquely defined) and* $\bar{s}_i \in S_i$ *such that*

(12) $$w_i(s) + v_i^k(s) = u_i^k(s^k \backslash \bar{s}_i^k) + g_i^k(s^k \backslash \bar{s}_i^k, v_i(s)), \; k \in I'.$$

Then $w_i(s) = f_i^0(s \backslash \bar{s}_i) \geqslant f_i^0(s \backslash r_i)$ *for each* $r_i \in S_i$.

Proof. Denoting by $\vec{w}_i(s)$ the constant sequence of l_∞ with general term $w_i(s)$ we can write

$$h(s \backslash \bar{s}_i, \vec{w}_i(s) + v_i(s)) = h(s \backslash \bar{s}_i, u_i(s \backslash \bar{s}_i) + g_i(s \backslash \bar{s}_i, v_i(s)))$$

From (9) (2), (3) and the linearity of h follows that

$$w_i(s) + h(s \backslash \bar{s}_i, v_i(s)) = f_i^0(s \backslash \bar{s}_i) + h(s \backslash_i, v_i(s)),$$

hence

$$w_i(s) = f_i^0(s \backslash \bar{s}_i).$$

Further, let r_i be arbitrary in S_i. Then

$$w_i(s) + v_i^k(s) \geqslant u_i^k(s^k \backslash r_i^k) + g_i^k(s^k \backslash r_i^k, v_i(s)).$$

Again, from (9), (2), (4) and the monotony of h it follows that $w_i(s) \geqslant f_i^0(s \backslash r_i)$.

Proof of Theorem 2. Define $\psi: S \to S$, $\psi(s) = (\psi_i^k(s))_{i \in I_n, \, k \in I'}$,

$$\psi_i^k(s) = \frac{1}{1 + \sum_{x_i^k} q_i^k(s; x_i^k)} (s_i^k + q_i^k(s)), \quad s \in S,$$

where $q_i^k(s) = (q_i^k(s; x_i^k))_{x_i^k \in x_i^k}$, and for each $x_i^k \in X_i^k$ the mapping $q_i^k(., x_i^k): S \to R$ is defined by

$$q_i^k(s; x_i^k) = \max \{0, \; u_k^i(s^k \backslash x_i^k) + g_i^k(s^k \backslash x_i^k, v_i(s)) - v_i^k(s) - f_i^0(s)\}$$

$v_i(s)$ together with $w_i(s)$ is the continuous solution of the system (11)).

From the fixed point theorem of Schauder-Tihonov follows the existence of an $\bar{s} \in S$ such that $\bar{s} = \psi(\bar{s})$.

Let us prove that $q_i^k(\bar{s}; x_i^k) \leqslant 0$ for all $x_i^k \in X_i^k$, $i \in I_n$, $k \in I'$. Indeed, if we admit that there are $i \in I_n$ and $k \in I'$ such that $\bar{X}_i^k = \{x_i^k \in X_i^k | g_i^k(\bar{s}; x_i^k) > > 0\} \neq \varnothing$, then we can write $u_i^k(\bar{s}^k \setminus x_i^k) + g_i^k(\bar{s}^k \setminus x_i^k, v_i(\bar{s})) > v_i^k(\bar{s}) + f_i^0(\bar{s})$ for every $x_i^k \in \bar{X}_i^k$.

On the other hand

$$\bar{s}_i^k = \sum_{x_i^k \in \bar{X}_i^k} \frac{q_i^k(\bar{s})}{1 + \sum_{x_i^k} q_i^k(\bar{s}; x_i^k)}$$

and the linearity of g_i^k implies

$$u_i^k(\bar{s}^k) + g_i^k(\bar{s}^k, v_i(\bar{s})) > v_i^k(\bar{s}) + f_i^0(\bar{s}),$$

therefore $w_i(\bar{s}) > f_i^0(\bar{s})$.

Then, for each $l \in I'$, $u_i^l(\bar{s}^l) + g_i^l(\bar{s}^l, v_i(\bar{s})) > v_i^l(\bar{s}) + f_i^0(\bar{s})$.

Hence, the monotony of h and (3) ensure us that

$$h(\bar{s}, u_i(\bar{s})) + h(\bar{s}, g_i(\bar{s}, v_i(\bar{s}))) > h(v_i(\bar{s})) + f_0^i(\bar{s}),$$

therefore $f_i^0(\bar{s}) > f_i^0(\bar{s})$, which is absurd.

Now we can write

$$f_i^0(\bar{s}) + v_i^k(\bar{s}) \geqslant \max_{x_i^k} [u_i^k(\bar{s}^k \setminus x_i^k) + g_i^k(\bar{s}^k \setminus x_i^k, v_i(\bar{s}))]$$

for every $i \in I_n$, $k \in I'$, $x_i^k \in X_i^k$ that is equivalent to

$$f_i^0(\bar{s}) + v_i^k(\bar{s}) \geqslant \max_{r_i^k} [u_i^k(\bar{s}^k \setminus r_i^k) + g_i^k(\bar{s}^k \setminus r_i^k, v_i(\bar{s}))]$$

for each $i \in I_n$, $k \in I'$. Or on account of Lemma 2 this implies that $f_i^0(\bar{s}) = = w_i(s) \geqslant f_i^0(\bar{s} \setminus s_i)$ for each $s_i \in S_i$, $i \in I_n$.

REFERENCES

1. Gillette, D., *Stochastic games with zero stop probabilities.* In: Dresher, M., Tucker, A.W Wolfe, P. (Eds.), *Contributions to the Theory of Games III.* Ann. of Math. Studies, No. 39, pp. 179–187. Princeton, Princeton Univ. Press, 1957.
2. Hoffman, A.J., Karp, R.M., *On nonterminating stochastic games.* Management Sci., **12**, 359–370 (1966).
3. Shapley, L.S., *Stochastic games.* Proc. Nat. Acad. Sci. U.S.A., **39**, 1095–1100 (1953).
4. Sobel, M.J., *Noncooperative stochastic games.* Ann. Math. Statist., 42, 1930–1935 (1971).
5. Ştefănescu, A., *Stochastic games: A unified theory.* Rev. Roumaine Math. Pures Appl., **29**, 183 – 193 (1984).

VOICU BOŞCAIU

On a general classification problem

The paper is an approach to some classification problems for which the set of diagnoses differs from the set of classes. Some theorems concerning classification procedures with minimum expected loss and some examples concerning the case of two overlapping classes are given.

There is a considerable literature dealing with the classification problems having the same sets of diagnoses and classes. This paper is an approach to a more general problem, namely the classification when the classes set (the set of classification categories) is different from the diagnoses set.

1. The general classification problem definition

Let $(\Omega, \mathcal{K}, p)$ be a probability space.

Definition 1. The *diagnoses set* is a finite partition of Ω, $D = \{d_i \in \mathcal{K}\}$. $p_d = p(d)$ is the *prior probability* of $d \in D$.

Definition 2. The *classes set* is a not empty subset, $C \subset \mathcal{K}(D) \setminus \{\emptyset\}$, of $\mathcal{K}(D)$, the σ-algebra generated by D. It is obvious that every class is a reunion of diagnoses.

Definition 3. The *classification criterion* is a random variable, $X\colon \Omega \to R^n$, whose distribution function F has density f. Let F_d and f_d be the conditional distribution function and conditional density of X given $d \in D$. The following relations are easily proved for $c \in \mathcal{K}(D)$:

(1) $$f = \sum_{d \in D} p_d f_d, \quad \sum_{d \in D} p_d = 1$$

(2) $$p_c f_c = \sum_{d \in D} a_d p_d f_d, \quad p_c = \sum_{d \in D} a_d p_d; \; a_d \in \{0,1\}$$

(3) $$f = p_c f_c + p_{\bar{c}} f_{\bar{c}}, \; p_c + p_{\bar{c}} = 1; \; \bar{c} = \Omega \setminus c.$$

Definition 4. The *C-classification procedure* is a partition of R^n, $R(C) = \{R_c | c \in C\}$. $y \in \Omega$ is classified in the class $c \in C$ iff $X(y) \in R_c$.

Definition 5. The *loss function* $L : \mathcal{H}(D) \times D \to \mathrm{R}$ has the following meaning: $L(c, d)$ is the "penalty paid" if an element having the diagnosis $d \in D$ is classified in the class $c \in C \subset \mathcal{H}(D) \setminus \{\emptyset\}$.

Definition 6. A *general classification problem* is described by a five-tuple

$$\mathcal{C} = (D, C, X, L, R(C)). \tag{4}$$

Definition 7. The *expected loss* of the general classification problem (4) is

$$G(\mathcal{C}) = \sum_{c \in C} \int_{R_c} \sum_{D \in d} L(c, d) p_d f_d(x)\, \mathrm{d}x. \tag{5}$$

If L is defined for every $c \in C$, $d \in D$ by

$$L(c, d) = \begin{cases} 0 & \text{if } d \subset c \\ 1 & \text{otherwise,} \end{cases} \tag{6}$$

$G(\mathcal{C})$ is the *probability of misclassification.*

Definition 8. A *general Bayes classification problem* is a general classification problem with a minimum expected loss.

Theorem 1. *The general classification problem $\mathcal{C}$ is a Bayes one if $R(C)$ is defined for every $c \in C$ by*

$$R_c = \left\{ x \in R^n | \min_{j \in C} \sum_{d \in D} L(j, d) p_d f_d(x) = \sum_{d \in D} L(c, d) p_d f_d(x) \right\}; \tag{7}$$

if for $c \neq s$ $R_c \cap R_s \neq \emptyset$, R_c should be replaced by $R_c \setminus R_s$ and this does not change the expected loss.

The proof is essentially that of Anderson [1] for the case of disjoint classes, taking into account the fact that for $R(C)$ defined by (7) we have

$$G(\mathcal{C}) = \int_{R^n} \min_{c \in C} \sum_{d \in D} L(c, d) p_d f_d(x)\, \mathrm{d}x. \tag{8}$$

Theorem 2. *Let $\mathcal{C}_i = (D, C_i, X, L, R(C_i))$, $i = 1, 2$, be two general Bayes classification problems. If $C_1 \subset C_2$, then $G(\mathcal{C}_1) \geqslant G(\mathcal{C}_2)$.*

The proof is obvious on account of (8).

Theorem 3. *Consider two partitions of Ω, $C_i \subset \mathcal{H}(\Omega)$ and the Bayes classification problems $\mathcal{C}_i = (C_i, C_i, X, L_i, R(C_i))$, where L is defined, taking into account (6), by $L_i(c, d) = i - \delta_{cd}$; $c, d \in C_i$, $i = 1, 2$. If C_1 is finer than C_2, then C_1 has a greater probability of misclassification than C_2.*

Proof. Let be $x \in R^n$ and $i \in C_1$ defined by $p_i f_i(x) = \max \{p_c f_c(x) \mid c \in C_1\}$. C_1 finer than C_2 implies the existence of $j \in C_2$, $i \subset j$ and moreover $j\bar{i} = j \setminus i \in \mathcal{H}(C_1)$. Therefore $p_j f_j(x) = p_i f_i(x) + p_{j\bar{i}} f_{j\bar{i}}(x)$ implies $p_j f_j(x) \geqslant p_i f_i(x)$ and then $\max \{p_c f_c(x) | c \in C_2\} \geqslant \max \{p_c f_c(x) | c \in C_1\}$. By integrating the last relation on R^n we get that the probability of correct classification for C_2 is greater or equal to the probability of correct classification for C_1.

Remarks. 1. If the classes set C does not cover Ω, $Q = \Omega \setminus \bigcup_{c \in C} c \neq \emptyset$ is a reunion of diagnoses. We have $Q = \bigcup_{d \in Z} d$, $Z \subset D$ and all Q elements will certainly be misclassified. If the "purity" of classes is not affected by Q elements, all the losses $L(c, d)$ must be equal for $c \in C$ and $d \in Z$. This is a usual case if the Q elements are not important or if they can be easily recognized by using an appropriate classification criterion (not X, of course).

2. If two nondisjoint classes $c, s \in C$ exist, $I = c \cap s \neq \varnothing$ is a reunion of diagnoses. We have $I = \bigcup_{d \in T} d$, $T \subset D$ and if there are no practical reasons for preferring to classify I elements in the class c or s it is necessary to have $L(c, d) = L(s, d)$ for every diagnosis $d \in T$.

3. If the classes set contains Ω, $\Omega \in C$, the classification procedure $R(C)$ contains a set R_0 such that if $X(y) \in R_0$, y remains unclassified. The existence of R_0 can be justified by high penalty for misclassification, small penalty for unclassification and a not "attractive" reward for correct classification.

4. A possible way of defining the loss function is to chose a positive penalty for misclassification or unclassification and a negative one (to diminish the loss) for a correct classification. For every $d \in D$, $c \in C$

$$L(c, d) \begin{cases} \leqslant 0 \text{ if } c \neq \Omega \text{ and } c \supset d \\ \geqslant 0 \text{ if } c = \Omega \text{ or } c \cap d = \varnothing \end{cases} \tag{9}$$

$(c \not\supset d)$ is equivalent to $c \cap d = \varnothing$ because $c \in \mathcal{K}(C)$).

5. Let $c \in C$. If (9) and there exists $s \in C$, $s \supset c$ so that for every $d \in D$ $L(c, d) \geqslant L(s, d)$, then, on account of Theorem 1, it is easy to find a Bayes procedure having $R_c = \varnothing$. But the above assumption is not a natural one: if for $d \in D$, $L(c, d) \geqslant L(s, d)$ and both sides are negative this means smaller or equal rewards for a stronger correct conclusion.

6. $C = D$ describes the usual case where the classes (categories) for classification are all possible diagnoses.

2. The classification problem in the case of two overlapping diagnoses

If $\pi_1, \pi_2 \subset \pi$ are two subsets of π (let us assume they consist of individuals of the given population π, which have the diseases "1" and "2") one has in general $\pi_1 \cap \pi_2 \neq \varnothing$, $\pi_1 \cup \pi_2 \neq \pi$. Because of Definition 1 the diagnoses set D must be a partition and then D cannot contain both π_1 and π_2. The diagnoses set is defined (this not being the unique possibility) by

$$D^\circ = \{\pi_{12} = \pi_1 \cap \pi_2,\ \pi_{1\bar{2}} = \pi_1 \setminus \pi_2,\ \pi_{\bar{1}2} = \pi_2 \setminus \pi_1,\ \pi_{\bar{1}} = \pi \setminus \pi_{\bar{1}} \cup \pi_2\}. \tag{10}$$

Beside Definitions 1—8, the following notation will be also used: $f_{\pi_i} = f_i$, $p_{\pi_i} = p_i$, $f_\pi = f$, $p_\pi = 1$, $R_\pi = R_0$, $L(\pi_i, \pi_j) = L(i, j)$, $L(\pi, \pi_i) = L(0, i)$. Equations (1), (2), (3) imply:

$$f = p_{12}f_{12} + p_{1\bar{2}}f_{1\bar{2}} + p_{\bar{1}2}f_{\bar{1}2} + p_{\bar{1}\bar{2}}f_{\bar{1}\bar{2}},\quad p_{12} + p_{1\bar{2}} + p_{\bar{1}2} + p_{\bar{1}\bar{2}} = 1 \tag{11}$$

$$p_1 f_1 = p_{12}f_{12} + p_{1\bar{2}}f_{1\bar{2}},\quad p_1 = p_{12} + p_{1\bar{2}} \tag{12}$$

and similar relations for $\pi_{\bar{1}}$, π_2, $\pi_{\bar{2}}$;

$$f = p_1 f_1 + p_{\bar{1}} f_{\bar{1}},\quad p_1 + p_{\bar{1}} = 1 \tag{13}$$

and a similar relation for π_2.

There are different possibilities to choose the classes set. Some of them will be shortly examined below.

Case 2.1. The classes set is $C^1 = \{\pi_1, \pi_2\}$.

If the loss function is defined taking into account Remarks 1 and 2, then $L(1, \bar{1}\bar{2}) = L(2, \bar{1}\bar{2})$ and $L(1, 12) = L(2, 12)$. The general classification problem (D°, C^1, X, L, R) has a Bayes procedure $R = \{R_1, R^n \setminus R_1\}$ defined by Theorem 1:

$$R_1 = \{x \in R^n \mid L(1, 12)p_{12}f_{12}(x) + L(1, 1\bar{2})p_{1\bar{2}}f_{1\bar{2}}(x) + L(1, \bar{1}2)p_{\bar{1}2}f_{\bar{1}2}(x) +$$

$$+ L(1, \bar{1}\bar{2})p_{\bar{1}\bar{2}}f_{\bar{1}\bar{2}}(x) \leqslant L(2, 12)p_{12}f_{12}(x) + L(2, 1\bar{2})p_{1\bar{2}}f_{1\bar{2}}(x) + L(2, \bar{1}2)p_{\bar{1}2}f_{\bar{1}2}(x) +$$

$$+ L(2, \bar{1}\bar{2})p_{\bar{1}\bar{2}}f_{\bar{1}\bar{2}}(x)\} = \{x \in R^n \mid [L(1, 1\bar{2}) - L(2, 1\bar{2})]p_{1\bar{2}}f_{1\bar{2}}(x) +$$

$$+ [L(1, \bar{1}2) - L(2, \bar{1}2)]p_{\bar{1}2}f_{\bar{1}2}(x) \leqslant 0\}.$$

Moreover, if $L(1, 1\bar{2}) = L(2, \bar{1}2) = -r$ and $L(1, \bar{1}2) = L(2, 1\bar{2}) = e$ $(r, e > 0)$, then using (12):

$$R_1 = \{x \in R^n \mid (r + e)(p_{\bar{1}2}f_{\bar{1}2}(x) - p_{1\bar{2}}f_{1\bar{2}}(x)) \leqslant 0\} =$$

$$= \{x \in R^n \mid p_{\bar{1}2}f_{\bar{1}2}(x) \leqslant p_{1\bar{2}}f_{1\bar{2}}(x)\} = \{x \in R^n \mid p_2f_2(x) \leqslant p_1f_1(x)\}.$$

This is the procedure of Anderson [1] for two disjoint classes which has the least misclassification probability. Thus we proved that the procedures for two overlapping classes and two disjoint classes are identical under some natural conditions.

Of course, a better choice of C^1 would be $\{\pi_1, \pi_2, \pi \setminus \pi_1 \cup \pi_2\}$, but that requires to know more than just f_1 and f_2.

Case 2.2. The classes set is $C^2 = \{\pi_1, \pi_{\bar{1}}, \pi_2, \pi_{\bar{2}}, \pi\}$.

The above definition of C^2 can provide information about the most probable class characterized by only one disease, but says nothing about the classes characterized by the two diseases, i.e.,

$$\pi_1 \cap \pi_2,\ \pi_1 \setminus \pi_2,\ \pi_2 \setminus \pi_1,\ \pi \setminus \pi_1 \cup \pi_2.$$

If the loss function L is such that all rewards for a correct classification are $-r$, all penalties for unclassification are u, all penalties for a misclassification are q $(u, q, r > 0)$, then the general classification problem (D°, C^2, X, L, R) has a Bayes procedure $R = \{R_1, R_{\bar{1}}, R_2, R_{\bar{2}}, R_0\}$ defined by Theorem 1:

$$R_i = \{x \in R^n \mid g_i(x) = \min \{g_j(x) \mid j \in \{1, \bar{1}, 2, \bar{2}, 0\}\}\} \tag{14}$$

for every $i \in \{1, \bar{1}, 2, \bar{2}, 0\}$, where, using (11) — (13):

$$g_1(x) = -rp_{12}f_{12}(x) - rp_{1\bar{2}}f_{1\bar{2}}(x) + qp_{\bar{1}2}f_{\bar{1}2}(x) + qp_{\bar{1}\bar{2}}f_{\bar{1}\bar{2}}(x) = -rp_1f_1(x) +$$

$$+ qp_{\bar{1}}f_{\bar{1}}(x) = qf(x) - (r + q)p_1f_1(x),\ g_2(x) = qf(x) - (r + q)p_2f_2(x),\ g_{\bar{1}}(x) =$$

$$= qf(x) - (r + q)p_{\bar{1}}f_{\bar{1}}(x),\ g_{\bar{2}}(x) = qf(x) - (r + q)p_{\bar{2}}f_{\bar{2}}(x),\ g_0(x) = uf(x).$$

It is easy to verify that the procedure defined by (14) remains unchanged if we define the functions g_i by

$$g_0 = [(q - u)/(g + r)]f,\ g_i = -p_i f_i \text{ for } i \in \{1, \bar{1}, 2, \bar{2}\}.$$

If $\pi_1 \cup \pi_2 = \pi$ and $\pi_1 \cap \pi_2 = \varnothing$ then $\pi_{\bar{1}} = \pi_2$, $\pi_{\bar{2}} = \pi_1$, $C^2 = \{\pi_1, \pi_2, \pi\}$ and $R = \{R_1, R_2, R_0\}$.

Case 2.3. The classes set is $C^3 = \{\pi_{12}, \pi_{1\bar{2}}, \pi_{\bar{1}2}, \pi_{\bar{1}\bar{2}}, \pi_1, \pi_{\bar{1}}, \pi_2, \pi_{\bar{2}}, \pi\}$.

Let $y \in \pi_{1\bar{2}}$. It can be correctly classified in four classes: $\pi_{1\bar{2}}, \pi_1, \pi_{\bar{2}}, \pi$. The classification procedure makes the choice between classes taking into account the loss function. If all rewards are zero ($L(1, 1) = L(1, 12) = L(1, 1\bar{2}) = L(\bar{1}, \bar{1}2) = 0$, etc.), $L(1, \bar{1}2) = L(12, \bar{1}2)$, $L(1, 1\bar{2}) = L(12, \bar{1}\bar{2})$, the assumptions of Remark 5 are fulfilled and therefore we can have $R_{12} = \varnothing$. This means that under the above assumptions and similar ones for $R_{1\bar{2}}, R_{\bar{1}2}, R_{\bar{1}\bar{2}}$ a Bayes procedure for the problem (D°, C^3, X, L, R) can have the structure $R = \{R_1, R_2, R_{\bar{1}}, R_{\bar{2}}, R_0\}$.

On the other hand, under the three assumptions below, the Bayes procedure can have $R_1 = R_{\bar{1}} = R_2 = R_{\bar{2}} = R_0 = \varnothing$ and therefore can be $R = \{R_{12}, R_{1\bar{2}}, R_{\bar{1}2}, R_{\bar{1}\bar{2}}\}$, which is identical to the Bayes classification problem with 4 disjoint classes defined by Anderson [1]. These assumptions are: a) (9) is verified, b) the rewards $L(i, ij)$ $i \neq j$ are small enough, c) the penalties $L(0, i)$ and $L(i, jk)$ $i \neq j, k$, $j \neq k$ are large enough for $i, j, k \in \{1, 2, \bar{1}, \bar{2}\}$.

If $\mathcal{C}_1 = (D^\circ, C^1, X, L, R^1)$, $\mathcal{C}_2 = (D^\circ, C^2, X, L, R^2)$, $\mathcal{C}_3 = (D^\circ, C^3, X, L, R^3)$ are general Bayes classification problems, Theorem 2 implies

$$G(\mathcal{C}_1) \geqslant G(\mathcal{C}_2) \geqslant G(\mathcal{C}_3).$$

This means that when one adds some new classes in a classification problem, the expected loss does not increase.

3. Conclusion

If one takes extremely care not to be wrong, it is better to adopt "doubtful" opinions instead of claiming certitudes. Even if one is wrong, which is also possible with a doubtful opinion, the mistake is smaller.

On the other hand, the most doubtful opinion is to say nothing, but, unfortunately, this also means to do nothing. Which should be "the best" proportion of "doubt" in an opinion? But what should one understand by "the best"? For the classification problems the answers could be to enrich the classes set, to extend the loss function to the new wider classes set and to minimize the expected loss.

Acknowledgements. I would like to thank prof. W. Schaafsma (Dept. of Mathematics, University of Groningen) for his goodwill of listening to me in Braşov and for his very useful comments concerning this subject.

REFERENCE

1. Anderson, T.W., *An Introduction to Multivariate Statistical Analysis*. New York, Wiley, 1958.

DOROTHEA MESSERER, A. NEISS[1], D. HOELZER[2]

Which factors influence the course of acute leukaemia in adults? A stochastic approach

Adult patients with acute leukaemia were treated in a German multi-centre trial. The patient's history can be thought of as being a series of different events (progression, remission, relapse etc.). The common practice of analysing each event separately is not adequate in our situation, because the clinicians were also interested in changes of one's patient course of the disease and especially in extracting prognostic factors thereby. Stochastic processes cope with these questions. Weiss and Zelen (1965) already applied a semi-Markov model to the uncensored data of patients with acute leukaemia, but Lagakos et al. (1978) improved the model by determining a general likelihood function for a set of partially censored observations. We applied the data of our study to this model, and compared the transitions probabilities and the sojourn distributions with respect to prognostic factors.

Introduction

In 1980 the government of the FRG started a program to initiate multi-centre clinical trials in the field of cancer, cardiovascular diseases, rheumatism and psychiatric illnesses. The so-called "Biometric Centers for Therapeutic Studies" which were founded by the government and which are closely connected with the local universities have thereby the responsibility to plan, to monitor and to analyse the clinical trials.

The first study with which we were concerned in Munich was an observational study in the treatment of acute leukaemia in adults. Patients were included in the study when the following criteria were met: 1. Unequivocal diagnosis of acute lymphoblastic (ALL) or acute undifferentiated leukaemia (AUL), not in remission, or in process of improvement, established by bone marrow aspiration. 2. Age between 15 and 65 years. 3. No prior therapy.

The therapeutic regimen was divided into four consecutive different parts [Hoelzer (1980): see Fig. 1]. Patients in whom the induction therapy was successfull continued with the consolidation therapy, reinduction therapy and again consolidation therapy lasting two years. 8 different drugs were given to the patients during induction therapy in a dosage which depended on the

[1] GIS, BZT, Pettenkoferstr. 35, 8000 München 2, Bundesrepublik Deutschland

[2] Universität Ulm, Zentrum für Innere Medizin, Steinhövelstr. 9, 7900 Ulm, Bundesrepublik Deutschland.

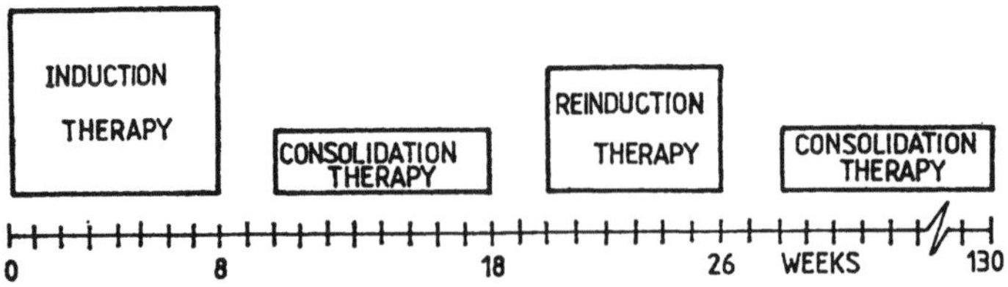

Fig. 1. ALL/AUL in adults: Therapeutic regimen.

patient's body surface area. This dosage had quite often to be reduced because of toxic side effects.

The study was started on October 1st, 1978, and 162 patients from 25 institutions were recruited until June, 1981. Their follow-up ended May 31st, 1982. This is one of the largest number of patients with ALL / AUL in adults in a clinical trial known until today in the literature.

The main question which arose in the discussions between clinicians and the Biometric Center was: which factors influence the course of the disease?

The model

The course of acute leukaemia in our study can be depicted in the following states: progression (P), partial remission or failure to respond (PR/F), complete remission (CR_I) without having been in PR/F, complete remission (CR_{II}) after having been in PR/F, relapse (R), death (D) and censored observation at the date set for analysis (C) (Fig. 2).

Individual courses of disease are for example: A patient dies shortly after start of therapy ($P \rightarrow D$). A patient enters complete remission at the first medical examination after start of therapy but relapses afterwards and is still under observation at the date set for analysis ($P \rightarrow CR_I \rightarrow R \rightarrow C$).

It is common practice in the evaluation of clinical trial data to analyse each of several endpoints separately (i.e. survival time, duration of remission). This approach, however, is not adequate in our situation, because there is no possibility to evaluate the patient's history in a more differential way. A therapeutic program with heavy side effects like this also requires the patient's informed consent. So it would be interesting to know the prognosis after each event in the patient's history to decide individually on the proceeding. A further argument is given by Lagakos (1978): "The simultaneous rather than separate analysis of several endpoints usually provides a better 'feel' for an individual patient's entire experience, and the knowledge that outcomes correspond to some probabilistic law can sometimes also be useful in designing new studies". An alternative approach to the separate analysis of individual events of interest is to model the patient experience as a multistate stochastic process.

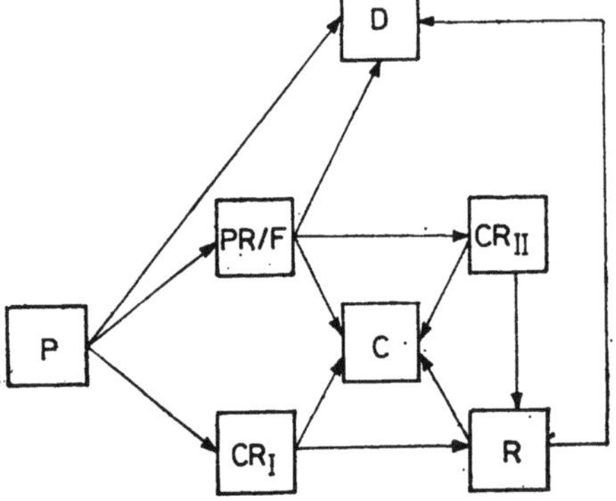

Fig. 2. The states of the model.

To decide which stochastic process is adequate in our situation we have to answer some questions.

(i) What do we know about the medical background to model the course of the disease?

(ii) Do we have to take into consideration any peculiarities in our data?

ad(i) We had no knowledge about the shapes of the sojourn distributions or about any dependencies between states in the patients' experiences. So no assumptions of these characteristics should be made.

ad(ii) Six different states should be allowed for. Because many patients were still under observation at the time of analysis, censored data should be considered.

The consequence was to choose a process which is neither restricted to a special type of sojourn distribution nor to a number of states, and there should be a statistical theory available to allow nonparametric estimation. A semi-Markov process was appropriate. It is a process in which state changes form an embedded Markov chain and sojourn times are independent with distributions depending only on adjoining states. The semi-Markov process assumes that transitions depend on time relative to patient whereas a Markov process needs to assume that the transitions in the state of a patient depend only on absolute time. The sojourn times follow exponential distributions in a Markov process. A semi-Markov process can be characterized in the following way , suppose that each subject begins in one of several states and that at each point in time will be in one of s possible states. The specific ordering of states is arbitrary.

Let Z_0 denote the initial state and Z_n the state corresponding to the patient's nth event. Then if T_n represents the sojourn time between the $(n-1)$st and nth event, the complete history H_m of a process having m events can be described as: $H_m = (Z_0, T_1, Z_1, \ldots, T_m, Z_m)$. The probability element associated with H_m has to be altered when patients' histories are censored. A censored history can be represented by taking $Z_m = s+1$ to denote that T_m is a censored sojourn time. The parameters which describe the process are: initial probabilities $\theta(i) = \mathrm{pr}(Z_0 = i)$, transition probabilities $\theta(i, j) = \mathrm{pr}(Z_{n+1} = j | Z_n = i)$, right-continuous survivorship functions

$$Q(t; i, j) = \mathrm{pr}(T_{n+1} > t | Z_n = i, Z_{n+1} = j)$$

In order to apply this to our set of data, we must estimate the unknown quantities $\theta(i)$, $\theta(i, j)$, $Q(t; i, j)$.

Therefore we used estimators described by Lagakos, Sommer and Zelen (1978), where the general likelihood function for a set of partially censored observations is determined and maximized nonparametrically.

The probability element for the complete of censored history H_m is

$$\theta(Z_0) \prod_{n=1}^{m-1} \{\theta(Z_{n-1}, Z_n)\ \mathrm{d}Q(t_n;\ Z_{n-1}, Z_n\ )) \times$$

$$\times \{\theta(Z_{m-1}, Z_m)\ \mathrm{d}Q(t_m; Z_{m-1}, Z_m)\}\, u(s - Z_m) \times$$

$$\times \left\{ \sum_{j=1}^{s} \theta(Z_{m-1}, j)\, Q(t_m; Z_{m-1}, j) \right\} u(Z_m - s - 1).$$

Here $\mathrm{d}Q(t; i. j)$ is the derivative of Q and $u(i) = 1$ if $i \geqslant 0$ and zero otherwise.

Two patient histories are for example:

(i) A patient dies shortly after start of therapy at time t_1:
history: $\{P, t_1, D\}$
probability element: $\theta(P) \cdot \{\theta(P, D)\ dQ(t_1; P, D)\}$

(ii) A patient enters complete remission at time t_1, relapses at time $t_1 + t_2$ and is censored at time $t_1 + t_2 + t_3$:
history: $\{P, t_1, CR_I, t_2, R, t_3, C\}$
probability element: $\theta(P)\ \{\theta(P, CR_I) dQ(t_1; P, CR_I)\} \times$

$$\times\ \{\theta(CR_I, R) dQ(t_2; CR_I, R) \times \{\theta(R, D)\ Q(t_3; R, D)\}.$$

To calculate the estimators for these parameters, we used a computer program by Muir and Messthaler, Technical University Munich. It calculates and plots the transition probabilities and the sojourn distributions as well as estimates for the variances. (The latter estimates are very time-consuming to calculate.)

To analyse the prognostic factors we considered them separately. We also pooled patients by combining factors in a way we thought is sensible from a medical point of wiev. Therefore we could compare matrices of transition probabilities and sets of sojourn distributions between subgroups.

Results

The main factors to be examined are presented in Table 1. The factors found to be of prognostic significance are age, hepatosplenomegaly and the simultaneous presence of the three main symptoms: fever, infection and bleeding. No dependencies could be shown between these factors.

Table 1

Description of prognostic factors

age		enlarged lymphnodes	
$\leqslant 35$	121	yes	91
> 35	41	no	71
diagnosis		white blood count	
ALL	96	$\leqslant 50.000$	39
AUL	66	> 50.000	123
hepatosplenomegaly		fever, infection, bleeding	
yes	44	all symptoms	111
no	118	not all symptoms	51

Here are some details. The estimated transition probabilities $\hat{\theta}(P, D)$ with associated standard errors differ considerably between the subgroup age $\leqslant 35$ with 0.08 ± 0.03 and the subgroup age > 35 with 0.18 ± 0.06 (Table 2).

Table 2

Transition probabilities in subgroups AGE ⩽ 35 and AGE > 35

(with standard errors)

	AGE ⩽ 35, N = 121 P	PR/F	CR_I	CR_{II}	R	D
P	0	0.30 0.04	0.62 0.05	0	0	0.08 0.03
PR/F	–	0	0	0.55 0.08	0	0.35 0.08
CR_I	–	–	0	0	1	0
CR_{II}	–	–	–	0	1	0
R	–	–	–	–	0	1

	AGE > 35, N = 41 P	PR/F	CR_I	CR_{II}	R	D
P	0	0.33 0.07	0,50 0,08	0	0	0.18 0.06
PR/F	–	0	0	0.64 0.14	0	0.36 0.14
CR_I	–	–	0	0	1	0
CR_{II}	–	–	–	0	1	0
R	–	–	–	–	0	1

Patients ⩽35 of age enter CR_I more frequently without having been in PR/F than the others whereas $\hat{\theta}$(P, PR/F), $\hat{\theta}$(PR/F, CR_{II}) and $\hat{\theta}$(PR/F, D) are comparable in both groups. The sojourn distributions $Q(t;$ CR_I, R) show a big advantage for patients ⩽ 35 of age, but $\hat{Q}(t;$ CR_{II}, R) show no difference between the groups (Figs. 3, 4). Our conclusion is that there is a better prognosis for patients ⩽ 35 in state P with respect to the probability of entering CR as well as dying and the remission duration.

After having been in PR/F it does not seem to make any difference what age the patient is. After having relapsed, however, the median sojourn time until death is 99 weeks for patients ⩽ 35 of age and 68 weeks for the others (Fig. 5).

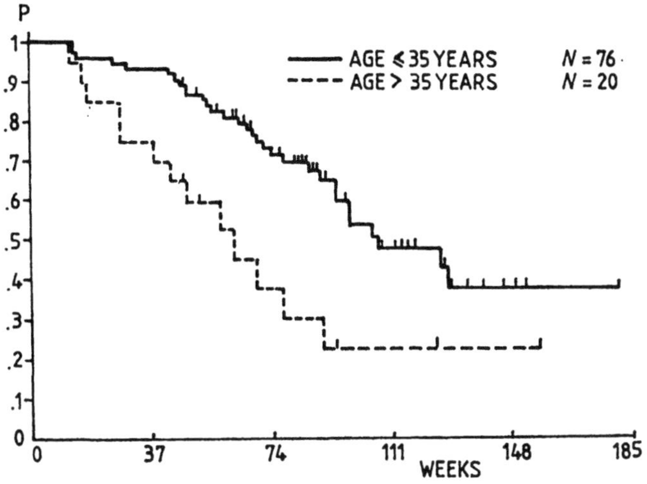

Fig. 3. Sojourn distributions $\hat{Q}(t;$ CR_I, R)

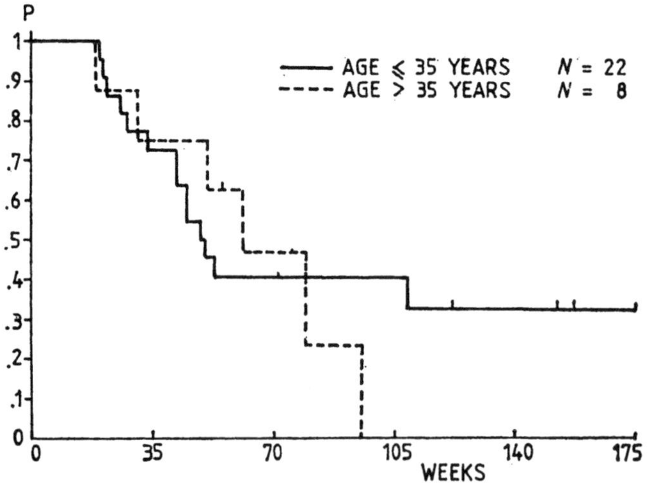

Fig. 4. Sojourn distributions $\hat{Q}(t;\ CR_{II},\ R)$.

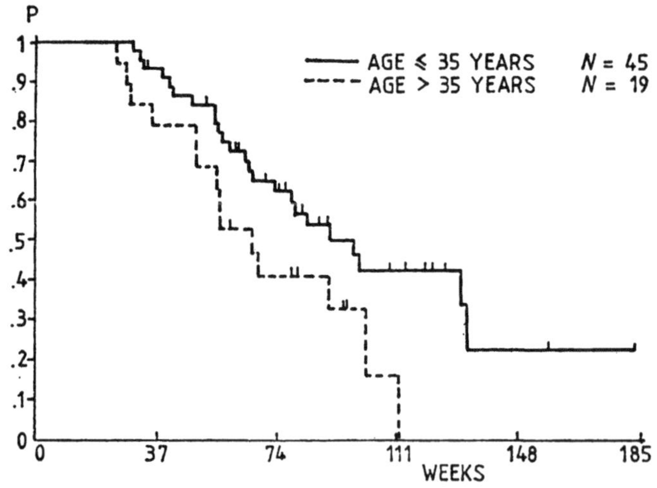

Fig. 5. Sojourn distributions $\hat{Q}(t;\ R,\ D)$.

Discussion

Weiss and Zelen (1965) have applied a semi-Markov process to data on acute leukaemia, but they assumed sojourn times followed gamma distributions and did not allow for censored data. We extended the application by using a method which considers censored observations and where the general likelihood function is maximized nonparametrically.

Weiss and Zelen applied their model on data of one heterogeneous group whereas we tried to extract prognostic factors. The extracted factors have each been compared independently. A better approach would be to use the factors simultaneously. Kay (1982) copes with this problem. He proposes to apply the proportional hazard model on each transition of the multistate stochastic process separately with a slightly different definition of the risk set. So he takes prognostic factors into consideration, but a global analysis of the course of the disease is impossible.

The assumptions of the semi-Markov process were fulfilled in our model. To reach this aim we had to construct two states of CR.

Our further research plans are to estimate transition probabilities by logistic regressions. Then we will be able to consider some probabilistic law and prognostic factors simultaneously.

The medical results were interesting for discussion to the clinicians. This approach is a possibility to consider the entire course of the disease and to derive a basis for further individual treatment decisions. The results are also important fundamentals for planning new studies.

We are well aware that the results of an observational study can be influenced by confounding effects.

Acknowledgements. We thank E. Muir and W. Messthaler, Technical University of Munich, for the programming. (The program is available from the authors.)

REFERENCES

1. Hoelzer, D. et al., *Studienprotokoll der multizentrischen Therapiestudien über akute lymphatische Leukämie (ALL) und akute undifferenzierte (AUL) des Erwachsenen.* Ulm, 1980.
2. Kay, R., *The analysis of transition times in multistate stochastic processes using proportional hazard regression models.* Comm. Statist. A, **11**, 1743–1756 (1982).
3. Lagakos, S.W., *Stochastic models in clininal trials.* In: Scheurlen, H.R. et al (Eds.), *Clinical Trials in "Early Breast Cancer"*, pp. 123–142. Berlin, Springer, 1978.
4. Lagakos, S.W., Sommer, C.J., Zelen, M., *Semi-Markov models for partially censored data.* Biometrika, **65**, 311–317 (1978).
5. Thurmayr, R., Neiss, A., Sintermann, R., *Analyse des Krankheitsverlaufs bei Prostatakarzinompatienten.* In: Horbach, L., Duhme, C. (Hrsg.) *Nachsorge und Krankheitsverlaufs analyse*, pp. 82–87. Berlin, Springer, 1981.
6. Weiss, G.H., Zelen, M., *A semi-Markov model for clinical trials.* J. Appl. Probab., 2, 269–285 (1965).

T. POSTELNICU

Estimation of non-measurable state variables in biotechnical processes

1. Introduction

Biotechnology constitutes a new and challenging field for the application of systems and control engineering. It has been defined as any process or process step in which a biological agent or concept is used or any technology that is directly supportive of the biological process. There is at present a vast assortment of new biotechnically produced medical drugs and chemical products. Modern biotechnology also offers solutions to many current problems concerning energy food supply and environmental pollution. It is now possible artificially to change the genetic make-up of organisms in order to obtain progeny with desired properties. Computers are nowadays available for practically every purpose and various control algorithms aimed at controlling biotechnical processes have been developed and tested using statistical models [3].

2. Estimation of model parameters

Certain problems arise in connection with parameter estimation which are specific to biological processes incorporating living organisms:

— Model parameter does not remain constant over a long period;

— The reproductibility of experiments is often uncertain as it may be difficult to obtain the same environmental conditions and to prevent changes in the internal state of the organisms;

— The quality of measurements, as well as measurement errors, are difficult to define. Some state variables, e.g. substrate (s), biomass (x) and product (p) concentrations are mainly determined by laboratory analyses. The errors appearing are therefore usually systematic, due to human factors, rather than stochastic;

— All the state variables defined, especially in the structured models, cannot be determined at all with sufficient accuracy and at reasonable cost (undeterminable variables are e.g. the substrate pool of plants and the concentration of active biomass in the activated sludge process);

— The expense and duration of laboratory determinations limit sampling frequency; the experiments needed for parameter estimation are usually time-consuming because of large time constants and expensive because of the need for continuous supervision.

In microbiological studies the parameters are usually calculated using linear regression analysis after some data modification, e.g. utilizing Lineweaver-Burke plots [7]. Steady-state data are often used. In certain studies nonlinear parameters estimation methods have been applied. D'Ans, Gottlieb and Kokotovic [2] determined the parameters of a batch process by minimizing an error index consisting of the weighted absolute difference between measured and simulated values.

Problems concerning convergence and uniqueness of the parameter estimation methods have not been sufficiently analysed previously. Nonlinear parameter estimation problems were considered for the case of enzymatic reactions [8].

3. Estimation of non-measurable state variables

A great obstacle to the development and implementation of control algorithms at fermentation and waste water treatment plants is the lack of suitable sensors for various chemical and biological activities. Unfortunately such process variables as biomass (x), product (p) and substrate (s) concentrations are seldom measurable on-line but usually require time consuming and expensive laboratory analysis. Rapid analyzers such as enzyme analyzers are being developed. Moreover, these are not yet in common use as they are still rather expensive and their reliability is not good enough for real time operation.

To illustrate the difficulties associated with obtaining information concerning the state variables, may we mention that the duration of the laboratory analyses used in the experimental studies can take values from 20 minutes to 3 weeks.

The single substrate-single organism dynamic model can be described as follows:

$$\frac{dx}{dt} = G(t) - K_d(t)\, x(t)$$

$$\frac{ds}{dt} = -G(t) - R(t)$$

$$G(t) = \mu(t)\, x(t)$$

$$R(t) = \frac{1 - Y_G}{Y_G}\, \mu(t)\, x(t) - m x(t),$$

in which s is the growth limiting substrate, x is the biomass concentration, G is the growth rate, R the respiration rate, K_d the death rate coefficient and $\mu(t)$ is the specific growth rate. Y_G is the true growth yield and m the maintenance rate coefficient. By introducing the overall biomass substrate yield coefficient $Y_{x/s}$ defined as

$$\frac{1}{Y_{x/s}} = \frac{1}{Y_G} + \frac{m}{\mu(t)}, \tag{2}$$

the second equation (1) can be written

$$\frac{ds}{dt} = -\frac{1}{Y_{x/s}}\,\mu(t)\,x(t). \tag{3}$$

Equation (2) indicates the time-dependent nature of the overall yield coefficient $Y_{x/s}$ which was empirically observed in research related to different processes. The model (1) yields both microbial and plant growth processes [5].

In a constant environment in which no environmental conditions limit growth, the specific growth rate $\mu(t)$ can be expressed by the so-called Michaelis-Menten equation

$$\mu(t) = \frac{\mu_m s(t)}{K_s + s(t)} \tag{4}$$

which originates from enzyme kinetics [5], [9] and was first applied to microbial cultures by Monod [6].

But the general model of microbial growth processes can be expressed as follows:

$$\frac{dx_1}{dt} = \sum_i \pm q_i(t)\,x_i(t) + \mu(t)x_1(t) - K_a x_1(t),$$

$$\frac{ds_1}{dt} = \sum_i \pm q_i(t)s_i(t) - \frac{1}{Y_{x/s}}\,\mu(t)\,x_1(t),$$

$$\frac{dp_1}{dt} = \sum_i \pm q_i(t)\,p_i(t) + Y_{p/x}\,G(t) + \beta x_1(t) - P_a x_1(t),$$

$$\frac{dc}{dt} = \sum_i \pm q_i(t)\,c_i(t) + k_L a(F_A(t))\,[c_s(T(t)) - c(t)] - \mathrm{OUR}(t), \tag{5}$$

$$\mu(t) = \mu_m \prod_i \mu_i(u_i),$$

$$\mathrm{OUR}(t) = aG(t) + bx_1(t),$$

$$G(t) = \mu(t)\,x_1(t), \quad q_i(t) = F_i(t)/V,$$

in which c is the concentration of dissolved oxygen in the growth reactor. The dillution rate q_i is positive, ak_L is the oxygen transfer coefficient which depends on the air flow rate F_A, c_s is the temperature dependent (T) saturation value of dissolved oxygen, OUR is the oxygen utilization rate of the organisms and a and b are parameters connected with oxygen utilization for growth and maintenance respectively.

Some studies have been reported in which nonlinear estimation algorithms have been applied to microbial growth processes [1]. For the purpose of practical application to the processes investigated in this paper, all the known methods are too demanding with respect to the required measurements. For short experiments it might be possible to carry out the required analyses but the methods are not suitable for continuous use.

By considering the system (5), when the OUR(t) could be measured or calculated from measurable data, a simple recursive growth estimator can be obtained in the following way. Assume the estimation law

$$\hat{G}(t) = Y_c(t)\,\mathrm{OUR}(t), \tag{6}$$

$$\widehat{\mathrm{OUR}}(t) = a\,\hat{G}(t) + b\hat{x}_1(t), \tag{7}$$

in which $\hat{G}(t)$ is the estimated growth rate, $\widehat{\mathrm{OUR}}\,(t)$ the estimated and OUR(t) the measured oxygen utilization rate, $Y_c(t)$ the oxygen yield coefficient.

The estimation error

$$e(t) = \mathrm{OUR}(t) - \widehat{\mathrm{OUR}}(t) \tag{8}$$

is zero when

$$Y_c(t) = \frac{\hat{G}(t)}{a\hat{G}(t) + b\hat{x}_1(t)}, \tag{9}$$

which can be used for updating Y_c. The estimate $\hat{x}_1(t)$ of the biomass can then be obtained recursively by substituting (6) and (9) into the first equation of the system (5), which become

$$\frac{\mathrm{d}\hat{x}_1(t)}{\mathrm{d}t} = \left(-q_1(t) - K_a - \frac{b}{a}\right)\hat{x}_1(t) + \sum_i q_i(t)\,x_i(t) + \tag{10}$$

$$+ \frac{1}{a}\,\mathrm{OUR}\,(t).$$

The state estimator can be extended to yield age distributed models and requires measurements only of the OUR and the dilution rates q_i.

The estimator is independent of the structure and parameters of the specific growth rate μ. Y_c is an auxiliary variable which however, has some practical interest and is also useful when implementing the algorithm. This was the indirect state estimation using the OUR. Estimation of OUR requires knowledge of either the oxygen transfer rate (OTR), or of ak_L. The dependence of ak_L on F_A was found empirically to be linear and a recursive least square estimation algorithm can be used.

The state estimator, described by equations (6) — (10), was obtained by utilizing the empirically determined correlation between oxygen consumption and growth. A similar relationship exists between heat production H and growth [4].

In fact, when developing and improving the estimation algorithms aimed at state estimation of biotechnical processes special attention should be paid to the problem of updating the estimates by utilizing the available measurements.

Typical features for these processes are long sampling intervals and the delay between sampling instants and the laboratory analyses.

REFERENCES

1. Aborhey, S., Williamson, D., *State and parameter estimation of microbial growth processes.* Automatica – J. IFAC, **14**, 493–498 (1978).
2. D'Ans, G., Gottlieb, D., Kokotovic, P.V., *Optimal control of bacterial growth.* Automatica – J. IFAC, **8**, 729–736 (1972).
3. Harris, C.J., *Modelling, simulation and control of stochastic systems with applications in wastewater treatment.* Internat. J. Systems Sci., **8**, 393–411 (1977).
4. Holmberg, A., *A system engineering approach to biotechnical processes.* Acta Polytech. Scand. Math. Comput. Sci. Ser., **33**, 5–46 (1981).
5. Michaelis, L., Menten, M.L., *Die Kinetik der Invertinwirkung.* Biochemische Z., **49**, 333–369 (1913).
6. Monod, J., *The growth of bacterial cultures.* Ann. Rev. Microbiology, **3**, 371–394 (1949).
7. Pirt, S.J., *Principles of Microbe and Cell Cultivation.* London, Blackwell, 1975.
8. Postelnicu, T., *Model matematic pentru studiul enzimelor cu doi centri activi.* An. Univ. Bucureşti Mat.-Mec., **18**, *1*, 9–11 (1969).
9. Tanner, R.D., *An enzyme kinetic model for describing fermentation processes.* Biotechnology Bioengineering, **12**, 831–843 (1970).
10. Thornley, J.H.M., *Mathematical Models in Plant Physiology.* London, Academic Press, 1976.

LIST OF PARTICIPANTS

1. AJUPOV Shavkat Abdullaevich, Department of Functional Analysis, Institute of Mathematics, Acad. Sci. Uzbek SSR, Akademgorodok, ul. F. Hadjaeva 29, 700143 Tashkent — 143, USSR.
2. ALBERS Willem, Department of Mathematics, Technological University Twente, P.O. Box 217, 7500 AE Enschede, The Netherlands.
3. ALTĂR Moise, Laboratoarele catedrei de cibernetică economică, Academia de studii economice, Calea Dorobanţi 15—17, 71131 Bucureşti, România.
4. ANDONIE Răzvan, CCE—CIAT, str. Poenelor 5, 2200 Braşov, România.
5. BALLY Vlad, Centrul de statistică matematică, str. Ştirbei Vodă 174, 77104 Bucureşti, România.
6. BARBU Gheorghe, Institutul de învăţămînt superior Piteşti, Bd. R.S. România 84, 0300 Piteşti, România.
7. BAZ Dragomira, Academia de studii economice, Piaţa Romană 6, 79550 Bucureşti, România.
8. BĂDIN Valeriu, Academia de studii economice, Piaţa Romană 6, 79550 Bucureşti, România.
9. BĂLTOIU Iuliana, Institutul central pentru conducere şi informatică, Bd. Miciurin 8—10, 79626 Bucureşti, România.
10. BĂRBUCEANU Veronica, Institutul central pentru conducere şi informatică, Bd. Miciurin 8—10, 79626 Bucureşti, România.
11. BEREANU, Bernard, Centrul de statistică matematică, str. Ştirbei Vodă 174, 77104 Bucureşti, România.
12. BERECHET Adriana, Centrul de statistică matematică, str. Ştirbei Vodă 174, 77104 Bucureşti, România.
13. BERGTHALLER Cristian, Centrul de statistică matematică, str. Ştirbei Vcdă 174, 77104 Bucureşti, România.
14. BERRY Donald A., School of Statistics, 270 Vincent Hall, University of Minnesota, Minneapolis, Minnesota 55455, USA.
15. BERŞADSCHI David, Întreprinderea chimică Dudeşti, Bd. Ion Şulea 246, 78722 Bucureşti, România.
16. BERTIN Emile M.J., Mathematisch Institut der Rijksuniversiteit te Utrecht, Budapestlaan 6, 3508 TA Utrecht, Nederland.
17. BLAKE Louis H., The City University of New York, College of Staten Island, Mathematics Department, 130 Stuyvesant Place, Staten Island, New York 10301, USA.
18. BOB Constantin, Laboratoarele catedrei de cibernetică econcmică, Academia de studii economice, Calea Dorobanţi 15—17, 71131 Bucureşti, România.
19. BOCONCIOS Rodica, Academia de studii economice, Piaţa Romană 6, 79550 Bucureşti, România.
20. BOCŞAN Gheorghe, Facultatea de ştiinţe ale naturii, Universitatea din Timişoara, Bd. V. Pârvan 4, 1900 Timişoara, România.
21. BORDEA Ion, Policlinica „Alexandru Sahia", Calea Dorobanţi 32, 71132 Bucureşti, România.
22. BOŞCAIU Voicu, Centrul de statistică matematică, str. Ştirbei Vcdă 174, 77104 Bucureşti, România.
23. BOTEZ Mihai C., Colectivul de studiul sistemelor, Facultatea de matematică, Universitatea din Bucureşti, str. Academiei 14, 70109 Bucureşti, România.
24. BRADLEY Richard C., Department of Mathematics, Indiana University, Bloomington, Indiana 47405, USA.
25. BRECKNER Wolfgang, Facultatea de matematică, Universitatea „Babeş-Bolyai", str. M. Kogălniceanu 1, 3400 Cluj-Napoca, România.
26. BRESCAN Mihai, Institutul de petrol şi gaze Ploieşti, Bd. Republicii 3, 2000 Ploieşti, România.

27. BRYC Włodzimierz, Instytut Matematyki, Politechnika Warszawska, Plac Jedności Rabotniczej 1, 00—661 Warszawa, Polska.
28. BUICULESCU Mioara, Centrul de statistică matematică, str. Ştirbei Vodă 174, 77104 Bucureşti, România.
29. BUSUIOC Aristiţa, Institutul de meteorologie şi hidrologie, Şos. Bucureşti—Ploieşti 97, 79641 Bucureşti, România.
30. BUZOIANU Petrache, Inspectoratul general de stat pentru controlul calităţii produselor, str. Scaune 6, 70069 Bucureşti, România.
31. CHIŢESCU Ion, Facultatea de matematică, Universitatea din Bucureşti, str. Academiei 14, 70109 Bucureşti, România.
32. COATU Silvia, Centrul de statistică matematică, str. Ştirbei Vodă 174, 77104 Bucureşti, România.
33. CONSTANTIN Gheorghe, Facultatea de ştiinţe ale naturii, Universitatea din Timişoara, Bd. V. Pârvan 4, 1900 Timişoara, România.
34. COPACIU Violeta, Institutul de meteorologie şi hidrologie, Şos. Bucureşti—Ploieşti 97, 79641 Bucureşti, România.
35. CORBU Sergiu, Institutul politehnic Bucureşti, Splaiul Independenţei 313, 79590 Bucureşti, România.
36. COSTEA Gheorghe, Inspectoratul general de stat pentru controlul calităţii produselor, str. Scaune 6, 70069 Bucureşti, România.
37. CRISTEA Alexandru, Institutul de virusologie „Şt. S. Nicolau", Şos. Mihai Bravu 285, 79650 Bucureşti, România.
38. CSÖRGŐ Miklós, Department of Mathematics, Carleton University, Ottawa, Ontario, Canada, K1S 5B6.
39. CUCULESCU Ion, Facultatea de matematică, Universitatea din Bucureşti, str. Academiei 14, 70109 Bucureşti, România.
40. DEÁK István, Computer and Automation Institute, Hungarian Academy of Sciences, Kende u. 13—17, P.O. Box 63, H—1502 Budapest, Hungary.
41. DEHEUVELS Paul, Institut de Statistique des Universités de Paris, 4, Place Jussieu, T. 45—55 E. 3, 75230 Paris Cedex 05, France.
42. DEMETRESCU Mihai C., Academia de studii economice, Piaţa Romană 6, 79550 Bucureşti, România.
43. DINCĂ Vasile, Institutul central pentru conducere şi informatică, Bd. Miciurin 8—10, 79626 Bucureşti, România.
44. DRAGOMIRESCU Liviu, Laboratorul de antropologie, Institutul „Victor Babeş", Splaiul Independenţei 97—101, 79630 Bucureşti, România.
45. DRAGOMIRESCU Mihai, Centrul de statistică matematică, str. Ştirbei Vodă 174, 77104 Bucureşti, România.
46. DRĂGUŢ Maria, Centrul de statistică matematică, str. Ştirbei Vodă 174, 77104 Bucureşti, România.
47. DUGUÉ Daniel, Institut de Statistique des Universités de Paris, 4, Place Jussieu, T. 45—55 E. 3,75230 Paris Cedex 05, France.
48. DUMITRESCU Tudor, Laboratoarele catedrei de cibernetică economică, Academia de studii economice, Calea Dorobanţi 15—17, 71131 Bucureşti, România.
49. DUMITRESCU-BAD Monica, Facultatea de matematică, Universitatea din Bucureşti str. Academiei 14, 70109 Bucureşti, România.
50. EMBRECHTS Paul, Departement Wiskunde K.U.L., Celestijnenlaan 200 B, B 3030 Leuven (Heverlee), Belgium.
51. ENĂCHESCU Denis Melvin, Centrul de calcul al Universităţii din Bucureşti, str. Ştefan Furtună 125, 77116 Bucureşti, România.
52. ENCHEV Ognian, Centre of Mathematics, 7000 Rousse, P.O. Box 325, Bulgaria.
53. ESSEEN Carl Gustav, Department of Mathematics, Uppsala University, Thunbergsvägen 3, S-752 38 Uppsala, Sweden.
54. EVSTIGNEEV Igor V., CEMI, ul. Krasikova 32, Moskow 117418, USSR.
55. FARCAŞ Dan, Centrul de calcul şi statistică sanitară al Ministerului sănătăţii, str. Alecu Constantinescu 9, 79934 Bucureşti, România.
56. FAYOUMI Mohammad, Department of Computer Science, Yarmouk University, Irbid, Jordan.
57. FILIP Argentina, Academia de studii economice, Piaţa Romană 6, 79550 Bucureşti, România.
58. FILIP Liana Maria, Institutul de meteorologie şi hidrologie, Şos. Bucureşti—Ploieşti 97, 79641 Bucureşti, România.
59. FLEISCHMANN Klaus, Institut für Mathematik, Akademie der Wissenschaften der DDR, Mohrenstrasse 39, DDR-1086 Berlin, Deutsche Demokratische Republik.
60. FLOREA Ioan, Facultatea de ştiinţe economice, Universitatea „Babeş-Bolyai", Piaţa Ştefan cel Mare 1, 3400 Cluj-Napoca, România.

61. FLORICEL Ion, INCERC, Şos. Pantelimon 266, 79669 Bucureşti, România.
62. GABER Cornelia, Institutul de petrol şi gaze Ploieşti, Bd. Republicii 3, 2000 Ploieşti, România.
63. GABRIEL Richard, Centrul de statistică matematică, str. Ştirbei Vodă 174, 77104 Bucureşti, România.
64. GARCEA Ştefan, str. Alexandru cel Bun 1, 2200 Braşov, România.
65. GHEORGHE Florin Cristian, Centrul de statistică matematică, str. Ştirbei Vodă, 174, 77104 Bucureşti, România.
66. GOLEŢ Ioan, Institutul politehnic „Traian Vuia", Piaţa Horaţiu 1, 1900 Timişoara, România.
67. GONZACENCO Florin, Centrul de statistică matematică, str. Ştirbei Vodă 174, 77104 Bucureşti, România.
68. GRIGORESCU Şerban, Centrul de statistică matematică, str. Ştirbei Vodă 174, 77104 Bucureşti, România.
69. GUT Allan, Department of Mathematics, Uppsala University, Thunbergsvägen 3, S-752 38 Uppsala, Sweden.
70. HAMMER Peter L., Department of Combinatorics and Optimization, University of Waterloo, Waterloo, Ontario, Canada N2L 3G1.
71. HARLAMOV Boris P., LOMI, Fontanka 27, 191011 Leningrad, USSR.
72. HERA Giorgia, Laboratoarele catedrei de cibernetică economică, Academia de studii economice, Calea Dorobanţi 15—17, 71131 Bucureşti, România.
73. HERŞCOVICI Anda, Centrul de statistică matematică, str. Ştirbei Vodă 174, 77104 Bucureşti, România.
74. HINDERER Karl, Institut für Mathematische Statistik der Universität Karlsruhe, Englerstrasse 2, D-7500 Karlsruhe 1, Bundesrepublik Deutschland.
75. HOLM Sture Axel, Department of Mathematics, Chalmers University of Technology, S-412 96 Göteborg, Sweden.
76. HOLST Lars, Department of Mathematics, Uppsala University, Thunbergsvägen 3, S-752 38 Uppsala, Sweden.
77. HRISTEA Gabriela, Centrul de statistică matematică, str. Ştirbei Vodă 174, 77104 Bucureşti, România.
78. ILIESCU Dan, Centrul de statistică matematică, str. Ştirbei Vodă 174, 77104 Bucureşti, România.
79. IORDACHE Octavian, Institutul politehnic Bucureşti, Splaiul Independenţei 313, 79590 Bucureşti, România.
80. IOSIFESCU Marius, Centrul de statistică matematică, str. Ştirbei Vodă 174, 77104 Bucureşti, România.
81. ISAAC Richard, Department of Mathematics, Herbert H. Lehman College of the City University of New York, Bedford Park Boulevard West, Bronx, New York 10468, USA.
82. IVAN Constantin, Fachhochschule Niederrhein, Reinarzstrasse 49, 4150 Krefeld 1, Bundesrepublik Deutschland.
83. JACOB Caius, Facultatea de matematică, Universitatea din Bucureşti, str. Academiei 14, 70109 Bucureşti, România.
84. JAGERS Peter, Department of Mathematics, Chalmers University of Technology, S-412 96 Göteborg, Sweden.
85. JARUŠKOVÁ Daniela, Katedra Matematiky a Des. Geometrie, Stavební Fakulta ČVUT, Thákurova 7, Praha 6, ČSSR.
86. JESDINSKY Hans, Medizinische Einrichtungen der Universität Düsseldorf, Institut für Medizinische Statistik und Biomathematik, Moorenstrasse 5, 4000 Düsseldorf 1, Bundesrepublik Deutschland.
87. KELLERER Hans G., Mathematisches Institut der Universität München, Theresienstrasse 39, D-8000 München 81, Bundesrepublik Deutschland.
88. KLEIN HANEVELD Willem K., Department of Econometrics, University of Groningen, P.O. Box 800, 9700 AV Groningen, The Netherlands.
89. KONECNY Franz, Institut für Mathematik und Angewandte Statistik der Universität für Bodenkultur Wien, Gregor Mendel-Strasse 33, A-1180 Wien, Österreich.
90. KUROTSCHKA Victor G., Institut fur Mathematik III, Freie Universität Berlin, Arnimallee 2—6, 1000 Berlin 33.
91. LEHMANN Erich L., University of California, Department of Statistics, Statistical Laboratory, Berkeley, California 94720, USA.
92. LENZ Norbert, Johannes Gutenberg-Universität in Mainz, Fachbereich Mathematik, Saarstrasse 21, Postfach 3980, D 6500-Mainz, Bundesrepublik Deutschland.
93. LEONTE Aristide, Facultatea de ştiinţe ale naturii, Universitatea din Craiova, str. Al.I. Cuza 13, 1100 Craiova, România.
94. LEWIN Marica, Institutul politehnic Iaşi, str. 23 August 11, 6600 Iaşi, România.

95. LICEA Gabriela, Facultatea de matematică, Universitatea din Bucureşti, str. Academiei 14, 70109 Bucureşti, România.
96. LINDQVIST Bo, University of Trondheim, Norwegian Institute of Technology, Division of Mathematical Statistics, Alfred Getz vei 1, N-7034 Trondheim, Norway.
97. LUBAN Florica, Laboratoarele catedrei de cibernetică economică, Academia de studii economice, Calea Dorobanţi 15—17, 71131 Bucureşti, România.
98. LUPŞA Liana, Facultatea de matematică, Universitatea „Babeş-Bolyai", str. M. Kogălniceanu 1, 3400 Cluj-Napoca, România.
99. MARCOVICI Maria Petruţa, str. Iulius Fucik 1, 70231 Bucureşti, România.
100. MAREŞ Ileana, Institutul de meteorologie şi hidrologie, Şos. Bucureşti—Ploieşti 97, 79641 Bucureşti, România.
101. MARINESCU Cornel, Institutul de petrol şi gaze Ploieşti, Bd. Republicii 3, 2000 Ploieşti, România.
102. MARINESCU Emil, Centrul de statistică matematică, str. Ştirbei Vodă 174, 77104 Bucureşti, România.
103. MARUŞCIAC Ioan, Facultatea de matematică, Universitatea „Babeş-Bolyai", str. M. Kogălniceanu 1, 3400 Cluj-Napoca, România.
104. MĂRGĂRITESCU Eugen, Centrul de statistică matematică, str. Ştirbei Vodă 174, 77104 Bucureşti, România.
105. MEHLMANN Alexander, Institut für Ökonometrie und Operations Research der Technischen Universität Wien, Argentinierstrasse 8, A-1040 Wien, Österreich.
106. MESSERER Dorothea, Biometrischen Zentrum für Therapiestudien, Pettenkoferstrasse 35, D-8000 Munich 2, Bundesrepublik Deutschland.
107. MIERLEA Vasile, Institutul central pentru conducere şi informatică, Bd. Miciurin 8—10, 79626 Bucureşti, România.
108. MIETHE Norbert, Institut für Mathematik, Akademie der Wissenschaften der DDR Mohrenstrasse 39, DDR-1086 Berlin, Deutsche Demokratische Republik.
109. MIHĂIŢĂ Nicolae, Academia de studii economice, Piaţa Romană 6, 79550 Bucureşti, România.
110. MIHOC Ion, Facultatea de ştiinţe economice, Universitatea „Babeş-Bolyai", Piaţa Ştefan cel Mare 1, 3400 Cluj-Napoca, România.
111. MOGYORÓDI József, Department of Probability Theory, Eötvös Loránd University, Múzeum krt. 6—8, H-1088 Budapest, Hungary.
112. MOROZAN Teodor, Secţia de matematică, INCREST, Bd. Păcii 220, 79622 Bucureşti, România.
113. NÅSELL Ingemar, Royal Institute of Technology, Department of Mathematics, S-100 44 Stockholm 70, Sweden.
114. NICOLAE Traian, Centrul de statistică matematică, str. Ştirbei Vodă 174, 77104 Bucureşti, România.
115. NICULESCU Florentina Rodica, Centrul de calcul al Universităţii din Bucureşti, str. Ştefan Furtună 125, 77116 Bucureşti, România.
116. NICULESCU Gheorghe, Institutul de învăţămînt superior Constanţa, Bd. Lenin 124, 8700 Constanţa, România.
117. NICULESCU Ionel, Academia de studii economice, Piaţa Romană 6, 79550 Bucureşti, România.
118. NICULESCU Ştefan P., Centrul de statistică matematică, str. Ştirbei Vodă 174, 77104 Bucureşti, România.
119. NICULESCU-DINCĂ Carmen, ICPE, Bd. Tudor Vladimirescu 45—47, 79623 Bucureşti, România.
120. NICULESCU-DINCĂ Viorel, ICPE, Bd. Tudor Vladimirescu 45—47, 79623 Bucureşti, România.
121. OANCEA Elena, Facultatea de matematică, Universitatea „Babeş-Bolyai", str. M. Kogălniceanu 1, 3400 Cluj-Napoca, România.
122. OBREJA Giorgică, Centrul de statistică matematică, str. Ştirbei Vodă 174, 77104 Bucureşti, România.
123. ONICESCU Octav, Centrul de statistică matematică, str. Ştirbei Vodă 174, 77104 Bucureşti, România.
124. OPREA Ana, Facultatea de matematică, Universitatea din Bucureşti, str. Academiei 14, 70109 Bucureşti, România.
125. OPREA Miron, Institutul de petrol şi gaze Ploieşti, Bd. Republicii 3, 2000 Ploieşti, România.
126. OPRIŞ Gheorghe, Catedra de matematică, Institutul politehnic Cluj-Napoca, str. Emil Isac 15, 3400 Cluj-Napoca, România.
127. OPRIŞAN Gheorghe, Institutul politehnic Bucureşti, Splaiul Independenţei 313, 79590 Bucureşti, România.

128. ORMAN Gabriel, Facultatea de matematică, Universitatea din Braşov, str. Gh. Gheorghiu-Dej 29, 2200 Braşov, România.
129. PÂRVAN Nicolae Cristian, IJCCP Călăraşi, str. Prel. Bucureşti 1—3, 8500 Călăraşi, România.
130. PELIGRAD Magda, Worcester Polytechnic Institute, Department of Mathematics, Worcester, Massachusetts 01609, USA.
131. PESARIN Fortunato, Università degli Studi di Padova, Facoltà di Scienze Statistiche, Demografiche ed Attuariali, Via VIII Febbraio, 2, I-35122 Padova, Italia.
132. PESCARU Ion Victor, Institutul de meteorologie şi hidrologie, Şos. Bucureşti—Ploieşti 97, 79641 Bucureşti, România.
133. PETRE Nicolae, IJCCP Ialomiţa, Aleea Parcului 4, Bl. B 27, 8400 Slobozia, România.
134. PHILIPPOU Andreas N., University of Patras, Department of Mathematics, Chair B of Applied Mathematics, Patras, Greece.
135. PLACHKY Detlef, Institut für Mathematische Statistik, Universität Münster, Einsteinstrasse 62, D-4400 Münster, Bundesrepublik Deutschland.
136. PLUCIŃSKA Agnieszka, Instytut Matematyki, Politechnika Warszawska, Plac Jedności Rabotniczej 1, 00—661 Warszawa, Polska.
137. POPA Antuza Sabina, Institutul de petrol şi gaze, Ploieşti, Bd. Republicii 3, 2000 Ploieşti, România.
138. POPA Victor, Institutul central pentru conducere şi informatică, Bd. Miciurin 8—10, 79626 Bucureşti, România.
139. POPESCU Alexandru, Academia de studii economice, Piaţa Romană 6, 79550 Bucureşti, România.
140. POPESCU Gheorghe, Centrul de statistică matematică, str. Ştirbei Vodă 174, 77104 Bucureşti, România.
141. POPESCU Mihai Florin, Centrul de statistică matematică, str. Ştirbei Vodă 174, 77104 Bucureşti, România.
142. POSTELNICU Tiberiu, Centrul de statistică matematică, str. Ştirbei Vodă 174, 77104, Bucureşti, România.
143. POSTELNICU Viorica Maria, Centrul de statistică matematică, str. Ştirbei Vodă 174, 77104 Bucureşti, România.
144. PRABHU Narahari U., School of Operations Research and Industrial Engineering, Cornell University, Ithaca, New York 14853, USA.
145. PREDA Vasile, Facultatea de matematică, Universitatea din Bucureşti, str. Academiei 14, 70109 Bucureşti, România.
146. RADIAN Eugen, Şcoala generală nr. 127, str. Munţii Carpaţi 72, 79598 Bucureşti, România
147. RADU Viorel, Facultatea de ştiinţe ale naturii, Universitatea din Timişoara, Bd. V. Pârvan 4, 1900 Timişoara, România.
148. RAPCSÁK Tamás, Computer and Automation Institute, Hungarian Academy of Sciences, Kende u. 13—17, P.O., Box 63, H-1502 Budapest, Hungary.
149. RĂDULESCU Marius Octavian, Institutul central pentru conducere şi informatică, Bd. Miciurin 8—10, 79626 Bucureşti, România.
150. RĂUŢU Gheorghe, Centrul de statistică matematică, str. Ştirbei Vodă 174, 77104 Bucureşti, România.
151. ROMANOF Nicolae, Institutul de meteorologie şi hidrologie, Şos. Bucureşti—Ploieşti 97, 79641 Bucureşti, România.
152. RUSU Gheorghe, Centrul de calcul al consiliului culturii şi educaţiei socialiste, Piaţa Scînteii 1, 71341 Bucureşti, România.
153. RUSU Ion Dumitru, Institutul de cercetări metalurgice, str. Mehadia 39, 79618 Bucureşti, România.
154. RUXANDA Gheorghe, Academia de studii economice, Piaţa Romană 6, 79550 Bucureşti, România.
155. SAMUEL Judita, Centrul de statistică matematică, str. Ştirbei Vodă 174, 77104 Bucureşti, România.
156. SANDI Ana-Maria, Centrul de cercetări sociologice, Universitatea din Bucureşti, str. Academiei 14, 70109 Bucureşti, România.
157. SANDI Horea, INCERC, Şos. Pantelimon 266, 79669 Bucureşti, România.
158. SARKADI Károly, Mathematical Institute of the Hungarian Academy of Sciences, Reáltanoda u. 13—15, H-1053 Budapest, Hungary.
159. SĂCUIU Ion, Academia de studii economice, Piaţa Romană 6, 79550 Bucureşti, România.
160. SĂVULESCU Benone, Academia de studii economice, Piaţa Romană 6, 79550 Bucureşti, România.
161. SÂMBOAN Anca, Centrul de statistică matematică, str. Ştirbei Vodă 174, 77104 Bucureşti, România.
162. SÂNGEORZAN Livia, ICPAT Braşov, str. 13 Decembrie 94, 2200 Braşov, România.

163. SBURLAN Silviu, Institutul de învăţămînt superior Constanţa, Bd. Lenin 124, 8700 Constanţa, România.
164. SCHAAFSMA Willem, Rijkuniversiteit Groningen, Mathematisch Instituut, Postbus 800, 9700 AV Groningen, Nederland.
165. SCHNEIDER Berthold, Medizinische Hochschule Hannover, Department für Biometrie und Medizinische Informatik, Abt. Biometrie, Postfach 610180, D-3000 Hannover, Bundesrepublik Deutschland.
166. SEIFERT Burkhardt, Institut für Mathematik, Akademie der Wissenschaften der DDR, Mohrenstrasse 39, DDR-1086 Berlin, Deutsche Demokratische Republik.
167. SHAFFER Juliet Popper, University of California, Department of Statistics, Statistical Laboratory, Berkeley, California 94720, USA.
168. SINGER Boris, Centrul de statistică matematică, str. Ştirbei Vodă 174, 77104 Bucureşti, România.
169. SPĂTARU Aurel, Centrul de statistică matematică, str. Ştirbei Vodă 174, 77104 Bucureşti, România.
170. SPĂTARU Gheorghe, IJCCP Ialomiţa, Aleea Parcului 4, Bl. B 27, 8400 Slobozia, România.
171. STANCU-MINASIAN Ion, Laboratoarele catedrei de cibernetică economică, Academia de studii economice, Calea Dorobanţi 15—17, 71131 Bucureşti, România.
172. STATE Ioan, Centrul de cercetare ştiinţifică şi inginerie tehnologică pentru tractoare, Bd. 13 Decembrie 94, 2200 Braşov, România.
173. STEINBACH Marc, Institutul de medicină internă, Şos Ştefan cel Mare 21, 79807 Bucureşti, România.
174. STOICA Ana Nina, Institutul central pentru conducere şi informatică, Bd. Miciurin 8—10. 79626 Bucureşti, România.
175. STOICA Doina, Centrul de statistică matematică, str. Ştirbei Vodă 174, 77104 Bucureşti, România.
176. STOICA Ion Lucreţiu, Secţia de matematică, INCREST, Bd. Păcii 220, 79622 Bucureşti, România.
177. STOICA Marcel Dragoş, Academia de studii economice, Piaţa Romană 6, 79550 Bucureşti, România.
178. STOICAN Viorel, Institutul de petrol şi gaze Ploieşti, Bd. Republicii 3, 2000 Ploieşti, România.
179. STOYANOV Jordan M., Institute of Mathematics, Bulgarian Academy of Sciences, P.O. Box 373, 1090 Sofia, Bulgaria.
180. SZÁKATS Adalbert, Institutul naţional de metrologie, Şos. Vitan-Bîrzeşti 11, 79642, Bucureşti, România.
181. SZEKELY Gábor J., Department of Probability Theory, Eötvös Loránd University, Muzeum krt. 6—8, H-1088 Budapest, Hungary.
182. ŞERBAN Mihai, str. Andrei Mureşanu 15 A, 3400 Cluj-Napoca, România.
183. ŞERBU Mihaela, Centrul de statistică matematică, str. Ştirbei Vodă 174, 77104 Bucureşti, România.
184. ŞTEFĂNESCU Anton, Facultatea de matematică, Universitatea din Bucureşti, str. Academiei 14, 70109 Bucureşti, România.
185. ŞTEFĂNESCU Gheorghe, Secţia de matematică, INCREST, Bd. Păcii 220, 79622 Bucureşti, România.
186. ŞTEFĂNESCU Maria V., Laboratoarele catedrei de cibernetică economică, Academia de studii economice, Calea Dorobanţi 15—17, 71131 Bucureşti, România.
187. ŞTEFĂNESCU Ştefan, Centrul de calcul al Universităţii din Bucureşti, str. Ştefan Furtună 125, 77116 Bucureşti, România.
188. TAVITIAN Mihai, Centrul de cercetare ştiinţifică şi inginerie tehnologică pentru tractoare, Bd. 13 Decembrie 94, 2200 Braşov, România.
189. TĂNASE Vasile, Institutul politehnic Bucureşti, Splaiul Independenţei 313, 79590 Bucureşti, România.
190. THÖNI Hanspeter, Institut für Angewandte Mathematik und Statistik der Universität Hohenheim, P.O. Box 106/01100, D-7000 Stuttgart 70, Bundesrepublik Deutschland.
191. TOIA Adrian, Institutul de cercetări metalurgice, str. Mehadia 39, 79618 Bucureşti, România.
192. TOMESCU Rodica, Institutul politehnic Bucureşti, Splaiul Independenţei 313, 79590 Bucureşti, România.
193. TOMKINS Jim R., University of Regina, Department of Mathematics and Statistics, Regina, Saskatchewan, Canada S4S 0A2.
194. TUDOR Constantin, Facultatea de matematică, Universitatea din Bucureşti, str. Academiei 14, 70109 Bucureşti, România.
195. ŢIMPEA Sebastian, IJCCP Braşov, str. Cuza Vodă 12, 2200 Braşov, România.
196. URSIANU Emiliana, Centrul de statistică matematică, str. Ştirbei Vodă 174, 77104 Bucureşti, România.

197. URSIANU Radu, Institutul politehnic Bucureşti, Splaiul Independenţei 313, 79590 Bucureşti, România.

198. VASILIU Dan Petru, Academia de studii economice, Piaţa Rcmană 6, 79550 Bucureşti, România.

199. VASILIU Florin, Inspectoratul general de stat pentru controlul calităţii produselor, str. Scaune 6, 70069 Bucureşti, România.

200. VĂDUVA Ion, Facultatea de matematică, Universitatea din Bucureşti, str. Academiei 14, 70109 Bucureşti, România.

201. VINCZE István, Mathematical Institute of the Hungarian Academy of Sciences, Réaltanoda u. 13—15, H-1053 Budapest, Hungary.

202. VODĂ Viorel, Centrul de statistică matematică, str. Ştirbei Vodă 174, 77104 Bucureşti, România.

203. WEBER Siegfried, Johannes Gutenberg-Universität in Mainz, Fachbereich Mathematik, Saarstrasse 21, Postfach 3980, D-6500 Mainz, Bundesrepublik Deutschland.

204. ZAHARIA Corneliu, Institutul de virusologie „Şt. S. Nicolau", Şos. Mihai Bravu 285, 79650 Bucureşti, România.

205. ZAHARIA Gheorghe, Academia de studii economice, Piaţa Romană 6, 79550 Bucureşti, România.

206. ZBĂGANU Gheorghiţă, Centrul de statistică matematică, str. Ştirbei Vodă 174, 77104 Bucureşti, România.

Bei Fragen zur Produktsicherheit wenden Sie sich bitte an:
If you have any questions regarding product safety,
please contact:

Walter de Gruyter GmbH
Genthiner Straße 13
10785 Berlin
productsafety@degruyterbrill.com